Essential Concepts in Molecular Pathology

Essential Concepts in Molecular Pathology

Edited by

William B. Coleman, Ph.D.

Department of Pathology and Laboratory Medicine
UNC Lineberger Comprehensive Cancer Center
University of North Carolina School of Medicine
Chapel Hill, NC

Gregory J. Tsongalis, Ph.D.

Department of Pathology
Dartmouth Medical School
Dartmouth Hitchcock Medical Center
Norris Cotton Cancer Center
Lebanon, NH

ELSEVIER

AMSTERDAM • BOSTON • HEIDELBERG • LONDON
NEW YORK • OXFORD • PARIS • SAN DIEGO
SAN FRANCISCO • SINGAPORE • SYDNEY • TOKYO
Academic Press is an imprint of Elsevier

Cover images provided by William Coleman and Gregory Tsongalis.

Academic Press is an imprint of Elsevier
30 Corporate Drive, Suite 400, Burlington, MA 01803, USA
525 B Street, Suite 1900, San Diego, California 92101-4495, USA
84 Theobald's Road, London WC1X 8RR, UK

Notice

Medicine is an ever-changing field. Standard safety precautions must be followed, but as new research and clinical experience broaden our knowledge, changes in treatment and drug therapy may become necessary or appropriate. Readers are advised to check the most current product information provided by the manufacturer of each drug to be administered to verify the recommended dose, the method and duration of administrations, and contraindications. It is the responsibility of the treating physician, relying on experience and knowledge of the patient, to determine dosages and the best treatment for each individual patient. Neither the publisher nor the authors assume any liability for any injury and/or damage to persons or property arising from this publication.

Library of Congress Cataloging-in-Publication Data
APPLICATION SUBMITTED

British Library Cataloguing-in-Publication Data
A catalogue record for this book is available from the British Library.

ISBN: 978-0-12-374418-0
For information on all Academic Press publications
visit our Web site at www.elsevierdirect.com

Printed and bound by CPI Group (UK) Ltd, Croydon, CR0 4YY

Transferred to Digital Print 2012

Dedication

This textbook contains a concise presentation of essential concepts related to the molecular pathogenesis of human disease. Despite the succinct form of this material, this textbook represents the state-of-the-art and contains a wealth of information representing the culmination of innumerable small successes that emerged from the ceaseless pursuit of new knowledge by countless experimental pathologists working around the world on all aspects of human disease. Their ingenuity and hard work have dramatically advanced the field of molecular pathology over time and in particular in the last two decades. This book is a tribute to the dedication, diligence, and perseverance of the individuals who have contributed to the advancement of our understanding of the molecular basis of human disease. We dedicate *Essential Concepts in Molecular Pathology* to our colleagues in the field of experimental pathology and to the many pioneers in our field whose work continues to serve as the solid foundation for new discoveries related to human disease. In dedicating this book to our fellow experimental pathologists, we especially recognize the contributions of the graduate students, laboratory technicians, and postdoctoral fellows, whose efforts are so frequently taken for granted, whose accomplishments are so often unrecognized, and whose contributions are so quickly forgotten.

We also dedicate *Essential Concepts in Molecular Pathology* to the many people that have played crucial roles in our successes. We thank our many scientific colleagues, past and present, for their camaraderie, collegiality, and support. We especially thank our scientific mentors for their example of research excellence. We are truly thankful for the positive working relationships and friendships that we have with our faculty colleagues. We also thank our students for teaching us more than we may have taught them. We thank our parents for believing in higher education, for encouragement through the years, and for helping our dreams into reality. We thank our brothers and sisters, and extended families, for the many years of love, friendship, and tolerance. We thank our wives, Monty and Nancy, for their unqualified love, unselfish support of our endeavors, understanding of our work ethic, and appreciation for what we do. Lastly, we give a special thanks to our children, Tess, Sophie, Pete, and Zoe, for providing an unwavering bright spot in our lives, for their unbridled enthusiasm and boundless energy, for giving us a million reasons to take an occasional day off from work just to have fun.

William B. Coleman
Gregory J. Tsongalis

Contents

Contents

Chapter

6   The Human Genome: Implications for the
     Understanding of Human Disease 77

Ashley G. Rivenbark, Ph.D.

Chapter

7   The Human Transcriptome: Implications
     for the Understanding of Human Disease 89

Christine Sers, Ph.D., Wolfgang Kemmner, Ph.D., and
Reinhold Schäfer, Ph.D.

Chapter

8   The Human Epigenome: Implications for
     the Understanding of Human Disease 105

Maria Berdasco, Ph.D., and Manel Esteller, Ph.D.

Chapter

9   Clinical Proteomics and Molecular
     Pathology 113

Mattia Cremona, Virginia Espina, M.S., M.T., Alessandra
Luchini, Ph.D., Emanuel Petricoin, Ph.D., and Lance A.
Liotta, M.D., Ph.D.

Chapter

10   Integrative Systems Biology: Implications
     for the Understanding of Human
     Disease 125

M. Michael Barmada, Ph.D., and David C.
Whitcomb, M.D., Ph.D.

PART

III   Principles and Practice of
      Molecular Pathology

Chapter

11   Pathology: The Clinical Description of
     Human Disease 137

William K. Funkhouser, M.D., Ph.D.

Chapter

12   Understanding Molecular Pathogenesis: The
     Biological Basis of Human Disease and
     Implications for Improved Treatment of
     Human Disease 143

William B. Coleman, Ph.D., and Gregory J.
Tsongalis, Ph.D.

PART

IV   **Molecular Pathology of
     Human Disease**

Contents

PART

V Practice of Molecular Medicine

List of Contributors

Dara L. Aisner, M.D., Ph.D.
Department of Pathology and Laboratory Medicine,
University of Pennsylvania School of Medicine, Hospital of
the University of Pennsylvania, Philadelphia, PA, USA

M. Michael Barmada, Ph.D.
Department of Human Genetics, Graduate School of Public
Health, University of Pittsburgh, Pittsburgh, PA, USA

Philippe L. Bedard, M.D.
Translational Research Unit, Jules Bordet Institute
Université, Libre de Bruxelles, Brussels, Belgium

David O. Beenhouwer, M.D.
Department of Medicine, David Geffen School of Medicine
at University of California, and Division of Infectious
Diseases, Veterans Affairs Greater Los Angeles Healthcare
System, Los Angeles, CA, USA

Jaideep Behari, M.D., Ph.D.
Department of Medicine, Division of Gastroenterology,
Hepatology, and Nutrition, University of Pittsburgh School of
Medicine, Pittsburgh, PA, USA

Maria Berdasco, M.D.
Cancer Epigenetics and Biology Program, Catalan Institute
of Oncology, Barcelona, Catalonia, Spain

Carlise R. Bethel, Ph.D.
Sidney Kimmel Comprehensive Cancer Center, Department
of Pathology, and the Brady Urological Research Institute,
Johns Hopkins University School of Medicine, Baltimore,
MD, USA

Joseph R. Biggs, Ph.D.
Departments of Pathology and Biological Sciences,
University of California, San Diego, La Jolla, CA, USA

Grant C. Bullock, M.D., Ph.D.
Department of Pathology, University of Virginia Health
System, Charlottesville, VA, USA

Sheldon M. Campbell, M.D., Ph.D., F.C.A.P.
Department of Laboratory Medicine, Yale University School
of Medicine Pathology and Laboratory Medicine, VA
Connecticut Healthcare System, West Haven, CT, USA

Wai-Yee Chan, Ph.D.
Laboratory of Clinical Genomics, National Institute of Child
Health and Human Development, NIH, Bethesda, MD, and
Departments of Pediatrics, Biochemistry & Molecular
Biology, Georgetown University School of Medicine,
Washington, D.C., USA

William B. Coleman, Ph.D.
Department of Pathology and Laboratory Medicine,
Curriculum in Toxicology, Program in Translational
Medicine, UNC Lineberger Comprehensive Cancer Center,
University of North Carolina School of Medicine,
Chapel Hill, NC, USA

Mattia Cremona
George Mason University, Center for Applied Proteomics
and Molecular Medicine, Manassas, VA, USA

Angelo M. De Marzo, M.D., Ph.D.
Sidney Kimmel Comprehensive Cancer Center, Department
of Pathology, and the Brady Urological Research Institute,
Johns Hopkins University School of Medicine, Baltimore,
MD, USA

Phuong Dinh, M.D.
Translational Research Unit, Jules Bordet Institute
Université, Libre de Bruxelles, Brussels, Belgium

Vladislav Dolgachev, Ph.D.
Department of Pathology, University of Michigan Medical
School, Ann Arbor, MI, USA

Virginia Espina, M.S., M.T.
George Mason University, Center for Applied Proteomics
and Molecular Medicine, Manassas, VA, USA

Manel Esteller, Ph.D
Cancer Epigenetics and Biology Program, Catalan Institute
of Oncology, Barcelona, Catalonia, Spain

Carol F. Farver, M.D.
Director, Pulmonary Pathology, Vice-Chair for Education,
Pathology and Laboratory Medicine Institute, Department of
Anatomic Pathology, Cleveland Clinic Foundation,
Cleveland, OH, USA

William K. Funkhouser, M.D., Ph.D.
Department of Pathology and Laboratory Medicine,
University of North Carolina School of Medicine,
Chapel Hill, NC, USA

Avrum I. Gotlieb, M.D.C.M.
Department of Pathology, Toronto General Research
Institute, University Health Network, Department of
Laboratory Medicine and Pathobiology, University of
Toronto, Toronto, Ontario, Canada

Robert F. Hevner, Ph.D., M.D.
Department of Neurological Surgery, Division of
Neuropathology, Department of Pathology, University of
Washington School of Medicine, Harborview Medical
Center, Seattle, WA, USA

W. Edward Highsmith, Jr., Ph.D.
Molecular Genetics Laboratory, Department of
Laboratory Medicine and Pathology, Mayo Clinic, Rochester,
MN, USA

C. Dirk Keene, M.D., Ph.D.
Department of Neurological Surgery, Division of
Neuropathology, Department of Pathology, University of
Washington School of Medicine, Harborview Medical
Center, Seattle, WA, USA

Wolfgang Kemmner, Ph.D.
Department of Surgery and Surgical Oncology,
Robert-Rössle-Klinik Berlin, Charité – Universitätsmedizin
Berlin, Berlin, Germany

Nigel S. Key, M.D.
Department of Medicine, University of North Carolina
School of Medicine, Chapel Hill, NC, USA

Hong Kee Lee, Ph.D.
Department of Pathology, Dartmouth Medical School,
Dartmouth Hitchcock Medical Center, Norris Cotton Cancer
Center, Lebanon, NH, USA

Joel A. Lefferts, Ph.D.
Department of Pathology, Dartmouth Medical School,
Dartmouth Hitchcock Medical Center, Norris Cotton Cancer
Center, Lebanon, NH, USA

John J. Lemasters, M.D., Ph.D.
Center for Cell Death, Injury and Regeneration,
Departments of Pharmaceutical & Biomedical Sciences and
Biochemistry & Molecular Biology, Medical University of
South Carolina, Charleston, SC, USA

Markus M. Lerch, M.D
Department of Internal Medicine A, Ernst-Moritz-Arndt-
Universität Greifswald, Greifswald, Germany

Lance A. Liotta, Ph.D.
George Mason University, Center for Applied Proteomics
and Molecular Medicine, Manassas, VA, USA

Alessandra Luchini, Ph.D.
George Mason University, Center for Applied Proteomics
and Molecular Medicine, Manassas, VA, USA

Amber Chang Liu, M.Sc.
Department of Pathology, Toronto General Research
Institute, University Health Network, Department of
Laboratory Medicine and Pathobiology, University of
Toronto, Toronto, Ontario, Canada

Karen Lu, M.D.
Department of Gynecologic Oncology, University of Texas
M.D. Anderson Cancer Center, Houston, TX, USA

Nicholas W. Lukacs, Ph.D.
Professor of Pathology, Director Molecular and Cellular
Pathology Graduate Program, University of Michigan
Medical School, Ann Arbor, MI, USA

Alice D. Ma, M.D.
Department of Medicine, University of North Carolina
School of Medicine, Chapel Hill, NC, USA

Julia Mayerle, Ph.D.
Department of Internal Medicine A, Ernst-Moritz-Arndt-
Universität Greifswald, Greifswald, Germany

Kara A. Mensink, M.S., G.C.G.
Molecular Genetics Laboratory, Department of Laboratory
Medicine and Pathology, Mayo Clinic, Rochester, MN, USA

Samuel Chi-ho Mok, Ph.D.
Department of Gynecologic Oncology, University of
Texas, M.D. Anderson Cancer Center, Houston, TX,
USA

Satdarshan (Paul) Singh Monga, M.D.
Director-Division of Experimental Pathology, Associate
Professor of Pathology and Medicine, University of
Pittsburgh, School of Medicine Pittsburgh, PA, USA

Thomas J. Montine, M.D., Ph.D.
Division of Neuropathology, Department of Pathology,
University of Washington, Harborview Medical Center,
Seattle, WA, USA

Jason H. Moore, Ph.D.
Computational Genetics Laboratory, Norris-Cotton Cancer
Center, Departments of Genetics and Community and Family
Medicine, Dartmouth Medical School, Lebanon, NH, USA

Amy K. Mottl, M.D., M.P.H.
Division of Nephrology and Hypertension, Department of
Medicine, University of North Carolina School of Medicine,
Chapel Hill, NC, USA

Karl Munger, Ph.D.
Department of Medicine, Brigham and Women's Hospital,
Harvard Medical School, Boston, MA, USA

Zoltan Nagymanyoki, M.D., Ph.D.
Division of Gynecologic Oncology, Department of Obstetrics
and Gynecology, Brigham and Women's Hospital, Harvard
Medical School, Boston, MA, USA

William G. Nelson, M.D., Ph.D.
Sidney Kimmel Comprehensive Cancer Center, Department
of Oncology, and the Brady Urological Research Institute,
Johns Hopkins University School of Medicine, Baltimore,
MD, USA

Carla Nester, M.D., M.S.A.
Department of Internal Medicine, University of Iowa
Hospitals and Clinics, Iowa City, IA, USA

Margret D. Oethinger, M.D., Ph.D.
Director of Clinical Microbiology and Molecular Pathology
Providence Portland Medical Center, Portland, OR, USA

Alan L.-Y. Pang, Ph.D.
Laboratory of Clinical Genomics, National Institute of
Child Health and Human Development, NIH, Bethesda,
MD, USA

Emanuel F. Petricoin, III, Ph.D.
George Mason University, Center for Applied Proteomics
and Molecular Medicine, Manassas, VA, USA

Ashley G. Rivenbark, Ph.D.
UNC Lineberger Comprehensive Cancer Center,
Department of Biochemistry and Biophysics,
University of North Carolina at Chapel Hill,
Chapel Hill, NC, USA

C. Harker Rhodes, M.D., Ph.D.
Norris-Cotton Cancer Center, Department of Pathology,
Dartmouth Medical School, Lebanon, NH, USA

Tara C. Rubinas, M.D.
Department of Pathology and Laboratory Medicine,
University of North Carolina School of Medicine,
Chapel Hill, NC, USA

Reinhold Schafer, Ph.D.
Laboratory of Molecular Tumor Pathology, Charité –
Universitätsmedizin Berlin, Berlin, Germany

Matthias Sendler, Ph.D.
Department of Internal Medicine A, Ernst-Moritz-Arndt-
Universität Greifswald, Greifswald, Germany

Antonia R. Sepulveda, M.D., Ph.D.
Department of Pathology and Laboratory Medicine,
University of Pennsylvania School of Medicine, Hospital of
the University of Pennsylvania, Philadelphia, PA, USA

Christine Sers, Ph.D.
Laboratory of Molecular Tumor Pathology, Charité –
Universitätsmedizin Berlin, Berlin, Germany

Lawrence M. Silverman, Ph.D.
Department of Pathology, University of Virginia Health
System, Charlottesville, VA, USA

Natasa Snoj, M.D.
Translational Research Unit, Jules Bordet Institute
Université, Libre de Bruxelles, Brussels, Belgium

Joshua A. Sonnen, Ph.D.
Division of Neuropathology, Department of Pathology,
University of Washington, Harborview Medical Center,
Seattle, WA, USA

Christos Sotiriou, M.D., Ph.D.
Translational Research Unit, Jules Bordet Institute
Université, Libre de Bruxelles, Brussels, Belgium

Gregory J. Tsongalis, Ph.D.
Department of Pathology, Dartmouth Medical School,
Dartmouth Hitchcock Medical Center and Norris Cotton
Cancer Center, Lebanon, NH, USA

Vesarat Wessagowit, M.D., Ph.D.
The Institue of Dermatology, Rajvithi Phyathai, Bangkok,
Thailand

David C. Whitcomb, M.D., Ph.D.
Department of Human Genetics, Graduate School of Public
Health, University of Pittsburgh,
Division of Gastroenterology, Hepatology, and Nutrition,
Department of Medicine, University of Pittsburgh Medical
Center, Pittsburgh, PA, USA

Kwong-kwok Wong, Ph.D.
Department of Gynecologic Oncology, University of Texas
M.D. Anderson Cancer Center, Houston, TX, USA

Dani S. Zander, M.D.
Professor and Chair of Pathology, University Chair in
Pathology, Penn State Milton S. Hershey Medical Center/
Penn State University College of Medicine, Department of
Pathology, Hershey, PA, USA

Dong-Er Zhang, Ph.D.
Departments of Pathology and Biological Sciences,
University of California, San Diego, La Jolla, CA, USA

Preface

Pathology is the scientific study of the nature of disease and its causes, processes, development, and consequences. The field of pathology emerged from the application of the scientific method to the study of human disease. Thus, pathology as a discipline represents the complimentary intersection of medicine and basic science. Early pathologists were typically practicing physicians who described the various diseases that they treated and made observations related to factors that contributed to the development of these diseases. The description of disease evolved over time from gross observation to microscopic inspection of diseased tissues based upon the light microscope, and more recently to the ultrastructural analysis of disease with the advent of the electron microscope. As hospital-based and community-based registries of disease emerged, the ability of investigators to identify factors that cause disease and assign risk to specific types of exposures expanded to increase our knowledge of the epidemiology of disease. While descriptive pathology can be dated to the earliest written histories of medicine and the modern practice of diagnostic pathology dates back perhaps 200 years, the elucidation of mechanisms of disease and linkage of disease pathogenesis to specific causative factors emerged more recently from studies in experimental pathology. The field of experimental pathology embodies the conceptual foundation of early pathology – the application of the scientific method to the study of disease – and applies modern investigational tools of cell and molecular biology to advanced animal model systems and studies of human subjects. Whereas the molecular era of biological science began over 50 years ago, recent advances in our knowledge of molecular mechanisms of disease have propelled the field of molecular pathology. These advances were facilitated by significant improvements and new developments associated with the techniques and methodologies available to pose questions related to the molecular biology of normal and diseased states affecting cells, tissues, and organisms. Today, molecular pathology encompasses the investigation of the molecular mechanisms of disease and interfaces with translational medicine where new basic science discoveries form the basis for the development of new strategies for disease prevention, new therapeutic approaches and targeted therapies for the treatment of disease, and new diagnostic tools for disease diagnosis and prognostication.

With the remarkable pace of scientific discovery in the field of molecular pathology, basic scientists, clinical scientists, and physicians have a need for a source of information on the current state-of-the-art of our understanding of the molecular basis of human disease. More importantly, the complete and effective training of today's graduate students, medical students, postdoctoral fellows, medical residents, allied health students, and others, for careers related to the investigation and treatment of human disease requires textbooks that have been designed to reflect our current knowledge of the molecular mechanisms of disease pathogenesis, as well as emerging concepts related to translational medicine. Most pathology textbooks provide information related to diseases and disease processes from the perspective of description (what does it look like and what are its characteristics), risk factors, disease-causing agents, and to some extent, cellular mechanisms. However, most of these textbooks lack in-depth coverage of the molecular mechanisms of disease. The reason for this is primarily historical – most major forms of disease have been known for a long time, but the molecular basis of these diseases are not always known or have been elucidated only very recently. However, with rapid progress over time and improved understanding of the molecular basis of human disease the need emerged for new textbooks on the topic of molecular pathology, where molecular mechanisms represent the focus.

In this volume on *Essential Concepts in Molecular Pathology* we have assembled a group of experts to discuss the molecular basis and mechanisms of major human diseases and disease processes, presented in the context of traditional pathology, with implications for translational molecular medicine. *Essential Concepts in Molecular Pathology* is an abbreviated version of *Molecular Pathology: The Molecular Basis of Human Disease*, that contains several distinct features. Each chapter focuses on essential concepts related to a specific disease or disease process, rather than providing comprehensive coverage of the topic. Each chapter contains *key concepts*, which capture the essence of the topic covered. In place of long lists of references to the primary literature, each chapter provides a list of *suggested readings*, which include pertinent reviews and/or primary literature references that are deemed to be most important to the reader. This volume is intended to serve as a multi-use textbook that would be appropriate as a classroom teaching tool for medical students, biomedical graduate students, allied health students, advanced undergraduate students, and others. We anticipate that this book will be most useful for teaching students in courses where the full textbook is not needed, but the concepts included are integral to the course of study. This book might also be useful for students that are enrolled in courses that utilize a traditional pathology textbook as the primary text, but need the complementary concepts related to molecular pathogenesis of disease. Further, this textbook will be valuable for pathology residents and other postdoctoral fellows that desire to advance their understanding of molecular mechanisms of disease beyond what they learned in medical/graduate school, and as a reference book and self-teaching guide for practicing basic scientists and physician scientists that need to understand the molecular concepts, but do not require comprehensive coverage or complete detail. To be sure, our understanding of the many causes and molecular mechanisms that govern the development of human diseases is far from complete. Nevertheless, the amount of information related to these molecular mechanisms has increased tremendously in recent years and areas of thematic and conceptual consensus have emerged. We hope that *Essential Concepts in Molecular Pathology* will accomplish its purpose of providing students and researchers with a broad coverage of the essential concepts related to the molecular basis of major human diseases in the context of traditional pathology so as to stimulate new research aimed at furthering our understanding of these molecular mechanisms of human disease and advancing the theory and practice of molecular medicine.

William B. Coleman
Gregory J. Tsongalis

Foreword

Pathology is a *bridging discipline* between basic biological sciences and clinical medicine. Experimental pathologists apply the knowledge and tools developed in basic science disciplines including biochemistry, cell biology, physiology, and molecular biology to understand mechanisms of disease. Clinical pathologists integrate this basic mechanistic understanding of disease with clinical, anatomic, and biochemical information to diagnose disease in individual patients. In the 21st century, this integrated diagnosis of human disease is increasingly based on molecular markers and understanding of disease pathogenesis at the genetic level. This textbook provides fresh insight into the pathogenesis and treatment of disease based on the new discipline of molecular pathology.

Biomedical, clinical, and translational research is conducted by *interdisciplinary teams.* Team members classically have a primary knowledge base and tools in one discipline; however, they must also have the breadth of knowledge and curiosity to incorporate insights from other disciplines to understand, diagnose, and treat human disease. *Essential Concepts in Molecular Pathology* will provide students with a basic foundation in this discipline that will enable them to participate in emerging interdisciplinary research and its clinical applications in the future. For example, molecular pathologists work together with geneticists and ethicists in genetic screening of inherited diseases such as cystic fibrosis. Future research teams including diagnostic pathologists, microbiologists, and biomedical engineers will develop inexpensive, portable devices to diagnose emerging infectious diseases.

Pathologists are also leaders in a new medical paradigm in the 21st century-the practice and application of *personalized medicine* using individual patterns of gene and protein expression. This new diagnostic paradigm relies on bioinformatics and systems biology using genomic and proteomic technologies. *Personalized medicine* promises more accurate diagnosis of complex diseases and individualized therapeutic approaches that are currently being developed for breast, lung, and colon cancers. The practice of medicine in the 21st century will also require new insights into basic mechanisms of disease. In the post-genomic era, molecular pathologists are exploring epigenetic alterations associated with disease that are based on heritable changes in DNA and chromatin organization in the absence of DNA mutations. Molecular pathologists are collaborating with epidemiologists to identify molecular biomarkers reflecting prior environmental exposures or susceptibility to development of future disease. Biostatisticians and systems biologists will collaborate with pharmacologists and pathologists to develop novel therapeutic approaches for human disease. The ultimate goal of these diverse interdisciplinary teams is disease prevention through early recognition of disease susceptibility using molecular biomarkers with potential for early intervention to prevent neurodegenerative diseases, cancer, type 2 diabetes, and cardiovascular disease.

Welcome to the team!

Agnes B. Kane, M.D., Ph.D.
Professor and Chair, Department of Pathology and Laboratory Medicine, Director, NIEHS Training Program in Environmental Pathology, Co-Director, GAANN Interdisciplinary Training Grant in Applications and, Implications of Nanotechnology
The Warren Alpert Medical School of Brown University.

Acknowledgements

The editors would like to acknowledge the significant contributions of a number of people to the successful production of *Essential Concepts in Molecular Pathology*.

We would like to thank the individuals that contributed to the content of this volume. The remarkable coverage of the state-of-the-art in the molecular pathology of human disease would not have been possible without the hard work and diligent efforts of the 65 authors of the individual chapters. Many of these contributors are our long-time colleagues, collaborators, and friends, and they have contributed to other projects that we have directed. We appreciate their willingness to contribute once again to a project that we found worthy. We especially thank the contributors to this volume that were willing to work with us for the first time. We look forward to working with all of these authors again in the future. Each of these contributors provided us with an excellent treatment of their topic and we hope that they will be proud of their individual contributions to the textbook. Furthermore, we would like to give a special thanks to our colleagues that co-authored chapters with us for this textbook. There is no substitute for an excellent co-author when you are juggling the several responsibilities of concurrently editing and contributing to a textbook. Collectively, we can all be proud of this volume as it is proof that the whole can be greater than the sum of its parts.

We would also like to thank the many people that work for *Academic Press* and *Elsevier* that made this project possible. Many of these people we have not met and do not know, but we appreciate their efforts to bring this textbook to its completed form. Special thanks goes to three key people that made significant contributions to this project on the publishing side, and proved to be exceptionally competent and capable. Ms. Mara Conner (*Academic Press*, San Diego, CA) embraced the concept of this textbook when our ideas were not yet fully developed and encouraged us to pursue this project. She was receptive to the model for this textbook that we envisioned and worked closely with us to evolve the project into its final form. We thank her for providing excellent oversight (and for displaying optimistic patience) during the construction and editing of the textbook. Ms. Megan Wickline (*Academic Press*, San Diego, CA) provided excellent support to us throughout this project. As we interacted with our contributing authors, collected and edited manuscripts, and through production of the textbook, Megan assisted us greatly by being a constant reminder of deadlines, helping us with communication with the contributors, and generally providing support for details small and large, all of which proved to be critical. Ms. Christie Jozwiak (*Elsevier*, Burlington, MA) directed the production of the textbook. She worked with us closely to ensure the integrity of the content of the textbook as it moved from the edited manuscripts into their final form. Throughout the production process, Christie gave a tremendous amount of time and energy to the smallest of details. We thank her for her direct involvement with the production and also for directing her excellent production team. This was our second major project working with Mara, Megan, and Christie. It was a pleasure to work with them on this book. We hope that they enjoyed it as much as we did, and we look forward to working with them again soon.

William B. Coleman
Gregory J. Tsongalis

Essential Pathology — Mechanisms of Disease

1

Molecular Mechanisms of Cell Death

John J. Lemasters

INTRODUCTION

A common theme in disease is death of cells. In diseases ranging from stroke to alcoholic cirrhosis of the liver, death of individual cells leads to irreversible functional loss in whole organs and ultimately mortality. For such diseases, prevention of cell death becomes a basic therapeutic goal. By contrast in neoplasia, the purpose of chemotherapy is to kill proliferating cancer cells. For either therapeutic goal, understanding the mechanisms of cell death becomes paramount.

MODES OF CELL DEATH

Although many stresses and stimuli cause cell death, the mode of cell death typically follows one of two patterns. The first is necrosis, a pathological term referring to areas of dead cells within a tissue or organ. Necrosis is typically the result of an acute and usually profound metabolic disruption, such as ischemia/reperfusion and severe toxicant-induced damage. Since necrosis as observed in tissue sections is an outcome rather than a process, the term oncosis has been introduced to describe the process leading to necrotic cell death, but the term has yet to be widely adopted in the experimental literature. Here, the terms oncosis, oncotic necrosis, and necrotic cell death will be used synonymously to refer both to the outcome of cell death and the pathogenic events precipitating cell killing.

The second pattern is programmed cell death, most commonly manifested as apoptosis, a term derived from an ancient Greek word for the falling of leaves in the autumn. In apoptosis, specific stimuli initiate execution of well-defined pathways leading to orderly resorption of individual cells with minimal leakage of cellular components into the extracellular space and little inflammation. Whereas necrotic cell death occurs with abrupt onset after adenosine triphosphate (ATP)

depletion, apoptosis may take hours to go to completion and is an ATP-requiring process without a clearly distinguished point of no return. Although apoptosis and necrosis were initially considered separate and independent phenomena, an alternate view is emerging that apoptosis and necrosis can share initiating factors and signaling pathways to become extremes on a phenotypic continuum of necrapoptosis or aponecrosis.

STRUCTURAL FEATURES OF NECROSIS AND APOPTOSIS

Oncotic Necrosis

Cellular changes leading up to onset of necrotic cell death include formation of plasma membrane protrusions called blebs, mitochondrial swelling, dilatation of the endoplasmic reticulum (ER), dissociation of polysomes, and cellular swelling leading to rupture with release of intracellular contents (Table 1.1, Figure 1.1). After necrotic cell death, characteristic histological features of loss of cellular architecture, vacuolization, karyolysis, and increased eosinophilia soon become evident (Figure 1.2). Cell lysis evokes an inflammatory response, attracting neutrophils and monocytes to the dead tissue to dispose of the necrotic debris by phagocytosis and defend against infection (Figure 1.3). In organs like heart and brain with little regenerative capacity, healing occurs with scar formation, namely replacement of necrotic regions with fibroblasts and collagen, as well as other connective tissue components. In organs like the liver that have robust regenerative capacity, cell proliferation can replace areas of necrosis with completely normal tissue within a few days. The healed liver tissue shows with little or no residua of the necrotic event, but if regeneration fails, collagen deposition and fibrosis will occur instead to cause cirrhosis.

Table 1.1   Comparison of Necrosis and Apoptosis

Necrosis	Apoptosis
Accidental cell death	Controlled cell deletion
Contiguous regions of cells	Single cells separating from neighbors
Cell swelling	Cell shrinkage
Plasmalemmal blebs without organelles	Zeiotic blebs containing large organelles
Small chromatin aggregates	Nuclear condensation and lobulation
Random DNA degradation	Internucleosomal DNA degradation
Cell lysis with release of intracellular contents	Fragmentation into apoptotic bodies
Inflammation and scarring	Absence of inflammation and scarring
Mitochondrial swelling and dysfunction	Mitochondrial permeabilization
Phospholipase and protease activation	Caspase activation
ATP depletion and metabolic disruption	ATP and protein synthesis sustained
Cell death precipitated by plasma membrane rupture	Intact plasma membrane

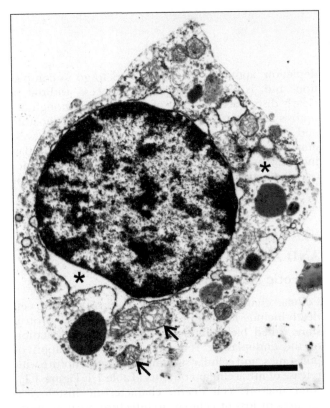

Figure 1.1   Electron microscopy of oncotic necrosis to a rat hepatic sinusoidal endothelial cell after ischemia/ reperfusion. Note cell rounding, mitochondrial swelling (arrows), rarefaction of cytosol, dilatation of the ER and the space between the nuclear membranes (*), chromatin condensation, and discontinuities in the plasma membrane. Bar is 2 μm.

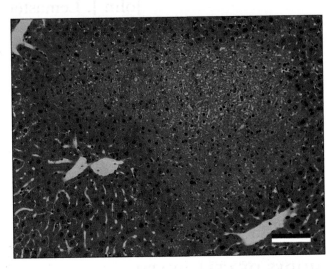

Figure 1.2   Histology of necrosis after hepatic ischemia/ reperfusion in a mouse. Note increased eosinophilia, loss of cellular architecture, and nuclear pyknosis and karyolysis. Bar is 50 μm.

Apoptosis

Unlike necrosis, which usually represents an accidental event in response to an imposed unphysiological stress, apoptosis is a process of physiological cell deletion that has an opposite role to mitosis in the regulation of cell populations. In apoptosis, cell death occurs with little release of intracellular contents, inflammation, and scar formation. Individual cells undergoing apoptosis separate from their neighbors and shrink rather than swell. Distinctive nuclear and cytoplasmic changes also occur, including chromatin condensation, nuclear lobulation and fragmentation, formation of numerous small cell surface blebs (zeiotic blebbing), and shedding of these blebs as apoptotic bodies that are phagocytosed by adjacent cells and macrophages for lysosomal degradation (Table 1.1, Figure 1.3). Characteristic biochemical changes also occur, typically activation of a cascade of cysteine-aspartate proteases, called caspases, leakage of proapoptotic proteins like cytochrome *c* from mitochondria into the cytosol, internucleosomal deoxyribonucleic acid (DNA) degradation, degradation of poly (ADP-ribose) polymerase (PARP), and movement of phosphatidyl serine to the exterior leaflet of the plasmalemmal lipid bilayer. Thus, apoptosis manifests a very different pattern of cell death than oncotic necrosis (Table 1.1, Figure 1.3).

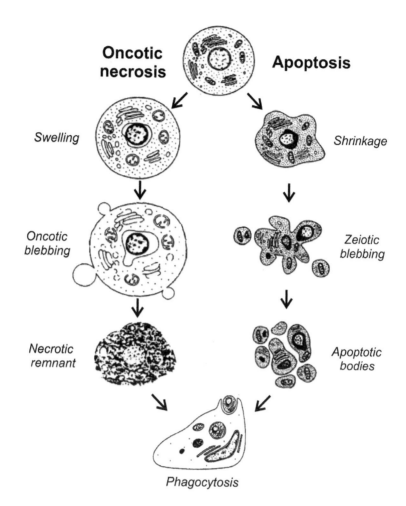

Figure 1.3   Scheme of necrosis and apoptosis. In oncotic necrosis, swelling leads to bleb rupture and release of intracellular constituents, which attract macrophages that clear the necrotic debris by phagocytosis. In apoptosis, cells shrink and form small zeiotic blebs that are shed as membrane-bound apoptotic bodies. Apoptotic bodies are phagocytosed by macrophages and adjacent cells. Adapted with permission from Van CS, Van Den BW. Morphological and biochemical aspects of apoptosis, oncosis and necrosis. *Anat Histol Embryol.* 2002;31(4):214–223.

CELLULAR AND MOLECULAR MECHANISMS UNDERLYING NECROTIC CELL DEATH

Metastable State Preceding Necrotic Cell Death

Cellular events culminating in necrotic cell death are somewhat variable from one cell type to another, but certain events occur regularly. As implied by the term oncosis, cellular swelling is a prominent feature of oncotic necrosis. In many cell types, swelling of 30–50% occurs early after ATP depletion associated with formation of blebs on the cell surface. These blebs contain cytosol and ER but exclude larger organelles like mitochondria and lysosomes. Mitochondrial swelling and dilatation of cisternae of ER and nuclear membranes accompany bleb formation (see Figure 1.1). After longer times, a metastable state develops, which is characterized by mitochondrial depolarization, lysosomal breakdown, ion dysregulation, and accelerated bleb formation with more rapid swelling. The metastable state lasts only a few minutes and culminates in rupture of a plasma membrane bleb. Bleb rupture leads to loss of metabolic intermediates such as those that reduce tetrazolium dyes, leakage of cytosolic enzymes like lactate dehydrogenase, uptake of dyes like trypan blue, and collapse of all electrical and ion gradients across the membrane.

Bleb rupture is the final irreversible event precipitating cell death, since removal of the instigating stress (e.g., reoxygenation of anoxic cells) leads to cell recovery prior to bleb rupture but not afterwards.

Mitochondrial Dysfunction and ATP Depletion

Ischemia as occurs in strokes and heart attacks is perhaps the most common cause of necrotic cell killing. In ischemia, oxygen deprivation prevents ATP formation by mitochondrial oxidative phosphorylation,

5

a process providing up to 95% of ATP utilized by highly aerobic tissues. As an alternative source of ATP, glycolysis partially replaces ATP production lost after mitochondrial dysfunction. Maintenance of as little as 15% or 20% of normal ATP then rescues cells from necrotic death. Glycolysis also protects against toxicity from oxidant chemicals, suggesting that mitochondria are also a primary target of cytotoxicity in oxidative stress. However, in pathological settings like ischemia, glycolytic substrates are rapidly exhausted.

Mitochondrial Uncoupling in Necrotic Cell Killing

Mitochondrial injury and dysfunction are progressive (Figure 1.4). Anoxia and inhibition with a toxicant like cyanide inhibit respiration to cause ATP depletion and ultimately necrotic cell death. Glycolysis can replace this ATP supply, although only partially in highly aerobic cells, to rescue cells from necrotic killing. However, when mitochondrial injury progresses to uncoupling (inner membrane permeability to hydrogen ions), accelerated ATP hydrolysis occurs that is catalyzed by the mitochondrial ATP synthase working in reverse. Since glycolytic ATP production cannot keep pace, ATP levels fall profoundly and necrotic cell death ensues. In the progression from respiratory inhibition to uncoupling, mitochondria become active agents promoting ATP depletion and cell death.

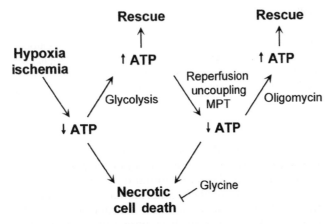

Figure 1.4 Progression of mitochondrial injury. Respiratory inhibition inhibits oxidative phosphorylation and leads to ATP depletion and necrotic cell death. Glycine blocks plasma membrane permeabilization causing necrotic cell death downstream of ATP depletion. Glycolysis restores ATP and prevents cell killing. Mitochondrial uncoupling as occurs after reperfusion due to the mitochondrial permeability transition (MPT) activates the mitochondrial ATPase to futilely hydrolyze glycolytic ATP, and protection against necrotic cell death is lost. By inhibiting the mitochondrial ATPase, oligomycin prevents ATP depletion and rescues cells from necrotic cell death if glycolytic substrate is present. With permission from Lemasters JJ, Qian T, He L, et al. Role of mitochondrial inner membrane permeabilization in necrotic cell death, apoptosis, and autophagy. *Antioxid Redox Signal*. 2002;4(5):769–781.

Mitochondrial Permeability Transition

In oxidative phosphorylation, respiration drives translocation of hydrogen ions out of mitochondria to create an electrochemical gradient composed of a negative inside membrane potential ($\Delta\Psi$) and an alkaline inside pH gradient (ΔpH). ATP synthesis is then linked to hydrogen ions returning down this electrochemical gradient through the mitochondrial ATP synthase. This chemiosmotic proton circuit requires the mitochondrial inner membrane to be impermeable to ions and charged metabolites.

In some pathophysiological settings, the mitochondrial inner membrane abruptly becomes nonselectively permeable to solutes of molecular weight up to about 1500 Da. Ca^{2+}, oxidative stress, and numerous reactive chemicals induce this mitochondrial permeability transition (MPT) whereas cyclosporin A and pH less than 7 inhibit it. The MPT causes mitochondrial depolarization, uncoupling, and large amplitude mitochondrial swelling driven by colloid osmotic forces. Opening of highly conductive permeability transition (PT) pores in the mitochondrial inner membrane underlies the MPT. Conductance through PT pores is so great that opening of a single PT pore may be sufficient to cause mitochondrial depolarization and swelling.

The composition of PT pores remains uncertain. In one model, PT pores are formed by ANT from the inner membrane, VDAC from the outer membrane, the cyclosporin A binding protein cyclophilin D (CypD) from the matrix, and possibly other proteins. Although once widely accepted, the validity of this model has been challenged by genetic knockout studies showing that the MPT still occurs in mitochondria that are deficient in ANT, VDAC, and CypD. An alternative model for the PT pore is that oxidative and other stresses damage membrane proteins that then misfold and aggregate to form PT pores in association with CypD and other molecular chaperones.

pH-dependent ischemia/reperfusion injury

Ischemia is an interruption of blood flow and hence oxygen supply to a tissue or organ. In ischemic tissue, anaerobic metabolism causes tissue pH to decrease by a unit or more. This naturally occurring acidosis of ischemia actually protects against onset of necrotic cell death.

Much of reperfusion injury is attributable to recovery of pH, since reoxygenation at low pH prevents cell killing entirely, whereas restoration of normal pH without reoxygenation produces similar cell killing as restoration of pH with reoxygenation, a so-called pH paradox. Cell killing in the pH paradox is linked specifically to intracellular pH and occurs independently of changes of cytosolic and extracellular free Na^+ and Ca^{2+}.

Role of the mitochondrial permeability transition in pH-dependent reperfusion injury

pH below 7 inhibits PT pores during ischemia. After reperfusion at normal pH, mitochondria repolarize initially. Subsequently and in parallel with recovery of

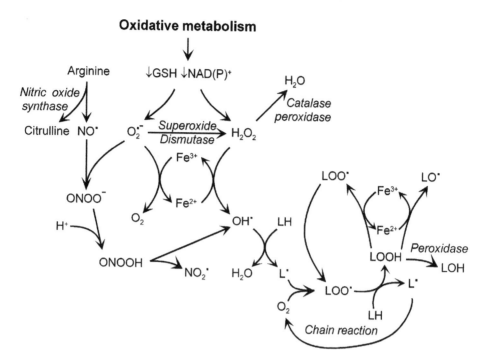

Figure 1.5 Iron-catalyzed free radical generation. Oxidative stress causes oxidation of GSH and NAD(P)H, important reductants in antioxidant defenses, promoting increased net formation of superoxide ($O_2^{\cdot-}$) and hydrogen peroxide (H_2O_2). Superoxide dismutase converts superoxide to hydrogen peroxide, which is further detoxified to water by catalase and peroxidases. In the iron-catalyzed Haber-Weiss reaction (or Fenton reaction), superoxide reduces ferric iron (Fe^{3+}) to ferrous iron (Fe^{2+}), which reacts with hydrogen peroxide to form the highly reactive hydroxyl radical ($OH^{\cdot}$). Hydroxyl radical reacts with lipids to form alkyl radicals ($L^{\cdot}$) that initiate an oxygen-dependent chain reaction which generates peroxyl radicals ($LOO^{\cdot}$) and lipid peroxides ($LOOH$). Iron also catalyzes a chain reaction generating alkoxyl radicals ($LO^{\cdot}$) and more peroxyl radicals. Nitric oxide synthase catalyzes formation of nitric oxide ($NO^{\cdot}$) from arginine. Nitric oxide reacts rapidly with superoxide to form unstable peroxynitrite anion ($ONOO^{-}$), which decomposes to nitrogen dioxide and hydroxyl radical. In addition to attacking lipids, these radicals also attack proteins and nucleic acids.

intracellular pH to neutrality, the MPT occurs. ATP depletion then follows, and necrotic cell death occurs. Reperfusion in the presence of PT pore blockers (e.g., cyclosporin A and its derivatives) prevents mitochondrial inner membrane permeabilization, depolarization, and cell killing. Notably, cyclosporin A protects when added only during the reperfusion phase, as now confirmed by decreased infarct size in patients receiving percutaneous coronary intervention (PCI) for ischemic heart disease. Thus, the MPT is the proximate cause of pH-dependent cell killing in ischemia/reperfusion injury.

Oxidative stress

Reactive oxygen species (ROS) and reactive nitrogen species (RNS), including superoxide, hydrogen peroxide, hydroxyl radical, and peroxynitrite, have long been implicated in cell injury leading to necrosis (Figure 1.5). Reperfusion after ischemia stimulates intramitochondrial ROS formation, onset of the MPT, and cell death. In neurons, excitotoxic stress with glutamate and N-methyl-D-aspartate (NMDA) receptor agonists also stimulates mitochondrial ROS formation, leading to the MPT and excitotoxic injury.

Iron potentiates injury in a variety of diseases and is an important catalyst for hydroxyl radical formation from superoxide and hydrogen peroxide (Figure 1.5).

During oxidative stress and hypoxia/ischemia, lysosomes rupture release chelatable (loosely bound) iron with consequent pro-oxidant cell damage. This iron is taken up into mitochondria by the mitochondrial calcium uniporter and helps catalyze mitochondrial ROS generation. Iron chelation with Desferal prevents this ROS formation and decreases cell death in oxidative stress and hypoxia/ischemia.

Other Stress Mechanisms Inducing Necrotic Cell Death

Poly (ADP-Ribose) Polymerase

Single strand breaks induced by ultraviolet (UV) light, ionizing radiation, and ROS (particularly hydroxyl radical and peroxynitrite) activate PARP. With excess DNA damage, PARP transfers ADP-ribose from NAD^+ to the strand breaks and elongates ADP-ribose polymers attached to the DNA. Consumption of the oxidized form of nicotinamide adenine dinucleotide (NAD^+) in this fashion leads to NAD^+ depletion, disruption of ATP-generation by glycolysis and oxidative phosphorylation, and ATP depletion-dependent cell death.

PARP-dependent necrosis is an example of programmed necrosis since PARP actively promotes a cell death-inducing pathway that otherwise would not occur. Necrotic cell death also frequently occurs when apoptosis is interrupted, as by caspase (cysteine-aspartate protease) inhibition. Such caspase independent cell death is the consequence of mitochondrial dysfunction or other metabolic disturbance.

Plasma membrane injury

An intact plasma membrane is essential for cell viability. Detergents and pore-forming agents like masto-paran from wasp venom defeat the barrier function of the plasma membrane and cause immediate cell death. Immune-mediated cell killing can act similarly. In particular, complement mediates formation of a membrane attack complex that in conjunction with antibody lyses cells. Complement component 9, an amphipathic molecule, inserts through the cell membrane, polymerizes, and forms a tubular channel visible by electron microscopy. Indeed, a single membrane attack complex may be sufficient to cause swelling and lysis of an individual erythrocyte.

PATHWAYS TO APOPTOSIS
Roles of Apoptosis in Biology

Apoptosis is an essential event in both the normal life of organisms and in pathobiology. In development, apoptosis sculpts and remodels tissues and organs, for example, by creating clefts in limb buds to form fingers and toes. Apoptosis is also responsible for reversion of hypertrophy to atrophy and immune surveillance-induced killing of preneoplastic cells and virally infected cells. Each of several organelles can give rise to signals initiating apoptotic cell killing. Often these signals converge on mitochondria as a common pathway to apoptotic cell death. In most apoptotic signaling, activation of caspases 3 or 7 from a family of caspases (Table 1.2) begins execution of the final and committed

Table 1.2   **Mammalian caspases** Caspases are evolutionarily conserved aspartate specific cysteine-dependent proteases that function in apoptotic and inflammatory signaling. Initiator caspases are involved in the initiation and propagation of apoptotic signaling, whereas effector caspases act on a wide variety of proteolytic substrates to induce the final and committed phase of apoptosis. Initiator and inflammatory caspases have large prodomains containing oligomerization motifs such as the caspase recruitment domain (CARD) and the DED. Effector caspases have short prodomains and are proteolytically activated by large prodomain caspases and other proteases. Proteolytic cleavage of procaspase precursors forms separate large and small subunits that assemble into active enzymes consisting of two large and two small subunits. Caspase activation occurs in multimeric complexes that typically consist of a platform protein that recruits procaspases either directly or by means of adaptors. Such caspase complexes include the apoptosome and the death-inducing signaling complex (DISC). Caspase 14 plays a role in terminal keratinocyte differentiation in cornified epithelium.

Initiator Caspases	Molecular Weight of Proenzyme (kDa)	Active subunits (kDa)	Prodomain	Amino Acid Target Sequence for Proteolysis
Caspase 2	51	19/12	Long with CARD	VDVAD
Caspase 8	55	18/11	Long with two DED	(L/V/D)E(T/V/I)D
Caspase 9	45	17/10	Long with CARD	(L/V/I)EHD
Caspase 10	55	17/12	Long with two DED	(I/V/L)EXD
Caspase 12	50	20/10	Long with CARD	ATAD
Effector Caspases				
Caspase 3	32	17/12	Short	DE(V/I)D
Caspase 6	34	18/11	Short	(T/V/I)E(H/V/I)D
Caspase 7	35	20/12	Short	DE(V/I)D
Inflammatory Caspases				
Caspase 1	45	20/10	Long with CARD	(W/Y/F)EHD
Caspase 4	43	20/10	Long with CARD	(W/L)EHD
Caspase 5	48	20/10	Long with CARD	(W/L/F)EHD
Caspase 11	42	20/10	Long with CARD	(V/I/P/L)EHD
Other Caspases				
Caspase 14	42	20/10	Short	(W/I)E(T/H)D

phase of apoptotic cell death. Caspase 3/7 has many targets. Degradation of the nuclear lamina and cytokeratins contributes to nuclear remodeling, chromatin condensation, and cell rounding. Degradation of endonuclease inhibitors activates endonucleases to cause internucleosomal DNA cleavage. The resulting DNA fragments have lengths in multiples of 190 base pairs, the nucleosome to nucleosome repeat distance. Additionally, caspase activation leads to cell shrinkage, phosphatidyl serine externalization on the plasma membrane, and formation of numerous small surface blebs (zeiosis). Unlike necrotic blebs, these zeiotic blebs contain membranous organelles and are shed as apoptotic bodies. However, not all apoptotic changes

depend on caspase 3/7 activation. For example, release of apoptosis-inducing factor (AIF) from mitochondria and its translocation to the nucleus promotes DNA degradation in a caspase 3-independent fashion.

Pathways leading to activation of caspase 3 and related effector caspases like caspase 7 are complex and quite variable between cells and specific apoptosis-instigating stimuli, and each major cellular structure can originate its own set of unique signals to induce apoptosis (Figure 1.6). Proapoptotic signals are often associated with specific damage or perturbation to the organelle involved. Consequently, cells choose death by apoptosis rather than life with organelle damage.

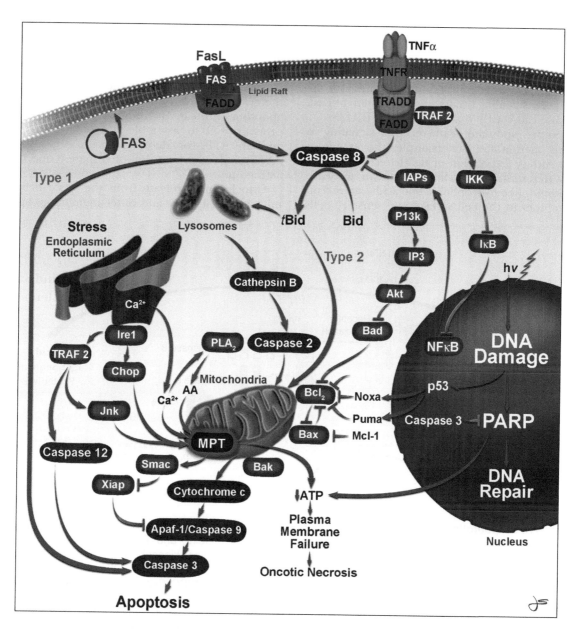

Figure 1.6 Scheme of apoptotic signaling from organelles. Adapted with permission from Lemasters JJ. Dying a thousand deaths: Redundant pathways from different organelles to apoptosis and necrosis. *Gastroenterology.* 2005; 129(1):351–360.

Plasma Membrane

The plasma membrane is the target of many receptor-mediated signals. In particular, death ligands (e.g., tumor necrosis factor α, or TNFα; Fas ligand; tumor necrosis factor-related apoptosis-inducing ligand, or TRAIL) acting through their corresponding receptors (TNF receptor 1, or TNFR1; Fas; death receptor 4/5, or DR4/5) initiate activation of apoptotic pathways. For example, binding of TNFα to TNFR1 leads to receptor trimerization and formation of a complex (Complex I) through association of adapter proteins (e.g., receptor interacting protein-1, or RIP1, and TNF receptor-associated death domain protein, or TRADD). Subsequently Complex II, or death-inducing signaling complex (DISC), forms through association with Fas-associated protein with death domain (FADD) and pro-caspase 8, which are internalized. Pro-caspase 8 becomes activated and in turn proteolytically activates other downstream effectors (Figure 1.7). In Type I signaling, caspase 8 activates caspase 3 directly, whereas in Type II signaling, caspase 3 cleaves Bid (novel BH3 domain-only death agonist) to truncated Bid (tBid) to activate a mitochondrial pathway to apoptosis. Similar signaling occurs after association of FasL with Fas (also called CD95) and TRAIL with DR4/5.

Many events modulate death receptor signaling in the plasma membrane. For example, the extent of gene and surface expression of death receptors is an important determinant in cellular sensitivity to death ligands. Stimuli like hydrophobic bile acids can recruit death receptors to the cell surface and sensitize cells to death-inducing stimuli. Surface recruitment of death receptors may also lead to self-activation even in the absence of ligand. Death receptors localize to lipid rafts containing cholesterol and sphingomyelin. After death receptor activation, sphingomyelin hydrolysis occurs, which promotes raft coalescence and formation of molecular platforms that cluster signal transducer components of DISC. Glycosphingolipids, such as ganglioside GD3, also integrate into DISCs to promote apoptosis.

MITOCHONDRIA

Cytochrome *c* release

Bid is a Bcl2 homology 3 (BH3) only domain member of the B-cell lymphoma-2 (Bcl2) family that includes both pro- and antiapoptotic proteins (Figure 1.8). tBid formed after caspase 8 activation translocates to mitochondria where it interacts with either Bak (Bcl2 homologous antagonist/killer) or Bax (a conserved homolog that heterodimerizes with Bcl2), two other proapoptotic Bcl2 family members, to induce cytochrome *c* release through the outer membrane into the cytosol. Cytochrome *c* in the cytosol interacts with apoptotic protease activating factor-1 (Apaf-1) and procaspase 9 to assemble haptomeric apoptosomes and an ATP (or deoxyadenosine triphosphate, or dATP)-dependent cascade of caspase 9 and caspase 3 activation.

Cytochrome *c* release from the space between the mitochondrial inner and outer membranes appears to

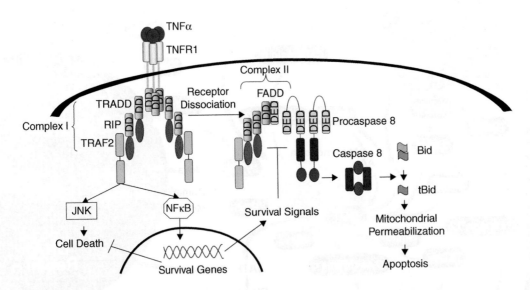

Figure 1.7 TNFα apoptotic signaling. TNFα binds to its receptor, TNFR1, and Complex I forms composed of TRADD (TNFR-associated protein with death domain), RIP (receptor-interacting protein), and TRAF-2 (TNF-associated factor-2). Complex I activates NFκB (nuclear factor kappa B) and JNK (c-jun N-terminal kinase). NFκB activates transcription of survival genes, including antiapoptotic inhibitor of apoptosis proteins (IAPs), antiapoptotic Bcl-XL, and inducible nitric oxide synthase. Complex I then undergoes ligand-dissociated internalization to form DISC Complex II. Complex II recruits FADD (Fas-associated death domain) via interactions between conserved death domains (DD) and activates procaspase 8 through interaction with death effector domains (DED). Active caspase 8 cleaves Bid to tBid, which translocates to mitochondria leading to mitochondrial permeabilization, cytochrome *c* release, and apoptosis. Adapted with permission from Malhi H, Gores GJ, Lemasters JJ. Apoptosis and necrosis in the liver: A tale of two deaths? *Hepatology*. 2006;43(2 Suppl 1):S31–S44.

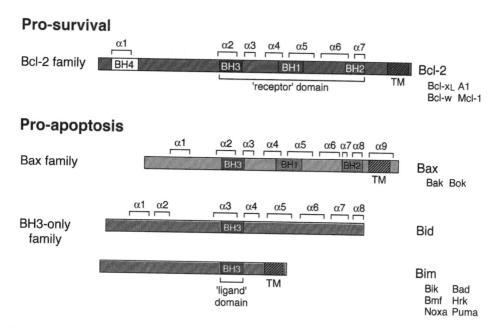

Figure 1.8   Bcl2 family proteins. BH1–4 are highly conserved domains among the Bcl2 family members. Also shown are α-helical regions. Except for A1 and BH3 only proteins, Bcl2 family members have carboxy-terminal hydrophobic domains to aid association with intracellular membranes. Reproduced with permission from Cory S, Adams JM. The Bcl2 family: Regulators of the cellular life-or-death switch. *Nat Rev Cancer.* 2002;2(9):647–656.

occur via formation of specific pores in the mitochondrial outer membrane. Except for the requirement for either Bak or Bax, the molecular composition and properties of cytochrome *c* release channels remain poorly understood. Alternatively, cytochrome *c* release can occur as a consequence of the MPT due to large amplitude mitochondrial swelling and rupture of the outer membrane.

After the MPT, progression to apoptosis or necrosis depends on other factors. If the MPT occurs rapidly and affects most mitochondria of a cell, as happens after severe oxidative stress and ischemia/reperfusion, a precipitous fall of ATP (and dATP) will occur that actually blocks apoptotic signaling by inhibiting ATP-requiring caspase 9/3 activation. With ATP depletion, oncotic necrosis ensues. However, when alternative sources for ATP generation are present (e.g., glycolysis), then necrosis is prevented and caspase 9/3 becomes activated and caspase-dependent apoptosis occurs instead (Figure 1.9). Crosstalk between apoptosis and necrosis also occurs in other ways. For example, after TNFα binding to its receptor, recruitment of RIP1 to TNFR1 can activate NADPH oxidase leading to superoxide generation, resulting in oncotic necrosis rather than apoptosis.

Regulation of the Mitochondrial Pathway to Apoptosis

Mitochondrial pathways to apoptosis vary depending on expression of procaspases, Apaf-1, and other proteins. Some neurons do not respond to cytochrome *c* with caspase activation and apoptosis, which may be linked to lack of Apaf-1 expression. Antiapoptotic Bcl2

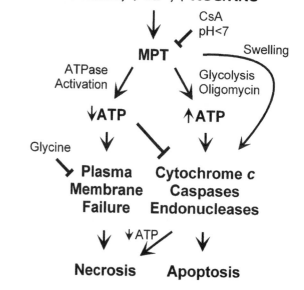

Figure 1.9   Shared pathways to apoptosis and necrosis.

proteins, like Bcl2, Bcl extra long (Bcl-xL), and myeloid cell leukemia sequence 1 (Mcl-1), block apoptosis and are frequently overexpressed in cancer cells (Figure 1.8). Antiapoptotic Bcl2 family members form heterodimers with proapoptotic family members like Bax and Bak, to prevent the latter from oligomerizing into cytochrome *c* release channels.

Inhibitors of apoptosis proteins (IAPs), including X-linked inhibitor of apoptosis protein (XIAP), cellular IAP1 (c-IAP1), cellular IAP2 (c-IAP2), and survivin, oppose apoptotic signaling by inhibiting caspase activation. Many IAPs can recruit E2 ubiquitin-conjugating enzymes and catalyse the transfer of ubiquitin onto target proteins, leading to proteosomal degradation. Some IAPs inhibit apoptotic pathways upstream of mitochondria at caspase 8, whereas others like XIAP inhibit caspase 9/3 activation downstream of mitochondrial cytochrome c release. Additional proteins like Smac suppress the action of IAPs, providing an "inhibitor of the inhibitor" effect promoting apoptosis. Smac is a mitochondrial intermembrane protein that is released with cytochrome c. Smac inhibits XIAP and promotes apoptotic signaling after mitochondrial signaling. Thus, high Smac to XIAP ratios favor caspase 3 activation after cytochrome c release. Other proapoptotic proteins released from the mitochondrial intermembrane space during apoptotic signaling include AIF (a flavoprotein oxidoreductase that promotes DNA degradation and chromatin condensation), endonuclease G (a DNA degrading enzyme), and HtrA2/Omi (a serine protease that degrades IAPs). Early in apoptosis, fragmentation of larger filamentous mitochondria into smaller more spherical structures typically occurs. Such fission seems to promote apoptotic signaling.

Antiapoptotic Survival Pathways

Ligand binding to death receptors can also activate antiapoptotic signaling to prevent activation of apoptotic death programs. Binding of the adapter protein, TNFR-associated factor 2 (TRAF2), to death receptors activates IκB kinase (IKK), which in turn phosphorylates IκB, an endogenous inhibitor of nuclear factor κB (NFκB), leading to proteosomal IκB degradation. IκB degradation relieves inhibition of NFκB and allows NFκB to activate expression of anti-apoptotic genes, including IAPs, Bcl-xL, inducible nitric oxide synthase (iNOS), and other survival factors. Nitric oxide from iNOS produces cGMP-dependent suppression of the MPT, as well as S-nitrosation and inhibition of caspases. In many models, apoptosis after death receptor ligation occurs only when NFκB ignaling is blocked, as after inhibition of proteosomes or protein synthesis.

The phosphoinositide-3-kinase (PI3) kinase/protooncogene product of the viral oncogene v-akt (Akt) pathway is another source of antiapoptotic signaling. When phosphoinositide 3-kinase (PI3 kinase) is activated by binding of insulin, insulin-like growth factor (IGF), and various other growth factors to their receptors, phosphatidylinositol trisphosphate (PIP3) is formed that activates Akt/protein kinase B, a serine/threonine protein kinase. One consequence is the phosphorylation and inactivation of Bad (heterodimeric partner for Bcl-xL), a proapoptotic Bcl2 family member, but other antiapoptotic targets of PI3 kinase/Akt signaling also exist. In cell lines, withdrawal of serum or specific growth factors typically induces apoptosis due to suppression of the PI3 kinase/Akt survival pathway.

NUCLEUS

In the so-called extrinsic pathway, death receptors initiate apoptosis by either a Type I (nonmitochondrial) or Type II (mitochondrial) caspase activation sequence. In the intrinsic pathway, by contrast, events in the nucleus activate apoptotic signaling, such as DNA damage caused by ultraviolet or ionizing (gamma) irradiation. DNA damage leads to activation of the p53 nuclear transcription factor and expression of genes for apoptosis and/or cell-cycle arrest, especially the proapoptotic Bcl2 family members PUMA, NOXA, and Bax for apoptosis, and p21 for cell-cycle arrest, especially the proapoptotic Bcl2 family members p53 upregulated modulator of apoptosis (PUMA), NOXA and Bax for apoptosis, and 21 kDa promoter (p21) for cell-cycle arrest (Figures 1.6 and 1.8). PUMA, NOXA, and Bax translocate to mitochondria to induce cytochrome c release by similar mechanisms as discussed previously for the extrinsic pathway. To escape p53-dependent induction of apoptosis, many tumors, especially those from the gastrointestinal tract, have loss of function mutations for p53.

DNA damage also activates PARP. With moderate activation, PARP helps mend DNA strand breaks, but with strong activation PARP depletes NAD^+ and compromises ATP generation to induce necrotic cell death. Caspase 3 proteolytically degrades PARP to prevent this pathway to necrosis. Thus, DNA damage can lead to either necrosis or apoptosis depending on which occurs more quickly—PARP activation and ATP depletion, or caspase 3 activation and PARP degradation.

ENDOPLASMIC RETICULUM

The ER also gives rise to proapoptotic signals. Oxidative stress and other perturbations can inhibit ER calcium pumps to induce calcium release into the cytosol. Uptake of this calcium into mitochondria from the cytosol may then induce a Ca^{2+}-dependent MPT and subsequent apoptotic or necrotic cell killing (Figure 1.6). ER calcium release into the cytosol can also activate phospholipase A2 and the formation of arachidonic acid, another promoter of the MPT.

ER calcium depletion also disturbs the proper folding of newly synthesized proteins inside ER cisternae to cause ER stress and the unfolded protein response (UPR). Blockers of glycosylation, inhibitors of ER protein processing and secretion, various toxicants, and synthesis of mutant proteins can also cause ER stress. Calcium-binding chaperones, including glucose-regulated protein-78 (GRP78) and glucose-regulated protein-94 (GRP94), mediate detection of unfolded and misfolded proteins. In the absence of unfolded proteins, GRP78 inhibits specific sensors of ER stress, but in the presence of unfolded proteins GRP78 translocates from the sensors to the unfolded proteins to cause sensor activation by disinhibition. The main sensors of ER stress are RNA-activated protein kinase (PKR), PKR-like ER kinase (PERK), type 1 ER transmembrane protein kinase (IRE1), and activating transcription factor 6 (ATF6). PKR and PERK are protein kinases whose activation leads to phosphorylation of eukaryotic initiation factor-2α (eIF-2α). Phosphorylation of eIF-2α suppresses ER protein synthesis, a negative feedback that can relieve the unfolding stress. IRE1 is both a protein kinase and a riboendonuclease that initiates splicing of a preformed mRNA encoding X-box-binding protein 1 (XBP) into an active form. ATF6 is another transcription factor that translocates to the Golgi after ER stress where proteases process ATF6 to an amino-terminal fragment that is taken up into the nucleus. Together IRE1 and ATF6 increase gene expression of chaperones and other proteins to alleviate the unfolding stress.

A strong and persistent UPR induces IRE1- and ATF6-dependent expression of C/EBP homologous protein (CHOP) and continued activation of IRE1 to initiate apoptotic signaling (Figure 1.6). Association of TRAF2 with activated IRE1 leads to activation of caspase 12 and JNK. Caspase 12 activates caspase 3 directly, whereas JNK and CHOP promote mitochondrial cytochrome c release as a pathway to caspase 3 activation.

LYSOSOMES

Lysosomes and the associated process of autophagy (self-digestion) are another source of proapoptotic signals. So-called autophagic cell death is characterized by an abundance of autophagic vacuoles in dying cells and is especially prominent in involuting tissues, such as post-lactation mammary gland. In autophagy, isolation membranes (also called phagophores) envelop and sequester portions of cytoplasm to form double membrane autophagosomes. Autophagosomes fuse with lysosomes and late endosomes to form autolysosomes. The process of autophagy acts to remove and degrade cellular constituents, an appropriate action for a tissue undergoing involution. Originally considered to be random, much evidence suggests that autophagy can be selective for specific organelles, especially if they are damaged. For example, stresses inducing the MPT seem to signal autophagy of mitochondria.

Whether or not autophagy promotes or prevents cell death is controversial. In some circumstances, suppression of expression of certain autophagy genes decreases apoptosis. Under other conditions, autophagy protects against cell death. When autophagic processing and lysosomal degradation are disrupted, cathepsins and other hydrolases can be released from autolysosomes to initiate mitochondrial permeabilization and caspase activation. Cathepsin B is released from lysosomes (or related structures such as late endosomes) during TNFα signaling to augment death receptor-mediated apoptosis and contribute to mitochondrial release of cytochrome c. In addition, lysosomal extracts cleave Bid to tBid, and cathepsin D, another lysosomal protease, activates Bax.

Necrapoptosis/Aponecrosis

In many and possibly most instances of apoptosis, mitochondrial permeabilization with release of cytochrome c is a final common pathway leading to a final and committed phase. At higher levels of stimulation, the same factors that induce apoptosis frequently also cause ATP depletion and a necrotic mode of cell death. Such necrotic cell killing is a consequence of mitochondrial dysfunction. Such shared pathways leading to different modes of cell death constitute necrapoptosis (or aponecrosis). In general, apoptosis is a better outcome for the organism since apoptosis promotes orderly resorption of dying cells, whereas necrotic cell death releases cellular constituents into the extracellular space to induce an inflammatory release that can extend tissue injury. Thus, an admixture of necrosis and apoptosis occurs in many pathophysiological settings.

CONCLUDING REMARK

Apoptosis and necrosis are prominent events in pathogenesis. An understanding of cell death mechanisms forms the basis for effective interventions to either prevent cell death as a cause of disease or promote cell death in cancer chemotherapy.

ACKNOWLEDGMENTS

This work was supported, in part, by Grants DK37034, DK73336, DK70844, DK070195, and AA016011 from the National Institutes of Health.

KEY CONCEPTS

- A common theme in disease is the life and death of cells. In diseases like stroke and heart attacks, death of individual cells leads to irreversible functional loss, whereas in cancer the goal of chemotherapy is to kill proliferating tumor cells. The mode of cell death typically follows one of two patterns: necrosis and apoptosis.
- Necrosis is the consequence of metabolic disruption with ATP depletion and is characterized by cellular swelling leading to plasma membrane rupture with release of intracellular contents. Apoptosis is a form of programmed cell death that causes orderly resorption of individual cells initiated by well-defined ATP-requiring pathways involving activation of proteases called caspases.
- In some pathophysiological settings, the mitochondrial inner membrane abruptly becomes permeable to solutes up to 1500 Da. This mitochondrial permeability transition causes uncoupling of oxidative phosphorylation, ATP depletion, mitochondrial swelling, and cytochrome c release that can lead to both necrosis and apoptosis.

- Each of several organelles gives rise to signals initiating apoptotic cell killing. Often these signals converge on mitochondria to cause cytochrome *c* release and Apaf-1-dependent caspase 9 and 3 activation as a final common pathway to apoptotic cell death.
- Death ligands like TNFα and Fas ligand activate their corresponding receptors in the plasma membrane to initiate caspase signaling cascades and the mitochondrial pathway to cell death. Inhibitor of apoptosis proteins (IAPs) oppose apoptotic signaling by inhibiting caspase activation.
- DNA damage activates p53, a nuclear transcription factor, and expression of proapoptotic Bcl2 family members like PUMA, NOXA, and Bax that translocate to mitochondria to induce cytochrome *c* release. Many tumors have loss of function mutations for p53 to escape p53-dependent apoptosis.
- Accumulation of unfolded/misfolded proteins in the ER causes ER stress. Initially, ER stress increases expression of molecular chaperones with inhibition of other protein synthesis to alleviate the unfolding stress. With prolonged ER stress, apoptotic pathways are activated. Lysosomes and the associated process of autophagy (self-digestion) are yet another source of proapoptotic signals. Some consider autophagic cell death as a separate category of programmed cell death.
- Apoptosis and necrosis can share common signaling pathways to be extreme end points on a phenotypic continuum of necrapoptosis or aponecrosis.

SUGGESTED READINGS

1. Colin J, Gaumer S, Guenal I, et al. Mitochondria, Bcl-2 family proteins and apoptosomes: Of worms, flies and men. *Front Biosci.* 2009;14:4127–4137.
2. Halestrap AP. What is the mitochondrial permeability transition pore? *J Mol Cell Cardiol.* 2009;46(6):821–831.
3. Kerr JF, Wyllie AH, Currie AR. Apoptosis: A basic biological phenomenon with wide-ranging implications in tissue kinetics. *Br J Cancer.* 1972;26(4):239–257.
4. Kim I, Xu W, Reed JC. Cell death and endoplasmic reticulum stress: Disease relevance and therapeutic opportunities. *Nat Rev Drug Discov.* 2008;7(12):1013–1030.
5. Krammer PH, Arnold R, Lavrik IN. Life and death in peripheral T cells. *Nat Rev Immunol.* 2007;7(7):532–542.
6. Kurz T, Terman A, Brunk UT. Autophagy, ageing and apoptosis: The role of oxidative stress and lysosomal iron. *Arch Biochem Biophys.* 2007;462(2):220–230.
7. Lamkanfi M, Festjens N, Declercq W, et al. Caspases in cell survival, proliferation and differentiation. *Cell Death Differ.* 2007;14(1): 44–55.
8. Lemasters JJ. Dying a thousand deaths: Redundant pathways from different organelles to apoptosis and necrosis. *Gastroenterology.* 2005;129(1):351–360.
9. Majno G, Joris I. Apoptosis, oncosis, and necrosis. An overview of cell death. *Am J Pathol.* 1995;146(1):3–15.
10. Malhi H, Gores GJ. Cellular and molecular mechanisms of liver injury. *Gastroenterology.* 2008;134(6):1641–1654.
11. Miyamoto S, Murphy AN, Brown JH. Akt mediated mitochondrial protection in the heart: Metabolic and survival pathways to the rescue. *J Bioenerg Biomembr.* 2009;41(2):169–180.
12. Mizushima N, Levine B, Cuervo AM, et al. Autophagy fights disease through cellular self-digestion. *Nature.* 2008;451(7182): 1069–1075.
13. Ow YL, Green DR, Hao Z, et al. Cytochrome c: Functions beyond respiration. *Nat Rev Mol Cell Biol.* 2008;9(7):532–542.
14. Piot C, Croisille P, Staat P, et al. Effect of cyclosporine on reperfusion injury in acute myocardial infarction. *N Engl J Med.* 2008;359(5):473–481.
15. Riedl SJ, Salvesen GS. The apoptosome: Signalling platform of cell death. *Nat Rev Mol Cell Biol.* 2007;8(5):405–413.
16. Song G, Ouyang G, Bao S. The activation of Akt/PKB signaling pathway and cell survival. *J Cell Mol Med.* 2005;9(1):59–71.
17. Sun B, Karin M. NF-kappaB signaling, liver disease and hepatoprotective agents. *Oncogene.* 2008;27(48):6228–6244.
18. Uchiyama A, Kim JS, Kon K, et al. Translocation of iron from lysosomes into mitochondria is a key event during oxidative stress-induced hepatocellular injury. *Hepatology.* 2008;48(5): 1644–1654.
19. Vaux DL, Silke J. IAPs, RINGs and ubiquitylation. *Nat Rev Mol Cell Biol.* 2005;6(4):287–297.
20. Yip KW, Reed JC. Bcl-2 family proteins and cancer. *Oncogene.* 2008;27(50):6398–6406.

2

Acute and Chronic Inflammation Induces Disease Pathogenesis

Vladislav Dolgachev . Nicholas W. Lukacs

INTRODUCTION

The recognition of pathogenic insults can be accomplished by a number of mechanisms that function to initiate inflammatory responses and mediate clearance of invading pathogens. This initial response when functioning optimally will lead to a minimal leukocyte accumulation and activation for the clearance of the inciting agent and have little effect on homeostatic function. However, often the inciting agent elicits a very strong inflammatory response, either due to host recognition systems or due to the agent's ability to damage host tissue. Thus, the host innate immune system mediates the damage and tissue destruction in an attempt to clear the inciting agent from the system. No matter, these initial acute responses can have long-term and even irreversible effects on tissue function. If the initial responses are not sufficient to facilitate the clearance of the foreign pathogen or material, the response shifts toward a more complex and efficient process mediated by lymphocyte populations that respond to specific residues displayed by the foreign material. Normally, these responses are coordinated and only minimally alter physiological function of the tissue. However, in unregulated responses the initial reaction can become acutely catastrophic, leading to local or even systemic damage to the tissue or organs, resulting in degradation of normal physiological function. Alternatively, the failure to regulate the response or clear the inciting agent could lead to chronic and progressively more pathogenic responses. Each of these potentially devastating responses has specific and often overlapping mechanisms that have been identified and lead to the damage within tissue spaces. A series of events take place during both acute and chronic inflammation that lead to the accumulation of leukocytes and damage to the local environment.

LEUKOCYTE ADHESION, MIGRATION, AND ACTIVATION

Endothelial Cell Expression of Adhesion Molecules

The initial phase of the inflammatory response is characterized by a rapid leukocyte migration into the affected tissue. Upon activation of the endothelium by inflammatory mediators, upregulation of a series of adhesion molecules is initiated that leads to the reversible binding of leukocytes to the activated endothelium. The initial adhesion is mediated by E and P selectins that facilitate slowing of leukocytes from circulatory flow by mediating rolling of the leukocytes on the activated endothelium. The selectin-mediated interaction with the activated endothelium potentiates the likelihood of the leukocyte to be further activated by endothelial-expressed chemokines, which mediate G-protein-coupled receptor (GPCR)-induced activation. If the rolling leukocytes encounter a chemokine signal and an additional set of adhesion molecules is also expressed, such as intracellular adhesion molecule-1 (ICAM-1) and vascular cell adhesion molecule-1 (VCAM-1), the leukocytes firmly adhere to the activated endothelium. The mechanism of chemokine-induced adhesion of the leukocyte is dependent on actin reorganization and a confirmational change of the β-integrins on the surface of the leukocytes. Subsequently, the firm adhesion allows leukocytes to spread along the endothelium and to begin the process of extravasating into the inflamed tissue following chemoattractant gradients that guide the leukocyte to the site of inflammation. Each of these events has been thoroughly examined over the past several years and has resulted in a better-defined process of coordinated events that lead the leukocyte from the vessel lumen into the inflamed tissue.

The transition from leukocyte rolling to firm adhesion depends on several distinct events to occur in the rolling leukocyte. First, the integrin needs to be modified through a G protein-mediated signaling event enabling a conformational change that exposes the binding site for the specific adhesion molecule. Second, the density of adhesion molecule expression needs to be high enough to allow the leukocyte to spread along the activated endothelium and appropriate integrin clustering on the leukocyte surface. Finally, it appears that a phenomenon known as outside-in signaling (recently reviewed) is also necessary for strengthening the adhesive interactions through several important signaling events that include FGR and HCK, two SRC-like protein tyrosine kinases (PTKs). Together, these coordinated events facilitate preparation of leukocytes for extravasation through the endothelium into the inflamed tissue.

Transendothelial cell migration of leukocytes requires that numerous potential obstacles be managed. After firmly adhering to the activated endothelium, leukocytes appear to spread and crawl along the border until they reach an endothelial cell junction that has been appropriately "opened" by the inflamed environment. While it has not been completely established, it appears that endothelial cell junctions that support transmigration of leukocytes express higher levels of adhesion molecules that allow a haptotatic gradient for the crawling cells to traverse through. This paracellular route of migration is a favored and well-supported mechanism that is optimized by tissue-expressed chemoattractants for mediating the crawling into the junctional region without harming the endothelial cell border. A number of molecules have been implicated in this route of migration, but PECAM1 has been the most thoroughly studied and appears to be functionally required for the process with targeted expression at the endothelial cell junction region. Another protein, junction adhesion molecule-A (JAM-A), has also been shown to be associated with migration of cells through the tight junctions of endothelial layer of vessels and is found on the surface of several leukocyte populations including PMNs. It appears that PECAM1 and JAM-A are utilized in a sequential manner to allow movement through the endothelial barrier.

The final obstacle for the leukocyte to traverse prior to entering into the tissue from the vessel is the basement membrane. The model that has been proposed over the years suggests that metalloproteinases (MMPs) are activated to degrade the basement membrane extracellular matrix (ECM), enabling leukocytes to penetrate toward the site of inflammation. While evidence *in vitro* suggests that matrix degradation is necessary and that MMPs are required, it has not been clearly identified how the basement membrane is traversed by leukocytes without substantial damage to the integrity of the vessel wall.

Chemoattractants

Over the past several years researchers have identified multiple families of chemoattractants that can participate in the extravasation of leukocytes. Perhaps the most readily accessible mediator class during inflammation is the complement system. These proteins are found in circulation or can be generated *de novo* upon cellular stimulation. Upon activation, bacterial products or immune complexes (as previously reviewed) through the alternative or classical pathways mediate cleavage of C3 and/or C5 into C3a and C5a that can provide an immediate and effective chemoattractant to induce neutrophil and monocyte activation. The role of C3a as an anaphylactic agent illustrates the importance of this early activation event on mast cell biology. In addition, C5a stimulates neutrophil oxidative metabolism, granule discharge, and adhesiveness to vascular endothelium. Interestingly, C5a activates endothelial cells via C5aR to induce expression of P-selectin that can further increase local inflammatory events. C3a lacks these latter activities. Altogether, these functions of C3a and C5a indicate that they are potent inflammatory mediators. While these chemoattractant molecules have previously been well described, recent literature has provided additional evidence that has rekindled excitement toward targeting these factors for therapeutic intervention.

A second mediator system that is involved in early and immediate leukocyte migration is the leukotrienes, a class of lipid mediators that are preformed in mast cells or are quickly generated through the efficient arachidonic acid pathway induced by 5-LO. In particular, leukotriene B4 (LTB4) has especially been implicated in the early induction of neutrophil migration, but also can generate long-term problems during inflammation. LTB4 can be rapidly synthesized by phagocytic cells (PMNs and macrophages) following stimulation with bacterial LPS or other pathogen products. Furthermore, the LTB4 receptor has been implicated in recruitment of T lymphocytes that mediate chronic inflammatory diseases, including rodent models of asthma and arthritis, as well as in transplantation rejection models. In particular, the LTB4 receptor BLT1 has been implicated in preferential recruitment of Th2 type T-lymphocytes during allergic responses. Thus, besides their potent function as a neutrophil chemoattractant in acute inflammatory events, LTB4 and BLT1 also play important roles in chronic immune reactions.

Chemokines represent a large and well-characterized family of chemoattractants composed of over 50 polypeptide molecules that are expressed in numerous acute and chronic immune responses. Given the number of individual molecules contained in the chemokine family, there is confusion of function. The determination of what leukocyte populations are recruited to a particular tissue during a response is dictated by the chemokine ligands that are induced and the specific receptors that are displayed on subsets of leukocytes. This latter aspect can be best observed during acute inflammatory responses, such as in bacterial infections, when the cellular infiltrate is primarily neutrophils and it is the production of CxCR binding chemokines that mediate the process. Likewise, when more insidious pathogens are present and the acute inflammatory mechanisms are not able to control the infectious process, immune cytokines, such as IFN and IL-4, tend to drive the production of chemokines that facilitate the recruitment of mononuclear cells, macrophages, and lymphocytes to the site of infection. This allows a more sophisticated immune response to develop for clearance of the pathogen. Thus, although there are numerous chemokines being produced during any single response, the overall profile of the response may be directed for recruitment of cells that are most appropriate to deal with the particular stimuli. In addition to their ability to bind to cellular receptors, chemokines are also able to bind to gly-cosaminoglycans (GAGs). Unlike with complement and lipid chemoattractants, this allows chemokines to accumulate within tissue and on endothelium for long periods without being washed away or otherwise cleared through various biological processes. Chemokines are important at the endothelial border where they mediate firm adhesion of leukocytes undergoing selectin-associated rolling to activate their β-integrins to the activated endothelial cells and subsequently direct migration of these cells to the site of inflammation. Finally, at higher concentrations (such as those found at the site of inflammation), chemokines induce leukocyte activation for effector function (for instance, degranulation). Thus, the progressive movement of leukocytes from the endothelial cell border in activated vessels through their arrival at the site of inflammation relies on the coordinated expression and interaction of chemokines and adhesion molecules.

ACUTE INFLAMMATION AND DISEASE PATHOGENESIS

The initiation of a rapid innate immune response to invading pathogens is essential to inhibit the colonization of microorganisms or to sequester toxic and noxious substances. Once an infection is established, pathogenic bacteria have the capability to multiply and expand at a rate that can surpass the ability of the host to clear and destroy the bacteria. A number of mechanisms have developed to inhibit the establishment of pathogenic bacteria in tissues. The primary mechanism is activation of edema and local fluid release to flood the affected tissue, along with early activation of the complement system in response to bacterial components, resulting in cleavage of C3 and C5. These early inflammatory mediators provide a relatively effective and rapid initiation of PMN and mononuclear phagocyte infiltration to sites of infection. The recruited phagocytic cells engulf invading pathogenic bacteria and quickly activate to begin producing LTB4 as well as early response cytokines, such as IL-1 and TNF, that enhance phagocytosis and killing. The early response cytokines subsequently activate resident cell populations to produce other important mediators of inflammation, such as IL-6 and IL-8 (CxCL8), and promote cytokine cascades that lead to continued leukocyte migration and activation. These early events are critical for regulation of the intensity of the inflammatory response as well as effective containment of the

pathogens and foreign substances. This multipronged approach to the activation of the inflammatory response and inhibition of the pathogen expansion is normally tightly regulated. However, in situations in which the inflammatory stimuli are intense, such as in a bolus dose of bacteria, severe trauma, or in burn victims, the acute inflammatory response can become dangerously unregulated. In these types of situations when mediators are produced in an unregulated manner, the host/patient can quickly become subjected to a systemic inflammatory response even though the initial insult may be quite localized. This is a result of mediator production (especially TNF and IL-1) that is systemically delivered to multiple organs, creating an overproduction of leukocyte chemoattractants in distal organs and inducing inflammatory cell influx. These types of responses can quickly damage target organs including liver, lung, and kidney. In this form of septic response, the overwhelming PMN recruitment and activation to multiple organs can lead to tissue damage and organ dysfunction.

The ensuing cytokine storm that develops in the affected tissues results from a cascade of cytokine and chemokine production, leading to uncontrolled leukocyte infiltration and activation that damages the tissue leading to organ dysfunction. While these events can affect any tissue of the organism, the liver and lung appear to be primary targets due to their relatively high numbers of resident macrophage populations that can quickly respond to the inflammatory cytokine signals. In the lung, the development of acute respiratory distress syndrome (ARDS) is often observed in patients experiencing a septic insult. Although early research focused on TNF and IL-1 as lead targets to combat these responses, clinical trials using specific inhibitors, along with more clinically relevant animal models, have not shown any benefit to blocking these central inflammatory mediators in acute diseases, such as sepsis. These failures are likely due to the fact that the early response cytokines are produced and cleared prior to the induction of the most severe aspects of disease. Thus, by the time these mediators were targeted in the sick patient, their detrimental inflammatory function has already been performed.

In the case of viral infections, the system must deal with clearance in a different manner since the ability to recognize and phagocytize virus particles is not reasonable due to their size. Instead, the system is geared to recognize the organisms once a target cell has been infected with the virus. One of the most effective early means of blocking the spread is through the immediate production of type I interferon (IFN). This class of mediators facilitates the blockade of spread by both altering the metabolism of infected cells and by promoting the production of additional antiviral factors in uninfected cells to reduce the chance of successful viral assembly and further spread. While the antiviral effects of type I IFNs were initially identified many years ago, researchers continue to attempt to fully understand the mechanisms that are initiated by this class of mediators.

PATTERN RECOGNITION RECEPTORS AND INFLAMMATORY RESPONSES

Toll-Like Receptors

The initiation of acute inflammation and the progression of chronic disease are often fueled by infectious agents that provide strong stimuli to the host. These responses evolved to be beneficial for the rapid recognition of pathogenic motifs that are not normally present in the host during homeostatic circumstances. These pathogen recognition systems form the basis for our innate immune system, which is rapidly activated to destroy pathogens prior to their colonization. However, at the same time the overactivation of this system can contribute to significant pathology in the host. While there are now a number of diverse families of pattern recognition systems, the best-characterized family is the toll-like receptors (TLR). The Toll system was first discovered in *Drosophila* as a crucial part of the antifungal defense for the organism. The TLR family in mammalian species consists of transmembrane receptors that reside either on the cell surface or within the endosome and that characteristically consist of leucine-rich repeats (LRR) for motif recognition and an intracytoplasmic region for signal transduction (Figure 2.1). Cellular activation signals are transmitted by TLRs via cytoplasmic adapter molecules that initiate a cascade of now well-defined activation pathways including NFkB, IRF3, IRF7, as well as a link to MAPK pathways. These activation pathways provide strong stimuli that alert the host with "danger signals" that allow effective immune cell activation.

One of the first molecules in this family that was identified was toll-like receptor 4 (TLR4), which primarily recognizes lipopolysaccharide (LPS; also known as endotoxin), a component of the cell wall of gram-negative bacteria. The TLR4 activation pathway is unique among TLRs as it signals via multiple adaptor proteins, including MyD88, TRIF, and MD-2, making it the most dynamic TLR within the family. It is the TLR4 pathway that is likely the most prominent during sepsis for the strong systemic activation during acute inflammatory responses.

Other TLR family members recognize distinct and now better-defined factors that allow the immediate activation of the innate immune system and subsequent signaling of the adaptive immune responses. TLR2 appears to have the most diverse range of molecules that are recognized directly including peptidoglycan, mycoplasm lipopeptide, a number of fungal antigens, as well as a growing number of carbohydrate residues on parasitic, fungal, and bacterial moities. In addition, TLR2 can heterodimerize with TLR1 and TLR6 to further expand its recognition capabilities. TLR5 specifically recognizes flagellin and is therefore important for recognizing both gram-negative and gram-positive bacteria. While the previously described TLRs are expressed on the cell membrane, a number of TLRs are predominantly expressed in endosomic membrane compartments of innate immune cells, including TLR3, TLR7, TLR8, and TLR9. These pathogen recognition receptors (PRR) are involved in recognition of nucleic acid motifs including dsRNA (TLR3), ssRNA (TLR7 and TLR8), and unmethylated CpG DNA (TLR9). Together, these TLRs function in the recognition of viral and bacterial pathogens that enter the cell via receptor-mediated endocytosis or that are actively phagocytized. All of these TLRs exclusively utilize the MyD88 adaptor pathway for activation, except TLR3, which uses TRIF. Thus, these pathways are important for the initiation of innate cytokines, including TNF, IL-12, and type I IFN, as outlined in Figure 2.1. However, the activation of antigen-presenting cells via the TLR pathways is also extremely important for integrating acute inflammatory events mediated by the innate immune system with acquired immunity and therefore also implicates TLR activation with chronic inflammation.

Cytoplasmic Sensors of Pathogens

While the TLR proteins have been the best characterized, it is now evident that they are not the only molecules that are important for recognition of pathogenic insults. One of the early observations that surround the TLR recognition system was the fact that they are expressed either on the surface membrane or on the endoplasmic membranes. However, if pathogens infect directly into the cytoplasm or escape endosomal degradation pathways, the host cell must have the ability to recognize and deal with the cytoplasmic insult. Cells have developed a number of additional recognition systems to specifically identify pathogen products in the cytoplasm. Similar to the TLR system, Nod-like receptors (NLR; also known as caterpillar proteins) are able to recognize specific pathogen patterns that are distinct from host sequences. Nod1 and Nod2 sense bacterial molecules produced by the synthesis and/or

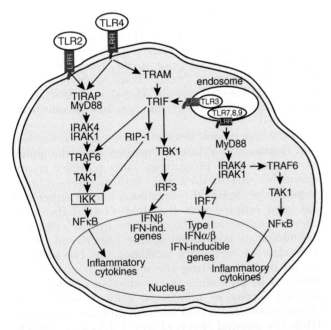

Figure 2.1   TLR activation leads to induction of diverse inflammatory, chemotactic, and activating cytokine production.

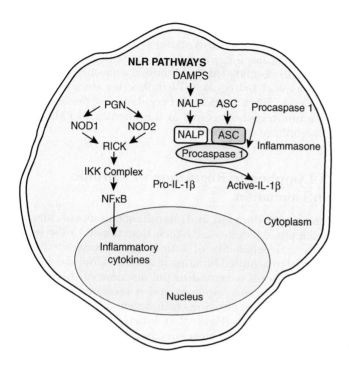

Figure 2.2 Cytoplasmic pathogen receptors allow activation of cytokines responsible for the upregulation of inflammatory processes.

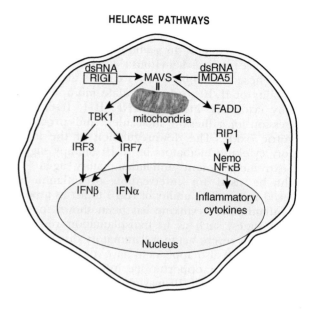

Figure 2.3 Helicase proteins induce early response and activating cytokines by recognition of dsRNA.

degradation of peptidoglycans (PGN) (Figure 2.2). Specifically, Nod1 recognizes PGNs that contain meso-diaminopimelic acid produced by gram-negative bacteria and some gram-positive bacteria, while Nod2 recognizes muramyl dipeptide that is found in nearly all PGNs. Other members of this family include NALPs (NACHT-, LRR-, and pyrin-domain-containing proteins), IPAF (ICE-protease activating factor), and NAIPs (neuronal apoptosis inhibitor proteins). An interesting aspect of NLRs is that they contain a conserved caspase associated receptor domain (CARD) that was initially related to proteins involved in programmed cell death or apoptosis. These proteins specifically activate cells through a complex of proteins known as the inflammasome. The central protein of the inflammasome is caspase 1, which is bound by the CARD of the NLRs. A simplified model of the NLR activation pathway suggests that binding to caspase 1 leads to processing of pro-IL-1b and pro-IL-18 to their active and released forms. Interestingly, these same activation pathways can lead to programmed cell death, possibly depending on the activation state of the responding cell and/or the intensity of the NLR signal. In addition to the inflammasome-mediated activation of proinflammatory cytokine release, NOD proteins have also been shown to induce NFkB and MAPK activation through a RICK/RIP2 signaling pathway. This opens up a number of additional gene activation events via this bacterial-induced pathway.

More recent investigations have identified two helicase proteins that have the ability to recognize dsRNA, RIG-I (retinoic acid-inducible gene) and MDA5 (melanoma differentiation-associated gene). The activation of the protein products of either of these genes in the cytoplasm leads to an immediate activation of type I

IFNs (Figure 2.3). The signaling pathway that RIG-I utilizes was initially surprising since it also utilizes its CARD to interact with a mitochondria-associated protein, MAVS. This interaction leads to a scaffold involving TRAF-3, TBK-1, and IRF-3 activation. While this pathway continues to be defined, it appears to be very important in several forms of viral infection. Interestingly, various viruses may differentially activate the two helicases, and this differential activation may define the type and intensity of the ensuing immune responses. Overall, the number of proteins that have the ability to recognize dsRNA not only demonstrates redundancy in the system but also indicates the importance of being able to detect this specific PAMP that is a clear sign of a productive viral infection.

Regulation of Acute Inflammatory Responses

The most successful therapy to date for controlling inflammation has been the use of steroidal compounds that nonspecifically inhibit the production of many inflammatory cytokines. The continued dependence on this strategy, although often effective, demonstrates our lack of complete understanding of the mechanisms that control inflammatory responses. A number of well-described regulators may be suited for management of acute inflammatory responses. Several anti-inflammatory mediators have been investigated, including IL-10, TGFβ, IL-1 receptor antagonist (IL-1ra), as well as IL-4 and IL-13. Perhaps the most attractive anti-inflammatory cytokine with broad-spectrum activity is IL-10. IL-10 is predominantly produced by macrophage populations, Th2-type lymphocytes, and B-cells, but can be produced by airway epithelial cells and by several types of tumor cell populations. The function of IL-10 appears to be important during normal physiological events, as IL-10 gene knockout mice develop lethal inflammatory bowel dysfunction. The importance of this cytokine has been demonstrated in models of endotoxemia and sepsis that neutralized IL-10 during the acute phase and led to increased lethality. In addition, administration of IL-10 to mice protects them from a lethal endotoxin challenge. These latter observations can be attributed to the ability of IL-10 to downregulate multiple inflammatory cytokines, including TNF, IL-1, IL-6, IFN-γ, expression of adhesion molecules, and production of nitric oxide. The downregulation of the inflammatory cytokine mediators by IL-10 therapy suggests an extremely potent anti-inflammatory agent that might be used for intervention of inflammation-induced injury. The ability of IL-10 to act as a potent anti-inflammatory cytokine has been shown in other disease states, such as in transplantation responses. However, in severe acute inflammatory diseases, such as sepsis, IL-10 also appears to have a role in promoting secondary or opportunistic infections due to its inhibitory activity. IL-10 may also have a role in promoting end-stage disease if not properly regulated. Thus, its role as a therapeutic has been questioned.

CHRONIC INFLAMMATION AND ACQUIRED IMMUNE RESPONSES

Perhaps one of the most difficult and important aspects of disease pathogenesis to regulate is when and how to turn off an immune/inflammatory response. Clearly, pathogen clearance is a primary focus for our immune system, and leukocyte accumulation and activation are a critical event that must be coordinated with the continued presence of pathogens. PRR activation is a critical recognition system that not only activates important cytokine and chemokine pathways for increasing leukocyte function, but also initiates critical antigen presentation cell (APC) functions to optimize lymphocyte activation for pathogen clearance. However, uncontrolled or inefficient immune responses can lead to continual inflammatory cell recruitment and tissue damage that, if persistent and unregulated, can result in organ dysfunction. There are numerous pathogen- and non-pathogen-related diseases that have been classically regarded as being caused by chronic inflammation, including rheumatoid arthritis (RA), chronic obstructive pulmonary disease (COPD), asthma, mycobacterial diseases, multiple sclerosis (MS), and viral hepatitis, among others. However, more recently additional diseases that have not been traditionally grouped with those associated with chronic inflammation have now been recognized to have defects in regulation of inflammation. These disease states include atherosclerosis, numerous obesity-related diseases, as well as various cancers. Chronic inflammation may be driven by non-antigenic stimuli that are persistent and cannot be effectively cleared, such as in the case of silicosis. In addition to the persistence of the inflammatory response during chronic inflammatory disorders, there is also a shift in the cellular composition of the leukocyte populations that accumulate. While neutrophils and macrophages may continue to be the end-stage effector cells that mediate the damage, a significant component of the inflammatory responses now comprises lymphocyte infiltration. The presence of activated T- and B-lymphocytes likely indicates the presence of a persistent antigen that induces cell-mediated and humoral immune responses. It is critical to regulate T-lymphocytes since they are central to the activation and regulation of the acquired immune response, as well as the intensity of PMN and macrophage activation.

T-Lymphocyte Regulation of Chronic Inflammation

The nature, duration, and intensity of episodes of chronic inflammatory events are largely determined by the presence and persistence of antigen that is recognized and cleared by acquired immune responses. Thus, the regulation of T-cells is central to the outcome of the inflammatory/immune responses and is mediated through a combination of cytokine environment and transcription factor regulation (Figure 2.4). When a pathogenic insult is encountered, the most effective immune response is a cell-mediated Th1-type response, which is induced by IL-12 with Stat4 activation and characterized by IFN production along with T-bet transcription factor expression. However, long term this immune response can be devastating to the host, as unregulated it can rapidly destroy local tissue and organ function. Thus, the immune response must be modulated and begin to shift to a less harmful response for the tissue, which in T-cells is regulated by IL-4 production and STAT6 activation leading to GATA3 transcription factor expression. While this shift in responses does not represent a sudden switch, rather a gradual transition, the long-term consequences of chronic inflammation are often a result of a combined cytokine phenotype that leads to altered macrophage function and continual tissue damage. One of the aspects of the Th2 response that can be detrimental is the shift toward tissue remodeling designed to promote both restoration of function and host protection.

Clearly, regulation of both the Th1- and Th2-type immune responses is central to resolving inflammatory responses and limiting damage once the inciting agent has been removed. An area that has held a significant level of interest has been the differentiation of T regulatory (Treg) cells during the development of chronic responses. This cell subset has been divided into several subpopulations, including natural Treg cells and inducible Treg cells (iTreg). Natural Treg cells develop in the thymus and are essential for control of autoimmune diseases, whereas inducible iTreg cells develop following an antigen-specific activation event and appear to function to modulate an ongoing response. In addition, Treg cells can also be subdivided based on the mechanism of inhibition that they use, such as production of IL-10 and/or TGFβ or use of CTLA-4. Some common ground has been forged in these cell populations based on the expression of Foxp3 transcription factor, although apparently not all Treg populations express this protein. Support for the importance of Foxp3+ Treg cells comes from studies with mice missing this factor and in humans who have mutations in FOXP3, both

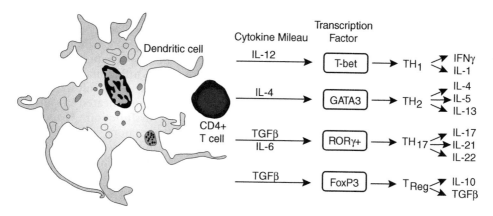

Figure 2.4 **The cytokine environment during antigen presentation controls T-cell differentiation by regulation of specific transcription factor activation.**

of which develop multiorgan autoimmune diseases. Thus, no matter the nomenclature, this cell population appears to be centrally important for the regulation of immune responses, and defects in this pathway may lead to chronic disease phenotypes.

More recent investigations have further enhanced our understanding of the role of T-cell subsets in chronic disease with the explosion of data that has described a newer subset of T-cells (Th17) that characteristically produce IL-17. This subset of T-cells has only recently begun to be understood during chronic inflammatory diseases. IL-17-producing cells were first described as an important component of antibacterial immunity. Subsequently, the Th17 subset has been identified as having a central role in the severity of autoimmune responses, in cancer, in transplantation immunology, as well as in infectious diseases. Interestingly, the critical aspect of whether T-cells will differentiate into a Th17 cell depends on the expression RORγt transcription factor. Similar to the Treg cells, the differentiation of these cells depends on exposure to TGFβ, but is additionally dependent on IL-6 or other STAT3 signals along with RORγt. The differentiation of Th17 cells also appears to be enhanced by IL-23, an IL-12 family cytokine that is upregulated in APC populations upon TLR signaling. As outlined in Figure 2.4, T-cell activation and cytokine phenotypes depend on the distinct transcription factor expressed for the activation of the particular T-cell subset. Clearly, these subsets and the cytokines that they produce dictate the outcome of a chronic response not only based on the cascade of mediators that they induce but also by the leukocyte subsets that are used as end-stage effectors during the responses.

B-Lymphocyte and Antibody Responses

The pathogenic role of the humoral immune system has been implicated in a number of chronic disease phenotypes, including allergic responses, autoimmune diseases, arthritis, vasculitis, and any other disease where immune complexes are deposited into tissues. Antibodies produced by B-lymphocytes are a primary goal of the acquired immune response to combat infectious organisms at mucosal surfaces, in the circulation, and within tissues of the host. To be effective, antibodies need to have the ability to bind to specific antigens on the surface of pathogen and through their Fc portion to facilitate phagocytosis by macrophages and PMNs for clearance and complement fixation for targeted killing of the microorganism by the lytic pathway. However, these features can also lead to detrimental aspects and tissue damage due to inappropriate activation of inflammatory effector cell populations. This is often a problem in autoimmune diseases such as systemic lupus erythematosus (SLE) and rheumatoid arthritis, where a wide array of antibodies directed against self-antigens is formed. In SLE, immune complexes form and are often deposited in the skin, driving a local inflammatory response and vasculitis. Even more serious for the host is the deposition of immune complexes in the kidney, leading to glomerulonephritis and possible severe kidney disease. Autoantibodies directed against tissue antigens can also induce damage due to FcR cross-linking on phagocytic cells including PMN, macrophages, and NK cells. This is often a central mechanism for initiating local inflammation and damage within autoimmune responses, such as with joints of RA patients with autoantibodies directed against matrix proteins.

The induction of allergic responses in developed nations has been steadily increasing for the past three decades. Incidence of food, airborne, and industrial allergies has a significant impact on the development of chronic diseases in the skin, lung, and gut, including atopic dermatitis, inflammatory bowel disease (IBD), and asthma. The production of IgE leading to mast cell and basophil activation is the central mechanism that regulates the induction of these diseases. As the antibody isotype produced by the B-cell is governed by the T-lymphocyte response and production of specific cytokines, determining mechanisms that regulate T-cells during allergic diseases has been central to research in these fields.

Exacerbation of Chronic Diseases

While much of the research in this field has centered on understanding the factors involved and defining targets for therapy of chronic disease, less research has focused on what exacerbates and/or extends the severity of these diseases. This becomes much more difficult when the mechanisms and causative agents that initiate the chronic response are not clearly identified or may be heterogeneous within the patient population. No matter the disease, it appears that a common initiating factor for the exacerbation is an infectious stimulus, bacterial or viral. The reality is that any stimulus that initiates a strong inflammatory signal has the ability to break the established maintenance and reinitiate the chronic response in individuals with an underlying disease. This is most often manifested in diseases where an antigenic response is the underlying cause of the chronic disease and the antigen is environmental or host available, such as allergic asthma or autoimmunity. The strong activation of immune cells locally within the affected tissue would provide the reactivation of a well-regulated immune response. The mediators that are upregulated, IL-12 and/or IL-23, in DC might dictate the type of effector response that is initiated, such as Th1 and/or Th17, respectively.

A common activation pathway for these responses is the use of molecules that can quickly and effectively recognize pathogens, such as the TLR family members. While these molecules are clearly expressed on immune cell populations and facilitate an effective host response, it may be their inappropriate expression on nonimmune cells, such as epithelial cells and fibroblasts, which presents the host with the most detrimental response. The expression of TLRs on nonimmune cells within chronic lesions would predispose these cells to strong infectious stimuli that initiate (or reinitiate) an inflammatory response through the activation and expression of cytokines and inflammatory mediators. While it has not been clearly established, a number of studies have indicated that TLR expression on nonimmune cells in chronic lesions is upregulated and would presumably predispose these tissues to hyperstimulation during infectious insults. This alone could cause the exacerbation of a chronic response without any other specific stimuli. The continual reactivation of tissue inflammation with infectious insults, such as in the lung and gut, could provide the mechanism for tissue damage and potentially remodeling that over time could lead to gradual but continuous organ dysfunction. The addition of other antigenic stimuli, such as an auto-antigen, would further enhance the damaging responses and lead to more severe and accelerated disease phenotypes along with end-stage disease that often accompanies tissue remodeling and ineffective repair.

TISSUE REMODELING DURING ACUTE AND CHRONIC INFLAMMATORY DISEASE

Repair of damaged tissues is a fundamental feature of biological systems and properly regulated has little harmful effect on normal organ function. Abnormal healing and repair, however, can lead to severe problems in the function of organs, and in some cases

the perpetuation of the remodeling and repair can result in end-stage disease. Damage to tissues can result from various acute or chronic stimuli, including infections, autoimmune reactions, or mechanical injury. In some cases acute inflammatory reactions, such as ARDS in septic patients, can result in a rapid and devastating disorder that is complicated by significant lung fibrosis and eventual dysfunction. However, more common chronic inflammatory disorders of organ systems (including pulmonary fibrosis, systemic sclerosis, liver cirrhosis, cardiovascular disease, progressive kidney disease) and the joints (such as rheumatoid arthritis and osteoarthritis) are a major cause of morbidity and mortality and enormous burden on healthcare systems. A common feature of these diseases is the destruction and remodeling of extracellular matrix (ECM) that has a significant effect on tissue structure and function. A delicate balance between deposition of ECM by myofibroblasts and ECM degradation by tissue leukocytes determines the tissue restructuring during repair processes, and proper function versus development of pathologic scarring. Chronic inflammation, tissue necrosis, and infection lead to persistent myofibroblast activation and excessive deposition of ECM (including collagen type I, collagen type III, fibronectin, elastin, proteoglycans, and lamin), which promote formation of a permanent fibrotic scar. Most chronic fibrotic disorders have a persistent irritant that stimulates production of proteolytic enzymes, growth factors, fibrogenic cytokines, and chemokines. Together, they orchestrate excessive deposition of connective tissue and a progressive destruction of normal tissue organization and function (Figure 2.5).

Profibrogenic Cytokines and Growth Factors Involved in Fibrotic Tissue Remodeling

Alterations in the balance of cytokines can lead to pathological changes, abnormal tissue repair, and tissue fibrosis. The most well-studied cytokines involved

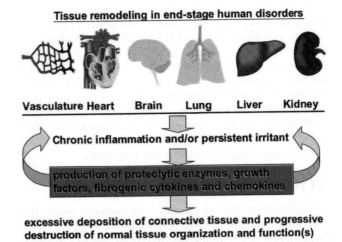

Tissue remodeling in end-stage human disorders

Vasculature Heart   Brain   Lung   Liver   Kidney

Chronic inflammation and/or persistent irritant

production of proteolytic enzymes, growth factors, fibrogenic cytokines and chemokines

excessive deposition of connective tissue and progressive destruction of normal tissue organization and function(s)

Figure 2.5 The persistent production of cytokines, fibrogenic growth factors, and proteolytic enzymes can result in tissue remodeling and eventual organ dysfunction.

in these processes include transforming growth factor β (TGFβ), tumor necrosis factor-α (TNFα), platelet-derived growth factor (PDGF), basic fibroblast growth factor (BFGF), monocyte chemoattractant protein-1 (MCP-1), macrophage inflammatory protein-1-α (MIP-1-α), and interleukin-1 (IL-1), IL-13, and IL-8.

TGFβ

TGFβ is one of the most well-studied profibrotic cytokines. Upregulation of TGFβ1 has been associated with pathological fibrotic processes in many organs (Table 2.1), such as chronic obstructive pulmonary disease, cataract formation, systemic sclerosis, renal fibrosis, heart failure, and many others. The TGFβ family represents a group of multifunctional cytokines that includes at least five known isoforms, three of which are expressed by mammalian cells (TGFβ1–3). TGFβ-induced effects are mainly mediated by signaling via the TGFβ receptors. TGFβ isoforms are known to induce the expression of ECM proteins (such as collagen I, collagen III, and collagen V; fibronectin; and a number of glycoproteins and proteoglycans normally associated with development) in mesenchymal cells, and to stimulate the production of protease inhibitors that prevent enzymatic breakdown of the ECM. In addition, some ECM proteins (such as fibronectin) are known to be chemoattractants for fibroblasts and are released in increased amounts by fibroblasts and epithelial cells in response to TGFβ. Elevated TGFβ expression in affected organs correlates with abnormal

connective tissue deposition observed during the beginning of fibrotic diseases. For example, fibroblasts derived from lungs of idiopathic pulmonary fibrosis patients show an enhanced synthetic activity in response to growth factors, whereas normal fibroblasts show a predominantly proliferative response. Thus, while TGFβ is necessary for normal repair processes, its overexpression plays a pivotal role in deposition of ECM and end-stage disease.

TNFα

Another cytokine that has often been linked to chronic remodeling diseases is tumor necrosis factor alpha. TNFα is a well-known early response cytokine that is key in initiating responses and is rapidly expressed in response to many kinds of stress (mechanical injury, burns, irradiation, viruses, bacteria). However, the role of TNFα is much more complex than simply serving as a trigger of cytokine cascades. TNFα is a proinflammatory mediator that is involved in the extracellular matrix network, as shown in the healing infarct and collagen synthesis by cardiac fibroblasts. The pathologic events in asthma also correlate with increased TNFα production both *in vivo* and *in vitro* in cellular isolates from asthmatic patients. The most convincing evidence that TNFα plays a role in pathophysiology of asthma was observed in normal subjects receiving inhaled recombinant TNFα. These studies demonstrate a significant increase in airway hyper-responsiveness and decreases in FEV1 within those subjects receiving TNFα compared to the placebo group linked to its ability to change extracellular matrix and upregulate pathways specific for leukocyte recruitment. A direct role of TNFα in fibrogenesis was recently demonstrated using human epithelioid dermal microvascular endothelial cell cultures. Exposure of those cells to TNFα for 20 days induced permanent transformation into myofibroblasts. Similar transformation events following chronic inflammatory stimulation *in vivo* may explain one source of myofibroblasts in skin fibrogenesis.

Rheumatoid arthritis (RA) represents a chronic disease that highlights the importance of TNFα. TNFα-dependent cytokine cascades were identified in the *in vitro* culture of synovium from joints of patients with RA and led to studies of TNFα blockage in experimental animal models of RA. Using a collagen-induced human RA model in DBA/J mice, researchers showed that anti-TNF antibodies ameliorate arthritis and reduce joint damage. This led to the use of TNF blockade in human RA. With the success of anti-TNF treatment in RA, this approach was tested in a number of other chronic disorders, including inflammatory bowel disease, asthma, and graft-versus-host disease (GVD) during bone marrow transplantation. Successful targeting of TNF has been observed in a number of chronic diseases where anti-TNF therapy has been approved for use and is now also being examined in numerous additional inflammatory, infectious, and neoplastic diseases (Table 2.2). However, anti-TNF therapy is

Table 2.1 TGFβ Contributes to Fibrosis of These Diseases by Excessive Matrix Accumulation

Organ	Disease
Eye	Graves ophthalmopathy
	Conjunctival cicatrization
Lung	Pulmonary fibrosis
	Pulmonary sarcoids
Heart	Cardiac fibrosis, cardiomyopathy
Liver	Cirrhosis
	Primary biliary cirrhosis
Kidney	Glomerulosclerosis
	Interstitial fibrosis
Pancreas	Chronic or fibrosing pancreatitis
Skin	Hypertrophic scar
	Keloids
	Scleroderma
Subcutaneous tissue	Dupuytren's contracture
Endometrium	Endometriosis
Peritoneum	Sclerosing peritonitis
	Postsurgical adhesion
Retroperitoneum	Retroperitoneal fibrosis
Bone	Renal osteodystrophy
Muscle	Polymyositis, dermatomyositis
	Muscular dystrophy
	Eosinophilia-myalgia
Bone marrow	Myelofibrosis
Neuroendocrine	Carcinoid

Table 2.2 Targeting of TNFα in Chronic Inflammatory Disorders

Approved for Use	Not Completed Trials and Pilot Studies	Clinical Failures
- Rheumatoid arthritis	- Ulcerative colitis	- Congestive heart failure
- Juvenile rheumatoid arthritis	- Behçet syndrome	- Multiple sclerosis
- Crohn's disease	- Vasculitis (small and large vessel)	- Chronic obstructive pulmonary disease
- Psoriatic arthritis	- Glomerulonephritis	- Sjögren syndrome
- Ankylosing spondylitis	- Systemic lupus erythematosus	- Wegener granulomatosis
- Psoriasis	- Joint prosthesis loosening	
	- Hepatitis	
	- Polymyositis	
	- Systemic sclerosis	
	- Amyloidosis	
	- Sarcoidosis	
	- Ovarian cancer	
	- Steroid-resistant asthma	
	- Refractory uveitis	

not appropriate in all diseases, as there have been a number of failed clinical trials, including those using anti-TNF treatment for MS and congestive heart failure. In addition, a potential side effect of using anti-TNF therapy is the increased susceptibility to infectious organisms.

KEY CONCEPTS

- Acute and chronic inflammation are coordinated responses that rely on early cytokine production, adhesion molecule expression, and chemoattractant directed migration of leukocytes.
- Pathogen recognition is facilitated by distinct families of molecules designed to "alert" the immune system and respond to the insult prior to colonization and mediate clearance.
- The nature of the innate immune response early in disease dictates the characteristics and severity of chronic inflammatory responses.
- The activation and phenotype of the CD4+ T helper lymphocyte has a primary role in dictating the development and features of chronic inflammation.
- Exacerbation of chronic disease can most often be mediated by infectious organisms that initiate strong innate immune responses that subsequently trigger the acquired immune system.
- Chronic inflammation often leads to tissue remodeling that can cause long-term dysfunction and increased disease severity.

SUGGESTED READINGS

1. Bacchetta R, Passerini L, Gambineri E, et al. Defective regulatory and effector T cell functions in patients with FOXP3 mutations. *J Clin Invest.* 2006;116:1713–1722.

2. Beutler B, Jiang Z, Georgel P, et al. Genetic analysis of host resistance: Toll-like receptor signaling and immunity at large. *Annu Rev Immunol.* 2006;24:353–389.

3. Chatila T. The regulatory T cell transcriptosome: E pluribus unum. *Immunity.* 2007;27:693–695.

4. DeLisser HM, Newman PJ, Albelda SM. Molecular and functional aspects of PECAM-1/CD31. *Immunol Today.* 1994;15:490–495.

5. Ivanov II, McKenzie BS, Zhou L, et al. The orphan nuclear receptor RORgammat directs the differentiation program of proinflammatory IL-17+ T helper cells. *Cell.* 2006;126:1121–1133.

6. Kanneganti TD, Lamkanfi M, Nunez G. Intracellular NOD-like receptors in host defense and disease. *Immunity.* 2007;27:549–559.

7. Kaplan MH, Grusby MJ. Regulation of T helper cell differentiation by STAT molecules. *J Leukoc Biol.* 1998;64:2–5.

8. Kato H, Takeuchi O, Sato S, et al. Differential roles of MDA5 and RIG-I helicases in the recognition of RNA viruses. *Nature.* 2006;441:101–105.

9. Kufer TA, Sansonetti PJ. Sensing of bacteria. NOD a lonely job. *Curr Opin Microbiol.* 2007;10:62–69.

10. Laudanna C, Kim JY, Constantin G, et al. Rapid leukocyte integrin activation by chemokines. *Immunol Rev.* 2002;186:37–46.

11. McEver RP, Cummings RD. Role of PSGL-1 binding to selectins in leukocyte recruitment. *J Clin Invest.* 1997;100:S97–S103.

12. McGeachy MJ, Cua DJ. Th17 cell differentiation: The long and winding road. *Immunity.* 2008;28:445–453.

13. Paust S, Cantor H. Regulatory T cells and autoimmune disease. *Immunol Rev.* 2005;204:195–207.

14. Rot A, Von Andrian UH. Chemokines in innate and adaptive host defense: Basic chemokinese grammar for immune cells. *Annu Rev Immunol.* 2004;22:891–928.

15. Sato M, Muragaki Y, Saika S, et al. Targeted disruption of TGF-beta1/Smad3 signaling protects against renal tubulointerstitial fibrosis induced by unilateral ureteral obstruction. *J Clin Invest.* 2003;112:1486–1494.

16. Takeda K, Kaisho T, Akira S. Toll-like receptors. *Annu Rev Immunol.* 2003;21:335–376.

17. Weaver CT, Harrington LE, Mangan PR, et al. Th17: An effector CD4 T cell lineage with regulatory T cell ties. *Immunity.* 2006;24:677–688.

18. Williams RO, Feldmann M, Maini RN. Anti-tumor necrosis factor ameliorates joint disease in murine collagen-induced arthritis. *Proc Natl Acad Sci USA.* 1992;89:9784–9788.

19. Yoneyama M, Kikuchi M, Natsukawa T, et al. The RNA helicase RIG-I has an essential function in double-stranded RNA-induced innate antiviral responses. *Nature Immunol.* 2004;5:730–737.

3

Infection and Host Response

Margret D. Oethinger . Sheldon M. Campbell

MICROBES AND HOSTS—BALANCE OF POWER?

Disease is one of the major driving forces of evolution. Humans have a generation time of roughly 20 years, and even small mammals reproduce in weeks to months. In contrast, microbial generation times range from minutes to days. Thus, microbes evolve hundreds to thousands of times more rapidly than their vertebrate hosts.

Large multicellular creatures represent concentrated, extremely rich nutrient sources for microbes. Therefore, the survival of multicellular creatures requires that they have sufficient defenses to prevent easy invasion and consumption. Recent advances in basic immunology illustrate the breadth and depth of the adaptations that have evolved to protect multicellular organisms from microbial invasion. However, the wondrous complexity and power of mammalian host defenses serve only as a backdrop to the even more astonishing complexity of microbial strategies for evading them. Humans (and other multicellular organisms) survive only because the microbes let us live.

In most cases, the pathologies induced by microbial pathogens primarily serve to aid microbial spreading to new hosts. Thus, coughing, sneezing, and diarrhea are all mechanisms for microbial spread, and the distress caused to the host is merely incidental. We will illustrate microbial adaptations and major host response mechanisms using examples of five microbes that are exposed to, but circumvent, those responses.

THE STRUCTURE OF THE IMMUNE RESPONSE

The response to invading microbes consists of three major arms: (i) the innate immune system, which recognizes pathogens and cellular damage; (ii) adaptive immunity, which mounts a pathogen-specific response; and (iii) effector mechanisms, directed by both innate and adaptive mechanisms, which inactivate pathogens

(listed in Table 3.1). These divisions are arbitrary, and nearly every cell in the body participates to a degree in all three functions; there really is no clear division between the immune system proper and the rest of the body. Of course, cells such as lymphocytes, phagocytes, and dendritic cells are much more deeply committed to defense against pathogens than most other specialized cell types.

Two major categories of molecules are recognized by the innate immune system: (i) microbial components and (ii) markers of tissue damage or death. Microbial molecules recognized by the innate immune system include peptidoglycan, lipopolysaccharide, and double-stranded RNA. These Pathogen-associated molecular patterns (PAMPs) are detected by pattern recognition receptors (PRRs) on host cells. Markers of tissue damage and death recognized by the innate immune system include tissue factor and other markers of cellular distress. The first-described and most important class of PRRs are the toll-like receptors (TLRs). Triggering PRRs in turn leads to a cascade of events which invoke the other two functions of the immune response. Depending on the tissue site, types of microbial structures, and category of cellular distress recognized, cells of the adaptive immune system and a broad range of effector mechanisms are recruited. Cellular recruitment is largely mediated by chemokines (peptide messengers that modulate immune cellular responses) and nonpeptide inflammatory mediators (such as prostaglandins).

Antigens processed by phagocytic and nonphagocytic cells are presented to lymphocytes, which then mount an adaptive immune response. The adaptive immune response is embodied in (i) T-lymphocytes, which regulate immune responses, invoke powerful effector mechanisms, and participate directly in cytotoxic effector responses; and (ii) B-lymphocytes, which produce antibodies. Antibodies are both direct effectors of the immune response and mediators of innate and adaptive immunity. Antibodies directly neutralize some organisms, but also invoke and enhance further effector mechanisms by opsonizing microbes to direct their ingestion by phagocytes and by initiating complement activity.

Table 3.1 Host Effector Mechanisms

Name	Properties	Effector Mechanisms
Soluble Effectors		
Complement system	Proteolytic cascade, activated by antibody, directly by microbial components, or via PRRs.	Direct destruction of pathogens via pore formation. Recruit inflammatory cells. Enhance phagocytosis and killing.
Coagulation system	Proteolytic cascade, activated by tissue and vascular damage.	Prevents blood loss. Bars access to bloodstream. Proinflammatory.
Kinin system	Proteolytic cascade triggered by tissue damage.	Proinflammatory. Causes pain response. Increases vascular permeability to allow increased access by plasma proteins.
Antibodies	Antigen-specific proteins produced by B-cells. Recognize a broad range of antigens.	Directly neutralize pathogens. Activate complement. Opsonize pathogens to enhance phagocytosis and killing.
Cellular Effectors		
Monocyte/ Macrophage	Have PRRs to recognize pathogens; activated by specific T-cells and chemokines.	Phagocytosis and microbial killing via multiple mechanisms. Antigen presentation; macrophages are the classic Antigen Presenting Cells (APCs).
Dendritic cell	Ingest large amounts of extracellular fluid; migrate to lymph node to present antigen to naïve T-lymphocytes.	Antigen uptake, transport, and presentation to T-lymphocytes. Initiate adaptive immune response.
Neutrophil	Have PRRs to recognize pathogens, activated antibody and complement.	Phagocytosis and microbial killing via multiple mechanisms APC.
Eosinophil	Recognize antibody-coated parasites.	Killing of multicellular pathogens.
Basophil/Mast cell	Associated with IgE-mediated responses.	Release of granules containing histamine and other mediators of anaphylaxis.
NK-cell	Lymphocyte lacking antigen-specific reactivity; recognize PAMPs of intracellular pathogens, activated by chemokines and by membrane proteins of infected cells.	Induce death of infected cells via membrane pores and induced apoptosis.
B-lymphocyte	Recognize antigens presented by APCs; regulated by T-cells and chemokines.	Produce antibody.
T-lymphocyte	Recognize antigens presented by APCs; regulate major portions of both adaptive and innate immunity.	Directly kill infected cells via membrane pores and induced apoptosis. Activate macrophages. Many other functions.

The essence of the adaptive immune response is somatic genetic variation, which produces diverse, antigen-specific molecules (antibodies and T-cell receptors). Each lymphocyte produces only a single receptor or antibody. Lymphocytes are elaborately selected to eliminate self-reactive molecules and to favor cells making receptors or antibodies to pathogens. While this process is complex, time-consuming, and wasteful (many lymphocytes are eliminated for each clone which survives), the specificity of the adaptive immune response makes it a central component of the mammalian defense system.

A huge range of effector mechanisms limit and eliminate infection either by direct antimicrobial activity, or by creating physical or chemical barriers to microbial proliferation and spread. A partial list of effectors is provided in Table 3.1.

REGULATION OF IMMUNITY

Because inflammatory responses are metabolically costly and capable of causing enormous damage to tissues, the immune system is tightly regulated. Soluble effector systems, such as the complement and coagulation cascades, have inhibitors which usually confine these responses to the area where the initiating stimulus

occurs. Cellular effectors are activated and inhibited via both chemokines and direct signaling via adhesion molecules and direct ligand-receptor interactions with regulatory cells, mostly T-lymphocytes. Adaptive immune responses undergo elaborate screening. Self-reactive cells are screened out, and cells reactive to current infectious challenges are activated and proliferate. In turn, the adaptive immune system directs the activities of the innate immune system. Antibodies activate complement and direct phagocytosis, and T-cells activate macrophages and secrete chemokines that modulate innate cellular responses.

PATHOGEN STRATEGIES

To evade the flexible and powerful system of host defenses, successful pathogens have evolved complex strategies. As examples, we have selected five pathogens: (i) the African trypanosomes (*Trypanosoma brucei* species), bloodstream-dwelling protists that evade antibody and complement via a remarkable strategy for generating antigenic diversity; (ii) *Staphylococcus aureus*, which employs a variety of strategies to evade and overload innate and adaptive immune responses; (iii) *Mycobacterium tuberculosis*, an intracellular bacterium which actually proliferates inside an effector immune cell, the macrophage, by manipulating the phagosome-lysosome trafficking of its vacuole; (iv) herpes simplex virus, a complex DNA virus which successfully disrupts intracellular mechanisms of viral control; and (v) human immunodeficiency virus, a small RNA virus that turns the immune system on itself and generates enormous molecular diversity during infection of a single host to evade and subvert the immune response over a period of years to decades. While these five organisms hardly demonstrate the incredible range of pathogen strategies of pathogenesis, they will allow for the discussion of the major aspects of immune function, in the context of meeting infectious challenges.

THE AFRICAN TRYPANOSOME AND ANTIBODY DIVERSITY: DUELING GENOMES

Blood is a tissue with many functions, many mediated by the cellular components of the bood. However, the acellular portion of the blood (the plasma) contains a host of molecules involved in defense. These include the proteins involved in coagulation, a variety of regulatory cytokines, and dedicated antimicrobial molecules such as antibodies and complement. Complement was initially described as a component of plasma that enhanced the antibacterial powers of antibodies. Hence, this activity that *complemented* the activity of antibodies became known as complement. Complement can be activated by antibodies (the classical pathway) or by interaction with specific molecular signatures. Once activated, the proteases of the complement system cleave targets, many of which are also proteases, continuing the cascade of activation and propagation of the response. The activated components of complement opsonize pathogens to enhance phagocytosis, attract immune cells, serve as co-receptors to enhance adaptive responses, and directly damage some bacteria by forming a membrane attack complex.

Antibodies are a major arm of the adaptive immune response. Antibodies function to activate complement, direct effector cells to the pathogen, neutralize, and sequester microbes. Thus, the bloodstream is an extremely hostile environment for microbes. Yet the agents of African sleeping sickness, *Trypanosoma brucei (b.) rhodesiense* and *T. b. gambiae*, establish and maintain extracellular, bloodstream infections that can last weeks, months, and occasionally years.

Generation of Antibody Diversity: Many Ways of Changing

An antigenic stimulus is required to induce B-lymphocytes to produce antibody, but the mere presence of antigen is insufficient to induce a robust response. Co-stimulation, either by helper T-cells or by particular antigens is required. Binding of complement to antigens activates co-receptors which markedly enhance the response. A complex series of stimulation, selection,

and differentiation events result in maturation of the B-cell into a mature antibody-producing plasma cell.

Structurally, antibodies are divided into variable and constant regions. The constant regions determine the antibody class: IgD, IgM, IgG, IgA, or IgE. The different classes of antibodies have different effector functions and different destinations. The genome is not large enough to contain a separate gene for each antibody needed to respond to any antigen that the host might encounter. Thus, antibody diversity is generated by five different mechanisms (Figure 3.1): (i) the inherent diversity of the variable-region sequences, (ii) the genetic recombination of those regions into functional immunoglobulins, (iii) the combination of different heavy and light chain variable regions to form a functional antigen-binding site, (iv) junctional diversity introduced during the joining process, and (v) somatic hypermutation of the V-region in activated B-cells.

The genome contains substantial diversity of variable regions. Table 3.2 lists the numbers and combinations. In theory, the genetic diversity of the variable-region genes and their combinations can produce 1.9×10^6 different antigen-binding sites. In practice, it is likely that many of these combinations are useless, unstable, or even recognize self-antigens, so the true diversity of pathogen recognition sequences available from simple recombination of germ-line elements is much lower. During the genetic recombination events which join the variable regions, additional diversity is generated at the junctions. In combination, these mechanisms add diversity in a semi-random way. Finally, somatic hypermutation of the variable regions occurs during proliferation of activated B-cells. During maturation, the antibody serves as the B-cell antigen receptor. The processes of stimulation, co-stimulation, and clonal deletion eliminate self-reactive cells and select for high-affinity receptors. The mechanisms for generating diversity, and for selecting high-affinity receptors combine to produce an immune response of high specificity and of increasing efficacy.

Trypanosoma brucei and Evasion of the Antibody Response: Diversity Responds to Diversity

African trypanosomes are unicellular, flagellate parasites carried by the tsetse fly (*Glossina* species). The fly injects the infectious metacyclic form of the parasite into the host. Subsequently, the parasite invades the subcutaneous tissues, then the regional lymph nodes, and finally the bloodstream. One of the salient characteristics of the infectious trypanosome is a homogeneous glycophosphatidylinositol (GPI)-linked surface protein called the variant surface glycoprotein (VSG). VSG is proinflammatory. As an antigen, VSG is recognized by B-cells and T-cells, and an antibody response is effectively generated. Most (but not all) trypanosomes are destroyed by antibody, complement, and phagocytosis.

A subpopulation of organisms manages to change its coat protein to a new VSG that is structurally similar but

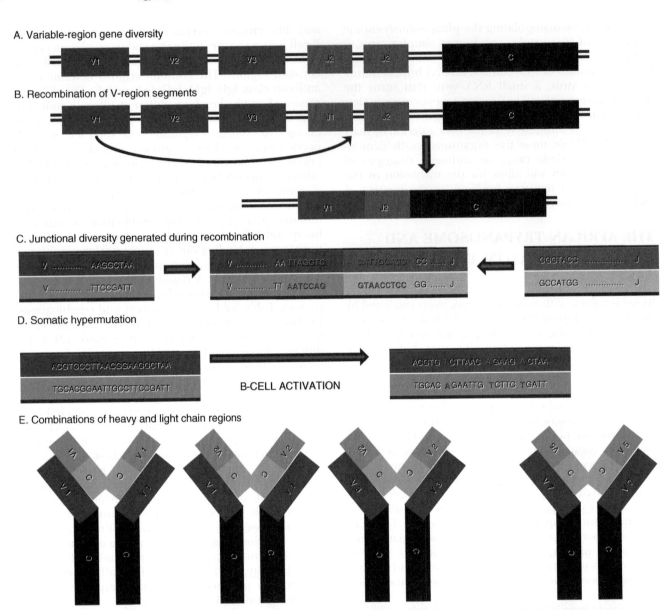

A. Variable-region gene diversity

B. Recombination of V-region segments

C. Junctional diversity generated during recombination

D. Somatic hypermutation

B-CELL ACTIVATION

E. Combinations of heavy and light chain regions

Figure 3.1   Mechanisms of generating antibody diversity. C regions and sequences are in shades of purple; J regions and sequences are in shades of green, and V regions and sequences are in shades of red. Altered or mutated sequences are in yellow. Designations of sequences (V1, etc.) are arbitrary and not meant to represent the actual arrangement of specific elements. (A) The inherent germline diversity of V and J regions provides some recognition diversity. (B) The combinations of V and J regions (V, D, and J in heavy-chains) provide additional diversity. (C) The V-J junctions undergo semirandom alterations during recombination, generating more variants. (D) In activated B-cells, the variable regions are hypermutated. (E) V regions of both light and heavy chains combine to form the antigen-recognition zone of the antibody. They can combine in different ways to provide still more variety of antigen recognition.

Table 3.2   Variable-region Gene Diversity

Immunoglobulin Class	Region	# of Genes
λ Light chains	V	30
120 total combinations	J	4
κ Light chains	V	40
200 total combinations	J	5
Heavy chains	V	40
6000 total combinations	D	25
	J	6

Overall; 1.9 × 10⁶ combinations

antigenically distinct from the original VSG. Again, the immune system responds to the antigen by producing an antibody. Once again, most of the flagellates are destroyed, but a few produce still another variation of the coat protein and the cycle continues. In experimental trypanosome infections, hundreds of cycles have been observed. In patients, parasitemia can persist for months or even years. This ability to change surface proteins as rapidly as the host immune system can generate new antibodies is the fruit of a set of genetic mechanisms hauntingly similar to the mechanisms that generate the antibody diversity the parasite is successfully evading.

On a genomic level, the trypanosome contains hundreds of VSG genes and pseudogenes, only one of which is expressed in a particular cell at a particular time. VSGs are transcribed as parts of polycistronic mRNAs from telomeric regions of the chromosome known as expression sites (ES). There are roughly 20 ES per cell. Most of the VSG genes not part of an ES (called silent VSGs) are found at the telomeres of roughly 100 minichromosomes of 50–150 kb each. These minichromosomes most likely evolved specifically to expand the accessible VSG repertoire of the organism. Finally, a few VSG genes and large numbers of pseudogenes (which are truncated, contain frameshift or in-frame stop-codon mutations, or lack the biochemical properties of expressed VSGs) are found in tandem-repeat clusters in subtelomeric locations. The structure of the VSG genes is depicted in Figure 3.2A.

There is only one active ES per cell, but several mechanisms can lead to expression of a new VSG, as depicted in Figure 3.2B. These mechanisms include (i) activation of a new ES (there are roughly 20 per genome) with inactivation of the original ES *in situ*, (ii) recombination of a VSG gene into the active ES via homologous recombination and telomere exchange, and (iii) segmental gene conversion of a portion or portions of VSG gene or genes into the ES VSG. Complex chimeric VSG containing elements of one or more silent VSG genes or pseudogenes may be produced. The ES promoter appears to be constitutively active and unregulated. Control of gene expression is mediated through post-transcriptional RNA processing and elongation. The active ES is located in a specialized nuclear region known as the expression site body. The molecular mechanisms underlying activation and inactivation of a given ES are not yet understood, but are hypothesized to involve a competition for transcription factors.

If switching between VSGs were simply random, one would expect an initial wave of parasites and then a second wave containing all the possible VSG variants which would overwhelm the host. However, this does not occur. There appears to be a hierarchy of switching mechanisms, so the more probable switching events occur early in infection, and VSGs generated by less probable mechanisms occur later in the infection. Perhaps not coincidentally, these mechanisms also contribute to variation over historical and evolutionary time of the parasite's VSG repertoire, as new sequences are assembled from the diverse genetic repertoire of potential VSG elements.

Trypanosomes are not the only organisms to utilize antigen switching or antigenic variation to evade immune responses. *Borrelia burgdorferi*, *Plasmodium*, *Neisseria gonorrhoeae*, and other organisms have various mechanisms of changing their antigenic constituents, but the mechanisms employed by the African trypanosome are the most spectacular.

STAPHYLOCOCCUS AUREUS: THE EXTRACELLULAR BATTLEGROUND

Staphyloccus aureus (S. aureus) is a gram-positive extracellular bacterium that is part of our commensal flora, living on the mucosal surfaces of humans and other mammals. It is a versatile pathogen in both community-acquired and hospital-acquired infections that range from superficial infections of skin and soft tissue to potentially life-threatening systemic disease. The first lines of defense against *S. aureus* are the recognition molecules and effector cells of the innate immune system; but *S. aureus* engages a multitude of mechanisms to subvert the innate immune response of the host.

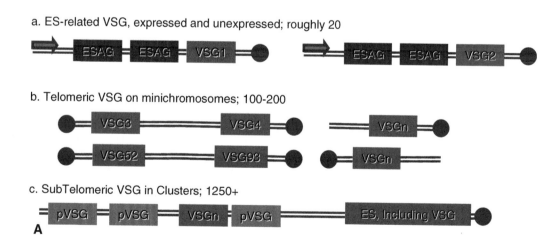

a. ES-related VSG, expressed and unexpressed; roughly 20

b. Telomeric VSG on minichromosomes; 100-200

c. SubTelomeric VSG in Clusters; 1250+

Figure 3.2A Mechanisms for generating variant surface glycoprotein diversity in trypanosomes: VSG genome structure. VSG sequences are in shades of red, others are purple. Silent VSG genes are dark red; expressed VSG genes are bright red, and VSG pseudogenes are pink. The large dots at the end of the chromosome represent telomeres. Green arrows are VSG promoters. ESAG are Expression Site Associated Genes, non-VSG genes, which are part of the polycistronic transcript driven by the VSG promoter. Designations of sequences (VSG1, etc.) are arbitrary and not meant to represent the actual arrangement of specific elements.

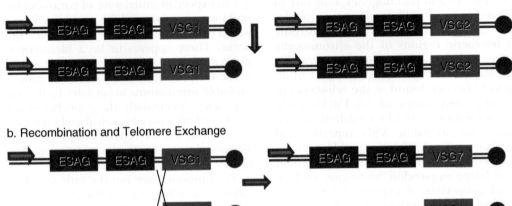

a. ES Switching – Post-transcriptional

b. Recombination and Telomere Exchange

c. Gene Conversion Events

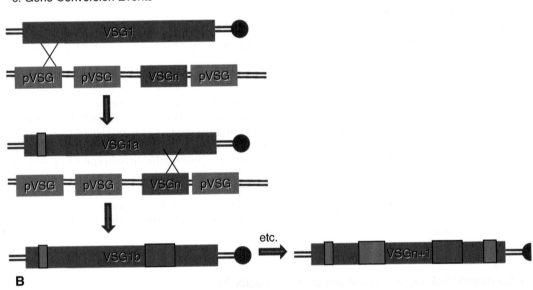

B

Figure 3.2B   Mechanisms for generating variant surface glycoprotein diversity in trypanosomes: Expressing new VSG. VSG sequences are in shades of red, others are purple. Silent VSG genes are dark red; expressed VSG genes are bright red, and VSG pseudogenes are pink. The large dots at the end of the chromosome represent telomeres. Green arrows are VSG promoters. The Xs represent recombination or gene conversion events. ESAG are Expression Site Associated Genes, non-VSG genes, which are part of the polycistronic transcript driven by the VSG promoter. Designations of sequences (VSG1, etc.) are arbitrary and not meant to represent the actual arrangement of specific elements. a. Post-transcriptional regulation causes different VSGs, located in alternative telomeric ESs, to be expressed. b. Recombination can switch a VSG gene from a minichromosome or other telomere to an ES. c. Gene conversion events can alter the sequence of VSGs located at ESs or elsewhere, drawing upon the sequence diversity not only of the silent VSGs but also of the VSG pseudogene pool.

The Innate Immune System: Recognition of Pathogens

After *S. aureus* breaches intact skin or mucosal lining, which constitutes the border by which the human body shields inside from outside, it first encounters resident macrophages. These are long-lived phagocytic cells that reside in tissue and participate in both innate and adaptive immunity. The macrophage expresses several receptors that are specific for bacterial constituents, such as LPS (endotoxin) receptors, TLR (see Table 3.3), mannose receptors, complement receptor C3R, glucan receptors, and scavenger receptors. After bacteria or bacterial constituents (such as lipopolysaccharide, peptidoglycan, or free bacterial DNA with CpG-rich oligonucleotide sequences) bind to their receptors, the macrophage engulfs them. The phagosome fuses with the lysosome to form the phagolysosome, degradative enzymes and antimicrobial substances are released, and the content of the phagolysosome is digested. The fragments are then presented to the adaptive immune system via MHC class II molecules.

Table 3.3 Recognition of Microbial Products Through Toll-like Receptors

Receptor	Ligands	Microorganisms Recognized	Notes
TLR-2 (TLR-1, -6) heterodimers	Peptidoglycan, bacterial lipoprotein and lipopeptide, porins, yeast mannan, lipoarabinomannan, glycophosphatidyl-inositol anchors	Gram-positive bacteria, mycobacteria, *Neisseria*, yeast, trypanosomes	Carried on macrophages
TLR-3 homodimer	Double-stranded RNA	Viral RNAs	
TLR-4 homodimer	Lipopolysaccharide (LPS)	Gram-negative bacteria	Carried on macrophages
TLR-5 homodimer	Flagellin	Gram-negative bacteria	Carried on intestinal epithelium; interacts directly with ligand
TLR-9 homodimer	DNA with unmethylated CpG motifs	Bacteria	Intracellular receptor

Activation of TLRs kicks off a common intracellular signal transduction pathway that involves an adaptor protein called MyD88 and the interleukin-1 receptor associated kinase (IRAK) complex. Ultimately, the transcription factor NFκB translocates to the nucleus of the macrophage and induces expression of inflammatory cytokines. Important cytokines that are secreted by macrophages in response to bacterial products include IL-1, IL-6, tumor necrosis factor α (TNFα), the chemokine CXCL8, and IL-12. These molecules have powerful effects and start off the local inflammatory response (Figure 3.3). A critical task of the macrophages is to

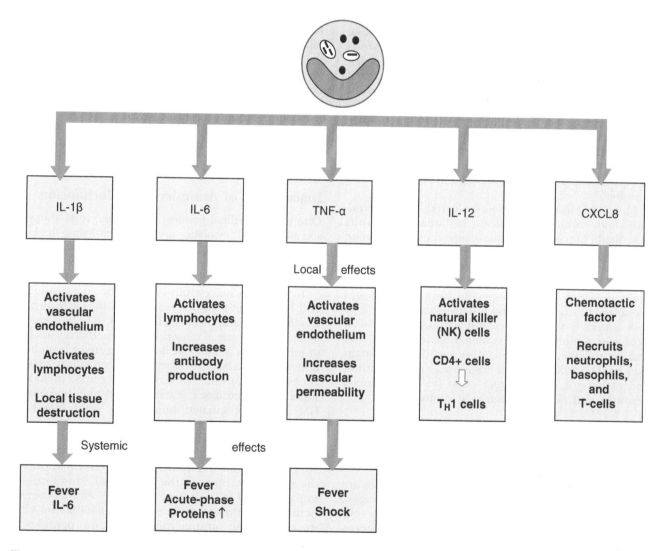

Figure 3.3 Chemokines secreted by macrophages in response to bacterial challenge. Chemokines secreted by macrophages have both local and systemic effects, which mobilize defenses to infection, but may have unfortunate consequences as well.

recruit neutrophils to the site of infection in an attempt to keep the infection localized.

As killers of bacteria, macrophages pale in comparison with neutrophils, which are short-lived, dedicated phagocytes circulating in the blood, awaiting a call from the macrophage to enter infected tissue. They are plentifully supplied with antimicrobial substances and mechanisms, stored in their granules and elsewhere, and are themselves programmed to die, on average within days after being released from the bone marrow. Neutrophils have receptors on their surfaces for inflammatory mediators derived both from the human host (CXCL8 from activated macrophages, C3a and C5a during complement activation) and from the bacteria themselves (for instance, bacteria-specific formylated peptides). Neutrophils home to sites of infection along the gradient of these chemoattractants.

The complement system is also part of the innate immune system. In the presence of microorganisms, the spontaneous low-level hydrolysis of complement factor C3 to iC3 is increased and leads to activation of C3, which ultimately leads to deposition of C3b on the microbial surface. This represents the alternative pathway of complement activation, and serves two purposes. First, C3b covalently bound to microbial surfaces tags them for more efficient phagocytosis since phagocytic cells have complement receptors. Second, the accumulation of C3b on the bacterial cell surface changes the specificity of the C3 convertase of the alternative pathway, Bb-C3b, to cleave C5, which then harbors the terminal complement proteins to form the lytic membrane attack complex consisting of C5b-C9. Gram-negative bacteria, but not gram-positive bacteria such as S. aureus, are successfully lysed by the pore-forming membrane attack complex.

Most infections are efficiently cleared by the ubiquitous and induced responses of innate immunity described. Once an infectious agent escapes innate mechanisms and spreads from the point of entry, it faces the adaptive immune response that is characterized by an extensive process in the draining lymph node in which pathogen-specific lymphocyte clones are selected, expanded, and differentiated.

S. aureus is such a successful pathogen because it expresses a multitude of virulence genes (Table 3.4) that act together to evade and subvert the three main axes of the innate immune responses: (i) recruitment and actions of inflammatory cells, (ii) antimicrobial peptides, and (iii) complement activation.

Inhibition of Inflammatory Cell Recruitment and Phagocytosis

While the macrophage and other primary defenses are sending messages about the presence of a pathogen, which recruit neutrophils and other inflammatory cells to the site of infection, S. aureus is busy blocking, scrambling, or subverting those messages. S. aureus contains an arsenal of antiadhesive and antimigratory proteins that specifically interfere with every step of host inflammatory cell recruitment. Staphylococcal chemotaxis inhibitory protein of S. aureus (CHIPS), present in approximately 60% of S. aureus strains, blocks neutrophil stimulation and chemotaxis by competing with the physiologic ligands at the complement receptor C5aR and formylated peptide receptor (FPR). Once near the site of infection, neutrophils must leave blood vessels through the vascular endothelium to reach the site of infection. This process starts with the rolling of leukocytes on activated endothelial cells by sticking to P-selectin expressed on the endothelial cell surface. Staphylococcal superantigen-like protein-5 (SSL-5) blocks this interaction. The next steps are adhesion to and transmigration through the endothelium, mediated in part by ICAM-1, the intercellular adhesion molecule-1 expressed on endothelial cells. S. aureus answers with production of extracellular adherence protein (Eap) that binds to ICAM-1, thereby interfering with extravasation of neutrophils at the site of infection.

Once a neutrophil manages to get close to an S. aureus cell, the bacterium still has means to evade phagocytosis. S. aureus expresses surface-associated antiopsonic proteins and a polysaccharide capsule that compromise efficient phagocytosis by neutrophils (Figure 3.4a). Protein A is a wall-anchored protein of S. aureus that binds the Fc portion of IgG and coats the surface of the bacterium with IgG molecules that are in the incorrect orientation to be recognized by the neutrophil Fc receptor (Figure 3.4c). Clumping factor A (ClfA) is a fibrinogen-binding protein present on the surface of S. aureus that binds to fibrinogen and coats the surface of the bacterial cells with fibrinogen molecules, additionally complicating the recognition process (Figure 3.4e).

Inactivation of Antimicrobial Mechanisms

One of the cardinal features of S. aureus is its ability to secrete several cytolytic toxins (hemolysins, leukocidins—see Table 3.4) that damage the membranes of host cells. They contribute to the development of abscesses with pus formation by direct killing of neutrophils.

If S. aureus is successfully engulfed by a neutrophil, its end has not yet come. It is well endowed with surface modifications and other mechanisms to help it survive in the phagosome. Transcriptional microarray analysis of mRNA from S. aureus following ingestion by neutrophils revealed a large number of differentially regulated genes. Many known stress-response genes, including superoxide-dismutases, catalase, and the leukotoxin Hlg, were upregulated immediately after ingestion. S. aureus is able to interfere with endosome fusion and the release of antimicrobial substances. Two superoxide dismutase enzymes help S. aureus to avoid the lethal effects of oxygen free radicals that are formed during the respiratory burst of the neutrophil. Modifications to the cell wall teichoic acid and other cell wall components change the cell surface charge such that the affinity of cationic, antimicrobial defensin peptides is reduced. Staphylokinase binds defensin peptides, and the extracellular metalloprotease aureolysin cleaves and inactivates certain defensin peptides.

Table 3.4 Examples of Virulence Factors Responsible for Immune Evasion by *Staphylococcus aureus* (Modified after Chavakis, T, Preissner KT, Herrmann M. The anti-inflammatory activities of *Staphylococcus aureus*. TRENDS in Immunology, 2007;28:408-418)

Name of Factor	Abbreviation	Function	Interference with Host Response
Anti-inflammatory Peptides			
Chemotaxis inhibitory protein of S. aureus	CHIPS	Binds to C5aR and formylated protein receptor (FPR)	Blocks chemotaxis
Staphylococcal complement inhibitor	SCIN	Stabilizes C2a-C4b and Bb-C3b convertases	Inhibits complement
Toxins			
Staphylococcal superantigen-like protein-5	SSL-5	Binds to P-selectin glycoprotein ligand-1 (PSGL-1)	Inhibits neutrophil recruitment
Staphylococcal superantigen-like protein-7	SSL-7	Binds to complement C5; binds to IgA	Inhibits complement
β-hemolysin	Hlb	Lysis of cytokine-containing cells	Cytotoxicity
γ-hemolysin	Hlg	Lysis of erythrocytes and leukocytes	Cytotoxicity
Panton-Valentine leukocidin	PVL	Stimulates and lyses neutrophils and macrophages	Change in gene expression of staphylococcal proteins; important in necrotizing pneumonia
Leukocidins D, E, M	LukD, LukE, LukM	Lysis of erythrocytes and leukocytes	Cytotoxicity
Exotoxins with superantigen activity enterotoxins	Se	Food poisoning when ingested; septic shock when systemic	Bridge MHC-II-TCR without antigen presentation; confer nonspecific T-cell activation and/or T-cell anergy; downregulate chemokine receptors
Toxic shock syndrome toxin-1	TSST-1		
Secreted Expanded Repertoire Adhesive Molecules			
Coagulase	Coa	Activates prothrombin and binds fibrin	Antiphagocytic
Extracellular adherence protein	Eap	Binds to endothelial cell membrane molecules, binds to ICAM-1 and T-cell receptors	Blocks neutrophil and T-cell recruitment; inhibits T-cell proliferation
Extracellular fibrinogen binding protein	Efb	Binds to fibrinogen; binds to complement factor C3 and inhibits its deposition on the bacterial cell surface	Inhibits complement activation beyond C3b, thereby blocking opsonophagocytosis; binds to platelets and blocks fibrinogen-induced platelet aggregation
Microbial Surface Components Recognizing Adhesive Matrix Molecules			
Clumping factor A	ClfA	Binds to fibrinogen	Antiphagocytic
S. aureus protein A	Spa	Binds to Fc portion of IgG and TNF receptor 1	Antiopsonic, antiphagocytic; modulates TNF signaling
Extracellular Enzymes			
Catalase	CatA	Inactivates free hydrogen peroxide	Required for survival, persistence, and nasal colonization
Staphylokinase	Sak	Plasminogen activator	Antidefensin; cleaves IgG and complement factors
Capsular Polysaccharides			
Capsular polysaccharide types 1, 5, and 8	CPS 1, CPS 5, CPS 8	Masks complement C3 deposition	Antiphagocytic effect

Inhibition of Complement Activation: You Can't Tag Me!

The prerequisite for complement activation is cleaving C3 into a soluble C3a and covalent attachment of C3b to the surface of *S. aureus*. This is either carried out by the C3 convertase C4bC2a (classical and lectin pathway) or C3bBb (alternative pathway). *S. aureus* secretes a protein called *Staphylococcus* complement inhibitor (SCIN) that stabilizes both C3 convertases and renders them less active. Similarly, the extracellular fibrinogen-binding protein Efb blocks C3 deposition on the bacterial surface (Table 3.5d). The effect is reduced opsonization and hence reduced phagocytosis. However, *S. aureus* not only prevents complement factor deposition, but is also capable of eliminating bound C3b and IgG through a very clever mechanism. Host plasminogen that is attached to the bacterial cell surface is activated by the *S. aureus* enzyme staphylokinase to plasmin, which then cleaves surface-bound C3b and IgG, resulting in reduced phagocytosis by neutrophils (Table 3.5b).

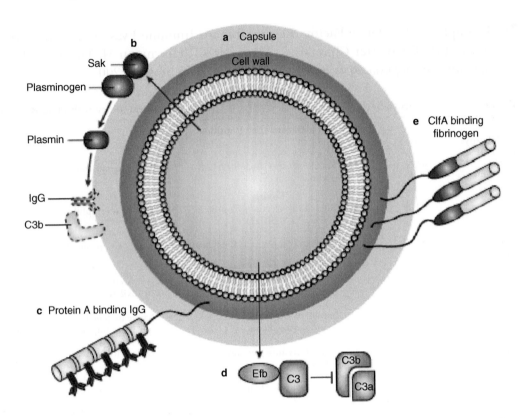

Figure 3.4   Mechanisms by which *Stapylococcus aureus* evades opsonophagocytosis. The figure illustrates **(a)** the capsular polysaccharide, which can compromise neutrophil access to bound complement and antibody; **(b)** the extracellular staphylokinase (Sak), which activates cell-bound plasminogen and cleaves IgG and C3b; **(c)** protein A with 5 immunoglobulin G (IgG) Fc-binding domains; **(d)** fibrinogen-binding protein (EfB), which binds complement factor C3 and blocks its deposition on the bacterial cell surface. Complement activation beyond C3b attachment is prevented, thereby inhibiting opsonization. **(e)** Clumping factor A (ClfA), which binds the γ-chain of fibrinogen. Reprinted with permission from Nature Publishing Group, *Nature Reviews Microbiology*, 2005;3:952.

Staphylococcal Toxins and Superantigens: Turning the Inflammatory Response on the Host

One of the most serious, life-threatening infections with *S. aureus* is toxic shock syndrome. It is caused by secreted exoenzymes and exotoxins (Table 3.4). Enterotoxins cause a fairly common, short-lived, benign gastroenteritis (food poisoning) when ingested, but act as a superantigen in systemic infections. The potent immunostimulatory properties of superantigens are a direct result of their simultaneous interaction with the V_b domain of the T-cell receptor and the MHC class II molecules on the surface of an antigen-presenting cell. Superantigens derive their name from the fact that they are able to polyclonally activate a large fraction of the T-cell population (2–20%) at picomolar concentrations, compared to a normal antigen-induced T-cell response where 0.001–0.0001% of the body's T-cells are activated. They bind to the variable part of the β chain of the T-cell receptor and to MHC class II molecules present on antigen-presenting cells without the need to be presented by antigen-presenting cells and cause an immune response that is not specific to any particular epitope on the superantigen. The cross-linking of MHC II molecule and TCR induces a signaling pathway that leads to proliferation of T-cells and a massive release

of cytokines. This systemic cytokine storm causes extravasation of plasma and protein, resulting in decreased blood volume and low blood pressure. Activation of the coagulation cascade leads to disseminated intravascular coagulation (DIC), which further compromises perfusion of end organs and eventually results in multi-organ failure and death. It is ironic that in toxic shock most of the deleterious effects on the host tissue are not related to actions of the bacteria, but the exaggerated host immune response. After all, it is usually not in the interest of *S. aureus* to kill the host.

MYCOBACTERIUM TUBERCULOSIS AND THE MACROPHAGE

Mycobacterium tuberculosis is an extremely successful pathogen and is one of the most important causes of worldwide morbidity and premature death. Spread by the aerosol route from person to person, organisms in droplet nuclei are deposited in the alveoli, where they encounter—and enter—their first immunological barrier, the alveolar macrophages.

Many pathogens utilize the intracellular compartment to evade host responses. While the intracellular lifestyle avoids many host defenses, such as complement and neutralizing antibody, other mechanisms of immunity operate intracellularly. In order to thrive intracellularly, pathogens must enter the cell. Utilization of the phagocytic pathway of entry carries the risk of fusion with lysosomes and destruction. Utilization of other pathways is more energy-intensive for the microbe, limits the potential for the parasite to exploit cellular mechanisms of transport and trafficking of endosomes, and may expose the microbe to cytoplasmic pattern-recognition receptors which will activate other defenses. After entering the cell, pathogens must survive and reproduce within whichever compartment is entered. This may require inhibition of lysosomal fusion, surviving constitutive intracellular inhibitory or killing mechanisms, transport of nutrients, and exploiting other aspects of host cell physiology. Next, intracellular pathogens must limit exposure to the cell-mediated immune system. Professional phagocytes are specifically equipped to present antigen via MHC class II, but most nucleated cells can present antigen via MHC class I pathways, which are specifically designed to initiate immune responses to intracellular pathogens by stimulating cytotoxic CD8+ T-cells. Finally, intracellular pathogens must prevent premature destruction of the host cell and pass successfully to another.

While tubercle bacilli are quite capable of extracellular growth and proliferation, in the early stages of infection this does not occur for long, since alveolar macrophages rapidly phagocytose them. The initial interaction between *M. tuberculosis* and the macrophage takes place in the absence of adaptive immunity. The mycobacterial surface appears to contain a rich array of TLR agonists that drive uptake of the organism and recruitment of inflammatory cells to the site of infection. For most microbes, that would be the end of the line; shortly after phagocytosis phagosome-lysosome fusion occurs, regulated by an elaborate array of proteins and glycolipids that control trafficking of membrane-bound organelles, and a variety of killing mechanisms are invoked. Acid pH, reactive oxygen and nitrogen intermediates, degradative enzymes, and toxic peptides kill and digest the invading pathogens. However, *Mycobacterium tuberculosis* survives and reproduces within the macrophage. Infected macrophages, though initially incapable of killing mycobacteria, nonetheless secrete chemokines that recruit other inflammatory cells to the area.

Mycobacterium and Macrophage: The Pathogen Chooses Its Destiny

The mycobacterial manipulation of the macrophage begins on uptake, and some of the mechanisms utilized by the pathogen are shown in Figure 3.5. In phagocytic cells, different receptors drive different processing of the phagosome. Thus, pathogens can control their intracellular fate by choosing the receptor that recognizes them. In the case of mycobacteria, multiple receptors may be involved, and the details of their interaction

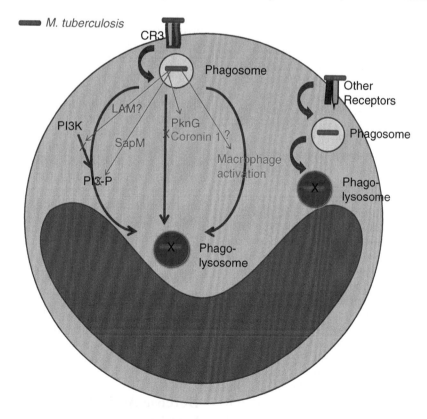

Figure 3.5 *Mycobacterium tuberculosis* **and the macrophage.** Gray arrows are host endosome processing pathways; the pink lines are mycobacterial mechanisms. Fusion of mycobacteria into mature phagolysosomes usually leads to death of the organism, so mycobacteria select their endocytotic pathway and interfere with mechanisms designed to result in phagosome-lysosome fusion. PI3K = phosphatidylinositol-3-kinase; PI3P = phosphatidylinositol-3-phosphate; LAM = lipoarabinomannan.

in vivo are not yet known. However, one of the major receptors is complement receptor type 3 (CR3), a pathway which prevents macrophage activation, in contrast with phagocytosis driven by other receptors. Once internalized in a phagosome, mycobacterial molecules actively interfere with phagosome-lysosome fusion, and cellular processes regulated by phosphorylation and dephosphorylation signals are controlled through modification of normal signals.

Despite all these mechanisms, *M. tuberculosis* have very little ability to block phagosome-lysosome fusion in *activated* macrophages stimulated via cytokines and TLRs. Probably the most important cytokines in activating macrophages to kill mycobacteria are interferon-γ and TNFα. Both in animal models and in humans, suppression of these pathways by drugs or mutations results in vulnerability to tuberculosis. A number of *M. tuberculosis* components inhibit elements of the activation pathway. LAM and its glycosylated derivatives can modulate signaling pathways initiated by interferon-γ and by TLRs and can block mitogen-activated protein kinase (MAPK) pathways within the network of activities that lead to activation. However, the macrophage is not entirely a passive host to the tubercle bacillus. The most effective response to this infection involves secretion of chemokines to recruit the adaptive immune system and other inflammatory cells and to present mycobacterial antigen to T-cells.

The Adaptive Response to *M. tuberculosis*: Containment and the Granuloma

Antigens from intracellular pathogens located in the endosomal compartment are primarily presented on MHC class II molecules and recognized by CD4 T-cells. This is not an exclusive arrangement since some antigen from the endosomal compartment is transported to the cytosol and presented to CD8 T-cells via MHC class I. Processing of antigen for class II requires acid pH and active acid proteases typically found in lysosomes; evidently not all *M. tuberculosis* succeed in inhibiting phagosome maturation and fusion.

CD4 T-cells mature into several populations of effector T-cells after extensive differentiation and selection. The best-known division is into Th$_1$ and Th$_2$ type cells. Th$_2$-type CD4 T-cells express cytokines that activate and induce class-switching in B-cells, resulting in an immune response centered around antibody production; opsonization; and handling of extracellular bacteria, viruses, and parasites. In contrast, the major activity of Th$_1$-type cells is to activate macrophages via interferon-γ, IL-2, TNFα, and other cytokines.

Differentiation of CD4 T-cells into the Th$_1$ and Th$_2$ lineages is controlled by the cytokines they encounter during the early stages of activation. Interferon-γ, secreted by macrophages and dendritic cells, induces differentiation toward the Th$_1$ phenotype. Since interferon-γ is a major cytokine produced by Th$_1$-type cells, the Th$_1$-type immune response is self-enforcing and tends to be stable unless other influences perturb the balance.

Activated macrophages are capable of killing ingested mycobacteria. The lesion that results from

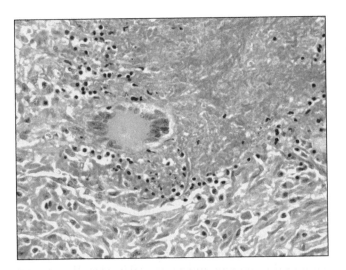

Figure 3.6   A mycobacterial granuloma. H&E stained sections of a mycobacterial granuloma. Central necrosis and an inflammatory response consisting of macrophages, lymphocytes, and fibroblasts are apparent. A large multinucleated giant cell, characteristic of the granulomatous reaction, is also present.

the effective immune response to *M. tuberculosis* infection is the granuloma, shown in Figure 3.6.

The granuloma structure is effective in containing, but typically not in eradicating, the mycobacteria. Persisting bacteria may remain viable for the life of the host, either within the necrotic center of the granuloma or in dynamic equilibrium within the inflammatory region of the lesion, or both. Under conditions of waning or suppressed immunity, the persisting mycobacteria can proliferate, spread, cause disease, and also escape, typically via the airborne coughed-out route, to a new host. *Mycobacterium tuberculosis* exploits the intracellular compartment to maintain itself in the host. Cell-mediated immune responses leading to macrophage activation, directed primarily by CD4 T-cells, lead to an at least partially protective response. A rather delicate balance of factors works in the interest of the pathogen. It is able to maintain itself for prolonged periods of time in a single host, awaiting an opportunity for transmission.

HERPES SIMPLEX VIRUS: TAKING OVER

Herpes simplex virus (HSV) types 1 and 2 are ubiquitous DNA viruses that cause a broad spectrum of disease, from painful oral and genital lesions to life-threatening brain and systemic infections. Once infected, a person usually harbors latent HSV for life.

Defense Against Viruses: Subversion and Sacrifice

Viruses present unique challenges to the immune system. Viruses enter host cells and then utilize the host protein-synthetic and other apparatus to assemble new viral particles. Entry into cells is rarely by phagocytosis until a mature immune response produces opsonized viral particles. Instead, viruses recognize and enter host cells via pathways that place them directly into the cytoplasm. Mechanisms by which viruses damage cells include direct cell lysis, inhibition of cell metabolism via diversion of resources and metabolic derangements, cell lysis or fusion, neoplastic transformation, and T-cell-mediated cellular injury.

In response to viral infection, the immune system has mechanisms for recognizing and managing viruses. There are pattern-recognition receptors that recognize the molecular signatures of viruses. In particular, TLR-3 recognizes double-stranded RNA, TLR-7 recognizes single-stranded RNA, and TLR-9 recognizes unmethylated CpG-containing DNA. The mannose-binding protein CD 206 recognizes some viral glycoproteins, including one HSV protein. Double-stranded RNA is also recognized by the cytoplasmic proteins RIG-1 and MDA-5. Recognition of viral components by TLR-3, RIG-1, or MDA-5 causes activation of the interferon-regulatory factors IRF3 and IRF7 and the nuclear translocation of regulatory factor NFκB, which induces a range of proinflammatory mechanisms, most importantly, production of the antiviral chemokines interferon (IFN) α and β.

IFNα/β induces a series of events that form a primary defense against viral infections. Via a series of protein phosphorylation events triggered by the Janus-family kinase linked to the interferon receptor, a number of antiviral activities are induced. These include (i) 2′–5′ oligoadenylate synthetase, an enzyme that produces 2′–5′-linked polyadenylates, which activate ribonuclease L to digest viral RNAs; and (ii) dsRNA-dependent protein kinase R (PKR), which modifies eukaryotic initiation factor 2α, leading to arrest of translation. In addition to the intracellular activities of IFNα/β which target viral replication, other activities include upregulation of IFNα/β synthesis (a positive feedback loop); (ii) alteration of the proteasome, the protein degradation system of the cell, to

favor production of peptides for presentation via MHC class I; (iii) upregulation of MHC class I expression and antigen presentation; and (iv) activation of macrophages, dendritic cells, and NK cells. Cytoplasmic antigens, such as most viral peptides, are exported from the cytosol to the endoplasmic reticulum. The heterodimeric proteins responsible for transport are called Transporters associated with Antigen Processing 1 and 2 (TAP1 and TAP2). TAP1 and TAP2 are induced by interferons, as are the MHC class I molecules which bind the antigen. After transport to the plasma membrane, MHC class I presents antigen to cytotoxic CD8 T-cells.

The final defense of virally infected cells is apoptosis. Apoptosis is an energy-requiring, deliberate, highly regulated process. Two major signaling pathways trigger apoptosis: (i) via extrinsic death receptors, such as the TNFα receptor, and (ii) via intracellular signals. In each case, a series of proteases called caspases are activated and destroy the critical infrastructure of the cell in a systematic way.

Herpes Simplex Virus on the High Wire: A Delicate Balancing Act

HSV has a large genome for a virus, consisting of ~150 kb, with at least 74 genes. While fewer than half the genes are required for replication in cell cultures, viruses isolated from human hosts almost always have the full complement. Those accessory genes not required for growth *in vitro* are mainly involved in evading or inhibiting host responses. A protein known as vhs (*viral host shutoff*) is an RNAse that degrades mRNA. Cellular responses to viral infection typically involve activation of response genes, and vhs globally inhibits such responses.

Because interferons play such a central role in defense against viral infection, a number of HSV proteins inhibit components of the IFNα/β response system (Table 3.5). HSV also acts to reduce presentation of antigen to the adaptive immune system.

Table 3.5 Interferon Actions and HSV Reactions

Mechanism	Effect	HSV Response
Activities That Inhibit Viral Gene Expression		
Activation of ribonuclease L	Digest viral RNAs	ICP0 inhibits ribonuclease L
ds-RNA-dependent phosphorylation of ribosomal initiation factor (PKR)	Arrest protein synthesis	Block the kinase responsible for phosphorylation; increase activity of phosphorylase, which restores activity
Activities That Enhance Inflammatory Responses		
Alteration of proteasome to favor production of peptides for class I MHC	Increase presentation of antigen to adaptive immune system	Unknown
Upregulation of MHC class I and associate mechanisms	Increase presentation of antigen to adaptive immune system	Block TAP transport of antigen, which in turn limits externalization of MHC class I
Activation of antigen-presenting and effector cells	Accelerate antibody and cell-mediated immune responses; induce apoptosis in infected cells	Infection of these cell types leads to downregulation of response elements, especially in dendritic cells
Upregulation of interferon synthesis	Positive-feedback loop to limit infectability of nearby cells	vhs globally inhibits host gene expression; ICP0 blocks multiple transduction mechanisms of IFN signaling

ICP47 blocks entry of peptides to the ER via binding to TAP. The activities that inhibit interferon actions also inhibit the interferon-mediated increase in MHC class I expression. In addition, HSV exerts broad inhibitory activities when it invades dendritic cells, inducing downregulation of co-stimulatory surface proteins, adhesion molecules, and class I MHC. Since dendritic cells play a central role in antigen presentation and control of the adaptive response, this inhibition, mediated in part by the US3 protein kinase, most likely slows the response to HSV infection.

HSV proteins even exert control over apoptosis. Infection with HSV initially makes cells resistant to apoptosis by either the extrinsic or intrinsic pathways. A number of viral proteins seem to be involved, and the complete pathway has yet to be determined. However, later in infection in some cell types, apoptosis is induced by HSV. The pathogen appears to create a delicate balance between inhibition of apoptosis early in infection, prior to production of virions, and induction of apoptosis late in the infective cycle.

HSV is an extremely prudent pathogen. It fails to completely inhibit the immune response, and local control of HSV infection is achieved relatively rapidly, usually with minimal lasting damage. Before this occurs, however, the virus has invaded neurons and entered a latent stage of infection, with only a small number of genes being transcribed at a low level. Neurons express low levels of MHC class I, which is further downregulated by HSV. Periodically, viral replication is turned on and viral particles are transported down the axon to its terminal near a mucosal surface, where the virus can invade epithelial cells and initiate a lesion, with more opportunities for transmission to a new host.

HIV: THE IMMUNE GUERILLA

By evolution, our immune system has developed several strategies to fight viral infections. In most chronic viral infections, both virus-specific T helper cells and cytotoxic T-lymphocytes (CTL) are required to effectively eliminate an infected cell. In turn, viruses have evolved numerous ways to evade the host immune system. Since the early 1980s, a new infectious disease of epidemic proportion has successfully emerged and spread around the globe: acquired immune deficiency syndrome (AIDS).

For more than two decades, the human immunodeficiency virus (HIV) has infected millions of people worldwide each year, mainly through mucosal transmission during unprotected sexual intercourse. In 2007, an estimated 33 million people lived with HIV globally, 2.5 million people were newly diagnosed with HIV infection, and 2.1 million patients died from AIDS. Since 1981, more than 25 million people have died from AIDS as a result of HIV infection (http://www.unaids.org/).

HIV is special in that this virus not only evades the immune response, but directly attacks the very effector cells that play a pivotal role in the fight against viruses, namely T-lymphocytes, macrophages, and dendritic cells. One of the paradoxes of HIV infection is that the virus elicits a broad immune response that is not completely protective, while it causes immune dysfunction on several levels. HIV infection is rarely eliminated by the immune system, but continues for many years and slowly progresses to AIDS and death if left untreated. Since the discovery of HIV in 1981, there has been an explosion of research aimed at deciphering the mechanism of infection, understanding why it cannot be controlled by our immune system, and at developing an effective vaccine.

Structure and Transmission of HIV—Small But Deadly

The human immunodeficiency virus is a human retrovirus belonging to the lentivirus group. Two genetically distinct forms exist, HIV-1 and HIV-2, but they cause similar syndromes and elicit similar host responses. HIV is an RNA virus that utilizes reverse transcriptase (RT) and other enzymes to convert its genome from RNA into an integrated proviral DNA. Its viral core contains the major capsid protein p24, nucleocapsid proteins, two copies of viral RNA, and three viral enzymes (protease, reverse transcriptase, and integrase). The viral particle is covered by a lipid bilayer that is derived from the host cell membrane. Two glycoproteins protrude from the surface: glycoproteins (gp)120 and gp41, which are critical for HIV infection of cells (Figure 3.7). In contrast with the 150 kb HSV genome, HIV has to accomplish all its tasks with a genome of only 9.8 kb.

The mode of transmission of HIV is mainly through close contact such as sexual intercourse: viral particles contained in semen enter the new host via microscopic lesions. Parenteral transmission, through blood products, sharing needles among drug users, or vertical transmission from mother to baby, is also an important route of transmission.

Invasion of Cells by HIV: Into the Lion's Den

The high affinity receptor that is used by HIV to enter host cells is the CD4 receptor, hence the major target for HIV is lymphoid tissue—more specifically, CD4+ T lymphocytes, macrophages, and dendritic cells; the very cells most deeply committed to dealing with viral infections (Figure 3.8). The first encounter between HIV and the naïve host takes place in the mucosa and draining lymph node. Dendritic cells play an important role in the infectious process. They are not only primary target cells, but also powerful professional antigen-presenting cells that are infected either directly or via capture of virus on their stellate processes. They can present antigens via MHC class I and class II molecules, stimulating both T-helper and CTL responses.

The presence of CD4 on host cells is not sufficient to mediate infection. The receptor needs to be accompanied by the presence of one of two chemokine receptors used as co-receptors: either CXCR4 or

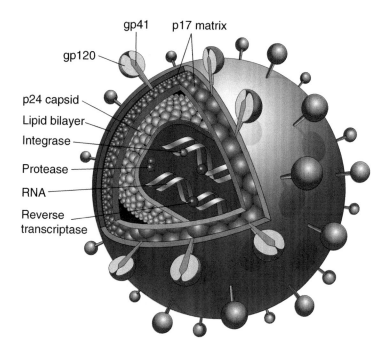

gp41   p17 matrix
gp120
p24 capsid
Lipid bilayer
Integrase
Protease
RNA
Reverse transcriptase

Figure 3.7   The structure of the HIV virion. Schematic illustration of an HIV virion. The viral particle is covered by a lipid bilayer that is derived from the host cell. Reprinted with permission from Elsevier Saunders. Robbins and Cotran: *Pathologic Basis of Disease*, 7th edition, copyright 2004, page 247.

CCR5. R4 viruses utilize CXCR4 as co-receptor which is expressed on lymphocytes, but not on macrophages. Hence, these viruses are called lymphocyte-tropic or T-tropic viruses. R5 viruses utilize CCR5 as a co-receptor, which is expressed on monocytes/macrophages, lymphocytes, and dendritic cells. R5 viruses are called macrophage-tropic or M-tropic viruses despite the fact that they can infect several cell types. Viruses that can use both CXCR4 and CCR5 as co-receptors are called dual tropic viruses. In the early phase of HIV infection, R5 (M-tropic) viruses dominate, but over the course of the infection the tropism often changes due to mutations in the viral genome, and R4 (lymphocyte-tropic) viruses increase in numbers.

The initial step in infection of any of the CD4+ cells is the binding of gp120 to CD4 molecules, which leads to a conformational change of the viral protein that now recognizes the co-receptor CCR5 or CXCR4. This interaction then triggers conformational change of gp41, which is noncovalently bound to pg120, and fusion of the viral bilayer with the host cell membrane. The HIV genome enters the host cell and reverse transcribes its RNA genome into cDNA (proviral DNA). In quiescent host cells, HIV cDNA may remain in the cytoplasm in linear form. In dividing host cells, the cDNA enters the nucleus and is then integrated in the host genome. In the case of the infected T-cell, proviral DNA may be transcribed, virions formed in the cytoplasm, and complete viral particles bud from the cell membrane. If there is extensive viral production (productive infection), the host cell dies. Alternatively, the HIV genome may remain silent, either in the cytoplasm or integrated as provirus into human

chromosomes, for months or even years (latent infection). Since macrophages and dendritic cells are relatively resistant to the cytopathic effect of HIV, they are likely important reservoirs of infection.

Clinically, the patient is asymptomatic or has flu-like symptoms during this first phase of HIV infection, also called the acute HIV syndrome. Approximately 40–90% of patients develop self-limiting symptoms (sore throat, myalgias, fever, weight loss, and a rash) 3–6 weeks after infection. This phase is characterized by widespread seeding of the lymphoid tissues, loss of activated CD4+ T-cells, and the highest level of viremia at any time during infection—unfortunately with high infectivity exactly when the infection is usually undiagnosed. The initial infection is readily controlled by the development of an HIV-specific cytotoxic T-lymphocyte (CTL) response and humoral response with antibodies raised against the envelope glycoproteins (Figure 3.8), and the patient again becomes asymptomatic. However, the antibody response is ineffective at neutralizing the virus. Thus, the role that antibodies play in controlling HIV disease is unclear. The level of viremia and the viral load in the lymphoid tissue at the end of the acute HIV syndrome define the so-called set point, which differs between individual patients and has prognostic implications. It is the result of a multifactorial process that is not yet clearly understood.

During the following clinical latency phase (also called the middle or chronic phase of HIV infection), the immune system is relatively intact. However, individual T-cells throughout the body, when activated by antigen contact or HIV itself, release intact virions and undergo apoptosis. Thus, it is misleading to talk about a latent infection in the context of HIV, since the definition of latency in the viral world implies a

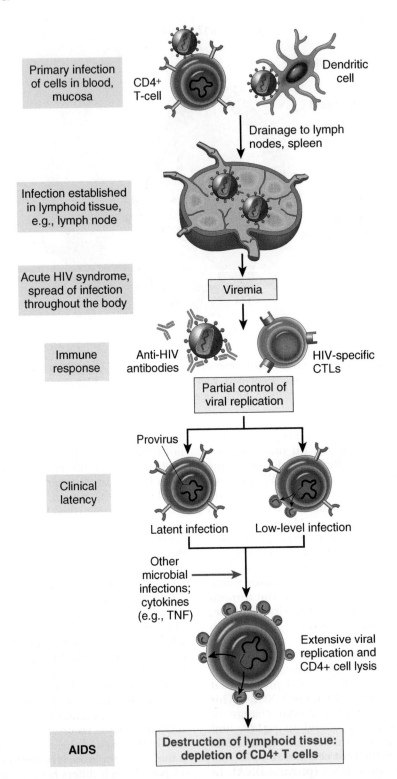

Figure 3.8 Pathogenesis of HIV-1 infection. Pathogenesis of HIV-1 infection. Initially, HIV-1 infects T-cells and macrophages directly or is carried to these cells by Langerhans cells. Viral replication in the regional lymph nodes leads to viremia and widespread seeding of lymphoid tissue. The viremia is controlled by the host immune response, and the patient then enters a phase of clinical latency. During this phase, viral replication in both T-cells and macrophages continues unabated, but there is some immune containment of virus. Ultimately, CD4+ cell numbers decline due to productive infection and other mechanisms, and the patient develops clinical symptoms of full-blown AIDS. Reprinted with permission from Elsevier Saunders. Robbins and Cotran: *Pathologic Basis of Disease*, 7th edition, copyright 2004, page 248.

lack of viral replication. In the case of HIV, there is continuous HIV replication, predominantly in the lymphoid tissues, which may last for years. This means that HIV infection lacks a phase of true microbiological latency. Indeed, the latent phase of HIV infection is a dynamic competition between the actively replicating virus and the immune system.

In a twist of fate, the life cycle in latently infected T-cells comes to completion (and usually leads to cell lysis) at the very moment when the T-cell is needed most—upon activation. On the molecular level this is achieved by sharing the transcription factor NFκB. After T-cells are activated by antigen or cytokines (such as TNFα, IL-1), signal transduction results in translocation of NFκB into the nucleus and upregulation of the expression of several cytokines. Flanking regions of the HIV genome also contain similar NFκB sites that are triggered by the same signal transduction molecule. Thus, the physiologic response of the T-cell stimulates virus production and leads ultimately to cell lysis and death of the infected cell. In addition, CD4+ T-cell loss is caused by mechanisms other than the direct cytopathic effect of the virus. Infected T-cells are killed by CTL cells that recognize HIV antigen presented on their cell surface. Even uninfected CD4+ T-cells, so-called innocent bystanders, are killed. Chronic activation of the immune system starts them down the pathway to apoptosis (programmed cell death), in this case activation-induced.

The last phase of HIV infection is progression to full-blown AIDS. A vicious cycle of increasingly productive viremia, loss of CD4+ cells, increased susceptibility to opportunistic infections, further immune activation, and progression of cell destruction develops. The clinical picture is characterized by a breakdown of host defense, a dramatic increase in circulating virus, and clinical disease. Patients present usually with long-lasting fever, fatigue, weight loss, and diarrhea. The onset of certain opportunistic infections such as invasive candidiasis, mycobacteriosis, or pneumocystosis (Figure 3.9), secondary neoplasms, or HIV-associated encephalitis marks the beginning of AIDS. Prior to the highly active antiretroviral therapy (HAART) era, AIDS was a death sentence, but treatment with several antiviral drugs has changed the fate of HIV-infected patients greatly.

Care has to be taken to avoid the development of drug resistance, which is due to the extreme plasticity of the HIV genome. Even early on during the infection, mutations are frequent due to error-prone replication by HIV and the structural flexibility of the viral envelope. Mutation and variation both in viral antigens and in viral physiology play an important role in the pathogenesis of HIV disease. This is also one reason why the quest for an HIV vaccine has remained elusive.

The devastating clinical course of AIDS and the unique pathological features of HIV infection, with significant viremia persisting for years, demonstrate the essential role of the CD4 T-cell in adaptive immunity (Table 3.6). While most CD4 T-cells have relatively

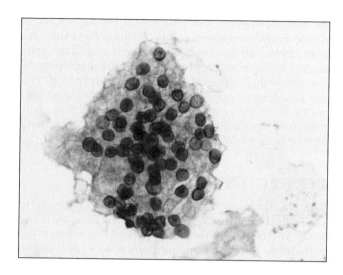

Figure 3.9 Opportunistic pathogen in AIDS. Cluster of *Pneumocystis jirovecii* cysts stained with toluidine blue in bronchoalveolar lavage of an HIV-positive patient (oil immersion, magnification 1000x).

Table 3.6 Immune Dysfunction in AIDS (Modified after Mori I and Nishiyama Y. Accessory genes define the relationship between the herpes simplex virus and its host. *Microbes and Infection.* 2006;8;2556–2562)

Altered Monocyte/Macrophage Functions

Decreased chemotaxis and phagocytosis
Decreased HLA class II antigen expression
Decreased antigen presentation capacity
Increased secretion of IL-1, IL-6, and TNFα

Altered T-cell Functions *In Vivo*

Preferential loss of memory T-cells
Susceptibility to opportunistic infections
Susceptibility to neoplasms

Altered T-cell Functions *In Vitro*

Decreased proliferate response to antigens
Decreased specific cytotoxicity
Decreased helper function for B-cell Ig synthesis
Decreased IL-2 and IFN-γ production

Polyclonal B-cell Activation

Hypergammaglobulinemia and circulating immune complexes
Inability to mount antibody response to new antigen
Refractoriness to normal B-cell activation *in vitro*

modest effector function, they are the central regulators of the immune response. CD4 T-cells are essential for maturation and development of B-cells and CD8+ cytotoxic T-cells, as well as for activation of macrophages. Patients with late-stage HIV infection become vulnerable to a host of opportunistic pathogens because they suffer from deteriorating antibody production, cell-mediated immunity, and decreased production and altered function of every kind of immune cell.

PERSPECTIVES

The five pathogens discussed here scarcely begin to cover the breadth of microbial interactions with the host. Microbes, for the most part, do not enter the forbidding interior of mammalian hosts with a single, general-purpose toxin or strategy for causing disease. Instead, they come equipped with a variety of very specific molecular tools, each with a particular target, which allow them to survive, replicate, and be transmitted to a new host.

It's difficult to overestimate the subtlety of the interactions of pathogens with the host on the molecular level. Pathogens rarely eradicate an immune response. Rather, because of cross-talk and redundancy on a molecular level between elements of the immune system, they attenuate, misdirect, and delay the host response. Overall, the complexity and flexibility of the interaction benefit both the host and the pathogen. The pathogen benefits because, even if it is ultimately eliminated from the host, it survives and proliferates long enough and well enough to be transmitted. The host benefits because a temperate response is less likely to result in severe collateral tissue damage. In addition, if host responses were so rigid and forceful that the pathogen was forced to kill the host or die itself, the microbes, ultimately, would win, due to their rapid evolution. The mammalian immune system, for all its extraordinary complexity and power, is a compromise between metabolic cost and efficacy, between elimination and containment of pathogens, and is able to live with what it cannot destroy.

KEY CONCEPTS

- Pathogens evolve rapidly to evade and manipulate host defenses to maximize the spread and transmission of the microbe.
- The African trypanosome has evolved into an intricate genetic apparatus for generating surface antigenic diversity to outstrip the ability of the immune system to generate antibody diversity.
- Many virulence factors, such as anti-inflammatory peptides, toxins, or adhesive molecules, allow *Staphylococcus aureus* to evade the innate immune system in the extracellular space.
- In staphylococcal toxic shock syndrome, host tissue is damaged in response to an exaggerated immune system, not directly by *Staphylococcus aureus*.

- *Mycobacterium tuberculosis* invades the macrophage and manipulates the cellular response to allow it to escape destruction and proliferate intracellularly.
- The immune response that produces a tuberculous granuloma contains but does not eliminate infection, which persists and may reactivate when immunity wanes.
- Herpes simplex virus devotes a substantial fraction of its genome to mechanisms for attenuating and misdirecting the immune response to viral infection, including intracellular pattern recognition, interferon-mediated mechanisms, and apoptosis.
- HIV infection is special in that this virus not only evades the immune response, but directly attacks and destroys the very effector cells that try to destroy it: CD4+ T-cells, macrophages, and dendritic cells.
- There is no true latency in HIV infection: the virus continuously replicates in lymphoid tissue.
- Destruction of CD4+ T-cells by HIV causes profound immune dysfunction, highlighting the central regulatory role that these cells play in our immune system.

SUGGESTED READINGS

1. Abbas AK. Diseases of immunity. In: Kumar V, Abbas A, Fausto N, eds. *Robbins and Cotran Pathologic Basis of Disease*. Philadelphia: Elsevier Saunders; 2004:245–258.
2. Chavakis T, Preissner KT, Herrmann M. The anti-inflammatory activities of *Staphylococcus aureus*. *TRENDS Immunology*. 2007; 28:408–418.
3. Cunningham AL, Diefenbach RJ, Miranda-Saksena M, et al. The cycle of herpes simplex virus infection: virus transport and immune control. *Journal of Infectious Disease*. 2006;194; S11–18.
4. Foster TJ. Immune evasion by staphylococci. *Nature Reviews*. 2005;3:948–958.
5. Gill N, Davies EJ, and Ashkar AA. The role of Toll-like receptor ligands/agonists in protection against genital HSV-2 infection. *American Journal of Reproductive Immunology*. 2008;59; 35–43.
6. HIV-1. UNAIDS. 2007 AIDS epidemic update. 2008. http://www.unaids.org/en/ accessed July 11, 2008.
7. Hladik F, Sakchalathorn P, Ballweber L, et al. Initial events in establishing vaginal entry and infection by human immunodeficiency virus type-1. *Immunity*. 2007;26:257–270.
8. Houben EN, Nguyen L, Pieters J. Interaction of pathogenic mycobacteria with the host immune system. *Current Opinion Microbiology*. 2006;9:76–85.
9. Johnston MI, Fauci AS. An HIV vaccine—Evolving concepts. *N Engl J Med*. 2007;356:2073–2081.
10. Mguyen ML and Blaho JA. Apóptosis during herpes simplex virus infection. *Advances in Virus Research*. 2007;69;67–97.
11. Mori I and Nishiyama Y. Accessory genes define the relationship between the herpes simplex virus and its host. *Microbes and Infection*. 2006;8;2556–2562.
12. Murphy K, Travers P, Walport M. *Janeway's Immunobiology*. 7th ed. New York: Garland Science; 2008.
13. Parham P. *The Immune System*. New York: Garland Science; 2005.
14. Pays E. Regulation of antigen gene expression in *Trypanosoma brucei*. *Trends Parasitology*. 2005;21:517–520.
15. Pays E, Vanhamme L, Perez-Morga D. Antigenic variation in *Trypanosoma brucei*: Facts, challenges and mysteries. *Current Opinion Microbiology*. 2004;7:369–374.

16. Pieters J. *Mycobacterium tuberculosis* and the macrophage: Maintaining a balance. *Cell Host Microbe.* 2008;3:399–407.
17. Russell DG. Who puts the tubercle in tuberculosis? *Nature Reviews Microbiology.* 2007;5:39–47.
18. Taylor JE, Rudenko G. Switching trypanosome coats: What's in the wardrobe? *Trends Genetics.* 2006;22:614–620.
19. Voyich JM, Braughton KR, Sturdevant DE, et al. Insights into mechanisms used by *Staphylococcus aureus* to avoid destruction by human neutrophils. *J Immunology.* 2005;175:3907–3919.
20. Walker BD, Burton DR. Toward an AIDS vaccine. *Science.* 2008;320:760–764.

4

Neoplasia

William B. Coleman . Tara C. Rubinas

INTRODUCTION

Cancer does not represent a single disease. Rather, cancer is a collection of myriad diseases with as many different manifestations as there are tissues and cell types in the human body, involving innumerable endogenous or exogenous carcinogenic agents and various etiological mechanisms. What all of these disease states share in common are certain biological properties of the cells that compose the tumors, including unregulated (clonal) cell growth, impaired cellular differentiation, invasiveness, and metastatic potential. It is now recognized that cancer, in its simplest form, is a genetic disease or, more precisely, a disease of abnormal gene expression. Recent research efforts have revealed that different forms of cancer share common molecular mechanisms governing uncontrolled cellular proliferation, involving loss, mutation, or dysregulation of genes that positively and negatively regulate cell proliferation, migration, and differentiation (generally classified as proto-oncogenes and tumor suppressor genes). The molecular mechanisms associated with neoplastic transformation and tumorigenesis of specific cell types is beyond the scope of this chapter. Rather, in the discussion that follows, we will introduce essential concepts related to neoplastic disease as a foundation for more detailed treatment of the molecular carcinogenesis of major cancer types provided elsewhere in this book.

CANCER STATISTICS AND EPIDEMIOLOGY

Cancer Incidence

Cancer is an important public health concern in the United States and worldwide. Due to the lack of nationwide cancer registries for all countries, the exact numbers of the various forms of cancer occurring in the world populations are unknown. Nevertheless, monitoring of long-range trends in cancer incidence and mortality among different populations is important for investigations of cancer etiology. Given the long latency for formation of a clinically detectable neoplasm (up to 20–30 years) following initiation of the carcinogenic process (exposure to carcinogenic agent), current trends in cancer incidence probably reflect exposures that occurred many years (and possibly decades) before. Thus, correlative analysis of current trends in cancer incidence with recent trends in occupational, habitual, and environmental exposures to known or suspect carcinogens can provide clues to cancer etiology. Other factors that influence cancer incidence include the size and average age of the affected population. The average age at the time of cancer diagnosis for all tumor sites is approximately 67 years. As a higher percentage of the population reaches age 60, the general incidence of cancer will increase proportionally. Thus, as the life expectancy of the human population increases due to reductions in other causes of premature death (due to infectious and cardiovascular diseases), the average risk of developing cancer will increase.

General Trends in Cancer Incidence

The American Cancer Society estimates that 1,437,180 new cases of invasive cancer were diagnosed in the United States in 2008. This number of new cancer cases reflects 745,180 male (52%) and 692,000 female cancer cases (48%). This estimate does not include carcinoma *in situ* occurring at any site other than in the urinary bladder and does not include basal and squamous cell carcinomas of the skin. In fact, basal and squamous cell carcinomas of the skin represent the most frequently occurring neoplasms in the United States, with an estimated occurrence of >1 million total cases in 2008. Likewise, carcinoma *in situ* represents a significant number of new cancer cases with 67,770 newly diagnosed breast carcinomas *in situ* and 54,020 new cases of melanoma carcinoma *in situ*.

Estimated site-specific cancer incidence for both sexes combined are shown in Figure 4.1. Cancers of the reproductive organs represent the largest group of newly diagnosed cancers in 2008 with 274,150 new

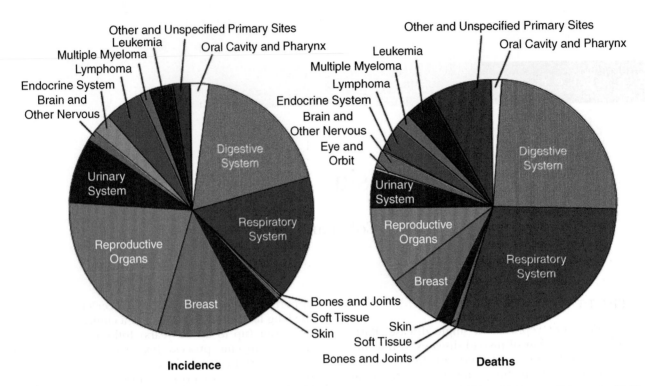

Figure 4.1   Cancer incidence and mortality by site for both sexes (United States, 2008). Cancers of the reproductive organs include those affecting the prostate, uterine corpus, ovary, uterine cervix, vulva, vagina, testis, penis, and other organs of the male and female genital systems. Cancers of the digestive system include those affecting esophagus, stomach, small intestine, colon, rectum, anus, liver, gallbladder, pancreas, and other digestive organs. Cancers of the respiratory system include those affecting the lung, bronchus, larynx, and other respiratory organs.

cases. This group of cancers includes prostate (186,320 cases), uterine corpus (40,100 cases), ovary (21,650 cases), and uterine cervix (11,070 cases), in addition to other organs of the genital system. The next most frequently occurring tumors originated in the digestive tract (271,290 cases), respiratory system (232,270 cases), and breast (184,450 cases). The majority of digestive system tumors involved colon (108,070 cases), rectum (40,740 cases), pancreas (37,680 cases), stomach (21,500 cases), liver and intrahepatic bile duct (21,370 cases), and esophagus (16,470 cases), in addition to the other digestive system organs. Most new cases of cancer involving the respiratory system affected the lung and bronchus (215,020 cases), with the remaining cases affecting the larynx or other components of the respiratory system. Other sites with significant cancer burden include the urinary system (125,490 cases), lymphomas (74,340 cases), skin (67,720 cases), leukemias (44,270 cases), and the oral cavity and pharynx (35,310 cases).

Among men, cancers of the prostate, respiratory system, and digestive system occur most frequently. Together, these cancers account for 62% of all cancers diagnosed in men. Prostate is the leading site, accounting for 186,320 cases and 25% of cancers diagnosed in men. Among women, cancers of the breast, respiratory system, and digestive system occur most frequently. Cancers at these sites combine to account for 59% of all cancers diagnosed in women. Breast is the leading site for tumors affecting women, accounting for

182,460 cases and 26% of cancers diagnosed in women.

General Trends in Cancer Mortality in the United States

Mortality attributable to invasive cancers produced 565,650 cancer deaths in 2008. This reflects 294,120 male deaths (52%) and 271,530 female deaths (48%). Estimated numbers of cancer deaths by site for both sexes are shown in Figure 4.1. The leading cause of cancer death involves tumors of the respiratory system (166,280 deaths), the majority of which are neoplasms of the lung and bronchus (161,840 deaths). The second leading cause of cancer deaths involve tumors of the digestive system (135,130 deaths), most of which are tumors of the colorectum (49,960 deaths), pancreas (34,290 deaths), stomach (10,880 deaths), liver and intrahepatic bile duct (18,410 deaths), and esophagus (14,280 deaths). Together, tumors of the respiratory and digestive systems account for 53% of cancer deaths.

Trends in cancer mortality among men and women mirror cancer incidence. Cancers of the prostate, lung and bronchus, and colorectum represent the three leading sites for cancer incidence and mortality among men. In a similar fashion, cancers of the breast, lung and bronchus, and colorectum represent the leading sites for cancer incidence and mortality among women. While cancers of the prostate and breast represent the leading sites for new cancer diagnoses among men and women, the majority of deaths in both sexes are related to cancers of the lung and bronchus. Tumors of the lung and bronchus are responsible for 31% of all cancer deaths among men and 26% of all cancer deaths among women. The age-adjusted death rate for lung cancer among men has increased dramatically over the last 60–70 years, while the death rates for other cancers have remained relatively stable. The lung cancer death rate for women has increased in an equally dramatic fashion since about 1960, becoming the leading cause of female cancer death in the mid-1980s after surpassing the death rate for breast cancer.

Global Cancer Incidence and Mortality

The IARC GLOBOCAN project estimates that 10,862,496 new cancer cases were diagnosed worldwide in 2002. This number of new cases represents 5,801,839 male cases (53%) and 5,060,657 female cases (47%). Mortality attributed to for the same year produced 6,723,887 deaths worldwide. This reflects 3,795,991 male deaths (56%) and 2,927,896 female deaths (44%). The leading sites for cancer incidence and mortality worldwide in 2002 included tumors of the lung, stomach, prostate, breast, colorectum, esophagus, and liver. Lung cancer accounted for the most new cases and the most deaths during this period of time, with 1,352,132 cases and 1,178,918 deaths for both sexes combined. The leading sites for cancer incidence among males included lung (965,241 cases), prostate (679,023 cases), stomach (603,419 cases), colorectum (550,465 cases), and liver (442,119 cases). Combined, cancers at these sites account for 44% of all cases among men. The leading causes of cancer death among men included tumors of the lung (848,132 deaths), stomach (446,052 deaths), liver (416,882 deaths), colorectum (278,446 deaths), and esophagus (261,162 deaths). Deaths from these cancers account for 59% of all male cancer deaths. The leading sites for cancer incidence among females included breast (1,151,298 cases), cervix uteri (493,243 cases), colorectum (472,687 cases), lung (386,891 cases), and stomach (330,518 cases). Combined, cancers at these sites account for 56% of cancer cases among women. The leading causes of cancer death among females directly mirrors the leading causes of cancer incidence: breast (410,712 deaths), lung (330,786 deaths), cervix uteri (273,505 deaths), stomach (254,297 deaths), and colorectum (250,532 deaths). Combined, these sites account for 52% of female cancer deaths.

Risk Factors for the Development of Cancer

Risk factors for cancer increase the chance that an individual will develop neoplastic disease. Individuals who have risk factors for cancer development are more likely to develop the disease at some point in their lives than the general population (lacking the same risk factors). However, having one or more risk factors does not necessarily mean that a person will develop cancer. It follows that some people with recognized risk factors for cancer will never develop the disease, while others lacking apparent risk factors will develop neoplastic disease. While certain risk factors are clearly associated with the development of neoplastic disease, making a direct linkage from a risk factor to causation of the disease remains very difficult and often impossible. Some risk factors for cancer development can be modified, while others cannot. For instance, cessation of cigarette smoking reduces the chance that an individual will develop cancer of the lung. In contrast, a woman with an inherited mutation in the *BRCA1* gene carries an elevated lifetime risk of developing breast cancer. Some of the major risk factors that contribute to cancer development include (i) age, (ii) race, (iii) gender, (iv) family history, (v) infectious agents, (vi) environmental exposures, (vii) occupational exposures, and (viii) lifestyle exposures.

Age, Race, and Gender as Risk Factors for Cancer Development

Cancer is predominantly a disease of old age. In fact, most malignant neoplasms are diagnosed in patients over the age of 65. According to the National Cancer Institute SEER Statistics (http://seer.cancer.gov/index.html) the median age at diagnosis for cancer of the lung is 71 (<2% occur in people <45 years old), the median age for cancer of the prostate is 68 (<10% occur in men <55 years old), the median age for cancer of the breast is 61 (<2% occur in women <35 years old), the median age for cancer of the colorectum is 71 (<5% occur in people <45 years old), the median age for cancer of the liver is 65 (<5% occur in people <45 years old), the median age for cancer of the ovary is 63 (<5% occur in women <35 years old), and the median age of melanoma is 59 (<10% occur in people <35 years old). In the case of prostate cancer, 86% occur in men over the age of 65, and 99.5% occur in men over 50. Likewise, 97% of prostate cancer deaths occur in men over the age of 65. In contrast, female breast cancer occurs much more frequently in younger individuals. Nonetheless, 63% of cases occur in women over the age of 65, and 88% occur in women over the age of 50. A notable exception to this relationship between advanced age and cancer incidence involves some forms of leukemia and other cancers of childhood. Acute lymphocytic leukemia (ALL) occurs with a bimodal distribution, with highest incidence among individuals less than 20 years of age, and a second peak of increased incidence among individuals of advanced age. The majority of ALL cases are diagnosed in children, with 40% of cases diagnosed in children under the age of 15. Despite the prevalence of this disease in childhood, a significant number of adults are affected. In fact, 32% of ALL cases are diagnosed in individuals over 65. In contrast to ALL, the other major forms of leukemia demonstrate the usual pattern of age-dependence observed with solid tumors, with large numbers of cases in older segments of the population.

Cancer incidence and mortality can vary tremendously with race and ethnicity. In the United States, African Americans and Caucasians are more likely to develop cancer than individuals of other races or ethnicity. African American men demonstrate a cancer incidence for all sites combined of 664 cases per 100,000 population, and Caucasian men exhibit a cancer rate of 557 cases per 100,000 population. In contrast, Native American men show the lowest cancer incidence among populations of the United States with 321 cases per 100,000 population for all sites combined. African American women demonstrate a cancer incidence for all sites combined of 397 cases per 100,000 population, and Caucasian women exhibit a cancer rate of 424 cases per 100,000 population, while Native American women show the lowest cancer incidence with 282 cases per 100,000 population. These differences in race-related cancer incidence rates can be magnified when site-specific cancers are considered. The overall incidence of prostate cancer in the United States (for all men) is 163 cases per 100,000 men. African American men have a significantly higher incidence rate (249 cases per 100,000 men) compared to Caucasian men (157 cases per 100,000). In contrast, Native American men have a significantly lower incidence of prostate cancer (73 cases per 100,000 men). The mechanisms that account for these differences are not known, but may be related to genetic factors or differences in various physiological factors (for instance, androgen hormone levels). Similar to the cancer incidence rates, mortality due to cancer is higher among African Americans (322 and 189 per 100,000 population for men and women) and Caucasians (235 and 161 deaths per 100,000 population for men and women) than other populations, including Asian/Pacific Islanders, American Indians, and Hispanics. Factors that are known to contribute to racial differences in cancer-related mortality include (i) differences in exposures (for instance, smoking prevalence), (ii) access to regular cancer screening (for breast, cervical, and colon cancers), and (iii) timely diagnosis and treatment.

Gender is clearly a risk factor for cancers affecting certain tissues such as breast and prostate where there are major differences between men and women. However, there are numerous other examples of cancers that appear to develop preferentially in men or women, and/or where factors related to gender increase risk. For instance, liver cancer affects men more often than women. The ratio of male to female incidence in the United States is 2:1, and worldwide is 2.4:1. However, in high incidence regions, the male to female incidence ratio can be as high as 8:1. This observation suggests that sex hormones and/or their receptors may play a significant role in the development of primary liver tumors. Some investigators have suggested that hepatocellular carcinomas overexpress androgen receptors, and that androgens are important in the promotion of abnormal liver cell proliferation. Others have suggested that the male predominance

of liver cancer is related to the tendency for men to drink and smoke more heavily than women, and are more likely to develop cirrhosis.

Family History as a Risk Factor for Cancer Development

Familial cancers have been described for most major organ systems, including colon, breast, ovary, and skin. Hereditary cancers are typically characterized by (i) early age at onset, (ii) neoplasms arising in first degree relatives of the index case, and (iii) in many cases, multiple or bilateral tumors. Epidemiologic evidence has consistently pointed to family history as a strong and independent predictor of breast cancer risk. Thus, women with a first-degree relative (mother or sister) who has been diagnosed with breast cancer are at elevated risk for the disease. The breast cancer susceptibility genes BRCA1, BRCA2, and p53 account for the majority of inherited breast cancers. Five to 10% of breast cancers occurring in the United States each year are related to genetic predisposition. Despite the recognition of multiple risk factors for development of breast cancer, 50% of affected women have no identifiable risk factors other than being female and aging. In many cases, the genetic predisposition to cancer development may be related to small but measurable risks associated with genetic variations (polymorphic variations) at multiple loci.

Infectious Agents as Risk Factors for Cancer Development

Liver cancers are frequently associated with hepatitis virus infection and related conditions. Both hepatitis B virus (HBV) and hepatitis C virus (HCV) are associated with development of hepatocellular carcinoma. Primary liver cancers are usually associated with chronic hepatitis, and 60–80% of hepatocellular carcinomas occurring worldwide develop in cirrhotic livers, most commonly nonalcoholic posthepatitic cirrhosis. However, hepatitis virus infection is not thought to be directly carcinogenic. Rather, HBV and/or HCV infection produces hepatocyte necrosis and regeneration (cell proliferation), which makes the liver susceptible to endogenous and exogenous carcinogens. Thus, preneoplastic nodules in the liver tend to occur in regenerative nodules of the injured liver (chronic hepatitis or cirrhosis). In certain geographic areas (such as China), large portions of the population are concurrently exposed to the hepatocarcinogen aflatoxin B_1 and HBV, which increases their relative risk for development of liver cancer.

Cervical cancer is the second most common cancer of women worldwide, with 493,243 cases and 273,505 deaths in 2002. Human papillomavirus infection (in particular HPV16 and HPV18) are specifically associated with development of cervical cancer (Figure 4.2). Case-control and prospective epidemiological studies have shown that HPV infection precedes high-grade

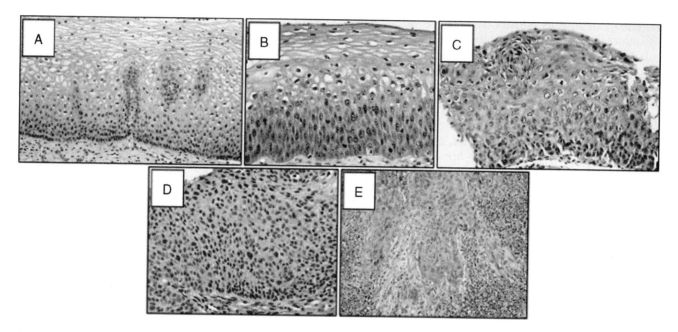

Figure 4.2 Progression of dysplasia in the cervix. (A) Normal squamous epithelium of the cervix. **(B)** Low-grade squamous dysplasia. **(C)** Moderate squamous dysplasia. The dysplastic squamous cells and mitotic figures involve the lower two-thirds of the epithelium. **(D)** High-grade squamous dysplasia. **(E)** Invasive squamous cell carcinoma.

dysplasia and invasive cancer, and represents the strongest independent risk factor for the development of cervical cancer. Molecular analyses of cervical cancers suggest that HPV plays a key role in cervical carcinogenesis since >90% of cervical cancer biopsies contain DNA sequences of high-risk HPV types.

Epstein-Barr virus (EBV) is a ubiquitous virus that infects >90% of the human population. Primary infection with EBV usually occurs in childhood, and in most cases there is no clinical course. However, in a subset of infected individuals, primary EBV infection can result in infectious mononucleosis. EBV causes a latent infection in lymphoid cells, which persists for life in a small subpopulation of B-lymphocytes. Although EBV has a very high transforming potential *in vitro*, it rarely causes tumors in humans, suggesting that EBV-infected cells are subject to continuous surveillance by cytotoxic T-cells. However, EBV infection is also closely associated with the development of lymphoid and epithelial malignancies in immunocompetent hosts. The human neoplasm that is classically associated with EBV infection is Burkitt's lymphoma. Other cancers that are associated with EBV infection include nasopharyngeal cancer, Hodgkin disease, and certain cancers of the stomach.

Environmental and Occupational Exposures as Risk Factors for Cancer Development

Environmental exposures represent significant risk factors for certain forms of cancer. The most well-studied hepatocarcinogen is a natural chemical carcinogen known as aflatoxin B_1 that is produced by the *Aspergillus flavus* mold. This mold grows on rice or other grains (including corn) that are stored without refrigeration

in hot and humid parts of the world. Ingestion of food that is contaminated with *Aspergillus flavus* mold results in exposure to potentially high levels of aflatoxin B_1. Aflatoxin B_1 is a potent, direct-acting liver carcinogen in humans, and chronic exposure leads inevitably to development of hepatocellular carcinoma.

Another naturally occurring carcinogen is the radioactive gas radon, which has been suggested to increase the risk of lung cancer development. This gas is ubiquitous in the earth's atmosphere, creating the opportunity for exposure of vast numbers of people. However, passive exposure to the background levels of radon found in domestic dwellings and other enclosures is not sufficiently high to increase lung cancer risk. High-level radon exposure has been documented among miners working in uranium, iron, zinc, tin, and fluorspar mines. These workers show an excess of lung cancer (compared to nonminers) that varies depending on the radon concentration encountered in the ambient air of the specific mine.

Although once heavily studied and thought to be a major mechanism of human cancer induction, exposure to chemical carcinogens does not represent an important risk factor for most of the general population. Nevertheless, several chemicals, complex chemical mixtures, industrial processes, and/or therapeutic agents have been associated with development of malignant neoplasms in exposed human populations. These exposures may include therapeutic exposure to the radioactive compounds (such as thorium dioxide or Thorotrast for the radiological imaging of blood vessels) and occupational exposures to certain industrial chemicals (such as vinyl chloride monomer, asbestos, bis[chloromethyl] ether, or chromium). In addition, certain food preservatives (such as nitrites) are of significant concern as potential carcinogens. Nitrites can produce nitrosylation of amines in various foodstuffs, resulting in formation of nitrosamine compounds, which are suspected to have carcinogenic potential in humans.

Lifestyle Exposures as Risk Factors for Cancer Development

Cancers of the lung, mouth, larynx, bladder, kidney, cervix, esophagus, and pancreas are related to consumption of tobacco products, including cigarettes, cigars, chewing tobacco, and snuff. Cigarette smoking alone is the suggested cause for one-third of all cancer deaths. Several lines of evidence strongly link cigarette smoking to lung cancer. Smokers have a significantly increased risk (11-fold to 22-fold) for development of lung cancer compared to nonsmokers, and cessation of smoking decreases the risk for lung cancer. Furthermore, heavy smokers exhibit a greater risk than light smokers, suggesting a dose-response relationship between cigarette consumption and lung cancer risk. Numerous mutagenic and carcinogenic substances have been identified as constituents of the particulate and vapor phases of cigarette smoke, including benzo-[a]pyrene, dibenza[a]anthracene, nickel, cadmium, polonium, urethane, formaldehyde, nitrogen oxides, and nitrosodiethylamine. There is also evidence that smoking combined with certain environmental (or occupational) exposures results in potentiation of lung cancer risk. Urban smokers exhibit a significantly higher incidence of lung cancer than smokers from rural areas, suggesting a possible role for air pollution in development of lung cancer.

Excessive alcohol consumption has been associated with increased risk of certain forms of cancer. Given that the liver is a target for alcohol-induced damage, it is not surprising that chronic alcohol consumption is associated with an elevated risk for primary liver cancer. However, it is important to note that whereas heavy sustained alcohol consumption is associated with risk of liver cancer, moderate consumption of alcohol is not. Alcohol is not directly carcinogenic to the liver; rather it is thought that the chronic liver damage produced by sustained alcohol consumption (hepatitis and cirrhosis) may contribute secondarily to liver tumor formation. For some other major cancer sites (such as lung), the role of alcohol consumption as a co-factor in cancer development is not clear.

All of the major forms of skin cancer (basal cell carcinoma, squamous cell carcinoma, and malignant melanoma) have been linked to sunlight exposure. The carcinogenic agent in sunlight that accounts for the neoplastic transformation of skin cells is ultraviolet (UV) radiation. Basal cell carcinoma is a malignant neoplasm of the basal cells of the epidermis that occurs predominantly in areas of sun-damaged skin. Thus, sun bathing and sun tanning using artificial UV light sources represent significant lifestyle risk factors for development of these tumors. Basal cell carcinoma is now diagnosed in some people at very young ages (second or third decade of life), reflecting increased exposures to UV irradiation early in life.

Squamous cell carcinoma is a malignant neoplasm of the keratinizing cells of the epidermis. As with basal cell carcinoma, extensive exposure to UV irradiation is the most important risk factor for development of this tumor. Likewise, development of malignant melanoma occurs most frequently in fair-skinned individuals and is associated to some extent with exposure to UV irradiation. This accounts for the observation that Caucasians develop malignant melanoma at a much higher rate than individuals of other races and ethnicity.

CLASSIFICATION OF NEOPLASTIC DISEASES

The word neoplasia is derived from the Greek words meaning "condition of new growth." The term tumor is commonly used to refer to a neoplasm. Tumor literally means "a swelling." In the early 1950s, R. A. Willis provided a description of neoplasm that we still utilize today: "A neoplasm is an abnormal mass of tissue the growth of which exceeds and is uncoordinated with that of the normal tissues and persists in the same manner after the cessation of the stimuli which evoked the change." Kinzler and Vogelstein define cancers to represent tumors that have acquired the ability to invade the surrounding normal tissues. This definition highlights one of the most important distinguishing factors in the classification of neoplasms—the distinction between benign and malignant tumors. Further subclassification of malignant neoplasms draws distinctions to (i) cancers of childhood versus cancers that primarily affect adults, (ii) solid tumors versus hematopoietic neoplasms, and (iii) hereditary cancers versus sporadic neoplasms.

Development of neoplastic disease is a multistep process through which cells acquire increasingly abnormal proliferative and invasive behaviors. Neoplasia also represents a unique form of genetic disease, characterized by the accumulation of multiple somatic mutations in a population of cells undergoing neoplastic transformation. Genetic and epigenetic lesions represent integral parts of the processes of neoplastic transformation, tumorigenesis, and tumor progression. Several forms of molecular alteration have been described in human cancers, including gene amplifications, deletions, insertions, rearrangements, and point mutations. In many cases specific genetic lesions have been identified that are associated with neoplastic transformation and/or tumor progression in a particular tissue or cell type. Epigenetic alterations (epimutations) in neoplastic disease include genome-wide hypomethylation of DNA (possibly resulting in induction of oncogene expression), gene-specific hypermethylation events (resulting in silencing of tumor suppressor genes), other changes in chromatin packaging, and aberrant post-transcriptional regulation of gene expression (related to abnormal microRNA expression). Statistical analyses of age-specific mortality rates for different forms of human cancer predict that multiple mutations or epimutations in specific target genes are required for the genesis and outgrowth of most clinically diagnosable neoplasms. It has been suggested that tumors grow through a process of clonal expansion driven by mutation or epimutations, where the first mutation/epimutation leads to limited expansion of progeny of a single cell, and each subsequent mutation/epimutation gives rise to a new clonal outgrowth with greater proliferative potential. The idea that carcinogenesis is a multistep process is supported by morphologic observations of the transitions between premalignant (benign) cell growth and malignant tumors. In colorectal cancer, the transition from benign lesion to malignant neoplasm can be easily documented and occurs in discernible stages, including benign adenoma, carcinoma *in situ*, invasive carcinoma, and eventually local and distant metastasis (Figure 4.3). Moreover, specific genetic alterations have been shown to correlate with each of these well-defined histopathologic stages of tumor development and progression. However, it is important to recognize that it is the accumulation of multiple genetic alterations in affected cells, and not necessarily the order in which these changes accumulate, that determines tumor formation and progression.

Both benign and malignant neoplasms are composed of (i) neoplastic cells that form the parenchyma, and (ii) the host-derived non-neoplastic stroma that is composed of connective tissue, blood vessels, and other cells. The tumor stroma serves a critical function in support of the growth of the neoplasm by providing a blood supply for oxygen and nutrients. In nearly all cases, the parenchymal cells determine the biological behavior (and clinical course) of the neoplasm.

Benign Neoplasms

The classification of neoplasms into benign and malignant categories is based on a judgment of the potential clinical behavior of the tumor. This judgment is based primarily on observations of the cellular features of the neoplasm, the growth pattern, and various clinical findings. Benign neoplasms are characterized by features that suggest a lack of aggressiveness. The most important characteristic of benign neoplasms is the absence of local invasiveness (Figure 4.4). Thus, while these neoplasms grow and expand, they do not invade

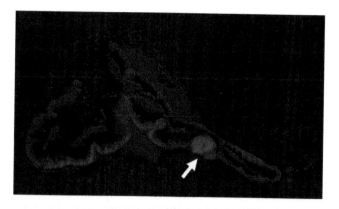

Figure 4.4   Adrenal adenoma. The normal cortex of the adrenal gland is yellow-gold in appearance. The white arrow points to a well-delineated, round lesion arising in the adrenal cortex. It does not appear to invade into the adjacent tissue. Courtesy of Kirsten Boland, MHS, PA(ASCP).

locally or spread to secondary tissue sites (remain localized). Since benign neoplasms remain localized, they are often amenable to surgical removal. However, benign neoplasms can cause adverse effects in the patient. Problems associated with benign neoplasms depend on (i) the size of the tumor, (ii) the location of the tumor, and (iii) secondary consequences related to presence of the neoplasm. Many benign neoplasms attain large size, impinge on important structures (like nerves or blood vessels), resulting in various types of local effects. Consider a few examples. Many/most brain tumors are considered benign by virtue of the fact that they do not invade locally or produce distant metastases. However, as expanding space-filling lesions, these neoplasms can cause severe effects on the host due to the application of pressure to nearby aspects of the brain or brainstem. For instance, a benign meningioma can cause cardiac and respiratory arrest by compressing the medulla. Likewise, hemangiomas represent a benign neoplastic lesion of blood vessels which creates a blood-filled cavity. Some hemangiomas (such as those affecting the liver) can become large and frequently impinge on the capsule of the organ. Lesions of this sort are subject to rupture, producing life-threatening bleeding in the patient.

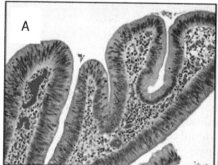
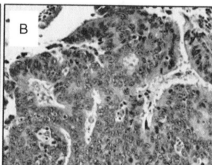
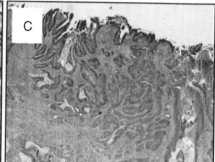

Figure 4.3   Progression of neoplastic transformation in the colon. (A) Low-grade glandular dysplasia. **(B)** High-grade glandular dysplasia. **(C)** Invasive adenocarcinoma.

Benign neoplasms are named by attaching the suffix *-oma* to the cell type from which the tumor originates. Thus, a benign neoplasm of fibrous tissue is termed a fibroma, a benign neoplasm of cartilaginous tissue is termed a chondroma, a benign neoplasm of osteoid tissue is termed an osteoma, a benign neoplasm arising from lipocytes is termed a lipoma, and a benign neoplasm arising from smooth muscle cells is termed a leiomyoma. The nomenclature for benign epithelial tumors is more complex. These neoplasms are classified either on the basis of their microscopic or macroscopic pattern, or according to their cells of origin. Thus, a benign epithelial neoplasm producing glandular patterns or a tumor arising from glandular cells is termed an adenoma, a benign epithelial neoplasm growing on any surface that produces microscopic or macroscopic finger-like fronds is termed a papilloma, a benign epithelial neoplasm that projects above a mucosal surface to produce a macroscopically visible structure is termed a polyp, and cystadenoma refers to hollow cystic masses.

Malignant Neoplasms

Malignant neoplasms are collectively known as cancers. Malignant neoplasms display aggressive characteristics, can invade and destroy adjacent tissues, and spread to distant sites (metastasize). These features of invasiveness distinguish malignant from benign neoplasms (Figure 4.5). Adverse effects associated with malignant neoplasms are generally associated with tumor burden on the host once the cancer has spread throughout the body. Specific adverse effects can arise from (i) the size and location of the primary tumor, (ii) consequences of local invasion and spread from the primary site, and (iii) consequences associated with tumor colonization of tissue sites distant to the primary tumor. Most commonly, the cause of death associated with malignant neoplasms can be attributed

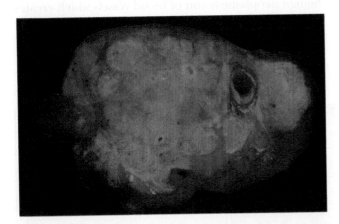

Figure 4.5 Pancreatic adenocarcinoma. This is a cut section of a pancreatic tumor. It is poorly delineated (meaning it is difficult to identify the exact borders of the lesion) and appears to be infiltrating into adjacent adipose tissue (patchy bright yellow areas along the bottom of the specimen). Courtesy of Kirsten Boland, MHS, PA(ASCP).

to the metastatic spread of the tumor. Common sites of metastasis for malignant epithelial neoplasms include the lungs, liver, bone, and brain. The lung is the most common site for cancer metastasis (involving most cancer types and primary sites), which may be accomplished by hematogenous spread (through the blood), lymphatic spread, or by direct invasion. Likewise, cancers of the breast, lung, and colon are well known to spread to the liver, but cancers associated with any site in the body (including leukemias and lymphomas) can colonize the liver. Metastatic cancer found in the lungs or liver is characterized by the presence of multiple cancer nodules that can replace large percentages of the normal tissue, and in liver can produce marked hepatomegaly. Adverse effects associated with lung metastasis include respiratory insufficiency and/or failure. Likewise, patient death related to liver metastasis results from various manifestations of liver insufficiency and/or failure.

The nomenclature for malignant neoplasms is very similar to that for benign neoplasms. Malignant neoplasms arising in mesenchymal tissues or its derivatives are called sarcomas. Sarcomas are designated by their histogenesis. Thus, a malignant neoplasm of fibrous tissue is termed a fibrosarcoma, a malignant neoplasm originating in cartilaginous tissue is termed a chondrosarcoma, a malignant neoplasm arising from lipocytes is termed a liposarcoma, a malignant neoplasm of osteoblasts is termed an osteosarcoma, a malignant neoplasm arising in blood vessels is termed an angiosarcoma, and a malignant neoplasm arising from smooth muscle cells is termed a leiomyosarcoma. Malignant epithelial neoplasms are called carcinomas. Carcinomas can be subclassified as adenocarcinoma and squamous cell carcinoma. Adenocarcinoma describes a malignant neoplasm in which the neoplastic cells grow in a glandular pattern. Squamous cell carcinoma describes a malignant neoplasm with a microscopic pattern that resembles stratified squamous epithelium. In each case, the nomenclature for a given tumor will specify the organ system of origin for the neoplasm (for instance, colonic adenocarcinoma or squamous cell carcinoma of the skin). In some cases, malignant neoplasms grow in an undifferentiated pattern that is inconsistent with a classification of adenocarcinoma or squamous cell carcinoma. In these cases, the neoplasm is termed a poorly differentiated carcinoma.

Mixed Cell Neoplasms

Some neoplastic cells undergo divergent differentiation during tumor formation, giving rise to tumors of mixed cell type. Examples of mixed cell tumors include that of salivary gland origin and breast fibroadenoma. These neoplasms are composed of epithelial components dispersed throughout a fibromyxoid stroma that may contain cartilage or bone.

In contrast to mixed cell tumors where all of the cellular components of the neoplasm are believed to derive from the same germ layer, teratomas contain recognizable mature or immature cells or tissues representative of more than one germ-cell layer, and sometimes all three. Teratomas typically originate from totipotential cells such as those found in the ovary and testis. The totipotent cells that give rise to teratomas have the capacity to differentiate into any cell type and can give rise to all of the tissues found in the adult body. It follows that teratomas are commonly composed of various tissue elements that can be recognized, including skin, hair, bone, cartilage, tooth structures, and others. Teratomas are further classified as benign (also referred to as mature) or malignant (also referred to as immature) (Figure 4.6). As the designation implies, mature teratomas contain well-formed tissue elements that appear normal, but arise in the abnormal context of the neoplasm. In contrast, immature teratomas are composed of abundant poorly differentiated primitive cells (blastema).

Confusing Terminology in Cancer Nomenclature

Some malignant neoplasms are conventionally referred to using terms that are suggestive of benign neoplasms based on the usual nomenclature for naming tumors. For example, lymphoma is a malignant neoplasm of lymphoid tissue, mesothelioma is a malignant neoplasm of the mesothelium, melanoma is a malignant neoplasm arising from melanocytes, and seminoma is a malignant neoplasm of the testicular epithelium. In each of these examples that refers to a specific tumor type, the name given implies a benign neoplasm even though all of these are malignant neoplasms. Likewise, hepatocellular carcinomas are often called hepatomas. Unfortunately, these tumor designations are well established in medical terminology and are unlikely to be corrected.

Preneoplastic Lesions

Neoplastic disease develops in patients over long periods of time and typically is preceded by development of one or more preneoplastic lesions. Well-characterized preneoplastic lesions include (i) metaplasia, (ii) hyperplasia, and (iii) dysplasia.

Metaplasia represents a reversible change in tissues characterized by substitution of one adult cell type (epithelial or mesenchymal) by another adult cell type (Figure 4.7). Metaplasia is a reactive condition, reflecting an adaptive replacement of cells that are sensitive to stress by cells that are resistant to the adverse conditions encountered by the tissue. In cigarette smokers, columnar to squamous epithelial metaplasia occurs in the respiratory tract in response to chronic irritation caused by inhalation of cigarette smoke. The ciliated columnar epithelial cells of the normal trachea and bronchi become replaced by stratified squamous epithelial cells in a pattern that may be focal or more widely distributed. Squamous metaplasia in the respiratory tract is accompanied by loss of function secondary to loss of the ciliated epithelial cells. In addition, development of squamous cell carcinoma of the lung may originate in focal areas of squamous metaplasia. Barrett's esophagus is an example of squamous to columnar metaplasia in response to refluxed gastric acid. The adaptive change is from stratified squamous epithelial cells to intestinal-like columnar epithelial cells, which are resistant to the effects of the gastric acid. Barrett's esophagus is frequently the site of development of esophageal adenocarcinomas.

Hyperplasia reflects an increase in the number of cells in an organ or tissue, typically resulting in an increased volume (or size) of the affected organ or tissue (Figure 4.8). Hyperplasia can be physiological or pathological. Physiological hyperplasia can be classified as (i) hormonal hyperplasia or (ii) compensatory hyperplasia. Most forms of pathological hyperplasia result from excessive (abnormal) hormonal or growth factor stimulation. Benign prostatic hyperplasia (BPH) is a commonly occurring condition among older men that is typical of pathological hyperplasia. In BPH, abnormal stimulation of the prostate tissue by androgen hormones results in a benign proliferation resulting in hypertrophy of the prostate gland. The abnormal cell proliferation that occurs in hyperplasia is controlled to the extent that upon cessation of the stimulus

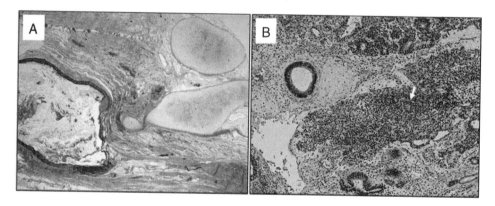

Figure 4.6 Neoplasms of mixed cell type. (A) Mature teratoma. This example demonstrates mature cartilage (right) and skin (left). **(B)** Immature teratoma. This type of teratoma contains areas of primitive-appearing hyperchromatic cells, known as blastema. There is focal rosette formation (white arrow).

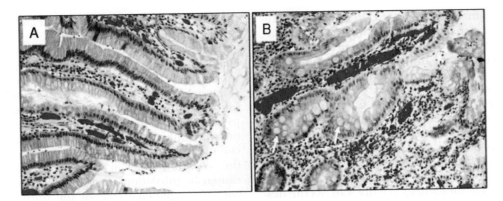

Figure 4.7   Metaplastic change in the stomach. (A) Normal glandular epithelium of the stomach. **(B)** Intestinal metaplasia. This image displays glandular epithelium of the stomach with goblet cells (white arrows).

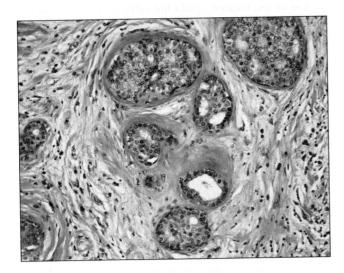

Figure 4.8   Epithelial hyperplasia of the breast. Normal mammary ducts are lined by a single layer of cuboidal epithelial cells (duct second from the center bottom). The lumens of the remaining ducts are filled with bland-appearing epithelium cells, representing prominent ductal epithelial hyperplasia.

(elimination of growth factor or hormone) cell proliferation will halt and the hyperplastic tissue will regress. Hyperplasia can precede neoplastic transformation, and many hyperplastic conditions are associated with elevated risk for development of cancer. For example, patients with endometrial hyperplasia are at increased risk for endometrial cancer.

Dysplasia is a proliferative lesion that is characterized by a loss in the uniformity of individual cells in a tissue and loss in the architectural orientation of the cells in a tissue. Thus, dysplasia can be simply described as a condition of disorderly but nonneoplastic cellular proliferation. Dysplastic cells show many alterations that are suggestive of their preneoplastic character, including cellular pleomorphism, hyperchromatic nuclei, high nuclear-to-cytoplasmic ratio, and increased numbers of mitotic figures (Figure 4.9). Lesions characterized by extensive dysplastic changes involving the entire thickness of the epithelium but remaining confined within the normal tissue are classified as carcinoma *in situ* (Figure 4.10). In many cases, dysplasia and/or carcinoma *in situ* are considered immediate precursors of invasive cancers.

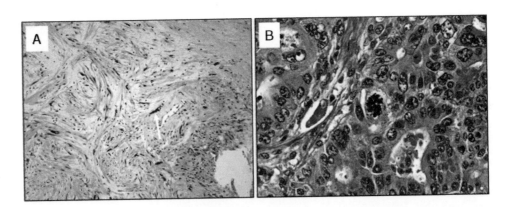

Figure 4.9   Nuclear pleomorphism and abnormal mitotic figures in neoplastic cells. (A) Pleomorphism. This is an example of leiomyosarcoma. Several of the malignant stromal cells are very large and different in shape from neighboring cells. **(B)** Mitotic figure (center). Malignant neoplasms often have an increased number of mitotic figures.

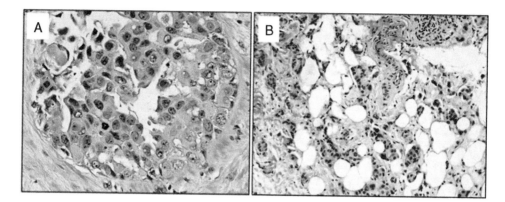

Figure 4.10   Progression of tumorigenesis in the breast. (A) Ductal carcinoma *in situ* (DCIS). This neoplastic process is *in situ* carcinoma because the dysplastic cells are confined to the lumen of the duct and have not invaded into the surrounding tissue. **(B)** Invasive adenocarcinoma of the breast. Glands lined by dysplastic cells haphazardly infiltrate into adipose tissue.

Cancers of Childhood

In 2008, an estimated 10,730 new cases of cancer and 1,490 cancer-associated deaths occurred in children under 15 years old. The major childhood cancers include leukemia (33% of all childhood cancers), brain tumors and other neoplasms of the nervous system (21% of all childhood cancers), neuroblastoma (7% of all childhood cancer), Wilms' tumor (5% of all childhood cancer), Non-Hodgkin lymphoma (4% of all childhood cancer), rhabdomyosarcoma (3.5% of all childhood cancer), retinoblastoma (3% of all childhood cancer), osteosarcoma (3% of all childhood cancer), and Ewing's sarcoma (1.5% of all childhood cancer). Acute lymphocytic leukemia is the most common childhood leukemia, representing >80% of all childhood leukemias. The incidence of acute lymphocytic leukemia peaks in children 2–5 years old. Neuroblastoma is a malignant neoplasm of the sympathetic nervous system and is the most frequently occurring neoplasm among infants, with peak incidence among children <1 year old (Figure 4.11). In fact, 40% of neuroblastomas are diagnosed in the first 3 months of life. Retinoblastoma is a tumor that originates in the retina of the eye. This neoplasm affects children as well as adults. Retinoblastoma occurring in children is typically associated with a genetic mechanism involving mutation of the *Rb1* gene. Wilms' tumor tends to occur in children <10 years old, with greatest incidence in children <5 years old. Wilms' tumor (also known as nephroblastoma) is the most commonly occurring pediatric kidney tumor. Astrocytomas represent the most frequently occurring brain tumor of children (52% of all childhood brain tumors), with ependymoma (9%), primitive neuroectodermal tumors (21%), and other gliomas (15%) representing most of the balance. Ependymomas and primitive neuroectodermal tumors occur most often in younger children (<5 years old), while astrocytomas are diagnosed with approximately the same frequency in children between birth and 15 years old. Rhabdomyosarcoma is the most common soft tissue sarcoma of children, typically occurring in children <10 years old. Osteosarcoma and Ewing's sarcoma are the most

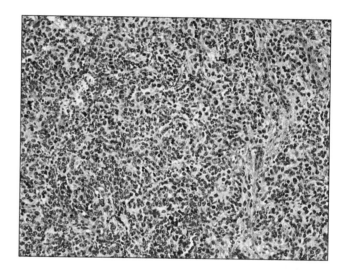

Figure 4.11   Childhood neuroblastoma. Neuroblastoma is considered one of the so-called small, round, blue cell tumors, as it is composed of small, hyperchromatic, monomorphic cells with scant cytoplasm.

commonly occurring bone tumors among children. There is a bimodal age distribution of osteosarcoma incidence, with peaks in early adolescence and in adults >65 years old. Like osteosarcoma, Ewing's sarcoma is a disease primarily of childhood and young adults, with highest incidence among children in their teenage years.

Hematopoietic Neoplasms

The majority of human cancers can be classified as solid tumors (grouped as carcinomas or sarcomas). The exceptions to this classification include the malignant neoplasms of hematopoietic origin, including lymphoma, myeloma, and leukemia (Figure 4.12). In 2008, 138,530 new malignant neoplasms of hematopoietic origin were diagnosed, representing approximately 9.5% of all new cancers. These include 74,340 cases of lymphoma, 19,920 cases of myeloma, and 44,270 cases of leukemia. Non-Hodgkin's lymphoma

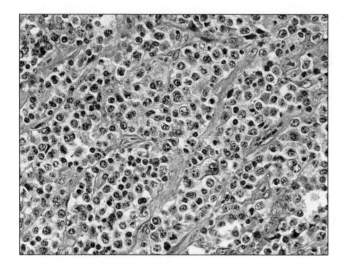

Figure 4.12 Lymphoma. This image of lymphoma demonstrates sheets of large cells with prominent nucleoli.

occurs much more frequently than Hodgkin's lymphoma and represents 89% of all lymphomas. The four major forms of leukemia (acute lymphocytic leukemia, chronic lymphocytic leukemia, acute myeloid leukemia, and chronic myeloid leukemia) account for 87% of all leukemias. The two most prevalent forms of leukemia are chronic lymphocytic leukemia and acute myeloid leukemia, which combine to account for 64% of all cases of leukemia.

Hereditary Cancers

A number of familial cancer syndromes and hereditary cancers have been recognized and characterized. Several rare genetic disorders involving dysfunctional DNA repair pathways are associated with elevated risk for cancer development. These disorders include xeroderma pigmentosum, ataxia telangiectasia, Bloom's syndrome, and Fanconi anemia. Individuals affected by these conditions are prone to development of various malignancies when exposed to specific DNA damaging agents.

Colorectal cancer is a fairly common disease worldwide, and particularly in populations from Western nations. A substantial fraction of colorectal cancers exhibit a genetic component, and several familial colorectal cancer syndromes are recognized, including familial adenomatous polyposis (FAP) and hereditary nonpolyposis colon cancer (HNPCC). Genes associated with each of these conditions have been identified and characterized. Of these familial colorectal cancer syndromes, HNPCC has been determined to be related to defective DNA repair. Tumors associated with HNPCC exhibit a unique form of genomic instability, which represents a unique mechanism for a genome-wide tendency for instability in short repeat sequences (microsatellites). The molecular defect responsible for microsatellite instability in HNPCC involves the genes that encode proteins required for normal mismatch repair.

Patients with Li-Fraumeni syndrome develop various types of neoplasms, including breast cancer, soft tissue sarcomas and osteosarcomas, brain tumors, leukemias, and several others. Cancer susceptibility among individuals with Li-Fraumeni syndrome follows an autosomal dominant pattern of inheritance and is highly penetrant (90% by age 70), but many neoplasms develop early in life. It is now known that Li-Fraumeni syndrome is associated with germline mutations in the *p53* tumor suppressor gene.

Familial melanoma is associated with (i) a family history of melanoma, (ii) the presence of large numbers of common or atypical nevi, (iii) a history of primary melanoma or other (nonmelanoma) skin cancers, (iv) immunosuppression, (v) susceptibility to sunburn, or (vi) a history of blistering sunburn. Two highly penetrant melanoma susceptibility genes have been identified: *CDKN2A* (which encodes cyclin-dependent kinase inhibitor 2A) and *CDK4* (which encodes cyclin-dependent kinase 4). Germline inactivating mutations of the *CDKN2A* gene are the most common cause of inherited susceptibility to melanoma, while mutations of *CDK4* occur much more rarely. The *CDKN2A* gene encodes two important cell-cycle regulatory proteins: $p16^{INK4A}$ and $p14^{ARF}$. Ongoing research is focused on the identification of low-penetrance melanoma susceptibility genes that confer a lower melanoma risk with more frequent variations. For instance, specific variants of the *MC1R* and the *OCA2* genes have been demonstrated to confer an increase in melanoma risk.

CHARACTERISTICS OF BENIGN AND MALIGNANT NEOPLASMS

Four fundamental features are particularly important when comparing and contrasting the characteristics of benign and malignant neoplasms: (i) cellular differentiation and anaplasia, (ii) rate of growth, (iii) presence of local invasion, and (iv) metastasis.

Cellular Differentiation and Anaplasia

The extent of cellular differentiation describes the degree to which the neoplastic cells resemble their normal counterparts based on morphology and function. Benign neoplasms are typically composed of well-differentiated cells that closely resemble normal cells. In well-differentiated benign neoplasms, cellular proliferation rates are low and mitoses are infrequent. However, when observed, mitotic figures appear normal. In contrast to benign tumors, malignant neoplasms exhibit an extremely wide range of cell differentiation. Many malignant neoplasms are very well differentiated, but it is not uncommon for these tumors to lack differentiated features and/or to appear completely undifferentiated.

In general, there is a direct relationship between the degree of cellular differentiation and the functional capabilities of the cells that compose the neoplasm. Thus, benign neoplasms (as well as some well-differentiated malignant neoplasms) of the endocrine glands can elaborate hormones characteristic of their origin. Likewise, well-differentiated squamous cell carcinomas (at various tissue sites) elaborate keratin (giving rise to histologically recognizable keratin pearls), and well-differentiated hepatocellular carcinomas synthesize bile salts. In contrast, many malignant neoplasms express genes and produce proteins or hormones that would not be expected from the cells of origin of the cancer. Some cancers synthesize fetal proteins that are not expressed by comparable cell types in adults. For instance, many hepatocellular carcinomas express α-fetoprotein, which is not expressed in mature hepatocytes of the adult liver. Further, malignant neoplasms of nonendocrine origin can excrete ectopic hormones producing various paraneoplastic syndromes. Certain lung cancers produce antidiuretic hormone (inducing hyponatremia in the patient), adrenocorticotropic hormone (resulting in Cushing syndrome), parathyroid-like hormone or calcitonin (both of which are implicated in hypercalcemia), gonadotropins (causing gynecomastia), serotonin and bradykinin (associated with carcinoid syndrome), or others.

Malignant neoplasms that are composed of undifferentiated cells are said to be anaplastic. Lack of cellular differentiation (or anaplasia) is considered a hallmark of cancer. The term anaplasia means "to form backward," which implies dedifferentiation (or loss of the structural and functional differentiation) of normal cells during tumorigenesis. In general, malignant neoplasms that are composed of anaplastic cells (typically rapidly growing) are unlikely to have specialized functional activities. Anaplastic cells tend to exhibit marked nuclear and cellular pleomorphism (extreme variation in nuclear or cell size and shape). The nuclei observed in anaplastic cells are typically hyperchromatic (darkly staining) and large, resulting in altered nuclear-to-cytoplasmic ratios (which may approach 1:1 instead of 1:4 or 1:6 as observed in normal cells). Very often malignant neoplasms contain giant cells (relative to the size of neighboring cells in the neoplasm), and these cells will contain an abnormally large nucleus or will be multinucleated. The nuclear pleomorphism observed in anaplastic cells of malignant neoplasms is characterized by nuclei that are highly variable in size and shape. These nuclei exhibit chromatin that appears coarse and clumped. Furthermore, the nucleoli may be very large relative to that observed in the nuclei of normal cells, possibly reflecting the extent of transcriptional activities taking place in these highly active cells. Given that malignant neoplasms often exhibit high rates of cell proliferation, numerous mitotic figures may be seen, and these mitotic figures are often abnormal. Typically, anaplastic cells will fail to organize into recognizable tissue patterns. This lack of cellular orientation reflects loss of normal cellular polarity, as well as a failure of normal structures to form.

Rate of Growth

In general, benign neoplasms grow more slowly than malignant neoplasms. However, there are exceptions to this rule. For instance, some benign neoplasms will exhibit changes in growth rate in response to hormonal stimulation or in response to alterations in blood supply. Leiomyomas are benign neoplasms of the uterus that originate in smooth muscle and are significantly influenced by circulating levels of estrogens. Thus, these neoplasms may display an elevated cell proliferation and increase rapidly in size in response to the hormonal changes seen in pregnancy. Once hormonal levels normalize (after childbirth), the lack of sufficient hormone levels to sustain high rates of cellular proliferation results in greatly diminished growth of the neoplasm. Subsequently, with the elimination of hormones related to menopause, these tumors may become fibrocalcific. Despite variations in growth rate among neoplasms and some physiological exceptions, most benign neoplasms proliferate slowly over time and increase in size slowly.

The rate of cell proliferations (and lesion growth) of malignant neoplasms generally correlates with the extent of cellular differentiation of the cells that compose the tumor. Thus, poorly differentiated neoplasms exhibit high rates of cell proliferation and tumor growth, and well-differentiated neoplasms tend to grow more slowly. Many or most malignant neoplasms grow relatively slowly and at a constant rate. However, there may be several patterns of growth among malignant neoplasms. Some of these lesions grow slowly for long periods of time before entering into a phase characterized by more rapid expansion. In this case, the rapid expansion phase probably reflects the emergence of a more aggressive subclone of neoplastic cells. Most malignant neoplasms progressively enlarge over time in relation to their cellular growth rate. Rapidly growing malignant neoplasms tend to contain a central area of necrosis that develops secondary to ischemia related to inadequate blood supply (Figure 4.13). Since the tumor blood supply is derived from normal tissues at the site of the neoplasm, the formation of new blood vessels to supply the

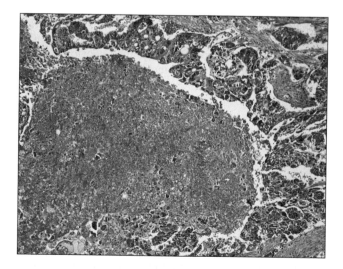

Figure 4.13   Tumor necrosis. Necrotic (dead tissue) debris is present within the center of the malignant glands in this adenocarcinoma.

expanding neoplasm may lag behind the proliferation of the neoplastic cells, resulting in inadequate supply of oxygen and other nutrients to the tumor mass.

Presence of Local Invasion

Benign neoplasms, by definition, remain localized at their site of origin. These neoplasms do not have the capacity to infiltrate surrounding tissues, invade locally, or metastasize to distant sites. Benign neoplasms typically have smooth borders, are sharply demarcated from the normal tissue at the tumor site, and are frequently encapsulated by a fibrous capsule that forms a barrier between the neoplastic cells and the host tissue. However, not all benign neoplasms have a capsule, and the lack of a capsule around a neoplasm does not indicate that the neoplasm is malignant. The capsule of a benign neoplasm is primarily the product of the elaboration of tumor stroma. Benign liver adenomas represent an example of a benign neoplasm that commonly exhibits a capsule. Uterine leiomyomas do not infiltrate adjacent normal tissues and are typically discretely demarcated from the surrounding smooth muscle by a zone of compressed and attenuated normal myometrium, but these neoplasms do not elaborate a capsule.

The growth of malignant neoplasms is characterized by progressive infiltration, invasion, destruction, and penetration of surrounding normal (non-neoplastic) tissues (Figure 4.5 and Figure 4.10B). Malignant neoplasms do not form well-developed capsules. However, some slow-growing malignant neoplasms appear histologically to be encased in stroma resembling a capsule. The invasive (malignant) nature of these neoplasms is revealed by close microscopic examination, which reveals penetration of the margins of the stroma by neoplastic cells and invasion of adjacent tissue structures. The infiltrative or locally invasive nature of the growth of malignant neoplasms requires that broad margins of non-neoplastic tissue must be resected during surgical excision of a tumor to ensure complete removal of all neoplastic cells. Local invasiveness and infiltration of adjacent tissue structures represent features that are strongly suggestive of a malignant neoplasm. These features represent the most reliable predictor of malignant behavior next to the development of distant metastasis.

Metastasis

The term metastasis describes the development of secondary neoplastic lesions at distant tissue locations that are separated from the primary site of a malignant neoplasm. The clinical finding of metastasis provides a definitive classification of a primary neoplasm as malignant. A significant percentage of patients with newly diagnosed malignant neoplasms exhibit clinically evident metastases at the time of diagnosis (~30%), and others have occult metastases at the time of diagnosis (~20%). Some malignant neoplasms demonstrate a propensity for metastasis. In general terms, malignant neoplasms that are anaplastic (less well differentiated) and larger are more likely to metastasize. However,

there are numerous exceptions; extremely small cancers may metastasize and have poor prognosis, while some large tumors may remain localized and not produce metastases. However, not all malignant neoplasms exhibit a strong tendency for metastasis. For instance, basal cell carcinomas of the skin are highly invasive at the primary site of the tumor and only rarely form distant metastases. Likewise, many malignant neoplasms of the brain are highly invasive at their primary sites but rarely give rise to distant metastatic lesions.

Malignant neoplasms metastasize and spread to distant sites in the patient through (i) direct invasion or seeding within body cavities, (ii) spread through the lymphatic system, or (iii) hematogenous spread via the blood. Direct invasion or spreading by seeding occurs when a malignant neoplasm invades a natural body cavity. This pathway of tumor spread is characteristic of malignant neoplasms of the ovary. These cancers often widely disseminate across peritoneal surfaces, producing a significant tumor burden without invasion of the underlying parenchyma of the abdominal organs. In this example, the ovarian carcinoma cells demonstrate an ability to establish at new sites (previously uninvolved peritoneal surfaces), but lack the capacity to invade into new tissue sites.

Malignant epithelial neoplasms (carcinomas) tend to metastasize through the lymphatic system (Figure 4.14). Hence, lymph node involvement represents an important factor in the staging of many cancers. However, given the numerous anatomic interconnections between the lymphatic and vascular systems, it is possible that some of these cancers may disseminate through either or both pathways. While enlargement of lymph nodes near a primary neoplasm may arouse suspicion of lymph node involvement and metastatic spread of the primary malignant neoplasm, simple lymph node swelling does not accurately predict spread of neoplastic cells. Patterns of lymph node seeding by metastatic tumor cells mainly depend on the site of the primary malignant neoplasm and the natural pathways of lymphatic drainage from that tissue site. Malignant neoplasms of the breast that arise in the upper outer quadrant of the breast will spread initially to axillary lymph nodes, while malignant tumors arising in the medial region of the breast may spread through the chest wall into the lymph nodes along the internal mammary artery. Seeding of the supraclavicular and infraclavicular lymph nodes by metastatic breast cancer cells can occur subsequent to the initial spread despite the initial route of tumor cell movement. In some cases, breast cancer cells appear to traverse lymphatic vessels without colonizing the lymph nodes that are immediately proximal to the site of the primary neoplasm, but do implant in lymph nodes that are subsequently encountered (producing metastatic lesions that are referred to as skip metastases). Given the complexity and potentially complicated pattern of lymphatic spread by metastatic breast cancer, it is not surprising that sentinel node biopsy has become a useful procedure for surgical staging of this malignant neoplasm. A sentinel lymph node is defined as the first lymph node in a regional lymphatic basin that receives lymphatic drainage from the

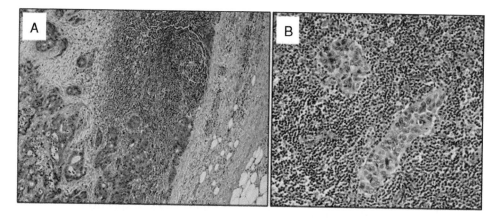

Figure 4.14   Metastatic spread of malignant cells through the lymphatic system. (A) Metastatic adenocarcinoma to a lymph node, low power. Lymph node tissue is present along the upper portion of the image, and infiltrating glands of adenocarcinoma are noted at the lower aspect of the image. **(B)** Metastatic carcinoma to a lymph node, high power. Two nests of plump malignant epithelial cells are present within the sea of smaller lymphocytes of the lymph node.

site of a primary neoplasm. It can be identified by injection of colored dyes (or radiolabeled tracers) into the site of the primary tumor with monitoring of the movement of the dye to the downstream lymph nodes. Histopathological evaluation of the sentinel lymph node provides a reliable indication of the extent of spread of a breast cancer and can be used to plan treatment.

Dissemination of a malignant neoplasm through the blood is referred to as hematogenous spread. Hematogenous spread of malignant mesenchymal neoplasms (sarcomas) occurs commonly, but this pathway is also readily utilized by malignant epithelial neoplasms (carcinomas). Invasion of the vascular system by neoplastic cells can involve either the arterial venous circulation (and associated vessels). However, the arteries are penetrated less readily than are veins, making venous involvement more likely. In the process of venous invasion by a malignant neoplasm, the neoplastic cells in the blood follow the normal pattern of venous blood flow draining the primary site of the neoplasm. It is well recognized that the liver and lungs represent the most frequently involved secondary sites in hematogenous spread of malignant neoplasms. This is due to the fact that all portal venous blood drainage flows to the liver and all caval blood drainage flows to the lungs. Numerous clinical and experimental observations combine to suggest that the anatomic location of the primary neoplasm and the natural pathways of venous blood flow from that site do not completely explain the patterns of metastatic disease observed with many forms of cancer. Thus, the so-called seed and soil hypothesis was developed to explain the tendency of certain cancers to preferentially spread to certain tissue sites. For example, prostatic carcinoma preferentially metastasizes to bone, bronchogenic carcinomas tend to spread to the adrenals and the brain, and neuroblastomas are prone to colonize the liver and bones. These observations suggest that factors intrinsic to the neoplastic cells and their secondary sites of involvement determine the ability of cancer cells to efficiently colonize a given tissue.

CLINICAL ASPECTS OF NEOPLASIA

Cancer-Associated Pain

Pain is frequently associated with malignant neoplasms, particularly in advanced disease. This cancer-associated pain is often difficult to treat and unrelenting, producing major challenges for management and typically requiring the use of narcotic drugs. The cause of the pain can be related to destruction of tissue by the neoplasm, infection, stretching of internal organs (due to tumor involvement), pressure (from an expanding neoplasm), or obstruction (secondary to tumor impingement). Destruction of bone tissue by metastatic cancer results in characteristic bone pain.

Cachexia

Cachexia refers to a progressive loss of body fat and lean body mass, accompanied by profound weakness, anorexia, and anemia. This condition affects many cancer patients with advanced disease. In fact, very often the size and extent of spread of the malignant neoplasm correlates with the severity of cachexia. However, the cachexic condition of the patient is not directly related to the nutritional needs of the neoplasm. Rather, cancer cachexia results from the action of soluble factors (such as cytokines) produced by the neoplasm. Cancer patients very often experience anorexia (loss of appetite), which can result in reduced caloric intake. Nevertheless, cancer patients often display increased basal metabolic rate and expend high numbers of calories in spite of reduced food caloric intake. It follows that as caloric needs increase (or remain high) in the patient, body fat reserves and lean tissue mass are consumed to meet the energy needs of the individual. Cancer cachexia cannot be treated effectively, although strategies for management of this aspect of cancer are being investigated. At the present time, the most effective treatment of cancer-associated cachexia is removal of the underlying cause (the malignant neoplasm).

Paraneoplastic Syndromes

Paraneoplastic syndromes refer to groupings of symptoms that occur in patients with malignant neoplasms that cannot be readily explained by local invasion or distant metastasis of the tumor, or the elaboration of hormones indigenous to the tissue of origin of the neoplasm. Paraneoplastic syndromes occur in 10–15% of cancer patients. These paraneoplastic syndromes present challenges to management of the cancer patient, potentially leading to significant clinical problems that affect quality of life, or contributing to potentially lethal complications. However, the symptoms associated with paraneoplastic syndromes may also represent an early manifestation of an occult neoplasm, presenting an opportunity for cancer detection and diagnosis. There are a number of different paraneoplastic syndromes that are associated with many different tumors. Well-characterized examples of paraneoplastic syndromes include (i) hypercalcemia, (ii) Cushing syndrome, and (iii) hypercoagulability (Trousseau syndrome). Hypercalcemia refers to a condition related to elevated plasma calcium concentrations. This condition occurs in 20–30% of patients with metastatic cancer and is the most common paraneoplastic syndrome. Hypercalcemia affects multiple organ systems, can be life-threatening, and predicts a poor outcome for the patient. Cushing syndrome is caused by ectopic secretion of corticotropin (hypercortisolism) and produces several major effects, including hyperglycemia, hyperkalemia, hypertension, and muscle weakness. Trousseau syndrome is characterized by hypercoagulopathy and produces the major effect of venous thrombosis of the deep veins in the cancer patient.

Grading and Staging of Cancer

To accurately predict the clinical course of a neoplasm and the probable outcome for the patient, clinicians make observations related to the histological aggressiveness of the cells that compose the neoplasm and the apparent extent and spread of the disease. This analysis yields a clinical description of the patient that reflects a score for the grade and stage of the malignant neoplasm.

The histological grade of a malignant neoplasm is based on (i) the degree of cellular differentiation of the neoplastic cells that compose the lesion and (ii) an estimate of the growth rate of the neoplasm (based on mitotic index). In general, histologic grade I refers to neoplasms that display 75–100% differentiation, grade II reflects 50–75% differentiation, grade III refers to 25–50% differentiation, and grade IV describes tumors with <25% differentiation. Current methods also take into consideration the mitotic activity of the neoplastic cells, the degree of infiltration of adjacent tissues, and the amount of tumor stroma that is present.

Staging of malignant neoplasms is based on (i) the size of the primary lesion, (ii) the extent of spread to regional lymph nodes, and (iii) the presence or absence of distant metastases. Effective staging of neoplastic disease relies on clinical assessment of the patient, radiographic examination (using CT, MRI, or other technologies), and in some cases surgical exploration. The major convention for staging of malignant neoplasms is known as the TNM system. In this system, the T refers to the primary tumor (with T1, T2, T3, and T4 reflecting increasing size of the primary tumor), the N refers to regional lymph node involvement (with N0, N1, N2, and N3 reflecting progressively advancing node involvement), and M refers to the presence or absence of distant metastases (with M0 and M1 reflecting absence and presence of distant metastases, respectively).

KEY CONCEPTS

- Cancer represents a major health problem in the United States and worldwide. In 2008, approximately 1.5 million new cases of invasive cancer were diagnosed in the United States, as well as over a million cases of common skin cancers, and there were 550,000 cancer-associated deaths. In 2002, nearly 11 million new cases of invasive cancer were diagnosed worldwide and over 6.5 million cancer-associated deaths occurred.

- The development of cancer is a multistep, multifactorial process that occurs over a long period of time, through which cells acquire increasingly abnormal proliferative and invasive behaviors. Major risk factors that contribute to cancer development include: (1) age, (2) race, (3) gender, (4) family history, (5) infectious agents, (6) environmental exposures, (7) occupational exposures, and (8) lifestyle exposures. These risk factors may interact with genetic polymorphisms and other genetic determinants to drive the development of cancer.

- The division of neoplastic diseases into benign and malignant categories is extremely important. Both benign and malignant neoplasms are composed of neoplastic cells and host-derived nonneoplastic stroma (composed of connective tissue and blood vessels). The classification of neoplasms into benign and malignant categories is based on a judgment of the potential clinical behavior of the tumor. Benign neoplasms are characterized by features that suggest a lack of aggressiveness. The most important characteristic of benign neoplasms is the absence of local invasiveness. Malignant neoplasms are collectively known as cancers. Malignant neoplasms display aggressive characteristics, can invade and destroy adjacent tissues, and spread to distant sites (metastasize).

- Neoplasms grow through a process of clonal expansion driven by mutation and/or epimutation, where the first mutation/epimutation leads to limited expansion of progeny of a single cell, and each subsequent mutation/epimutation gives rise to a new clonal outgrowth with greater proliferative potential. The idea that carcinogenesis is a multistep process is supported by morphologic observations of the transitions between premalignant (benign) cell growth and malignant tumors. For example, in colorectal cancer this transition from benign to malignant neoplasm can be easily documented and occurs in discernible stages, including benign adenoma, carcinoma *in situ*, invasive carcinoma, and eventually local and distant metastasis. The stepwise development of neoplastic disease provides opportunity for detection of developing disease, diagnosis at a definable stage, and appropriate therapeutic intervention.

- Staging of malignant neoplasms is based on (1) the size of the primary lesion, (2) the extent of spread to regional lymph nodes, and (3) the presence or absence of distant metastases. The major convention for staging of malignant neoplasms is the TNM system, where the T refers to the primary tumor (with T1, T2, T3, and T4 reflecting increasing size of the primary tumor), the N refers to regional lymph node involvement (with N0, N1, N2, and N3 reflecting progressively advancing node involvement), and M refers to the presence or absence of distant metastases (with M0 and M1 reflecting absence and presence of distant metastases, respectively). Accurate staging of clinical neoplasms is important for management of the disease (application of appropriate treatment) and prediction of patient outcomes (prognosis).

SUGGESTED READINGS

1. Fearon ER, Vogelstein B. A genetic model for colorectal tumorigenesis. *Cell.* 1990;61:759–767.

2. Gayther SA, Pharoah PD, Ponder BA. The genetics of inherited breast cancer. *J Mammary Gland Biol Neoplasia.* 1998;3:365–376.

3. Hagan JP, Croce CM. MicroRNAs in carcinogenesis. *Cytogenet Genome Res.* 2007;118:252–259.

4. Hankey BF, Gloeckler Ries LA, Miller AB, et al. Overview. In: *Cancer Statistics Review 1973–1989.* NIH Publication Number 92–2789 ed. 1992;I.1–17.

5. Heimann R, Hellman S. Aging, progression, and phenotype in breast cancer. *J Clin Oncol.* 1998;16:2686–2692.

6. Herranz M, Esteller M. DNA methylation and histone modifications in patients with cancer: Potential prognostic and therapeutic targets. *Methods Mol Biol.* 2007;361:25–62.

7. Irigaray P, Newby JA, Clapp R, et al. Lifestyle-related factors and environmental agents causing cancer: An overview. *Biomed Pharmacother.* 2007;61:640–658.

8. Jemal A, Siegel R, Ward E, et al. Cancer statistics, 2008. *CA Cancer J Clin.* 2008;58:71–96.

9. Kinzler KW, Vogelstein B. Introduction. In: Vogelstein B, Kinzler KW, eds. *The Genetic Basis of Human Cancer.* 2nd ed. New York: McGraw-Hill; 2002:3–6.

10. Marchioli CC, Graziano SL. Paraneoplastic syndromes associated with small cell lung cancer. *Chest Surg Clin N Am.* 1997;7:65–80.

11. Parkin DM, Bray F, Ferlay J, et al. Global cancer statistics, 2002. *CA Cancer J Clin.* 2005;55:74–108.

12. Piris A, Mihm Jr MC. Mechanisms of metastasis: Seed and soil. *Cancer Treat Res.* 2007;135:119–127.

13. Singh Ranger G, Mokbel K. The evolving role of sentinel lymph node biopsy for breast cancer. *Eur J Surg Oncol.* 2003;29:423–425.

14. Stewart AF. Clinical practice. Hypercalcemia associated with cancer. *N Engl J Med.* 2005;352:373–379.

15. Tisdale MJ. Loss of skeletal muscle in cancer: Biochemical mechanisms. *Front Biosci.* 2001;6:D164–D174.

16. Willis RA. *The Spread of Tumors in the Human Body.* London: Butterworths; 1952.

17. Wogan GN, Hecht SS, Felton JS, et al. Environmental and chemical carcinogenesis. *Semin Cancer Biol.* 2004;14:473–486.

18. Zeigler JL, Buonaguro FM. Infectious agents and human malignancies. *Front Biosci.* 2009;14:3455–3464.

Concepts in Molecular Biology and Genetics

5

Basic Concepts in Human Molecular Genetics

Kara A. Mensink . W. Edward Highsmith, Jr.

INTRODUCTION

Molecular diagnostics is the branch of laboratory medicine or clinical pathology that utilizes the techniques of molecular biology to diagnose disease, predict disease course, select treatments, and monitor the effectiveness of therapies. Molecular diagnostics is associated with virtually all clinical specialties and is a vital adjunct to several areas of clinical and laboratory medicine, but is most predominantly aligned with infectious disease, oncology, and genetics. The subject of this chapter is molecular genetics, which is concerned with the analysis of human nucleic acids as they relate to disease.

Since the completion of the first working draft of the human genome sequence in 2000 and the completion of the polished sequence in 2003, progress in molecular genetics has been swift and shows no signs of abating. Relatively few gene tests were clinically available in the late 1990s, whereas over 1000 are available today. Further, molecular genetic testing has proven useful and robust enough to expand into population-based screening. Molecular testing serves as the final confirmatory test for several disorders included as part of expanded newborn screening programs, and in 2003, the *American Colleges of Medical Genetics and Obstetrics and Gynecology* recommended that population-based carrier screening for cystic fibrosis using molecular testing be implemented in the United States.

Molecular genetics as a discipline and as a clinical laboratory service does not exist in a vacuum. Rather, it is intimately tied to molecular and cell biology and the central paradigm of molecular biology—that genes code for proteins. Thus, it is through the analysis of genes that insight into the genesis of protein malfunction can be achieved. Such examination specifically entails an assessment of how the DNA sequence of a gene compares with its wild-type or normal sequence. Ultimately, protein malfunctions related to gene mutations lead to organ dysfunction and disease states.

MOLECULAR STRUCTURE OF DNA

The human genome is composed of 3 billion base pairs of DNA. This is not present as one continuous piece of double-stranded DNA, but is distributed among 22 pairs of autosomal chromosomes and 2 sex chromosomes. The DNA is associated with a large number of proteins (histones and others) that serve regulatory functions and package the genetic material into these large chromosomal units. Chromosomes range in size from the 33.4 Mb of chromosome 22 to the 263 Mb of chromosome 1, which is the largest chromosome. Along the length of each chromosome, DNA is organized into linear domains consisting of genes (primarily nonrepetitive DNA), repetitive elements, and apparently functionless regions, much like beads on a string. Approximately half of the human genome consists of repetitive DNA, while the other half consists of nonrepetitive sequence. Nonrepetitive DNA includes regulatory sequences, intronic sequence, and protein coding (exon) sequence. Protein coding regions account for a relatively small fraction of genes within the human genome. In fact, it is estimated that only 6% of the human genome consists of protein coding, nonrepetitive DNA.

Genes are found among the nonrepetitive DNA in the genome. Genes code for specific protein chains, each with a specific function in cell physiology. A gene is composed of regulatory elements, which determine where, when, and how a gene is transcribed and coding regions, which are broken into segments, termed exons (expressed sequences). An example of a regulatory element is the promoter, which is the site where gene transcription is initiated. The exons are separated by noncoding regions of DNA called introns (intervening sequences). The average gene is about 2.7 kb (2700 bp) of DNA in length.

Chemically, genes are composed of 2-deoxyribonucleic acid (DNA). DNA is a linear, nonbranching polymer of nucleotides. Repeating ribose and phosphate subunits form a backbone; and attached to each of the ribose moieties is a purine (adenine, guanine) or pyrimidine (thymine or cytosine) base. Following standard nomenclature for the naming of ring containing compounds, the nitrogenous bases have their various carbon and heteroatom components numbered 1–6 (for the pyrimidines) or 1–9 (for the purines) and the ribose positions are indicated by numbers $1'$–$5'$. The bases are attached to the ribose subunits at the $1'$ position of the sugar molecule. The ribose subunits are joined by phosphodiester linkages between the $5'$ position of one ribose to the $3'$ position of the next. Thus, the molecule is not symmetrical and there is directionality implicit in a DNA strand. There is a $5'$ end of a DNA strand and a $3'$ end. Two DNA strands bind together to form the familiar double helical structure of double-stranded DNA. In order for a double helix to be stable, there must be a complementary base on the opposite strand for every base on a strand of DNA. The complementary pairs of bases are adenosine and thymine (A:T) and guanine and cytosine (G:C). The two strands join in an antiparallel fashion (one strand runs $5'$ to $3'$ and the other $3'$ to $5'$). The ribose sugars form the scaffolding for the complementary nitrogenous bases connected by hydrogen bonds on the inside of the molecule. The DNA double helix is dynamic, and the weak hydrogen bonding between complementary bases allows for the DNA strands to easily denature and reassociate with themselves. In the laboratory, the process of separating (denaturing) double-stranded DNA and then allowing the complementary single strands of DNA to reassociate and return to a double-stranded configuration is called hybridization. The basis of many of the laboratory techniques central to molecular diagnostics hinge on hybridization and the remarkable specificity of a nonrepetitive sequence of bases that make up a single strand of DNA to bind to its complementary sequence and no other. *In vivo*, the denaturing and reassociation of double stranded DNA is inherent to the process of gene transcription.

MODES OF INHERITANCE

A detailed family history provides the foundation for genetic diagnosis and risk assessment. Visually recorded using standardized symbols and nomenclature, the pedigree provides the tool by which inheritance patterns are elucidated, and subsequent risk assessment is calculated. In addition to diagnosis and risk assessment, the pedigree is a powerful research tool, aiding in the discovery of new genes and helping to better understand the phenotypic expression of genes already discovered. Observations made from controlled monohybrid and dihybrid crosses of peapod plants in the 1860s formed the foundation for Gregor Mendel's landmark laws of heredity that still govern basic pedigree interpretation today. Since that time, much has been learned and the study of inheritance is now far more complex than Mendel himself may have imagined. This section of the chapter reviews modes of inheritance and factors that may influence pedigree interpretation.

Mendelian Inheritance

The concepts that two copies of a gene segregate from each other (law of segregation) and are transmitted unaltered (particulate theory of inheritance) from parents to their offspring help to explain the concepts of dominant and recessive traits. When the presence of one copy of a particular allele results in phenotypic expression of a particular trait, the trait is dominant. When two copies of a particular allele must be present for the phenotypic expression of a trait, the trait is recessive. Note that it is the phenotypic expression that is described as dominant or recessive, not the allele or gene itself. Thus, patterns of inheritance are distinguished by where the gene resides within the genome (autosome or sex chromosome) and whether or not phenotypic expression occurs in the heterozygous or homozygous state. Traditionally recognized mendelian patterns of inheritance include autosomal dominant, autosomal recessive, X-linked recessive, X-linked dominant, and Y-linked (holandric). When each of these inheritance patterns is represented in a pedigree diagram, distinguishing features can be visually recognized (Table 5.1).

Autosomal Dominant Inheritance

Autosomal dominant inheritance is designated when no difference in phenotypic expression is observed between heterozygous and homozygous genotypes. Visually, the autosomal dominant pedigree shows multiple affected generations in a vertical pattern, an equal distribution of males and females affected, and both males and females transmit the phenotype (including males transmitting the phenotype to other males). Typically, dominant disorders occur when a mutation confers an inappropriate activity on a gene product. Examples include Huntington disease and other polyglutamine disorders where expansion of a triplet repeat with in a polyglutamine tract causes cellular toxicity, or familial amyloidosis where mutant transthyretin protein is relatively unstable and deposits as amyloid in tissues. However, some dominant disorders, such as those involved in the majority of the inherited cancer syndromes, occur with inheritance of a single copy of a gene where the mutant copy has not acquired a novel, pathogenic function, but is inactivated. Further, the great majority of cells and tissues carrying single copies of these mutant genes are functionally normal. The resolution of this apparent paradox came when, after observing familial cases of bilateral retinoblastoma (RB) and comparing those cases with sporadic (nonfamilial) cases of unilateral RB, Knudson proposed that two hits or mutational events were needed for the initiation of tumor growth. In the case of familial RB, a germline mutation in one tumor suppressor allele was postulated, and there was a much higher probability of tumor initiation because the individual was born with one hit (mutation). Therefore, somatic mutations that hit or render nonfunctional the remaining normal allele would be tumorigenic. In contrast, in a normal

Table 5.1   Mendelian Inheritance Patterns

Inheritance Pattern	Example Pedigree	Clinical Example
Autosomal dominant		Huntington disease Myotonic dystrophy Retinoblastoma Lynch syndrome Neurofibromatosis I TTR associated-amyloidosis And many others
Autosomal recessive		Cystic fibrosis Galactosemia Autosomal recessive (AR) deafness AR epidermolysis bullosa Tay-Sachs disease Klippel-Feil syndrome And many others
X-linked recessive		Duchenne muscular dystrophy Hemophilia A X-linked ichthyosis X-linked mental retardation Opitz syndrome Emery-Dreifuss muscular dystrophy And many others
X-linked dominant		Vitamin D-resistant rickets Coffin-Lowry syndrome And others
Y-linked (Holandric)		Hairy ears Y-linked deafness Very few others

individual (one not carrying a mutant *RB* gene in the germline), that same somatic event would not lead to the initiation of a tumor because one functional allele would remain. Tumor initiation takes place in a normal individual only if two somatic events occur at the same locus. As tumor initiation is not observed with a single abnormal allele, it is said that tumor suppressor genes act as recessive alleles at the cellular level, but as dominant disorders at the organism level. Additional examples of dominant disorders associated with inactivating mutations in tumor suppresser genes include Lynch syndrome (inactivation of one of the mismatch repair genes *MLH1*, *MSH2*, or *MSH6*) or familial breast cancer (inactivation of *BRCA1* or *BRCA2*).

Autosomal Recessive Inheritance

Autosomal recessive inheritance is designated when phenotypic expression is observed only when both copies of a gene are inactivated or mutated. Visually, the autosomal recessive pedigree typically shows a horizontal pattern where multiple affected individuals can be observed within the same sibship, and an equal number of males and females are affected. In instances of autosomal recessive inheritance, each parent of an affected individual has a heterozygous genotype composed of one copy of the mutated gene and one copy of the normal/functional gene. When a pedigree is analyzed, individuals who must be genetic carriers of the disorder in question, such as parents of an affected child, are termed obligate carriers. Other individuals in the pedigree may be at risk for being carriers. The risk to be a carrier is defined by each individual's position in the pedigree relative to affected individuals, or known carriers. For example, a sibling of an obligate carrier of sickle cell anemia has a 50% probability of being a sickle cell carrier, while a first cousin of a cystic fibrosis patient has a 25% chance of being a carrier.

Typically in recessive disorders, having only a single copy of a mutant gene is insufficient for manifestation of disease. Alternatively stated, one copy is enough for normal cellular and tissue function. Examples of recessive disorders include enzyme deficiencies, such as galactosemia (galactose-1-phosphate uridyl transferase) or phenylketonuria (phenylalanine hydroxylase), or deficiency of transport proteins, such as cystic fibrosis (CFTR). Consanguinous mating, or mating between related individuals, increases the risk of autosomal recessive phenotypic expression for certain genes because the proportion of shared genes among offspring is increased.

X-Linked Recessive Inheritance

X-linked recessive inheritance is designated when phenotypic expression is observed predominantly in males of unaffected, heterozygous mothers. All female offspring of affected males are obligate carriers. Visually, the pedigree typically shows a horizontal pattern of affected individuals with no instance of direct male-to-male transmission. However, males may transmit the disorder to a grandson through carrier female daughters.

It is not uncommon for X-linked recessive disorders to appear in a family such that before a certain generation the disease is not apparent, but is observed to be segregating in the family after that generation. This phenomenon is due to new mutations appearing *de novo* in an individual. This was explained by the American geneticist Haldane, and his theory is referred to as the Haldane hypothesis. If the reproductive fitness of a male affected with an X-linked recessive disorder is low or nil, then in a population one-third of all affected X chromosomes will be removed from the gene pool every generation. An example of decreased reproductive fitness among males is Duchenne muscular dystrophy. If the incidence of the disease is constant, then one-third of cases must be due to mutations arising *de novo* in a family.

X-Linked Dominant Inheritance

X-linked dominant inheritance is designated when phenotypic expression is observed predominantly in females (ratio of about 2:1) and all daughters of affected males are affected and none of the sons of affected males are affected. Visually, the pedigree typically shows a vertical pattern of affected individuals, with no instance of direct male-to-male transmission. X-linked dominant conditions are substantially less common than X-linked recessive disorders. An example is X-linked, vitamin D-resistant rickets, which is caused by mutations in the *PHEX* (phosphate-regulating endopeptidase homolog, X-linked) gene located at Xp22.

X-Linked Dominant Male Lethal Inheritance

X-linked dominant male lethal inheritance is designated when phenotypic expression is observed only in females. Visually, the pedigree typically shows a vertical pattern with an increased rate of spontaneous abortion and where approximately 50% of the daughters from affected mothers are also affected. Although the great majority of cases of Rett syndrome are not familial (but sporadic), familial cases have been described and would be classified as an X-linked dominant male lethal disorder. Rett syndrome is caused by mutations in the *MeCP2* gene (methyl CpG binding protein 2). Rett syndrome is a neurodevelopmental disorder characterized by arrested development between 6 and 18 months of age, regression of acquired skills, loss of speech, stereotypical hand movements, microcephaly, seizures, and mental retardation. Affected males rarely survive to term, and the majority of affected females do not reproduce.

Y-Linked or Holandric Inheritance

Y-linked or Holandric inheritance is designated when phenotypic expression is observed only in males with a Y chromosome. Visually, the pedigree shows only male-to-male transmission. Hairy ears are an example of a Y-linked trait. Few disease states have been shown to be Y-linked. However, there is one report of a multigenerational Chinese family with Y-linked deafness.

Non-Mendelian Inheritance

Epigenetic Inheritance—Imprinting

When the phenotypic expression of a gene is essentially silenced dependent on the gender of the transmitting parent, the gene is referred to as imprinted. The phenomenon of imprinting renders the affected genes functionally haploid. In the case of imprinted genes, the functional haploid state disadvantages the imprinted gene because the gene is more susceptible to adverse effects of uniparental disomy, recessive mutations, and epigenetic (like DNA methylation-dependent gene silencing) defects. Visually, pedigrees that represent imprinting may appear similar to autosomal recessive or sporadic pedigrees and show a horizontal pattern. Imprinting disorders may also appear autosomal dominant and show a grandparental effect in the case of imprinting center mutations. Males and females are equally affected, and transmission is dependent on the gender of a parent. The two most well-known imprinting disorders are Prader-Willi and Angelman syndrome. Prader-Willi syndrome is caused by an absence of paternally contributed 15q11–13 (PWS/AS) region, whereas Angelman syndrome is caused by an absence of maternal contribution at the same locus. In the case of PWS, lack of paternally contributed genes at 15q11–13 (regardless of mechanism) results in unmethylated and overexpressed genes in this region. The same is true for Angelman syndrome. However, it is a lack of maternally contributed genes that causes the phenotype in this instance. The clinical phenotype of each disorder is distinct, but both are associated with mental retardation.

Inheritance Through Mitochondrial DNA

The inheritance of mitochondrial disease is complicated by the fact that mitochondrial disease can be either the result of mutations in nuclear DNA (nDNA) and thereby subject to the Mendelian forms of inheritance described previously or the result of mutations in organelle-specific mitochondrial DNA (mtDNA). Since the mitochondrial genome is maternally inherited, pedigrees demonstrating mitochondrial inheritance show an affected mother with all of her offspring (male and female) affected. The common phenomenon of heteroplasmy, where mtDNA mutations are present in only a portion of the mitochondria within a cell, can make laboratory analysis and clinical assessment difficult. It is estimated that only 10–25% of all mitochondrial disease is the result of maternally inherited mutations in the mitochondrial genome. Therefore, mitochondrial disease should not always be equated with mitochondrial inheritance.

Multifactorial Inheritance

Sorting out whether a particular phenotype is predominantly the result of inherited genetic variation, environmental influence, or some combination therein can be difficult. When the combined effects of both inherited and environmental factors cause disease, the disorder is said to exhibit multifactorial inheritance. Multifactorial inheritance is associated with most, if not all cases of complex, common disease (cancer, heart disease, asthma, autism, mental illness, and others). Typically, multiple loci or multiple genes are associated with the same complex disease phenotype. Such genetic heterogeneity works additively, such that the net effect of multiple mutations in multiple genes exacerbates and/or detracts from a particular clinical phenotype.

Sporadic Inheritance

Sporadic inheritance, where only one isolated case occurs within a family, is the most common pedigree pattern observed in clinical practice. Chromosomal abnormalities and new dominant mutations typically demonstrate sporadic inheritance. It is easy to imagine how autosomal recessive and X-linked recessive disorders can often appear sporadic, especially in situations where family size is small or clinical knowledge about extended family is limited. Thus, both Mendelian and non-Mendelian explanations for sporadic inheritance, each with its own recurrence risks, can apply. As a result, clinicians tend to refer to isolated cases as apparently sporadic rather than absolutely sporadic. Since noninherited disorders are associated with virtually negligible recurrence risk as compared to those exhibiting Mendelian inheritance, those associated with chromosomal abnormalities, and those associated with new dominant mutations, it is important to make every effort to distinguish apparently sporadic cases from truly sporadic ones. However, it is often not possible to make this determination and recurrence risk can be narrowed only to a broad range encompassing all possibilities.

Differences in Phenotypic Expression Can Complicate Pedigree Analysis

The occurrence of reduced penetrance, variable expressivity, anticipation, and gender influence or limitation can confound pedigree analysis. The clinical subtlety and nuance associated with each phenomenon can impact recognition of the correct inheritance pattern (usually autosomal dominant, but not always) and result in overlooked or even incorrect diagnoses. Further, accurate recurrence risk is dependent on correct diagnosis and pedigree assessment.

Genetic Penetrance

The penetrance of a genetic disorder is measured by evaluating how often a particular phenotype occurs given a particular genotype or vice versa. Some disorders show 100% penetrance, where all individuals with a particular genotype express disease, while others show reduced penetrance, such that a proportion of individuals with a particular genotype never develop any features (even mild) of the associated clinical phenotype. Thus, penetrance is the probability that any phenotypic effects resulting from a particular genotype will occur. Certain factors are known to influence the gene penetrance for specific disorders. For example, phenotypic expression of a particular phenotype may be modified by age, termed age-related penetrance. Sometimes, as age increases, penetrance increases. For example, only 25% of individuals with a specific Huntington disease genotype (41 repeats) exhibit symptoms at age 50, while 75% exhibit symptoms at age 65. Although less common, penetrance can also decrease with age. Gender-related penetrance has been observed in cases of hereditary hemochromatosis where some females with a particular HH genotype show no evidence of iron accumulation in contrast with their affected male siblings who are known to have the identical genotype. Reduced penetrance can sometimes obscure an autosomal dominant inheritance pattern because, while some family members may have affected offspring, they themselves are not affected due to reduced penetrance of the disorder.

Sex-Influenced Disorders

Sex-influenced disorders are disorders that demonstrate gender-related penetrance. When the probability of phenotypic expression is more likely given a specific gender, the disorder is said to be sex-influenced. BRCA2-related hereditary breast/ovarian cancer (HBOC) and APOE4-associated late onset familial Alzheimer disease are sex-influenced disorders. BRCA2-related HBOC is an autosomal dominant disorder associated most predominantly with increased risk for breast and/or ovarian cancer. Although less common than breast or ovarian cancer, BRCA2 carriers may also be at increased risk for several other cancers including neoplasms of the skin, prostate, pancreas, larynx, esophagus, colon, stomach, gallbladder, bile duct, and hematopoietic system. In cases of HBOC caused by BRCA2 mutation, about 6% of males as opposed to 86% of females are expected to develop breast cancer by age 70. With respect to APOE4-associated late onset familial Alzheimer disease, women who are heterozygous for APOE4 alleles are at 2-fold increased risk to develop late onset familial Alzheimer disease as compared to males with the same genotype.

Sex-Limited Disorders

Sex-limited disorders refer to autosomal disorders that are nonpenetrant for a particular gender. Male limited precocious puberty is one example. Males heterozygous for mutations in the LCGR gene located on chromosome 2 exhibit this phenotype, but females with the same genotype do not. Very few sex-limited disorders have been documented.

Variable Expressivity

Variable expressivity refers to the difference in severity of disease among affected individuals, both between related and unrelated individuals. It is important to note that even between related individuals (with the same genotype) variable expressivity occurs. Variable expressivity is distinct from penetrance because it implies a degree of affectedness, not whether or not the individual is affected at all. The majority of inherited disease demonstrates some degree of variable expressivity. Variable expressivity can complicate pedigree analysis because individuals with subtle clinical manifestations can be mistaken for unaffected individuals. Neurofibromatosis type I is an autosomal dominant neurocutaneous disorder that affects 1/3000 individuals. There is a high degree of variable expressivity observed within family members that carry the same NF1 mutation. Some affected individuals with the same mutation may show only a few café-au-lait macules of the skin, while others may be more severely affected with large invasive plexiform neurofibromas and hundreds of cutaneous and subcutaneous neurofibromas.

Pleiotropy

Pleiotropy refers to disorders where multiple, seemingly unrelated organ systems are affected. For example, one individual in a pedigree may exhibit cardiac arrhythmia, whereas another individual with the same disorder in either the same or different pedigree shows muscle weakness and deafness. Since the manifestations of disease are so vastly and usually inexplicably different, disorders that show a high degree of pleiotropy are often difficult to diagnose. As a group, mitochondrial disorders typically show a high degree of pleiotropy, as any organ system can be affected, to almost any degree, with any age of onset.

Anticipation

A disorder shows anticipation when an earlier age of onset or increased disease severity occurs in successive generations. Anticipation is predominantly associated with neurodegenerative trinucleotide repeat disorders (spinocerebellar ataxias, Huntington disease, myotonic dystrophy, etc.). In such cases, the number of trinucleotide repeats expands through generations and is correlated with severity of disease and age of onset. However, not all disorders that exhibit anticipation are trinucleotide repeat disorders. Dyskeratosis congenita-Scoggins type characterized by nail dystrophy, skin hyperpigmentation, and mucosal leukoplakia shows anticipation via a mechanism of progressive telomere shortening in successive generations. While the mechanism remains unclear, anticipation is observed in families with a specific TTR gene mutation (V30M) associated with amyloidosis.

Other Factors That Complicate Pedigree Analysis

Genetic Mosaicism

Mosaicism occurs when two or more genetically distinct cell lines are derived from a single zygote. The timing of the post-zygotic event(s) and tissues involved determine the clinical consequence and help to distinguish one type of mosaicism from another. Gonosomal mosaicism occurs early in embryonic development and is more likely to involve gonadal tissue and result in phenotypic expression. The clinical effects are often milder for mosaic individuals where only a proportion of cells carry a particular mutation, as compared to those who inherit germline mutations where all cells are affected. When mosaicism is confined to gonadal tissue, there are usually no clinical consequences to the gonadal mosaic individual. However, such individuals are at higher risk for having affected offspring. Thus, since gonosomal mosaic parents have some proportion of mutant germ cells, they can (and do) have nonmosaic, affected offspring. There is no practical way to exclude the possibility of gonadal mosaicism or effectively test for it. This can cause a dilemma with respect to providing accurate recurrence risks to families. Gonadal mosaicism has been found to be more common for certain disorders and some empiric risk estimates have been determined. For example, Duchenne muscular dystrophy, an X-linked disorder, has an empiric risk for gonadal mosaicism of 10–30%. This means that even when a mother of an affected boy tests negative for a *DMD* gene mutation, a male who inherits the same X chromosome as an affected sibling has a 10–30% chance of being affected.

Consanguinity

Consanguinity is both a social and genetic concept. Generally, it refers to marriage or a reproductive relationship between two closely related individuals. The degree of relatedness between two individuals defines the proportion of genes shared between them. The offspring of consanguineous couples are at increased risk for autosomal recessive disorders due to their increased risk for homozygosity by descent. A frequent way that consanguinity can complicate pedigree analysis is when a provider is unaware of consanguinity at the time they are evaluating the pedigree and what appears to be an autosomal dominant inheritance pattern is associated with an autosomal recessive disease phenotype.

Preferential Marriage Between Affected Individuals

Increased reproductive risk can be the result of preferential marriage between affected individuals. It is not uncommon for similarly affected individuals to attend the same school (for example, deaf high schools) or make connections at support groups (for example, Little People of America). And a proportion of such relationships may develop such that an affected couple may decide to start a family. Such selective mating can increase the likelihood of pseudodominance within a pedigree because the mating environment is selected such that an autosomal recessive disorder appears more frequently than expected. An increased risk for autosomal dominant disorders is also present. In cases where both reproductive partners are affected with the same autosomal dominant condition recurrence risk ranges from 66% (when homozygous dominant inheritance is not compatible with life) to 75%.

Other Considerations for Pedigree Construction and Interpretation

Pedigree construction requires equal amounts of skill, science, and art. Those obtaining them must have a strong base of medical genetic knowledge, so as to know the important questions to ask and construct the pedigree correctly. In addition, careful attention must be paid to establishing trust, navigating social relationships, educating, and communicating effectively. Family histories are deeply personal, and the psychosocial impact of the required informational gathering can be significant. Further, inaccurate information can result in misinterpretation and ultimately misdiagnosis. Whenever possible, reported diagnoses must be confirmed with medical records. The effort this requires should not be minimized, as privacy and confidentiality must be upheld for all family members throughout the process.

Clinical molecular genetics seeks to identify genetic variation and to determine whether or not the observed genetic variation has a phenotypic effect. Certainly, the latter cannot be accomplished without astute and thorough clinical evaluation and family history. Even an apparently negative family history is an important one that can guide test selection and result interpretation. In addition, the impact of pedigree analysis on genomic research is formidable. As a result of detailed pedigree assessment, numerous genes have been discovered, genotype:phenotype correlations elucidated, natural history knowledge obtained, and certainly inheritance patterns revealed.

CENTRAL DOGMA AND RATIONALE FOR GENETIC TESTING

The clinical relevance of molecular genetics is fundamentally rooted in the central paradigm of molecular biology: genes encode proteins. Genes are the blueprint for the proteins that form the macromolecules of cellular structure and function. Cells, their respective functions, and the interactions between them translate to the observable characteristics, or clinical phenotype, of an organism. Endogenous and exogenous molecular, cellular, and organismal environments also play an important role in influencing clinical phenotype. So, the expression of DNA at the molecular level coupled with environmental effects leads to more tangible morphological and physiological traits at the level of the organism. However, organisms do not exist in isolation. Each organism functions as part of a population within a larger species and external environment. A species, and the organisms within it, is subject to evolutionary forces, including natural selection, genetic drift, and gene flow. Such forces ultimately impose, overlook, propagate, or extinguish genetic variation. The dynamic relationships between genetic variation, proteins, cells, organisms, populations, and environment(s) connect genetic laboratories to clinical practice, as evaluating for genetic variation (molecular genetics, cytogenetics) and/or its biochemical consequence (biochemical genetics) provides an explanation and/or causative evidence for clinical phenotype and diagnosis.

Diagnostic and Predictive Molecular Testing

The clinical applications of molecular genetic testing can be generalized into two groups based on whether the clinical information sought is intended for diagnostic or predictive purposes. Occasionally overlap between diagnostic and predictive testing occurs. Although most commonly performed for the purpose of diagnosing a disorder in a symptomatic individual, diagnostic testing can also be informative for presymptomatic at-risk individuals. The degree of gene penetrance must be known in order for this diagnostic yet predictive testing to impart clinical value. Penetrance does not necessarily have to be 100% to be useful to the patient and/or family; for example, *BRCA1* mutations are associated with a lifetime risk of approximately 60–80% for the development of breast cancer.

The molecular genetic test for Huntington disease (HD) illustrates how the same test can be used to determine diagnosis for affected individuals and to predict affected status for as yet unaffected individuals. If testing is performed on a 25-year-old asymptomatic individual known to be at 50% risk for HD and the result is consistent with a repeat expansion known to be fully penetrant by age 47, the result of predictive testing is consistent with a diagnosis of HD during the presymptomatic period. Predictive diagnostic testing can be more difficult to interpret in cases where gene penetrance is not so absolute. For example, if the same 25-year-old asymptomatic individual at 50% risk for HD was found to have an expansion mutation in the reduced penetrance range (36–39 repeats), the ultimate diagnosis is not so absolute.

The second broad group of molecular genetic tests includes those performed for the purpose of revising an already known risk. Predictive molecular testing typically employs molecular screening tests to more accurately determine the individual and familial/reproductive risks for an individual that is already a member of a high-risk population. Typically, a targeted mutation analysis method is used. For example, for Caucasian individuals of Northern European ancestry and no family history of cystic fibrosis (CF), the risk for being a heterozygous carrier is 1/25. After such an individual is screened for the 23 mutations in the *CFTR* gene that account for approximately 90% of CF alleles in the Northern European population, a negative screen result decreases that risk slightly over 10-fold, to 1/265. It is important to recognize that while many predictive molecular screening tests are focused on evaluating at-risk individuals for autosomal recessive carrier status, other subgroups of predictive screens help to distinguish germline from somatic disease or revise prognosis or risk related to complex disease based on presence or absence of disease-associated SNPs. For example, colon cancer is typically a sporadic disease that has a genetic component but does not typically follow a simple Mendelian inheritance pattern of a single-gene disorder. However, a small fraction, approximately 5% of cases, do indeed follow a Mendelian inheritance pattern and are due to inherited mutations in single genes. It is clear that identification of these families can have enormous importance for family members because individuals who are shown to carry the familial mutation can greatly benefit from enhanced monitoring and prophylactic measures. Similarly, family members that are shown not to carry the familial mutation are freed from the need for intensive monitoring and are returned to the same risk as the general population for the development of colon cancer. Unfortunately, simple pedigree analysis is seldom sufficient to identify such families because colon cancer is not a particularly rare condition, and it is not uncommon for multiple family members, who often share many environmental risk factors, to develop sporadic colon cancer. In addition, the penetrance of the disorder may not be complete, thereby making it more difficult to recognize a specific inheritance pattern. Some inherited colon cancer syndromes, such as familial adenomatous polyposis colon cancer (FAP) have a distinctive phenotype (many thousands of colonic polyps) so ascertainment of families is usually straightforward. However, other syndromes, such as Lynch syndrome (or HNPCC), which is due to mutations in the mismatch repair (MMR) pathway, cannot be distinguished from sporadic colon cancer using clinical or pathological criteria. One might investigate all of the relevant MMR genes in cases suggestive of Lynch syndrome, early age of onset (<50 years), or familial clustering, but such testing (which could involve whole gene sequencing and deletion analysis for up to three large genes) could be very expensive. Further, selection of potential cases on clinical grounds and family history often has a relatively low yield. Because tumors from Lynch syndrome patients are defective in MMR, they exhibit a type of genomic instability known as microsatellite instability. Since approximately 30% of sporadic tumors have microsatellite instability, screening tumor specimens from individuals at risk for Lynch syndrome for loss of expression for mismatch repair proteins and microsatellite stability can eliminate approximately 70% of colon cancer patients from a Lynch syndrome diagnostic algorithm and greatly reduce cost. If microsatellite instability is present in a tumor from an at-risk individual, germline testing can subsequently be performed. If microsatellite instability is not present, suspicion for an underlying germline defect is low.

Benefits of Molecular Testing

Psychosocial benefits of a confirmed molecular genetic diagnosis may include (i) reduced anxiety associated with a known versus unknown diagnosis, (ii) reduced anxiety if the diagnosis confirmed is considered by the patient to be less severe among those being considered for patient, (iii) reduced anxiety associated with a cease in the diagnostic odyssey that many patients with rare disorders experience (multiple medical consults, procedures, and laboratory tests associated with a continued search for a diagnosis). In addition, psychosocial benefits may accrue from implementation of a more individualized and, in some cases, preventive medical management approach. For individuals undergoing presymptomatic testing, benefits may also include a sense of empowerment, regardless of their test result and a sense of relief if they test negative. Knowledge of one's risk for having children with a genetic disorder also assists with family planning, with individuals and couples being able to access genetic counseling and prenatal diagnosis.

Clinical benefits of a molecular diagnosis often include the ability for the care provider to recommend a preventive medicine and treatment plan based on the known natural history of a particular disorder. Genotype:phenotype correlations have been established for some particular mutations/disorders, such that a more individualized medical approach to care with respect to severity of disease, expected age of onset for presymptomatic cases, and increased risk for certain associated complications can be determined. So, the genotype result may give care providers and their patients information that could lead to more individualized medical management. Also, when a molecular diagnosis is confirmed, predictive testing options become available for at-risk family members. Consider an individual at 50% risk for FAP undergoing presymptomatic testing. Identification of the causative mutation for this individual allows for early intervention by screening and prophylactic colectomy, and informative presymptomatic testing options for at-risk family members. A negative test result directs implementation of a more appropriate, less aggressive screening strategy.

Risks Associated with Molecular Testing

Risks and limitations of genetic testing should always be reviewed and openly discussed with patients as part of the informed consent process prior to testing. Risks associated with molecular genetic testing are most often psychological and financial. Limitations of molecular genetic testing are usually related to confounding results, interpretive restrictions, or imperfections of the method used.

Although it can be of profound benefit, the knowledge of a molecular genetic test result can also be a risk, regardless of whether the result confirms the presence of disease. A positive test result can be devastating, and a true negative result can sometimes invoke survivor guilt. Consider two siblings who undergo presymptomatic testing where one sibling tests positive for a life-threatening disorder and the other tests negative. Both siblings will likely experience psychological repercussions of testing. Such psychological risks must be discussed before sample collection and may influence an individual's decision to undergo testing. Financial risks may include inability to obtain life

insurance or certain types of health insurance should an individual test positive. However, the newly enacted Genetic Information Nondiscrimination Act (GINA) should reduce these risks. Many insurance plans do not cover the costs of genetic testing or will cover only part of the expense. Given that many molecular genetic tests are expensive due to the high costs of the technology used and the highly skilled personnel required to process and interpret the sample, personal financial cost to the patient can be substantial.

Limitations of molecular genetic testing should also be discussed with the patient as part of the informed consent process prior to sample collection. Molecular genetic testing is often misunderstood as perfectly decisive. While new technologies and detection rates are continuously improving, not all mutations are identified such that the risk for a false negative result is always a possibility. An even more common problem involves the identification of alterations whose medical or functional significance is not clear. These genetic alterations are termed variants of uncertain significance and can be especially complicated to interpret. Although rare, laboratory errors such as performing the wrong test or mislabeling samples can also occur. Disease and test-specific limitations of molecular testing are truly method, disease, and case specific, and it would be impossible to address each of them here.

Considerations for Selection of a Molecular Test

Selecting an appropriate molecular genetic test is dependent on the purpose for testing, the clinical information known, the sample(s) and testing methods available, and the clinical information sought. Molecular screening tests usually involve methods that investigate for common mutations (for example, targeted mutation detection by RFLP), whereas diagnostic testing methods are typically more comprehensive (for example, DNA sequencing). The molecular methods used for the purpose of revising a known risk can sometimes be the same, but are often different than those used for diagnostic purposes. When evaluating the method to be used, the expected detection rate for individuals who are classically affected with the disorder in question and the clinical context of the patient being tested should be considered. To maximize the informative value of presymptomatic testing, in most cases, one must know the familial mutation(s). Practically, this translates into the necessity that an affected individual should be tested before presymptomatic testing is performed on at-risk family members. Preferred testing algorithms developed by expert clinicians and laboratorians are especially useful, though ultimately each clinical situation is different and should be considered within its own unique context. An increasing number of molecular genetics laboratory directors are employing genetic counselors that act as a liaison between the ordering provider and the laboratory to serve as a resource for the identification of case-specific benefits, risks, and limitations to testing, as well as to assist with test selection, case coordination, and interpretation of results.

ALLELIC HETEROGENEITY AND CHOICE OF ANALYTICAL METHODOLOGY

The great majority of analyses performed in the clinical molecular genetics laboratory are based on the polymerase chain reaction (PCR). PCR is a technique, developed by Kary Mullis in 1984 (then at Cetus Corp.), for the rapid, *in vitro* amplification of specific DNA sequences. The rapid introduction of PCR into research and later into clinical laboratory practice has revolutionized the practice of molecular biology. In 1993, Dr. Mullis was awarded the Nobel Prize in Chemistry for his achievement.

Knowledge of the sequence of the region of DNA flanking the area of interest is required for PCR. Two synthetic oligodeoxynucleotides (primers), typically 20 to 30 bases in length, are prepared (or purchased) such that one of the primers is complementary to an area on one strand of the target DNA 5′ to the sequences to be amplified, and the other primer is complementary to the opposite strand of the target DNA, again 5′ to the region to be amplified. To perform the amplification, one places the sample DNA in a tube along with a large molar excess of the two primers, all four deoxynucleotide triphosphates (dNTPs), buffer, magnesium ion, and a thermostable DNA polymerase. Successive rounds of heating to 93–95°C to denature the DNA, cooling to 50–60°C to allow annealing of the oligonucleotides, and heating to 72°C (the temperature optimum for the DNA polymerase isolated from *Thermus aquaticus*) result in synthesis of the DNA that lies between the two primers. The amount of amplified DNA being synthesized doubles (approximately) with every temperature cycle. The amount of DNA produced is exponential with respect to cycle number. After 30 cycles of denaturation, annealing, extension, 2^{30}, or approximately 10^9, copies of the DNA sequences lying between the two primers will have been generated. In a typical experiment starting with 20–100 ng of human DNA, 30 cycles of amplification will produce enough DNA from a single copy gene to be visualized on an ethidium bromide stained gel. As each cycle takes 2–5 minutes, amplification of a specific sequence can easily be accomplished in several hours. After amplification, the DNA can be analyzed by one of several techniques, depending on the specific problem.

Specific Versus Scanning Methods

Analytical methods in molecular genetics can be grouped into two broad categories: mutation detection techniques, which are used to investigate the actual base sequence at a particular locus, and quantitative methods, in which PCR-based techniques are used to quantify specific nucleic acid sequences. Mutation detection strategies can be further grouped into specific or scanning techniques.

Specific Mutation Detection

Specific mutation detection entails straightforward, and largely routine, procedures that can be used to analyze DNA samples for previously identified mutations using an assay designed for maximum specificity. This approach targets known mutations in potentially large cohorts of patients or small panels of specific mutations in disorders characterized by one or a few common alleles. Results from these types of analyses may confirm or establish clinical diagnoses. Furthermore, in families at risk for a particular genetic disease, specific or targeted mutation detection allows for rapid screening of an entire family for the mutation identified in the proband (the first member of a family to be diagnosed with a genetic disorder), thereby permitting accurate carrier determinations that may aid reproductive decisions. Rapid testing of large numbers of patients permits an assessment of the frequency of a mutation among disease-causing alleles, thereby determining which mutations are most prevalent in different patient populations and guiding the creation of effective clinical mutation testing panels. Examples of genetic disorders that are characterized by low allelic heterogeneity and are most often investigated using specific mutation detection methods include hypercoagulable states due to Factor V Leiden or prothrombin mutations, hemochromatosis, galactosemia, and alpha-1-antitrypsin deficiency. Although cystic fibrosis has a high degree of allelic heterogeneity (over 1500 mutations identified), carrier screening is typically done with a panel of 23–100 mutations, which detect approximately 90% of mutations in the target population of Northern European Caucasians.

The specific mutation detection methods can themselves be divided into those that utilize electrophoretic- or hybridization-based methods. Both types of platforms are robust, and in experienced hands yield reproducible results. Both types of systems are in widespread use in clinical and research laboratories. One criterion for choice between these general platforms is the cost incurred per sample analyzed. In the authors' experience, when the number of samples to be analyzed at one time (samples per batch) is low, electrophoretic methods are often the most cost effective to develop, validate, and implement. However, when the number of samples per batch is larger (greater than 8–12 samples), then the hybridization-based techniques, many of which can be adapted to 96-well microplate formats or real-time, are often more cost effective.

Mutation Scanning Approaches

Mutation scanning methods interrogate DNA fragments for all sequence variants present. By definition, these strategies are not predicated on specificity for specific alleles, but are designed for highly sensitive detection for all possible variants. In principle, all sequence variants present will be detected without regard to advance knowledge of their pathogenic consequences. Once evidence for a sequence variant is found, the sample must be sequenced to determine its molecular nature. The advantage of using a scanning method followed by sequencing of only positive PCR products is that the scanning methods are typically less costly to perform than DNA sequencing. Although a number of mutation scanning methods have been developed, including single-strand conformation polymorphism (SSCP), heteroduplex analysis (HA), conformation-specific gel electrophoresis (CSGE), thermal gradient gel electrophoresis (TGGE), and melt curve analysis, they have been almost completely replaced by what is considered the gold standard mutation scanning method—DNA sequencing. There are a number of disease-associated genes that have high allelic heterogeneity, or very few recurrent mutations in the population, that are typically addressed for diagnostic purposes by whole gene sequencing, including *BRCA1* and *BRCA2*, the mismatch repair genes; *MSH2, MLH1,* and *MSH6; CFTR* (for diagnostic, nonscreening applications); biotinidase (*BTD*); and medium chain acylCoA dehydrogenase (*ACADM*). Only when they are combined with appropriate genetic data and *in vitro* functional studies can investigators distinguish disease-causing mutations from polymorphisms without clinical consequence. In the research laboratory, mutation screening is a critical and obligatory final step toward identifying genes that underlie genetic disease. In the clinical laboratory, these methods are applied toward the detection of mutations in diseases marked by significant allelic heterogeneity. As the number of laboratories offering whole-gene sequencing assays for an increasing number of genes grows, the amount of variation in coding regions is beginning to be understood to be significantly greater than previously thought. This has the consequence that obtaining a previously unknown sequence variation in a patient sample is not uncommon. The interpretation of such results is challenging and is not a solved problem.

Interpretation of Molecular Testing Results

Of the three types of coding region mutations caused by single nucleotide changes, two are often relatively straightforward to interpret. It is generally assumed that nonsense mutations (or indels giving rise to an in-frame stop codon) are deleterious and are likely to be associated with a disease phenotype. Similarly, silent mutations are most often assumed to be benign. Exceptions exist, of course; silent mutations occurring at the first or last bases of an exon may influence RNA splicing. In addition, silent mutations may interrupt an exonic splice enhancer, again leading to altered splicing. An example of the disruption of an exonic splice enhancer is found in spinal muscular atrophy (SMA). The great majority of *SMA* is caused by deletion of exon 7 of the telomeric copy of the survival motor neuron gene *(SNMt)*. There exists a very highly homologous gene, the centromeric copy *(SMNc)* that has only 5 nucleotide changes relative to *SNMt*. Why is the presence of this gene, which is structurally normal in almost all cases of SMA, not sufficient to prevent neuronal death even if the telomeric copies are mutated? One of the nucleotide differences between *SNMc* and

SNMt is a C to T change in the centromeric copy. Although a silent mutation from an amino acid standpoint (both sequences code for Valine), the T allele is not recognized as an exonic splice enhancer. Thus, the *SMNc* gene transcript also lacks exon 7 and is unable to compensate for the lack of the *SNMt* gene.

The interpretation of missense changes is challenging. Many examples (affecting many different genes) exist in which missense changes are either pathogenic or benign. The distinction typically requires the examination of multiple families carrying a given missense mutation, and/or functional studies of recombinant, mutant protein. When a novel missense change is encountered in a clinical laboratory setting, these studies are not available. Thus, novel missense changes are typically referred to as variants of uncertain significance (VUS).

There are two schools of thought with respect to how VUS should be reported. One school holds that unless the laboratory can give a clean interpretation and offer documentation as to whether a given variant is known to be pathogenic or benign, the report should simply indicate that a VUS was detected. Thus, the contribution of the genetic test to the management of the patient is nil; it is as if the test were not performed (and cannot be performed). Clearly, the advantage in this approach is that one is not tempted to overinterpret the results, potentially leading to an incorrect medical decision. The disadvantage is the frustration on the part of the patient (and healthcare provider) that a rather expensive test (typically) has been performed and no useful information was obtained. The other school of thought holds that the laboratory should use all the tools available and when possible make a probabilistic statement as to the potential effect of the variant. The advantage to this approach is that the final decision as to how the result will be used in guiding patient care remains with the patient and his/her healthcare provider. The clear disadvantage is the possibility that the result provided may lead to incorrect medical management. Because of this, interpretation of VUS should be done very carefully.

The *insilico* characterization of missense VUS changes is still in its infancy, and much more work needs to be done in this area. As the era of whole genome sequencing rapidly approaches, the urgency of the need to characterize novel changes is increasing.

CONCLUSION

Molecular genetics utilizes the laboratory tools of molecular biology to relate changes in the structure and sequence of human genes to functional changes in protein function, and ultimately to health and disease. New technology, such as is being developed for the $1000 genome project, promises to greatly increase the reach and scope of molecular genetics. Indeed, some subspecialties, such as biochemical and cytogenetics may ultimately merge with molecular genetics and offer the medical community a more comprehensive and integrated approach to understanding the role of our genomic variation in health and disease. However, the interpretations of results from the clinical molecular genetics laboratory will always be rooted in the fundamentals of molecular and cell biology and in the central paradigm—that genes encode proteins. It will be from these roots that modern, personalized medicine will grow.

KEY CONCEPTS

- Deoxyribonucleic acid, DNA, is a double-stranded molecule consisting of two antiparallel polymers composed of ribonucleosides linked together by phosphodiester bonds. Weak hydrogen bonding between complementary bases allows for easy denaturing and reassociation of double-stranded DNA.

- Recognizing and correctly identifying the mode of inheritance of a disorder in a family provide powerful clues that can assist in establishing a genetic diagnosis and understanding risk to family members.

- Genetic testing typically is sought for either diagnostic or predictive purposes.

- Risks and limitations of genetic testing should be discussed openly with the patient as part of the informed consent process *prior* to the initiation of testing.

- The selection of either a mutation scanning technology or a specific mutation detection technique generally depends upon the allelic heterogeneity for which the disorder is being tested.

SUGGESTED READINGS

1. Bennett RL, Steinhaus KA, Uhrich SB, et al. Recommendations for standardized human pedigree nomenclature. Pedigree Standardization Task Force of the National Society of Genetic Counselors. *Am J Hum Genetics.* 1995;56:745–752.

2. Bruns DE, ER Ashwood, CA Burtis, eds. *Fundamentals of Molecular Diagnostics.* Saunders; 2007.

3. Chan PA, Duraisamy S, Miller PJ, et al. Interpreting missense variants: Comparing computational methods in human disease genes CDKN2A, MLH1, MSH2, MECP2, and tyrosinase (TYR). *Hum Mutat.* 2007;28:683–693.

4. Coleman WB, Tsongalis GJ, eds. *Molecular Diagnostics for the Clinical Laboratorian.* Humana Press; 2005.

5. Gödde R, Akkad DA, Arning L, et al. Electrophoresis of DNA in human genetic diagnostics—State-of-the-art, alternatives and future prospects. *Electrophoresis.* 2006;27:939–946.

6. Leonard DGB, ed. *Molecular Pathology in Clinical Practice.* Springer; 2006.

7. Lyon E, Wittwer CT. Light cycler technology in molecular diagnostics. *J Mol Diagn.* 2009;11(2):93–101.

8. Mais DD, Lowery-Nordberg M, eds. *Quick Compendium of Molecular Pathology (ASCP Quick Compendium).* American Society for Clinical Pathology; 2008.

9. Mardis ER. New strategies and emerging technologies for massively parallel sequencing: Applications in medical research. *Genome Med.* 2009;1:40.

10. Nikolausz M, Chatzinotas A, Táncsics A, et al. The single-nucleotide primer extension (SNuPE) method for the multiplex detection of various DNA sequences: from detection of point mutations to microbial ecology. *Biochem Soc Trans.* 2009;37:454–459.

11. Potts DM. *Queen Victoria's Gene.* Sutton Publishing, 1999.

12. Sykes B. *The Seven Daughters of Eve: The Science That Reveals Our Genetic Ancestry.* W.W. Norton & Co.; 2002.

13. Taylor CF, Taylor GR. Current and emerging techniques for diagnostic mutation detection: An overview of methods for mutation detection. *Methods Mol Med.* 2004;92:9–44.

14. Tchernitchko D, Goossens M, Wajcman H, et al. In silico prediction of the deleterious effect of a mutation: Proceed with caution in clinical genetics. *Clin Chem.* 2004;50:1974–1978.

15. Tsongalis GJ, Silverman LM. Molecular diagnostics. A historical perspective. *Clin Chim Acta.* 2006;369:188–192.

16. Tubbs RO, Stoler MH, eds. *Cell and Tissue Based Molecular Pathology: A Volume in the Foundations in Diagnostic Pathology Series.* Churchill Livingstone; 2008.

17. Wittwer CT. High resolution melting analysis: Advancements and limitations. *Hum Mutat.* 2009;30:857–859.

18. Yu B, Sawyer NA, Chiu C, et al. DNA mutation detection using denaturing high-performance liquid chromatography (DHPLC). *Curr Protoc Hum Genet.* 2006.

6

The Human Genome: Implications for the Understanding of Human Disease

Ashley G. Rivenbark

INTRODUCTION

Genetics is the study of cells, individuals, heredity, variation, and the population within each organism. The modern science of genetics started in the mid-19th century with the work of Gregor Mendel when he observed that organisms inherit traits in a discrete manner—later called genes. In a matter of about five decades, perhaps the most far reaching endeavor that the field of genetic research has ever attempted was accomplished in the sequencing of the human genome. Applied research in genetics has produced many benefits, including the recognition of the molecular basis of human genetic disorders and cancer. For example, many genetic diseases have been discovered as a result of a single mutation or a specific chromosomal rearrangement and are now understood at the molecular level, including, but not limited to, sickle-cell anemia, hemophilia, cystic fibrosis, Duchenne muscular dystrophy, Tay-Sachs disease, Down syndrome, Li-Fraumeni syndrome, Wilms' tumor, Prader-Willi syndrome, Angelman's syndrome, and many metabolic disorders. This would not have been accomplished in a timely manner for many of these diseases if it were not for the entire sequence of the human genome produced by the Human Genome Project in 2003. This chapter will touch in summary on the structure and organization of the human genome, the overview of the Human Genome Project, and the impact it has made on the identification of disease-related genes, as well as the sources of variation in the human genome, types of genetic diseases, and human malignancies we have come to understand based on the sequencing of the human genome. This chapter is not comprehensive but covers essential concepts and observations and, I hope, gives the reader an understanding that there has never been a more thrilling time to be immersed in the study of human genetics.

STRUCTURE AND ORGANIZATION OF THE HUMAN GENOME

In the 1940s, deoxyribonucleic acid (or DNA) was shown to be the genetic material of all living organisms. In 1953, James D. Watson and Francis H. C. Crick discovered the double helical structure of DNA. This discovery single-handedly revolutionized molecular biology and biological sciences. Many advances in genetics followed, and researchers established that the units of genetic information (genes) encoded information for the synthesis of enzymes/proteins. The complete set of genetic information or DNA instructions is referred to as a *genome*.

DNA Carries Genetic Information

The genetic blueprint of life occurs in the form of DNA, which is faithfully packaged within the nucleus of each cell in our body. In order to understand genetic disease, we need to first examine the structure of DNA. DNA is quite simple chemically; it is composed of phosphate, deoxyribose sugar, and four nitrogenous bases. These nucleotide bases in DNA are adenosine (A), thymine (T), cytosine (C), and guanine (G). The A is always paired to the T and the C is always paired to the G. Each triplet combination of these nucleotides makes up our genetic code and constitutes a code word. These words or bases are strung together to make genes, which instructs the cell how to make a specific protein. There are approximately 20,000–25,000 genes in the human genome that are transcribed to ribonucleic acid (RNA), and then translated to produce tens of thousands of proteins. DNA regions that are found and expressed in the mRNA sequence are called exons, and the DNA sequences that are not found in the final mRNA product are called introns. In addition, there are regions of intergenic DNA located between functional genes. For a certain phenotype or observable characteristic, a gene can give rise to several different combinations called alleles. An individual's genotype is made up of different alleles that arise from both parents. Alleles can be dominant or recessive. The dominant allele will result in a certain phenotype if only one copy is present, but there must be two copies of a recessive allele to result in the same phenotype. If a genotype (or pair of alleles) of an individual is two dominant alleles or two recessive alleles, the individual is said to be homozygous, and if an individual inherits two different alleles (one dominant and one recessive), then that individual is heterozygous. Ultimately, one gene can give rise to multiple transcripts that give rise to multiple proteins with different functions.

The genomes of any two people are more than 99% similar; therefore, the small fraction of the genome that varies among humans is very important. These variations of DNA are what make humans unique. However, variations in DNA can occur in the form of genetic mutations in which a base is missing or changed. This results in an aberrant protein and can lead to disease. Through studies of the genetic variation of humans, it is hoped that we can gain insight into phenotypic variation and disease susceptibility. In addition, it is thought that DNA structure plays a role in certain human genetic diseases. Certain trinucleotide (CTG and CCG) repeat sequences have been found in genes whose aberrant expression leads to disease. The severity of the disease is associated with the number of repeats; diseased individuals have greater than 50 repeats, whereas normal individuals have very few repeats.

General Structure of the Human Genome

DNA packaging, and how it gets organized in the nucleus of a cell, turns out to be a job for a class of proteins called histones. These histones can be altered by a number of chemical modifications, which have been shown to regulate the accessibility of the underlying DNA. The core nucleosome particle is composed of 147 base pairs of DNA wrapped around an octomer of four core histone proteins. These nucleosomes fold into 30 nm chromatin fibers, which are the components that make up a chromosome. The human genome is 3×10^9 base pairs or a length of about 1 meter, which compacts into a nucleus that is only 10^{-5} meters in diameter. Regulation of chromatin

has profound consequences for the cell, as the ability to open and close the environment in which DNA is packaged is the primary mechanism by which the genes encoded within the DNA get expressed into proteins. The structure of chromatin is now well understood, but how chromatin is packaged into a chromosome is not. Chromosomes are clearly visible with dyes that react with DNA, which can then be visualized under a primitive light microscope. The word *chromosome* is derived from Greek and describes a colored body, which reflects the ability to visualize dense regions. Dense, compact regions of chromosomes are referred to as heterochromatin consisting of mostly untranscribed and inactive DNA. Regions called euchromatin are less compact and consist of more highly transcribed genes. The genetic code that is inherited is in the DNA sequence, although the way in which the DNA is packaged into chromatin plays an important role in controlling and organizing the information that the DNA holds. When packaged into chromatin, some information is accessible and some is not, which depends on chemical modifications to the chromatin proteins (histones). Chromatin is dynamic, and the accessible regions of DNA change during human development or different disease states. This process of altering gene expression in a stable, heritable manner without changing the DNA code is referred to as epigenetics.

Chromosomal Organization of the Human Genome

Our genome contains 46 chromosomes with 22 autosomal pairs and two sex chromosomes. These chromosomes differ about 4-fold in size from chromosome 1 to chromosome 21, which is largest to smallest, respectively. Each of the 46 chromosomes in human cells contains a centromere (central region) and telomere (ends of the chromosome) composed of genes (2%), regulatory elements (1–2%), noncoding DNA (50%), which includes chromosome structural elements, replication origins, repetitive elements, and other sequences (45%). At the end of the 19th century it was accepted by numerous researchers that chromosomes formed the basis of inherited traits. There are approximately 60 trillion cells in the human body, which all originate from a single fertilized cell. The cells in the body undergo cell division, or mitosis, in which the chromosomes are condensed and genomic DNA is faithfully replicated. The nomenclature used to define the segments on a chromosome was determined by G-banding chromosomal staining, where the mitotic chromosomes were digested with trypsin and followed with Giemsa staining, which stains centromeric regions. The short arm region of the chromosome (usually displayed above the centromeres) is referred to as the p arm (for instance, 17p) and the long arm region of the chromosome (displayed below the centromeres) is called the q arm (for instance, 13q) with each band having a number associated with it. Chromosomal banding studies using Giemsa staining have shown that heterochromatin comprises 17–20% of the human chromosome consists of different families of alpha satellite DNAs and other higher order repeats.

Subchromosomal Organization of Human DNA

DNA features

Once the human genome was sequenced, it was recognized that our genome consists of a significant number of different repetitive DNA sequences. Several classes of repetitive DNA exist in the human genome, including Alu repeats, mammalian interspersed repeat (MIR), medium reiteration (MER), long terminal repeat (LTR), and long interspersed nucleotide elements (LINE). Repetitive DNA such as short interspersed nuclear fragments (SINEs), including MIRs and LINEs, are dispersed throughout the genome, whereas satellite DNAs are clustered in discrete areas (centromeres). Repetitive sequences have been recently proposed to be involved in genome compaction. Genomic regions of satellite DNA are condensed throughout the cell cycle, and there is evidence that LINEs are involved in X-chromosome condensation. Single nucleotide polymorphisms (SNPs) or a single nucleotide base change occurs by random and independent mutations, and in a high degree of variability among chromosomes.

CpG islands appear in approximately 50% of human genes and are located preferentially at the promoter region of genes, flanking the transcription start site. It has been estimated that there are around 30,000 to 45,000 total CpG islands in the human genome. A CpG island is defined as a region with greater than 200 base pairs with a G+C percentage that is greater than 50% with an observed/expected CpG ratio that is greater than 0.6. CpG dinucleotides are sites for DNA methylation and in turn can downregulate gene expression. DNA methylation has been shown to be important during gene imprinting and tissue-specific gene expression. Gene inactivation by aberrant DNA methylation has been correlated with cancer in many different cell types. In addition, the identification of CpG islands throughout the genome can help predict promoter regions for human genes.

Gene structure

Now that the human genome has been sequenced, one of the next steps is to utilize genomic tools to obtain a picture of how DNA is targeted by transcription factors and cofactors to lead to gene expression. These proteins (transcription factors and cofactors) control whether a gene is on or off. Transcription factor binding sites are thought to contain conserved sequence motifs of 6–20 base pairs. Transcription factor binding proteins bind to *cis*-acting elements including promoters, enhancers, silencers, splicing regulators, chromosome boundary elements, insulator elements, and locus control regions to control gene expression that regulates cell development and fate. The goal of the Encyclopedia of DNA Elements (ENCODE) Project is to identify and define all of these sequences in the human genome. Nuclease-hypersensitive sites are regions of DNA that interact with transcription factors in the chromatin environment *in vivo*. Trans-acting factors bind chromatin at DNAse1-hypersensitive sites (DHSs), which occur at accessible chromatin regions. Interestingly, CpG islands are associated with DHSs that are either constitutive or tissue-specific.

OVERVIEW OF THE HUMAN GENOME PROJECT

Decoding the DNA sequence that is made up of 3 billion base pairs was highly anticipated throughout the scientific and nonscientific community alike. The contribution of the Human Genome Project (HGP) to scientific research has undeniably contributed significantly toward understanding the causation of human disease, and the interaction between the environment and heritable traits defining human conditions. The sequencing of the human genome was first proposed by Robert Sinsheimer (chancellor of the University of California at Santa Cruz) in 1985. This idea was met with some critiques from the scientific community, many thinking the idea was premature and crazy. However, in 1988, Nobel Laureate James Watson gave the HGP a significant boost when he began to lead a National Institutes of Health (NIH) component of the project after joint funding from the NIH and the Department of Energy (DOE). In 1990, the HGP was officially initiated and was proposed to take 15 years, with a budget of $3 billion. The first 5 years of the HGP under the direction of James Watson was determined to map the genetic and physical features of the human genome, and his comment was "only once would I have the opportunity to let my scientific life encompass the path from the double helix to the 3 billion steps of the human genome." The managers of the HGP were Francis Collins at the NIH, Michael Morgan at The Wellcome Trust, and Aristides Patrinos at the DOE. Although the United States made the largest contribution to the HGP, it was an international effort with contributions from Britain, France, Germany, Canada, China, and Japan. Several species of bacteria and yeast had been completely sequenced in 1996, and this progress spurred the attempt at sequencing the human genome on to a more pilot scale. Eight years into the project in 1998, the plan included a sequencing facility to be built that would help sequence the human genome in only a 3-year period, ahead of schedule. The HGP agreed to release all sequences to the public and on June 26, 2000, a working draft of the human genome became available. Almost 3 years later on April 14, 2003, the HGP accomplished its ultimate goal and announced that the sequencing of the human genome

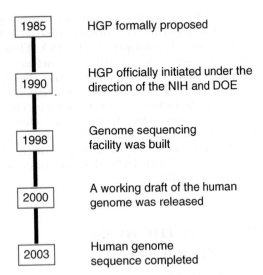

1985	HGP formally proposed
1990	HGP officially initiated under the direction of the NIH and DOE
1998	Genome sequencing facility was built
2000	A working draft of the human genome was released
2003	Human genome sequence completed

Figure 6.1   Human Genome Project timeline. The vertical timeline emphasizes the major years of the Human Genome Project (HGP) boxed in green with a summary of what occurred in that year to the right of the date.

was completed. See Figure 6.1 for a timeline depiction highlighting the course of events leading up to the completion of the human genome. Remarkably, this announcement occurred almost 50 years to the date of Watson and Crick's influential publication of the DNA double helix. Therefore, genomic science has rapidly gone from the identification of the structure of DNA to the sequencing of the human genome (and many other organisms) in a span of 50 years.

The Human Genome Project's Objectives and Strategy

The primary goal of the HGP was to obtain the complete DNA sequence of the human genome by 2005. Through use of a whole-genome random shotgun method and a whole-genome assembly, along with a regional chromosome assembly, and through combination of sequence data from Celera (a private sector company that agreed to help sequence the human genome for profitable purposes) and the publicly funded genome center, a 2.91 billion base pair consensus sequence was derived from the DNA of 5 individuals. At first the whole-genome shotgun approach proposed in 1997 by Weber and Mayers for the sequencing of the human genome was not well received. However, at that time (almost 8 years into the sequencing of the human genome), only 5% of the genome sequence had been completed, and it was clear that the goal of finishing by 2005 was unattainable. At that time PE Biosystems (now Applied Biosystems) developed a sequencer called the ABI PRISM 3700 DNA analyzer, which was going to be a part of Celera. Now with the ability to sequence with an automated, high-throughput capillary DNA sequencer, as well as new developments in tracking for whole-genome assembly, the chosen test case of the whole-genome

assembly on a eukaryotic genome was *Drosophila melanogaster.* The *Drosophila* genome, comprising 120 Mb of euchromatic DNA, was sequenced over a 1-year period.

The HGP enrolled 21 donors and collected approximately 130 mL of blood from males and females from a variety of ethnic backgrounds. From the 21 donors, 5 were chosen, including two males and three females: two Caucasians, one Hispanic Mexican, one Asian Chinese, and one African American. In order for the shotgun sequencing method to be fully utilized, the plasmid DNA libraries needed to be uniform in size, nonchimeric, and representative of the whole genome (rather than randomly representing the genome). Therefore, DNA from each donor was inserted in either a 2 Kb, 10 Kb, or 50 Kb plasmid library.

Human Genome Project Findings and Current Status

The HGP findings were extensive and exciting, as expected. Therefore, only highlighted findings related to the understanding of human disease are reported here. A complete and detailed analysis was published by Venter et al. In the wake of the human genome sequence, there was considerable acceleration in the success of the identification of genes that were important for the development of disease. In 1990, fewer than 10 genes had been identified by positional cloning, but by 1997 that number grew to more than 100 genes.

The HGP defined 26,383 genes with confidence using a unique rule-based system called Otto. Regions of sequence that were likely gene boundaries were matched up with BLAST and partitioned by Otto, and grouped into bins of related sequence that may define a gene. Known genes were then matched to the corresponding cDNA and were annotated as a predicted transcript. However, the genome sequence has variations and frameshifts, and it was not always possible to predict a transcript that agrees 100%. Therefore, if a transcript matched the genome assembly for at least 50% of its length at greater than 92% identity, then the region was annotated by Otto. It was predicted that an average gene in the human genome is approximately 27,874 bases. These variations in the human genome are being cataloged and will provide clues for the risk and diagnosis of common genetic diseases. More than 2 million single nucleotide polymorphisms have been identified. DNA arrays are now being employed to study the gene expression patterns of as many as 10,000 genes at one time. This analysis will help researchers understand the gene expression patterns between normal tissues and diseased tissues.

The HGP examined the genome for regions that were gene-rich and regions that were devoid of genes. A gene-poor region was defined as a region greater than 500 Kb lacking an open reading frame. Under these conditions, about 20% of the genome contains gene-poor regions, and they were not evenly distributed throughout the genome. Gene-poor chromosomes were 4, 13, 18, X, reflecting 27.5% gene-free regions of the total 492 Mb; and gene-rich chromosomes were 17, 19, 22, having only 12% gene-free regions within their 171 Mb. The next few years following the sequencing of the human genome were spent closing the sequencing gaps of all the chromosomes. Chromosomes 21 and 22 were completed first.

The HGP correlated CpG islands with gene start sites of computationally annotated genome transcripts and the entire human genome sequence. The HGP compared the variation of the CpG island computation with Larsen et al. and used two different thresholds of CG dinucleotide likelihood, including the original ratio of 0.6. The analysis showed a strong correlation between first coding exons and CpG islands. Genome-wide repeat elements were examined by the HGP. They observed that approximately 35% of the human genome was composed of different repeat elements, with chromosome 19 having 57% repeat density, the highest repeat density as well as the highest gene density. Gene density and Alu repeat elements exhibit an association, whereas this was not observed with the other classes of repeat elements.

The human genome sequence and the variations contained within must be utilized to identify the gene target of all hereditary diseases. However, this is an exciting but huge task; nevertheless, much large-scale gene duplication was identified in the HGP. These included duplications that were known to be associated with proteins involved in disease such as bleeding disorders, developmental disease, and cardiovascular conduction abnormalities. The duplications were located throughout the genome. However, there were gene families that were scattered in blocks within the genome, such as the olfactory receptor family. Chromosome 2 contains two very large duplications that are shared by two different chromosomes, 14 and 12. The first duplicated region is a block of 33 proteins spread in 8 different regions spanning 20 Mb of 2p, and these genes are also found on chromosome 14 spanning 63 Mb. The second duplication is on 2q and chromosome 12. This duplication includes two of the four known Homeotic (*Hox*) gene clusters, and the other two *Hox* gene clusters are also seen as duplications on two different chromosomes. These *Hox* genes play a fundamental role in controlling embryonic development, X-inactivation, and renewal of stem cells. According to the HGP, SNPs occur frequently at about 1 per 1200–1500 bp, but only less than 1% affect protein assembly and function. These analyses were based on the potential of an SNP to impact protein function based on SNPs that are located within predicted gene coding regions. Interestingly, the frequency of SNPs is highest in intronic regions, followed by intergenic regions, and then exonic regions.

Sequences of known proteins were compared to predicted proteins by the HGP. Analysis demonstrated that out of the predicted 26,588 proteins, 12,809 (41%) of the gene products could not be classified and were termed proteins with unknown function. The remainder of the proteins were classified into broad groups based on at least two lines of evidence. Importantly, the molecular functions of the majority

of the predicted proteins by the HGP are transcription factors and proteins that regulate nucleic acid metabolism. Many other proteins were receptors, kinases, and hydrolases, as well as proto-oncogenes, and proteins involved in signal transduction, cell-cycle regulators, and proteins that modulate kinase, G protein, and phosphatase activity. Large-scale analysis for characterizing proteins (proteomics) by their structure, function, modifications, localization, and interactions are being accomplished and utilized to gain understanding of their role in disease and cell differentiation.

A big challenge now that the human genome is sequenced is to understand how the DNA code is transcribed into biological processes that determine cell development and fate. The human genome sequence is only the first level of understanding and all functions of genes and the factors that regulate them must be defined. For example, in a disease such as diabetes mellitus, there may be up to 10 genes that result in an increased risk for the development of this very common disorder. It is exciting to think that small molecule drugs are being designed to block or stimulate a certain pathway. For example, the main etiology for chronic myelogenous leukemia is a translocation between chromosomes 9 and 22, and a drug was designed to inhibit the kinase activity of the bcr-abl kinase, which is the protein that is produced as a result of this translocation. This understanding will help explain how these underlying molecular processes when disturbed can lead to human genetic diseases and cancer. Francis S. Collins and Victor A. McKusick have predicted that "by the year 2010, it is expected that predictive genetic tests will be available for as many as a dozen common conditions, allowing individuals who wish to know this information to learn their individual susceptibilities and to take steps to reduce those risks for which interventions are or will be available." This is an exciting possibility, and as research escalates to help provide more preventative medicine and more early treatment options, we are getting closer to personalized medicine.

IMPACT OF THE HUMAN GENOME PROJECT ON THE IDENTIFICATION OF DISEASE-RELATED GENES

The identification of disease-related genes has been quite a laborious process. However, several strategies have been employed by using the information that is known about a candidate gene, including the knowledge of the protein/enzyme involved in the disease, the location of the gene within a chromosomal region, and a known animal model of the human disease in question. Linkage analysis, microsatellite markers, large DNA fragment-cloning techniques, and expressed sequence tags (ESTs) are important in the identification of genes responsible for human diseases. The HGP has without a doubt facilitated the strategies used for the identification of disease-related genes. The human DNA sequence has provided the template by which mutations are identified as well as cDNA, and genomic source data in order to provide numerous candidate genes for future studies. Presently, there are more genes than disease phenotypes, which have helped identify many if not all single gene disorders. The challenge now is to identify genes that are involved in polygenic disorders such as diabetes, hypertension, and most cancers.

Positional Gene Cloning

Positional cloning is used to determine the location of a gene without the understanding of its function and isolating the gene starting from the knowledge of its physical location in the human genome. The progress of positional cloning moved slowly at first because of its laborious nature with methods that required chromosome walking and the identification of expressed sequences. In 1986, Stuart Orkin and colleagues first reported their success in positional cloning the X-linked gene for chronic granulomatous disease. Usually, the first step in positional cloning is linkage analysis of the disorder in disease-prone families to determine chromosomal location, and then subsequent isolation and testing of genes for mutations that are segregating with the chromosomal location of this disorder. On average there may be 20 to 50 genes in this chromosomal location, and the gene contributing to the disease can be segregated often on the basis of some presumptions of the disorder in question. Positional cloning has been used to identify a number of inherited gene disorders as well as human cancers (Table 6.1).

Functional Gene Cloning

In functional cloning the protein is known and a gene is isolated based on fundamental knowledge and/or function of the protein product causing the human disease without information known about chromosomal location. The amino acid sequence and/or the antibodies available are used to determine the gene coding sequence. A cDNA library can then be screened with an oligonucleotide probe (antibody or degenerate oligonucleotides) based on the nucleotide sequence of the gene, and polymerase chain reaction (PCR) can amplify the cDNA using oligonucleotides from the amino acid sequence. Functional cloning has been used to identify genes causing human diseases such as phenylketonuria and sickle cell anemia. However, our fundamental understanding of human disease is lacking, and this gene discovery tool is not really available often.

Candidate Gene Approach

In the candidate gene approach, the cloning of a specific gene depends on having some functional information about the disease and relies on the availability of information on genes that had been previously isolated. This may not be the best way to clone genes

Table 6.1	Inherited Disease-Related Genes Identified by Positional Cloning

Disease	Year
Chronic granulomatous disease	1986
Duchenne muscular dystrophy	1986
Retinoblastoma	1986
Cystic fibrosis	1989
Wilms' tumor	1990
Neurofibromatosis type 1	1990
Testis determining factor	1990
Choroideremia	1990
Fragile X syndrome	1991
Familial polyposis coli	1991
Kallmann syndrome	1991
Aniridia	1991
Myotonic dystrophy	1992
Lowe syndrome	1992
Norrie syndrome	1992
Menkes disease	1993
X-linked agammaglobulinemia	1993
Glycerol kinase deficiency	1993
Adrenoleukodystrophy	1993
Neurofibromatosis	1993
Huntington disease	1993
von Hippel-Lindau disease	1993
Spinocerebellar ataxia I	1993
Lissencephaly	1993
Wilson disease	1993
Tuberous disease	1993
McLeod syndrome	1994
Polycystic kidney disease	1994
Dentatorubral pallidoluysian atrophy	1994
Fragile X "E"	1994
Achondroplasia	1994
Wiskott-Aldrich syndrome	1994
Early onset breast/ovarian cancer	1994
Diastrophic dysplasia	1994
Aarskog-Scott syndrome	1994
Congenital adrenal hypoplasia	1994
Emery-Dreifuss muscular dystrophy	1994
Machado-Joseph disease	1994
Spinal muscular atrophy	1995
Chondrodysplasia punctata	1995
Limb-girdle muscular dystrophy	1995
Ocular albinism	1995

because an informed guess is made about the kind of protein that may be responsible for the human disorder. Missense mutations in the *p53* genes were cloned using the candidate gene approach and were shown to be the cause of Li-Fraumeni syndrome, an inherited cancer disorder.

Positional Candidate Gene Approach

The positional candidate approach relies on the sequence of the human genome in that disease-related genes have been mapped to the correct chromosomal location and a survey of the sequence of that region is used to identify genes that are good candidates for cloning and testing. Candidate genes are analyzed by comparing the amino acid sequence of the genes to that of proteins with known functions and then studied in affected individuals in order to determine what gene(s) is responsible for the genetic disorder. The gene responsible for Marfan syndrome, an autosomal dominant disorder of connective tissue, was mapped using the positional candidate approach. Marfan syndrome was mapped to 15q by linkage analysis as well as the fibrillin gene. When DNA from patients with Marfan syndrome was analyzed, mutations were found in the fibrillin gene. Another example of genes identified by the positional candidate approach is four genes found in the mismatch repair process that had been previously implicated in hereditary nonpolyposis colon cancer.

SOURCES OF VARIATION IN THE HUMAN GENOME

The DNA sequence between humans is 99.9% identical. Therefore, it is important to examine the sequence variation between individuals to gain insight into phenotypic variation, as well as disease susceptibility. Single nucleotide polymorphisms (SNPs); short tandem repeats; micro/minisatellites; and less than 1Kb insertions, deletions, inversions, and duplications are responsible for most of the genetic variation in the human population (Figure 6.2). These genome variations can give rise to diseases through a gain or loss of dosage-sensitive genes. Through the sequencing of the human genome, new techniques such as genome-scanning arrays and comparative DNA sequence analysis have been developed to examine the composition of the human genome. These technologies have been important in finding copy-number variants or segments of DNA that are 1 Kb or larger, including insertions, deletions, and duplications. Genomic disorders are influenced by the genome architecture around

the recombination event and share a common mechanism for genomic rearrangement, that is, nonallelic homologous recombination or ectopic homologous recombination between low-copy repeats that flank the rearranged DNA segment. Inversions are created by nonallelic homologous recombination events that occur between inverted low-copy repeats, whereas a nonallelic homologous recombination by direct low-copy repeats results in a duplication or deletion. In addition, nucleotide substitutions and point mutations cause alterations in protein sequence and can result in disease.

TYPES OF GENETIC DISEASES

Genetic Diseases Associated with Gene Inversions

Structural variants have been identified in the general population to be the cause of genetic disease in the offspring of parents who exhibit certain DNA inversions. In patients with *Williams-Beuren syndrome*, there is a 1.5 Mb inversion at 7q11.23 that occurs in approximately one-third of the patients' parents with a 5% frequency of this inversion in the general population; this syndrome has an incidence of 1/20,000–50,000. An inversion that is 4 Mb at 15q12 is associated with *Angelman's syndrome*, and about half of the parents of these patients have this variation as well as 9% of the general population; this syndrome has an incidence of 1/10,000–20,000. There are diseases in which inversions found in the patients affected have not been detected in the general population. Patients with *hemophilia A* have a 400 Kb inversion in intron 22 in the *factor VIII* gene, and two copies are located 400 Kb telomeric in an inverted orientation; this nonallelic

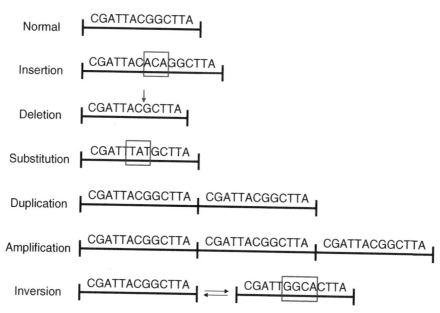

Figure 6.2 DNA rearrangements. Structural variations in the human genome result from rearrangements of DNA ranging from gross changes to single nucleotide alterations. The letters are used to depict nucleotides in a DNA sequence and lines are used to depict a gene. Each DNA rearrangement is outlined by a red box. The red arrow represents a nucleotide deletion.

homologous recombination event results in inactivation of the *factor VIII* gene. In addition, a smaller inversion in the emerin gene in *Emery-Dreifuss muscular dystrophy* has been identified. *Hunter syndrome* is an X-linked dominant disorder. Nonallelic homologous recombination between the iduronate 2-sulphatase gene (*IDS*) and an *IDS* pseudogene generates a genomic inversion resulting in a disruption of the functional *IDS* gene occurring in approximately 13% of *Hunter syndrome* patients. Within the Japanese population, fathers of *Soto syndrome* patients, a microdeletion syndrome, carry a 1.9 Mb inversion variant at 5q35 that predisposes their offspring to this disease. Constitutional translocations in the human genome can be mediated by a polymorphic inversion at olfactory-receptor gene clusters at loci 4p16 and 8p23, which occur at frequencies of 12.5% and 26%, respectively. Heterozygous carriers of these translocations exhibit no phenotypic characteristics, whereas their offspring who inherit these translocations show phenotypes from mild dysmorphic features to *Wolf-Hirschhorn syndrome*, which is characterized by growth defects and severe mental retardation. These examples signify the importance in continuing to characterize inversions within the human genome in the general population in order to examine the risk these variations have on the carriers' offspring.

Genetic Diseases Associated with Gene Deletions

Genomic disorders can be responsible for commonly occurring diseases. For instance, *α-thalassemia* affects 5–40% of the population in Africa and 40–80% in South Asia, and results from a homologous deletion of an approximately 4 Kb fragment that is flanked by two α-globin genes on 16q13.3. A nonallelic homologous recombination event between these two copies of the α-globin genes results in the deletion of one functional copy. Red and green pigment genes are located on Xq23, and individuals who have normal color vision have one copy of the red pigment gene and one or more copies of the green pigment gene. In *red-green color blindness*, which affects 4–5% of males, deletions or fusions caused by nonallelic homologous recombination occur. In patients with *incontinentia pigmenti*, an 11 Kb deletion occurs by nonallelic homologous recombination between two low-copy repeats with one in the diseased gene (*NEMO*) and one 4 Kb downstream of the gene. *Hereditary neuropathy with liability to pressure palsy* (HNPP) is a common autosomal dominant neurological disorder that is caused by a 1.4 Mb deletion of a genomic fragment on 17p12. The gene *NF1* that encodes for *neurofibromatosis type 1* is located on 17q11.2 and a 1.5 Mb deletion encompassing this gene accounts for 5–22% of patients with this disease. Patients with *DiGeorge syndrome/velocardiofacial syndrome* (DGS/VCFS) can exhibit a 3 Mb deletion within a region-specific repeat unit, LCR22 that is flanked by LCR22A and D, or a 1.5 Mb deletion that is flanked by LCR22A and B located on chromosome 22q11.2. Patients with this congenital disease experience recurrent infection, heart defects, and known facial features. *Smith-Magenis syndrome* (SMS) effects 1/25,000 individuals and is caused by a 4 Mb deletion on several loci contained within chromosome 17 and depending on the loci involved determines the severity of mental retardation exhibited by the patient.

Genetic Diseases Associated with Gene Duplications

Charcot-Marie-Tooth disease (CMT) is an inherited autosomal dominant trait that occurs in about 1/25,000 individuals and is characterized by atrophy of the muscles in the legs, progressing over time to the hands, forearms, and feet. There are two clinical classes of CMT: type I and type II. In CMT type I (CMT1A) 75% of individuals have a duplication in one of the peripheral myelin protein 22 (*PMP22*) genes. Duplication on both chromosomes at 17p12 produces a severe form of CMT1A where essentially there are four copies of the *PMP22* gene. A central nervous system disorder affecting the myelin sheath covering the nerve fibers in the brain is called *Pelizaeus-Merzbacher disease*. The majority of patients with this disease have a duplication of the proteolipid protein gene (*PLP1*) which is found on Xq21–22.

GENETIC DISEASES AND CANCER

The HGP and the completion of the human genome sequence have made a huge impact for the practice of medicine and molecular genetic research. The human genome sequence has helped make advances in the development of designer drugs targeting molecular pathways that disrupt diseases caused by single genes or a complex array of gene products. It is beyond the scope of this chapter to provide a comprehensive review of genetic diseases and/or the genetic causes of cancer. However, included are a few examples of how the HGP has advanced our understanding of human disease and will continue to make an impact forever.

Cystic Fibrosis (CF)

The cystic fibrosis transmembrane regulator (*CFTR*) gene is located on chromosome 7q31.2 and was the first gene to be identified by the HGP. Researchers defined the *CFTR* gene by the positional cloning approach. The function of the CFTR protein is to regulate chloride secretion and the inhibition of sodium absorption across the cell membrane. Approximately 1547 mutations of the *CFTR* gene have been described, with the most common mutation as a 3–base pair DNA deletion that results in a loss of the amino acid phenylalanine at position 508 occurring in 66% of CF patients. Out of these mutations only 23 have been shown directly to cause sufficient loss of *CFTR* to confer CF disease, and these mutations are seen in 85% of the diseased population. Interestingly, two or more

CFTR mutations can be located in *trans* on two separate chromosomes and this will confer CF. However, if the mutations are found in *cis* on the same chromosome, this is not associated with the disease. Unfortunately, this distinction between *cis* and *trans* on chromosomes is not made in most commercial laboratories. In addition, the different mutations will confer different phenotypic responses to CF with some resulting in milder forms of the disease. Approximately 9.7% of genotyped individuals in the Cystic Fibrosis Foundation Patient Registry have at least one unidentified mutation, but the majority (90%) of *CFTR* mutations can be picked up by regular screening methods. The discovery of the *CFTR* gene has given researchers a better understanding of the etiology, the genetic bases, and the pathobiology of CF.

CF is an autosomal recessive disease that occurs in approximately 1 in 3500 newborns, is the most common lethal inherited genetic disease among the Caucasian population, and affects almost 30,000 Americans. Treatment advances for patients with CF have increased the survival age from mid-teens in the 1970s, to late-twenties to thirties in the 1990s, to more than 36 years old today. A CF diagnosis is based on several clinical characteristics, a familial history of CF or a positive CF newborn screening test, and a mutation in the *CFTR* gene and/or protein. Newborn screening has been implemented in 40 states currently, and one hopes that by the year 2010 it will be available in all states. CF is a disease that is caused by the improper regulation of the ion channels between the cell cytoplasm and the surrounding fluid, resulting in the inability of the exocrine epithelial cells to transport fluid and electrolytes in and out of the cells. CF patients cannot effectively clear inhaled bacteria and have an abnormal accumulation of viscous, dehydrated mucus, and because of this, an excessive inflammatory response to pathogens.

Francis Collins summarized how important the HGP has been in understanding genetic diseases and used CF as an example:

> *Cystic fibrosis has become the paradigm for the study of genetic diseases and indeed, for the medicine of the future. The notion that it is possible to identify genes whose structure and function are unknown and to use that information to understand given disease and develop "designer" therapies is becoming the central paradigm of biomedical research, and cystic fibrosis is the disease that leads that charge.*

Phenylketonuria (PKU)

Mutations in the phenylalanine hydroxylase gene (*PAH*) encoding the protein L-phenylalanine hydroxylase causes a mental retardation disease called phenylketonuria (PKU). The inability to hydrolyze phenylalanine to tyrosine can lead to hyperphenylalaninemia and if untreated has a toxic effect on the brain. In the 1980s, the *PAH* gene was cloned and sequenced and mapped to chromosome 12, region 12q23.2. The PKU phenotype is not a simple disease, nor does it have a simple explanation. As with many other genetic diseases, each patient with PKU has to be treated differently. With the help of the HGP, understanding PKU and its resulting phenotype has been extremely beneficial. The locus for *PAH* covers 1.5 Mb of DNA with SNPs, repeat sequences, polymorphisms, and *cis* control elements embedded in the sequence, as well as harboring five other genes, and thus providing for a wide range of disease-causing mutations. The *PAH* gene is expressed in the liver and kidney.

PKU caused a paradigm shift of attitudes about genetic disease by becoming one of the first disorders to show a treatment effect. PKU is an autosomal recessive inherited disease, causing mental retardation; a mousy odor; light pigmentation; peculiarities of sitting, standing, and walking; as well as eczema and epilepsy. The average incidence of PKU in the United States is 1 in 8000. PKU is one of the first genetic diseases to have an effective rational therapy. PKU can be identified with a biochemical test in newborns and can be treated by a phenylalanine-free, tyrosine-supplemented diet, which permits normal or near-normal cognitive development. In the adolescent and adult patients, it was difficult until recently to adopt the diet recommendations of PKU based on deficiencies in both organoleptic properties and nutrient content in the food. Fortunately, many diet deficiencies are being overcome, and diagnosis is occurring earlier so that patients start the recommended diets sooner, and are more aggressive throughout life.

Breast Cancer

The majority of hereditary breast and ovarian cancers are caused by mutations in the breast cancer-predisposing gene 1 or 2 (*BRCA1* or *BRCA2*). *BRCA1* was found by candidate gene approach in 1991, and *BRCA2* was located by linkage analysis and positional cloning in 1995 using familial breast cancer pedigrees with multiple cases of breast cancer in many generations. *BRCA1* is located on 17q21 encoding an 1863-amino acid polypeptide, and *BRCA2* is found on chromosome 13q12-13 encoding 3418 amino acids. *BRCA1* has been implicated in cell-cycle regulation, chromatin remodeling, protein ubiquitylation, and both proteins are involved in DNA repair. In the Ashkenazi Jewish population, there are founder mutations that occur at specific locations in *BRCA1* (185delAG and 5382insC) and *BRCA2* (617delT), but most mutations occur anywhere along the gene, including frameshift or nonsense mutations as well as deletions or duplications. DNA-based methods have been recently employed to conduct analysis of both *BRCA1* and *BRCA2* for the presence of genomic rearrangements. The prevalence of genomic rearrangements in *BRCA1* is higher than that of *BRCA2*, accounting for 8–19% of the total mutations in *BRCA1* and 0–11% in *BRCA2* mutations.

Breast cancer affects one in eight women in the United States, and a woman born in the United States has an average lifetime risk of 13% for developing breast cancer. Familial breast cancer is associated with 10–20% of all breast cancer cases. Mutations in the *BRCA1* and *BRCA2* genes in women have a 60–80% increase of developing breast cancer. In addition, women who carry a mutation in the *BRCA1* gene have a 15–60% lifetime risk for ovarian cancer, which is a much more increased risk when compared to a mutation in the *BRCA2* gene (10–27%). Women with *BRCA2* mutations tend to develop ovarian cancer after age 50. Women who are under the age of 50 and are *BRCA1* mutation carriers have a 57% chance of being diagnosed with breast cancer, and only a 28% chance of developing breast cancer if the *BRCA2* gene is mutated. Interestingly, men who harbor a *BRCA2* mutation are estimated to have a 6% chance of being diagnosed with breast cancer. *BRCA1* breast tumors are found to be more poorly differentiated, whereas *BRCA2* tumors tend to be more high-grade tumors when compared to sporadic breast tumors (nonhereditary). *BRCA1* breast tumors when compared to sporadic breast tumors are frequently negative for estrogen receptor, progesterone receptor, and HER-2/Neu overexpression. It is recommended for women with *BRCA* mutations to begin monthly breast self-examinations at the age of 18 and clinical breast examinations and annual mammograms beginning at age 25.

Nonpolyposis Colorectal Cancer (HNPCC)

Hereditary nonpolyposis colorectal cancer (HNPCC) is caused by mutations in the mismatch repair (MMR) genes *MLH1*, *MSH2*, *MSH6*, and *PMS2*. A hypermutation phenotype was discovered in 1993 in families with HNPCC similar to that observed in MMR-deficient bacteria and yeasts. Linkage analysis and positional cloning in HNPCC families subsequently identified *MSH2* and *MLH1* genes, and mutations in these genes account for 60–80% of HNPCC diagnosis. Additionally, the MMR genes *PMS2* and *MSH6* were associated with HNPCC. A higher risk of colorectal cancer occurs in *MSH2* and *MLH1* mutation carriers as compared to *MSH6* or *PMS2* mutation carriers. The *MSH2* and *MSH6* genes are located on chromosomes 2p22-p21 and 2p16, respectively. *MLH1* is found on 3p21.3, and *PMS2* is located on chromosome 7p22.

There are approximately 160,000 new cases of colorectal cancer diagnosed in the United States each year, with HNPCC accounting for 2–7% of diagnosed colorectal cancer, affecting about 1 in 200 individuals. The average age of HNPCC diagnosis is 44 years old. HNPCC can also be called Lynch syndrome in honor of Dr. Henry T. Lynch, professor at Creighton University Medical Center. HNPCC is an autosomal dominant trait and exhibits phenotypic characteristics of less than 100 colonic polyps and early onset of multiple tumors in the colon. The Amsterdam criteria were established for the clinical designation of a family with HNPCC: (i) three or more relatives with colon cancer, one of them must be a first degree relative (parent, child, sibling) of the other two, (ii) at least two affected generations, (iii) one or more members of a family must develop colon cancer before the age of 50, and (iv) familial adenomatous polyposis (FAP) should be excluded from the diagnosis.

PERSPECTIVES

Millions of people around the world waited and watched for the completed human genome sequence to be released with the expectation that it would benefit humankind. Decades ago it was not anticipated that genomic disorders would represent such a common cause of human genetic disease. Now from this perspective humans are the best model organisms that we have in order to study the human genome, disease, and its associated phenotypes. Currently, there is a huge amount of sequence information that has been generated from genome sequencing projects, including vertebrates and nonvertebrates. One objective of the human genome sequence is to derive medical benefit from analyzing the DNA sequence of humans. It is undeniable that genomic science will begin to unlock more of the mysteries of complex hereditary factors in heart disease, cancer, diabetes, schizophrenia, and many more. Genetic tests have become available for individuals who have a strong family history or are more susceptible to a particular disorder, such as breast cancer or colorectal cancer. Healthcare professionals will become practitioners of genomic medicine as more genetic information about common illnesses is available and healthy individuals want to protect themselves from illness. Clinicians will have to grasp the understanding and advances of molecular genetics, and a group of physicians, nurses, and other clinicians called *The National Coalition for Health Professional Education in Genetics* has been organized to help prepare for the genomics era. Within the next decade, it is exciting to think that designer drugs will be available for diabetes mellitus, hypertension, mental illness, and many other genetic disorders. In addition, it is likely that all tumors could have a molecular fingerprint associated with them and the promise of individualized medicine by tailoring prescribing practices and management to that person's unique molecular profile. Also within a decade or two, it may be possible to sequence the genome of an individual human with minimal laboratory cost (maybe $1000). If this becomes reality, we can imagine the possibilities for scientific research, clinical care, treatment options, and overall dramatically changing the face of medicine. Francis Collins, one of the pioneers of the Human Genome Project, stated it best when he said, "if the past 50 years of biology is any indication of the future, the best is certainly yet to come."

KEY CONCEPTS

- In 1953, James D. Watson and Francis H. C. Crick discovered the double helical structure of DNA. This discovery single-handedly revolutionized molecular biology and biological sciences.

- The genomes of any two people are more than 99% similar; therefore the small fraction of the genome that varies among humans is very important. These variations of DNA are what make humans unique. However, variations in DNA can occur in the form of genetic mutations in which a base is missing or changed. This results in an aberrant protein and can lead to disease.
- The contribution of the Human Genome Project (HGP) to scientific research has undeniably contributed significantly toward understanding the causation of human disease, and the interaction between the environment and heritable traits defining human conditions.
- In the wake of the human genome sequence there was considerable acceleration in the success of the identification of genes that were important for the development of disease. In 1990, fewer than 10 genes had been identified by positional cloning, but by 1997 that number grew to more than 100 genes.
- Single nucleotide polymorphisms (SNPs), short tandem repeats, micro/minisatellites, and less than 1 Kb insertions, deletions, inversions, and duplications are responsible for most of the genetic variation in the human population. These genome variations can give rise to diseases through a gain or loss of dosage sensitive genes. Through the sequencing of the human genome, new techniques such as genome-scanning arrays and comparative DNA sequence analysis have been developed to examine the composition of the human genome.
- The human genome sequence has helped make advances in the development of designer drugs targeting molecular pathways that disrupt diseases caused by single genes or a complex array of gene products.
- The cystic fibrosis transmembrane regulator (*CFTR*) gene is located on chromosome 7q31.2 and was the first gene to be identified by the HGP.
- The majority of hereditary breast and ovarian cancers are caused by mutations in the breast cancer-predisposing gene 1 or 2 (*BRCA1* or *BRCA2*). *BRCA1* was found by candidate gene approach in 1991 and *BRCA2* was located by linkage analysis and positional cloning in 1995 using familial breast cancer pedigrees with multiple cases of breast cancer in many generations.
- Hereditary nonpolyposis colorectal cancer (HNPCC) is caused by mutations in the mismatch repair (MMR) genes *MLH1, MSH2, MSH6,* and *PMS2*. A hypermutation phenotype was discovered in 1993 in families with HNPCC similar to that observed in MMR deficient bacteria and yeasts. Linkage analysis and positional cloning in HNPCC families subsequently identified *MSH2* and *MLH1* genes, and mutations in these genes account for 60–80% of HNPCC diagnosis.
- Within the next decade it is exciting to think that designer drugs will be available for diabetes mellitus, hypertension, mental illness, and many other genetic disorders. In addition, it is likely that all tumors could have a molecular fingerprint associated with them and the promise of individualized medicine by tailoring prescribing practices and management to that person's unique molecular profile.

SUGGESTED READINGS

1. Aaltonen LA, Peltomaki P, Leach FS, et al. Clues to the pathogenesis of familial colorectal cancer. *Science.* 1993;260: 812–816.
2. Collins FS. Positional cloning moves from perdition to traditional. *Nat Genet.* 1995;9:347–350.
3. Collins FS, McKusick VA. Implications of the Human Genome Project for medical science. *JAMA.* 2001;285:540–544.
4. Collins FS, Morgan M, Patrinos A. The Human Genome Project: Lessons from large-scale biology. *Science.* 2003;300: 286–290.
5. Farrell PM, Rosenstein BJ, White TB, et al. Guidelines for diagnosis of cystic fibrosis in newborns through older adults: Cystic Fibrosis Foundation consensus report. *J Pediatr.* 2008;153: S4–S14.
6. Feuk L, Carson AR, Scherer SW. Structural variation in the human genome. *Nat Rev Genet.* 2006;7:85–97.
7. Fishel R, Lescoe MK, Rao MR, et al. The human mutator gene homolog MSH2 and its association with hereditary nonpolyposis colon cancer. *Cell.* 1993;75:1027–1038.
8. Guttmacher AE, Collins FS. Welcome to the genomic era. *N Engl J Med.* 2003;349:996–998.
9. Higgs DR, Vernimmen D, Hughes J, et al. Using genomics to study how chromatin influences gene expression. *Annu Rev Genomics Hum Genet.* 2007;8:299–325.
10. Kerem B, Rommens JM, Buchanan JA, et al. Identification of the cystic fibrosis gene: Genetic analysis. *Science.* 1989;245: 1073–1080.
11. Maeshima K, Eltsov M. Packaging the genome: The structure of mitotic chromosomes. *J Biochem.* 2008;143:145–153.
12. Miki Y, Swensen J, Shattuck-Eidens D, et al. A strong candidate for the breast and ovarian cancer susceptibility gene BRCA1. *Science.* 1994;266:66–71.
13. Narod SA, Foulkes WD. BRCA1 and BRCA2: 1994 and beyond. *Nat Rev Cancer.* 2004;4:665–676.
14. Nicolaides NC, Papadopoulos N, Liu B, et al. Mutations of two PMS homologues in hereditary nonpolyposis colon cancer. *Nature.* 1994;371:75–80.
15. Plaschke J, Engel C, Kruger S, et al. Lower incidence of colorectal cancer and later age of disease onset in 27 families with pathogenic MSH6 germline mutations compared with families with MLH1 or MSH2 mutations: The German Hereditary Nonpolyposis Colorectal Cancer Consortium. *J Clin Oncol.* 2004;22: 4486–4494.
16. Scriver CR. The PAH gene, phenylketonuria, and a paradigm shift. *Hum Mutat.* 2007;28:831–845.
17. Sebat J, Lakshmi B, Troge J, et al. Large-scale copy number polymorphism in the human genome. *Science.* 2004;305: 525–528.
18. Solinas-Toldo S, Lampel S, Stilgenbauer S, et al. Matrix-based comparative genomic hybridization: Biochips to screen for genomic imbalances. *Genes Chromosomes Cancer.* 1997;20:399–407.
19. Venter JC, Adams MD, Myers EW, et al. The sequence of the human genome. *Science.* 2001;291:1304–1351.
20. Wilson JM, Wilson CB. Cystic fibrosis: Pointing to the medicine of the future. In: *Highlights (Selected Proceedings from the Tenth Annual North American Cystic Fibrosis Conference, October 24–27, 1996).* 1997: 1–2.

The Human Transcriptome: Implications for the Understanding of Human Disease

Christine Sers . Wolfgang Kemmner . Reinhold Schäfer

INTRODUCTION

The most fascinating aspect of transcriptomics is that the entire set of messenger RNA (mRNA) molecules or transcripts produced in a population of cells or in tissues can be analyzed simultaneously. The present microarray technology produces devices equivalent to the size of a stamp for gene expression profiling. Analyzing the transcriptome is a challenging task, since the mRNA content of a biological entity is heterogeneous and can vary substantially. The abundance of individual transcripts varies from a few copies to hundreds or thousands of copies per cell. The kinds and copy numbers of individual transcripts expressed at a given time depend on the developmental stage, on external conditions, and environmental stimuli. Quantitative and qualitative alterations of mRNAs can be directly linked to the molecular mechanism of disease or reflect the downstream consequences of these disease processes.

This chapter outlines the methodological prerequisites for transcriptome analysis and describes typical applications in molecular cell biology and pathology. During the last three decades, technology development and experimental approaches aiming at mRNA analysis were significantly fueled by molecular cancer research. One of the main reasons for progress in this area was the availability of relevant cell lines that could be propagated indefinitely and served as reproducible sources of RNA and of sufficient quantities of normal and diseased tissues. A strong motivation lay in the demand for distinguishing as many transcripts as possible in normal and tumorigenic cells to understand cancer-specific alterations in gene expression. While early work along these lines was mostly related to pathogenesis, more recent applications deal with diagnostic issues such as tumor outcome, prognosis, and therapy response prediction.

GENE EXPRESSION PROFILING: THE SEARCH FOR CANDIDATE GENES INVOLVED IN PATHOGENESIS

To date, microarray-based expression profiling is accepted as the gold standard in transcriptome analysis. Before microarrays were available for most researchers in sufficient quantity and quality (as well as affordable at reasonable costs), alternative techniques were instrumental in answering questions related to the quantity of transcripts expressed ubiquitously, and to identifying tissue-specific expression patterns and candidate genes related to disease.

Early Gene Expression Profiling Studies

Intriguingly, the question of how many transcription units distinguish normal from tumor cells, a question that is expected to be the domain of transcriptomics using microarrays, was addressed nearly at the same time when techniques in molecular biology permitted the identification and thorough analysis of individual mRNAs. In 1977 and 1980, researchers described the northern blot technique for transferring electrophoretically separated RNA from an agarose gel to paper strips, the coupling of the RNA to the paper surface and the detection of specific RNA bands by hybridization with ^{32}P-labeled DNA probes followed by autoradiography for the first time. A report published in 1980 provided evidence for the complexity of cellular transformation at the RNA level, when scientists had studied the RNA pool of chicken embryonic breast muscle cells infected with Rous sarcoma virus (RSV). The authors compared the hybridization kinetics of nuclear RNA preparations from normal and RSV-transformed cells, respectively, incubated in solution with tracer amounts of labeled single-copy chicken DNA. Based on the assumption that an average transcription unit is about 10 times larger than its corresponding mRNA, the authors concluded that the observed increase in the number of stable transcription products in transformed cells relative to normal cells was equivalent to approximately 1,000 transcription units. Several years later, other scientists used a more sophisticated approach for contrasting mRNA patterns of cellular material obtained from colon tumor biopsies. The researchers took advantage of the molecular cloning techniques that allowed the establishment of a set of complementary DNAs (cDNAs) obtained by reverse transcription of mRNA. The reference cDNA library of some 4,000 clones represented abundant and middle abundant RNA sequences. Replicas of the library were then hybridized to ^{32}P-labeled cDNA probes synthesized from polyadenylated RNA from small biopsies obtained from normal and neoplastic intestinal mucosa. The comparison of normal colonic mucosa with carcinomas showed expression alterations of ~7% of the cloned sequences and was extrapolated to the entire, yet unknown, set of transcripts. The number of alterations was smaller between normal mucosa and benign adenomas indicating that transcriptional changes accumulate during cancer progression.

cDNA Libraries and Data Mining

Further advances in deciphering cancer-related transcripts were driven by increased efforts in cDNA cloning, and sequence analysis. Collections of cDNAs were obtained from various normal and diseased tissues, as well as from reference cell lines. The functional characterization of transcribed sequences progressed at the same time. However, due to the complexity of gene function in biological systems, functional information lagged significantly behind sequence information. The large cDNA collections deposited in expression databases often provided only partial sequence information. The corresponding cDNAs known to be expressed in various tissues or cell types analyzed were designated expressed sequence tags (or ESTs). With increasing entries into these EST catalogues, it became feasible to merge overlapping partial sequences and eventually to define full-length open-reading frames (ORFs). As a practical consequence of the global gene expression information provided by cDNA/EST databases, an approach termed the electronic northern became feasible. The electronic northern analysis facilitated prediction of expression changes between normal and diseased tissues. Extensive mining of EST databases using stringent statistical tests permitted identification of candidate genes whose altered (stimulated or reduced) expression correlated with the disease state.

cDNA Subtraction

The data mining approach was limited by the existing sequence information and the available gene annotations. To circumvent this bias, researchers established several elegant methods that permitted enrichment of mRNA sequences (or cDNAs) associated with special experimental conditions or with particular cellular features. The methods established were cDNA subtraction, differential display PCR (DD), representational difference analysis, and serial analysis of gene expression (SAGE).

In general, cDNA subtraction is a method for separating cDNA molecules that distinguish related cDNA samples, for instance, prepared by reverse transcription of mRNA from normal precursor cells and related neoplastically transformed cells. The basis of subtraction is that cDNAs prepared from two different cell types to be compared are rendered single-stranded, subsequently mixed, and incubated to allow annealing of sequences common to both cell species. These sequences will hybridize, while sequences unique to one of the cells will stay single-stranded. In the classical subtraction approach, single-stranded and double-stranded cDNAs were separated by hydroxylapatite chromatography. Subsequently, the unique cDNA fragments are cloned and sequenced. The major drawback of this method is that the enrichment of differentially expressed sequences usually does not exceed a factor of 100, that abundant mRNAs (cDNAs) are over-represented due to the lack of normalization, and that rare transcripts are not detected at all. These inherent disadvantages were overcome by development of a method called suppression subtractive hybridization (SSH), a PCR-based subtraction method that combines normalization and subtraction into a single procedure. Differential amplification of unique cDNA fragments is achieved by ligating different primers to each restricted cDNA originating from the cell types to be compared prior to the annealing step and the PCR. The normalization step equalizes the abundance of cDNA fragments within the target population, and the subtraction step excludes sequences that are common to the cell populations being contrasted. Using this method, the probability of recovering differentially expressed cDNAs of low abundance is largely increased (by a factor of 1000 or more). RDA is a technique that combines subtractive hybridization with PCR-mediated kinetic enrichment for the detection of differences between two complex genomes or transcriptomes.

Differential Display PCR

Differential display PCR is a method to separate and clone individual mRNAs that are differentially expressed by means of the polymerase chain reaction. A set of oligonucleotide primers is used, one being anchored to the polyadenylated tail of a subset of mRNAs, the other being short and arbitrary in sequence to allow annealing at different sites relative to the first primer. The mRNA subpopulations defined by the primer pairs are amplified after reverse transcription and the products resolved (displayed) on DNA sequencing gels. Differential display visualizes mRNA compositions of cells by displaying subsets of mRNAs as short cDNAs. The beauty of this approach is that many samples can be run in parallel to reveal differences in mRNA composition. The differentially expressed cDNA fragments can be recovered by cloning techniques.

Serial Analysis of Gene Expression

While the previously discussed approaches directly aim at identifying important differences between closely related cell types, the key element of the SAGE method is to represent all transcripts in a given cell type in a quantitative manner. The basic principle of SAGE is that short nucleotide sequence tags of 10 to 14 base pairs contain sufficient information to uniquely identify transcripts. Moreover, concatenation of these short sequence tags permits an efficient analysis of transcripts serially by sequencing of multiple tags within a single cloned element. More recent variants of the method are based on longer sequence tags and integrate microarray technology. Two years after the initial publication of the method, Johns Hopkins University researchers for the first time reported on gene expression profiles in normal and cancer cells based on SAGE. The total number of transcripts varied from approximately 14,000 to 20,000 between cell populations. Most transcripts (86%) were expressed at fewer than 5 copies per cell; however, the bulk of the mRNA mass consisted of more abundant transcripts (more than 5 copies per cell). The relative expression levels of transcripts were determined by dividing the number of tags observed in tumor and normal tissue. Most transcripts were expressed at similar levels. However, 548 of 14,000 to 20,000 transcripts were overrepresented or underrepresented in tumor versus normal cells. The average difference in expression for these transcripts was 15-fold. About 20% of them were less than 3-fold different. The authors also addressed the issue of whether cultured cell lines, frequently used in molecular cancer research, display gene expression patterns that mimic those found in the organ microenvironment. Interestingly, 72% of transcripts expressed at reduced levels in cancer specimens were also expressed at lower levels in cell lines. Likewise, 43% of transcripts exhibiting elevated expression in cancers were also upregulated in cell lines. Useful links and SAGE databases can be found at http://www.sagenet.org/.

A procedure very similar to SAGE is used to study cellular microRNAs (miRNAs), which are short ∼22-nucleotide segments of RNA that have been found to play an important role in gene regulation. Small RNAs are isolated, linkers are added to each of them, and the RNA is converted to cDNA. Afterwards the linkers containing internal restriction sites are digested with the appropriate restriction enzyme and the sticky ends are concatamerized. The concatamers are ligated into plasmid vectors and cloned, followed by sequencing. In this way, the expression levels of miRNA can be quantitatively assessed by counting the number of times they are present.

TRANSCRIPTOME ANALYSIS BASED ON MICROARRAYS: TECHNICAL PREREQUISITES

Microarray technology was pioneered by Pat Brown and colleagues at Stanford University. The researchers were not only the first to use microarrays for studying biological questions, but also described the necessary technical devices in detail. In this way, they contributed to the fast spread of the technology and supported its industry-based propagation. The central element of the technique is that DNA molecules, cDNA fragments or oligonucleotides, are arrayed and immobilized at defined positions on a solid support or matrix. The probes are hybridized with complementary and fluorochrome-labeled RNA or DNA molecules (targets) derived from biological specimens such as cells, tissues, or blood. The hybridization intensity/fluorochrome staining intensity obtained within the position of the probe is equivalent to the abundance of the corresponding nucleotide sequence in the complex mixture of targets. The different kinds of microarrays available today are distinguished by the number, density, design and size of nucleotide probes, the way of chip manufacturing, and the experimental protocols for target hybridization.

During surgical removal of tissue specimens, one of the first steps is the ligation of the arterial blood supply. From this moment on, the tissue, e.g., a malignant tumor, is exposed to hypoxia at body temperature. The duration between ligation and final tissue removal can vary considerably and may not be reduced to a standardized interval under clinical conditions. Logistical constraints may lead to further considerable delay before the tissue is finally transferred to −80°C. Thus, this lengthy process might lead to a considerable extent of RNA-degradation.

Frozen tissue samples are dissected, fixed on glass-slides, and stained. Histological characterization is essential to show the tissue composition, for example, the percentage of the cell types of interest, necrotic areas with degraded RNA, or fatty tissue from which RNA-extraction is difficult. One way to accomplish this goal is laser-microdissection of tissue areas of interest; for instance, only areas with carcinoma cells or stromal material only. Microdissection of 5×10^6 mm^2 per specimen yields about 10–20 ng RNA in about 2 hours working time. RNA-yield and quality are checked by electrophoresis. The RNA is used for synthesis of the labeled sample nucleic acid, mostly cDNA or aRNA, which is quality-checked again. The last step is hybridization of the fluorescently labeled sample nucleic acid to the probe DNA on the microarray (Figure 7.1).

Solid microarray supports typically are glass slides or silicon surfaces. Probes are often covalently linked to a

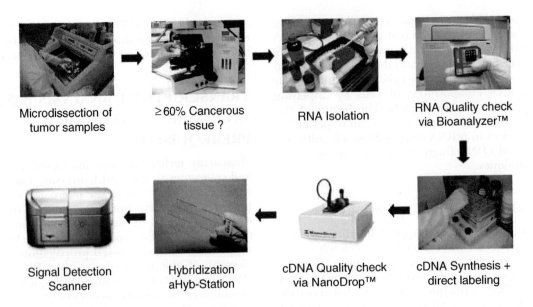

Figure 7.1   Laboratory workflow of a typical microarray experiment.

chemical matrix (e.g., via epoxy-silane or amino-silane). The probes are dispensed either by contact spotting or applied as micro-droplets via techniques resembling ink jet technologies used in printing. One of the industrial suppliers (Affymetrix Inc.) produces arrays produced using photolithographic methods as in silicone chip production.

During surgical removal of tissue specimens, one of the first steps is the ligation of the arterial blood supply. From this moment on, the tissue, e.g., a malignant tumor, is exposed to hypoxia at body temperature. The duration between ligation and final tissue removal can vary considerably and may not be reduced to a standardized interval under clinical conditions. Logistical constraints may lead to further considerable homogeneous material.

For hybridization to the probe nucleic acids on the microarray, a labeled sample nucleic acid is needed. RNA extracted from the sample material can be used for synthesis of labeled complementary DNA without amplification of the sample RNA, or for the production of antisense-RNA. The process of aRNA-synthesis allows high amplification of the sample material, which is of relevance if only low amounts of sample material are available.

Common amplification procedures utilize bacteriophage T7, T3 or SP6 polymerases to transcribe RNA from DNA templates. The DNA template requires an appropriate polymerase binding site in its sequence that is located upstream of the region to be transcribed. A complex of this approximately 20 base pair binding sequence linked to an oligo-T sequence is incorporated into the cDNA by reverse transcription of the sample RNA (first strand synthesis). RNA then is degraded by RNase-treatment and a second strand is fabricated by DNA-polymerase. The cDNA now becomes the template strand for the T7-RNA-polymerase producing RNA (aRNA) in antisense direction compared to the orientation of the template RNA. The whole procedure can be repeated a second time leading to 1000-fold or more amplification of the RNA. By including labeled nucleotides in the in-vitro transcription reaction, one can incorporate labels into the synthesized RNA.

Hybridization of the target nucleic acid molecule to the probe DNA on the chip is most commonly detected and quantified by fluorescence-based detection. This requires a target molecule labeled with a fluorophore such as Cy3 or Cy5. The aim of the procedure is to determine the relative abundance of the target molecule within the sample. Spotting of the probe molecules to the miroarray surface is a critical step. To account for spot to spot variations due to differing amounts of probe molecules in the spots, two-colour experiments are used and the ratio of the two fluorophores in any single spot is determined. In a typical two-colour experiment, RNA is extracted for instance from tumor tissue and neighbouring normal tissue. The RNA samples are labelled with different fluorophores, for instance tumor RNA with a red fluorophore, normal RNA with a green one. Both samples are hybridized together on the microarray. If the spot then appears in red, this means higher expression of that gene in tumor tissue compared to normal tissue. If the spot looks green, higher expression in normal tissue compared to tumor tissue has been detected. Yellow means equal expression in tumor versus normal tissue (Figure 7.2).

MICROARRAYS: APPLICATIONS IN BASIC RESEARCH AND TRANSLATIONAL MEDICINE

An Early Example for Microarray-Based Gene Expression Profiling Aimed at Understanding Metabolism

One of the early applications of microarray technology was published in 1997 by Pat Brown's group at Stanford University. The Brown group used a microarray

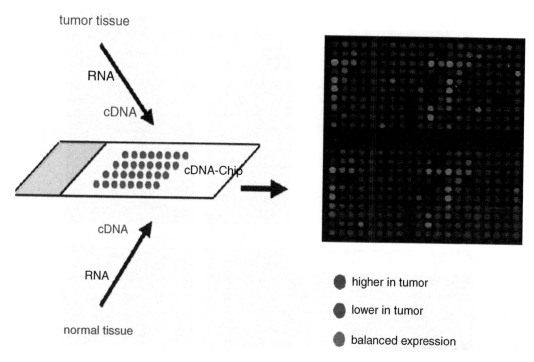

Figure 7.2 Two-color microarray experiment.

representing 6,600 yeast genes to study a process known as diauxic shift. In glucose-rich medium, yeast cells generate energy by fermentation and convert the substrate glucose to acetaldehyde, which is then reduced to ethanol by alcohol dehydrogenase. When glucose is consumed, cells switch from fermentation to respiration and utilize the produced ethanol as a carbon source to generate glycogen. To study gene activity during this process, the researchers labeled cDNA obtained from cells before reaching exponential growth phase with the red fluorescent dye Cy3 as a reference. RNA was prepared at several time points during growth phase and substrate shift, reverse transcribed into cDNA, and labeled with the green fluorescent dye Cy5. Then the Cy5-labeled cDNA (RNA) targets and the Cy3-labeled reference were hybridized to the arrays, and the relative intensities of Cy3 versus Cy5 were measured for each time point. With increasing yeast cell growth indicated by enhancement of the optical density of the cultures, the number of differentially expressed genes increased, as did the level of differential expression indicated by the intensity of red and green staining (Figure 7.3). While in sparse culture, only 0.3% of the genes were altered and the maximal difference in expression was 2.7-fold, 30% of the genes were altered at the final time point of the experiment. More than 300 genes exhibited a differential expression of more than 4-fold. This experiment confirmed that alterations of expression can be efficiently determined in a time-resolved manner by microarray analysis. In addition, a number of genes that had not been characterized, approximately 400 at that time, could potentially play a role in the diauxic shift, growth control, and energy generation. In summary, these candidate genes were placed into a potential functional framework. This became one of the major goals of microarray experiments in subsequent microarray

studies, not only in yeast but also in mammalian systems including human cells and tissues.

In the yeast microarray experiment, the Stanford researchers went one step further and asked the question, if co-expressed genes are regulated in a similar fashion. Several distinct gene clusters comprising elements that exhibit the same expression pattern of upregulation or downregulation over time were identified. When the gene promoters of the co-expressed genes were analyzed, common regulatory sequences were recovered. For example, all but one gene (*IDp2*) contained a regulatory element named CSRE—carbon source responsive element (Figure 7.4). The CSRE is required to activate transcription of the genes involved in gluconeogenesis and the glyoxylate cycle in yeast. And indeed, all of the genes found in this cluster play a role in the glyoxylate cycle (*MLS1, IDP2, ICL1*), in the conversion of acetate to acetyl-CoA (*ACR1*), and in the production of fructose-6-phosphate (*FBP1*). In summary, the basic conclusions from the yeast experiment were (i) similar function is associated with co-regulation, (ii) co-regulation provides a way to define novel functional modules, (iii) co-regulation provides a way to define potential functions for unknown genes, (iv) co-regulation is based on similar transcriptional regulatory factors, and (v) co-regulation is a basis for the identification of regulatory mechanisms.

Elucidating the Transcriptional Basis of the Serum Response in Human Cells

Diploid human fibroblasts, like most other cell cultures, require the presence of serum growth factors in their culture medium. Routinely, these factors are supplied by adding fetal calf serum to the culture medium.

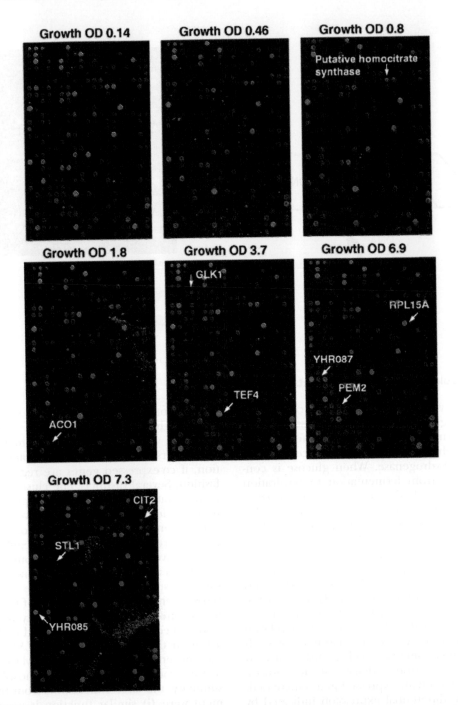

Figure 7.3   Gene expression changes associated with increased culture density over time. In each of the arrays used to analyze gene expression during the diauxic shift, red spots represent genes that were induced relative to the initial time point, and green spots represent genes that were repressed. Note that distinct sets of genes are induced and repressed in the different experiments. Cell density as measured by optical density (OD) at 600 nm was used to monitor the growth of the culture. Reproduced with permission from AAAS, *Science* 1997;278:680–686.

Cultured cells can be made quiescent by serum deprivation. When fetal calf serum is added to such cells, they quickly resume cell-cycle progression and proliferation. This cellular reaction is called the serum response, which was chosen as another early example to demonstrate the power of microarray analysis. This time the Stanford researchers obtained RNA from serum-starved cultures and prepared target cDNA labeled with Cy3. RNA from all other time points following serum stimulation was used to prepare Cy5-labeled targets. In two-color microarray experiments, the targets were hybridized to a human cDNA array representing 8600 human sequences. About 4000 of them were known human genes, 2000 sequences were related to these annotated genes, while the remaining genes were ESTs without known function. Figure 7.5 shows a subset of genes ($n = 517$) whose expression changed up to 8-fold during serum stimulation.

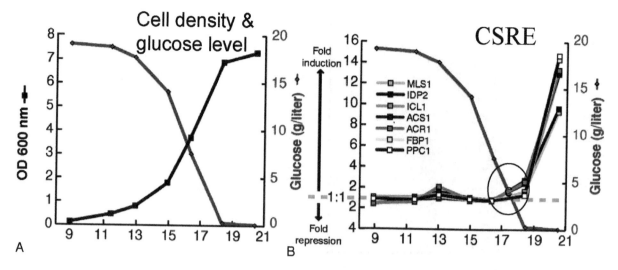

Figure 7.4   Analysis of regulatory modules within the promoters of co-regulated genes associated with the diauxic shift.
(**A**) Growth curve of yeast cells shown as increasing optical density (black line) upon glucose consumption (red line). (**B**) Induction of a group of genes carrying a carbon source element (CSRE) within their promoters. The decreasing glucose level (red line) allows determination of a threshold for the onset of gene expression (grey and black lines) mediated by the CSRE. Reproduced by permission from AAAS, *Science* 1997;278:680–686.

The transcriptional response toward serum stimulation was observed within 15 minutes. The genes can be divided into several clusters, which exhibit a common regulatory scheme. Some clusters show a characteristic pattern of upregulation followed by downregulation (cluster C) or the reverse pattern (clusters E and B). Based on current knowledge in molecular cell biology and data mining for gene functions, the Stanford researchers performed a functional gene clustering. This analysis was done at a time before the Gene Ontology became available. One functional cluster of co-regulated genes comprised transcription factors including the ones known to be involved in the immediate early gene response, permitting rapid responses without the need for protein synthesis. Another cluster included phosphatases. Their functional relevance was not known at the time of the analysis. Today it is well established that the phosphatases limit signaling kinase activity, which is rapidly stimulated upon growth stimulation, by negative feedback. Not surprisingly, the researchers recovered genes encoding cell-cycle regulatory proteins. Inhibitory genes were quickly downregulated, paving the way for re-entry of the serum-starved cells into the cell cycle. With a short delay, cell-cycle stimulatory genes were upregulated, among them cyclin D1 and DNA topo-isomerase, which is required for chromosome segregation at mitosis. A more surprising feature of the analysis was the appearance of genes with known functions in wound healing. This referred not only to genes whose products function intracellularly, but also to genes whose products play a role in remodeling clot structure and the extracellular matrix, as well as in intercellular signaling. While previous studies had aimed at elucidating intracellular events in wound healing, gene expression profiling of the serum response indicated the relevance of extracellular events during the first 24 hours in this process.

Microarray Applications in Cancer Pathogenesis and Diagnosis

A recent PubMed search revealed that the majority of microarray and gene expression profiling studies in medicine are devoted to some aspect of cancer. Cancer studies far outnumber similar studies in cardiovascular diseases, neurodegenerative diseases, infection, inflammation, and other diseases (Table 7.1). Therefore, we have chosen some prominent applications of microarrays in the field of cancer as paradigms to demonstrate the power of transcriptome analysis.

To elucidate the mechanisms of tumorigenesis and metastasis, particularly to study the complexity of the underlying processes, researchers frequently use microarrays. Cancer classification based on microarray studies aims at identifying characteristics beyond anatomical site and histopathology. Outcome prediction tries to overcome the limitations of current diagnostic procedures by establishing gene-based criteria to indicate and predict tumor prognosis and therapy response, even for individual cancer patients. Basically, there are three types of microarray-based approaches: (i) class comparison, (ii) class discovery, and (iii) class prediction. Using class comparison, one tries to compare the expression profiles of two (or more) predefined classes. For example, two tissue samples, normal versus malignant cells or tissues, different developmental stages, or cells treated with drugs under different conditions. Using class discovery, one tries to identify novel subtypes within an apparently homogenous population. In this case, microarray analysis is used to identify features that cannot be distinguished by other available tools. The starting point usually is a homogenous group of specimens, in which a concealed proportion behaves aberrantly or exhibits invisible or unknown features. The problem of cancer treatment

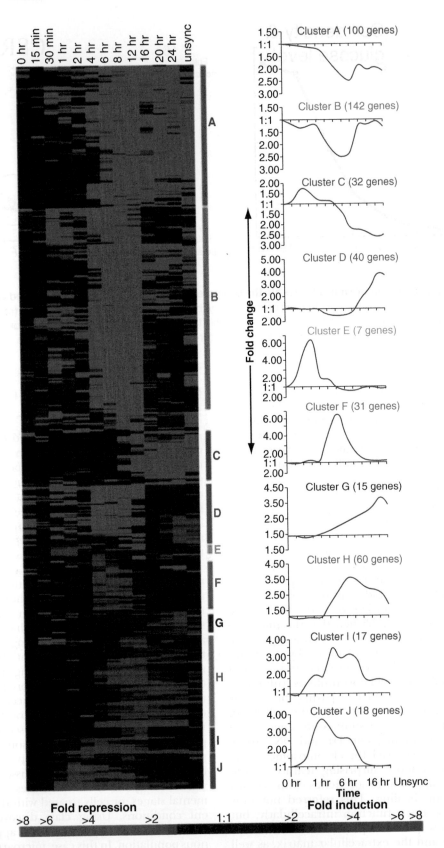

Figure 7.5   Hierarchical clustering of genes induced or repressed during serum response in human fibroblasts. Ten gene clusters (**A–J**) harboring 517 genes, which show significant alterations in gene expression over time, are depicted. For each gene, the ratio of mRNA levels in fibroblasts at the indicated time intervals after serum stimulation compared to their level in the serum-deprived (time zero) fibroblasts is represented by a color code, according to the scale for fold-induction and fold-repression shown at the bottom. The diagram at the right of each cluster depicts the overall tendency of the gene expression pattern within this cluster. The term *unsync* denotes exponentially growing cells. Reproduced with permission from AAAS, *Science* 1999;283:83–87.

Table 7.1	Number of Published Microarray and Gene Expression Profiling Applications in Research. Results of a PubMed Search Dated June 29, 2009, Using Single Keywords and Combinations of Two Keywords Without Limits to Publication Years.

| | 2nd Keyword | | |
1st Keyword	None	Gene Expression Profiling	Microarray
None	—	70,227	31,158
Pharmacology	4,377,665	15,897	6,852
Diseases	3,712,501	16,125	7,287
Cancer	*2,279,276*	*22,323*	*10,375*
Pathology	1,916,497	15,388	6,978
Cardiovascular diseases	*1,526,007*	*1,974*	*751*
Development	1,345,130	15,789	6,667
Immunology	1,137,843	6,339	2,592
Infection	*939,977*	*3,428*	*1,554*
Nutrition	195,387	668	307
Drug development	302,361	3,382	1,483
Inflammation	*302,474*	*2,608*	*1,231*
Neurodegenerative diseases	*162,823*	*807*	*328*
Toxicology	87,300	1,136	529

falls into this category, since patients who are stratified into treatment groups according to standard histopathological criteria often respond differently to therapy. We will see that microarray studies can help to successfully address this urgent clinical problem. Class prediction means to find a set of features that are predictive for a certain, predefined class. This is perhaps the most sophisticated type of microarray application. It is usually based on class discovery, followed by the idea to establish a classifier. A classifier is a set of features, like genes, proteins, micro-RNAs that are surrogate markers for a certain class. This is the common approach to identify predictive gene sets or gene signatures that can predict clinical outcome or therapy response.

Identification of Hidden Subtypes Within Apparently Homogenous Cancers

The group of T. Sorlie identified 456 genes out of 8000 genes on a microarray that discriminated between tumor subclasses in a cohort of 65 tumors from 42 breast cancer patients. Gene expression patterns of breast carcinomas helped to distinguish tumor subclasses with clinical implications. Using hierarchical clustering, the researchers distinguished five distinct tumor groups characterized by their gene expression pattern: the basal epithelial cancer type, the luminal epithelial cancer types A–C, a group displaying expression of the breast cancer oncogene *ERBB2* (*HER2*), and a group without any known feature. There was yet another group showing features of normal breast epithelial cells (Figure 7.6). In the next step of the analysis, the researchers addressed the question as to whether these different groups are characterized by distinct clinical parameters. Therefore, they compared the groups by certain statistical methods, among others by univariate statistical analysis, for either overall survival or relapse-free survival monitored for up

to 4 years (Figure 7.7). The patient groups that were *ERBB2*-positive or were characterized as basal epithelial breast tumors had the shortest survival times. While this information was not new for the *ERBB2*-positive tumors, the basal epithelial breast cancers belong to a novel group with an obviously bad prognosis. One characteristic of this tumor type is the high frequency of *TP53* mutations. The tumor suppressor gene *TP53*, well known as the guardian of the genome, is lost or mutated in more than 50% of all advanced human cancers, and might be responsible for the bad prognosis. There was also a difference in clinical outcome between the luminal-type breast cancers. Most strikingly, luminal A tumors exhibited a very good outcome at least within 4 years, while luminal B or luminal C tumors were intermediate. In conclusion, this study opened the door to further screen many tumors for gene signatures indicative of the clinical performance of breast cancer patients. With respect to cancer treatment, the most important issue is to find gene sets predictive for the susceptibility or resistance to therapy, particularly to chemotherapy, and to clinical outcome in the absence of other conventional indicators.

Gene Expression Profiling Can Predict Clinical Outcome of Breast Cancer

Breast cancer patients with the same stage of disease exhibit markedly different treatment responses and overall outcome. However, histopathological assessment of these cancers does not have sufficient power to discriminate which patients will perform well versus those that will not. The strongest predictors for metastases (such as lymph node status and histological grade) fail to classify accurately breast tumors according to their clinical behavior. None of the signatures of breast cancer gene expression reported to date allow for patient-tailored therapy strategies. The study published by van't

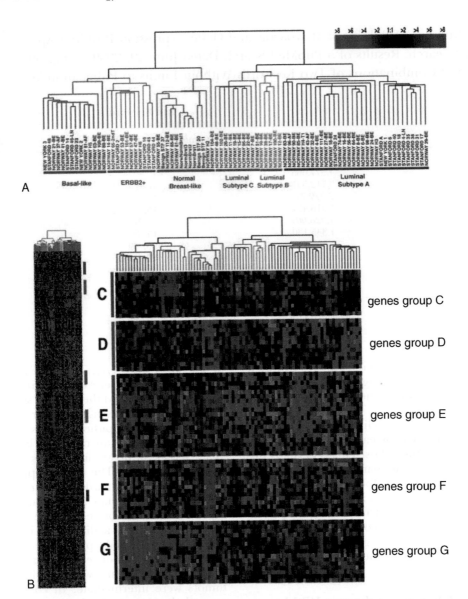

Figure 7.6   Differential breast cancer gene expression. Gene expression patterns of 85 experimental samples (78 carcinomas, 3 benign tumors, 4 normal tissues) analyzed by hierarchical clustering using a set of 476 cDNA clones. (**A**) Tumor specimens were divided into 6 subtypes based on their differences in gene expression: luminal subtype A, dark blue; luminal subtype B, yellow; luminal subtype C, light blue; normal breast-like, green; basal-like, red; and ERBB2+, pink. (**B**) The full cluster diagram obtained after two-dimensional clustering of tumors and genes. The colored bars on the right represent the characteristic gene groups named C to G and are shown enlarged in the right part of the graph: (**C**) ERBB2 amplification cluster, (**D**) novel unknown cluster, (**E**) basal epithelial cell-enriched cluster, (**F**) normal breast epithelial-like cluster, (**G**) luminal epithelial gene cluster containing ER (estrogen receptor). Reproduced with permission from the National Academy of Sciences USA, *Proc Natl Acad Sci U S A* 2001;98:10869–10874.

Veer et al. in the Netherlands has pioneered gene array-based breast cancer diagnostics. The study was based on a well-characterized cohort of breast cancer patients ($n = 117$). This included 78 sporadic primary invasive ductal and lobular breast carcinomas of less than 5 cm in size. The tumor stages were T1 or T2, nodal status N0 (without axilliary metastases), patient age <55 years at diagnosis without a history of previous malignancies. The patients received surgical treatment followed by radiotherapy, but no adjuvant chemotherapy (except for 5 patients). The follow-up period of the patient cohort was 5 years. Tissue samples contained more than

50% tumor cells by pathological inspection; estrogen receptor (ER) and progesterone receptor (PR) status were known. The cohort was supplemented by 20 hereditary tumors carrying *BRCA1/BRCA2* mutations that were of similar histology to the sporadic cancers. Target RNA/cDNA was labeled and hybridized to an oligonucleotide array representing more than 24,000 human sequences and more than 1000 control sequences. The reference target used in this system was a pooled cRNA derived from an RNA mixture of all patients. This means that gene expression of each sample was determined relative to the pool of all

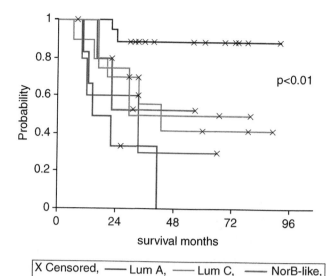

Figure 7.7 Survival analysis (Kaplan-Meier plot) of patient groups distinguished according to gene expression profiling. The Y-axis shows the survival probability for each individual group; the X-axis represents the time scale according to patient follow-up data. All groups identified by gene expression profiling are shown. Luminal type A, dark blue; luminal type B, yellow; luminal type C, light blue; normal type, green; ERBB2-like type, pink; and basal type, red. Patients with ERBB2-like or basal type tumors had the shortest survival times; luminal-type A patients had the best prognosis. All others showed an intermediate probability and were not clearly distinguishable. Reproduced with permission from the National Academy of Sciences USA, *Proc Natl Acad Sci U S A* 2001;98:10869–10874.

samples. The hybridizations were performed in duplicate and ~5000 genes appeared significantly regulated more than 2-fold with a *p*-value of less than 0.01.

In a supervised classification procedure, the researchers from the Netherlands used the gene expression profiles obtained from the sporadic tumors only. In the first step of the classification procedure, the ~5000 genes that were significantly regulated in more than 3 of 78 tumors were selected from the 25,000 genes represented on the array. The correlation of each gene expression profile with the clinical outcome of patients was calculated, and 231 genes were found to be significantly associated with disease progression. In the second step, the 231 informative genes were rank-ordered according to their correlation coefficient. In the third step, the number of genes in this preliminary prognosis classifier was optimized by cross-validation, particularly by the leave-one-out procedure. The final result was a signature of 70 genes, which predict the clinical outcome—distant metastasis within 5 years—with an accuracy of 83% (Figure 7.8 A, B). This means that of 78 patients, 65 were assigned to the right category, poor prognosis or good prognosis. Five patients with poor prognosis and 8 patients with good prognosis were misclassified. Van't Veer et al. used an independent set of 19 lymph-node negative

breast tumors to validate their classifier (Figure 7.8 B, C). This time, 2 of 19 patients were assigned to the wrong group. Thus, the classifier predictive of a short interval to distant metastases (poor prognosis signature) in patients without tumor cells in local lymph nodes at diagnosis (lymph node negative patients) showed a similar performance on this test set of tumors as compared to the training set.

Today, three gene expression-based prognostic breast cancer tests have been licensed for use. These are MammaPrint (Agendia BV, Amsterdam, the Netherlands; based on the work described above), Oncotype DX (Genomic Health, Redwood City, California), and H/I (AvariaDX, Carlsbad, California). However, a recent comparative study showed that for all tests offered, the relationship of predicted to observed risk in different patient populations and their incremental contribution over conventional predictors, optimal implementation, and relevance to patients receiving current therapies need further study. A particular caveat on the currently available predictors was also provided in a paper published in 2005. The authors re-evaluated data from 8 different microarray-based studies with more than 800 tumor samples. The results suggested that the list of genes identified as predictors was highly unstable and that the molecular signatures strongly depended on the selection of patients in the training set. Notably, 5 of 7 studies re-evaluated did not classify patients better than chance.

From Gene Expression Signatures to Simple Gene Predictors

In 1999, a group of scientists in the United States assembled a specialized microarray representing genes preferentially expressed in lymphoid cells. The so-called lympho-chip harbored more than 17,000 cDNA probes derived from libraries specific for germinal center B-cells, diffuse large B-cell lymphoma (DLBCL), follicular lymphoma, mantle cell lymphoma, chronic lymphatic leukemia (CLL), genes induced or repressed in T-cell or B-cell activation, supplemented by lymphocyte-specific genes and cancer genes. The consortium interrogated these chips using targets prepared from normal cells and tumors to define signatures for the different immune cell types, under different conditions and developmental stages. Particularly, the researchers analyzed the most prevalent adult lymphomas using the lympho-chip. They identified signatures for distinct types of diffuse large B-cell lymphoma (DLBCL) exhibiting a bad prognosis, follicular lymphoma (FL) exhibiting a low proliferation rate, and for chronic lymphatic leukemia (CLL) with slow progression (>20 years). Clustering analysis placed the CLL and FL profiles close to those of resting B-cells, while genes of the so-called proliferation signature were weakly expressed in these tumors. DLBCL, the highly proliferative, more aggressive disease, had higher expression levels of proliferation-associated genes. An additional signature characterized germinal center B-cells, which was clearly different from the resting blood B-cells and from the

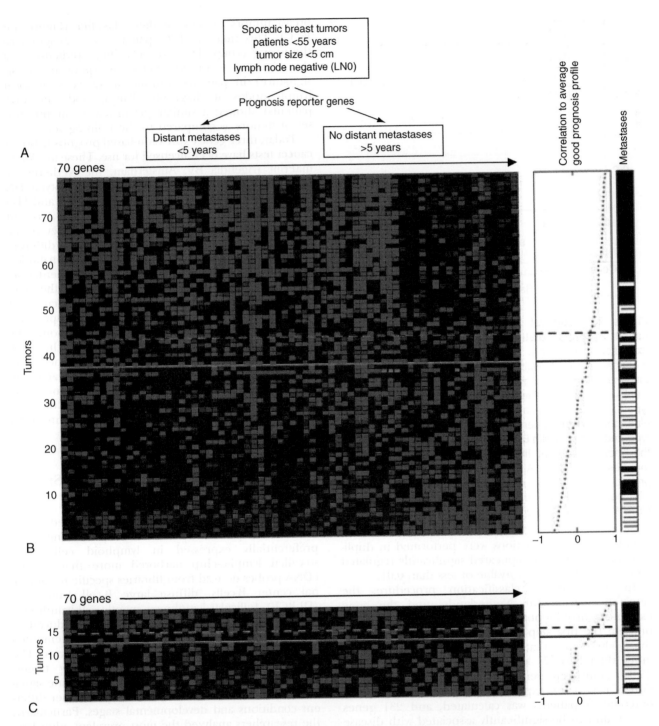

Figure 7.8 Identification of the prognostic breast cancer gene set using a supervised approach. The 231 genes identified as being most significantly correlated to disease outcome were used to recluster, as described in the text. Each row represents a tumor and each column a gene. The genes are ordered according to their correlation coefficient with the two prognostic groups. The tumors are ordered according to their correlation to the average profile of the good prognosis group. The solid line marks the prognostic classifier showing optimal accuracy; the dashed line marks the classifier showing optimized sensitivity. Patients above the dashed line have a good prognosis signature, while patients below the dashed line have a poor prognosis signature. The metastasis status for each patient is shown on the right. White bars indicate patients who developed distant metastases within 5 years after the primary diagnosis; black indicates disease-free patients. Reprinted with permission from van't Veer et al., *Nature* 2002;415:530-536.

in vitro activated B-cells. This indicated that that germinal center B-cells represent a distinct stage of B-cells and do not simply resemble activated B-cells located in the lymph node.

When the scientists reclustered all DLBCL cases, particularly considering the genes that define the germinal center B-cells, they could clearly separate two different subclasses of DLBCL. One of them strictly showed the signature of the germinal center B-cells, while the other one was clearly distinct. These data suggested that a certain class of DLBCL was derived from germinal center B-cells and retained its differentiation signature even after malignant transformation. By investigating the genes exclusively expressed in either of the DLBCL types and reclustering, the authors defined two signatures representative of either the germinal center-type (GC-like) and what they called the activated-type DLBCL (Figure 7.9). Analysis of the clinical follow-up showed that the GC-like tumors have a much better prognosis than the activated type of DLBCL (Figure 7.10). When the authors compared the microarray-based classification to the standard classifiers that define high and low clinical risk, there was obviously no significant classification progress (Figure 7.10B). However, when the low-risk patients initially classified conventionally are further stratified by subgrouping them into the GC and activated-type DLBCL types, the molecular classifier was superior.

Several further microarray studies confirmed that gene signatures were associated with clinical outcome of diffuse B-cell lymphoma. However, among these studies there were disparities with regard to the number and the nature of informative genes. A recent study tried to circumvent the technical and bioinformatic issues of microarray analysis by using quantitative real-time polymerase-chain reaction. Scientists from Miami and Stanford studied the expression of 36 genes that had previously been reported to be of predictive value among 66 lymphoma patients. The prediction of survival could be based on only 6 genes. This result opens the interesting perspective that selecting informative genes that have been filtered through genome-wide microarray studies may permit the application of conventional methods in the future and may obviate microarray applications in routine clinical testing.

PERSPECTIVES

Microarrays have developed into an indispensable tool for transcriptome analysis in basic research, translational studies, and clinical investigations. In experimental pathology, gene expression activities under various conditions can be assessed at an unprecedented quantity, speed, and precision. Commercial microarray platforms exhibit a high degree of standardization allowing service laboratories and academic core facilities to offer the technology to users from industry and academia, respectively, who do not have the means to develop their own specific expertise in this field. Together with other '–omics' technologies, transcriptomics will be an essential component

in worldwide efforts to understand normalcy and disease at the systems level. Already now, transcriptomic approaches are a standard strategy for data collection in systems biology and systems medicine.

In the clinical situation, the current instabilities of predictive gene signatures will probably be scrutinized by enforcing standard operating procedures and efforts aiming at the general standardization of diagnostic approaches, as was the case in the optimization period of microarray technology. The strong need for predictive markers in the clinic, the issue of personalized medicine, and the requirement to study the effects of old and novel drugs at the genome level are expected to increase the use of microarray technologies even further. Alternative high-throughput approaches such as proteomic profiling combined with mass spectroscopy or deep sequencing will probably not be regarded as competitive approaches. Rather, these techniques will further increase our knowledge on complex biological phenomena and pathogenic mechanisms. New types of microarrays have become available that allow analysis of alternative splicing at the level of the transcriptome, as well as to analyze the expression of microRNAs, a novel class of gene expression regulators in development, normal physiology, and disease. Rapid progress will also be made in understanding the molecular basis for the transcriptional alterations that can be assessed by microarrays, by combining chromatin immuno-precipitation (ChIP) and microarray analysis (ChIP-on-chip). Last but not least, efforts are being made to develop chip technologies that permit a truly quantitative estimation of mRNA expression. It is tempting to speculate that these novel chip technologies will gradually replace the currently available microarrays, facilitate transcriptome analysis with even higher precision, and obviate extensive validation (based on real-time PCR, immunohistochemistry, or other methods) and quantification procedures. After all, the race is open for deciphering the protein and RNA master regulators of the transcriptome.

KEY CONCEPTS

- Microarray-based analysis of gene expression permits assessment of normal mRNA levels and disease-related alterations thereof in biological systems such as cells, tissues, organs on a genome-wide scale. In contrast to other approaches used to study gene expression of multiple genes simultaneously such as serial analysis of gene expression, microarray based analysis is supported by manufactured standardized platforms.
- Rigid statistical assessments of microarray data and the use of sufficiently high numbers of specimens are obligatory due to the inherent problem of multiple testing.
- Gene expression patterns (often called gene signatures, molecular portraits, or gene profiles) are widely used to deduce information on cellular pathways that are active, or specifically altered, in physiological states and disease (e.g., in cancer).

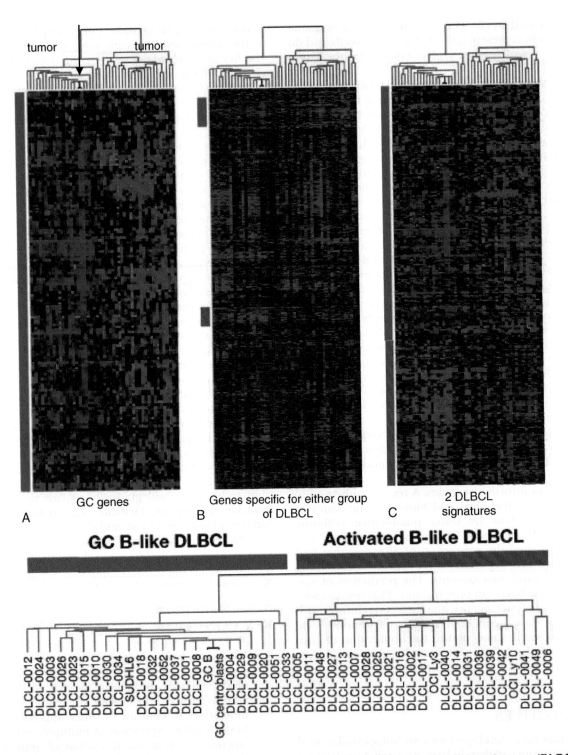

Figure 7.9 Gene signatures representing germinal center (GC)-like diffuse large B-cell lymphomas (DLBCL) and activated B-cell-like DLBCL. (A) Genes characteristic for normal germinal center B-cells were used to cluster the tumor samples. This process defines two distinct classes of B-cell lymphomas: GC-like DLBCL and activated B-cell-like DLBCL. **(B)** Genes that were selectively expressed either in GC-like DLBCL (yellow bar) or activated B-cell-like DLBCL (blue bar) were identified in the tumor samples. **(C)** Result of hierarchical clustering that generated GC-like and activated B-cell-like DLBCL gene signatures. Reprinted with permission from Alizadeh et al., *Nature* 2000;403:503–511.

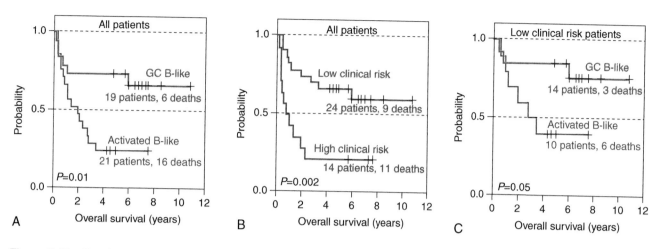

Figure 7.10 Survival analysis of diffuse large B-cell lymphoma patients distinguishable according to gene expression profiling, conventional clinical criteria, and a combination of both sets of criteria. (**A**) DLBCL patients grouped on the basis of gene expression profiling. The GC-like (germinal center-like) and the activated B-cell-like show clearly different survival probabilities. (**B**) DLBCL patients grouped according to the International Prognostic Index (IPI) form two groups with clearly different survival, independent of gene expression profiling. Low clinical risk patients (IPI score 0–2) and high clinical risk patients (IPI score 3–5) are plotted separately. (**C**) Low clinical risk DLBCL patients (IPI score 0–2) shown in **B** were grouped on the basis of their gene expression profiles and exhibited two distinct groups with different survival probabilities. Reprinted with permission from Alizadeh et al., *Nature* 2000;403:503–511.

This information enables the nomination of candidate genes (or sets of genes) that may play a regulatory role in the processes under study or of genes that possibly execute them.

- In molecular pathology, gene expression patterns were identified that correlate with disease state or clinical outcome, independent of classifications based on conventional approaches such as morphology, histology, and clinical parameters. Disease entities indistinguishable by histopathological criteria were defined by expression profiling as well.

- For example, gene signatures related to lymphoma subtypes and to malignant breast cancer were identified. A number of signatures related to prediction of prognosis and response to therapy are already in clinical use.

- Although microarray-based profiling exhibits a tremendous resolution in terms of gene number or transcriptome coverage, expression profiling essentially generates correlative information only. Particularly, gene signatures diagnostic of disease prognosis and therapy response have to be thoroughly validated through unbiased analysis of large cohorts of specimens. Regarding the causal relationship between gene expression alterations and the cellular phenotypes under study, independent functional experiments have to be performed.

- The initial applications of microarray analysis targeted mRNA expression. Meanwhile, the prognostic value of micro-RNA based profiles has been recognized.

- In systems-based analysis, microarray profiling (together with other "–omics" technologies) has become an indispensible tool for assessing the impact of pharmacological or genetic perturbations on the transcriptome.

SUGGESTED READINGS

1. Alizadeh AA, Eisen MB, Davis RE, et al. Distinct types of diffuse large B-cell lymphoma identified by gene expression profiling. *Nature.* 403:503–511.
2. Ashburner M, Ball CA, Blake JA, et al. Gene ontology: Tool for the unification of biology. The Gene Ontology Consortium. *Nat Genet.* 2000;25:25–29.
3. DeRisi JL, Iyer VR, Brown PO. Exploring the metabolic and genetic control of gene expression on a genomic scale. *Science.* 1997;278:680–686.
4. Dupuy A, Simon RM. Critical review of published microarray studies for cancer outcome and guidelines on statistical analysis and reporting. *J Natl Cancer Inst.* 2007;99:147–157.
5. Eisen MB, Spellman PT, Brown PO, et al. Cluster analysis and display of genome-wide expression patterns. *Proc Natl Acad Sci U S A.* 1998;95:14863–14868.
6. Iyer VR, Eisen MB, Ross DT, et al. The transcriptional program in the response of human fibroblasts to serum. *Science.* 1999; 283:83–87.
7. Schena M, Shalon D, Davis RW, et al. Quantitative monitoring of gene expression patterns with a complementary DNA microarray. *Science.* 1995;370:467–470.
8. Sorlie T, Perou CM, Tibshirani R, et al. Gene expression patterns of breast carcinomas distinguish tumor subclasses with clinical implications. *Proc Natl Acad Sci U S A.* 2001;98:10869–10874.
9. Van't Veer LJ, Dai H, van de Vijer MJ, et al. Gene expression profiling predicts clinical outcome of breast cancer. *Nature.* 2002;415:530–536.
10. Walther A, Johnstone E, Swanton C, et al. Genetic prognostic and predictive markers in colorectal cancer. *Nat Rev Cancer.* 2009;9:489–499.

8

The Human Epigenome: Implications for the Understanding of Human Disease

Maria Berdasco ▪ Manel Esteller

INTRODUCTION

Epigenetic processes, defined as the heritable patterns of gene expression that do not involve changes in the sequence of the genome, and their effects on gene repression are increasingly understood to be such a way of modulating phenotype transmission and development. The patterns of DNA methylation, histone modifications, microRNAs, and several chromatin-related proteins of sick cells usually differ from those of healthy cells, highlighting the importance of epigenetic regulation in most human pathologies. The aim of the present review is to provide an overview of how epigenetic factors contribute to the development of human diseases such as abnormal imprinting-causative pathologies, cancer malignancies, as well as autoimmune, cardiovascular, and neurological disorders. These studies have provided extensive information about the mechanisms that contribute to the phenotype of human diseases, but also provided opportunities for therapy.

EPIGENETIC REGULATION OF THE GENOME

The Human Epigenome Project

To date, small-scale studies of specific epigenetic marks have provided limited information about the regulation of genes from different pathways, for example, the hypermethylation-dependent silencing of tumor suppressor genes in cancer or the mutational inactivation of *MeCP2* in Rett patients. However, we need to develop an understanding of these processes

that is based on a broader perspective. A range of matters remains to be resolved, such as the relationships between the epigenetic players (the epigenetic code) and how the environment and/or aging modulate the epigenetic marks. Much of this could be achieved by analyzing the epigenetic patterns on a genome-wide scale, an approach that has at last become possible thanks to recent technological advances. For example, comprehensive DNA-methylation maps (called the methylome) could be assessed by combining the methyl-DIP strategy with tilling or promoter microarray analyses. In a similar manner, the use of the chromatin immunoprecipitation technique followed by genomic microarray hybridization (ChIP-on-chip) has begun to provide extensive maps of histone modifications. Although the groundbreaking discoveries in the field of human disease were initially performed in cancer cells, the characterization of the error-bearing epigenomes underlying other disorders, such as neurological, cardiovascular, and immunological pathologies, has only just begun. The fact that epigenetic aberrations control the function of the human genome and contribute to normal and pathological states justifies carrying out a comprehensive human epigenome project. The goal of the Human Epigenome Project is "to identify all the chemical changes and relationships among chromatin constituents that provide function to the DNA code." This "will allow a fuller understanding of normal development, aging, abnormal gene control in cancer, and other diseases as well as the role of the environment in human health." It is important to bear in mind that there is no single epigenome, but rather many different ones that are characteristic of normal and diverse human disorders, so it is essential to define the chosen starting material. The information extracted from the whole-genome assays will help us understand the role of the epigenetic marks and could have translational research benefits for diagnosis, prognostic, and therapeutic treatment. Defining human epigenomes associated with human disorders will help select patients who are likely to benefit from epigenomic therapies or prevention strategies, determine their efficacy and specificity, and lead to the identification of surrogate markers and end points of its effects. The aim of the present review is to provide an overview of how epigenetic factors contribute to the development of human diseases such as abnormal imprinting-causative pathologies, cancer malignancies, as well as autoimmune, cardiovascular, and neurological disorders with aberrant epigenetic profiles.

GENOMIC IMPRINTING

Epigenetic Regulation of Imprinted Genes

Genomic imprinting is a genetic phenomenon by which epigenetic chromosomal modifications drive differential gene expression according to the parent-of-origin. Expression is exclusively due either to the allele inherited from the mother (such as the *H19* and *CDKN1C* genes) or to that inherited from the father (such as *IGF2*). It is an inheritance process that is independent of the classical Mendelian model. Most imprinted genes, which have been identified in insects, mammals, and flowering plants, are involved in the establishment and maintenance of particular phases of development. Nucleus transplantation experiments in mouse zygotes carrying reciprocal translocations carried out in the early 1980s suggested that imprinting may be fundamental to mammalian development. Assays confirmed that normal gene expression and development in mice require the contribution of both maternal and paternal alleles. However, it was not until 1991 that the first imprinted genes, insulin-like growth factor 2 (*IGF2*) and its receptor (*IGF2R*), were identified. Since then, 83 imprinted genes have been identified in mice and humans, about one-third of which are imprinted in both species. It has recently been predicted that 600 genes have a high probability of being imprinted in the mouse genome, and a similar genome-wide analysis predicts humans to have about half as many imprinted genes. The molecular mechanisms underlying genomic imprinting are poorly understood. As imprinting is a dynamic process and the profile of imprinted genes varies during development, regulation must be epigenetic. DNA methylation has been widely described as the major mechanism involved in the control of genes subjected to imprinting. One model for this regulation is based on the cluster organization of imprinted genes. This structure within clusters allows them to share common regulatory elements, such as noncoding RNAs and differentially methylated regions (DMRs). DMRs are up to several kilobases in size, rich in CpG dinucleotides (such as CpG islands), and may contain repetitive sequences. DNA methylation of DMRs is thought to interact with histone

modifications and other chromatin proteins to regulate parental allele-specific expression of imprinted genes. Furthermore, the aforementioned regulatory elements usually control the imprinting of more than one gene, giving rise to imprinting control regions (ICRs). This cluster organization, observed in 80% of imprinted genes, and the specific DNA methylation patterns associated with DMRs are two of the main characteristics of imprinted genes. Deletions or aberrations in DNA methylation of ICRs lead to loss of imprinting (LOI) and inappropriate parental gene expression. Imprinted genes have diverse roles in growth and cellular proliferation, and specific patterns of genomic imprinting are established in somatic and germline cells. Imprinting is erased in germline cells, and reprogramming involving a *de novo* methyltransferase is necessary to ensure sex-specific gene expression in the individual. Methyl groups are incorporated into most ICRs in oocytes, although only some ICRs are methylated during spermatogenesis. After fertilization, the specific methylation profiles of ICRs must be maintained during development in order to mediate the allelic expression of imprinted genes. In the primordial germline cells of the developed individual, these imprinting marks must be freshly erased by DNA demethylation to allow the subsequent establishment of new oocyte-specific and sperm-specific imprints. As a consequence, mammalian imprinting can be described as a development-dependent cycle based on germline establishment, somatic maintenance, and erasure.

Imprinted Genes and Human Genetic Diseases

Since expression of imprinted genes is monoallelic, and thereby functionally haploid, there is no protection from recessive mutations that the normal diploid genetic complement would provide. For this reason, genetic and epigenetic aberrations in imprinted genes are linked to a wide range of diseases. Modulation of perinatal growth and human pregnancy has played a central role in the evolution of imprinting, and many of the diseases associated with imprinted genes involve some disorders of embryogenesis. This is the case of the hydatidiform mole disorder, where all nuclear genes are inherited from the father. In most cases, this androgenesis arises when an anuclear egg is fertilized by a single sperm, after which all the chromosomes and genes are duplicated. However, fertilization by two haploid sperm (diandric diploidy) may occasionally occur. Most cases are sporadic and androgenetic, but recurrent hydatidiform mole has biparental inheritance with disrupted DNA methylation of DMRs at imprinted loci. In contrast, the disorder of ovarian dermoid cysts arises from the spontaneous activation of an ovarian oocyte that leads to the duplication of the maternal genome. These abnormalities suggest that normal human development is possible only when the paternal and maternal genomes are correctly transmitted. Parent-of-origin effects involved in behavioral and brain disorders have been widely reported at the prenatal and postnatal stages of development.

A postnatal growth retardation syndrome associated with the *MEST* gene expression is an illustrative example. The effect of introducing a targeted deletion into the coding sequence of the mouse *MEST* gene strongly depends on the paternal allele. When the deletion is paternally derived, *Mest+/−* mice are viable and fertile, but mutant mice show growth retardation and high mortality. *Mest−/+* animals, with a maternally derived deletion, show none of these effects. This suggests that the phenotypic consequences of this mutation are detected only through paternal inheritance and are the result of imprinting. There is also evidence that some imprinting effects are associated with increased susceptibility to cancer. Absence of expression of a tumor suppressor gene could be the result of LOI or uniparental disomy (UPD) in imprinted genes. Conversely, LOI or UPD of an imprinted gene that promotes cell proliferation (an oncogene) may allow gene expression to be inappropriately increased. These aberrations could have a widespread effect if the aberrant imprinting occurs in an ICR, resulting in the epigenetic dysregulation of multiple imprinted oncogenes and/or tumor suppressor genes. The most common genetic disorders associated with imprinting aberrations are described next.

Prader-Willi Syndrome and Angelman Syndrome

Prader-Willi syndrome (PWS; OMIM #176270) and Angelman syndrome (AS; OMIM #105830) are very rare genetic disorders with autosomal dominant inheritance in which gene expression depends on parental origin. The clinical features of both syndromes are quite similar; they are neurological disorders with mental retardation and developmental aberrations. PWS is characterized by diminished fetal activity, feeding difficulties, obesity, muscular hypotonia, mental retardation, poor physical coordination, short stature, hypogonadism, and small hands and feet, among other traits. AS is characterized by mental retardation, movement or balance disorder, characteristic abnormal behaviors, increased sensitivity to heat, absent or little speech, and epilepsy. The prevalence of the two syndromes is not accurately known, but is estimated to be between 1 in 12,000 and 1 in 15,000 live births, respectively. Most cases of PWS are caused by the deficiency of the paternal copies of the imprinted genes on chromosome 15 located in the 15q11–q13 region, while AS affects maternally imprinted genes in the same region. This deficiency could be due to the deletion of the 15q11–q13 region (3–4 Mb), parental uniparental disomy of chromosome 15, or imprinting defects. The imprinted domain on human chromosome 15q11–q13 is regulated by an ICR that is responsible for establishing the imprinting in the gametes and for maintaining the patterns during the embryonic phases. ICR regulates differential DNA methylation and chromatin structure, and in consequence, differential gene expression affecting the two parental alleles. The ICR in 15q11–q13 appears to have a bipartite structure; one part seems to be responsible for the control of paternal expression and the other for maternal gene expression. PWS is in effect a contiguous gene syndrome

resulting from deficiency of the paternal copies of the imprinted *SNRF/SNRPN* gene, the necdin gene, and possibly other genes. It has been estimated that the region could contain more than 30 genes, so PWS probably results from a stochastic partial inactivation of important genes. While PWS appears to be more closely related to deficiencies caused by chromatin aberrations affecting several genes, AS is associated with mutations in single genes. For instance, the most common genetic defect leading to AS is a ~4 Mb maternal deletion in chromosomal region 15q11–13, which causes an absence of *UBE3A* expression in the maternally imprinted brain regions. Mutations in the gene encoding the ubiquitin-protein ligase E3A (*UBE3A*) have been identified in 25% of AS patients. The *UBE3A* gene is present on both the maternal and paternal chromosomes, but differs in its pattern of methylation. Paternal silencing of the *UBE3A* gene occurs in a brain-region-specific manner, with the maternal allele being active almost exclusively in the Purkinje cells, hippocampus, and cerebellum. Another maternally expressed gene, *ATP10C* (aminophospholipid-transporting ATPase), is also located in this region and has been implicated in AS. Like *UBE3A*, it exhibits imprinted, preferential maternal expression in human brain.

Beckwith-Wiedemann Syndrome

Beckwith-Wiedemann syndrome (BWS; OMIM #130650) is a well-characterized human disease involving imprinted genes that are epigenetically regulated, and studies of BWS patients have contributed much to the understanding of normal imprinting. BWS is a rare genetic or epigenetic overgrowth syndrome with an estimated prevalence of about 1 in 15,000 births and a high mortality rate in the newborn (about 20%). Neonatal patients are mainly characterized by exomphalos, macroglossia, and gigantism, but other symptoms may occur, such as organomegaly, adrenocortical cytomegaly, hemihypertrophy, and neonatal hypoglycemia. There is also an increased risk of developing specific tumors, such as Wilms' tumor and hepatoblastoma. The imprinted domain BWS-related genes are located on 11p15 and are regulated by a bipartite ICR. Two clusters of imprinting genes have been described: (i) *H19/IGF2* (imprinted, maternally expressed, untranslated mRNA/insulin-like growth factor 2); and (ii) *p57KIP2* (a cyclin-dependent kinase inhibitor), *TSSC3* (a pleckstrin homology-like domain), *SLC22A1* (an organic cation transporter), *KvLQT1* (a voltage-gated potassium channel), and *LIT1* (*KCNQ1* overlapping transcript 1). Both clusters are regulated by two DMRs differentially methylated regions: DMR1, which is responsible for *H19/IGF2* control and is methylated on the paternal but not the maternal allele, and DMR2, which is located upstream of *LIT1* and is normally methylated on the maternal but not the paternal allele. BWS can appear as a consequence of two separate mechanisms. First, some patients are characterized by UPD, which consists of the complete genetic replacement of the maternal allele region with a second paternal copy, and/or LOI affecting the *IGF2*-containing region and/or the *LIT1* gene, which causes a switch in the epigenotype of the *H19/IGF2* subdomain or the *p57/KvLQT1/LIT1* subdomain, respectively. When UPD affects *IGF2*, it yields a double dose of this autocrine factor, resulting in tissue overgrowth and increased cancer risk. The LOI mechanism involves aberrant methylation of the maternal *H19* DMR. Second, maternal replacements of the allele and localized abnormalities of allele-specific chromatin modification on *p57KIP2* (also known as *CDKN1C*) or *LIT1* could also contribute to the BWS phenotype. In conclusion, BWS is a model for the hierarchical organization of epigenetic regulation in progressively larger domains. Additionally, mutations in NSD1 (*nuclear receptor-binding SET domain-containing protein 1*), the major cause of the Sotos overgrowth syndrome, have been described in BWS patients, demonstrating the role this gene plays in imprinting the chromosome 11p15 region.

CANCER EPIGENETICS

Cancer encompasses a fundamentally heterogeneous group of disorders affecting different biological processes and is caused by abnormal gene/pathway function arising from specific alterations in the genome. Initially, cancer was thought to be solely a consequence of genetic changes in key tumor suppressor genes and oncogenes that regulate cell proliferation, DNA repair, cell differentiation, and other homeostatic functions. However, recent research suggests that these alterations could also be due to epigenetic disruption. The study of epigenetic mechanisms in cancer, such as DNA methylation, histone modification, nucleosome positioning, and microRNA expression, has provided extensive information about the mechanisms that contribute to the neoplastic phenotype through the regulation of expression of genes critical to transformation pathways. These alterations and their involvement in tumor development are briefly reviewed in the following sections.

DNA Hypomethylation in Cancer Cells

The low level of DNA methylation in tumors compared with that in their normal-tissue counterparts was one of the first epigenetic alterations to be found in human cancer. It has been estimated that 3–6% of all cytosines are methylated in normal human DNA, although cancer cell genomes are usually hypomethylated with malignant cells, featuring 20–60% less genomic 5-methylcytosine than their normal counterparts. Global hypomethylation in cancer cells is generally due to decreased methylation in CpGs dispersed throughout repetitive sequences, which account for 20–30% of the human genome, as well as in the coding regions and introns of genes. A chromosome-wide and large-promoter-specific study of DNA methylation in a colorectal cancer cell line using the methyl-DIP approach has revealed extensive hypomethylated genomic regions located in gene-poor areas. Importantly, the degree of hypomethylation of genomic DNA increases as the lesion progresses from a benign cellular proliferation to an invasive cancer. From a functional point of view, hypomethylation in cancer cells is associated with a number of adverse outcomes, including chromosome instability, activation of transposable elements, and LOI (Figure 8.1). Decreased methylation of repetitive sequences in the satellite DNA of the pericentric region of chromosomes is associated with increased chromosomal rearrangements, mitotic recombination, and aneuploidy. Intragenomic endo-parasitic DNA, such as L1 (long interspersed nuclear elements) and Alu (recombinogenic sequence) repeats, are silenced in somatic cells and become reactivated in human cancer. Deregulated transposons could cause transcriptional deregulation, insertional mutations, DNA breaks, and an increased frequency of recombination, contributing to genome disorganization, expression changes, and chromosomal instability. DNA methylation underlies the control of several imprinted genes, so the effect on the loss of imprinting must also be considered. Wilms' tumor, a nephroblastoma that typically occurs in children, is the best-characterized imprinting effect associated with increased susceptibility to cancer. Other changes in the expression of imprinted genes caused by changes in methylation have been demonstrated in malignancies such as osteosarcoma, hepatocellular carcinoma, and bladder cancer. Finally, DNA methylation acts a mechanism for controlling cellular differentiation, allowing the expression only of tissue-specific and housekeeping genes in somatic differentiated cells. It is possible that some tissue-specific genes became reactivated in cancer in a hypomethylation-dependent manner. Activation of *PAX2*, a gene that encodes a transcription factor involved in proliferation and other important cell activities, and *let-7a-3*, an miRNA gene, have been implicated in endometrial and colon cancer.

Hypermethylation of Tumor Suppressor Genes

Aberrations in DNA methylation patterns of the CpG islands in the promoter regions of tumor suppressor genes are accepted as being a common feature of human cancer (Figure 8.1). The initial discovery of silencing was performed in the promoter of the retinoblastoma (*Rb*) tumor suppressor gene, but hypermethylation of genes like *VHL* (associated with von Hippel-Lindau disease), *p16INK4a, 8-11 hMLH1* (a homologue of *Escherichia coli* MutL), and *BRCA1* (breast-cancer susceptibility gene 1) has also been described. The presence of CpG island promoter hypermethylation affects genes from a wide range of cellular pathways, such as cell cycle, DNA repair, toxic catabolism, cell adherence, apoptosis, and angiogenesis, among others, and may occur at various stages in the development of cancer. In recent years, a CpG-island hypermethylation profile of human primary tumors has emerged, which shows that the CpG island hypermethylation profiles of tumor suppressor genes are specific to the cancer type. Each tumor type can be assigned a specific, defining DNA "hypermethylome," rather like a

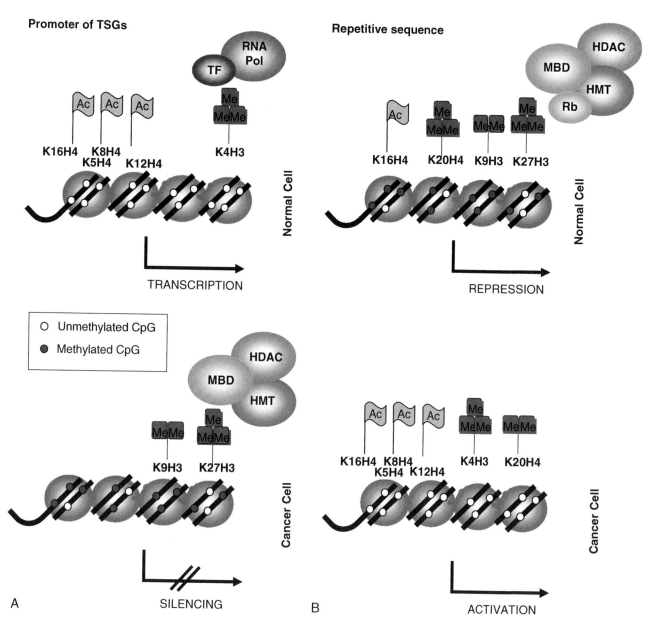

Figure 8.1   A model for the disruption of histone modifications and DNA methylation patterns in cancer cells. Nucleosome arrays are located in the context of genomic regions that include (**A**) promoters of tumor suppressor genes (TSGs), (**B**) repetitive sequences in heterochromatin regions. Nucleosomes consisting of two copies of histones H2A, H2B, H3, and H4 are represented as gray cylinders. DNA (black lines) is wrapped around each nucleosome. In normal cells, TSG promoter regions are unmethylated and enriched in histone modification marks associated with active transcription, such as acetylation of histone H4 (at lysine K5, K8, K12, and K16), or trimethylation of histone H3 (at lysine K4). Transcription machinery recognizes these active marks and transcription of TSGs is allowed. In the same cells, the repetitive genomic DNA is silenced due to the high degree of DNA methylation and histone-repressive marks: histone H4 is densely trimethylated at lysine 20, and histone H3 is dimethylated at lysine 9 and trimethylated at lysine k27. This epigenetic profile is disrupted in transformed cells. TSG promoters are silenced by the loss of the histone-active marks and gain of promoter hypermethylation. Repetitive sequences are activated by replacement of the repressive marks, leading to activation of endoparasitic sequences, genomic instability, or loss of imprinting.

physiological or cytogenetic marker. These marks of epigenetic inactivation occur not only in sporadic tumors but also in inherited cancer syndromes, in which hypermethylation may be the second lesion in Knudson's two-hit model of cancer development. It is expected that improvements in genome-wide epigenomic studies

will increase the number of hypermethylated tumor suppressor genes in a broad spectrum of tumors. However, to date 100–400 instances of gene-specific methylation have been noted in a given tumor. Sometimes the epigenetic alteration of a tumor suppressor gene has genetic consequences, for example, when a DNA repair gene (*hMLH1, BRCA1, MGMT,* or Werner's syndrome gene) is silenced by promoter methylation and functionally blocked.

Histone Modifications of Cancer Cells

The histone modification network is very complex. Histone modifications can occur in various histone proteins (including H2B, H3, H4) and variants (such as H3.3) and affect different histone residues (including lysine, arginine, serine). Several chemical groups (methyl, acetyl, phosphate) may be added in different degrees (for example, monomethylation, dimethylation, or trimethylation). Furthermore, the significance of each modification depends on the organism, the biological process, and the chromatin-genomic region. Due to this diversity of permutations and combinations, little is known about the patterns of histone modification disruption in human tumors. Recent results have shown that the CpG promoter-hypermethylation event in tumor suppressor genes in cancer cells is associated with a particular combination of histone markers: deacetylation of histones H3 and H4, loss of H3K4 trimethylation, and gain of H3K9 methylation and H3K27 trimethylation (Figure 8.1). Increased acetylated histones H3 and H4, and H3/K4 at the *p21* and *p16* transcription start sites after genistein induction (an isoflavone found in the soybean with tumor suppressor properties), in the absence of *p21* promoter methylation, have been reported. The association between DNA methylation and histone modification aberrations in cancer also occurs at the global level. In human and mouse tumors, histone H4 undergoes a loss of monoacetylated and trimethylated lysines 16 and 20, respectively, especially in the repetitive DNA sequences. Subsequent studies showed that loss of trimethylation at H4K20 is involved in disrupting heterochromatic domains and may reduce the response to DNA damage of cancer cells. Immunohistochemical staining of primary prostatectomy tissue samples revealed that patterns of H3 and H4 were predictors of clinical outcome independently of tumor stage, preoperative prostate-specific antigen levels, and capsule invasion in prostate cancer.

Epigenetic Regulation of microRNAs in Cancer

miRNA expression patterns must be tightly regulated during development in a tissue-specific manner and play important roles in cell proliferation, apoptosis, and differentiation. miRNA expression profiles differ between normal and tumor tissues, and also among tumor types, whereby some microRNAs are downregulated in cancer (like tumor suppressor genes). This comparison suggests that miRNAs could be silenced by epigenetic mechanisms. DNA methylation has been shown to be the regulatory mechanism in at least two microRNAs, miR-127 and miR-124a. miR-127, which negatively regulates the protooncogene *BCL6* (B-Cell Lymphoma 6), is usually expressed in normal cells but is silenced by DNA methylation in cancer cells. Similarly, the hypermethylation-dependent silencing of miR-124a results in an increase in expression of cyclin D-kinase 6 oncogene (*CDK6*), and is recognized as a common feature of a wide range of tumors.

Aberrations in Histone-Modifier Enzymes

Aberrations in the epigenetic profiles, with respect to DNA methylation and histone modifications, could also be a consequence of genetic disruption of the epigenetic machinery. A preliminary set of genes involved in epigenetic modifications with mutations in cancer cells but not in the normal counterparts has been found. A list of genes involved in epigenetic modifications that are disrupted in human cancer is presented in Table 8.1. Although alterations in the levels of DNMTs and MBD-containing proteins are commonly observed in human tumors, no genetic lesion has been described in the DNA-methylation machinery in cancer cells. The picture is different for histone modifier enzymes. In leukemia and sarcoma, chromosomal translocations that involve histone-modifier genes, such as histone acetyltransferases (HATs) [such as cyclic AMP response-element-binding protein (CREB)-binding protein-monocytic leukemia zinc finger (*CBP-MOZ*)] and histone methyltransferases (HMTs) [such as mixed-lineage leukemia 1 (*MLL1*)], nuclear-receptor binding SET domain protein 1 (*NSD1*), and nuclear-receptor binding SET-domain protein 3 (*NSD3*), create aberrant fusion proteins. In solid tumors, both HMT genes such as *EZH2*, mixed-lineage leukemia 2 (*MLL2*), or *NSD3*, and a demethylase [Jumonji domain-containing protein 2C (*JMJD2C/GASC1*)] are known to be amplified. Genetic aberrations also disrupt expression of histone deacetylases, such as histone deacetylase 2 (*HDAC2*), which could be affected by mutational frameshift inactivation in colon cancer, and chromatin remodeling proteins, such as *HLTF* (helicase-like transcription factor), *BRG1* (Brahma-related gene 1), and other components of the SWI/SNF family of proteins.

HUMAN DISORDERS ASSOCIATED WITH EPIGENETICS

Aberrant Epigenetic Profiles Underlying Immunological, Cardiovascular, Neurological, and Metabolic Disorders

The patterns of DNA methylation, histone modifications, microRNAs, and several chromatin-related proteins of sick cells usually differ from those of healthy cells, highlighting the importance of epigenetic regulation in most human pathologies. Most of our knowledge regarding human epigenetic diseases was first obtained from cancer cells, but nowadays there is increasing interest in understanding the role of epigenetic modifications in the etiology of human disease. Classical autoimmune disorders, such as systemic lupus erythematosus (SLE), an autoimmune disease characterized by the production of a variety of antibodies against nuclear components and which

Table 8.1   Disruption of Genes Involved in DNA Methylation and Histone Modifications in Cancer

Gene	Alteration	Tumor Type
Alterations affecting DNA methylation enzymes (DNMTs)		
DNMT1	Overexpression	Various
DNMT3b	Overexpression	Various
Alterations involving Methyl-CpG-binding proteins (MBPs)		
MeCP2	Overexpression, rare mutations	Various
MBD1	Overexpression, rare mutations	Various
MBD2	Overexpression, rare mutations	Various
MBD3	Overexpression, rare mutations	Various
MBD4	Mutations in microsatellite instable tumors	Colon, stomach, endometrium
Alterations disrupting histone acetyltransferases (HATs)		
p300	Mutations in microsatellite instable tumors	Colon, stomach, endometrium
CBP	Mutations, translocations, deletions	Colon, stomach, endometrium, lung, leukemia
pCAF	Rare mutations	Colon
MOZ	Translocations	Hematological malignancies
MORF	Translocations	Hematological malignancies, leiomyomata
Alterations disrupting histone deacetylases (HDACs)		
HDAC1	Aberrant expression	Various
HDAC2	Aberrant expression, mutations in microsatellite instable tumors	Various
Alterations affecting histone methyltransferases (HMTs)		
MLL1	Translocation	Hematological malignancies
MLL2	Gene amplification	Glioma, pancreas
MLL3	Deletion	Leukemia
NSD1	Translocation	Leukemia
EZH2	Gene amplification, overexpression	Various
RIZ1	Promoter CpG-island hypermethylation	Various
Alterations affecting histone demethylases		
GASC1	Gene amplification	Squamous cell carcinoma

Adapted from Esteller M. Cancer epigenomics: DNA methylomes and histone-modification maps. *Nat Rev Genet.* 2007;8:286–298.

causes inflammation and injury of multiple organs, and rheumatoid arthritis, a chronic systemic autoimmune disorder which primarily causes inflammation and destruction of the joints, are characterized by massive genomic hypomethylation. This decrease in DNA methylation levels is highly reminiscent of the global demethylation observed in the DNA of tumor cells compared with their normal tissue counterparts. How this hypomethylation in T-cells induces SLE is not well understood. It has been proposed that demethylation induces overexpression of integrin adhesive receptors and leads to an autoreactive response. Identification of the full set of genes deregulated by DNA hypomethylation could help to explain these immunological disorders and will also enable the development of effective therapies to cure SLE. More pathologies with epigenetic regulation of immunology have been reported, including the epigenetic silencing of the ABO histo-blood group genes, the silencing of human leukocyte antigen (HLA) class I antigens, and the melanoma antigen-encoding gene (MAGE) family. It is known that *MAGE* gene expression is epigenetically repressed by promoter CpG methylation in most cells, but *MAGE* genes may be expressed in various tumor types via CpG demethylation and can act as antigens that are recognized by cytolytic T-lymphocytes. Alterations of specific genomic DNA methylation levels have been described not only in the fields of oncology and immunology but also in a wide range of biomedical and scientific fields. In neurology, for example, mutations in

MeCP2, which encodes methyl CpG binding protein 2, cause most cases of Rett syndrome (RTT) and autism. Future research needs to determine whether aberrant DNA methylation is important in the complex etiology of other frequent neurological pathologies, such as schizophrenia and Alzheimer's disease. Beyond this, DNA methylation changes are also known to be involved in cardiovascular disease, the biggest killer in Western countries. For example, aberrant CpG island hypermethylation has been described in atherosclerotic lesions. Since DNA methylation and histone modifications are mechanistically linked, it is likely that different changes in DNA methylation are associated with changes in histone modifications in human diseases. To date, we have been largely ignorant of how these histone modification markers are disrupted in human diseases. A preview of the patterns of histone modifications and their cellular location has only been described for cancer malignancies. For other human pathologies, our knowledge of the alterations in histone modification patterns comes from the use of epigenetic drugs. Therapeutic assays have demonstrated that histone deacetylase (HDAC) inhibitors can improve deficits in synaptic plasticity, cognition, and stress-related behaviors in a wide range of neurological and psychiatric disorders, including Huntington's and Parkinson's diseases, anxiety and mood disorders, and Rubinstein-Taybi and Rett syndromes. Abnormal histone modification patterns associated with specific gene expression have also been described in lupus CD4+ T-cells, and HDAC inhibitors are able to reverse gene expression significantly.

Genetic Aberrations Involving Epigenetic Genes

Genetic alterations of genes coding for enzymes that mediate chromatin structure could result in a loss of adequate regulation of chromatin compaction, and finally, the deregulation of gene transcription and inappropriate protein expression. Although the consequences in cancer malignancies have been widely described, in this section we extend the review to include other genetic diseases involving the function of several enzymes of the epigenetic machinery. The phenotype of these diseases also helps to clarify the role of various chromatin proteins in cell proliferation and differentiation. These include disorders arising from alterations in chromatin remodeling factors, alterations of the components of the DNA methylation machinery, and aberrations disturbing histone modifiers.

KEY CONCEPTS

- Epigenetics: The heritable patterns of gene expression that do not involve changes in the sequence of the genome.
- Epigenome: The overall epigenetic state of a cell.
- Genome imprinting: The epigenetic marking of a locus on the basis of parental origin, which results in monoallelic gene expression.
- Chromatin: The complex of DNA and protein that composes chromosomes. Chromatin packages DNA into a volume that fits into the nucleus, allows mitosis and meiosis, and controls gene expression. Changes in chromatin structure are affected by DNA methylation and histone modifications.
- DNA methylation: The addition of a methyl group to DNA at the 5-carbon of the cytosine pyrimidine ring that precedes a guanine.
- Histone modifications: A set of reactions that introduce functional chemical groups into the histone tails. Posttranslational modifications of the histone tails include methylation, acetylation, phosphorylation, ubiquitination, sumoylation, citrullination, and ADP-ribosylation.

SUGGESTED READINGS

1. Agrelo R, Cheng WH, Setien F, et al. Epigenetic inactivation of the premature aging Werner syndrome gene in human cancer. *Proc Natl Acad Sci U S A.* 2006;103:8822–8827.
2. Bernstein BE, Kamal M, Lindblad-Toh K, et al. Genomic maps and comparative analysis of histone modifications in human and mouse. *Cell.* 2005;120:169–181.
3. Eden A, Gaudet F, Waghmare A, et al. Chromosomal instability and tumors promoted by DNA hypomethylation. *Science.* 2003;300:455.
4. Esteller M, Corn PG, Baylin SB, et al. A gene hypermethylation profile of human cancer. *Cancer Res.* 2001;61:3225–3229.
5. Esteller M. The necessity of a human epigenome project. *Carcinogenesis.* 2006;27:1121–1125.
6. Esteller M. Cancer epigenomics: DNA methylomes and histone-modification maps. *Nat Rev Genet.* 2007;8:286–298.
7. Feinberg AP, Tycko B. The history of cancer epigenetics. *Nat Rev Cancer.* 2004;4:143–153.
8. Feinberg AP. Phenotypic plasticity and the epigenetics of human disease. *Nature.* 2007;447:433–440.
9. Fraga MF, Ballestar E, Villar-Garea A, et al. Loss of acetylation at Lys16 and trimethylation at Lys20 of histone H4 is a common hallmark of human cancer. *Nat Genet.* 2005;37:391–400.
10. He L, Hannon GJ. MicroRNAs: Small RNAs with a big role in gene regulation. *Nat Rev Genet.* 2004;5:522–531.
11. Herman JG, Baylin SB. Gene silencing in cancer in association with promoter hypermethylation. *N Engl J Med.* 2003;349:2042–2054.
12. Jones PA, Baylin SB. The epigenomics of cancer. *Cell.* 2007;128:683–692.
13. Jones PA, Martienssen R. A blueprint for a Human Epigenome Project: the AACR Human Epigenome Workshop. *Cancer Res.* 2005;65:11241–11246.
14. Kurdistani SK, Tavazoie S, Grunstein M. Mapping global histone acetylation patterns to gene expression. *Cell.* 2004;117:721–733.
15. Lefebvre L, Viville S, Barton SC, et al. Abnormal maternal behaviour and growth retardation associated with loss of the imprinted gene Mest. *Nat Genet.* 1998;20:163–169.
16. Lujambio A, Esteller M. CpG island hypermethylation of tumor suppressor microRNAs in human cancer. *Cell Cycle.* 2007;6:1455–1459.
17. Ropero S, Fraga MF, Ballestar E, et al. A truncating mutation of HDAC2 in human cancers confers resistance to histone deacetylase inhibition. *Nat Genet.* 2006;38:566–569.
18. Seligson DB, Horvath S, Shi T, et al. Global histone modification patterns predict risk of prostate cancer recurrence. *Nature.* 2005;435:1262–1266.
19. Ubeda F, Wilkins JF. Imprinted genes and human disease: An evolutionary perspective. *Adv Exp Med Biol.* 2008;626:101–115.
20. Zhang X, Yazaki J, Sundaresan A, et al. Genome-wide high-resolution mapping and functional analysis of DNA methylation in arabidopsis. *Cell.* 2006;126:1189–1201.

9

Clinical Proteomics and Molecular Pathology

Mattia Cremona . Virginia Espina . Alessandra Luchini . Emanuel Petricoin . Lance A. Liotta

UNDERSTANDING CANCER AT THE MOLECULAR LEVEL: AN EVOLVING FRONTIER

Genomic and proteomic research is launching the next era of cancer molecular medicine. Molecular expression profiles can uncover clues to functionally important molecules in the development of human disease and generate information to subclassify human tumors and tailor a treatment to the individual patient. The next revolution is the synthesis of proteomic information into functional pathways and circuits in cells and tissues. Such synthesis must take into account the dynamic state of protein post-translational modifications and protein–protein or protein–DNA/RNA interactions, cross-talk between signal pathways, feedback regulation within cells, between cells, and between tissues. This higher level of functional understanding will be the basis for true rational therapeutic design that specifically targets the molecular lesions underlying human disease.

The rapid progress in molecular medicine is largely due to new insights emanating from data generated by emerging technologies. This chapter summarizes new technologies in the exploding field of proteomics. These technologies hold promise for the early cancer diagnostics from the drop of a patient's blood, to the molecular dissection of a patient's individual tumor cells, to the development of individualized molecularly targeted therapies.

MICRODISSECTION TECHNOLOGY BRINGS MOLECULAR ANALYSIS TO THE TISSUE LEVEL

Molecular analysis of pure cell populations in their native tissue environment is necessary to understand the microecology of the disease process. Accomplishing this goal is much more difficult than just grinding up a piece of tissue and applying the extracted molecules to a panel of assays. The reason is that tissues are complicated three-dimensional structures composed of large numbers of different types of interacting cell populations. The cell subpopulation of interest may constitute a tiny fraction of the total tissue volume. For example, if the goal is to analyze the genetic changes in the malignant cells in a biopsy, this subpopulation is frequently located in microscopic regions occupying less than 5% of the tissue volume. After the computer adage "garbage in, garbage out," if the extract of a complex tissue is analyzed using a sophisticated technology, the output will be severely compromised if the input material is contaminated by the wrong cells. Culturing cell populations from fresh tissue is one approach to reducing contamination. However, cultured cells may not accurately represent the molecular events taking place in the actual tissue from which they were derived. Assuming methods are successful to isolate and grow the tissue cells of interest, the gene expression pattern of the cultured cells is influenced by the culture environment and can be quite different from the genes expressed in the native tissue state. The reason is that the cultured cells are separated from the tissue elements that regulate gene expression, such as soluble factors, extracellular matrix molecules, and cell–cell communication. Thus, the problem of cellular heterogeneity has been a significant barrier to the molecular analysis of normal and diseased tissue. In fact analysis of critical gene expression and protein patterns in normal developing and diseased tissue progression requires the separation of a microscopic homogeneous cellular subpopulation from its complex tissue milieu. This problem can now be overcome by new developments in the field of laser tissue microdissection.

Laser capture microdissection (LCM) has been developed to provide scientists with a fast and dependable method of capturing and preserving specific cells from tissue, under direct microscopic visualization. With the ease of procuring a homogeneous population of cells from a complex tissue using the LCM, the approaches to molecular analysis of pathologic processes are significantly enhanced. The mRNA from microdissected cancer lesions has been used as the starting material to produce cDNA libraries, microchip microarrays, differential display, and other techniques to find new genes or mutations.

The development of LCM allows investigators to determine specific protein expression patterns from tissues of individual patients. Using multiplex analysis, investigators can correlate the pattern of expressed genes and post-translationally modified proteins with the histopathology and response to treatment. Microdissection can be used to study the interactions between cellular subtypes in the organ or tissue microenvironment. Efficient coupling of LCM of serial tissue sections with multiplex molecular analysis techniques is leading to sensitive and quantitative methods to visualize three-dimensional interactions between morphologic elements of the tissue. For example, it will be possible to trace the gene expression pattern and quantitate protein signaling activation state along the length of a prostate gland or breast duct to examine the progression of neoplastic development. The end goal is the integration of molecular biology with tissue morphogenesis and pathology.

Beyond Functional Genomics to Cancer Proteomics

Whereas DNA is an information archive, proteins do all the work of the cell. The existence of a given DNA sequence does not guarantee the synthesis of a corresponding protein. Moreover, protein complexity and versatility stem from context-dependent post-translational processes, such as phosphorylation, sulfation, or glycosylation. Nucleic acid profiling (including microRNA) does not provide information about how proteins link together into networks and functional machines in the cell. In fact, the activation of a protein signal pathway, causing a cell to migrate, die, or initiate division, immediately take place before any changes occur in DNA/RNA gene expression. Consequently, the technology to drive the molecular medicine revolution from the correlation to the causality phase is emerging from protein analytic methods.

The term *proteome*, which denotes all the proteins expressed by a genome, was first coined in late 1994 at the Sienna two-dimensional (2D) gel electrophoresis meeting. Proteomics is the next step after genomics. A major goal of investigators in this exciting field is to assemble a complete library of all proteins. To date, only a small percentage of the proteome has been cataloged. Although a number of new technologies are being introduced for high-throughput protein characterization and discovery, the traditional method of protein identification continues to be 2D gel electrophoresis. When a mixture of proteins is applied to the 2D gel, individual proteins in the mixture are separated out into signature locations on the display, depending on their individual size and charge. The protein spot can be procured from the gel, and a partial amino acid sequence can be read using mass spectrometry. An experimental 2D gel image can be captured and overlaid digitally with known archived 2D gels. In this way it is possible to immediately highlight proteins that are differentially abundant in one state versus another (for instance, tumor versus normal or before and after hormone treatment).

Two-dimensional gels have traditionally required large amounts of protein-starting material equivalent to millions of cells. Thus, their application has been limited to cultured cells or ground-up heterogeneous tissue. Not unexpectedly, this approach does not provide an accurate picture of the proteins that are in use by cells in real tissue.

While individualized treatments have been used in medicine for years, advances in cancer treatment have now generated a need to more precisely define and identify patients that will derive the most benefit from new-targeted agents. Molecular profiling using gene expression arrays has shown considerable potential for the classification of patient populations in all of these respects. Nevertheless, transcript profiling, by itself, provides an incomplete picture of the ongoing molecular network for a number of clinically important reasons. First, gene transcript levels have not been found to correlate significantly with protein expression or the functional (often phosphorylated) forms of the encoded proteins. RNA transcripts also provide little information about protein–protein interactions and the state of the cellular signaling pathways. Finally, most current therapeutics are directed at protein targets, and these targets are often protein kinases and/or their substrates. The human kinome, or full complement of kinases encoded by the human genome, comprises the molecular networks and signaling pathways of the cell. The activation state of these proteins and these networks fluctuate constantly depending on the cellular microenvironment. Consequently, the source material for molecular profiling studies needs to shift from *in vitro* models to the use of actual diseased human tissue. Technologies which can broadly profile and assess the activity of the human kinome in a real biological context will provide a rich source of new molecular informations critical for the realization of patient-tailored therapy (Figure 9.1).

Protein Microarray Tools to Guide Patient-Tailored Therapy

Theoretically, the most efficient way to identify patients who will respond to a given therapy is to determine, prior to treatment initiation, which potential signaling pathways are truly activated in each patient. Ideally, this would come from analysis of tissue material taken from the patient through biopsy procurement. In general, previous traditional proteomic technologies such as 2D gel electrophoresis have significant limitations when they are applied to very small

PROTEIN MICROARRAYS

Circuit Mapping in the Tissue Microenvironment

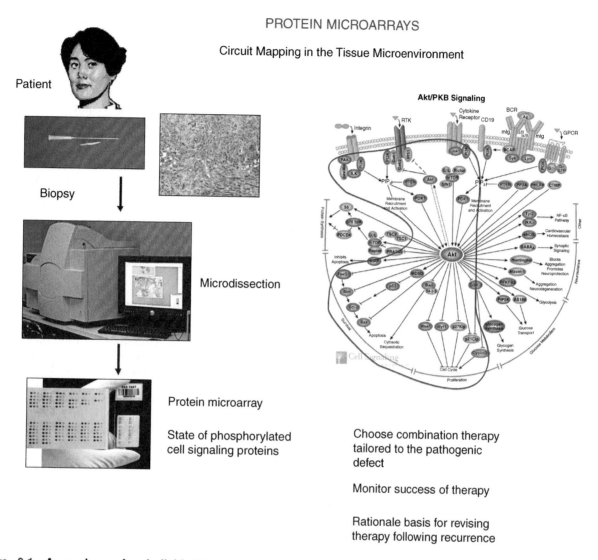

Figure 9.1   A roadmap for individualized cancer therapy. Following biopsy or needle aspiration, and laser microdissection, signal pathway analysis is performed using protein microarrays for phosphoproteomic analysis, and RNA transcript arrays. The specific signaling portrait becomes the basis of a patient-tailored therapeutic regime. Therapeutic assesment is obtained by follow-up biopsy, and the molecular portrait of signaling events is reassessed to determine if therapeutic selection should be modified further.

tissue samples, such as biopsy specimens where only a few thousand cells may be procured. Protein microarrays represent an emerging technology that can address the limitations of previous measurement platforms and are quickly becoming powerful tools for drug discovery, biomarker identification, and signal transduction profiling of cellular material. The advantage of protein microarrays lies in their ability to provide a map of known cellular signaling proteins that can reflect, in general, the state of information flow through protein networks in individual specimens. Identification of critical nodes, or interactions, within the network is a potential starting point for drug development and/or the design of individual therapy regimens. Protein microarrays that examine protein–protein recognition events (phosphorylation) in a global, high-throughput manner can be used to profile the working state of cellular signal pathways in a

manner not possible with gene arrays. Protein microarrays may be used to monitor changes in protein phosphorylation over time, before and after treatment, between disease and nondisease states and responders versus nonresponders, allowing one to infer the activity levels of the proteins in a particular pathway in real-time to tailor treatment to each patient's cellular circuitry.

The application of this technology to clinical molecular diagnostics will be greatly enhanced by increasing numbers of high-quality antibodies that are specific for the modification or activation state of target proteins within key pathways. Antibody specificity is particularly critical, given the complex array of biological proteins at vastly different concentrations contained in cell lysates. Very often, the final number of actual tumor cells microdissected or from biopsy tissues for analysis may be as low as a few thousand. Assuming that the proteins of interest, and their phosphorylated counterparts, exist in low abundance, the total concentration of analyte proteins in the sample will be very low. Newer generations of protein microarrays with highly sensitive and specific antibodies are now able to achieve adequate levels of sensitivity for analysis of clinical specimens containing fewer than a few thousand cells.

At a basic level, protein microarrays are composed of a series of immobilized spots. Each spot contains a homogeneous or heterogeneous bait molecule. The array is queried with (i) a probe (labeled antibody or ligand) or (ii) an unknown biological sample (for instance, cell lysate or serum sample) containing analytes of interest. When the query molecules are tagged directly or indirectly with a signal-generating moiety, a pattern of positive and negative spots is generated. For each spot, the intensity of the signal is proportional to the quantity of applied query molecules bound to the bait molecules. An image of the spot pattern is captured, analyzed, and interpreted.

Protein microarray formats fall into two major classes, forward phase arrays (FPAs) and reverse phase arrays (RPMAs), depending on whether the analyte(s) of interest is captured from solution phase or bound to the solid phase. In FPAs, capture molecules are immobilized onto the substratum and act as the bait molecule. In the FPA format, each array is incubated with one test sample (for instance, a cellular lysate from one treatment condition or serum sample from disease/control patients), and multiple analytes are measured at once. Examples of their use in cancer research include the identification of changes in protein levels following treatment of colon cancer cells with ionizing radiation, identification of serum protein biomarkers for bladder cancer diagnosis and outcome stratification, and prostate cancer diagnosis. Despite their great potential, antibody array use is limited currently by the availability of well-characterized antibodies. A second obstacle to routine use of antibody arrays surrounds detection methods for bound analyte on the array. Current options include the use of specific antibodies recognizing distinct analyte epitopes from the capture antibodies (similar to a traditional sandwich-type ELISA), or the direct labeling of the analytes used for probing the array, both of which present distinct technical challenges.

In contrast to the FPA format, the RPMA format immobilizes an individual test sample in each array spot, such that an array is composed of hundreds of different patient samples or cellular lysates. Though not limited to clinical applications, the RPMA format provides the opportunity to screen clinical samples that are available in very limited quantities, such as biopsy specimens. Because human tissues are composed of hundreds of interacting cell populations, RPMAs coupled with LCM provide a unique opportunity for discovering changes in the cellular proteome that reflect the cellular microenvironment. The RPMA format is capable of extremely sensitive analyte detection with detection levels approaching attogram amounts of a given protein and variances of less than 10%. The sensitivity of detection for the RPMAs is such that low abundance phosphorylated protein isoforms can be measured from a spotted lysate representing fewer than 10 cell equivalents. This level of sensitivity combined with analytical robustness is critical if the starting input material is only a few hundred cells from a biopsy specimen. Since the reverse phase array technology requires only one antibody for each analyte, it provides a facile way for broad profiling of pathways where hundreds of phosphospecific analytes can be measured concomitantly. Most importantly, the reverse phase array has significantly higher sensitivity than bead arrays or ELISA such that broad screening of molecular networks can be achieved from tissue specimens routinely procured. A number of studies illustrate the utility of reverse phase protein microarrays for the analysis of human tissues and demonstrate the potential for the technology.

Combination Therapies

There is increasing evidence demonstrating the promise and potential of combination therapies combining conventional treatments such as chemotherapy or radiotherapy and molecular-targeted therapeutics such as erlotinib (Tarceva®, Roche) and trastuzumab (Herceptin®, Roche) that interfere with kinase activity and protein–protein interactions in specific deregulated pathways. However, strategies that target multiple interconnected proteins within a signaling pathway have not been explored to the same extent. The view of individual therapeutic targets can be expanded to that of rational targeting of the entire deregulated molecular network, extending both inside and outside the cancer cell. Mathematical modeling of network-targeted therapeutic strategies has revealed that attenuation of downstream signals can be enhanced significantly when multiple upstream nodes or processes are inhibited with small molecule inhibitors compared with inhibition of a single upstream node. Also, inhibition of multiple nodes within a signaling cascade allows reduction of downstream signaling to desired levels with smaller doses of the necessary targeted drugs. While therapeutic strategies incorporating these lower dosages could lead to reduced toxicities and a broadened spectrum of available drugs, it must be recognized that testing these interacting drug modalities will necessitate clinical trials of complex design.

Ultimately, targeting, response assessment, and therapeutic monitoring will be individualized, and will reflect the subtle pre-therapy and post-therapy changes at the proteomic level, as well as the protein signaling cascade systems between individuals. The ability to visualize these interconnections both inside and outside a cell could have a profound effect on how we view biology, and can enable the realization of the recent emphasis on personalized combinatorial molecular medicine.

For lung cancer, targeting of the EGFR tyrosine kinase with small molecule inhibitors has received significant attention. Gefitinib (Iressa®, AstraZeneca) and erlotinib have shown significant clinical benefit in specific subsets of patients. Promising phase I trial results with gefitinib demonstrated that the drug decreased levels of activated EGFR and mitogen-activated protein kinase in post-treatment skin biopsies indicating that the intended target was being inhibited. Subsequent international phase II trials in patients with progressive non-small-cell lung cancer showed that only a small minority of patients respond durably to these EGFR tyrosine kinase inhibitors. Further analysis of this small subset led to the sequencing and identification of *EGFR* mutations that are associated with sensitivity to these drugs. The presence of *EGFR* mutations correlates remarkably well with the identified correlative clinical parameters.

To date, the differences in signaling between wild-type and mutant EGFRs are poorly understood. The EGFR signaling network can be activated in a number of ways: mutation of the receptor, overexpression of the receptor, and mutation of downstream kinases (such as phosphatidylinositol 3-kinase) are just a few examples. This suggests that mutation analysis alone should not serve as the sole criterion for identification and treatment selection, and that further studies incorporating proteomic profiling of tissue may be beneficial in identifying additional patients who will benefit from EGFR tyrosine kinase inhibitor treatment. Proteomic profiling of EGFR-related signaling activity in preclinical and *in vitro* experiments as well as in clinical specimens could provide useful information for characterizing drug responses. Reverse-phase array technology is well suited to assess the signaling differences between mutant and wild-type cells, and these studies are currently under way. Instead of single measurements of selected analytes such as EGFR and ErbB2 levels, future pathology reports can be envisioned to include a phosphoproteomic portrait of the functional state of many relevant specific downstream end points, and entire classes of signaling pathways as a guide for therapeutic decision making and prognosis.

Molecular profiling of the proteins and signaling pathways produced by the tumor microenvironment, host, and peripheral circulation hold great promise in effective selection of therapeutic targets and patient stratification. For many of the more common sporadic cancers, there is significant heterogeneity in cell signaling, tissue behavior, and susceptibility to chemotherapy. Proteomic analysis is particularly useful in this area given the ability to study multiple pathways simultaneously. Cataloging of abnormal signaling pathways for large numbers of specimens will provide the data necessary for a rationally based formulation of combination therapy that presumably would be more effective than monotherapy and help to minimize the issues of tumor heterogeneity. The promise of proteomic-based profiling, different from gene transcript profiling alone, is that the resulting prognostic signatures are derived from drug targets (such as activated kinases), not genes, so the pathway analysis provides a direction for therapeutic mitigation. Thus, phosphoproteomic pathway analysis becomes both a diagnostic/prognostic signature as well as a guide to therapeutic intervention.

Protein biomarker stability in tissue: a critical unmet need

The promise of tissue protein biomarkers to provide revolutionary diagnostic and therapeutic information will never be realized unless the problem of tissue protein biomarker instability is recognized, studied, and solved. There is a critical need to develop standardized protocols and novel technologies that can be used in the routine clinical setting for seamless collection and immediate preservation of tissue biomarker proteins, particularly those that have been post-translationally modified, such as phosphoproteins. While molecular profiling offers tremendous promise to change the practice of oncology, the fidelity of the data obtained from a diagnostic assay applied to tissue must be monitored and ensured; otherwise, a clinical decision may be based on incorrect molecular data. Under the current standard of care, tissue is procured for pathologic examination in three main settings: (i) surgery in a hospital-based operating room, (ii) biopsy conducted in an outpatient clinic, and (iii) image-directed needle biopsies or needle aspirates conducted in a radiologic suite. Based on current reseach standards, tissue must be snap-frozen in order to perform proteomic studies. In the real world of a busy clinical setting, it will be impossible to immediately preserve procured tissue in liquid nitrogen. Moreover, the time delay from patient excision to pathologic examination and molecular analysis is often not recorded and may vary from 30 minutes to many hours depending on the time of day, the length of the procedure, and the number of concurrent cases.

There are two categories of variable time periods that define the stability intervals for human tissue procurement. The first variable time period is the post-excision delay time (or EDT). EDT is the time from the moment that tissue is excised from the patient and becomes available *ex vivo* for analysis and processing to the time that the specimen is placed in a stabilized state (immersed in fixative or snap-frozen in liquid nitrogen). Given the complexity of patient-care settings, during the EDT the tissue may reside at room temperature in the operating room or on the pathologist's cutting board, or it may be refrigerated in a specimen container. The second variable time period is the processing delay time (PDT) that is the period between the immersion of the tissue in a preservative solution or the storage in a freezer and the processing for molecular analysis. In addition to the uncertainty about the length of these two time intervals, a host of known and unknown variables can influence the stability of tissue molecules during these time periods (such as the temperature fluctuations prior to fixation or freezing). Even if a strict protocol is followed, there is no ultimate assurance that processing variables are free from compromise up to the time that the molecular profile data are collected. Even if a strict protocol is followed, there is no ultimate assurance that processing variables are free from compromise up to the time that the molecular profile data are collected.

Protein stability is unrelated to RNA transcript stability

Several studies have been conducted concerning the stability of RNA in tissue *ex vivo*. While this information is applicable to gene array profiling, it has no bearing on protein stability in general or phosphoprotein stability specifically. Chemical conditions favoring protein stability may be completely different from those for RNA stability. This is true for the following reasons.

a. Gene transcript array data cannot reflect the posttranslational state of a protein.
b. Gene transcript array data cannot accurately reflect the activated state of a protein signal pathway or the phosphorylated state of a kinase substrate.
c. Gene array data often do not quantitatively correlate with protein expression for signal proteins and cytokines.

Recognition that the tissue is alive and reactive following procurement

While investigators have worried about the effects of vascular clamping and anesthesia, prior to excision, a much more significant and underappreciated issue is the fact that excised tissue is alive and reacting to *ex vivo* stresses. The instant a tissue biopsy is removed from a patient, the cells within the tissue react and adapt to the absence of vascular perfusion, ischemia, hypoxia, acidosis, accumulation of cellular waste, absence of electrolytes, and temperature changes. In as little as 30 minutes post-excision, drastic changes can occur in the protein signaling pathways of the biopsy tissue as the tissue remains in the operating room suite or on the pathologist's cutting board. It would be expected that a large surge of stress-related, hypoxia-related, and wound repair-related protein signal pathway proteins and transcription factors will be induced in the tissue. Over time the levels of candidate proteomic markers (or RNA species) would be expected to widely fluctuate upward and downward. This will significantly distort the molecular signature of the tissue compared to the state of the markers *in vivo*. Moreover, the degree of *ex vivo* fluctuation could be quite different between tissue types and influenced by the pathological microenvironment.

Formalin Fixation May Be Unsuitable for Quantitative Protein Biomarker Analysis in Tissue

Although it is now possible to extract proteins from formalin-fixed tissue, because of the long period required to formalin tissue fixation, the procedure may be not optimal for phosphoprotein analysis. For tissue placed directly in formalin, the standard procedure for the past 100 years, the formalin penetration rate is 0.1 mm/hr, so the cellular molecules in the depth of the tissue will have significantly degraded by the time formalin permeates the tissue. Formalin cross-linking, the formation of methylene bridges between amide

groups of protein, blocks analyte epitopes, as well as decreases the yield of proteins extracted from the tissue. Since the dimensions of the tissue and the depth of the block that are samples are unknown variables, formalin fixation would be expected to cause significant variability in protein and phosphoprotein stability for molecular diagnostics. Phosphorylation and dephosphorylation of structural and regulatory proteins are major intracellular control mechanisms. Protein kinases transfer a phosphate from ATP to a specific protein, typically via serine, threonine, or tyrosine residues. Phosphatases remove the phosphoryl group and restore the protein to its original dephosphorylated state. Hence, the phosphorylation-dephosphorylation cycle can be regarded as a molecular on-off switch. At any point in time within the cellular microenvironment, the phosphorylated state of a protein is a function of the local stoichiometry of associated kinases and phosphatases specific for the phosphorylated residue. During the *ex vivo* time period, if the cell remains alive, it is conceivable that phosphorylation of certain proteins may transiently change. A variety of chemical-based and protein-based inhibitors of phosphatases exist. Thus, there is adequate chemistry knowlege to design rational stabilizers for the preservation of phosphoprotein stability without freezing.

SERUM PROTEOMICS: AN EMERGING LANDSCAPE FOR EARLY STAGE CANCER DETECTION

The recognition that cancer is a product of the proteomic tissue microenvironment and involves communication networks has important implications. First, it shifts the emphasis away from therapeutic targets being directed solely against individual molecules within pathways and focuses the effort on targeting nodes in multiple pathways inside and outside the cancer cell that cooperate to orchestrate the malignant phenotype. Second, the tumor-host communication system may involve unique enzymatic events and sharing of growth factors. Consequently, the microenvironment of the tumor-host interaction could be a source for biomarkers that could ultimately be shed into the serum proteome.

Application of Serum Proteomics to Early Diagnosis

Cancer is too often diagnosed and treated too late, when the tumor cells have already invaded and metastasized. At this stage, therapeutic modalities are limited in their success. Detecting cancers at their earliest stages, even in the premalignant state, means that current or future treatment modalities might have a higher likelihood of a true cure. For example, ovarian cancer is usually treated when it is at an advanced stage. The resulting 5-year survival rate is 35% to 40% for patients with late-stage disease who receive the best possible surgical and chemotherapeutic intervention. By contrast, if ovarian cancer is detected at an early stage conventional therapy produces a high rate (95%) of 5-year survival. Thus, early detection, by itself, could have a profound effect on the successful treatment of this disease. A clinically useful biomarker should be measurable in a readily accessible body fluid, such as serum, urine, or saliva. Clinical proteomic methods are especially well suited to discovering such biomarkers. Serum or plasma has been the preferred medium for discovery because this fluid is a protein-rich information reservoir that contains the traces of what has been encountered by the blood during its constant perfusion and percolation throughout the body. Biomarker discovery is moving away from the idealized single cancer specific biomarker. Despite decades of effort, single biomarkers have not been found that can reach an acceptable level of specificity and sensitivity required for routine clinical use for the detection or monitoring of the most common cancers. Most investigators believe that this is due to the patient-to-patient molecular heterogeneity of tumors. Taking a cue from gene arrays, the hope is that panels of tens to hundreds of protein and peptide markers may transcend the heterogeneity to generate a higher level of diagnostic specificity. The overall specificity and sensitivity required for clinical use depends entirely on the intended use of the marker(s). Markers for general population screening for rare diseases may have to approach 100% specificity to be accepted. On the other hand, markers that are used for high-risk screening or for relapse monitoring can have much lower specificity but require high sensitivity.

The low molecular weight (LMW) range of the serum proteome (generally defined here as peptides less than 50,000 Daltons) is called the peptidome due to the abundance of protein peptides and fragments. While some dismissed the peptidome as noise, others have proposed that just the opposite is the case—it may contain a rich, untapped source of disease-specific diagnostic information. Tissue proteins that are normally too large to passively diffuse through the endothelium into the circulation can still be represented as fragments of the parent molecule.

Actually, the early studies revealed an apparent abundance of disease-specific information in the circulatory peptidome of patients and the disease-specific ions appear to be fragments of large molecules either from endogenous high-abundant proteins or low abundance cellular and tissue proteins. The appreciation of a fragment peptide as a new analyte isoform in and of itself has now given way to recent general optimism about the diagnostic potential of the peptidome.

The Peptidome: A Recording of the Tissue Microenvironment

Cancer is a product of the tissue microenvironment (Figure 9.2). While normal cellular processes (and the peptide content generated by these processes) are also a manifestation of the tissue microenvironment, the tumor microenvironment, through the process of aberrant cell growth, cellular invasion, and altered immune system function, represents a unique constellation of enzymatic (e.g., kinases, phosphatases) and protease activity (e.g., matrix metalloproteases), resulting in changed stoichiometry of molecules within the peptidome itself compared to the normal milieu. An added benefit for cancer biomarkers is the leaky nature of newly formed blood vessels and the increased hydrostatic pressure within tumors. This pathological physiology

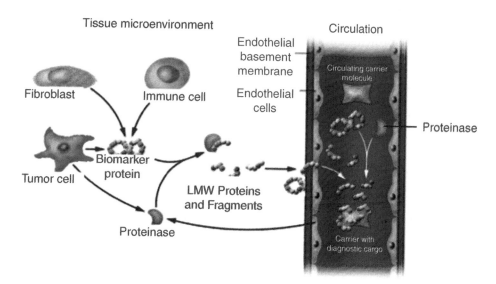

Biomarker Cascades Generated in the Tissue Microenvironment

Figure 9.2   The peptidome hypothesis: Circulating peptides and protein fragments are shed from all cell types in the tissue microenvironment. Proteolytic cascades within the tissue generate fragments that diffuse into the circulation. The identity and cleavage pattern of the peptides provide two dimensions of diagnostic information.

would tend to push molecules from the tumor interstitium into the circulation (Figure 9.2). As cells die within the microenvironment, they will shed the degraded products. The mode of death, apoptosis versus necrosis, would be expected to generate different classes of degraded cellular constituents. As a consequence, the blood peptidome may reflect ongoing recordings of the molecular cascade of communication taking place in the tissue microenvironment (Figure 9.2). Combinations of peptidome markers representing the specific interactions of the tumor tissue microenvironment at the enzymatic level can achieve a higher specificity and a higher sensitivity for early stage cancers. This optimism is in part based on the concept that the biomarkers are derived from a population of cells that comprise a volume that is greater than just the small precancerous lesion itself. In this way, the peptidome can potentially supersede individual single biomarkers and transcend the issues of tumor and population heterogeneity.

Physiologic Roadblocks to Biomarker Discovery

Candidate biomarkers are expected to exist in very low concentration, have the potential to be rapidly excreted, and must be separated from high abundance blood proteins, such as albumin, which exist in a billion-fold excess. Early stage disease lesions such as premalignant cancer may arise within a tissue volume less than 0.10 mL. Assuming all the putative biomarkers emanating from this volume are uniformly dispersed within the entire blood volume of 5000 mL, then the dilution factor will be 50,000. We can also reasonably hypothesize that the most physiologically relevant proteins specific for the disease constitute a minor subpopulation of the cellular proteome. Consequently, the greatest challenge to biomarker discovery is the isolation of very rare candidate proteins within a highly concentrated complex mixture of blood proteins massively dominated by seemingly nonrelevant proteins. Because of the low abundance of the biomarkers, analytical sensitivity is the first challenge for biomarker discovery and also routine measurement. During the discovery phase, it is likely that large plasma or serum volumes, including pooled samples, can be available for analysis. In contrast, once a candidate marker is taken forward to clinical testing, the volume of blood available for an individual patient's assay may be less than one mL. When one takes all of these factors into consideration, the analytical platform used to measure the candidate marker must have a detection sensitivity sufficient to reliably detect marker concentrations in the subfemtomolar or attomolar concentration.

Requirement for New Classes of Diagnostic Technology

Cancer biomarkers of the future are predicted to emerge from multiplexed measurement of patterns of specific size fragments of known parent molecules. This will require new generations of immunoassay-based technology that can determine both the identity and exact size of the biomarker. Conventional immunoassay platforms such as antibody arrays and bead capture arrays cannot effectively measure panels of fragment analytes. The reason is that immunoassays, by their very definition, rely on antibody-based capture and detection. An antibody-based assay cannot distinguish the parent molecule from its clipped fragments since the antibody recognizes its cognate epitope in both the parent and fragment molecule. Immuno-MS is one possibility: using this two-dimensional immunoassay technology, based on the amino acid sequence of the peptide fragment, a miniaturized affinity antibody column, perhaps in a multiplexed microwell format, is first used to capture all species of molecules that contain the antibody recognition site. Next, the captured fragments are eluted off the antibody column directly into an MS (such as a MALDI-TOF), which can provide an extremely accurate mass determination of the entire population of captured peptides. Thus, in only two automated steps, a panel of peptide fragments derived from a known parent molecule can be rapidly sorted and tabulated. The result is an immediate read-out of identities and specific fragment sizes of a given biomarker candidate, and both dimensions of information are captured concomitantly.

Reduction of Bias in the Discovery Phase of Peptide Biomarkers

The composition of the circulation is a true mirror of ongoing cellular and organ system function. While this means that subsets of the blood peptidome can potentially reflect subtle disease events in small tissue volume, it also means that the peptidome is constantly fluctuating due to ongoing daily physiologic events. Epidemiologists and clinical chemists fear that the level of individual blood-borne biomarkers can be greatly influenced by a variety of non-disease-related epidemiologic factors and normal physiologic conditions. Thus, the promise of the specificity of the peptidome is counterbalanced by the sensitivity to interfering factors. Consequently, as with any biomarker discovery (single, or as a panel), great care is needed to reduce sample bias during the discovery and validation phase of peptidome biomarker translational research.

Methods for Discovering and Validating Candidate Protein Biomarkers

Researchers can choose from a series of separation, chromatography, electrophoresis, and mass spectrometry-based methodologies useful for discovering the low molecular weight peptidome. Each methodology has advantages and disadvantages. Mass spectrometry profiling technology is useful for rapidly obtaining an ion fingerprint of a test fluid sample. SELDI-TOF is a MALDI-TOF-related research system that is relatively sensitive compared to other MS and can rapidly read out peptide ion signatures derived from a small sample volume. The disadvantages of SELDI are that the resolution of the instrumentation is usually quite low (many peptides can exist within each ion peak) and the ion peaks of interest cannot be directly identified by mass spectrometry-based sequence determination. However, SELDI-TOF, as with other more common MALDI and ESI instruments, cannot routinely detect proteins in the low abundance range (<4 pg/ml range), which almost all immunoassays can reach with ease. An intermediate system is represented by a solid particle or bead capture systems. These methods are used to harvest peptides by fairly nonspecific hydrophobic binding. The harvested peptides eluted from the beads can be profiled by MALDI or ESI MS, followed by MS sequencing and identification of abundant peaks. Native carrier proteins such as albumin constitute an endogenous resident *in vivo* affinity chromatography system for harvesting peptides. Carrier protein harvesting is a facile method for obtaining high-resolution ion profiles. The disadvantages of the bead and carrier protein capture systems are low throughput and minimal prefractionation.

Various methods exist to prefractionate a complex LMW peptidome mixture prior to sequencing. However, given the new finding that a great deal of LMW information appears to exist prebound to albumin and other high-abundant proteins, and that many investigations for biomarker discovery begin with depleting the blood of these high-abundance proteins, a great deal of caution is warranted in using this approach for LMW candidate discovery. An alternate approach would be to denature the input material and dissociate the LMW information from the carrier molecules, and then separate, isolate, and enrich for the LMW archive. Thus, a researcher planning an LMW peptidome biomarker discovery project must carefully select the proper sample preparation method tailored to the volume of the available samples, the resolution desired, the required throughput, and the need for identification of the candidate biomarker peptides.

A recommended staging process for biomarker discovery and validation is described above. The starting point for rigorous analysis is the development of a discovery study set consisting of a population of serum or plasma samples from patients who have (i) histologically verified cancer; (ii) benign or inflammatory non-neoplastic disease; and (iii) unaffected, apparently healthy controls or hospital controls, depending on the intended use. The issue of including specimens for initial discovery from patients with no evidence of cancer but with inflammatory conditions, reactive disease, and benign disorders is of critical importance to ensure that specific markers are enriched for from the outset. This issue is critical for cancer research, especially since the disease almost always occurs in the background of inflammatory processes that are part of the disease pathogenesis itself. The peptidome, as a mirror of the ongoing physiology of the entire individual, may be especially sensitive to these

processes, which is why care must be taken to at least minimize the chance that nonspecific markers are selected. In fact, the reduction of the potential upfront bias is critical prior to undertaking the discovery phase of the research.

The next stage is LMW peptidome fractionation, separation, isolation and enrichment, concentration, and mass spectrometry-based identification. At this stage it is essential that iterative and repetitive MS-based analysis and MS/MS sequencing be conducted on each sample. Candidate peptides identified repetitively over many iterations within a sample and within a study set have a higher likelihood of being correct. The researcher ends up with a list of candidate diagnostic markers that are judged to be differentially abundant in the cancer versus the control populations. The next step is to find or make specific antibodies or other ligands for each candidate peptide marker. After each antibody is validated for specificity using a reference analyte, the antibody can then be used to validate the existence of the predicted peptide marker in the disease and nondiseased discovery set samples. Multiple Reaction Monitoring (MRM) is an emerging generation of MS technology which offers the potential to quantify and identify peptides with such high confidence that antibody validation may not be required. The goal is to develop a panel of candidate peptide biomarkers along with measurement reagents that are independent of the analytical technology that will ultimately be used in the clinical lab. Clinical validation of the candidate biomarkers starts with ensuring the sensitivity and precision of the measurement platform. Once the measurement platform is proven to be reliable and reproducible, then the clinical validation can proceed.

The final and most critical stage of research clinical validation is blinded testing of the biomarker panel using independent (not used in discovery), large clinical study sets that are ideally drawn from at least three geographically separate locations. The required size of these test sets for adequate statistical powering depends on both the performance of the peptide analyte panel in the platform validation phase and the intended use of the analyte in the clinic. It is important to emphasize that sensitivity and specificity in an experimental test population does not translate to the positive predictive value that would be seen if the putative test is used routinely in the clinic. The true positive predictive value is a function of the indicated use and the prevalence of the cancer (or other disease condition within the target population). The percentage of expected cancer cases in a population of patients at high genetic risk for cancer is higher than the general population. Consequently, the probability of false positives in the latter population would be much higher. Finally, the information content of the peptidome will never be fully realized unless blood collection protocols and reference sets are standardized, new instrumentation for measuring panels of specific fragments are proven to be reproducible and sensitive, and extensive clinical trial validation is conducted under full CAP/CLIA regulatory guidelines.

Frontiers of Nanotechnology and Medicine

Nanotechnology will have a significant impact on early diagnosis and targeted drug delivery. The development of inorganic nanoparticles that bind specific tumor markers that exist at very low concentrations in serum may be able to be used as serum harvesting agents. In the future, patients may be injected with such nanoparticles that seek out and bind tumor or disease markers of interest. Once the nanoparticles have bound their targets, they can be "harvested" from the serum to enable diagnosis or to monitor disease progression.

Some investigators have created core shell hydrogel nanoparticles to address these fundamental roadblocks to biomarker purification and preservation. These smart nanoparticles conduct enrichment and encapsulation of selected classes of proteins and peptides from complex mixtures of biomolecules such as plasma, purify them away from endogenous high-abundance proteins such as albumin, and protect them from degradation during subsequent sample handling. Particles incubated with serum trap the target analytes and are isolated by centrifugation. Candidate biomarkers are then released from particles by means of elution buffers. The ratio of the volume of elution buffer to the original starting solution establishes the concentration amplification factor. This concentration step is a fundamental point for biomarker discovery and measurement because it provides a means to effectively raise the concentration of rare biomarkers that become the input for a measurement system such as an immunoassay platform or mass spectrometry. With the synthesis of peptide carriers designed to transport drug therapies to specific vascular beds, treatments can be designed to treat specific tissues or tumor sites, leaving nonmalignant tissues unaffected. The use of homing nanoparticles, or semiconductor quantum dots, may be an important part of imaging diagnostics, in addition to drug delivery, in the health care of the future.

Future of Cancer Clinical Proteomics

The pathologist of the future will detect early manifestations of disease using proteomic patterns of body-fluid samples and will provide the primary physician a diagnosis based on proteomic signal pathway network signatures as a complement to histopathology. He or she will be able to dissect a patient's individual tumor molecularly, identifying the specific regulatory pathways that are deranged in the cell cycle, differentiation, apoptosis, and invasion and metastasis. Based on this knowledge, recommendations will be made for an individualized selection of therapeutic combinations of molecularly targeted agents that best strike the entire disease-specific protein network of the tumor. The pathologist and the diagnostic imaging physician will assist the clinical team to perform real-time assessment of therapeutic efficacy and toxicity. Proteomic and genomic analysis of recurrent tumor lesions could be the basis for rational redirection of therapy because it could reveal changes in the diseased protein network that are associated with drug resistance. The paradigm shift will directly affect clinical practice, as it has an impact on all of the crucial elements of patient care and management.

KEY CONCEPTS

- Microdissection, the extraction of a microscopic homogeneous cellular subpopulation from its complex native tissue environment, is necessary to a reliable molecular analysis of critical gene expression and protein patterns.
- Protein complexity and versatility stem from context-dependent posttranslational processes. Nucleic acid profiling (including microRNA) does not provide information about how proteins link together into networks and functional machines in the cell. In fact, the activation of a protein signal pathway, causing a cell to migrate, die, or initiate division, can immediately take place before any changes occur in DNA/RNA gene expression.
- Cancer is a product of the tissue microenvironment. Although normal cellular processes are also a manifestation of the tissue microenvironment, the tumor microenvironment represents a unique constellation of enzymatic and protease activity compared to the normal milieu.
- There is increasing evidence demonstrating the promise and potential of combination therapies combining conventional treatments such as chemotherapy or radiotherapy and molecular-targeted therapeutics.
- The promise of tissue protein biomarkers to provide revolutionary diagnostic and therapeutic information will never be realized unless the problem of tissue protein biomarker instability is recognized, studied, and solved.

SUGGESTED READINGS

1. Erickson HS, Gillespie JW, Emmert-Buck MR. Tissue microdissection. *Methods Mol Biol.* 2008;424:433–448.
2. Espina V, Edmiston KH, Heiby M, et al. A portrait of tissue phosphoprotein stability in the clinical tissue procurement process. *Mol Cell Proteomics.* 2008;7(10):1998–2018.
3. Espina V, Wulfkuhle JD, Calvert VS, et al. Laser capture microdissection. *Nature Protocols.* 2006;1:586–603.
4. Flaherty KT. The future of tyrosine kinase inhibitors: Single agent or combination. *Curr Oncol Rep.* 2008;10(3):264–270.
5. Gil PR, Parak WJ. Composite nanoparticles take aim at cancer. *ACS Nano.* 2008;2(11):2200–2205.
6. Hait WN, Hambley TW. Targeted cancer therapeutics. *Cancer Res.* 2009;69(4):1263–1267.
7. Issaq H, Veenstra T. Two-dimensional polyacrylamide gel electrophoresis (2D-PAGE): Advances and perspectives. *Biotechniques.* 2008;44(5):698–700.
8. Joyce JA, Pollard JW. Microenvironmental regulation of metastasis. *Nat Rev Cancer.* 2009;9(4):239–252.
9. Kiehntopf M, Siegmund R, Deufel T. Use of SELDI-TOF mass spectrometry for identification of new biomarkers: Potential and limitations. *Clin Chem Lab Med.* 2007;45(11):1435–1449.

10. Kulasingam V, Diamandis EP. Strategies for discovering novel cancer biomarkers through utilization of emerging technologies. *Nat Clin Pract Oncol.* 2008;5(10):588–599.

11. Luchini A, Geho DH, Bishop B, et al. Smart hydrogel particles: Biomarker harvesting: One-step affinity purification, size exclusion, and protection against degradation. *Nano Lett.* 2008; 8:350–361.

12. Mbeunkui F, Johann Jr DJ. Cancer and the tumor microenvironment: A review of an essential relationship. *Cancer Chemother Pharmacol.* 2009;63(4):571–582.

13. Overdevest JB, Theodorescu D, Lee JK. Utilizing the molecular gateway: The path to personalized cancer management. *Clin Chem.* 2009;55(4):684–697.

14. Petricoin EF, Belluco C, Araujo RP, et al. The blood peptidome: A higher dimension of information content for cancer biomarker discovery. *Nat Rev Cancer.* 2006;6(12):961–967.

15. Preisinger C, von Kriegsheim A, Matallanas D, et al. Proteomics and phosphoproteomics for the mapping of cellular signalling networks. *Proteomics.* 2008;8(21):4402–4415.

16. Speer R, Wulfkuhle J, Espina V, et al. Molecular network analysis using reverse phase protein microarrays for patient tailored therapy. *Adv Exp Med Biol.* 2008;610:177–186.

17. VanMeter A, Signore M, Pierobon M, et al. Reverse-phase protein microarrays: Application to biomarker discovery and translational medicine. *Expert Rev Mol Diagn.* 2007;7(5):625–633.

18. Villanueva J, Shaffer DR, Philip J, et al. Differential exoprotease activities confer tumor-specific serum peptidome patterns. *J Clin Invest.* 2006;116:271–284.

19. Yap TA, Carden CP, Kaye SB. Beyond chemotherapy: Targeted therapies in ovarian cancer. *Nat Rev Cancer.* 2009;9(3):167–181.

20. Zhou M, Veenstra T. Mass spectrometry: m/z 1983–2008. *Biotechniques.* 2008;44(5):667–668, 670.

Integrative Systems Biology: Implications for the Understanding of Human Disease

M. Michael Barmada . David C. Whitcomb

INTRODUCTION

Networks have recently generated considerable interest in biological sciences, not because of any change in our basic knowledge about them, but because of key technological changes that have enhanced our ability to interrogate them and thereby develop descriptive models allowing prediction of their behavior. This shift in ability has fueled new perspectives on how to apply the scientific method to biomedical science, emphasizing integration instead of reduction. Systems (or network) biology focuses on the investigation of complex interactions in biological systems. Much of the promise of systems biology in biomedicine revolves around its potential to explain disease as a perturbation of a normal system. Many disease syndromes include elements from multiple tissue-specific and systemic systems. For example, chronic pancreatitis is defined by variable amounts of maldigestion (loss of pancreatic acinar cell function), diminished bicarbonate secretion (loss of duct cell function), diabetes mellitus (loss of islet cell function), inflammation (immune system), fibrosis (regenerative systems), pain (nervous system), and cancer risk (DNA repair and cell-cycle regulation systems). Since genetic and environmental factors affect each of these systems in different ways, it is not surprising that dysfunction of these systems in response to stress or injury will be variable between patients. However, it is also clear that each of these systems interacts with each other. Careful modeling of these different relationships within the phenome (the set of all phenotypes of an organism) or within the interactome (the set of all interactions of an organism) can greatly increase the power

to identify functional genomic variation important in disease phenotypes and lead to a better understanding of how to readjust the system to return to a normal state.

At its most basic, systems biology comprises three challenges: (i) how to generate a sufficient quantity of data to analyze variability in networks, (ii) how to properly integrate data from multiple disparate sources into a usable corpus of knowledge, and (iii) how to use that corpus of knowledge to model a component system and optimize it. Model construction allows predictions of the behavior of a system and, ultimately, an understanding of how to change that behavior in predictable ways. The promise of this approach is vast because the exact definition of a system is not fixed—anything from a particular biochemical pathway up to the level of an entire organism can be considered a system in biomedical applications. In this regard, population-level biosciences, such as epidemiology and population genetics, have long practiced a form of systems biology, but their focus has always been on a collection of individuals as the system of interest. Current systems biology approaches model systems of interest to clinical outcomes, such as metabolic pathways, individual cells, and organs. Modeling at this level has the appeal of allowing translation of results from decades of population-based laboratory research to be applied to individualized clinical medicine.

The field of systems biology borrows from control theory, which deals with the behavior of dynamic systems (systems that fluctuate in a dependent fashion). Control theory proposes the idea that systems can best be modeled as a cycle of controllers which modify inputs to produce the desired outputs, and sensors which provide feedback to the system, producing a steady-state system in which equilibrium is achieved by balancing the input and output concentrations and the feedback signals. In a similar fashion, systems biology includes the definition and measurement of the components of a system, formulation of a model, and the systematic perturbation (either genetically or environmentally) of and remeasurement of the system. The experimentally observed responses are then compared with those predicted by the model, and new perturbation experiments are designed and performed to distinguish between multiple or competing models. This cycle of test-model-retest is repeated until the final model predicts the reaction of the system under a broad range of perturbations.

Systems Biology as a Paradigm Shift

Although systems biology is itself a new discipline, the study of systems in biology is not new. Early studies of enzyme kinetics employed a similar cycle of testing and modeling, followed by retesting of new hypotheses. In a similar fashion, modeling of neurophysiological processes like the propagation of action potentials along a neuron are illustrative of early forays into mathematical modeling of cellular processes. However, like all next-generation methodological shifts, systems biology involves a rethinking of basic principles—a return to a more basic understanding or a more simplistic framework in which to generate hypotheses. Traditional science follows a bottom-up paradigm—that is, break a problem down into its individual components (reductionism), learn everything there is to know about those individual components, and then integrate the information together to get information about a system. Systems biology can work within this paradigm, creating models for individual components, and then integrating the models (creating, in essence, models of models, to explain the behavior of the original system). However, this approach requires that knowledge of the individual components is sufficient to explain the behavior of the system. Apart from a few exceptions, current reductionist research models have not been successful at explaining the large-scale behavior of biological systems. To this end, systems biology endorses a more top-down approach, allowing modeling to occur at a higher level of organization (not at the level of the components, but at the level of the organ or the individual). In this point of view, a better understanding of the system as a whole can create a better understanding of the components. "Systems biology ... is about putting together rather than taking apart, integration rather than reduction. It requires that we develop ways of thinking about integration that are as rigorous as our reductionist programmes, but different." Or, put more simply: "You [can] study each part of a Boeing 777 aircraft and describe how it functions, but that still wouldn't tell you how the airplane flies." This has led to the recognition that new paradigms may be necessary to understand and model systems with multiple interacting quantitative components.

DATA GENERATION

Central to the success of any modeling endeavor is the generation of large quantities of data. Because of the increased interest in systems biology, the theme of the late 1990s and early 21st century reflected exactly this necessity, exemplified by the fact that several papers were published on methods of dealing with the so-called "data deluge." Genomic technologies were the first to enter the high-throughput realm and subsequently the first to produce large quantities of high-quality data. Coupled with a concurrent explosion in computing power, high-throughput data generation made realistic modeling feasible. New technologies produced during this time vastly increased the ability to query an entire system at once and led to the advent of the major international efforts of the early 21st century, such as the Human Genome Project (the effort to sequence the entire complement of human chromosomes). The HapMap Project (the effort to catalog common genetic variation across multiple human ethnic populations), the ENCODE Project—(an effort to resequence particular portions of the human genome in multiple individuals from different ethnic populations to explore common and rare genetic variation), and the 1000 Genomes Project (an effort to completely sequence 1000 human genomes). Specifically, microarray technology coupled with advances in mass spectroscopy, combinatorial chemistry, and robotics created an explosion of data and unique visualizations of cellular processes.

Microarrays

Microarray technology grew out of a confluence of trends and technologies. On the experimental level, microarrays are most similar to Southern blotting, where fragmented DNA is attached to a substrate and then probed (hybridized) with a known gene or fragment to identify complementary sequences. The trend through the 1990s was to increase capacity in individual experiments, allowing the analysis of more and more samples (or more and more markers) in a single experiment. With the advent of paper-blotting techniques (allowing DNA molecules to be immobilized or spotted on special paper, then hybridized), and subsequently glass-blotting techniques, coupled with advances in robotic pipetting (allowing smaller and smaller volumes of liquid to be spotted, in greater and greater densities), microarrays became feasible. Initial microarray experiments began with small numbers of immobilized probes, owing mostly to limited knowledge about the genes that make up a genome. This was quickly followed by arrays with hundreds, thousands, tens of thousands, and now (currently) millions of probes as our understanding of the components of genomes grew (owing, in large part, to early efforts to identify and catalog all expressed genes in organisms and to large-scale efforts like the Human Genome and ENCODE projects).

Microarrays have now found use in multiple experimental venues. Most commonly, the molecules being immobilized on the array are DNA molecules, allowing measurement of expression levels or detection of polymorphisms (such as single-nucleotide polymorphisms or SNPs). A DNA microarray consists of thousands of microscopic spots (or features), each containing a specific DNA sequence. Each feature is typically labeled with a fluorescent tag and then used as probes in hybridization experiments with a DNA or RNA sample (the target), which is labeled with a complementary fluorescent label. Hybridization is quantified by fluorescence-based scanning of the array, allowing determination of the relative abundance of nucleic acid sequences in the target by characterization of the relative abundance of each fluorophore (Figure 10.1).

Transcriptomics

Transcriptomics is the study of relative RNA transcript abundances, using microarray technologies. Chips that are specialized for this purpose are known as RNA microarrays and are typically prepared with a library of transcripts of known origin (representative tags for a known complement of genes, for example—the current human RNA arrays contain approximately 60,000 probes, representative of a majority of known RNA species from the 20,000 or so human genes). These are then interrogated with RNA (typically reverse transcribed into cDNA) from two different samples, labeled with different dyes (commonly green and red dyes). This allows the relative abundance (in one sample versus the other) of each RNA transcript to be assessed. Experiments of this type are routinely used to determine which RNAs are upregulated or downregulated in a disease sample versus a normal sample. However, note that the utility of the information generated from transcriptomics studies is highly variable, as transcription levels are potentially influenced by a wide range of factors, including disease phenotypes. To be of general use, transcriptomic data must be generated under a wide variety of conditions and compared, to eliminate the trivial sources of variation.

Genotyping

One of the earliest explosions of data on a whole-genome scale came from early genetic linkage studies. The first whole-genome genetic linkage scans were performed in the early 1990s using panels of ~350 genetic markers scattered throughout the autosomal chromosomes, as well as the X-chromosome. Subsequent advances in mapping techniques and marker discovery increased the number of markers in the genetic maps, and included both sex chromosomes, as well as the mitochondrial genome. Current genotyping arrays now contain a mixture of simple polymorphic markers (SNPs) and so-called copy-number probes, allowing the assessment of copy-number changes across the genome, at a density which provides excellent coverage for most ethnic groups (~1,000,000 SNPs and nearly that many copy number probes, meaning each array carries almost 2 million features). It is expected that chip densities will continue to grow, allowing for higher-throughput genotyping. However, at the same time rapid low-cost sequencing technologies are becoming available that may well allow for affordable whole-genome sequence-based assessment of genomic variation, replacing any need for increased array densities.

Other Omic Disciplines

Recognizing that many properties of biological systems are emergent, or a result of complex interactions between components that are not understandable at the level of individual system components, scientists in the 21st century turned to analysis of different levels of organization in an organism. Each level of organization had to be given an appropriate name. For example, since the genome is the entire set of genes of an organism, genomics is the name given to the study of the whole set of genes of a biological system. Various methods, such as genotyping, sequencing (examining the linear sequence of DNA) and transcriptomics (examining the expression of all genes in an organism) can be used to interrogate the genome. Likewise, given that the proteome is the collection of proteins expressed in a system, proteomics is the name given to the study of the proteome. Proteomic methods include traditional techniques such as mass spectroscopy—identifying compounds based on the mass-charge ratio of ionized particles. But just as microarrays are a modification of traditional genomic techniques for high-throughput experiments, proteomics modifies the techniques of mass spectroscopy in a similar fashion—creating experimental and informatics pipelines that allow a sample (like a blood draw) to be partitioned into various fractions, have those fractions examined via mass spectroscopy, and have the results of the mass spectroscopy experiments compared to other mass

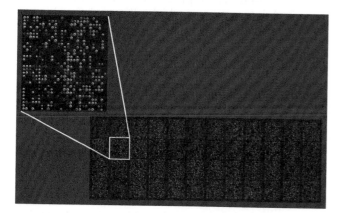

Figure 10.1   Example of an approximately 40,000 probe spotted oligonucleotide microarray with enlarged inset to show detail (Image from Wikimedia Commons).

spectroscopy profiles to allow for identification of individual components.

Other omic disciplines are similarly named—the general rule being that the respective discipline arises from the study of the associated "ome" (or set). This gives rise to disciplines such as metabolomics—the study of the entire range of metabolites taking part in a biological process; interactomics—the study of the complete set of interactions between proteins or between these and other molecules; localizomics—the study of the localization of transcripts, proteins, and other molecules; and phenomics—the study of the complete set of phenotypes of a given organism. These various "omes" can be organized hierarchically, as demonstrated in Figure 10.2, based on the relationships recognized from years of biological experimentation. As with genomics and proteomics, experimental procedures for these other "omics" disciplines are adaptations of traditional methods (such as microscopy for localizomics, or single metabolite measurements for metabolomics) for high-throughput protocols.

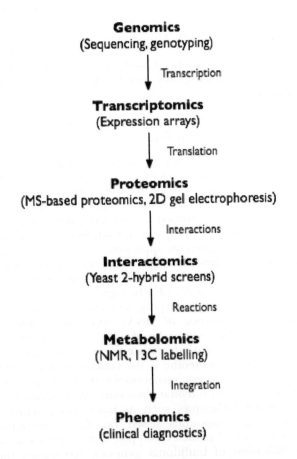

Figure 10.2   Omics technologies gather data on numerous levels. Various "omics" fields are displayed here along with the laboratory techniques used to generate the data, as well as the relationship of data from one level to another. Adapted from Fischer HP. Towards quantitative biology: Integration of biological information to elucidate disease pathways and drug discovery. *Biotechnol Annu Rev.* 2005;11:1–68.

DATA INTEGRATION

While data generation is certainly of paramount importance for systems biology, because the technologies that generate the data have their own unique (and often proprietary) formats for storing the data, systems biologists must also concern themselves with integrating data that span multiple resources. Since 1998, the number of recognized online databases related to biological information has increased 10-fold (from just under 100 in 1998 to over 1000 in 2008). Resources like the Bioinformatics Links Directory (http://bioinformatics.ca/links_directory/) extend this further by collecting links to molecular resources and tools as well as databases, and currently list 2300 links. However, the problem is more than just quantity. Issues of data quality, lack of standards, lack of interfaces allowing for integration, and longevity (or lack thereof) continue to plague online biological data resources, and make cross-resource querying or integration difficult. Nonetheless, tools continue to appear to assist in the integration of data from disparate sources. Using technology adopted from the burgeoning Semantic Web (or Web 2.0) projects, biologists and bioinformaticists are able to capture instances of data sufficient for systems biology scale efforts.

Semantic Web Technologies

Many of the problems inherent to integration of biological data resources are similar to those being faced by the larger community of World Wide Web users. The Semantic Web is a vision for how to have computers infer information relating one web page (or element) to another. It is an extension of the current web protocols (primarily the HyperText Transport Protocol, or HTTP), which allows for meaning to be imbedded together with content, such that automated agents can make associations between data without needing user input. In essence, creating the Semantic Web involves recasting the information in the World Wide Web (which currently stores relationships in the form of hyperlinks, which can link anything to anything, and so do not represent information content or meaning) into a format which allows relationships to be represented. The eXtensible Markup Language (or XML) is an early example of this type of recasting—allowing content to be stored with representative tags that describe the content. Resource Description Framework (or RDF) builds on XML by using triples (subject-predicate-object) to represent the information in XML tags (or in hyperlinks), and defining a standard set (or schema) of RDF triples to describe a particular object. Each part of a triple names a resource using either a Uniform Resource Identifier (URI) or Uniform Resource Locator (URL), or a literal. The advantage of this format is that RDF schemas are predefined so that meaning can be imbedded in the definition, or by using a hierarchy or ontology, describing the relationship within and between schemas. Also, since the RDF triple can contain a location as well as attribute-value pair, the components of a schema do not need to be located in the same place. Thus, if we have a schema representing a web page in which the sections (header, footer, left panel, right panel, title, and others) are defined by RDF triples, these web pages can easily integrate data from multiple sources.

The representation of information by RDF allows the location of data and data resources to be independent of the location of user interfaces or analytic resources. In essence, this allows the integration of data from multiple repositories and the querying of the integrated data. Coupled with the concept of RDF schemas to accurately describe knowledge about objects and ontologies to organize the relationship of objects to one another, the use of RDF can solve several of the problems mentioned for data integration (namely, the lack of standards and the lack of interfaces for integration). Another current trend—the use of wikis (a website that allows collaborative editing of its content by its users)—addresses the problem of data quality by allowing users to mark data as reliable or not, or by allowing users to update out-of-date or incorrect data.

MODELING SYSTEMS

Once the appropriate types of data have been generated, and the various sources of data collected and integrated, the resulting information is turned into knowledge by interpreting what the data actually mean, and how they address questions that need to be answered. Data modeling is used to understand the relationships important in defining the system. As noted previously, systems biology draws heavily from control theory, which itself is derived from mathematical modeling of physical systems. Although the mathematical derivation of control problems is complex, the fundamental concepts governing the formulation of control models are more intuitive and relatively limited. Three basic concepts can be thought of as central to forming control models: (i) the need for control (regulation or feedback), (ii) the need for fluctuation, and (iii) the need for optimization. A simple control model is presented in Figure 10.3.

Key characteristics of this model include a controller (responsible for the conversion between input and output) and sensor (responsible for determining the degree and direction of the feedback)—each of which can be the result of a single or multiple elements. An initial model of this form can often be derived from an initial data generation step, which can measure a baseline set of conditions for the system and also provide an idea of the components of the system. For example, an expression network (groups of genes that are co-regulated in some fashion) that regulates a biochemical pathway can be thought of as a control model. A single transcriptomics experiment can give you information on genes that are potentially co-regulated (sets of genes that are overexpressed compared to control, and so which might be providing the function of a controller), as well as information about potential sensors (genes that differ in response—one being overexpressed, the other underexpressed). However, the problem with a single experiment (or a single snapshot of the transcriptome) is that the information derived is not sufficient to disentangle the true positives (elements that truly should be components of the model) from the false positives (random variation which transiently mimics the behavior of the model). This reflects the need for fluctuation. By perturbing (changing an aspect of the system to produce a predictable outcome) and remeasuring the system, additional confidence regarding the true positives can be achieved assuming the new measurements reflect the predictions of the model. If not, another model which explains the previous data measurements is selected, and another test is devised. This cycle continues until the final model accurately represents the data measured in all tests. In this way, the most optimized model is selected (Figure 10.4).

Lastly, although not immediately obvious from the preceding discussions, data modeling also requires large amounts of computational power. Due to the size of the high-throughput data sets currently in use (expression data on 30,000 genes for multiple time points; SNP genotypes for millions of markers; occurrence of all known protein-protein interactions; computational prediction and annotation of all known protein-DNA binding sites; and others), the advanced mathematical and visualization frameworks currently in use, and the iterative nature of analysis, large amounts of computing power are necessary, often pushing the limits of even parallelized or grid-enabled computational clusters. Continued growth in computing power will be necessary to support the

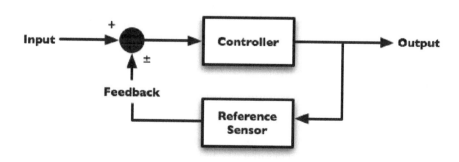

Figure 10.3   Example of a control module. This module represents a simple feedback loop, with output from the sensor either upregulating or downregulating the process that converts the input into output.

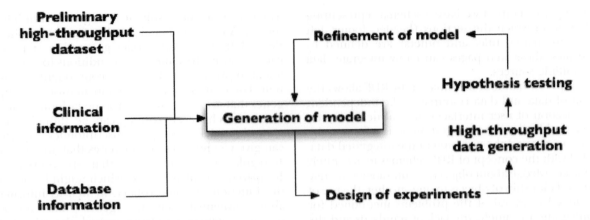

Figure 10.4   The iterative nature of systems biology research. Note that every refinement of the model needs additional data generation for retesting of specific hypotheses. Adapted from Studer SM, Kaminski N. Towards systems biology of human pulmonary fibrosis. *Proc Am Thorac Soc.* 2007;4:85–91.

ongoing use of systems biology as the models in use become more and more elaborate and incorporate information from greater numbers of high-throughput methodologies.

IMPLICATIONS FOR UNDERSTANDING DISEASE

The simplicity and early success of the germ theory of disease, in which a single pathological agent was responsible for a specific disease with characteristic signs and symptoms, lulled physicians into thinking that all disease would follow a similar pattern. This reductionist-like approach in medicine allowed physicians and scientists from different disciplines to triangulate on the same target using different methods and perspectives. It has identified numerous components of individual systems—many of which are reused over and over, leading to a modular view of system components—and has provided numerous insights into human disease. Despite these advances, the reductionist approach in medicine has been less successful in identifying the many complex interactions between disease components, and in explaining how system properties (like disease phenotypes) emerge—a concept which becomes important when considering manipulation of systems to produce predictable and desirable outcomes, as required for preventative medicine. As such, new paradigms are required, likewise requiring new approaches and new methods.

One such new paradigm is the advent of personalized medicine, which can be thought of as an application of systems biology-type thinking to medicine. Personalized medicine represents a transition from population-based thinking (describing risks on a population level) to an individual-based approach (describing risks for an individual based on his or her personal genomic/proteomic/metabolomic/phenomic profiles). Reductionist approaches have successfully identified many of the biomarkers required to define disease states

in complex disorders on a population level (that is attributing population risks to various changes), but have not been able to describe how individual exposures or biomarkers (or combinations of these components) interact to create the disease state in individuals. To achieve this transition, systemic models that explain the function of systems (cells, organs, etc.) are needed, from which the implications of changes in particular exposures or biomarkers (genetic changes, proteomic changes, and others) can be understood on the system level.

Redefining Human Diseases

The first major hurdle in transitioning from allopathic medicine to personalized medicine is to redefine common human diseases. In allopathic medicine, diseases are typically defined by characteristic signs and symptoms that occur together and meet accepted criteria. Chronic inflammatory diseases are defined by the location of the inflammation, the persistence of inflammation, and the presence of tissue destruction or scarring. These definitions indicate that the specific mechanism causing a specific organ to develop and to sustain an inflammatory reaction is not known. Furthermore, it has been impossible to identify the cause when using the traditional scientific methods used to identifying a single, causative infectious agent. Instead, the possibility that disease affecting one or more organs can occur through any one of multiple pathways, and that each pathway has multiple steps and regulatory components, and that multiple effects on multiple systems and multiple environmental exposures may be required before disease is manifest must be embraced to transition to personalized medicine. Indeed, physicians recognized that diseases located in specific organs share some common systemic features such as the type of inflammatory response (autoimmune or fibrosing) or chronic pain since they use a limited number of anti-inflammatory or pain medications for a variety of diseases. Additionally, studies that have looked at the clustering of disease phenotypes based on their representative genomics (associated gene/SNP polymorphisms or expression profiles) have demonstrated that even disease phenotypes we once thought were roughly homogeneous (for instance, disease subtypes like Crohn's disease with ileal involvement in Caucasian-only populations) are in fact associated with different genetic loci (heterogeneous). Personalized medicine must deconvolute systemic and tissue-specific pathways and reconstruct them in a way that leads to precise, patient-specific treatments.

The Transition to Personalized Medicine

The effect of approaching complex inflammatory disorders as a single disease rather than as a complex process is illustrated through the comparison of multiple small studies by meta-analysis. While some of the variance in estimated effect sizes from small genetic studies can be attributed to random chance, the possibility also exists that populations from which the samples were taken were not equivalent, and that different etiologies will lead to the same end-stage signs and symptoms through different, parallel pathways. In cases where a candidate gene is critical to some pathological pathway, but not others, the wide variance between small genetic studies may reflect the fraction of subjects that progress to disease through that specific gene-associated pathway.

Evidence of this effect was recently demonstrated in an evaluation of reports on the effect of the pancreatic secretory trypsin inhibitor gene (SPINK1) N34S polymorphism in chronic pancreatitis. A model was developed to test the hypothesis that alcohol, which is known to be associated with chronic pancreatitis and fibrosis via the collagen-producing pancreatic stellate cell (PSC), drives pancreatic fibrosis through a recurrent trypsin activation pathway as seen in hereditary pancreatitis, or through a trypsin-independent pathway, as illustrated in Figure 10.5.

We identified 24 separate genetic association studies of the effect of the SPINK N34S mutation on chronic pancreatitis with effect sizes (odds ratio, OR) reported to be between nonsignificant to ~80. Using

meta-analysis, we determined an overall effect of the SPINK1 pN34S on risk of chronic pancreatitis to be high (OR 11.00; 95% CI: 7.59–15.93), but with significant heterogeneity. Subdividing the patients into four groups based in proximal etiological factors (alcohol, tropical region, family history, and idiopathic), we found that the effect of SPINK1 pN34S on alcohol (OR 4.98, 95% CI: 3.16–7.85) was significantly smaller than idiopathic chronic pancreatitis (OR 14.97, 95% CI: 9.09–24.67) or tropical chronic pancreatitis (OR 19.15, 95% CI: 8.83–41.56). Thus, we conclude that alcohol acts through Factor A-type pathway, while tropical chronic pancreatitis and idiopathic chronic pancreatitis act through trypsin-associated pathways. The fact that a higher percentage of alcoholic patients than the control populations had SPINK1 mutations may also mean that some idiopathic patients were misdiagnosed as being alcoholics since the threshold for alcohol-associated risk was previously unknown. It could also mean that alcohol accelerates the progression from recurrent acute pancreatitis to chronic pancreatitis as demonstrated in animal models. Thus, these results are important for understanding the etiology and progression of alcoholic chronic pancreatitis, but have broader implications for understanding the presence and effects of heterogeneity of pathways in complex disorders.

The chronic pancreatitis model also illustrates a major problem for transitioning to personalized medicine. Very large association studies require the recruitment of subjects from many large medical centers in different regions and different countries. This approach will tend to obscure mechanism-based heterogeneity and converge on only the most common features of the disease (population-level effects), even though other factors may have a stronger biological effect in a limited number of patients, while being irrelevant in others. This argues for more detailed phenotyping of study populations (phenomics) and careful analysis of context-dependent effects (interactomics). The application of systems biology to this data would allow for a better understanding of the various pathways leading to disease states, and so a better understanding of the possible interactions (or context-dependent effects) that accurately predict disease in a single individual.

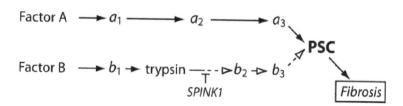

Figure 10.5 Hypothesis of etiology-defined pathways to pancreatic fibrosis. Hypothetical influence diagram illustrating pathological pathways linking proximal factor (Factor A and B) to PSC (pancreatic stellate cell) and fibrosis through multiple steps (a1, a2, a3). Etiological factors of type B activate trypsinogen to trypsin, and therefore their pathological pathway to the PSC can be interrupted by SPINK1. Etiological factors of type A are independent of trypsin, and therefore will not be influenced by variations in SPINK1 expression or function.

Applications of Systems Biology to Medicine

The realization of personalized medicine requires not only a more systems-like perspective regarding risk factors, it also requires a rethinking of existing classifications, which are largely derived from observation and reductionist approach (common observable phenotypes should be the result of common underlying factors). As an example of this rethinking, a recent study undertook the reclassification of the disease phenome (the space of all associations between disease phenotypes) using the explosion of current information on genetic disease associations. Disease-associated genes were clustered based on information about gene-gene or protein-protein interactions (from transcriptomics or interactomics studies) and superimposed with a representation of the organ system of the associated disease phenotype. The resulting network graph (Figure 10.6) demonstrates the association of different genes together in modules, many of which represent associations of proteins in macromolecular complexes (protein-protein interactions), or association of proteins in metabolic or regulatory pathways.

This view of the network of disease genes can at the same time inform us about novel associations we might not have been aware of (such as PAX6 in the ophthalmological cluster, or PTEN and KIT in the cancer cluster), and also direct us toward pathways (clusters) that might be of most benefit in understanding the network more completely (that is, those that are involved in multiple disease states or that have the most connections to them). This network-centric view of disease can also inform clinical medicine in terms of the choice of medications (indicating that medications used in one disease might also function appropriately in another, based on clustering of associated genetic factors).

DISCUSSION

We have outlined many of the reasons that a systems biology-based approach is needed for understanding complex disorders. Once the key components are organized into a logical series, then formal modeling of the disease process can be applied, tested, additional experimental data added, the system calibrated and retested. The optimal model or models are yet to be determined experimentally for any system, but with the continued rapid increase in high-throughput data generation and the continued increase in computational power, large-scale integrations and models can be achieved. The challenge will be to accurately anticipate the information necessary to be included in the model, as accurate assessment of the context is essential to the appropriate understanding of the effect of individual variation and the integration of that understanding into clinical practice. It is not enough to know that certain changes affect a phenotype (like disease risk) in a population, as these overall effects are typically very small (on the order

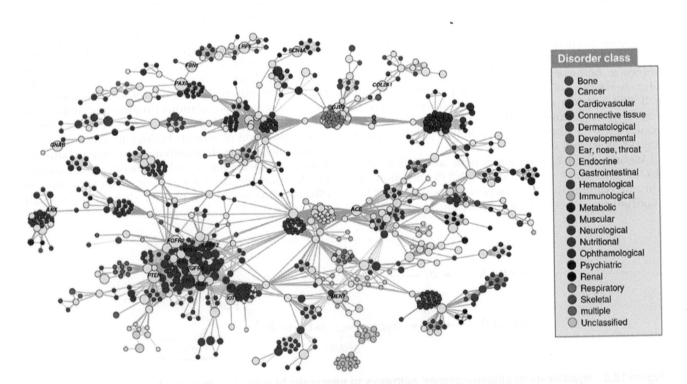

Figure 10.6 Disease gene network. Each node is a single gene, and any two genes are connected if implicated in the same disorder. In this network map, the size of each node is proportional to the number of specific disorders in which the gene is implicated. Reproduced with permission from Loscalzo J, Kohane I, Barabási AL. Human disease classification in the postgenomic era: A complex systems approach to human pathobiology. *Mol Syst Biol.* 2007;3:124.

of an odds ratio of 1.1 to 1.2). However, with a proper understanding of the context in which each variant is important, appropriate medical interventions can be implemented.

KEY CONCEPTS

- Systems biology focuses on the investigation of complex interactions (or networks) in biological systems.
- Three challenges face researchers interested in systems biology: (a) how to generate sufficient amounts of data for modeling of whole systems; (b) how to integrate data from multiple data sources; and (c) how to use integrated data to model a system and optimize the model.
- Various "omics" disciplines (genomics, proteomics, transcriptomics) have grown out of technological advances associated with the Human Genome Project and other large-scale biological projects of the early 21st century. Microarray technologies, in particular, have greatly expanded the quantity and quality of data generation for modeling purposes.
- Semantic web technologies, including XML and RDF, coupled with a greater understanding of the importance of ontologies, have provided much needed data integration capabilities.
- Control theory can be used to derive general frameworks for model construction, in which effects are organized into either regulatory (controlling) or monitoring (feedback) components, and in which derived models are tested via

a cycle of prediction-perturbation-measurement, allowing hypothesis tests to be conducted and leading to model refinements and the generation of more appropriate models.

- Implications of systems biology for medicine include the development of a better understanding of diseases and their associations with one another. As disease models are built that explain conglomerations of diseases grouped by systems-level thinking (rather than by clinical measures), individual risk factors (i.e., components of the controller or feedback components of the model) become more easily identifiable, leading to better understanding of pathological mechanisms, and better drug and intervention targets.

SUGGESTED READINGS

1. Bader S, Kühner S, Gavin AC. Interaction networks for systems biology. *FEBS Lett.* 2008;582(8):1220–1224.
2. Fischer HP. Mathematical modeling of complex biological systems. *Alcohol Res Health.* 2008;31(1):49–59.
3. Loscalzo J, Kohane I, Barabási AL. Human disease classification in the postgenomic era: A complex systems approach to human pathobiology. *Mol Syst Biol.* 2007;3:124.
4. Makarow M, Hojgaard L, Ceulemans R. Advancing systems biology for medical applications. *Science Policy Briefing.* 2008;1–12.
5. Sieberts SK, Schadt EE. Moving toward a system genetics view of disease. *Mamm Genome.* 2007;18(6–7):389–401.
6. Studer SM, Kaminski N. Towards systems biology of human pulmonary fibrosis. *Proc Am Thorac Soc.* 2007;4(1):85–91.
7. Wierling C, Herwig R, Lehrach H. Resources, standards and tools for systems biology. *Brief Funct Genomic Proteomic.* 2007;6(3):240–251.

Principles and Practice of Molecular Pathology

11

Pathology: The Clinical Description of Human Disease

William K. Funkhouser

"... Future discoveries will not likely be made by morphologists ignorant of molecular biologic findings, or by biologists unaware or scornful of morphologic data, but by those willing and capable of integrating them through a team approach...."
Rosai J. *Rosai and Ackerman's Surgical Pathology.* Mosby; 2004.

CURRENT PRACTICE OF PATHOLOGY

Diseases can be distinguished from each other based on differences at the molecular, cellular, tissue, fluid chemistry, and/or individual organism level. One hundred and fifty years of attention to the morphologic and clinical correlates of diseases has led to sets of diagnostic criteria for the recognized diseases, as well as a reproducible nomenclature for rapid description of the changes associated with newly discovered diseases. Sets of genotypic and phenotypic abnormalities in the patient are used to determine a diagnosis, which then implies a predictable natural history and can be used to optimize therapy by comparison of outcomes among similarly afflicted individuals. The disease diagnosis becomes the management variable in clinical medicine, and management of the clinical manifestations of diseases is the basis for day-to-day activities in clinics and hospitals nationwide. The pathologist is responsible for integration of the data obtained at the clinical, gross, morphologic, and molecular levels, and for issuing a clear and logical statement of diagnosis.

Clinically, diseases present to front-line physicians as patients with sets of signs and symptoms. Symptoms are the patient's complaints of perceived abnormalities. Signs are detected by examination of the patient. The clinical team, including the pathologist, will work up the patient based on the possible causes of the signs and symptoms (the differential diagnosis). Depending on the differential diagnosis, the workup typically involves history-taking, physical examination, radiographic examination, fluid tests (blood, urine, sputum, stool), and possibly tissue biopsy.

Radiographically, abnormalities in abundance, density or chemical microenvironment of tissues allows distinction from surrounding normal tissues. Traditionally, the absorption of electromagnetic waves by tissues led to summation differences in exposure of silver salt photographic film. Tomographic approaches such as CT (1972) and NMR (1973) complemented summation radiology, allowing finely detailed visualization of internal anatomy in any plane of section. In the same era, ultrasound allowed visualization of tissue with density differences, such as a developing fetus or gallbladder stones. More recently, physiology of neoplasms can be screened with positron emission tomography (PET, 1977) for decay of short half-life isotopes such as fluorodeoxyglucose. Neoplasms with high metabolism can be distinguished physiologically from adjacent low-metabolism tissues, and can be localized with respect to normal tissues by pairing PET with standard CT. The result is an astonishingly useful means of identifying and localizing new space-occupying masses, assigning a risk for malignant behavior and, if malignant, screening for metastases in distant sites. This technique is revolutionizing the preoperative decision-making of clinical teams, and improves the likelihood that patients undergo resections of new mass lesions only when at risk for morbidity from malignant behavior or interference with normal function.

Pathologically, disease is diagnosed by determining whether the morphologic features match the set of diagnostic criteria previously described for each disease. Multivolume texts are devoted to the gross and microscopic diagnostic criteria used for diagnosis, prognosis, and prediction of response to therapy. Pathologists diagnose disease by generating a differential diagnosis, then finding the best fit for the clinical presentation, the radiographic appearance, and the pathologic (both clinical lab and morphologic) findings. Logically, the Venn diagram of the clinical, radiologic, and pathologic differential diagnoses should overlap. Unexpected features expand the differential diagnosis and may raise the possibility of previously undescribed diseases. For example, Legionnaire's disease, human immunodeficiency virus (HIV), Hantavirus pneumonia, and severe acute respiratory syndrome (SARS) are examples of newly described diagnoses during the last 30 years. The mental construct of etiology (cause), pathogenesis (progression), natural history (clinical outcome), and response to therapy is the standard approach for pathologists thinking about a disease. A disease may have one or more etiologies (initial causes, including toxins, mutagens, drugs, allergens, trauma, or genetic mutations). A disease is expected to follow a particular series of events in its development (pathogenesis), and to follow a particular clinical course (natural history). Disease can result in a temporary or lasting change in normal function, including patient death. Multiple diseases of different etiologies can affect a single organ, for example, infectious and neoplastic diseases involving the lung. Different diseases can derive from a single etiology, for example, emphysema, chronic bronchitis, and small cell lung carcinoma in long-term smokers. The same disease (for instance, emphysema of the lung) can derive from different etiologies (emphysema from α-1-antitrypsin deficiency or cigarette smoke).

Modern diagnostic pathology practice hinges on morphologic diagnosis, supplemented by histochemical stains, immunohistochemical stains, cytogenetics, and clinical laboratory findings, as well as the clinical and radiographic findings. Sections that meet all of these criteria are diagnostic for the disease. If some, but not all, of the criteria are present to make a definitive diagnosis, the pathologist must either equivocate or make an alternate diagnosis. Thus, a firm grasp of the diagnostic criteria and the instincts to rapidly create and sort through the differential diagnosis must be possessed by the diagnostic pathologist.

The pathologic diagnosis has to make sense, not only from the morphologic perspective, but from the clinical and radiographic vantage points as well. It is both legally risky and professionally erosive to make a clinically and pathologically impossible diagnosis. In the recent past, limited computer networking meant numerous phone calls to gather the relevant clinical and radiographic information to make an informed morphologic diagnosis. For example, certain diseases such as squamous and small cell carcinomas of the lung are extremely rare in nonsmokers. Thus, a small cell carcinoma in the lung of a nonsmoker merits screening for a nonpulmonary primary site. Fortunately for pathologists, computing and networking technologies now allow us access to preoperative clinical workups, radiographs/reports, clinical laboratory data, and prior pathology reports. All of these data protect pathologists by providing them with the relevant clinical and radiographic information, and protect patients by improving diagnostic accuracy. Just as research scientists "... ignore the literature at their peril...", diagnostic pathologists "... ignore the

presentation, past history, workup, prior biopsies, and radiographs at their peril. ..."

There are limitations to morphologic diagnosis by H&E stains. First, lineage of certain classes of neoplasms (including small round blue cell tumors, clear cell neoplasms, spindle cell neoplasms, and undifferentiated malignant neoplasms) is usually clarified by immunohistochemistry, frequently by cytogenetics (when performed), and sometimes by electron microscopy. Second, there are limitations inherent in a snapshot biopsy or resection. Thus, the etiology and pathogenesis can be obscure or indeterminate, and rates of growth, invasion, or timing of metastasis cannot be inferred. Third, the morphologic changes may not be specific for the underlying molecular abnormalities, particularly the rate-limiting (therapeutic target) step in the pathogenesis of a neoplasm. For example, *Ret* gain of function mutations in a medullary thyroid carcinoma will require DNA level screening to determine germline involvement, familial risk, and presence or absence of a therapeutic target. Fourth, the same morphologic appearance may be identical for two different diseases, each of which would be treated differently. For example, there is no morphologic evidence by H&E stain alone to distinguish host lymphoid response to hepatitis C viral (HCV) antigens from host lymphoid response to allo-HLA antigens in a liver allograft. This is obviously a major diagnostic challenge when the transplant was done for HCV-related cirrhosis, and the probability of recurrent HCV infection in the liver allograft is high.

Paraffin section immunohistochemistry has proven invaluable in neoplasm diagnosis for clarifying lineage, improving diagnostic accuracy, and guiding customized therapy. If neoplasms are poorly differentiated or undifferentiated, the lineage of the neoplasm may not be clear. For example, sheets of undifferentiated malignant neoplasm with prominent nucleoli could represent carcinoma, lymphoma, or melanoma. To clarify lineage, a panel of immunostains is performed for proteins that are expressed in some of the neoplasms, but not in others. Relative probabilities are then used to lend support (rule in) or exclude (rule out) particular diagnoses in the differential diagnosis of these several morphologically similar undifferentiated neoplasms. The second role is to make critical distinctions in diagnosis that cannot be accurately made by H&E alone. Examples of this would include demonstration of myoepithelial cell loss in invasive breast carcinoma but not in its mimic, sclerosing adenosis (Figure 11.1), or loss of basal cells in invasive prostate carcinoma (Figure 11.2). The third role of immunohistochemistry is to identify particular proteins, such as nuclear estrogen receptor (ER) (Figure 11.3) or the plasma membrane HER2 proteins (Figure 11.4), both of which can be targeted with inhibitors rather than generalized systemic chemotherapy. Morphology remains the gold standard in this diagnostic process, such that immunohistochemical data support or fail to support the H&E findings, not vice versa.

Probability and statistics are regular considerations in immunohistochemical interpretation, since very few antigens are tissue-specific or lineage-specific.

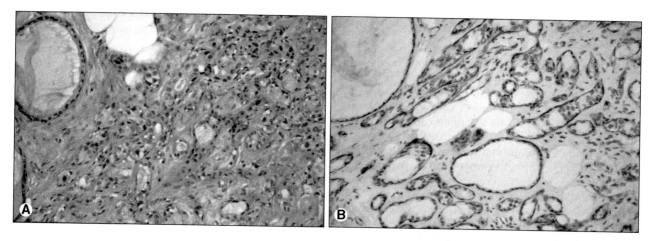

Figure 11.1 Sclerosing adenosis of breast. (A) H&E of sclerosing adenosis of breast. By H&E alone, the differential diagnosis includes infiltrating ductal carcinoma and sclerosing adenosis. **(B)** Actin immunostain of sclerosing adenosis of breast. Actin immunoreactivity around the tubules of interest supports a diagnosis of sclerosing adenosis and serves to exclude infiltrating carcinoma.

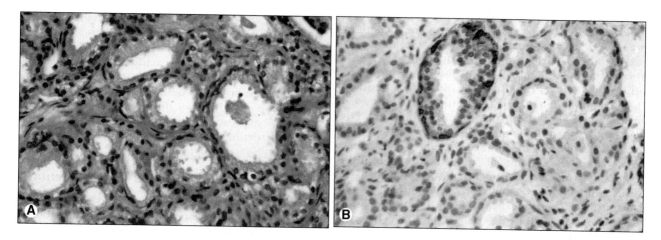

Figure 11.2 Invasive adenocarcinoma of prostate. (A) H&E of invasive adenocarcinoma of prostate. By H&E alone, the differential diagnosis includes invasive adenocarcinoma and adenosis. **(B)** High molecular weight cytokeratin immunostain of invasive adenocarcinoma of prostate. Loss of high molecular weight cytokeratin (34βE12) immunoreactivity around the glands of interest supports a diagnosis of invasive adenocarcinoma.

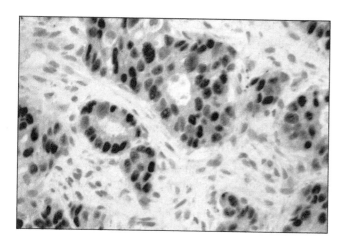

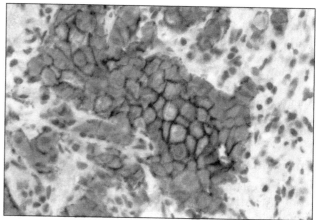

Figure 11.3 Estrogen receptor immunostain of breast carcinoma. Strong nuclear immunoreactivity for ER is noted, guiding use of ER inhibitor therapy.

Figure 11.4 HER2/c-erbB2 immunostain of breast carcinoma. Strong plasma membrane immunoreactivity for c-erbB2/HER2 is noted, guiding use of either anti-HER2 antibody or HER2 kinase inhibitor therapy.

Cytokeratin is positive in carcinomas, but also in synovial and epithelioid sarcomas. This example may imply aspects of the lineage of these two sarcomas that may be helpful in our categorization of these neoplasms. Another example would be the diagnosis of small cell carcinoma in the lung of a nonsmoker. Because lung primary small cell carcinoma is extremely uncommon, in non-smokers, this diagnosis would prompt the pathologist to inquire about screening results for other, nonpulmonary, primary sites. Likewise, immunohistochemistry results are always put into the context of the morphologic, clinical, and radiographic findings. For example, an undifferentiated CD30(+) neoplasm of the testis supports embryonal carcinoma primary in the testis, whereas a lymph node effaced by sclerotic bands with admixed CD30(+) Reed-Sternberg cells supports nodular sclerosing Hodgkin's disease.

A wealth of information is conveyed to a service pathologist in a tried-and-true H&E section. Analogous to the fact that a plain chest X-ray is the sum total of all densities in the beam path, the morphologic changes in diseased cells and tissues are the morphologic sum total of all of the disequilibria in the abnormal cells. For most neoplastic diseases, morphologic criteria are sufficient to predict the risk of invasion and metastasis (the malignant potential), the pattern of metastases, and the likely clinical outcomes. For example, the etiology and pathogenesis in small cell lung carcinoma can be inferred (cigarette smoking, with carcinogen-induced genetic mutations) and the outcome predicted (early metastasis to regional nodes and distant organs, with high probability of death within 5 years of diagnosis). New molecular data for both neoplastic and non-neoplastic diseases will most likely benefit unaffected individuals by estimating disease risk, and will most likely benefit patients by defining the molecular subset for morphologically defined diagnostic entities, thus guiding individualized therapy.

THE FUTURE OF DIAGNOSTIC PATHOLOGY

Diagnostic pathology will continue to use morphology and complementary data from protein (immunohistochemical) and nucleic acid (cytogenetics, *in situ* hybridization, DNA sequence, and RNA abundance) screening assays. New data will be integrated into the diagnostic process by reducing the cost and turnaround time of current technologies, and by development of new technologies, some of which are described.

Individual Identity

For transplant candidates, major histocompatibility complex (MHC, HLA in human) screening is evolving from cellular assays and serology toward sequencing of the alleles of the class I and II HLA loci. Rapid sequencing of these alleles in newborn cord blood would allow databasing of the population's haplotypes, facilitating perfect matches for required bone marrow or solid organ transplants.

Rapid Cytogenetics

Current uses of *in situ* hybridization to screen for viruses (such as EBV), light chain restriction (in B lymphomas), and copy number variation (for instance, HER2 gene amplification) demonstrate the benefit of *in situ* nucleic acid hybridization assays. It is possible that interphase FISH/CISH will become rapid enough to be used in the initial diagnostic workup of certain patients, including for sarcoma-specific translocations, ploidy analysis in hydatidiform moles, and gene amplification of receptor tyrosine kinase genes.

Rapid Nucleic Acid Sequence and RNA Abundance Screening

Current uses of nucleic acid screening for *bcr-abl* translocation, donor:recipient ratios after bone marrow transplant, microsatellite instability, quantitative viral load (for EBV, BK, CMV, and others), and single gene mutations (for *CFTR, Factor 2, α-1-antitrypsin*) demonstrate the benefit of nucleic acid screening in diagnosis and management. It is possible that each new neoplasm will be promptly defined as to ploidy, translocations, gene copy number differences, DNA mutations, and RNA expression cluster subset, allowing residual disease screening as well as individualized therapy.

Computer-Based Prognosis and Prediction

Current uses of morphology, immunohistochemistry, and molecular pathology demonstrate their benefit through improved diagnostic accuracy. However, diagnosis, extent of disease, and molecular subsets are currently imperfect estimators of prognosis and response to therapy. Relational databases which correlate an individual's demographic data, family history, concurrent diseases, morphologic features, immunophenotype, and molecular subset, and which integrate disease prevalence by age, sex, and ethnicity using Bayesian probabilities, should improve accuracy of prognosis and prediction of response to therapy. As risk correlates are developed, it is possible that healthy individuals will be screened and given risk estimates for development of different diseases.

Normal Ranges and Disease Risks by Ethnic Group

Current uses of normal ranges for serum chemistry assumes a similar bell-curve distribution across ages, sexes, and races. This may be true for most but not all analytes. Computer reference databases will likely generate normal ranges specific for the particular age/sex/ethnicity of individual patients. Similarly, familial risk for an inherited disease may vary by ethnic group, and this variation should be used in Bayesian calculations to define risk for unaffected at-risk family members.

Individual Metabolic Differences Relevant to Drug Metabolism

Current uses of liver and renal impairment to guide drug dosage demonstrate the benefit of using patient physiology to customize therapy. It is likely that individual differences in enzymatic metabolism of particular drugs (for instance, warfarin or tamoxifen) will be defined at the enzyme sequence level, and that gene haplotype data will be determined for new patients prior to receipt of these drugs.

Serum Biomarkers

Current uses of prostate-specific antigen (PSA) to screen for prostate carcinoma and its recurrence demonstrates the benefit of serum biomarkers in common neoplasms. It is likely that high-sensitivity screening of single and clustered serum analytes will lead to improved methods for early detection and persistence of neoplasms, autoimmune diseases, and infections.

CONCLUSION

Pathologists consider each disease to have a natural, mechanical, physicochemical basis. Each disease has an etiology (initial cause), a pathogenesis (stepwise progression), and a natural history with effects on normal function (clinical outcome). Pathologists collect the data needed to answer patients' and clinicians' questions, simply phrased as "what is it?" (diagnosis), "how it going to behave?" (prognosis), and "how do I treat it?" (prediction of response to therapy). Instincts and diagnostic criteria, as well as the optical, mechanical, chemical, and computing technologies described previously, are the basis for modern service pathology. As the human genome is deciphered, and as the complex interactions of cellular biochemistry are refined, risk of disease in unaffected individuals will be calculable, disease diagnosis will be increasingly accurate and prognostic, and molecular subsets of morphologically defined disease entities will be used to guide customized therapy for individual patients. It is a great time in history to be a pathologist.

KEY CONCEPTS

- Clinically, diseases present to front-line physicians as patients with sets of signs and symptoms. Symptoms are the patient's complaints of perceived abnormalities. Signs are detected by examination of the patient. The clinical team (including the pathologist) evaluate the patient based on the possible causes of the signs and symptoms (the differential diagnosis).
- Pathologically, disease is diagnosed by determining whether the morphologic features match the set of diagnostic criteria previously described for each disease. Pathologists diagnose disease by

generating a differential diagnosis, then finding the best fit for the clinical presentation, the radiographic appearance, and the pathologic (both clinical lab and morphologic) findings.

- Etiology describes the causes of a disease. One disease entity can have more than one etiology, and a single etiology can lead to more than one disease. For example, emphysema, chronic bronchitis, and small cell lung carcinoma can all occur in long-term smokers (different diseases derived from a single etiology). Likewise, the same disease (for instance, emphysema of the lung) can derive from different etiologies (emphysema from α-1-antitrypsin deficiency or cigarette smoke).
- The pathogenesis of a disease describes its stepwise progression after initiation in response to a specific etiologic factor (or factors). Pathogenesis can refer to the changes in the structure or function of an organism at the gross/clinical level, and it can refer to the stepwise molecular abnormalities leading to changes in cellular and tissue function.
- The natural history of a disease describes the expected course of disease, including chronicity, functional impairment, and survival. However, not all patients with a given disease will naturally follow the same disease course, so differences in patient outcome do not necessarily correspond to incorrect diagnosis. Variables that independently correlate with clinical outcome differences are called independent prognostic variables, and are assessed routinely in an effort to predict the natural history of the disease in the patient.
- Likewise, variables that independently correlate with response to therapy are called independent predictive variables, and are assessed routinely in an effort to optimize therapeutic response for each patient.

SUGGESTED READINGS

1. Coons AH, Kaplan MH. Localization of antigen in tissue cells; improvements in a method for the detection of antigen by means of fluorescent antibody. *J Exp Med.* 1950;91(1):1–13.
2. Crick FH, Barnett L, Brenner S, et al. General nature of the genetic code for proteins. *Nature.* 1961;192:1227–1232.
3. Early P, Huang H, Davis M, et al. An immunoglobulin heavy chain variable region gene is generated from three segments of DNA: VH, D and JH. *Cell.* 1980;19(4):981–992.
4. Gal AA. In search of the origins of modern surgical pathology. *Adv Anat Pathol.* 2001;8(1):1–13.
5. Gall JG, Pardue ML. Formation and detection of RNA-DNA hybrid molecules in cytological preparations. *Proc Natl Acad Sci U S A.* 1969;63(2):378–383.
6. Goeddel DV, Kleid DG, Bolivar F, et al. Expression in *Escherichia coli* of chemically synthesized genes for human insulin. *Proc Natl Acad Sci U S A.* 1979;76(1):106–110.
7. Holmes O. Contagiousness of puerperal fever. *New England Quarterly Journal of Medicine.* 1843;1:503–530.
8. Koch R. The etiology of tuberculosis. *Berliner Klinische Wochenschrift.* 1882;15:221–230.
9. Kohler G, Milstein C. Continuous cultures of fused cells secreting antibody of predefined specificity. *Nature.* 1975;256(5517):495–497.

10. Lander ES, Linton LM, Birren B, et al. Initial sequencing and analysis of the human genome. *Nature.* 2001;409(6822):860–921.

11. Medawar PB. The immunology of transplantation. *Harvey Lect.* 1956;(Series 52):144–176.

12. Mullis K, Faloona F, Scharf S, et al. Specific enzymatic amplification of DNA in vitro: the polymerase chain reaction. *Cold Spring Harb Symp Quant Biol.* 1986;51(Pt 1):263–273.

13. Nirenberg M, Caskey T, Marshall R, et al. The RNA code and protein synthesis. *Cold Spring Harb Symp Quant Biol.* 1966;31: 11–24.

14. Raju TN. Ignac Semmelweis and the etiology of fetal and neonatal sepsis. *J Perinatol.* 1999;19(4):307–310.

15. Ririe KM, Rasmussen RP, Wittwer CT. Product differentiation by analysis of DNA melting curves during the polymerase chain reaction. *Anal Biochem.* 1997;245(2):154–160.

16. Rosai J. The continuing role of morphology in the molecular age. *Mod Pathol.* 2001;14(3):258–260.

17. Sanger F, Nicklen S, Coulson AR. DNA sequencing with chain-terminating inhibitors. *Proc Natl Acad Sci U S A.* 1977;74(12): 5463–5467.

18. Smith LM, Sanders JZ, Kaiser RJ, et al. Fluorescence detection in automated DNA sequence analysis. *Nature.* 1986;321(6071): 674–679.

19. Turk JL. Rudolf Virchow—father of cellular pathology. *J R Soc Med.* 1993;86(12):688–689.

20. Venter JC, Adams MD, Myers EW, et al. The sequence of the human genome. *Science.* 2001;291(5507):1304–1351.

21. Watson JD, Crick FH. Molecular structure of nucleic acids; a structure for deoxyribose nucleic acid. *Nature.* 1953;171(4356): 737–738.

12

Understanding Molecular Pathogenesis: The Biological Basis of Human Disease and Implications for Improved Treatment of Human Disease

William B. Coleman . Gregory J. Tsongalis

INTRODUCTION

Disease has been a feature of the human existence since the beginning of time. Over time, our knowledge of science and medicine has expanded and with it our understanding of the biological basis of disease. In this regard, the biological basis of disease implies that more is understood about the disease than merely its clinical description or presentation. In the last several decades, we have moved from causative factors in disease to studies of molecular pathogenesis. Molecular pathogenesis takes into account the molecular alterations that occur in response to environmental insults and other contributing factors, to produce pathology. By developing a deep understanding of molecular pathogenesis, we will uncover the pathways that contribute to disease, either through loss of function or gain of function. By understanding the involvement of specific genes, proteins, and pathways, we will be better equipped to develop targeted therapies for specific diseases. Continued growth in our knowledge base with respect to underlying mechanisms of disease has resulted in unprecedented patient management strategies. Identification of genetic variants in genes once associated with the diagnosis of a disease process are now being re-evaluated as they may impact new therapeutic options.

In this chapter we describe three disease entities (Hepatitis C virus infection, acute myeloid leukemia, and cystic fibrosis) as examples of our increased understanding of the pathology represented by these diseases and how novel therapeutics are being introduced into clinical practice.

HEPATITIS C VIRUS INFECTION

Hepatitis C virus (HCV) infection represents the most common chronic viral infection in North America and Europe, and a common viral infection worldwide. In the United States' third National Health and Nutrition Examination Survey, it was estimated that 3.9 million people had detectable antibodies to HCV, indicating a prior exposure to the virus, and 75% of these individuals were positive for HCV RNA, suggesting an active infection. HCV infection has been found to be more common in certain populations, including prison inmates and homeless people, where the prevalence of infection may be as high as 40%. Worldwide, it is estimated that 340 million individuals are chronically infected with HCV.

Identification of the Hepatitis C Virus

HCV was first recognized in 1989 using recombinant technology to create peptides from an infectious serum that were then tested against serum from individuals with non-A, non-B hepatitis. This approach resulted in the isolation of a section of the HCV genome. Subsequently, the entire HCV genome was sequenced. HCV is a member of the family of flaviviridae. Flaviviruses are positive, single-stranded RNA viruses. The HCV genome encodes a gene for production of a single polypeptide chain of approximately 3000 amino acids. This polypeptide gives rise to a number of specific proteins. The Env proteins are among the most variable parts of the peptide chain and are associated with multiple molecular forms in a single infected person. The mutations affecting this portion of the HCV genome (and the endcoded Env protein) seem to be critical for escape of the virus from the host immune response. The HCV protein NS5a contains an interferon-response element. Evidence from several studies suggests that mutational variation in the HCV genome encoding this protein is associated with resistance to interferon, the main antiviral agent used in treatment of HCV. Other proteins encoded by the HCV genome include the NS3 region that codes for a protease and the NS5b region that codes for an RNA polymerase. Drugs that target the HCV protease or polymerase are now undergoing trials as therapeutic agents to treat HCV infection.

There are several HCV strains that differ significantly from each other. The nomenclature adopted to describe these HCV strains is based on division of the HCV RNA into three major levels: (i) genotypes, (ii) subtypes, and (iii) quasispecies. There are 6 recognized genotypes of HCV which are numbered from 1 to 6. Among these HCV genotypes there is <70% homology in the nucleotide sequence. HCV subtypes typically display 77–80% homology in nucleotide sequence, while quasispecies have >90% nucleotide sequence homology. Infection of an individual with HCV involves a single genotype and subtype (except in rare instances). However, infected individuals will carry many quasispecies of HCV because these RNA viruses do not contain a proofreading mechanism and acquired mutations in the HCV genome over time are common.

Risk Factors for Hepatitis C Virus Infection

There are a number of recognized risk factors for HCV. Among the most common risk factors for HCV infection are (i) the use of injection drugs and (ii) blood transfusion or organ transplant recipient before 1992. A significant percentage of people who used recreational injection drugs in the 1960s and 1970s became infected with HCV. Less commonly, HCV infection can be transmitted by dialysis, by needlestick injury, and through vertical transmission from an infected mother to her child. The likelihood of infection from needlestick injury or vertical transmission is estimated to be approximately 3–5%.

Hepatitis C Infection

The primary target cell type for HCV infection is the mature hepatocyte, although there is some evidence that infection can also occur in other cell types, particularly circulating mononuclear cells. Following the initial HCV infection, there is a latency period of 2–4 weeks before viral replication is detectable. In most cases, there is no clinical evidence of the infection even after viremia develops. In fact, only 10–30% of individuals with HCV infection will develop the clinical symptomatology of acute hepatitis. When acute HCV hepatitis develops, patients display symptoms of fever, loss of appetite, nausea, diarrhea, and specific liver symptoms, including discomfort and tenderness in the right upper abdomen, jaundice, dark urine, and pale-colored stools. Typically, these symptoms occur 2–3 months after the initial HCV infection and then gradually resolve over a period of several weeks. During this time, liver enzymes such as alanine aminotransferase (ALT) and aspartate aminotransferases (AST) are found at elevated levels in the blood, reflecting hepatocyte injury and death. In acute HCV hepatitis, these enzymes are typically increased from 10-fold to 40-fold the upper reference limit of normal. In the majority of individuals infected with HCV, there are no signs or symptoms that accompany the initial infection. In most of these cases, a chronic HCV infection develops, resulting in chronic hepatitis (ongoing inflammation in the liver). In general, chronic HCV infections can be clinically silent for many years without obvious symptomatology associated with the infection or liver injury, or produce only mild, nonspecific symptoms such as fatigue, loss of energy, and difficulty performing tasks that require concentration. The major end-stage diseases that result from chronic hepatitis include cirrhosis and hepatocellular carcinoma. It has been estimated that 20–30% of individuals with chronic HCV infection will progress to cirrhosis after 20 years of infection, although the fibrotic changes in the liver progress at different rates in different individuals. Cirrhosis due to HCV infection has now become the most common indication for liver transplantation in the United States.

Testing for Hepatitis C Virus Infection

Most clinical testing for HCV infection begins with detection of antibodies against HCV proteins. The sensitivity of the anti-HCV assay is reported to be in the range of 97–99% for detecting HCV infection. Most false negative results of the anti-HCV assay occur in the setting of immunosuppression, such as with human immunodeficiency virus (HIV) infection, or in renal failure. Anti-HCV antibodies are detectable after 10–11 weeks of infection (on average) using the second generation anti-HCV assays, but the third generation anti-HCV assays show improved sensitivity with positive detection of anti-HCV antibodies by 7–8 weeks after the initial infection. At the time of clinical presentation with acute HCV hepatitis, >40% of patients lack detectable anti-HCV. In the current clinical laboratory setting, the major method employed to determine the presence of active HCV infection is HCV RNA measurement. With acute HCV infection, HCV RNA becomes detectable 2–4 weeks after infection, and viral loads climb rapidly. Average HCV viral loads are approximately 2–3 million copies per mL. Qualitative assays are designed to determine the presence or absence of HCV RNA, without consideration of actual viral load. Two primary methodologies are used in this type of assay:

(i) reverse-transcriptase polymerase chain reaction (RT-PCR) and (ii) transcription-mediated amplification (TMA). The detection limit for assays of this type is <50 IU/mL. The approaches employed for qualitative determination of HCV RNA utilize a known amount of a synthetic standard to enable quantitative measurement of HCV RNA through comparison of the amounts of HCV amplified and the amount of standard amplified using a calibration curve. Determination of HCV viral load has become standard of care in evaluating patients before and during treatment for chronic HCV infection. Real-time PCR allows for reduced carryover amplification, more rapid detection of amplification, increased low-end sensitivity, and a wider dynamic range for detection and quantification (Figure 12.1).

Several techniques have been developed to determine the particular genotype and subtype of HCV causing infection in an individual infected patient. These techniques typically target the 5'-untranslated and/or core regions of the HCV genome which represent the most highly conserved regions. Because most amplification methods for HCV RNA also target the 5'-untranslated region, qualitative PCR methods can provide amplified RNA for use in determination of HCV genotype. The most widely used technique is a commercial line probe assay (Figure 12.1). In this assay, a large number of oligonucleotide sequences are immobilized on a membrane, incubated with amplified RNA, and then detected using a colorimetric reagent that detects areas of hybridization. The line probe assay enables recognition and identification of most HCV types and subtypes accurately, although there are several subtypes that cannot be distinguished from one another.

Clinical Course of Hepatitis C Virus Infection

Although anyone who is infected with HCV will experience an initial infection incident, in most cases this phase of the infection will be clinically silent without obvious symptoms. Acute infection with HCV is most likely to be detected when it occurs following a needlestick exposure

from a person with known HCV, or when the infection arises under other circumstances but produces symptomatic infection and jaundice (estimated to occur in less than one-third of all cases). There is some evidence to suggest that patients who develop clinical jaundice are actually more likely to clear the infection and not progress to chronic HCV hepatitis. During the initial incubation period approximately 2 weeks following infection, HCV RNA is either undetectable or can be detected only intermittently. Subsequently, there is a period of rapid increase in the amount of circulating HCV, with an estimated doubling time of <24 hours. HCV viral loads reach very high levels during this period of time, typically reaching values of 10^7 IU/mL and occasionally higher. Evidence of liver injury appears after an additional 1–2 months of HCV infection. This liver injury can be detected secondary to increased serum levels of ALT and AST. Approximately 40–50% of individuals with acute HCV infection that are clinically diagnosed are detected during this stage of infection, prior to the development of anti-HCV. By 7–8 weeks following infection, anti-HCV becomes detectable using the third generation immunoassays. However, detection of anti-HCV using the second generation immunoassay cannot be accomplished until 10–12 weeks following infection. At the time of this seroconversion, the HCV viral loads decrease, sometimes to undetectable levels. In most individuals who will progress to chronic hepatitis (and in some that eventually clear the infection) the HCV viral load remains detectable, but at reduced levels. In a person suspected of having acute HCV infection, the most reliable test for proving exposure is HCV RNA. Because of the high viral loads seen, either qualitative or quantitative assays would be acceptable for this purpose. Detectable HCV RNA in the absence of anti-HCV is strong evidence of recent HCV infection.

Treatment of Hepatitis C Infection

In contrast to other chronic viral infections such as those associated with hepatitis B virus or HIV, treatment has been successful in eradicating replicating HCV and

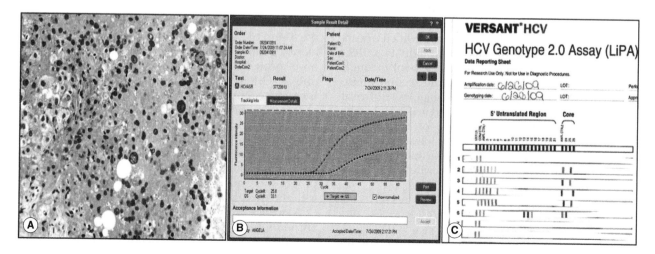

Figure 12.1 Testing for hepatitis C virus. (A) traditional H&E stained section of liver, (B) amplification curve from viral load testing using the Roche Taqman assay, and (C) HCV genotyping using the Siemens LiPA assay.

halting progression of liver damage. Interferon alpha-2 is the agent of choice for treatment of chronic HCV infection. There are currently two potential approaches to treatment of chronic HCV: (i) interferon alone or (ii) a combination of interferon plus ribavirin. While ribavirin is ineffective as a single agent for treating HCV, it increases the effectiveness of interferon. Application of ribavirin in combination with interferon increases the number of patients who respond to therapy by 2-fold to 3-fold. For many years, the only form of interferon available was standard dose interferon. Using standard dose interferon, large doses (typically 3 million units) were delivered to patients infected with HCV several times each week. The short half-life of this interferon produced widely fluctuating interferon levels in these patients, diminishing its therapeutic effectiveness. In 2001, a longer-acting form of interferon was approved for use in treating HCV infection. The longer-acting interferon was modified by attachment of polyethylene glycol (pegylated interferon), which resulted in increased half-life for the administered drug. Use of pegylated interferon results in sustained high levels of interferon in the patient, reducing the number of required administrations to a single injection each week. There was also an improvement in response rates among patients treated with the pegylated interferon. Currently, the preferred treatment for chronic HCV infection is the combination of pegylated interferon plus ribavirin.

Guided Treatment of Hepatitis C Virus

The appropriate duration of treatment for HCV infection varies depending on the HCV strain (genotype) that infects the patient. HCV genotypes 2 and 3 respond much better to standard treatment regimens. Thus, only 24 weeks of therapy are needed to achieve maximum benefit, compared to 48 weeks in persons infected with other HCV genotypes. In current clinical practice, treatment is offered to all patients with HCV infection except those with decompensated cirrhosis, where treatment may lead to worsening of the patient's condition. Once treatment is initiated, the most reliable means to determine efficacy is to evaluate the response by measuring HCV RNA. Successful treatment is associated with at least two different phases of viral clearance. The first phase, which occurs rapidly over the course of days, is thought to reflect HCV RNA clearance from a circulating pool through the antiviral effect of interferon. In the second phase of clearance, infected liver cells (the major site of viral replication) undergo cell turnover and are replaced by uninfected cells. The second phase of clearance is more variable in duration. First phase clearance is less specific for detecting success of antiviral treatment; therefore, it is necessary to evaluate whether second phase clearance has occurred.

Summary

HCV infection represents a relatively recently identified infectious agent that has a varied natural history from patient to patient. Intensive research efforts have characterized the phases of HCV infection and the clinical symptomatology of acute and chronic HCV infection. Through improved understanding of the biology of the HCV virus and its life cycle in the infected host, effective and sensitive diagnostic tests have been developed. Unlike some other chronic viral infections, HCV infection can be effectively treated using interferon in combination with ribavirin. However, it is now recognized that effective therapy of the patient depends on knowing the genotype of the HCV causing the infection. With continued advances in the understanding of the pathogenesis of HCV infection, new treatments and/or new modes of administration of known anti-HCV drugs will emerge that provide effective control of the viral infection with minimal adverse effects for the patient.

ACUTE MYELOID LEUKEMIA

The human leukemias have been classified as a distinct group of clinically and biologically heterogeneous disorders that are a result of genetic abnormalities that affect specific chromosomes and genes. The acute myeloid leukemias (AML) represent a major form of leukemia. AML is characterized by accumulation of neoplastic immature myeloid cells, consisting of $\geq 30\%$ myeloblasts in the blood or bone marrow and classified on the basis of their morphological and immunocytochemical features. AML can arise (i) *de novo*, (ii) in a setting of a pre-existing myelodysplasia, or (iii) secondary to chemotherapy for another disorder.

Chromosomal Abnormalities in Acute Myelogenous Leukemia

Various cytogenetic and/or molecular abnormalities have been associated with various types of AML. Chromosomal translocations are the most common form of genetic abnormality identified in acute leukemias. Typically, these translocations involve genes that encode proteins that function in transcription and differentiation pathways. As a result of chromosomal translocation, the genes proximal to the chromosome breakpoints are disrupted, and the 5′-segment of one gene is joined to the 3′-end of a second gene to form a novel fusion (chimeric) gene. When the chimeric gene is expressed, a novel protein product is produced from the chimeric mRNA. Other genetic alterations such as point mutations, gene amplifications and numerical gains or losses of chromosomes can also be identified in the acute leukemias. The clinical heterogeneity seen in AML may be due in part to differences in the number and nature of genetic abnormalities that occur in these cancers. However, these same molecular differences define various prognostic and therapeutic characteristics associated with the specific disorder in a given patient.

A major chromosomal translocation in AML involves chromosomes 15 and 17. This genetic abnormality, t(15;17)(q21;q21), occurs exclusively in acute promyelocytic leukemia (APL). APL represents approximately 5–13% of all *de novo* AMLs. The presence of the t(15;17) translocation consistently predicts responsiveness to a specific treatment utilizing all-*trans* retinoic acid (ATRA). Retinoic acid is a ligand for the retinoic acid receptor (RAR), which is involved in the t(15;17). ATRA is thought to overcome the block in myeloid cell maturation, allowing the neoplastic cells to mature (differentiate) and be eliminated. Approximately 75% of patients with APL present with a bleeding diathesis, usually the result of one or more processes including disseminated intravascular coagulation, increased fibrinolysis, and thrombocytopenia, and secondary to the release of procoagulants or tissue plasminogen activator from the granules of neoplastic promyelocytes. This bleeding diathesis may be exacerbated by standard cytoreductive chemotherapy. Two morphologic variants of APL have been described, typical (hypergranular) and microgranular, both of which carry the t(15;17) translocation. In the typical or hypergranular variant, the promyelocytes have numerous azurophilic cytoplasmic granules that often obscure the border between the cell nucleus and the cytoplasm. Cells with numerous Auer rods in bundles are common. In the microgranular type the promyelocytes contain numerous small cytoplasmic granules that are difficult to discern with the light microscope but are easily seen by electron microscopy.

Consequence of the t(15;17) Translocation in Acute Myelogenous Leukemia

The t(15;17) is a balanced and reciprocal translocation in which the *PML* (for promyelocytic leukemia) gene on chromosome 15 and the *RARα* gene on chromosome 17 are disrupted and fused to form a hybrid gene. The *PML-RARα* fusion gene, located on chromosome 15, encodes a chimeric mRNA and a novel protein. On the derivative chromosome 15, both the *PML* and *RARα* genes are oriented in a head-to-tail orientation. The function of the normal *PML* gene is poorly understood. However, the gene is ubiquitously expressed and encodes a protein that contains a dimerization domain and is characterized by an N-terminal region with two zinc-finger-like motifs (known as a ring and a B-box). Given its structural features, the PML protein is thought to be involved in DNA binding. Furthermore, the normal PML protein appears to have an essential role in cell proliferation. The *RARα* gene encodes a transcription factor that binds to DNA sequences in *cis*-acting retinoic acid-responsive elements. High-affinity DNA binding also requires heterodimerization with another family of proteins, the retinoic acid X receptors. The RARα protein contains transactivation, DNA binding, heterodimerization, and ligand binding domains. The normal RARα protein plays an important role in myeloid differentiation.

There are three major forms of the *PML-RARα* fusion gene, corresponding to different breakpoints in the *PML* gene (Figure 12.2). The breakpoint in the *RARα* gene occurs in the same general location in all cases, involving the sequences within intron 2. Approximately 40–50% of cases have a *PML* breakpoint in exon 6 (the so-called long form, termed *bcr1*), 40–50% of cases have

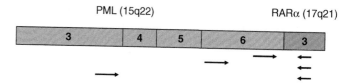

Figure 12.2   Schematic of real time PCR detection of the t(15;17) in AML. Arrows indicate primers used in amplification.

the *PML* breakpoint in exon 3 (the so-called short form, termed *bcr3*), and 5–10% of cases have a breakpoint in *PML* exon 6 that is variable (the so-called variable form, termed *bcr2*). In each form of the translocation, the PML-RARα fusion protein retains the 5′-DNA binding and dimerization domains of PML and the 3′-DNA binding, heterodimerization, and ligand (retinoic acid) binding domains of RARα. Recent studies indicate that the different forms of *PML-RARα* fusion mRNA correlate with clinical presentation or prognosis. In particular, the *bcr3* type of *PML-RARα* correlates with higher leukocyte counts at time of presentation. Both higher leukocyte counts and variant morphology are adverse prognostic findings, and the *bcr3* type of *PML-RARα* does not independently predict poorer disease-free survival.

Detection of the t(15;17) Translocation in Acute Myelogenous Leukemia

A number of methods may be used to detect the t(15;17) translocation. Conventional cytogenetic methods detect the t(15;17) in approximately 80–90% of APL cases at time of initial diagnosis. Suboptimal clinical specimens and poor quality metaphases explain a large subset of the negative results. Fluorescence *in situ* hybridization (FISH) is another useful method for detecting the t(15;17) in APL. Different methods employ probes specific for either chromosome 15 or chromosome 17 (or both), and commercial kits are available. Southern blot hybridization is another method to detect gene rearrangements that result from the t(15;17). The chromosomal breakpoints consistently involve the second intron of the *RARα* gene, and, therefore, probes derived from this region are the most often utilized. Virtually all cases of APL can be detected by Southern blot analysis using two or three genomic *RARα* probes. RT-PCR is a very convenient method for detecting the *PML-RARα* fusion transcripts. Primers have been designed to amplify the potential transcripts, and each type of transcript can be recognized (Figure 12.2). Results using this method are equivalent to or better than other methods at time of initial diagnosis.

Polyclonal and monoclonal antibodies reactive with the PML and RARα proteins have been generated, and immunohistochemical studies to assess the pattern of staining appear to be useful for diagnosis. Some investigators have studied a number of APLs, showing that the pattern of PML or RARα immunostaining correlates with the presence of the t(15;17). APL cells immunostaining for either PML or RARα reveals a microgranular pattern.

The fusion protein may prevent PML from forming normal oncogenic domains, since treatment with ATRA allows PML reorganization into these domains. For the diagnosis of residual disease or early relapse after therapy, conventional cytogenetic studies, Southern blot analysis, and immunohistochemical methods are limited by low sensitivity. Quantitative RT-PCR and FISH methods are very useful. The sensitivity and rapid turnaround time of RT-PCR make this method very useful for monitoring residual disease after therapy.

Summary

Acute promyelocytic leukemia is a distinct subtype of acute myeloid leukemia that is cytogenetically characterized by a balanced reciprocal translocation between chromosomes 15 and 17 [t(15;17)(q21;q21)], which results in a gene fusion involving *PML* and *RARα*. This disease is the most malignant form of acute leukemia with a severe bleeding tendency and a fatal course of only weeks in affected individuals. In the past, cytotoxic chemotherapy was the primary modality for treatment of APL, producing complete remission rates of 75–80% in newly diagnosed patients, a median duration of remission from 11–25 months, and only 35–45% of the patients were cured. However, with the introduction of all-trans retinoic acid (ATRA) in the treatment and optimization of the ATRA-based regimens, the complete remission rate increased to 90–95% and 5-year disease-free survival improved to 74%.

CYSTIC FIBROSIS

Cystic fibrosis (CF) is a clinically heterogeneous disease that exemplifies the many challenges of complex genetic diseases and the causative underlying mechanisms. CF is the most common lethal autosomal-recessive disease in individuals of European descent with a prevalence of 1:2500 to 1:3300 live births. While CF occurs most commonly in the Caucasian population, members of other racial and ethnic backgrounds are also at risk for this disease. In the United States, approximately 850 individuals are newly diagnosed on an annual basis, and 30,000 children and adults are affected. The majority of CF diagnoses are made in individuals who are less than one year of age (http://www.genetests.org/).

Cystic Fibrosis Transmembrane Conductance Regulator Gene

The Cystic Fibrosis Transmembrane Conductance Regulator (*CFTR*) gene is responsible for CF. This gene is large, spanning approximately 230 kb on chromosome 7q, and consists of 27 coding exons. The *CFTR* mRNA is 6.5 kb and encodes a CFTR membrane glycoprotein of 1480 amino acids with a mass of ~170,000 daltons. CFTR functions as a cAMP- regulated chloride channel in the apical membrane of epithelial cells. To date, over 1000 unique mutations in the CFTR gene have been described (Cystic Fibrosis Mutation Data Base, http://www.genet.sickkids .on.ca/cftr/). Hence, *CFTR* mutation testing in the molecular diagnostics laboratory examines multiple common mutation sites (Figure 12.3). The most common *CFTR* mutation is the deletion of phenylalanine at position 508 (ΔF508). This mutation affects 70% of patients worldwide. The allelic frequency of *CFTR* mutations varies by ethnic group. For example, the ΔF508 *CFTR* mutation is only present in 30% of the affected Ashkenazi Jewish population.

Diagnosis of Cystic Fibrosis

A diagnosis of CF in a symptomatic or at-risk patient is suggested by clinical presentation and confirmed by a sweat test. In the presence of clinical symptoms (such as recurrent respiratory infections), a sweat chloride above 60 mmol/L is diagnostic for CF. Although the results of this test are valid in a newborn as young as

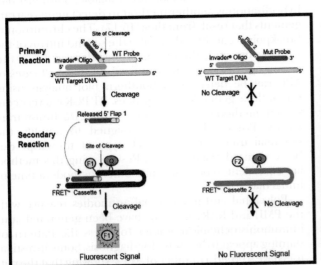

Figure 12.3 Cystic fibrosis mutation analysis using the Hologic CFTR Inplex Invader assay. A schematic diagram illustrates the chemistry involved in the Invader assay and the table represents data output from the 43 mutation panel assay.

24 hours, collecting a sufficient sweat sample from a baby younger than 3 or 4 weeks old is difficult. The sweat test can also confirm a diagnosis of CF in older children and adults, but is not useful for carrier detection. Mutations in the *CFTR* gene are grouped into six classes, including (i) Class I, characterized by defective protein synthesis where there is no CFTR protein at the apical membrane; (ii) Class II, characterized by abnormal/ defective processing and trafficking where there is no CFTR protein at the apical membrane; (iii) Class III, characterized by defective regulation where there is a normal amount of nonfunctional CFTR at the apical membrane; (iv) Class IV, characterized by decreased conductance where there is a normal amount of CFTR with some residual function at the apical membrane; (v) Class V, characterized by reduced or defective synthesis/trafficking where there is a decreased amount of functional CFTR at the apical membrane; and (vi) Class VI, characterized by decreased stability where there is a functional but unstable CFTR at the apical membrane. Of the CFTR mutations, classes I–III are the most common and are associated with pancreatic insufficiency. The ΔF508 CFTR mutation (which is most common worldwide) represents a class II mutation, with varying frequency between ethnic groups.

Abnormal Function of CFTR in Cystic Fibrosis

CFTR is a member of an ATP-binding cassette family with diverse functions such as ATP-dependent transmembrane pumping of large molecules, regulation of other membrane transporters, and ion conductance. Mutations in the *CFTR* gene can lead to an abnormal protein with loss or compromised function that results in defective electrolyte transport and faulty chloride ion transport in apical membrane epithelial cells affecting the respiratory tract, pancreas, intestine, male genital tract, hepatobiliary system, and the exocrine system, resulting in complex multisystem disease. The loss of CFTR-mediated anion conductance explains a variety of CF symptoms including elevated sweat chloride, due to a defect in salt absorption by the sweat ducts, and meconium ileus, a defect in fluid secretion by intestinal crypt cells. The malfunction of CFTR as a regulator of amiloride-sensitive epithelia Na+ channel leads to increased Na+ conductance in CF airways, which drives increased absorption of Cl− and water. Most of the symptoms associated with CF, such as meconium ileus, loss of pancreatic function, degeneration of the vas deferens, thickened cervical mucus, and failure of adrenergically mediated sweating are due to the role CFTR plays in Cl-driven fluid secretion.

CFTR is an anion channel that functions in the regulation of ion transport. It plays multiple roles in fluid and electrolyte transport, including salt absorption, fluid absorption, and anion-mediated fluid secretion. Defects in this protein lead to CF, the morbidity of which is initiated by a breach in host defenses and propagated by an inability to clear the resultant infections. Since inflammatory exacerbations precipitate

irreversible lung damage, the innate immune system plays an important role in the pathogenesis of CF. Respiratory epithelial cells containing the CFTR also provide a crucial environmental interface for a variety of inhaled insults. The local mucosal mechanism of defense involves mucociliary clearance that relies on the presence and constituents of airway surface liquid (ASL). The high salt in the ASL found in CF patients interferes with the natural antibiotics present in ASL such as defensins and lysozyme. Some investigators have categorized the role of CFTR in the pathogenesis of CF-related lung disease by dividing patients into two groups. The first describes defects in CFTR that result in altered salt and water concentrations of airway secretions. This then affects host defenses and creates a milieu for infection. The second is associated with CFTR deficiency that results in biologically and intrinsically abnormal respiratory epithelia. These abnormal epithelial cells fail as a mechanical barrier and enhance the presence of pathogenic bacteria by providing receptors and binding sites or failing to produce functional antimicrobials.

Much debate exists regarding the relative biological activity of antibacterial peptides such as beta-defensins and cathelicidins in human ASL and their role in the pathogenesis in CF-related lung disease. It is possible that the innate immune system provides a first line of host defense against microbial colonization by secreting defensins, small cationic antimicrobial peptides produced by epithelia. The innate antibiotics are thought to possess salt-sensitive bactericidal capabilities. Hence, these innate antibiotics demonstrate altered (impaired) function in the lungs of CF patients. Mannose-binding lectin represents another antimicrobial molecule that is present in ASL and is thought to be inactivated by high salt concentrations in the lungs of CF patients. Mannose-binding lectin, an acute phase serum protein produced in the liver, opsonizes bacteria and activates complement. Common variations in the mannose-binding lectin gene (MBL2) are associated with increased disease severity, increased risk of infection with *Burkholderia cepacia*, poor prognosis, and early death. The understanding that such naturally occurring peptide antibiotics exist has resulted in the pharmacologic development of these peptides for therapeutics.

Pathophysiology of Cystic Fibrosis

The occurrence of CF leads to clinical, gross, and histologic changes in various organ systems expressing abnormal CFTR, including the pancreas, respiratory, hepatobiliary, intestinal, and reproductive systems. In addition, pathologic changes have been observed in organ systems that do not express the *CFTR* gene (such as the rheumatologic and vascular systems). The current age of individuals affected with CF ranges from 0–74 years, and the predicted survival age for a newly diagnosed child is 33.4 years. The increasing age of survival of CF has led to increased manifestation of pulmonary and extrapulmonary disorders (gastrointestinal, hepatobiliary, vascular, and musculoskeletal) associated with the disease. The extent and severity of disease tend to correlate with the degree of CFTR function. Although all these organ systems are affected, the pulmonary changes are the most pronounced and the major cause of mortality in most cases.

Lung infection remains the leading cause of morbidity and mortality in CF patients. It is currently recognized that CF-related lung disease is the consequence of chronic pulmonary consolidation by the well-known opportunistic pathogens *Pseudomonas aeruginosa* (mucoid and nonmucoid), *Burkholderia cepacia*, *Staphylococcus aureus*, and *Haemophilus influenzae*. Morbidity and mortality due to persistent lung infection despite therapeutic advances focus attention toward the expanding microbiology of pulmonary colonizers. These increasingly prevalent flora include *Burkholderia cepacia* complex (genomovar I–IX), Methicillin-Resistant *Staphylococcus aureus* (MRSA), *Stenotrophomonas maltophilia*, *Achromobacter xylosoxidans*, *Mycobacterium abscessus*, *Mycobacterium-avium* complex, *Ralstonia* species, and *Pandoraea* species. Inflammatory exacerbation precipitates progressive irreversible lung damage, of which bronchiectases are the landmark changes. Bronchial mucous plugging facilitates colonization by microorganisms. Repetitive infections lead to bronchiolitis and bronchiectasis. Other pulmonary changes include interstitial fibrosis and bronchial squamous metaplasia. Often, subpleural bronchiectatic cavities develop and communicate with the subpleural space with resultant spontaneous secondary pneumothorax, the incidence of which increases later in life.

Exocrine pancreas insufficiency is present in the majority of patients with CF. This clinically manifests by failure to thrive, and fatty bulky stools owing to deficiency of pancreatic enzymes. However, pancreatic lesions vary greatly in severity, and the pancreas may be histologically normal in some patients who die in infancy. Early in the postnatal development of the pancreas, patients with CF have a deficiency of normal acinar development. Increased secretory material within the ducts and increased duct volume also contribute to progressive degradation and atrophy of pancreatic acini. These factors result in duct obstruction and progressive pancreatic pathology. Exocrine pancreatic disease appears to develop as a result of deficient ductal fluid secretion due to decreased anion secretion. Coupled to normal protein load derived from acinar cell secretion, this then leads to pancreatic protein hyperconcentration within the pancreatic ducts. The protein hyperconcentration increases susceptibility to precipitation and finally obstruction of the duct lumina. Hence, the characteristic lesion is cystic ductal dilation, atrophy of pancreatic acini, and severe parenchymal fibrosis.

The manifestation of CF in the hepatobiliary system is directly related to CFTR expression. The liver disease in CF is considered inherited liver disease due to impaired secretory function of the biliary epithelium. While defective CFTR may be expressed, males are more likely to be affected than females and the risk for developing liver disease is between 4% and 17% as assessed by yearly exams and biochemical testing. CFTR is expressed in epithelial cells of the biliary tract. Therefore, any or all cells of the biliary tree may be affected. While a variety of liver manifestations exist, including fatty infiltration (steatosis), common bile duct stenosis, sclerosing cholangitis, and gallbladder disease, the rare but characteristic liver lesion in CF is focal biliary cirrhosis, which develops in a minority of patients and is usually seen in older children and adults. With the increasing life expectancy in patients with CF, liver-related deaths have increased and may become one of the major causes of death in CF. The associated liver disease usually develops before or at puberty, is slowly progressive, and is frequently asymptomatic. There is negligible effect on nutritional status or severity of pulmonary involvement. Only a minority of patients go on to develop a clinically problematic liver disease with rapid progression. Abnormal bile composition and reduced bile flow ultimately lead to intrahepatic bile duct obstruction and focal biliary cirrhosis. Diagnosis of CF-associated liver disease is based on clinical exam findings, biochemical tests, and imaging techniques. Although liver biopsy is the gold standard for the diagnosis of most chronic liver diseases, only rarely is it employed in the diagnostic workup, mainly due to sampling error.

The gastrointestinal manifestations of cystic fibrosis are seen mainly in the neonatal period and include meconium ileus, distal intestinal obstruction syndrome (DIOS), fibrosing colonopathy, strictures, gastroesophageal reflux, rectal prolapse and constipation in later childhood. Throughout the intestines CFTR is the determinant of chloride concentration and secondary water loss into the intestinal lumen. Decreased water content results in viscous intestinal contents, with a 10–15% risk of developing meconium ileus in babies born with cystic fibrosis. This also accounts for DIOS and constipation in older children. DIOS (formerly meconium ileus equivalent) is a recurrent partial or complete obstruction of the intestine in patients with CF and pancreatic insufficiency.

Arthritis is a rare but recognized complication of cystic fibrosis that generally occurs in the second decade. Three types of joint diseases are described in patients with cystic fibrosis: (i) cystic fibrosis arthritis (CFA) or episodic arthritis (EA), (ii) hypertrophic pulmonary osteoarthropathy (HPOA), and (iii) co-existent or treatment-related arthritis. The most common form, episodic arthropathy, is characterized by episodic, self-limited polyarticular arthritis with no evidence of progression to joint damage. Histologic features are minimal with prominent blood vessels and interstitial edema occurring most commonly, or rarely lymphocytic inflammation.

Infertility is an inevitable consequence of cystic fibrosis in males occurring in >95% of patients and is due to congenital bilateral absence or atrophy of the vasa deferentia (CBAVD) and/or dilated or absent seminal vesicles. Spermatogenesis and potency remain normal. Mutations in the *CFTR* gene are present in up to 70% of the patients with CBAVD. Diagnosis of obstructive azoospermia may be diagnosed by semen analysis; however, it must be confirmed by testicular biopsy and no other reason for azoospermia. Fertility in females may be impaired due to dehydrated cervical mucus, but their reproductive function is normal. Advances in techniques such as microscopic epididymal sperm aspiration

(MESA) and intracytoplasmic sperm injection have allowed males with cystic fibrosis the ability to reproduce.

Summary

Cystic fibrosis is a complex multiorgan system disease that results from mutation in the CFTR gene. Advances in the understanding of the pathogenesis of this disease and related complications (such as recurrent lung infection) have led to improvement in diagnosis and treatment of affected individuals, resulting in improved life expectancy. With continued expansion of our understanding of the molecular pathogenesis of this disease and the variant manifestations of CF-related disorders, it is expected that new treatments will emerge that attempt to counteract or correct the pathologic consequences of *CFTR* mutation.

KEY CONCEPTS

- Molecular pathogenesis describes the molecular alterations that occur in response to environmental insults, exogenous exposures, genetic predispositions, and other contributing factors, to produce pathology. By developing a complete understanding of molecular pathogenesis, the pathways that contribute to disease through loss of function or gain of function will be elucidated. A greater understanding of the involvement of specific genes, proteins, and pathways in the causation of specific diseases will facilitate the development of targeted therapies for particular diseases.

- The natural history of HCV infection varies from patient to patient. Intensive research efforts have characterized the phases of HCV infection and the clinical symptomatology of acute and chronic HCV infection. Through improved understanding of the biology of the HCV virus and its life cycle in the infected host, effective and sensitive diagnostic tests have been developed. Unlike some other chronic viral infections, HCV infection can be effectively treated using interferon in combination with ribavirin. However, effective therapeutic treatment of the individual patient requires knowledge of the genotype of the HCV associated with the infection. With continued advances in the understanding of the pathogenesis of HCV infection, new treatments and/or new modes of administration of known anti-HCV drugs will emerge that provide effective control of the viral infection with minimal adverse effects for the patient.

- Acute promyelocytic leukemia (APL) is a distinct subtype of acute myeloid leukemia that is cytogenetically characterized by a balanced reciprocal translocation between chromosomes 15 and 17 [t(15;17)(q21;q21)], which results in a gene fusion involving *PML* and *RARα*. This disease is associated with a severe bleeding tendency and a fatal course of only weeks in affected individuals. Cytotoxic chemotherapy was once the primary modality for APL treatment, producing complete remission rates of 75–80% in newly diagnosed patients, a median duration of remission from 11 to 25 months, but only 35–45% of the patients were cured. However, with the introduction of all-*trans* retinoic acid (ATRA), the complete remission rate increased to 90–95% and five-year disease free survival improved to 74%.

- Cystic fibrosis is a complex multiorgan system disease that results from mutation in the *CFTR* gene. Advances in the understanding of the pathogenesis of this disease and related complications (such as recurrent lung infection) have led to improvement in diagnosis and treatment of affected individuals, resulting in improved life expectancy. With continued expansion of our understanding of the molecular pathogenesis of this disease and the variant manifestations of CF-related disorders, it is expected that new treatments will emerge that attempt to counteract or correct the pathologic consequences of *CFTR* mutation.

- Understanding of the molecular pathogenesis of disease creates opportunities for the development of new molecular diagnostics and targeted treatments that combine to improve the available modalities for treating affected individuals.

SUGGESTED READINGS

1. Alberti A, Boccato S, Vario A, et al. Therapy of acute hepatitis C. *Hepatology*. 2002;36:S195–S200.
2. Bukh J, Miller RH, Purcell RH. Genetic heterogeneity of hepatitis C virus: Quasispecies and genotypes. *Semin Liver Dis*. 1995; 15:41–63.
3. Cummings KJ, Lee SM, West ES, et al. Interferon and ribavirin vs interferon alone in the retreatment of chronic hepatitis C previously nonresponsive to interferon: A meta-analysis of randomized trials. *JAMA*. 2001;285:193–199.
4. Di Bisceglie AM, Hoofnagle JH. Optimal therapy of hepatitis C. *Hepatology*. 2002;36:S121–S127.
5. Gretch DR. Diagnostic tests for hepatitis C. *Hepatology*. 1997; 26:43S–47S.
6. Grignani F, Fagioli M, Alcalay M, et al. Acute promyelocytic leukemia: From genetics to treatment. *Blood*. 1994;83:10–25.
7. Honda M, Kaneko S, Sakai A, et al. Degree of diversity of hepatitis C virus quasispecies and progression of liver disease. *Hepatology* 1994;20:1144–1151.
8. Ideo G, Bellobuono A. New therapies for the treatment of chronic hepatitis C. *Curr Pharm Des*. 2002;8:959–966.
9. Milla PJ. Cystic fibrosis: Present and future. *Digestion*. 1998;59: 579–588.
10. Miller WH Jr, Kakizuka A, Frankel SR, et al. Reverse transcription polymerase chain reaction for the rearranged retinoic acid receptor alpha clarifies diagnosis and detects minimal residual disease in acute promyelocytic leukemia. *Proc Natl Acad Sci U S A*. 1992;89:2694–2698.
11. Rabbitts TH. Translocations, master genes, and differences between the origins of acute and chronic leukemias. *Cell*. 1991; 67:641–644.
12. Ratjen F, Doring G. Cystic fibrosis. *Lancet*. 2003;361:681–689.
13. Rosenstein BJ, Zeitlin PL. Cystic fibrosis. *Lancet*. 1998;351:277–282.
14. Seeff LB. Natural history of chronic hepatitis C. *Hepatology*. 2002; 36:S35–S46.

15. The Cystic Fibrosis Genotype-Phenotype Consortium. Correlation between genotype and phenotype in patients with cystic fibrosis. *N Engl J Med.* 1993;329:1308–1313.

16. Vankeerberghen A, Cuppens H, Cassiman JJ. The cystic fibrosis transmembrane conductance regulator: An intriguing protein with pleiotropic functions. *J Cyst Fibros.* 2002;1:13–29.

17. Vyas RC, Frankel SR, Agbor P, et al. Probing the pathobiology of response to all-trans retinoic acid in acute promyelocytic leukemia: Premature chromosome condensation/fluorescence in situ hybridization analysis. *Blood.* 1996;87:218–226.

18. Wang ZY, Chen Z. Acute promyelocytic leukemia: From highly fatal to highly curable. *Blood.* 2008;111:2505–2515.

19. Wine JJ. The genesis of cystic fibrosis lung disease. *J Clin Invest.* 1999;103:309–312.

20. Zielenski J. Genotype and phenotype in cystic fibrosis. *Respiration.* 2000;67:117–133.

Integration of Molecular and Cellular Pathogenesis: A Bioinformatics Approach

Jason H. Moore . C. Harker Rhodes

INTRODUCTION

Historically the touchstone of diagnostic pathology has been the histologic appearance of diseased and normal tissues when stained with conventional stains, usually hematoxylin and eosin. The microscopic appearance of these tissues reflects their cellular structures, which in turn is due to the differential expression of approximately 20,000 protein-coding genes in the human genome. The development of immunohistochemical techniques in the 1960s that allowed the histologic detection of specific proteins revolutionized both diagnostic pathology as well as the scientific study of disease. This differential expression of specific proteins, of course, is due in large part to changes at the DNA level, changes in transcription and mRNA splicing, and alterations in other aspects of the complex regulation of RNA metabolism and translation. With the sequencing of the human genome and the recent advances in both PCR-based and array-based technologies that allow the easy quantitation of specific RNA levels, the simultaneous detection of tens or hundreds of thousands of different genetic transcripts, and the DNA-based technologies for looking at genetic changes and changes in chromosomal organization, we are now on the verge of another, similar revolution. The goal of this chapter is to introduce bioinformatics as an important part of the modern molecular pathology research strategy.

The challenge is to integrate the tremendous wealth of information now available on the molecular changes associated with normal physiology and disease processes with the practice of pathology. In the practice of conventional H&E pathology the coordinate expression of thousands of genes results in the creation of distinctive microscopic appearances, which are recognized by trained pathologists and serve as the basis both for the diagnostic pathology and for the scientific study of disease. With the advent of immunohistochemistry these subtle and complex images were supplemented with information about the expression of a few specific proteins, one for each stain that is done. Now, as DNA- and RNA-based technologies are being introduced into the general practice of pathology two things are happening. First, information about specific genetic alterations or changes in specific RNA levels is being added to the information available about diseased tissue. The wealth of new insights provided by these techniques rivals that which became available when immunohistochemistry was introduced, but conceptually the integration of this information into the mainstream of pathology is straightforward and similar to what has been done in the past. But in addition and unlike the previous technologies, the nucleic-acid-based technologies can be incorporated into arrays that provide thousands or hundreds of thousands of individual pieces of information—genotype information, information about DNA methylation, data about gene copy number, or RNA expression levels. The analysis of this avalanche of data that can be generated from a single biopsy or autopsy specimen requires the techniques of bioinformatics if it is to be reduced to meaningful information.

For example, consider the information currently available from a biopsy of a high grade glioma such as a glioblastoma. The diagnosis is made today as it has been for decades based on the H&E appearance of the lesion—on the presence of endothelial proliferation and tumor necrosis in what is histologically a malignant glial lesion with the nuclear morphology typical of an astrocytic tumor. For many generations that was all the pathologist could tell the clinician about the tumor and it was all the clinician needed to know to select a therapy and treat the patient.

It had been recognized since the 1940s that these tumors fell into two distinct, but overlapping clinical categories: primary glioblastomas, which tended to be found in older patients with a shorter clinical course and appeared to arise as glioblastomas; and secondary glioblastomas, which were found in slightly younger patients and arose by progression from lower grade astrocytic tumors. But the distinction was of no therapeutic importance, and there were no pathologic correlates that could be used to distinguish a primary from secondary glioblastoma.

The use of immunohistochemical techniques did not change the situation for this particular tumor. Stains for glial fibrillary acidic protein could be used to demonstrate the astrocytic character of the lesion and markers for cycling cells as opposed to those in Go like the Ki-67 antigen, which could be used to demonstrate the relatively high proliferative index of the tumor, but they rarely provided clinically important new information.

Then in the mid 1990s studies using nucleic-acid-based technologies demonstrated that not only did primary and secondary glioblastomas have distinctive clinical histories, but on a molecular basis these histologically identical tumors were completely different with distinctive molecular signatures including frequent *EGFR* amplification and mutation in the primary tumors and *p53* gene mutations in the secondary GBMs. That distinction remained of limited clinical significance, but it helped establish the idea that in spite of the pathologist's inability to separate the glioblastomas into subtypes based on H&E histology they were a molecularly heterogeneous group of tumors. And now specific molecularly targeted therapies directed, for example, against tumors overexpressing EGFR are entering clinical trials.

Although these single marker, candidate gene studies of glioblastoma did a great deal to elucidate the molecular pathogenesis of these tumors and define molecularly distinct subsets of GBM, it was clear to most pathologists that the separation of GBM into primary and secondary tumors did not adequately capture the complexity of the situation. Within the last several years studies using RNA expression arrays to classify these tumors based on unsupervised clustering of thousands of mRNA levels have suggested that glioblastomas are perhaps best thought of as being of three types, those with "proneural," "proliferative," or "mesenchymal" molecular signatures. Other studies based on high-resolution copy number analysis using oligonucleotide-based array comparative genomic hybridization also have identified three subsets of GBM, one that seems to correspond to the classically defined primary GBM and two others that represent secondary GBMs. The integration of these studies into a single unified classification system remains to be done, but in the meantime other single-marker studies are providing clinically important information about these tumors. For example, the epigenetic silencing of the gene for the DNA-repair enzyme MGMT has been shown to influence the response of these tumors to conventional therapies and stratification of GBM

patients based on MGMT promoter methylation status is rapidly entering clinical practice. Similarly, the recognition of a subset of tumors that are histologically indistinguishable from other glioblastomas, but which, like oligodendrogliomas have loss of the short arm of chromosome 1 and/or the long arm of chromosome 19 and which have a better prognosis than the usual GBM, is becoming standard clinical practice.

As just illustrated, these examples demonstrate the potential for biotechnology to significantly impact our ability to use molecular pathology to understand disease processes. However, our ability to exploit these new technological resources will depend critically on our ability to make sense out of mountains of data collected for a set of pathology samples. The remainder of this chapter will introduce bioinformatics and the resources that are available to pathologists for making full use of genetics, genomics, and proteomics.

OVERVIEW OF BIOINFORMATICS

Bioinformatics is an interdisciplinary field that blends computer science and biostatistics with biomedical sciences such as epidemiology, genetics, genomics, and proteomics. Bioinformatics emerged as an important discipline shortly after the development of high-throughput DNA sequencing technologies in the 1970s. It was the momentum of the Human Genome Project that spurred the rapid rise of bioinformatics as a formal discipline. The word *bioinformatics* didn't start appearing in the biomedical literature until around 1990, but quickly caught on as the descriptor of this important new field. An important goal of bioinformatics is to facilitate the management, analysis, and interpretation of data from biological experiments and observational studies. Thus, much of bioinformatics can be categorized as database development and implementation, data analysis and data mining, and biological interpretation and inference. The goal of this chapter is to review each of these three areas and to provide some guidance on getting started with a bioinformatics approach to molecular pathology investigations of disease susceptibility.

The need to interpret information from whole-genome sequencing projects in the context of biological information acquired in decades of research studies prompted the establishment of the National Center for Biotechnology Information (NCBI) as a division of the National Library of Medicine (NLM) at the National Institutes of Health (NIH) in the United States in November of 1988. When the NCBI was established, it was charged with (1) creating automated systems for storing and analyzing knowledge about molecular biology, biochemistry and genetics; (2) performing research into advanced methods of computer-based information processing for analyzing the structure and function of biologically important molecules and compounds; (3) facilitating the use of databases and software by biotechnology researchers and medical care personnel; and (4) coordination of efforts to gather biotechnology information worldwide. Since 1988, the NCBI has fulfilled many of these goals and has delivered a set of databases and computational tools that are essential for modern biomedical research in a wide range of different disciplines including molecular epidemiology. The NCBI and other international efforts such as the European Bioinformatics Institute (EBI) that was established in 1992 have played a very important role in inspiring and motivating the establishment of research group and centers around the world that are dedicated to providing bioinformatics tools and expertise. Some of these tools and resources will be reviewed here.

DATABASE RESOURCES

One of the most important prestudy activities is the design and development of one or more databases that can accept, store, and manage molecular pathology data. There are eight steps for establishing an information management system for genetic studies. These are broadly applicable to many different kinds of studies. The first step is to develop the experimental plan for the clinical, demographic, sample, and molecular/laboratory information that will be collected. What are the specific needs for the database? The second step is to establish the information flow; that is, how does the information find its way from the clinic or laboratory to the database? The third step is to create a model for information storage. How are the data related? The fourth step is to determine the hardware and software requirements. How much data needs to be stored? How quickly will investigators need to access the data? What operating system will be used? Will a freely available database such as mySQL (http://www.mysql.com) serve the needs of the project or will a commercial dataset solution such as Oracle (http://www.oracle.com) be needed? The fifth step is to implement the database. The important consideration here is to define the database structure so that data integrity is maintained. The sixth step is to choose the user interface to the database. Is a web page portal to the data sufficient? The seventh step is to determine the security requirements. Do HIPAA regulations (http://www.hhs.gov/ocr/hipaa) need to be followed? Most databases need to be password-protected at a minimum. The eighth and final step outlined by Haynes and Blach is to select the software tools that will interface with the data for summary and analysis. Some of these tools will be reviewed as follows.

Although most investigators choose to develop and manage their own database for security and confidentiality reasons, there are an increasing number of public databases for depositing data and making it widely available to other investigators. The tradition of making data publicly available soon after it has been analyzed and published can largely be attributed to the community of investigators using gene expression microarrays. Microarrays represent one of the most revolutionary applications that derived from the knowledge of whole genome sequences. The extensive use of this technology has led to the need to store and search expression data for all the genes in the genome acquired in different genetic backgrounds or in different environmental conditions. This has resulted in a number of public databases such as the Stanford Microarray Database (http://genome-www5.stanford.edu), the Gene Expression Omnibus (http://www.ncbi.nlm.nih.gov/geo), ArrayExpress (http://www.ebi.ac.uk/arrayexpress), and others from which anyone can download data. The nearly universal acceptance of the data sharing culture in this area has yielded a number of useful tools that might not have been developed otherwise. The need for defining standards for the ontology and annotation of microarray experiments has led to proposals such as the Minimum Information about a Microarray Experiment (MIAME) (http://www.mged.org/Workgroups/MIAME/miame.html) that provided a standard that greatly facilitates the storage, retrieval, and sharing of data from microarray experiments. The MIAME standards provide an example for other types of data such as SNPs and protein mass spectrometry spectra. The success of the different databases depends on the availability of methods for easily depositing data and tools for searching the databases often after data normalization.

Despite the acceptance of data sharing in the genomics community, the same culture does not yet exist in molecular pathology. One of the few such examples is the Pharmacogenomics Knowledge Base of PharmGKB (http://www.pharmgkb.org). PharmGKB was established with funding from the NIH to store, manage, and make available molecular data in addition to phenotype data from pharmacogenetic and pharmacogenomic experiments and clinical studies. It is anticipated that similar databases for molecular epidemiology will appear and gain acceptance over the next few years as the NIH and various journals start to require data from public research be made available to the public.

In addition to the need for a database to store and manage molecular pathology data collected from experimental or observational studies, there are a number of database resources that can be very helpful for planning a study. A good starting point for database resources are those maintained at the NCBI (http://www.ncbi.nlm.nih.gov). Perhaps the most useful resource when planning a molecular pathology study is the Online Mendelian Inheritance in Man or OMIM database (http://www.ncbi.nlm.nih.gov/omim). OMIM is a catalog of human genes and genetic disorders with detailed summaries of the literature. The NCBI also maintains the PubMed literature database with more than 15 million indexed abstracts from published papers in more than 4700 life science journals. The PubMed Central database (http://www.pubmedcentral.nih.gov) quickly is becoming an indispensable tool with more than 400,000 full text papers from over 200 different journals. Rapid and free access to the complete text of published papers significantly enhances the planning, execution, and interpretation phases of any scientific study. The new Books database (http://www.ncbi.nlm.nih.gov/books) provides free access for the first time to electronic versions of many textbooks and other resources such as the NCBI Handbook, which serves as a guide to the resources that NCBI has to offer. This is a particularly important resource for students and investigators who need to learn a new discipline such as genomics. One of the oldest databases provided by the NCBI is the GenBank DNA sequence resource (http://www.ncbi.nlm.nih.gov/Genbank). DNA sequence data for many different organisms has been deposited in GenBank for more than two decades now, totaling more than 100 gigabases of data. GenBank is a common starting point for the design of PCR primers and other molecular assays that require specific knowledge of gene sequences. Curated information about genes, their chromosomal location, their function, their pathways, and so on can be accessed through the Entrez Gene database (http://www.ncbi.nlm.nih.gov/entrez/query.fcgi?db=gene), for example.

Important emerging databases include those that store and summarize DNA sequence variations. NCBI maintains the dbSNP (http://www.ncbi.nlm.nih.gov/projects/SNP/) database for single-nucleotide polymorphisms or SNPs. dbSNP provides a wide range of different information about SNPs including the flanking sequence primers, the position, the validation methods, and the frequency of the alleles in different populations. As with all NCBI databases it is possible to link to a number of other datasets such as PubMed and OMIM. The recently completed International Haplotype Map (HapMap) project documents genetic similarities and differences among different populations. Understanding the variability of SNPs and the linkage disequilibrium structure plays an important role in determining which SNPs to measure when planning a molecular epidemiology study. The International HapMap Consortium maintains an online database with all the data from the HapMap project (http://www.hapmap.org/thehapmap.html). Another useful database is the Allele Frequency Database or ALFRED (http://alfred.med.yale.edu/alfred/index.asp), which currently stores information on more than 3700 polymorphisms across 518 populations.

In addition to databases for storing raw data, there are a number of databases that retrieve and store knowledge in an accessible form. For example, the Kyoto Encyclopedia of Genes and Genomes (KEGG) database stores knowledge on genes and their pathways (http://www.genome.jp/keg). The Pathway component of KEGG currently stores knowledge on 42,937 pathways generated from 307 reference pathways. The Pathway component documents molecular interaction in pathways, whereas the Brite database stores knowledge on higher-order biological functions. One of the most useful knowledge sources is the Gene Ontology (GO) project that has created a controlled vocabulary to describe genes and gene products in any organism in terms of their biological processes, cellular components, and molecular functions (http://www.geneontology.org). GO descriptions and KEGG pathways both are captured and summarized in the NCBI databases. For example, the description of *p53* in Entrez Gene includes KEGG pathways such as cell cycle and apoptosis. It also includes GO descriptions such as protein binding and cell proliferation.

In general, a good place to start for information about available databases is the annual Database issue and the annual Web Server issue of the journal *Nucleic Acids Research*. These special issues include annual reports from many of the commonly used databases.

DATA ANALYSIS

Once the data are collected and stored in a database, an important goal of molecular pathology is to identify biomarkers or molecular/environmental predictors of disease end points. Statistical methods in bioinformatics provide a good starting point for the analysis of molecular pathology data. This can include commonly used methods such as t-tests, analysis of variance, linear regression, and logistic regression, for example, or may include more advanced data mining and machine learning methods such as cluster analysis or neural networks. Although many of these methods require special training in mathematics, statistics, or computer science, the good news is that most simple and advanced analysis methods are easily implemented in one or more freely available software packages. We briefly review several of these next.

Data Mining Using R

R is perhaps the one software package that everyone should have in their bioinformatics arsenal. R is an open-source and freely available programming language and data analysis and visualization environment that can be downloaded from http://www.r-project.org. According to the web page, R includes (1) an effective data handling and storage facility; (2) a suite of operators for calculations on arrays, in particular matrices; (3) a large, coherent, integrated collection of intermediate tools for data analysis; (4) graphical facilities for data analysis and display either on-screen or on hardcopy; and (5) a well-developed, simple, and effective programming language that includes conditionals, loops, user-defined recursive functions, and input and output facilities. A major strength of R is the enormous community of developers and users who ensure that just about any analysis method you need is available. This includes analysis packages such as Rgenetics (http://rgenetics.org) for basic genetic and epidemiologic analysis such as testing for deviations from Hardy-Weinberg equilibrium or haplotype estimation, epitools for basic epidemiology analysis (http://www.epitools.net), geneland for spatial genetic analysis (http://www.inapg.inra.fr/ens_rech/mathinfo/personnel/guillot/Geneland.html), and popgen for population genetics (http://cran.r-project.org/src/contrib/Descriptions/popgen.html). Perhaps the most useful contribution to R is the Bioconductor project (http://www.bioconductor.org). According to the Bioconductor web page, the goals of the project are to (1) provide access to a wide range of powerful statistical and graphical methods for the analysis of genomic data; (2) facilitate the integration of biological metadata (e.g., PubMed, GO) in the analysis of experimental data; (3) allow the rapid development of extensible, scalable, and interoperable software; (4) promote high-quality and reproducible research; and (5) provide training in computational and statistical methods for the analysis of genomic data.

There are numerous packages for machine learning and data mining that either are part of the base R software or can be easily added. For example, the neural package includes routines for neural network analysis (http://cran.r-project.org/src/contrib/Descriptions/neural.html). Others include arules for association rule mining (http://cran.r-project.org/src/contrib/Descriptions/arules.html), cluster for cluster analysis (http://cran.r-project.org/src/contrib/Descriptions/cluster.html), genalg for genetic algorithms (http://cran.r-project.org/src/contrib/Descriptions/genalg.html), som for self-organizing maps (http://cran.r-project.org/src/contrib/Descriptions/som.html), and tree for classification and regression trees (http://cran.r-project.org/src/contrib/Descriptions/tree.html). Many others are available. A full list of contributed packages for R can be found at http://cran.r-project.org/src/contrib/PACKAGES.html. The primary advantage of using R as your data mining software package is its power. However, the learning curve can be challenging at first. Fortunately, plenty of documentation is available on the web and in published books.

Data Mining Using Weka

One of the most mature, open-source, and freely available data mining software packages is Weka (http://www.cs.waikato.ac.nz/ml/weka). Weka is written in Java and thus will run in any operating system (e.g., Linux, Mac, Sun, Windows). Weka contains a comprehensive list of tools and methods for data processing, unsupervised and supervised classification, regression, clustering, association rule mining, and data visualization. Machine learning methods include classification trees, k-means cluster analysis, k-nearest neighbors, logistic regression, naïve Bayes, neural networks, self-organizing maps, and support vector machines, for example. Weka includes a number of additional tools such as search algorithms and analysis tools such as cross-validation and bootstrapping. A nice feature of Weka is that it can be run from the command line, making it possible to run the software from Perl or even R (see http://cran.r-project.org/src/contrib/Descriptions/RWeka.html). Weka includes an experimenter module that facilitates comparison of algorithms. It also includes a knowledge flow environment for visual layout of an analysis pipeline. This is a very powerful analysis package that is relatively easy to use. Further, there is a published book that explains many of the methods and the software.

Data Mining Using Orange

Orange is another open-source and freely available data mining software package (http://www.ailab.si/orange) that provides a number of data processing, data mining, and data visualization tools. What makes

Orange different and in some way preferable to other packages such as R is its intuitive visual programming interface. With orange, methods and tools are represented as icons that are selected and dropped into a window called the canvas. For example, an icon for loading a dataset can be selected along with an icon for visualizing the data table. The file load icon is then "wired" to the data table icon by drawing a line between them. Double-clicking on the file load icon allows the user to select a data file. Once loaded, the file is then automatically transferred by the "wire" to the data table icon. Double-clicking on the data table icon brings up a visual display of the data. Similarly, a classifier such as a classification tree can be selected and wired to the file icon. Double-clicking on the classification tree icon allows the user to select the settings for the analysis. Wiring the tree viewer icon then allows the user to view a graphical image of the classification tree inferred from the data. Orange facilitates high-level data mining with minimal knowledge of computer programming. A wide range of different data analysis tools are available. A strength of Orange is its visualization tools for multivariate data. Recent additions to Orange include tools for microarray analysis and genomics such as heat maps and GO analysis.

Interpreting Data Mining Results

Perhaps the greatest challenge of any statistical analysis or data mining exercise is interpreting the results. How does a high-dimensional statistical pattern derived from population-level data relate to biological processes that occur at the cellular level? This is an important question that is difficult to answer without a close working relationship between pathologists, for example, and statisticians and computer scientists. Fortunately, a number of emerging software packages are designed with this in mind. GenePattern (http://www.broad.mit.edu/cancer/software/genepattern/), for example, provides an integrated set of analysis tools and knowledge sources that facilitate this process. Other tools such as the Exploratory Visual Analysis (EVA) database and software (http://www.exploratoryvisualanalysis.org/) are designed specifically for integrating research results with biological knowledge from public databases in a framework designed for pathologists, for example. These tools and others will facilitate interpretation.

THE FUTURE

We have only scratched the surface of the numerous bioinformatics methods, databases, and software tools that are available to the pathology community. We have tried to highlight some of the important software resources such as Weka and Orange, which might not be covered in other reviews that focus on more traditional methods from biostatistics. Although there are an enormous number of bioinformatics resources today, the software landscape is changing rapidly as new technologies for high-throughput biology emerge. Over the next few years we will witness an explosion of novel bioinformatics tools for that analysis of genomewide association data and, more importantly, the joint analysis of SNP data with other types of data such as gene expression data and proteomics data. Each of these new data types and their associated research questions will require special bioinformatics tools and perhaps special hardware such as faster computers with bigger storage capacity and more memory. Some of these datasets will easily require 1 to 2 Gb or more of memory or more for analysis and could require as many as 100 processors or more to complete a data mining analysis in a reasonable amount of time. The challenge will be to scale our bioinformatics tools and hardware such that a genomewide SNP dataset can be processed as efficiently as we can process a candidate gene dataset with perhaps 20 SNPs today. Only then can molecular pathology truly arrive in the genomics age.

KEY CONCEPTS

- An important challenge for the practice of pathology is to integrate the tremendous wealth of information now available on the molecular changes associated with normal physiology and disease processes.
- Bioinformatics is an interdisciplinary field that blends computer science and biostatistics with biomedical sciences such as epidemiology, genetics, genomics, and proteomics.
- One of the most important pre-study activities is the design and development of one or more databases that can accept, store, and manage molecular pathology data.
- Once the data are collected and stored in a database, an important goal of molecular pathology is to use biostatistics and data mining to identify biomarkers or molecular/environmental predictors of disease end points.
- The challenge for the practice of pathology in the future will be to scale bioinformatics software and

hardware such that information from the entire genome or proteome, for example, can be harnessed in an efficient manner.

SUGGESTED READINGS

1. Ashburner M, Ball CA, Blake JA, et al. Gene ontology: Tool for the unification of biology. The Gene Ontology Consortium. *Nat Genet.* 2000;25:25–29.
2. Brazma A, Krestyaninova M, Sarkans U. Standards for systems biology. *Nat Rev Genet.* 2006;7:593–605.
3. Benson D, Boguski M, Lipman DJ, et al. The National Center for Biotechnology Information. *Genomics.* 1990;6:389–391.
4. Benson D, Lipman DJ, Ostell J. GenBank. *Nucleic Acids Res.* 1993;21:2963–2965.
5. Boguski MS. Bioinformatics. *Curr Opin Genet Dev.* 1994;4: 383–388.
6. Gene Ontology Consortium. The Gene Ontology (GO) project in 2006. *Nucleic Acids Res.* 2006;34:D322–D326.
7. Gentleman R, Carey VJ, Huber W, et al. *Bioinformatics and Computational Biology Solutions using R and Bioconductor.* New York, NY: Springer; 2005.
8. Hamosh A, Scott AF, Amberger J, et al. Online Mendelian Inheritance in Man (OMIM), a knowledgebase of human genes and genetic disorders. *Nucleic Acids Res.* 2005;33:D514–D517.
9. Hastie T, Tibshirani R, Friedman J. *The elements of statistical learning.* New York, NY: Springer; 2001.
10. Haynes C, Blach C. Information management. In: Haines JL, Pericak-Vance MA, eds. *Genetic analysis of complex disease.* Hoboken, NJ: Wiley; 2006.
11. International HapMap Consortium. A haplotype map of the human genome. *Nature.* 2005;437:1299–1320.
12. Ogata H, Goto S, Sato K, et al. KEGG: Kyoto encyclopedia of genes and genomes. *Nucleic Acids Res.* 1999;27:29–34.
13. Reimers M, Carey VJ. Bioconductor: An open source framework for bioinformatics and computational biology. *Methods Enzymol.* 2006;411:119–134.
14. Robinson C. The European Bioinformatics Institute (EBI)—Open for business. *Trends Biotechnol.* 1994;12:391–392.
15. Rosner, B. *Fundamentals of Biostatistics.* Duxbury Press; 2000.
16. Schena M, Shalon D, Davis RW, et al. Quantitative monitoring of gene expression patterns with a complementary DNA microarray. *Science.* 1995;270:467–470.
17. Sherlock G, Hernandez-Boussard T, Kasarskis A, et al. The Stanford microarray database. *Nucleic Acids Res.* 2001;29:152–155.
18. Sherry ST, Ward M, Sirotkin K. dbSNP-database for single nucleotide polymorphisms and other classes of minor genetic variation. *Genome Res.* 1999;9:677–679.
19. Venables WN, Ripley BD. *Modern Applied Statistics with S.* New York, NY: Springer; 2002.
20. Whitten IH, Frank E. *Data Mining.* Boston, MA: Elsevier; 2005.

Molecular Pathology of Human Disease

14

Molecular Basis of Cardiovascular Disease

Amber Chang Liu ▪ Avrum I. Gotlieb

GENERAL MOLECULAR PRINCIPLES OF CARDIOVASCULAR DISEASES

Several important concepts have emerged as the molecular biology of cardiovascular disease is investigated. (1) Cells of the cardiovascular system have unique properties (Table 14.1). (2) Cell function is regulated by the combined actions of specific molecules, some that promote and others that inhibit a cellular process. It is the balance between the bioactivity of all these molecules that dictates the function of the cell at any given moment in time. (3) Depending on conditions, the same molecule may both promote or inhibit a given cellular function, generally by directly or indirectly acting on signaling molecules that regulate different pathways. (4) Signaling pathways interact by sharing downstream molecules. (5) Microenvironments are important in autocrine and paracrine regulation of cardiovascular cell function, including across cell types. (6) Cell-extracellular matrix interactions are critical to normal physiology and pathogenesis of disease. (7) Physical forces regulate functions of cardiovascular cells that are important in both maintaining normal physiology and regulating pathogenesis of disease.

THE CELLS OF CARDIOVASCULAR ORGANS

Vascular Endothelial Cells

Vascular endothelial cells (ECs), which are embryologically derived from splanchnopleuric mesoderm, form a thromboresistant barrier on the surface of the vascular tree. The cells are quiescent but have the ability to proliferate once appropriate genes are activated in response to injury and/or disease. These cells are highly metabolically active and alter their function as their microenvironment changes. These cell functions are balanced between the regulation of physiological functions that maintain normal homeostasis and the endothelial dysfunction that is associated with pathobiology (Table 14.2). Genetic conditions result in several coagulopathies. A major role of ECs is to transduce hemodynamic shear stress from a physical force to a biochemical signal that regulates gene expression and/or protein secretion of bioactive agents. These shear stress activated-molecules include vasoactive compounds, extracellular matrix proteins and degradation enzymes, growth factors, and coagulation and inflammatory factors (Table 14.3).

Vascular Smooth Muscle Cells

Vascular SMCs form the cells of the media and maintain the matrix of the normal vascular wall. Smooth muscle cells are quiescent in this media. However, upon injury, the cells undergo phenotypic transformation to proliferating, secreting, and migrating cells with a capacity to become myofibroblasts and participate in repair. Smooth muscle cells may become foam cells through ingestion of lipids. They participate in autocrine and paracrine pathways, especially in interactions with macrophages and ECs. Smooth muscle cells are important regulators of vascular remodeling.

Valve Endothelial Cells

Valve endothelial cells (VECs) form a single cell layer of adherent cells that cover the surface of the valve. The cells are quiescent, but upon injury, they will proliferate to reconstitute the thromboresistant surface. VECs are heterogenous and show important differences when compared to vascular EC. Using microarray technology, VECs have been shown to differentially express 584 genes on the aortic side versus the ventricular side of

Table 14.1 The Cells of the Cardiovascular System

Heart
  Cardiac myocytes
  Cardiac interstitial fibroblasts
  Valve interstitial cells (VICs)
  Valve endothelial cells (VECs)
  Endothelial cells, smooth muscle cells, pericytes of blood vessels

Blood vessels
  Endothelial cells (ECs)
  Smooth muscle cells (SMCs)
  Pericytes
  Adventitial fibroblasts
  Endothelial cells, smooth muscle cells of vasa vasorum

Stem cells
  Endothelial progenitor cells (EPCs)
  Mesenchymal stem cells
    Bone marrow derived
    Tissue derived

Cells associated with disease
  Dendritic cells
  Macrophages/foam cells
  Lymphocytes
  Mast cells
  Giant cells

Table 14.2 Endothelial Function

Physiological Function	Endothelial Dysfunction
Platelet resistant	Platelet adhesion
Anticoagulation	Procoagulation
Fibrinolysis	Antifibrinolysis
Quiescent	Migration/proliferation
Leukocyte resistant	Leukocyte adhesion
Anti-inflammatory	Proinflammatory
Selective impermeability	Enhanced permeability
Quiescent SMC	SMC activation
Provasodilation	Provasoconstriction
Matrix stability	Matrix remodeling
Vessel stability	Angiogenesis

Table 14.3 Shear Stress-Regulated Factors in Endothelium

Vasoactive Compounds
- Angiotensin converting Enzyme (ACE)
- NO—endothelial Nitric Oxide Synthase (eNOS)
- NO—induced Nitric Oxide Synthase (iNOS)
- Prostacyclin
- Endothelin-1

ECM/ECM Degradation Enzymes
- Matrix Metalloproteinase-9 (MMP-9)
- Collagen XII
- Thrombospondin

Growth Factors
- Epidermal Growth Factor (EGF)
- Basic Fibroblast Growth Factor (bFGF)
- Granulocyte Monocyte-Colony Stimulating Factor (GM-CSF)
- Insulin-Like Growth Factor Binding Protein (IGFBP)

Coagulation/Fibrinolysis
- Thrombomodulin
- Tissue Factor
- Tissue Plasminogen Activator (tPA)
- Protease-Activated Receptor-1—thrombin receptor (PAR-1)

Inflammation Factors
- Monocyte chemoattractant protein (MCP-1)
- Vascular cell adhesion molecule (VCAM-1)
- Intercellular adhesion molecule (ICAM-1)
- E-Selectin

Others
- Extracellular superoxide dismutase (ecSOD)
- Sterol regulatory element binding protein (SREBP)
- Platelet/endothelial cell adhesion molecule (PECAM-1)

normal adult pig aortic valves. Several of these observed differences could help explain the vulnerability of the aortic side of the valve cusp to calcification in diseases such as calcific aortic stenosis (CAS). However, because calcification occurs within the valve tissue, it is likely that VECs may be playing more of a transducing role, regulating VIC function. Valvular ECs also show phenotypic differences in response to shear when compared to vascular ECs.

Valve Interstitial Cells

The term valve fibroblasts is still used in the literature. However, it should be abandoned and the term valve interstitial cells (VICs) should be used because these cells do have specific features which are context dependent to the heart valve and show differences when compared to fibroblasts in other tissues and organs.

The valve matrix contains VICs distributed in all three layers of the leaflet: the fibrosa, the spongiosa, and the ventricularis. Compartmentalization occurs at late gestation. However, between 20 and 39 weeks the valves only have a bilaminar structure. It is not known how remodeling of the valve into individual compartments occurs. It is clear that physical forces do play some role because the three layers seen in the adult architecture are not complete until early adulthood. Cell cultures of VICs have been characterized and have provided new information on the cell and molecular biology of these cells.

Five phenotypes best represent the VIC family of cells (Figure 14.1). Each phenotype exhibits specific sets of cellular functions essential in normal valve physiology and in pathobiologic conditions. Although these phenotypes may exhibit plasticity and convert from one form to another (Figure 14.1), characterizing VIC function by distinct phenotypes brings clarity to our understanding of the complex VIC biology and pathobiology by focusing investigations on the interaction of each specific VIC phenotype within the valve and the systemic environment in which it resides.

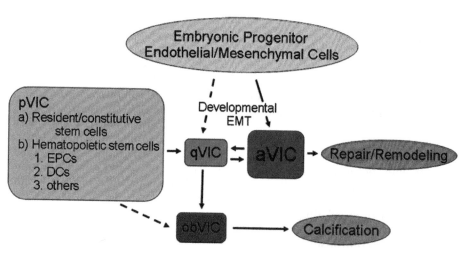

Figure 14.1   The current literature describes numerous VIC functions which can be conveniently organized into five phenotypes: Embryonic progenitor endothelial/mesenchymal cells, quiescent VICs (qVICs), activated VICs (aVICs), stem cell-derived progenitor VICs (pVICs), and osteoblastic VICs (obVICs). These represent specific sets of VIC functions in normal valve physiology and pathophysiology. Embryonic progenitor endothelial/mesenchymal cells undergo endothelial-mesenchymal transformation in fetal development to give rise to aVIC and/or qVICs resident in the normal heart valve. The VICs undergoing the transformation do have features of aVICs, including migration, proliferation, and matrix synthesis. When the heart valve is subjected to an insult, be it abnormal hemodynamic/mechanical stress or pathological injury, qVICs become activated, giving rise to aVICs, which participate in repair and remodeling of the valve. pVICs including bone marrow-derived cells, circulating cells, and resident valvular progenitor cells are another source of aVICs in the adult. The relationship between bone marrow, circulating, and resident pVICs is unknown. Under conditions promoting valve calcification, such as in the presence of osteogenic and chondrogenic factors, qVICs can undergo osteoblastic differentiation into obVICs. It is possible that obVICs are derived from pVICs. obVICs actively participate in the valve calcification process. Compartmentalizing VIC function into distinct phenotypes recognizes as well the transient behavior of VIC phenotypes. The hatched arrows depict possible transitions for which there is no solid evidence currently. Reprinted with permission from the American Society for Investigative Pathology, *Am J Pathol.* 2007;171:1407–1418.

Leukocytes

There are occasional macrophages and lymphocytes in the normal vessel wall, especially in the intima. Endothelial dysfunction due to injury and/or inflammation promotes monocytes to enter the wall, become activated macrophages, and promote vessel dysfunction and further injury. Macrophages may transform into foam cells. Polymorphonuclear leukocytes are prominent at the interface of necrotic and intact myocardium in a myocardial infarction and in the early stages of several vasculitides.

Vascular Progenitor/Stem Cells

In the last decade, human studies and experimental animal model and cell culture investigations have identified a variety of embryonic and adult-derived cell types that exhibit the potential for vascular or myocardial repair of injured and diseased tissue. Some of these cells are differentiated cells, such as skeletal myoblasts and cardiomyocytes, and others are multipotent embryonic stem cells and multipotent adult stem cells. Some of these cells can adopt both vascular and cardiomyocyte phenotypes. The stem or progenitor cells are usually rare within a population of cells, and specific techniques are required to isolate the cells and then expand the population, usually *ex vivo*. The cells are identified by specific markers. The therapeutic

potential of these cells following transplantation is considered to be due to paracrine effects since the cells do not readily expand *in vivo* following cell therapy and do not persist for long. The biology and pathobiology of these cells are still poorly understood, and the literature contains numerous controversies, primarily due to the fact that different methods of isolation and different sources of cells have been utilized experimentally and clinically.

Endothelial progenitor cells (EPCs) are a specialized subset of hematopoietic cells found in the adult bone marrow and peripheral circulation arising from hemangioblasts prenatally. They are phenotypically characterized by antigens including CD133, CD34, c-kit, VEGFR2, CD144, and Sca-1. EPCs are immature cells which have the capacity to proliferate, migrate, and differentiate into endothelial lineage cells, but have not yet acquired characteristics of mature ECs, including surface expression of vascular endothelial cadherin and von Willebrand factor and loss of CD133. The discovery of circulating EPCs in the adult changed the view that new blood vessel growth occurs exclusively by angiogenesis postnatally. The process of vasculogenesis in the developing embryo is thought to be re-employed in the adult when EPCs are mobilized and recruited to regions of neovascularization to form new blood vessels. Parallels in the regulatory steps of embryonic and adult vasculogenesis suggest that the underlying initiating stimulus and regulatory pathways may be conserved. EPCs have been the subject of intense experimental and clinical investigations due to their therapeutic potential in cardiovascular regeneration. Since endothelial damage and/or dysfunction may initiate atherosclerosis, bone marrow-derived and circulating EPCs may play an important role in re-establishing a normal endothelium and thus protecting the vessel wall from progression of disease.

The role of EPCs in determining the pathogenesis and prognosis of cardiovascular disease is under study although there are currently no means to precisely track the kinetics of bone marrow-derived EPCs for pathological neovascularization in humans. The cell number and migratory activity of circulating EPCs are decreased in patients with stable coronary artery disease compared to age-matched control subjects and is inversely correlated with the number of coronary risk factors in coronary artery disease patients. Proliferation of EPCs obtained from patients with type II diabetes was decreased by 48% compared with controls. EPCs from subjects at high risk for cardiovascular events had higher rates of *in vitro* senescence than cells from subjects at low risk. These clinical findings indicate that EPCs may be sensitive indicators of heightened risk of cardiovascular diseases.

EPC transplantation to promote collateral circulation is a new therapy offering hope to treat tissue ischemia. When *ex vivo*-expanded EPCs obtained from healthy human peripheral blood are intravenously administered into immunodeficient mice with hindlimb ischemia, the result is cell incorporation and *in situ* differentiation into EC lineage, as well as physiological evidence of enhancement of limb blood flow. *Ex vivo* expanded human EPCs intravenously administered into nude rats with acute myocardial infarction contributed to ischemic neovascularization following recruitment into the ischemic area. The EPC therapy also inhibited left ventricular fibrosis and preserved left ventricular function. Human EPC-derived cardiomyocytes and SMC were identified in the rat infarcted myocardium. Whether improvement was due to increased myocardial perfusion and/or increased cardiac muscle mass was not studied. On the basis of the promising outcomes in animal models, clinical applications of EPCs for ischemic diseases are in progress. Intramyocardial injection of autologous bone marrow EPCs into patients with chronic myocardial infarction at the time of coronary artery bypass grafting surgery improved left ventricular global function, improved left ventricular ejection fraction, as well as cardiac perfusion 3–9 months later. Benefits of intramyocardial transplantation in intractable angina in chronically ischemia myocardium is likely due to induction of neovascularization. However, the life of transplanted EPCs is often short, so many investigators are interpreting the improved function to be due to paracrine effects that the EPCs have on resident cells due to the secretion of bioactive molecules such as cytokines.

Endothelial damage is a well-known trigger of restenosis after percutaneous balloon angioplasty or stenting. EPC-based therapeutic strategies to improve the function and number of ECs following angioplasty and stenting

have gained considerable interest. Statins were found to inhibit restenosis in injured murine carotid arteries through enhanced mobilization and incorporation of bone marrow-derived EPCs for re-endothelialization. Implanting VEGF-2 coated stents in rabbit injured iliac arteries also inhibited restenosis through enhanced mobilization of bone marrow-derived EPCs. Thus, EPCs appear to contribute to re-endothelialization in damaged vessels and inhibit restenosis. Intravenous infusion of EPCs in a mouse model of carotid artery injury leads to incorporation of EPCs into the injured vessel wall and re-endothelialization, resulting in inhibition of neointimal hyperplasia. Recently, stainless steel stents coated with anti-CD34 antibody were developed to capture circulating EPCs onto the stent surface to augment re-endothelialization and prevent restenosis and thrombosis. Preliminary clinical studies are under way.

Cardiac Stem Cells

The adult heart is considered to be a postmitotic organ comprising fully differentiated cardiomyocytes which survive a lifetime without replenishment. An elevated number of immature cardiomyocytes, capable of mitotic division, have been identified at the infarct border zone of myocardial infarction in some investigations, whereas other studies failed to identify sufficient proliferating cells at sites of myocardial infarction. However, adult heart may contain a population of stem cells either entering from the circulation or resident *in situ* which possess regenerative properties. If the biology of these cells can be characterized, then they may become useful clinically.

ATHEROSCLEROSIS

Atherosclerosis is a chronic vascular disease initially developing in the intima of elastic and larger muscular arteries and characterized by the presence of fibro-inflammatory lipid plaques (atheromas), which grow in size to protrude into the vascular lumen and to involve the media of the artery. Focal plaque growth eventually leads to clinical disease characterized by the development of complicated plaques, lumenal stenosis, and focal weakening of vessel walls, especially the aorta. Clinically, atherosclerosis leads to local aneurysm and/or rupture, and to end-stage organ disease including ischemic heart disease, cerebral vascular disease, and peripheral vascular disease. Epidemiologic studies have identified environmental and genetic conditions that increase the risk of developing clinical atherosclerotic disease (Table 14.4). Currently, risk modification has become an important medical

Table 14.4	Risk Factors for Atherosclerosis
LDL receptor mutations	Obesity
Hyperlipidemias	Family history
Diabetes mellitus	Gender
Hypertension	Advancing age
Cigarette smoke	

approach to prevention and treatment of atherosclerosis. Biomarkers of endothelial dysfunction are being studied including VCAM-1, ICAM-1, ELAM-1, high sensitivity C-reactive protein, IL-1, IL-6, and TNFα. It is hoped that these markers will provide a greater predictive capability beyond the traditional risk factors to identify high-risk individuals.

The interplay between an individual's genetic disposition and the environment adopted by the individual may result in an imbalance between proatherogenic and antiatherogenic factors and processes that then leads to initiation and growth of the atherosclerotic plaque. At present, it is unlikely to identify a single atherogenic gene that explains pathogenesis. Instead, multiple genes (polygenic), including clusters of genes forming networks regulating specific cell functions, interact with the environment and with each other to promote atherogenesis. Recently, the use of high-density genotyping arrays have led to genome-wide association studies showing a strong association of coronary artery disease with a chromosomal locus on chromosome 9p21.3. These studies in subjects of Northern European origin identified four single nucleotide polymorphisms (SNPs), two associated with risk for coronary artery disease and two associated with risk for myocardial infarction. The pathobiologic role of the 9p21.3 locus is not understood at present.

The development of atherosclerosis may begin as early as the fetal stage, with the formation of intimal cell masses, or perhaps shortly after birth, when fatty streaks begin to evolve. However, the characteristic lesion, which is not initially clinically significant, requires as long as 20 to 30 years to form.

A convenient way to view pathogenesis is to propose that three stages can be identified over the course of the evolution of the clinical plaque in humans: (i) a plaque initiation and formation stage, (ii) a plaque adaptation stage, and (iii) a clinical stage. Biologically active molecules regulate a number of dynamic cellular functions. As more molecules are identified in the vessel wall and the plaque, their function is studied to determine their antiatherosclerotic (atheroprotective) or atherogenic (atherogenic) potential. In addition, a given molecule may have both atheroprotective and atherogenic effects. Much effort has gone into identifying genetic variants of specific genes that are described as candidate genes because they have been shown to be important in health and disease of the artery wall. Examples include ACE insertion/deletion, *APOE*, *APOE2, APOE3, APOE4,* and *MTHFR* C677T. These genetic association studies represent our best understanding at present, but most are not robust and further studies are needed. Caution is required in interpreting genetic association studies, especially single reports and reports that are difficult to confirm because there is variability in outcomes of several different studies.

Stage I: Plaque Initiation and Formation

The intimal lesions initially occur at vascular sites considered to be predisposed to plaque formation. Endothelial injury is an early event and may be due to several conditions including microorganisms, toxins, hyperlipidemia, hypertension, and immunologic events. Hemodynamic shear stress at branch points and curves may also induce endothelial dysfunction, a predisposition for plaque formation. The accumulation of subendothelial SMCs, as occurs in an intimal cell mass (eccentric intimal thickening, intimal cushion) at branch points and at other sites in certain vessels, particularly the coronary arteries, is considered a predisposing condition for plaque formation since it provides a readily available source of SMCs. It is thought that this intimal thickening is a physiological adaptation to mechanical forces. In humans, atherosclerotic lesions tend to occur at sites where shear stresses are low but fluctuate rapidly, such as at branch points. Low shear has been shown to induce cell adhesion molecules on the surface of endothelial cells to promote monocyte attachment. This is regulated by upregulation of vascular cell adhesion molecule (VCAM) on the surface of ECs. The leukocytes first roll along the endothelium mediated by P-selectin and E-selectin and then adhere due to chemokine-induced EC activation and integrin interactions with cell adhesion molecules. The leukocytes penetrate the endothelial barrier at interendothelial sites, regulated by platelet EC adhesion molecule (PECAM, CD31). Low shear also disrupts normal repair following endothelial injury, thus exposing a denuded endothelial surface to blood flow for longer periods of time. Hemodynamic forces induce gene expression of several biologically active molecules in ECs that are likely to promote atherosclerosis, including FGF-2, tissue factor (TF), plasminogen activator, and endothelin (Table 14.3). However, shear stress also induces gene expression of agents that are considered antiatherogenic, including NOS and PAI-1 (Table 14.3).

Lipid accumulation depends on disruption of the integrity of the endothelial barrier through disruption of cell-cell adhesion junctions, cell loss, and/or cell dysfunction. Low-density lipoproteins carry lipids into the intima. Monocyte/macrophages adhere to activated ECs and transmigrate into the intima bringing in lipids. Some macrophages become foam cells, due in part to the uptake of oxidized LDL via scavenger receptors, and undergo necrosis and release lipids. A change in the types of connective tissue and proteoglycans synthesized by the SMCs in the intima also renders these sites prone to lipid accumulation.

Macrophages, in addition to playing a central role by participating in lipid accumulation, secrete several types of cytokines and release growth factors, thereby promoting further accumulation of SMCs. Oxidized lipoproteins induce tissue damage and further macrophage accumulation. Macrophages secrete MCP-1, which promotes macrophage accumulation. Macrophages secrete reactive oxygen species. Monocyte/macrophages synthesize PDGF, FGF, TNF, IL-1, IL-6, interferon-γ (IFN-γ), and TGFβ, each of which can modulate the growth of SMCs and ECs. For example, IFN-γ and TGFβ inhibit cell proliferation and could account for the failure of EC regeneration to maintain an intact surface over the lesion as it protrudes into the lumen. Cytokines IL-1 and TNF stimulate endothelial cells to produce platelet-activating factor (PAF), TF, and plasminogen activator inhibitor (PAI). TF expression is also upregulated by oxidized lipids and by disruption of the fibrous cap. Thus, the normal anticoagulant vascular surface becomes a procoagulant one.

As the lesion progresses, small mural thrombi may develop on the damaged intimal surface, which has become prothrombotic. This stimulates the release of numerous molecules from adherent and activated platelets, including PDGF, which accelerates smooth muscle proliferation; TGFβ, which enhances the secretion of matrix components; and thrombin, ADP, and thromboxane, which promote further platelet activation. The thrombus grows as these molecules promote the prothrombotic state. Since thrombosis also initiates fibrinolysis and inhibition of factors in the coagulation pathway, the thrombus may alternatively lyse. Another scenario modulated in part by TGFβ is organization of the thrombus and incorporation into the plaque. Further growth of the thrombus is a function of the coagulation cascade, which may continue to be stimulated by cytokines and tissue factor.

The deeper parts of the thickened intima are poorly nourished. Hypoxia promotes HIF-1alpha translocation to the nucleus of SMCs and macrophages, which bind to the promoter-specific hypoxia response element, leading to the transcriptional activation of VEGF, and other target genes. Macrophages and SMCs undergo ischemic necrosis, as well as apoptosis. Cell death is also promoted by proteolytic enzymes released by macrophages and by tissue damage caused by oxidized LDL and other reactive oxygen species. VEGF initiates angiogenesis with new vessels forming in the plaque derived from the vasa vasorum. Some regard the presence of neovascularization as a condition that establishes permanency to the plaque and prevents significant regression.

The fibroinflammatory lipid plaque is formed, with a central necrotic core and a fibrous cap which separates the necrotic core from the blood in the lumen (Table 14.5). The plaque is heterogeneous with respect to inflammatory cell infiltration, lipid deposition, and matrix organization. TGFβ is an important regulator of plaque remodeling and extracellular matrix deposition. TGFβ increases several types of collagen, fibronectin, and proteoglycans. It inhibits proteolytic enzymes that promote matrix degradation and enhances expression of protease inhibitors.

The expression of HLA-DR antigens on both ECs and SMCs in plaques implies that these cells have undergone some type of immunological activation, perhaps in response to IFN-γ released by activated T-cells in the plaque. The presence of T-cells reflects an immune response that is important for the progression of atherosclerotic lesions. Possible antigens include oxidized LDL to which antibodies have been identified in the plaque.

Stage II: Adaptation Stage

As the plaque encroaches upon the lumen (in coronary arteries), the wall of the artery undergoes remodeling to maintain the original lumen size likely regulated by TGFβ. Once a plaque encroaches upon half the lumen, compensatory remodeling can no longer maintain normal patency, and the lumen of the artery becomes narrowed (stenosis). Hemodynamic shear stress is an important regulator of vessel wall remodeling acting through the mechanotransduction properties of the ECs. Shear stress activates the expression of a variety of genes that encode for proteins that promote remodeling such as MMPs, collagens, bFGF, TGFβ, and inflammatory factors. Smooth muscle cell turnover-proliferation and apoptosis, and matrix synthesis and degradation modulate remodeling of the vessel and the plaque in the face of atherosclerosis. The molecules that are important in matrix remodeling are metalloproteinases (MMP) and their inhibitors (TIMP). This compensatory remodeling is useful because it maintains patency and blood flow in the lumen. However, it may delay clinical diagnosis of the presence of atherosclerosis since the plaque may be clinically silent. At this stage it would be very useful to have a group of biomarkers that can reliably assess the extent of subclinical atherosclerosis present in the unsuspecting person.

Stage III: Clinical Stage

Plaque growth continues as the plaque encroaches into the lumen. Hemorrhage into a plaque due to leakage from the small fragile vessels of neovascularization may not necessarily result in actual rupture of the plaque but may still increase plaque size. Complications develop in the plaque, including surface erosion, ulceration, fissure formation, calcification, and aneurysm formation. Calcification is driven by chondrogenesis and osteogenesis, regulated in part by TGFβ, osteogenic progenitor cells, and bone-forming proteins. Activated mast cells are found at sites of erosion and may release proinflammatory mediators and cytokines. Continued plaque growth leads to severe stenosis or occlusion of the lumen. Plaque rupture, through the fibrous cap, and ensuing lumen

Table 14.5	Composition of Atherosclerotic Plaque

Cells	– Endothelial – Smooth Muscle – Macrophages – Foam – Lymphocytes (T-cells) – Giant cells
Matrix	– Collagens – Proteoglycans—biglycan, versican, perlecan – Elastin – Glycoproteins
Lipids and lipoproteins, cholesterol crystals	
Serum proteins	
Platelet and leukocyte products	
Necrotic debris	
Microvessels	
Hydroxyapatite crystals	

Table 14.6	Plaque Rupture

- Endothelial erosion, ulceration, fistula
- Thin fibrous cap
- Decreased smooth muscle cells in cap
- Inflammation—Macrophages
- Foam cells
- Hemodynamic shear stress
- Imbalance in matrix synthesis/degradation (metalloproteinases, tissue inhibitors of metalloproteinases)
- Nodular calcification

thrombosis and occlusion may precipitate acute catastrophic events in these advanced plaques, such as acute myocardial infarct. Table 14.6 describes risk factors for plaque rupture. Plaques causing less than 50% stenosis may also suddenly rupture. Investigations to discover biomarkers to identify patients with plaques at risk for rupture have not been successful to date.

ISCHEMIC HEART DISEASE

Ischemic heart disease (coronary heart disease) is described clinically as stable angina and acute coronary syndromes including unstable angina, non-ST elevation myocardial infarction, and ST-elevation myocardial infarction. Reliable biomarkers are available to identify cardiac damage, especially sensitive cardiac troponin assays. There have been investigations to validate putative risk factors of acute coronary syndromes, but candidate gene variants have not been well characterized. Those in the field are aware that new novel biomarkers will not enter the clinical arena unless they are shown to significantly improve diagnosis and provide clinicians with better tools to monitor and guide treatment than are currently available.

Ischemia of the myocardial cell leads to a series of intracellular structural and biochemical changes which begin almost immediately after onset and evolve over time. Initiation of ATP depletion begins within seconds. Initially the reaction of the cells results in reversible injury; however, by 20–30 minutes the myocardial cells become irreversibly injured and undergo necrosis. In the myocardium, cardiomyocytes in the subendocardium are most at risk for ischemia. Thus, irreversible injury occurs first in the subendocardium and progresses as a wave front toward the epicardium, resulting in a transmural myocardial infarct. Progression of necrosis will involve the full cardiac bed supplied by the occluded coronary artery and usually is complete by 6 hours. The repair of the infarcted myocardium follows a well-characterized sequence of necrosis, inflammation, granulation tissue, remodeling, and scar formation.

ANEURYSMS

Aneurysms may occur in any vessel. However, the thoracic aorta, the abdominal aorta, and the cerebral arteries are common sites. These aneurysms often result in rupture. Blood dissects along the long axis of the media resulting in a channel filled with blood. There is an intimal tear, the entry point, and often a distal exit tear back into the lumen. Dissection may involve aortic branches. Thoracic dissections are classified based on involvement of the ascending aorta: Type A (ascending aorta involved) and Type B (distal, sparing the ascending aorta). Associated conditions include atherosclerosis, hypertension, bicuspid aortic valve, and idiopathic aortic root dilation.

Medial degeneration characterized by fragmentation and loss of elastic fibers, accumulation of proteoglycans, and depletion of SMCs is a common nonspecific histologic finding in aneurysm or dissection. Degenerative medial change is associated with hyaline and hyperplastic arteriosclerosis of the adventitial vasa vasorum. In all cases of aneurysms and dissection whatever the histologic picture, vascularization is present in the areas of medial destruction in the form of thin-walled, widely patent vessels. Thus, the histopathology is not helpful in establishing the pathogenesis of thoracic aortic aneurysms and dissections. However, genetic information has linked several of these cases to genetic syndromes, including Marafan, Ehlers-Danlos (Type IV), and Loeys-Dietz syndromes, and filamin A mutations. In addition, these conditions may also occur as an inherited autosomal dominant condition with decreased penetrance and variable expression without the syndromes. Mutations have been identified in genes for fibrillin-1 (*FBN1*), TGFβ receptor 2 (*TGFβR2*), TGFβ receptor1 (*TGFβR1*), SMC specific β myosin (*MYH11*), and α-actin (*ACTA2*).

Studies suggest that both human and experimental mouse models show that an increase in TGFβ signaling is important in the pathogenesis of Marfan and Loeys-Dietz syndromes (see pages 172–173). Patients with single gene defects suggest that mutations disrupt the contractile functions of vascular SMCs, which leads to activation of the stress and stretch pathways of the SMC. It has been postulated that the stretch pathways promote increased levels of matrix metalloproteins (MMPs; especially MMP2 and MMP9) and proteoglycans, and promote proliferative agents such as IGF-1, TGFβ, and MIP1α and MIP1β. A recent review suggests that the progress in identifying genetic determinants for intracranial aneurysms is very limited. Genome-wide linkage studies have identified two loci on chromosomes 1p34.3-p36.13 and 7q11 with association with positional candidate genes, perlecan gene and elastin and collagen type 1 2A gene, respectively. The authors discuss the difficulties encountered in studies in the genetics of intracranial aneurysms.

The pathogenesis of abdominal aortic aneurysms is poorly understood. However, connective tissue degradation, inflammation, and loss of smooth muscle are characteristic features. Rupture is considered to be due to collagen degradation due to increased MMP2, MMP9, and cysteine collagenase. Recent studies identified MMP8; cysteine proteases cathepsin K, L, and S; and osteoclastic proton pump vH+-ATPase as important enzymes that are responsible for proteolysis of the medial and adventitial type I/III fibrillar collagen.

Others have focused on the role of cytokines in abdominal aortic aneurysms, including IL1β, TNFα, MCP-1, IL-8, and others. Utilizing expression profiling of a 42-cytokine protein array, it was shown that the aneurysm wall exhibits a specific cytokine profile of upregulated proinflammatory cytokines, chemokines, and growth factors. The descriptive analysis of end stage human tissue confirmed previous findings and also identified several new factors, including GCSF, MCSF, IL-13, and others.

VALVULAR HEART DISEASE

Mitral Valve Prolapse

Mitral valve prolapse (MVP) is a heart valve condition characterized by progressive thinning of the mitral leaflet tissue, causing leaflets to billow backward during ventricular contraction, prolapsing into the left atrium beyond their normal position of closure at the level of the mitral ring or annulus. The most common etiology of systolic mitral regurgitation in patients with severe mitral insufficiency referred for mitral valve surgery remains myxomatous degeneration. The myxomatous changes are seen as part of connective tissue syndromes or as primary valve disease.

The natural history of asymptomatic MVP is extremely heterogeneous. It can vary from benign, with a normal life expectancy, to adverse, with significant morbidity and mortality attributed to the development of valvular insufficiency. Fortunately, its complications, including heart failure, mitral regurgitation, bacterial endocarditis, thromboembolism, and atrial fibrillation, are extremely uncommon, affecting less than 3% of subjects with MVP. A cause of abrupt clinical deterioration is sudden chordal rupture due to attenuation and thinning of chordal tissue. When associated with systemic disease, like Marfan syndrome, the myxomatous degeneration is more extensive and involves other heart valves.

MVP valves show myxomatous degeneration with greatly increased type III collagen, some increase of type I and V collagens, and an accumulation of dermatan sulfate, a glycosaminoglycan, within the valve matrix. The accompanying loss in elastin and reduction in SMCs is similar to the histological changes in the valve cusps that were described previously in dissecting aneurysm and are often part of syndromes that involve valve and aorta.

Connective Tissue Disorders

MVP is a feature of many patients with Ehlers-Danlos syndrome and Marfan syndrome, which are linked to collagen and *fibrillin-1* mutations, respectively. MVP has also been documented to be more prevalent in patients with osteogenesis imperfecta and other collagen-related disorders. This association of MVP with inherited connective tissue disorders suggests that abnormalities in matrix proteins of the connective tissues are important in the etiology of MVP.

Ehler-Danlos syndrome is a rare and heterogeneous group of numerous connective tissue heritable disorders characterized by joint hypermobility, skin hyperextensibility, cardiac valvular defects, and tissue fragility. Ehler-Danlos syndrome type IV is associated with mutations in the *COL3A1* gene, which encodes type III procollagen.

Marfan syndrome is an autosomal dominant genetic disorder of the connective tissue associated with mutations in *fibrillin-1*, a major component of the microfibrils that form a sheath surrounding amorphous elastin. *Fibrillin-1* is essential for the proper formation of the extracellular matrix including the biogenesis and maintenance of elastic fibers. The extracellular matrix is critical for both the structural integrity of connective tissue but also serves as a reservoir for growth factors which are essential in the normal maintenance of valve structure and function as well as in response to injury and regulation of repair. The interaction of hemodynamic and mechanical forces with the genetically altered extracellular matrix is not well studied. However, it is likely that these physical forces are important determinants of valve dysfunction.

TGFβ Dysregulation

TGFβ, a 25 kDa protein that is a member of the TGFβ superfamily, is a well-known regulator of extracellular matrix deposition and remodeling (Figure 14.2). It is secreted by numerous cell types including heart valve interstitial cells, the predominant cell type in heart valves, with potent autocrine effects. It is known to promote differentiation of mesenchymal cells into myofibroblasts and to regulate multiple aspects of the myofibroblast phenotype through transcriptional activation of alpha-smooth muscle actin, collagens, matrix metalloproteinases, and other cytokines, such as connective tissue growth factor and basic fibroblast growth factor. TGFβ is secreted in a latent complex containing active TGFβ and latency-associated protein (LAP). This latent complex is tethered through latent TGFβ binding proteins (LTBPs) to matrix proteins to allow cells to tightly regulate TGFβ bioavailability and create special gradients in the cell microenvironment. Fibrillin-1 regulates TGFβ activation. Fibrillin-1 interacts with LTBPs to sequester latent TGFβ at specific locations in the matrix and stabilizes the inactive large latent complex (TGFβ, LAP, and LTBP), rendering it less prone to activation. Reduced or mutated fibrillin-1 leads to increased TGFβ activation and subsequent elevated levels of TGFβ signaling, resulting in cellular responses such as extensive degradation and remodeling of the extracellular matrix. In mouse models of Marfan syndrome with fibrillin-1 mutations, abnormal mitral valves show increased TGFβ activity leading to an MVP-like phenotype. This mitral valve abnormality can be rescued by perinatal administration of neutralizing antibodies to TGFβ.

The importance of the TGFβ pathway in MVP pathogenesis is further highlighted by the discovery of Loeys-Dietz syndrome, which has many similar clinical features to Marfan syndrome. Loeys-Dietz syndrome is caused by mutations in the genes encoding TGFβR1

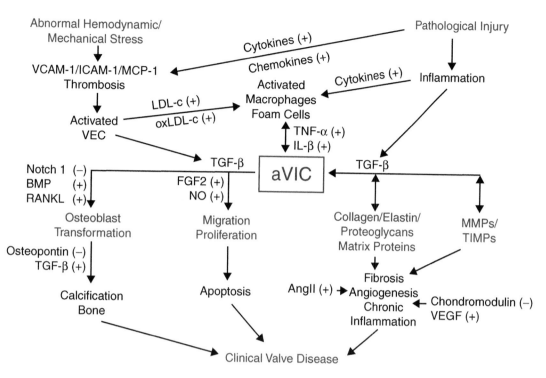

Figure 14.2 **The normal adult heart valve is well adapted to its physiological environment, able to withstand the unique hemodynamic/mechanical stresses under normal conditions.** Under conditions of pathological injury or abnormal hemodynamic/mechanical stresses, VICs become activated through activation of VECs and by inflammation and associated cytokine and chemokine signals. Macrophages will also be activated. aVICs increase matrix synthesis; upregulate expression of matrix remodeling enzymes; migrate, proliferate and undergo apoptosis; as well as undergo osteoblast transformation. These processes are regulated by a variety of factors, several secreted by the aVIC. If the aVICs continue to promote these cellular processes, angiogenesis, chronic inflammation, fibrosis, and calcification result, leading to progressive clinical valve disease. Reprinted with permission from the American Society for Investigative Pathology, *Am J Pathol.* 2007;171:1407–1418.

or TGFβR2. Losartan, an angiotensin II receptor antagonist that modulates the interaction of TGFβ with valves and vascular structures leading to blocking of TGFβ activity, can rescue the expression of the lethal aortic aneurysm in the mouse model of Marfan syndrome.

The newly discovered filamin A mutation responsible for X-linked valvulopathy with MVP-like phenotypes may also exert its effect through the interaction of *filamin A* with molecules in the TGFβ pathway. *Filamin A* is a ubiquitous phosphoprotein that cross-links actin filaments and links the actin cytoskeleton to the plasma membrane by interacting with transmembrane proteins such as β-integrins. *Filamin A* has also been implicated in regulating many cellular signaling pathways by acting as a scaffold for intracellular proteins involved in signal transduction. *Filamin A* may contribute to the development of myxomatous changes of the heart valves by augmenting TGFβ signaling through its interaction with Smad proteins such as Smad-2 and Smad-5. Thus, defective Smad-mediated TGFβ signaling due to *filamin A* mutations appears to underlie the pathogenesis of X-linked valvulopathy.

CARDIOMYOPATHIES

The impact of cell and molecular biology on cardiovascular disease is reflected in the recent reclassification of cardiomyopathies published by the *America Heart Association* and by the *European Society of Cardiology*. Cardiomyopathies are a heterogeneous group of human diseases of the myocardium with primary dysfunction of the cardiomyocytes, the etiology of which is either known or unknown. The former includes ischemia, hypertension, heart valve abnormalities, alcoholism, as well as systemic diseases such as diabetes, hemochromatosis, and amyloidosis. These are usually referred to by their etiology. Those of unknown etiology are classified into dilated, hypertrophic, restrictive, and arrhythmogenic right ventricular cardiomyopathy based on gross and microscopic features and clinical features. Recently, noncompaction cardiomyopathy has been considered either as a distinct cardiomyopathy or as a trait common to several types of cardiomyopathies. As the cell and molecular biology becomes known to us, many of those with unknown etiology turn out to be primary disorders of the myocardium due to one of hundreds of inherited mutations in genes encoding cytoskeletal or sarcomere proteins. Hypertrophic cardiomyopathy (HCM) is considered a disease of the sarcomere; dilated cardiomyopathy, a disease of sarcomere-sarcolemma; and arrhythmogenic right ventricular cardiomyopathy, a disease of desmosomes. Although overlaps in mutations in genes encoding proteins of the respective structures occur, the molecular pathogenesis of each cardiomyopathy is unique. However, in most cases it is not known how the specific mutation in the gene results in the dysfunction seen at the cellular level. To date, understanding the genetic causes and the molecular pathogenesis of these distinct cardiomyopathies has allowed improved classification and management of these diseases, as well as the development of better therapies. The *American Heart Association* classification takes molecular genetics under consideration, and the scope of the classification has been broadened to include ion channelopathies. Clinical genetic testing is available for several monogenic forms of cardiomyopathy. However, at the present time physicians need to discuss the benefits and limitations of such testing with patients and family members.

Molecular Genetics and Pathogenesis of Hypertrophic Cardiomyopathy

HCM is a commonly inherited cardiomyopathy characterized by marked thickening of the left ventricular wall in the absence of increased external load and with cardiomyocyte disarray and cardiac fibrosis. Clinical consequences of HCM include heart failure, arrhythmias, and sudden death. HCM is primarily inherited as an autosomal dominant trait, and in rare cases shows mitochondrial inheritance.

Sixty percent of HCM is due to more than 450 different mutations within 8 sarcomeric genes. Most HCM mutations occur in two genes, *MYH7* and *MYBPC3*, encoding the beta-myosin heavy chain and myosin binding protein C, respectively. Mutations in genes that encode cardiac troponin T and cardiac troponin I, essential myosin light chain, regulatory myosin light chain, alpha-tropomyosin, and cardiac actin are less common in HCM patients. Rare mutations in cardiac troponin C and alpha-myosin heavy chain also cause HCM. In addition, mutations in titin, muscle LIM protein, telethonin, and myozenin, which are Z-disc proteins forming a framework connecting sarcomere units to each other, can also cause HCM.

Despite the numerous genetic mutations leading to HCM, it is unclear whether the molecular pathogenesis of HCM is due to impaired sarcomere function or to a gain of new sarcomere function. Findings of mutant myosin fragments and skeletal muscles carrying HCM mutations show decreased actin sliding velocity, supporting the hypothesis that mutations in the myosin components in HCM cause a decrease in motor function of sarcomeres and a compensatory hypertrophic response of the myocardium. Models of HCM carrying mutations in myosin heavy chain, myosin binding protein C, and troponin T support the alternate hypothesis that mutant proteins result in abnormal sarcomere function rather than a decrease in motor function. These animal models show enhanced actin-activated myosin ATPase activity, increased force generation, and accelerated actin sliding velocity, which may be the molecular mechanisms that result in increased cardiac performance often evident in patients with HCM.

While the functional changes in sarcomeres in HCM remain unclear, the initial step in the pathogenesis of HCM is known to be incorporation of mutant protein with altered mechanical properties to the normal sarcomere. It is likely that the heterogeneity of mutant and wild-type structural proteins within the sarcomere uncouples the normal mechanical coordination between myosin heads and the associated enhanced ATPase activity, resulting in higher levels of energy consumption. This increased energy consumption may trigger hypertrophy, and combined with decreased energy supply as a result of impaired blood flow to the hypertrophied heart, myofibrillar disarray may result, leading to cardiomyocyte death and fibrosis.

Ca2+ homeostasis is disrupted very early in the pathogenesis of HCM. For example, cardiomyocytes from mice with alpha-myosin heavy chain mutations show significant reductions in sarcoplasmic reticulum Ca2+ content and reduced expression levels of proteins responsible for intracellular Ca2+ regulation such as the cardiac ryanodine receptor Ca2+-release channel, the sarcoplasmic reticulum Ca2+ storage protein calsequestrin-2, and the associated anchoring proteins, triadin and junction. A proposed mechanism is that sarcomeric release of Ca2+ at the end of systole is impaired, leading to Ca2+ accumulation within the sarcomere and Ca2+ release from the sarcoplasmic reticulum into the cytoplasm. Elevated cytoplasmic diastolic Ca2+ levels induce hypertrophic responses of cardiomyocytes through the calcineurin-NFAT signaling pathway. Increased intracellular Ca2+ results in calmodulin saturation and calcineurin activation, leading to subsequent dephosphorylation of the nuclear factor of activated T-cells (NFAT) transcription factor in the cytoplasm. Upon dephosphorylation, NFAT translocates to the nucleus, where it induces expression of growth-related genes for cardiomyocytes. This mechanism is supported by the evidence that transgenic mice overexpressing activated calcineurin demonstrate a profound hypertrophic response of cardiomyocytes, leading to a 2-fold to 3-fold increase in heart size, which rapidly progresses to dilated heart failure. Administration of the L-type Ca2+ channel inhibitor diltiazem corrects some of the Ca2+-related changes and prevents HCM 50% of the time.

Molecular Genetics and Pathogenesis of Dilated Cardiomyopathy

Dilated cardiomyopathy (DCM) is the most common form of cardiomyopathy, involving about 90% of all clinical cardiomyopathies. It occurs more frequently in men than in women. It is characterized by increased ventricular size, reduced ventricular contractility, and left or biventricular dilation. Clinical features include heart failure. Twenty to forty percent of DCM patients have familial forms of the disease, with autosomal dominant inheritance being most common. Autosomal recessive, X-linked, and mitochondrial inheritance of the disease is also found. The majority of mutations are in genes encoding either cytoskeletal or sarcomeric proteins. Mutations in the cytoskeletal proteins desmin, delta-sarcoglycan, and metavinculin lead to defects of force transmission, while mutations in the sarcomeric proteins lead to defects of force generation in the myocardium. Both of these conditions result in the DCM phenotype. In the majority of instances, the mechanism through which a loss or a defect in these proteins alters function is not well understood. However, these genetic mutations can be used as biomarkers to carry out risk stratification in order to improve management of cardiomyopathies. Further, these proteins may be potential therapeutic targets.

Cytoskeletal Defects

Mutations in 13 genes have been identified to date in autosomal dominant DCM, including desmins, delta-sarcoglycan, metavinculin, alpha-actinin-2, ZASP, actin, troponin T, beta-myosin heavy chain, titin, and myosin binding protein C. The majority of DCM mutations are in genes encoding the cytoskeletal proteins desmin, delta-sarcoglycan, and metavinculin.

Desmin is a cytoskeletal protein which forms intermediate filaments in cardiac, skeletal, and smooth muscle. Desmin is found at the intercalated discs and functions to attach and stabilize adjacent sarcomeres. Desmin networks also provide a scaffold allowing connections to adjacent sarcolemma and to sarcolemmal costameres.

Delta-sarcoglycan is a member of the sarcoglycan subcomplex in the DAPC. It is involved in stabilization of the cardiomyocyte sarcolemma as well as signal transduction. In the absence of delta-sarcoglycan, the remaining sarcoglycans (beta, gamma, sigma) cannot assemble properly in the endoplasmic reticulum. Mouse models of delta-sarcoglycan deficiency also sometimes demonstrate HCM instead of DCM. Some patients with delta-sarcoglycan mutations show a form of autosomal recessive limb girdle muscular dystrophy.

Patients with mutations in the cytoskeletal gene metavinculin encoding vinculin and its splice variant metavinculin also present with DCM. Vinculin is ubiquitously expressed and metavinculin is coexpressed with vinculin in cardiac, skeletal, and smooth muscle. It is localized to subsarcolemmal costameres in the heart, where they interact with alpha-actinin, talin, and gamma-actin to form a microfilamentous network linking cytoskeleton and sarcolemma. In addition, vinculin and metavinculin are present in adherens junctions and in intercalated discs and participate in cell-cell adhesion.

Sarcomeric Defects

Mutations in sarcomeric genes may produce DCM or HCM. In the former, sarcomeric mutations giving rise to autosomal dominant inheritance are mainly in genes encoding actin, alpha-tropomyosin, and troponins. Cardiac actin is a sarcomeric protein in sarcomeric thin filaments interacting with tropomyosin and troponin complexes. Mutations in sarcomeric thin filament proteins cardiac actin, alpha-tropomyosin, cardiac troponin T, and troponin I all give rise to DCM. In addition to mutations in thin filament proteins, mutations in the thick filament protein beta-myosin heavy chain also cause DCM. These mutations also perturb the actin-myosin interaction and force generation, as well as alter cross-bridge movement during myocardial contraction.

X-linked and mitochondrial inheritance are the rarer forms of inheritance for DCM. X-linked DCM is caused by mutations in the cytoskeletal protein dystrophin. Multiple mutations in the dystrophin gene identified in DCM patients are affecting the 5′ end of the gene, leading to defective promoter region or N-terminus of the protein. Cardiac actin binds to the N-terminus of dystrophin, linking the sarcomere to the sarcolemma to stabilize the contractile structure.

Complete loss of dystrophin protein or dystrophin with an abnormal N-terminus leads to defective force transmission in the myocardium and DCM. Patients with mitochondrial myopathy due to mitochondrial mutations of the protein MIDNA found in the mitochondrial respiratory chain and responsible for energy generation also present with DCM.

Recent evidence indicates that mutations in lamin A/C in Emery-Dreifuss muscular dystrophy are also associated with DCM. The lamins are proteins in the nucleoplasmic side of the inner nuclear membrane. Lamin A and C are expressed only in heart and skeletal muscle. The mechanisms through which lamin A/C are responsible for the development of DCM are being studied and may be related to their role in maintaining nuclear integrity.

Molecular Genetics and Pathogenesis of Arrhythmogenic Right Ventricular Cardiomyopathy

Arrhythmogenic right ventricular cardiomyopathy (ARVC) is characterized by right ventricular fibro-fatty replacement of myocardial tissue. ARVC presents with palpitations or syncope as a result of ventricular tachyarrhythmias and is an important cause of sudden death at young ages.

ARVC is familial in about 50% of cases and is mainly autosomal dominant with some autosomal recessive inheritance. Mutations in ARVC have been identified in desmosomal proteins such as plakophilin-2, desmoplakin, plakoglobin, desmoglein-2, and desmocolin-2. Patients carrying mutations in the *PKP2* gene encoding plakophilin-2, an important desmosomal protein linking cadherins to intermediate filaments, present with ARVC at an earlier age than those without *PKP2* mutations. To date, 50 independent *PKP2* mutations, the majority of which are truncation mutations, are found in 70% of familial cases of ARVC. Mutations in desmoplakin are identified in 6% of ARVC patients in North America, and are associated with the autosomal recessive inherited syndromes—Carvajal syndrome with DCM, wooly hair, and palmoplantar keratoderma, and Naxos-like disorder consisting of ARVC, wooly hair, and an epidermolytic skin disorder. The C-terminus of desmoplakin binds desmin to tightly anchor intermediate filaments, and C-terminal mutations of desmoplakin are associated with an early left ventricular involvement in ARVC, while N-terminus mutations of desmoplakin are associated with a predominantly right ventricular phenotype. Homozygous deletions in the plakoglobin gene are identified in patients with Naxos disease with autosomal recessive inherited ARVC and palmoplantar keratoderma and wooly hair. *In vitro* studies in ECs demonstrate that inhibition of plakoglobin expression induces cell-cell dissociation in response to shear stress. Plakoglobin null mouse embryos die from ventricular rupture when the myocardium undergoes increased mechanical stress in embryonic development. Ten percent of ARVC patients are identified with mutations in *DSG2* encoding desmoglein-2, and 5% of ARVC patients are identified with mutations in *DSC2* encoding desmocollin-2, both major components of desmosomal cadherins. A number of these mutations are splice-site mutations that lead to left ventricular involvement.

The molecular mechanisms through which desmosomal mutations lead to ARVC are currently believed to be related to defects in desmosome composition and function, abnormalities of intercalated discs, and defective Wnt/β-catenin signaling. Mutations in genes encoding components of the desmosomal complex lead to insufficient incorporation of the desmosomal proteins into the complex, absence of protein-protein interactions, and incorrect incorporation of these proteins into the complex. These result in disturbed formation or reduced numbers of functional desmosomes. Defects and disruptions in desmosomes also lead to the inability of desmosomes to protect other junctions in the intercalated discs from mechanical stress such as the adhesion junctions between cardiomyocytes leading to loss of cell-cell contacts and cardiomyocyte death. The destabilization of cell adhesion complexes further perturbs the kinetics of gap junctions with smaller and fewer gap junctions present and reduced localization of the gap junction protein connexin 43 at the intercalated discs. This results in heterogeneous conduction, a contributor to the characteristic arrhythmogenesis in ARVC.

The recent discovery that desmosomal components participate in the Wnt/β-catenin pathway involved in a variety of developmental processes including cell proliferation and differentiation help explain the histological findings of fibro-fatty tissue replacement of cardiomyocytes in ARVC. Desmosomal dysfunction results in nuclear translocation of the desmosomal protein plakoglobin, resulting in competition between plakoglobin and β-catenin. This leads to inhibition of Wnt/β-catenin signaling and a shift in cell fate from cardiomyocyte to adipocyte. Plakophilins are also localized to the nucleus, and may be involved in transcriptional regulation or in regulating β-catenin activity in Wnt-signaling. In all cases, defective desmosomes are believed to be unable to maintain tissue integrity under excessive mechanical stress, such as in the thinnest areas of the right ventricle. These include the right ventricular outflow tract, inflow tract, and the apex, which are most vulnerable to ARVC. Thus, it is postulated that patients with desmosomal protein mutations are predisposed to damage in these areas, leading to disruption and subsequent degeneration of cardiomyocytes followed by replacement by fibro-fatty tissue.

Molecular Genetics and Pathogenesis of Noncompaction Cardiomyopathy

Noncompaction cardiomyopathy (NCC) is a rare congenital cardiomyopathy. In NCC, myocardial development is hindered during embryogenesis beginning around 8 weeks post-conception. At this stage in development, the myocardium is sponge-like to maximize surface area and to allow perfusion of the myocardium from the left ventricular cavity. However, as the embryo grows, the myocardium compacts and matures. The myocardium of NCC patients fail to fully compact, leaving the myocardium with a spongiform appearance. Symptoms result from poor pumping performance of the heart and tachyarrhythmia, thromboembolisms, and sudden death.

NCC can be inherited in an autosomal dominant manner or through X-linked inheritance. The most common gene responsible for NCC is *TAZ*, an X-linked gene encoding taffazin, which is involved in the biosynthesis of cardiolipin, an essential component of the mitochondrial inner membrane. Mutations in *TAZ* also cause Barth syndrome, a metabolic condition with DCM, with or without noncompaction, neutropenia, skeletal myopathy, and 3-methylglutaconic aciduria. Mutations in cytoskeleton and sarcomere-related genes *DTNA* and *LDB3* can also give rise to NCC. *DTNA* encodes alpha-dystrobrevin, a dystrophin-associated protein involved in maintaining the structural integrity of the sarcolemma. *LDB3* encodes the sarcomeric Z-band protein, LIM domain binding 3 protein. Unfortunately, due to its rare occurrence, the molecular pathogenesis for NCC is not well understood.

Channelopathies

Congenital long QT syndromes, which affect about 1 in 3000 persons, are a potentially lethal group of cardiac conditions described by delayed repolarization of the myocardium and QT prolongation. This arrhythmogenic disorder is characterized by a significant increased risk of syncope, seizures, and sudden cardiac death. Since several mutations in genes that encode ion channels or their associated proteins account for about 80% of cases, postmortem genetic testing for sudden unexplained death in the young is very useful.

KEY CONCEPTS

- Basic cardiovascular research is essential to provide new knowledge that can have innovative and transformative clinical applications.
- Molecular pathology driven by genomics and proteomics is enhancing the understanding of pathogenesis, the diagnosis, and the treatment of cardiovascular diseases.
- Although some cardiovascular diseases are monogenic in nature due to alterations in a single gene, many cardiovascular diseases are likely to be polygenic making investigations more complex.
- Utilization of stem cells and progenitor cells offer promising therapeutic approaches to cardiovascular diseases, especially with respect to cellular regeneration and repair.
- Although many known genetic abnormalities have not yet been linked to specific cellular dysfunctions, the expectation is that molecular pathology will provide new avenues for diagnosis, treatment, prognosis, and prevention of cardiovascular disease.

SUGGESTED READINGS

1. Alcalai R, Seidman JG, Seidman CE. Genetic basis of hypertrophic cardiomyopathy: From bench to clinics. *J Cardiovasc Electrophysiol.* 2008;19:104–110.
2. Allaire E, Schneider F, Saucy F, et al. New insights in aetiopathogenesis of aortic disease. *Eur J Vasc Endovasc Surg.* 2009;37:531–537.
3. Cunningham KS, Gotlieb AI. The role of shear stress in the pathogenesis of atherosclerosis. *Lab Invest.* 2005;85:9–23.
4. Dejana E, Tournier-Lasserve E, Weinstein BM. The control of vascular integrity by endothelial cell junctions: Molecular basis and pathological implications. *Dev Cell.* 2009;16:209–221.
5. Elliott P, Andersson B, Arbustini E, et al. Classification of the cardiomyopathies: A position statement from the European Society of Cardiology working group on myocardial and pericardial disease. *Eur Heart J.* 2008;29:270–276.
6. Gotlieb AI. Blood vessels. In: Rubin T, Strayer DS, eds. *Rubin's Pathology.* 5th ed. Philadelphia: Lippincott Williams and Wilkins; 2007:387–426.
7. Hamsten A, Eriksson P. Identifying the susceptibility genes for coronary artery disease: From hyperbole through doubt to cautious optimism. *J Intern Med.* 2008;263:538–552.
8. Hirschhorn JN, Lohmueller K, Byrine E, et al. A comprehensive review of genetic association studies. *Genetic Med.* 2002;4:45–61.
9. Liu AC, Joag VR, Gotlieb AI. The emerging role of valve interstitial cell phenotypes in regulating heart valve pathobiology. *Am J Pathol.* 2007;171:1407–1418.
10. Loeys BL, Schwarze U, Holm T, et al. Aneurysm syndromes caused by mutations in the TGF-beta receptor. *N Engl J Med.* 2006;355:788–798.
11. Maron BJ, Towbin JA, Thiene G, et al. Contemporary definitions and classification of the cardiomyopathies: An American Heart Association Scientific Statement from the Council of Clinical Cardiology, Heart Failure and Transplantation Committee; Quality of Care and Outcomes Research and Functional Genomics and Translational Biology Interdisciplinary Working Groups; and Council on Epidemiology and Prevention. *Circulation.* 2006;113:1807–1816.
12. Milewicz DM, Guo D-C, Tran-Fadulu V, et al. Genetic basis of thoracic aortic aneurysms and dissections: Focus on smooth muscle cell contractile dysfunction. *Annu Rev Genomics Hum Genet.* 2008;9:283–302.
13. Ng CM, Cheng A, Myers LA, et al. TGF-β-dependent pathogenesis of mitral valve prolapsed in a mouse model of Marfan syndrome. *J Clin Invest.* 2004;114:1586–1592.
14. Olivotto I, Girolami F, Ackerman MJ, et al. Myofilament protein gene mutation screening and outcome of patients with hypertrophic cardiomyopathy. *Mayo Clin Proc.* 2008;83:630–638.
15. Singh NN, Ramji DP. The role of transforming growth factor-β in atherosclerosis. *Cytokine Growth Factor Rev.* 2006;17:487–499.
16. Tester DJ, Ackerman MJ. Postmortem long QT syndrome genetic testing for sudden unexplained death in the young. *J Am Coll Cardiol.* 2007;49:240–246.
17. Towbin JA, Bowles NE. Dilated cardiomyopathy: A tale of cytoskeletal proteins and beyond. *J Cardiovasc Electrophysiol.* 2006;17:919–926.
18. Van Gils JM, Zwaginga JJ, Hordijk PL. Molecular and functional interactions among monocytes, platelets, and endothelial cells and their relevance for cardiovascular diseases. *J Leukoc Biol.* 2009;85:195–204.
19. Wu KH, Liu YL, Zhou B, et al. Cellular therapy and myocardial tissue engineering: the role of adult stem and progenitor cells. *Eur J Cardiothorac Surg.* 2006;30:770–781.

15

Molecular Basis of Hemostatic and Thrombotic Diseases

Alice D. Ma . Nigel S. Key

INTRODUCTION AND OVERVIEW OF COAGULATION

Blood coagulation is the process whereby cells and soluble protein elements interact to form an intravascular blood clot. When this occurs in response to vessel injury, it is an important protective mechanism that functions to seal vascular bleeds, and thereby prevent excessive hemorrhage. This physiological process is generally referred to as hemostasis. However, in pathological situations, blood coagulation may be triggered by a variety of stimuli and lead to the formation of a maladaptive intravascular clot or thrombus that may obstruct blood flow to or from a critical organ, and/or embolize to a distal site through the circulatory system. This process is known as thrombosis or thromboembolism, and it may affect either the arterial or venous circulations. A comprehensive review of the molecular basis of all of the defects leading to disorders of hemostasis and thrombosis is beyond the scope of this chapter. We will instead focus on broad themes, particularly defects in the soluble coagulation factors, defects in platelet number or function, other defects leading to hemorrhage, and inherited defects predisposing to thrombosis.

Coagulation can be conceptualized as a series of steps occurring in overlapping sequence. Primary hemostasis refers to the interactions between the platelet and the injured vessel wall, culminating in formation of a platelet plug. The humoral phase of clotting (secondary hemostasis) encompasses a series of enzymatic reactions, resulting in a hemostatic fibrin plug. Finally, fibrinolysis and wound repair mechanisms are recruited to restore normal blood flow and vessel integrity. Each of these steps is carefully regulated, and defects in any of the main components or regulatory mechanisms can predispose to either hemorrhage or thrombosis. Depending on the nature of the defect, the hemorrhagic or thrombotic tendency can be either profound or subtle.

Primary hemostasis begins at the site of vascular injury, with platelets adhering to the subendothelium, utilizing interactions between molecules such as collagen and von Willebrand factor (VWF) in the vessel wall with glycoprotein and integrin receptors on the platelet surface. Specifically, the primary platelet receptor for subendothelial VWF under high shear conditions is the glycoprotein (GP) Ib/V/IX complex, while the primary platelet receptors for direct collagen binding are GP VI and the integrin $\alpha_2\beta_1$. Platelet spreading is followed by activation and release of granular components, and exposure to the cocktail of agonists exposed at a wounded vessel amplifies the process of platelet activation. Via a process known as inside-out signaling, the integrin $\alpha_{2b}\beta_3$ (also known as GP IIbIIIa) undergoes a conformational change in order to be able to bind fibrinogen, which cross-links adjacent platelets and leads to platelet aggregation. Activated GP IIbIIIa can also bind VWF under certain circumstances (such as higher shear rates). Secretion of granular contents is also triggered by external activating signals, further potentiating platelet activation. Lastly, the membrane surface of the platelet changes to serve as a scaffold for the series of enzymatic reactions that result in thrombin generation. This process is primarily dependent on the exposure of negatively charged phospholipids that are normally confined to the inner leaflet of the plasma membrane bilayer.

Our understanding of the process by which fibrin is ultimately generated by thrombin cleavage of soluble fibrinogen has undergone several iterations over the past half century. The waterfall (or cascade) model of coagulation was developed by two groups nearly simultaneously and included an extrinsic, intrinsic, and a common pathway leading to fibrin formation. In this model, the intrinsic pathway can be initiated by the action of kaolin, a negatively charged surface activator that interacts with the contact factors (prekallikrein or PK, high molecular weight kininogen or HMWK, and factor XII) to generate factor XIIa. Factor XIIa then cleaves and activates factor XI. Factor XIa activates factor IX to IXa, which cooperates with its cofactor, factor VIIIa, to form the tenase complex, which then proteolytically cleaves factor X to generate factor Xa. Factor Xa, along with its cofactor Va, forms the pro-thrombinase complex, which generates thrombin from its zymogen (factor II or prothrombin). The thrombin formed by this complex cleaves fibrinogen to allow it to polymerize into insoluble fibrin strands. Factor XIII is also activated by thrombin, and the role of factor XIIIa, a transglutaminase, is to cross-link fibrin strands to provide additional clot strength and stability. The extrinsic pathway requires tissue factor (TF) in complex with factor VIIa to form the tenase complex. Additionally, TF/VIIa can activate factor IX to IXa, serving as an alternate method of activating the intrinsic pathway (Figure 15.1).

While the cascade hypothesis explains the prothrombin time (PT) and the activated partial thromboplastin time (APTT) tests as they are performed *in vitro*, it fails to explain the relative intensity of the bleeding diathesis seen in individuals deficient in factors XI, IX, and VIII, as well as the lack of bleeding in those deficient in factor XII, HMWK, or PK. The first reconciliation of this paradox came with the discovery that the tissue factor/VIIa complex activates both factor X, as well as factor IX. More recently, a cell-based model of hemostasis was proposed to address these deficiencies, and to integrate the role of cell surface-bound coagulation reactions in hemostasis. In this model, a tissue factor-bearing cell such as an activated monocyte or fibroblast serves as the site for generation of a small amount of thrombin and factor IXa. The initial thrombin burst is quickly limited by the interaction of generated factor Xa with tissue factor pathway inhibitor (TFPI), a Kunitz-like inhibitor present in plasma and at cell surfaces. The binary complex of Xa-TFPI then forms an inhibitory quaternary complex with TF-VIIa. The small amount of thrombin generated by this initiation step is insufficient to cleave fibrinogen, but serves to activate platelets and cleave circulating factor VIII from its noncovalently associated VWF, allowing for the formation of VIIIa. The factor IXa formed on the TF-bearing cell cooperates with VIIIa to form the tenase complex on the surface of the activated platelet. The Xa thus formed interacts with the Va generated on the platelet surface to form the prothrombinase complex. This complex generates a large burst of thrombin that is now sufficient to cleave fibrinogen, activate factor XIII, and activate the thrombin activatable fibrinolysis inhibitor (TAFI), thus allowing for formation of a stable fibrin clot (Figure 15.2). In this model, XI binds to the surface of activated platelets where it is activated by thrombin. Factor XIa so formed boosts the activity of the tenase complex, but is not necessary for thrombin generation.

Fibrinolysis leads to clot dissolution, thereby restoring normal blood flow. Plasminogen is activated to plasmin by the action of either tissue plasminogen

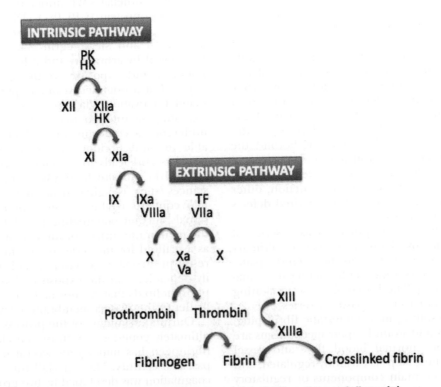

Figure 15.1   Classic view of hemostasis according to the "waterfall" or "cascade" model.

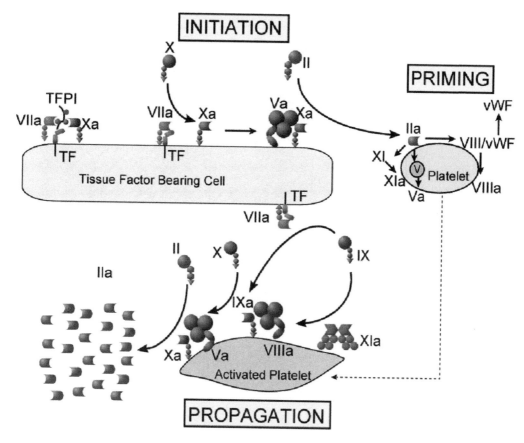

Figure 15.2 **The cell-based model of hemostasis.** Figure Courtesy of Dr. Dougald Monroe.

activator (t-PA) or urokinase plasminogen activator (u-PA). Both plasminogen and the plasminogen activators bind to free lysine residues exposed on fibrin, and in this way fibrin acts as the cofactor for its own destruction. Plasmin is capable of degrading both fibrin and fibrinogen and can thus dissolve both formed clot as well as its soluble precursor. However, non-fibrin-bound plasmin is inhibited by a number of circulating inhibitors, of which α_2-plasmin inhibitor (α_2-PI) is the most significant. Plasminogen activation is also inhibited by a number of molecules; chief among them is plasminogen activator inhibitor-1 (PAI-1), a serine protease inhibitor (SERPIN) that irreversibly binds to—and inactivates—the plasminogen activators. Lastly, cellular receptors act to localize and potentiate or clear plasmin and plasminogen activators.

DISORDERS OF SOLUBLE CLOTTING FACTORS

While classic hemophilia/hemophilia A (factor VIII deficiency) and Christmas disease/hemophilia B (factor IX deficiency) are the best known examples of the clotting factor deficiencies, the following overview will discuss each deficiency in the numerical order ascribed by the Roman numeral classification system (see Table 15.1). Note that fibrinogen is rarely referred to as factor I even though this is how it is

designated using strict nomenclature; similarly, tissue factor, a cell-bound transmembrane glycoprotein, is rarely referred to by its designated number (factor III). No tissue factor deficiency states have been described in humans. In addition, factor IV is reagent calcium, and there is no factor VI in the current scheme. While hemophilia A occurs in approximately 1:10,000 of the population, and hemophilia B in about 1:40,000, the remaining factor deficiencies (with the possible exception of factor XI in Ashkenazi Jewish populations) typically occur at a frequency between 1:500,000 and 1:2,000,000 of the general population, and are thus usually known collectively as the rare inherited bleeding disorders.

Rather than attempt to provide an exhaustive list of mutations affecting all coagulation proteins in this text, we have provided a number of useful websites containing described mutations in these proteins in Table 15.2. In addition, a summary of all described mutations in factors II, V, VII, X, XI, XIII, and combined V/VIII deficiency can be found at www.med. unc.edu/isth/mutations-databases/mutations_rare_bleeding_disorders.html. In addition, the Human Gene Mutation database, available at www.hgmd.org, contains detailed information on many genetic mutations in a variety of inherited disorders, including those affecting the coagulation pathways. Table 15.3 summarizes the essential biochemical data on each of the important inherited deficiency states.

Table 15.1   Clinical Features of Inherited Coagulation Factor Deficiency States

Defect	Inheritance Pattern	Bleeding Manifestations	Diagnostic Testing					Treatment
			PT	PTT	TCT	BT	
Fibrinogen abnormalities — Afibrinogenemia	Autosomal	Severe, but less so than severe hemophilia A and B	Infinite	Infinite	Infinite	Prolonged	Fibrinogen concentrate, Cryoprecipitate
Dysfibrinogenemia	Autosomal	Variable bleeding and/or clotting	Prolonged	Prolonged	Prolonged or shortened	Normal	Cryoprecipitate
Prothrombin deficiency	Autosomal	Varies with prothrombin levels	Prolonged	Prolonged	Normal	Normal	PCCs
Factor V deficiency	Autosomal	Mild–moderate	Prolonged	Prolonged	Normal	Prolonged	FFP, potential need for exchange transfusion
Factor VII deficiency	Autosomal	Moderate–severe	Prolonged	Normal	Normal	Normal	Recombinant-activated factor VII
Hemophilia A	X-linked recessive	Variable, depending on factor VIII level	Normal	Prolonged	Normal	Normal	Factor VIII concentrates, DDAVP in mild cases
Hemophilia B	X-linked recessive	Variable, depending on factor IX level	Normal	Prolonged	Normal	Normal	Factor IX concentrates
Factor X deficiency	Autosomal	Variable, depending on factor X level	Prolonged	Prolonged	Normal	Normal	Plasma or PCCs
Factor XI deficiency	Autosomal	Variable, but NOT dependent on factor XI levels	Normal	Prolonged	Normal	Normal	Plasma or recombinant-activated factor VII
Deficiency of Factor XII, prekallikrein, or high molecular weight kininogen	Autosomal	None	Normal	Prolonged	Normal	Normal	None needed
Factor XIII deficiency	Autosomal	Severe	Normal	Normal	Normal	Normal	Cryoprecipitate
Deficiency of alpha-2 plasmin inhibitor or plasminogen activator inhibitor-1	Autosomal	Severe	Normal	Normal	Normal	Normal	Antifibrinolytic agents (epsilon aminocaproic acid or tranexamic acid)

Table 15.2   Useful Websites for Documented Mutations in Individual Coagulation Factor Deficiency States

Fibrinogen	www.geht.org/databaseang/fibrinogen/
Prothrombin	www.coagMDB.org/
Factor V	www.lumc.nl/rep/cod/redirect/4010/ research/factor_v_gene.html
Factor VII	www.coagMDB.org/
Factor VIII	http://europium.csc.mrc.ac.uk/
Factor IX	www.kcl.ac.uk/ip/petergreen/ haemBdatabase.html; www.coagMDB .org/
Factor X	www.coagMDB.org/
Factor XI	www.FactorXI.org/
Factor XIII	www.f13-database.de/ (xhgmobrswxgori45zk5jre45)/index.aspx
VWF	www.vwf.group.shef.ac.uk/
Protein S	www.med.unc.edu/isth/ssc/ communications/plasma_coagulation/ proteins.htm
Protein C	www.coagMDB.org/
Antithrombin	www1.imperial.ac.uk/medicine/about/ divisions/is/haemo/coag/ antithrombin/

Fibrinogen Abnormalities

Fibrinogen abnormalities are inherited in an autosomal pattern and occur in two main patterns: hypo/afibrinogenemia and dysfibrinogenemia. Afibrinogenemia is a very rare disorder that occurs when any one of the three genes coding for the alpha, beta, or gamma chains that make up the fibrinogen molecule is mutated. If the mutation is sufficient to disrupt formation or secretion of any of the three chains, afibrinogenemia results. Afibrinogenemic patients have a severe bleeding disorder manifest by bleeding after trauma into subcutaneous and deeper tissues that may result in dissection. Bleeding from the umbilical stump at birth occurs frequently. Though hemarthroses do occur in these patients, they are less frequent than in the severe forms of hemophilia A and B. Less severe mutations in the fibrinogen chains, lead to reduced but measurable circulating levels of fibrinogen, or hypofibrinogenemia. Hypofibrinogenemic patients manifest a variable but concordant reduction in fibrinogen antigen and activity levels in plasma.

Dysfibrinogenemia is also rare but is more common than afibrinogenemia, with the majority of patients being heterozygous for the disorder. The dysfibrinogens are the result of missense, nonsense, or splice junction mutations. Several hundred mutations have been recorded, many of which result in neither a hemorrhagic nor thrombotic state. However, other dysfibrinogens are associated with bleeding episodes, while a few may be associated with venous or arterial thrombosis.

Prothrombin (Factor II) Deficiency

Inherited prothrombin deficiency is rare, with fewer than 50 distinct mutations being reported. It is an autosomal recessive disorder, and heterozygotes have no bleeding symptoms. Symptomatic patients may be homozygous or doubly heterozygous for causative mutations. By convention, patients with hypoprothrombinemia manifest a concordant reduction in circulating prothrombin antigen and functional activity, whereas those with hypodysprothrombinemia classically have a reduced concentration of prothrombin antigen with a discordantly low functional activity. Bleeding in affected patients varies from mild to

Table 15.3   Summary of Biochemical Features and Hemostatic Levels of Each Coagulation Factor

	Chromosome	MW (Da)	Number of Chains (Active)	$t\frac{1}{2}$ (Half-Life)	Plasma Concentrations	Number of Gla Domains	Number of Described Mutations	Prevalence of Deficiency State	Hemostatic Levels (% ref. Range)
FVII	13	50,000	2	4–6 hr	10 nM	10	>130	1/500,000	15–20%
FX	13	59,000		40–60 hr		11	~70	1/10⁶	15–20%
FII	11	72,000	2	3–4 d	10 µg/ml	10	~40	1/2×10⁶	20–30%
FV	1	330,000	2		20 nM	—	~30	1/10⁶	15–25%
FV/FVIII	18	330,000	2 (FV),	36 hr (FV),		—	>15	1/2×10⁶	15–20%
ERGIC 53	2	(FV)	3 (FVIII)	10–14 hr			(ERGIC)		
MCFD2		(FVIII)		(FVIII)			>5 (MCFD2)		
FXI	4	160,000	2	40–72 hr (52 hr)	30 nM (~5 µg/ mol)	—	>50	1/10⁶	15–20%
FXIII (2 genes)	6 (A chain), 1 (B chain)	320,000	4	11–14 d		—	>50	1/2×10⁶	2–5%
Fibrinogen (3 genes)	4 (α, β, and γ chains)	340,000	6	2–4 d	150–350 mg/dL	—	>300 (Dysfib) >30 (Hypofib)	1/10⁶	50 mg/dL

179

severe, depending on the functional prothrombin activity level. The complete absence of prothrombin probably leads to embryonic lethality.

Factor V Deficiency

FV is a large molecule (330 kDa) that shares significant structural homology with FVIII and ceruloplasmin. Factor V deficiency is an autosomal recessive disorder that results from mutations in the factor V gene. Heterozygotes are generally asymptomatic, while homozygotes or combined heterozygotes may have mild to moderately severe bleeding symptoms. While about 40 mutations in the factor V gene have been reported, they seem to occur less frequently than in genes for other clotting factors. Bleeding manifestations are similar to those seen in classic hemophilia, except that they tend to be milder, and hemarthroses are less common.

Factor VII Deficiency

Approximately 1% of the total factor VII in plasma circulates in the active serine protease form (FVIIa, 10–100 pmol/L). Controversy still exists as to which enzyme is responsible for basal activation of factor VII *in vivo*, although there is evidence implicating FIXa. In the absence of its cofactor TF, FVIIa is a very weak enzyme.

Factor VII deficiency is an autosomal recessive bleeding disorder that occurs in mild, moderate, and severe forms. It is generally considered to be the most common of the rare bleeding disorders, with a prevalence of 1:300,000 to 1:500,000 in most populations. More than 100 mutations in the gene for factor VII have been reported. Bleeding manifestations vary, but in severely affected patients, bleeding can be as severe as that seen in severe classic hemophilia and may include crippling hemarthroses. Factor VII levels of 10% of normal are probably sufficient to control most bleeding episodes, but in some scenarios, higher levels may be required for hemostasis. Some patients with almost no measurable factor VII activity may express very few hemorrhagic manifestations. Furthermore, there are reports of thrombotic events occurring in patients with FVII deficiency, such that it has been unclear whether distinct mutations may actually be prothrombotic. However, at present, registry analysis suggests that while factor VII deficiency does not seem to predispose to thrombosis, neither is it protective.

Hemophilia A and Hemophilia B (Classic Hemophilia and Christmas Disease)

Hemophilia A and B result from deficiencies of factor VIII and IX, respectively. Factor VIII and IX are necessary for the sustained generation of factor Xa (and ultimately thrombin) to form a normal hemostatic plug-in response to vascular injury. Hemophilia A and B are the only two soluble clotting factor deficiencies that

are inherited as X-linked recessive disorders. Several hundred distinct mutations in each gene have been reported. These mutations result in mild, moderate, and severe forms of hemophilia, and the clinical manifestations of hemophilia A and B are, for all practical purposes, indistinguishable. In the severe form (<1% basal activity), both disorders are characterized by recurrent hemarthroses that result in chronic crippling arthropathy unless treated by replacing the deficient factor on a scheduled prophylactic basis. Central nervous system hemorrhage is especially hazardous and remains one of the leading causes of death. Retroperitoneal hemorrhage and bleeding into the pharynx may also be life-threatening.

Factor X Deficiency

Like factor VII deficiency, factor X deficiency is inherited in an autosomal recessive fashion and can be mild, moderate, or severe. Numerous mutations have been recorded. Severely affected patients have symptoms similar to severe classic hemophilia, including hemarthroses and chronic crippling hemarthropathy. About one-third of affected individuals exhibit a type 1 deficiency, while the rest have a type 2 pattern, with factor X antigen levels that are preserved, or at least significantly above the level of factor X activity.

Factor XI Deficiency

Factor XI deficiency is an autosomal recessive disorder that commonly occurs in patients of Ashkenazi Jewish descent. In these communities, homozygotes may be as prevalent as 1:500, compared to a frequency of about 1:1,000,000 in most other populations. Among Jewish factor XI-deficient patients, three mutations account for most cases. Factor XI-deficient patients have normal levels of factor VIII and IX to form the tenase complex and normal levels of factors V and X to form the prothrombinase complex. Whether or not factor XI-deficient patients bleed may depend on differences in their ability to generate thrombin, and/or the ability to activate thrombin activatable fibrinolysis inhibitor (TAFI). TAFIa cleaves free lysine residues on fibrin, preventing fibrin-dependent activation of plasminogen by plasminogen activators. Thus, factor XIa has both procoagulant (factor IX activating) and antifibrinolytic activities.

Deficiencies of Factor XII, Prekallikrein (PK), and High Molecular Weight Kininogen (HK)

Deficiencies of factors XII, PK, and HK (the so-called contact factors) cause a marked prolongation of the APTT, but other screening tests of coagulation are normal. These defects are inherited in an autosomal recessive fashion. They are not associated with bleeding even after trauma or surgery, although the prolonged APTT may cause a great deal of consternation among those not familiar with these defects. A good history revealing the absence of bleeding in these patients and their family members despite a long APTT is the best indication that one is dealing with one of these defects. Recent data suggest that although patients with factor XII deficiency do not bleed excessively, they may be relatively protected from arterial thrombosis. The kallikrein-kinin system may be more physiologically important in inflammation, blood pressure regulation, and fibrinolysis than in hemostasis.

Factor XIII Deficiency

FXIII is activated by thrombin in the presence of calcium. Factor XIIIa is a plasma transglutaminase that covalently cross-links fibrin alpha and gamma chains through γ-glutamyl-ε-lysine bonds to form an impermeable fibrin clot. Although a clot may form in the absence of factor XIII and be held together by hydrogen bonds, this clot is excessively permeable and is easily dissolved by the fibrinolytic system. The clot formed in the absence of factor XIII does not form a normal framework for wound healing, and abnormal scar formation may occur.

Factor XIII consists of two A chains and two B chains. Factor XIII-A is synthesized in megakaryocytes, monocytes, and macrophages, whereas factor XIII-B is synthesized in hepatocytes. The complete molecule is an A_2B_2 tetramer with the A chains containing the active site and the B chains acting as a carrier for the A subunits. Platelet alpha granules contain A chains but not B chains. Factor XIII deficiency may result from mutations in the genes encoding either the A or B chains, with A chain mutations being more common. Autosomal genes govern hepatic synthesis of the factor, and the disease is expressed as a recessive disorder. Typically, only patients with severe deficiency of factor XIII ($<3–5\%$) are symptomatic.

Bleeding manifestations are generally severe, and hemorrhage can occur into any tissue. Umbilical stump bleeding in the neonatal period is common in factor XIII deficiency. Intracranial hemorrhage is also a relatively common manifestation that may mandate long-term prophylactic factor replacement.

Multiple Clotting Factor Deficiencies

The two most common multiple clotting factor deficiencies are a combined deficiency of factors V and VIII and a combined deficiency of the vitamin K-dependent factors (factors II, VII, IX, X, and Protein C and S).

A combined deficiency of factors V and VIII is inherited in an autosomal recessive fashion and can be distinguished from a combined inheritance of mild classic hemophilia and mild factor V deficiency by family studies or by genetic analysis. The disorder is due to defects in one of two genes: the *LMAN1* gene and a newly discovered gene called the multiple clotting factor deficiency 2 (*MCFD2*) gene. The products of both genes play a critical role in the transport of factors V and VIII from the endoplasmic reticulum to the Golgi apparatus and are necessary for normal secretion of these factors. The disorder results in a mild to moderate bleeding tendency with factor V and VIII levels ranging from 5–30% of normal.

Combined deficiencies of the vitamin K dependent factors can be due to defects in either the gene for vitamin K-dependent carboxylase or the gene for vitamin K epoxide reductase. These are autosomal recessive disorders that may be associated with severe deficiency of prothrombin; factor VII, IX, and X; as well as Protein C and S. The diagnosis must be distinguished from surreptitious ingestion of coumarin drugs (including rodenticides), which is an acquired disorder with bleeding manifestations of recent onset.

Von Willebrand Disease (VWD)

The most common hereditary bleeding disorder arises from abnormalities in von Willebrand factor (VWF). VWF occurs in plasma as multimers of a 240,000 Dalton subunit, with molecular weights ranging from 1 million to 20 million Daltons. The principal functions of VWF are to act as a carrier for clotting factor VIII and to mediate platelet adhesion to the injured vessel wall. The larger molecular weight multimers are the most effective at mediating platelet adhesion. VWF binds to glycoprotein Ib on the platelet surface and also to collagen in the vessel wall. VWF is particularly important for platelet adhesion under high shear stress vascular beds. It has been elegantly demonstrated that shear stress leads to unfolding of the globular collagen-bound VWF multimers, thereby uncovering the molecular domain responsible for binding to VWF. VWF also cross-links platelets via binding to glycoprotein IIb-IIIa.

There are three major types of von Willebrand disease (VWD): type 1, 2, and 3 (see Table 15.4). Type 1 is autosomal dominant and represents a partial quantitative deficiency of VWF. Generally, it is explained by reduced synthesis of VWF, and analysis of multimers in plasma reveals a global decrease in multimers of all sizes. More recently, it has been appreciated that the pathophysiology of some cases of type 1 VWF is accelerated clearance of VWF from plasma. The prototype of this phenotype is the so-called Vicenza R1205H mutation, which causes a moderate to severe variant of von Willebrand disease. Type 3 is an autosomal recessive severe quantitative deficiency, in which there is an absence of VWF. Type 2 VWD usually occurs as an autosomal dominant disorder with qualitative abnormalities in VWF function. Type 2 occurs in four major forms: 2A, 2B, 2N, and 2M. Types 2A and 2B are characterized by absence of the higher molecular weight multimers of VWF in plasma. Type 2B is also associated with thrombocytopenia as a result of a gain of function mutation resulting in a VWF molecule with higher affinity for the GP Ib receptor, thus enhancing platelet agglutination and accelerated clearance. Type 2M patients show reduced binding of VWF to GPIb, although they have normal VWF multimeric composition in plasma.

Table 15.4   Von Willebrand Disease Subtypes

| | | Inheritance Pattern | Bleeding Manifestations | Diagnostic Testing | | | | Treatment |
				VWF Ag	VWF Activity	Factor VIII Activity	VWF Multimers	
Type 1		Autosomal dominant	Generally mild	Low	Low	Low	Normal	DDAVP, factor VIII concentrates rich in VWF
Type 2	2A	Autosomal dominant	Mild–moderate	Low	Lower than antigen	Variable	Absent high molecular weight forms	Factor VIII concentrates rich in VWF
	2B	Autosomal dominant	Mild–moderate	Low	Lower than antigen	Variable	Absent high molecular weight forms	Factor VIII concentrates rich in VWF
	2N	Autosomal recessive	Mild	Normal	Normal	Low	Normal	Factor VIII concentrates rich in VWF
	2M	Autosomal dominant	Mild–moderate	Normal	Lower than antigen	Normal	Normal	DDAVP, factor VIII concentrates rich in VWF
Type 3		Autosomal recessive	Severe	Near absent	Near absent	Near absent	Absent	Factor VIII concentrates rich in VWF

Type 2N VWD is a rare autosomal recessive disorder arising from a mutation in the factor VIII binding site on the VWF molecule. Without the protection provided by VWF binding, plasma factor VIII levels are reduced because of a markedly decreased half-life. VWF multimers and antigen and activity levels may be normal, while the factor VIII levels are low enough to be confused with mild classic hemophilia. These two disorders can be distinguished by an ELISA-based factor VIII binding assay. Clinically affected patients are either homozygous for one of several gene mutations in the D' or D3 domain of mature VWF, or are combined heterozygous for a 2N mutation and a type 1 mutation.

Although VWD is a defect in a soluble clotting factor, bleeding in patients with this disorder is more similar to that produced by a defect in platelet number or function. The bleeding manifestations tend to be more of the "oozing and bruising" variety, with hematoma formation being rare. Bleeding in types 1 and 2 VWD is usually mild to moderate, although severe bleeding may occur with trauma and surgery. Some patients with type 1 VWD may be relatively asymptomatic. Table 15.4 lists the diagnostic features of the various types of VWD.

DISORDERS OF PLATELET NUMBER OR FUNCTION

Disorders of Platelet Production

Inherited disorders causing thrombocytopenia are a heterogeneous group of conditions. Some are associated with a profound thrombocytopathy, some are associated with other somatic changes, while others manifest thrombocytopenia only.

The MYH9-Associated Disorders

The May-Hegglin anomaly is the prototype of a family of disorders now known to be due to a defect in the *MYH9* gene. May-Hegglin anomaly, Sebastian syndrome, Fechtner syndrome, and Epstein syndrome are autosomal dominant macrothrombocytopenias that are distinguished by different combinations of clinical and laboratory signs, such as sensorineural hearing loss, cataract, nephritis, and polymorphonuclear inclusions known as Döhle-like bodies. Mutations in the *MYH9* gene encoding for the nonmuscle myosin heavy chain IIA (NMMHC-IIA) have been identified in all these syndromes.

Defects in Transcription Factors

Alterations in megakaryocyte development due to defective transcription factors underlie a large number of the familial thrombocytopenias. Derangements in development of other cell types as well as other somatic mutations can also occur. Mutations in *HOXA11* have been described in two unrelated families with bone marrow failure and skeletal defects.

The Paris-Trousseau syndrome is an autosomal dominant condition characterized by macrothrombocytopenia with giant alpha granules. It is caused by hemizygous loss of the *FLI1* gene due to deletion at 11q23. Lack of FLI1 protein leads to lack of platelet production due to arrested megakaryocyte development. The 11q23 deletion is also seen in patients with Jacobsen's syndrome who also have congenital heart disease, trigonocephaly, dysmorphic facies, mental retardation and multiple organ dysfunction as well as macrothrombocytes with abnormal alpha granules.

Mutations in *GATA-1* lead to the X-linked congenital dyserythropoietic anemia and thrombocytopenia syndrome. The platelets are large and exhibit defective collagen-induced aggregation. X-linked thrombocytopenia without anemia is due to mutations within *GATA-1* that disrupt FOG-1 (Friend of GATA) interactions while leaving DNA-binding intact. The acute megakaryoblastic anemia seen in conjunction with Down's syndrome can be associated with mutations in *GATA-1*.

Mutations in *RUNX1* lead to familial thrombocytopenic syndromes with a predisposition to development of acute myelogenous leukemia. *RUNX1* mutations cause an arrest in megakaryocyte development with an expanded population of progenitor cells. The platelets that are produced show defects in aggregation. The development of acute leukemia likely requires a second mutation within *RUNX1* or another gene.

Defects in Platelet Production

Congenital amegakaryocytic thrombocytopenia (CAMT) is due to defects in the *c-MPL* gene encoding the thrombopoietin receptor. Children born with this disorder have severe thrombocytopenia and may go on to develop deficiencies in other cell types.

Thrombocytopenia with absent radii (TAR) is a syndrome characterized by severe congenital thrombocytopenia along with absent or shortened radii. The platelets produced show abnormal aggregation. Although thrombopoietin levels are elevated, no defect in *c-MPL* has been identified, and abnormal intracellular signaling pathways are postulated as the cause of this rare disorder.

Perhaps the most common hereditary thrombocytopenia with small platelets is the Wiskott-Aldrich syndrome (WAS), a disorder associated with the triad of immune deficiency, eczema, and thrombocytopenia. This syndrome is X-linked and results from mutations in the gene for Wiskott-Aldrich syndrome protein (WASP). Platelets as well as T-lymphocytes show defective function, and clinical manifestations vary widely. As opposed to the macrothrombocytopenic defects, WAS platelets are small and defective in function.

Disorders of Platelet Function

Defects in Platelet Adhesion

Bernard Soulier Syndrome (BSS) is a severe bleeding disorder characterized by macrothrombocytopenia, decreased platelet adhesion to VWF, abnormal prothrombin consumption, and reduced platelet survival. Deficient platelet binding to subendothelial von Willebrand factor is due to abnormalities (either qualitative or quantitative) in the GP-Ib-IX-V complex. Mutations in GPIbα binding sites for P-selectin, thrombospondin-1, factor XI, factor XII, and high molecular weight kininogen may mediate variations in the phenotype seen. The product of four separate genes (*GPIBA, GPIBB, GP9,* and *GP5*) assemble within the megakaryocyte to form the GP-Ib-IX-V on the platelet surface. Defects in any of the genes

may lead to BSS. Classically, affected platelets from affected individuals aggregate normally in response to all agonists except ristocetin. A database of described causative mutations for BSS may be found at www.bernardsoulier.org/.

Platelet-type von Willebrand disease is due to a gain of function mutation such that plasma VWF binds spontaneously to platelets, and the platelets exhibit agglutination in response to low dose ristocetin. Mutations generally lie within the *GPIBA* gene. High molecular weight multimers of VWF bound to platelets are cleared from the circulation, which may result in bleeding. The phenotype is identical to that seen in type 2B VWD, in which the mutation lies within the VWF rather than its receptor. It can therefore be quite difficult to distinguish between type 2B VWD and platelet-type VWD. Gene sequencing of the VWF gene, the *GPIBA* gene, or both may be required.

Defects in Platelet Aggregation

Glanzmann thrombasthenia is a rare, autosomal recessive disorder characterized by absent platelet aggregation. It is due to absent or defective GP IIbIIIa on the platelet surface. Patients have severe mucocutaneous bleeding, which becomes refractory to platelet transfusions as alloantibodies to transfused platelets form. Though demonstration of absent platelet aggregation in response to all agonists (with the exception of ristocetin) will suggest the diagnosis, definitive diagnosis relies on showing absence of functional GPIIbIIIa on the platelet surface, either by flow cytometry or by electron microscopy using immuno-gold labeled fibrinogen imaging. A database of described mutations in the GPIIb and GPIIIa genes that may result in Glanzmann thrombasthenia may be found at http://sinaicentral.mssm.edu/intranet/research/glanzmann/menu. Acquired Glanzmann thrombasthenia has been described in patients who develop autoantibodies against GPIIbIIIa. These patients may have underlying immune thrombocytopenic purpura, but the severity of their bleeding is out of proportion to platelet number.

Some patients exhibit defects in aggregation in response to specific agonists. Platelets from these patients may show defects in either platelet receptors or in the downstream intracellular signaling pathways leading to activation. These disorders must be distinguished from the effects of drugs such as aspirin, whose ingestion can produce similar effects on platelet function.

Disorders of Platelet Secretion: The Storage Pool Diseases

Platelets contain two types of intracellular granules: alpha and delta (or dense) granules. Alpha granules contain proteins either synthesized within the megakaryocytes or endocytosed from the plasma, including fibrinogen, factor V, thrombospondin, platelet-derived growth factor, multimerin, fibronectin, factor XIII A chains, high molecular weight kininogen, and VWF among others. Their membrane contains molecules such as P-selectin and CD63 that are translocated to the outer plasma membrane after secretion and membrane fusion. Dense granules contain ATP and ADP as well as calcium and serotonin, and any deficiency of dense granules thus leads to a defective secondary wave of platelet aggregation.

Defects in Alpha Granules

The gray platelet syndrome (GPS) is an autosomal recessive condition that leads to a mild bleeding diathesis. It may be recognized by examination of a Wright-Giemsa stained peripheral blood smear showing platelets that appear gray without the usual red-staining granules. Electron microscopy showing a depletion of alpha granules is a more sensitive method to diagnose the syndrome. GPS is thus classified with the other platelet secretion defects, but may also be classified with the macrothrombocytopenias, since platelets may be slightly larger than usual, albeit not as large as those seen in the giant platelet disorders described previously. Furthermore, the platelet count is only moderately depressed, and bleeding symptoms are mild. Patients with GPS may also develop early onset myelofibrosis, a probable consequence of the impaired storage of growth factors such as PDGF.

The Quebec platelet disorder (QPD) is associated with a normal to slightly low platelet count with a mild bleeding disorder. Pathophysiologically, this is due to abnormal proteolysis of alpha granule proteins that appears to be mediated by ectopic intragranular production of urokinase. QPD was first recognized as a specific deficiency of platelet factor V associated with normal concentrations of plasma factor V. The platelets appear normal on peripheral blood smears under the light microscope, and diagnosis depends on showing decreased alpha granule proteins. It is inherited in an autosomal dominant fashion, but the specific genetic defect is unknown.

Defects in Dense Granules

The Hermansky-Pudlak syndrome is the association of delta storage pool deficiency with oculocutaneous albinism and increased ceroid in the reticuloendothelial system. There are several subtypes of the Hermansky-Pudlak syndrome resulting from several distinct mutations. The syndrome is inherited in an autosomal recessive pattern. Granulomatous colitis and pulmonary fibrosis are also part of the syndrome. Mutations in at least eight genes (*HPS-1* through *HPS-8*) lead to defects in HPS proteins responsible for organelle biosynthesis and protein trafficking. Described mutations accounting for the Hermansky-Pudlak syndrome are collated in the database available at http://liweilab.genetics.ac.cn/HPSD/.

The Chediak-Higashi syndrome is also associated with storage pool deficiency and is characterized by oculocutaneous albinism, neurologic abnormalities, immune deficiency with a tendency to infections, and giant inclusions in the cytoplasm of platelets and leukocytes. The disorder is rare, and bleeding manifestations are relatively mild. The syndrome is due to mutations in the LYST (lysosomal trafficking regulator) gene; these are listed in the Chediak-Higashi database at http://bioinf.uta.fi/LYSTbase/?content=pin/IDbases. Affected patients are homozygous, while heterozygotes are phenotypically normal.

The Scott Syndrome

In this disorder, platelets, when activated, cannot translocate phosphatidylserine from the inner to the outer platelet membrane when the flip-flop of the membrane leaflet occurs, presumably due to defects in the activity of the scramblase enzyme. Because of this defect, factors Xa and Va are unable to efficiently bind to the membrane to assemble the prothrombinase complex, and thrombin generation on the platelet surface is impaired. Scott syndrome is characterized by a mild bruising and bleeding tendency. It can be detected using flow cytometry with antibodies against annexin V which will show the defective microvesicle formation characteristic of this disorder.

Disorders of Platelet Destruction

Disorders of platelet destruction are too numerous to discuss here. Therefore, the discussion in the section will be limited to those disorders where the molecular pathogenesis is known at a greater level of detail. A more comprehensive description of disorders characterized by platelet destruction can be found in recent reviews.

Antibody-Mediated Platelet Destruction

Neonatal alloimmune thrombocytopenia (NAIT) is a bleeding disorder caused by transplacental transfer of maternal antibodies directed against fetal platelet antigens inherited from the father. In Caucasians, the antigens most frequently implicated include HPA-1a (PL^{A1}) and HPA-5b (Br^a). In Asians, HPA-4a and HPA-3a account for the majority of NAIT cases. NAIT occurs with a lower frequency in Caucasians than is expected by the incidence of HPA-1a negativity in the population, suggesting that other factors influence antibody development. Additionally, NAIT mediated by antibodies against HPA-1a is more clinically severe, perhaps because these antibodies may also block platelet aggregation, since HPA-1a is an antigen expressed on platelet GPIIIa. Mothers who are negative for the antigen in question can develop antiplatelet antibodies that cross the placenta, leading to severe fetal thrombocytopenia. Even the first child may be affected, and intracranial hemorrhage is a feared and devastating complication. Subsequent pregnancies have a near 100% rate of NAIT.

Post-transfusion purpura (PTP) is associated with thrombocytopenia resulting from a mismatch between platelet antigens. In this condition, patients previously sensitized against certain platelet antigens (the same ones that lead to NAIT) develop acute, severe thrombocytopenia 5–14 days after transfusion. Though packed red cells are most commonly associated with PTP, transfusion of any blood component may precipitate this disorder. These blood components contain platelet microparticles that express the offensive platelet antigen, leading to an anamnestic production of antibodies. However, paradoxically, these patients develop antibodies directed against their own platelets, either by fusion of the exogenous microparticles with their own platelets, or by a process in which exposure to foreign platelets leads to formation of autoantibodies.

Thrombotic Microangiopathies

Thrombotic thrombocytopenic purpura (TTP) is an acute disorder that usually presents in previously healthy subjects. It is highly lethal unless treated promptly, which generally entails plasma exchange with or without additional immunosuppression. In 1982, Moake made the seminal observation that the plasma of patients with TTP contained ultralarge (UL) multimers of VWF, which were absent in normal plasma. He hypothesized that TTP could be due to the absence of a protease or depolymerase responsible for cleaving the UL VWF multimers. The protease was identified in 1996 by the groups of Tsai and Furlan, its gene was cloned, and the enzyme named ADAMTS-13, when it was found to be a member of the "a disintegrin-like and metalloprotease with thrombospondin repeats" family of metalloproteases. ADAMTS-13 levels are found to be low in most patients with both familial and sporadic TTP, and an IgG auto-antibody inhibitor to ADAMTS-13 is found in a majority of (but not all) patients with sporadic TTP.

The hemolytic uremic syndrome (HUS) shares many clinical features with TTP, including microangiopathic hemolytic anemia, thrombocytopenia, and renal insufficiency. Renal findings are more prominent and neurologic findings less so. HUS is divided into diarrhea-associated HUS (D+HUS) and atypical (diarrhea-negative) HUS. Diarrhea-positive HUS is triggered by infection with a Shiga-toxin-producing bacteria and is much more commonly encountered in children. *Escherichia coli* O157:H7 is implicated in 80% of the cases, but other bacteria including other *E. coli* subtypes and *Shigella dysenteriae* serotype 1 can cause D+HUS. Shiga toxins bind to the glycosphingolipid receptor globotriaosylceramide (Gb3) on the surface of renal mesangial, glomerular, and tubular epithelial cells. Protein synthesis is impaired through inhibition of 60S ribosomes, and cell death occurs. Plasma from patients with HUS demonstrates markers of abnormal thrombin generation. As compared with TTP, ADAMTS-13 levels are typically normal in patients with HUS, and the fibrin microthrombi do not contain VWF strands.

Atypical HUS occurs in patients without a diarrheal prodrome. Underlying conditions, such as organ transplantation or exposure to certain drugs may be present. In as many as 30–50% of patients, mutations in one of three proteins involved in complement regulation occur. Factor H (CFH) and membrane cofactor protein (MCP or CD46) are regulators of complement factor I (CFI), which is a serine protease that cleaves and inactivates surface-bound C3b and C4b. Autoantibodies against these proteins have also been reported, suggesting that unregulated complement activation plays a role in the pathogenesis of HUS.

Heparin-Induced Thrombocytopenia

Heparin-induced thrombocytopenia (HIT) is a common iatrogenic thrombocytopenic disorder that can paradoxically lead to thrombosis. It occurs in 1–5% of patients treated with standard unfractionated heparin for at least 5 days and in <1% of those treated with low molecular weight heparin. Approximately 50% of patients develop venous and/or arterial thromboses. New thromboses develop in 25% of patients, amputations are required in 10%, and reported mortality rates are between 10% and 20%. The pathogenic autoantibodies that cause HIT are directed against neoepitopes on PF4 that are induced by heparin and other anionic glycosaminoglycans (GAGs). PF4 is an abundant protein stored in the alpha granules of platelets in complex with chondroitin sulfate (CS). Upon platelet activation, PF4/CS complexes are released and bind to the platelet surface. Heparin can displace CS, forming PF4/heparin complexes. Binding of IgG anti-PF4/heparin to the platelet leads to Fcγ receptor-mediated clearance of platelets but also leads to platelet activation and generation of procoagulant microparticles via FcγRIIA. PF4/heparin complexes also form on the surface of monocytes and endothelial cells, and antibody binding leads to tissue-factor driven thrombin generation and hence to clot formation. The PF4/heparin complexes are most antigenic when PF4 and heparin are present at equimolar concentrations, where they form ultralarge molecular complexes. Low molecular weight heparin forms these ultralarge complexes less efficiently and at concentrations that tend to be supratherapeutic, perhaps explaining the lower frequency of HIT in patients treated with LMWH as opposed to standard unfractionated heparin.

THROMBOPHILIA

An understanding of how the coagulation system is physiologically regulated is necessary when seeking to determine how it can become deranged. Thrombosis can result from excessive activation of coagulation and/or impaired endogenous regulation. This section will focus on the two major natural anticoagulant pathways that serve to inhibit thrombin generation: the protein C/S pathway and the antithrombin pathway.

The Protein C/S Pathway and Thrombosis

Protein C (PC) is a vitamin K-dependent protein that is activated by thrombin. When bound to the endothelial cell surface protein thrombomodulin, thrombin changes its substrate specificity, losing its ability to cleave fibrinogen and activate platelets. Instead, the thrombin-thrombomodulin complex proteolytically activates zymogen PC to form activated protein C (APC). Activated protein C and its cofactor protein S (another vitamin K-dependent protein) inactivate factors Va and VIIIa, thereby inhibiting thrombin generation. APC also downregulates inflammatory pathways and inhibits *p53*-mediated apoptosis of ischemic brain endothelium. The endothelial protein C receptor (EPCR), localized on the surface of endothelial cells, serves to bind PC and thereby enhance its activation by thrombin-thrombomodulin 5-fold. EPCR is also found in a soluble form in plasma, and its levels are enhanced in such multifocal conditions as disseminated intravascular coagulation and systemic lupus erythematosus. EPCR also binds to APC, shifting its substrate specificity to favor activation of the protease activated receptor-1 (PAR-1). This pathway thereby facilitates cross-talk between the coagulation system and inflammatory cell, endothelial, and platelet functions.

Heterozygous protein C deficiency is a recognized risk factor for venous thromboembolism, with an odds ratio of 6.5–8. Most of the causative mutations are of the type I variety, with a concordant decrease in activity and antigen. These mutations affect protein folding and lead to unstable molecules that are either poorly secreted or are degraded more rapidly. Type II defects lead to activity levels that are reduced disproportionately to the antigen levels and result in dysfunctional molecules with ineffective protein-protein interactions. Heterozygous protein C deficiency has a prevalence of 0.2–0.4% in the general population and approximately 4–5% of patients with confirmed deep venous thrombosis. Protein C-deficient individuals with personal and family histories of thrombosis may have a second thrombophilic defect, such as factor V Leiden, to account for the thrombotic tendency. Venous thromboembolic disease (VTE) occurs in 50% of heterozygous individuals in affected families by age 45, with half of the events being spontaneous in onset. Venous thrombosis at unusual sites (cerebral sinus and intra-abdominal) is a clinical hallmark. Arterial thrombosis is rare, though reported. Homozygous protein C deficiency with levels <1% generally presents with neonatal purpura fulminans and massive thrombosis in affected infants. Individuals with protein C deficiency are predisposed to develop warfarin skin necrosis when anticoagulated with vitamin K antagonists such as Coumadin. Since protein C has a much shorter half-life (8 hours) than the procoagulant vitamin K-dependent factors such as prothrombin and factor X (24–48 hours), a transient hypercoagulable state can occur in patients treated with vitamin K antagonists in the absence of an alternate bridging anticoagulant such as heparin. This risk is magnified in patients with underlying deficiency of either protein C or vitamin K.

Protein S is a vitamin K-dependent protein that is not a serine protease; rather, it acts as a cofactor for APC. In normal plasma, 60% of protein S is bound to C4b-binding protein (C4BP), and the remainder is present in the free form. Only the free form of protein S can function as the cofactor for APC. Protein S also exhibits anticoagulant activities independent of APC by binding to and inactivating factors Va, VIIIa, and Xa. Most recently, it has been shown that protein S is a cofactor for TFPI-mediated inactivation of tissue factor. Protein S deficiency exists in three forms: (i) type I has equal decrements of antigen and activity; (ii) type II has low activity but normal antigen levels, and (iii) type III shows low free protein S levels, with total protein S levels in the low to normal range. The odds ratio for VTE with protein S deficiency has been variably reported as 1.6, 2.4, 8.5, and 11.5. More than 50% of VTE events are unprovoked. Arterial thromboses occur at higher frequency, especially among smokers or those with other thrombotic risk factors. Laboratory testing needs to be interpreted with caution. Normal levels vary with age and gender, being lower in premenopausal women than men, with further reductions occurring as a result of estrogen therapy or pregnancy. Measured protein S activity can be falsely low in patients with inherited resistance to activated protein C. Acquired protein S deficiency occurs in a variety of conditions, including acute thrombosis, inflammation, liver disease, nephrotic syndrome, vitamin K deficiency, disseminated intravascular coagulation, and in association with the lupus anticoagulant. Antibodies to protein S can be seen in children with varicella or other viral illnesses.

Addition of APC to plasma normally causes a prolongation of the APTT. In 1993, Dahlback reported a series of thrombophilic families in which the plasma of the probands and their affected relatives exhibited resistance to APC, with much less prolongation of the APTT than would be expected. Mixing studies showed this defect to be due to a problem with factor V, and the genetic defect responsible for APC resistance was shown to be a mutation at the major cleavage site of APC on factor Va from arginine to glutamine (R506Q). This mutation, now known as factor V Leiden, after the Dutch city in which it was first discovered, is the most prevalent inherited mutation leading to thrombophilia. It is found in approximately 5% of Caucasian populations and is felt to be the result of a founder mutation in a single ancestor 21,000 to 34,000 years ago. The mutant factor Va is inactivated by APC 10-fold more slowly, thereby leading to excessive thrombin generation. Factor V Leiden is estimated to account for 20–25% of inherited thrombophilia. Approximately 90% of affected individuals do not suffer any venous thromboembolic events during their lifetime. On the other hand, homozygotes have an odds ratio for VTE of 50–100, and half of such individuals will have thromboses during their lives. Coronary artery thrombosis may also occur with greater frequency in young men and women with other risk factors, such as smoking. In general, however, factor V Leiden is not considered to be a major risk factor for arterial thrombosis. The risk for venous thrombosis in individuals with factor V Leiden is greatly magnified when other risk factors for thrombosis are present. These risks may be either genetic or acquired, including PC deficiency, PS deficiency, the prothrombin G20210 mutation, elevated levels of factor VIII, antiphospholipid antibodies, hyperhomocysteinemia, prolonged immobility, surgery, malignancy, pregnancy, or use of oral contraceptives. An acquired form of APC resistance may be caused by conditions other than factor V Leiden, including pregnancy, lupus anticoagulants, inflammation, and use of anticoagulants. Testing for APC resistance is best performed using factor V-deficient plasma, which will eliminate the preceding conditions. Genetic testing for factor V Leiden, generally using a PCR-based assay, is also available and is sensitive and specific for the disorder.

A mutation found in 1% of Caucasians is the second most frequent cause of inherited thrombophilia. A mutation in the 3'-untranslated region of the prothrombin gene (G20210A) results in elevated prothrombin synthesis. Thrombotic risk is probably a result of increased thrombin generation and/or decreased fibrinolysis mediated by enhanced activation of TAFI. The relative risk for first episode of VTE in heterozygotes is between 2 and 5.5, and 4–8% of patients presenting with their first VTE will be found to have this mutation. Homozygosity for the mutation appears to confer a higher risk of VTE. Venous clots in odd locations, as well as arterial clots, are found with increased frequency, especially in patients younger than 55, and especially in those with other thrombotic risk factors. PCR amplification of the pertinent region, followed by DNA sequencing is required for the diagnosis. Measurement of factor II levels is neither sensitive nor specific for the disorder.

Antithrombin Deficiency

Antithrombin (AT) is a SERPIN that inactivates thrombin and clotting factors Xa, IXa, and XIa by forming irreversible 1:1 complexes in reactions accelerated by glycosaminoglycans such as heparin or heparan sulfate on the surface of endothelial cells. Deficiency of antithrombin therefore results in potentiation of thrombosis. In type I deficiency, the antigen and activity levels are decreased in parallel, whereas in type II deficiency, a dysfunctional molecule is present. Type IIa mutations affect the active center of the inhibitor, which is responsible for complexing with the active site of the protease. Type IIb mutations target the heparin-binding site, and type IIc mutations are heterogeneous. Severe antithrombin deficiency with levels <5% is rare, resulting from one of several IIb mutations, and leads to severe recurrent arterial and venous thromboses. The odds ratio for venous thrombosis in heterozygotes is approximately 10–20. Lower extremity deep vein thrombosis is common, and clots in unusual sites have been reported. Clots tend to occur at a younger age, with 70% presenting before age 35, and 85% before age 50. Some patients with AT deficiency exhibit resistance to the anticoagulant effects of heparin. Other conditions associated with reduced levels of AT include treatment with heparin, acute thrombosis, disseminated intravascular coagulation, nephrotic syndrome, liver disease, treatment with the chemotherapeutic agent L-asparaginase, and pre-eclampsia.

KEY CONCEPTS

- Defects in hemostasis may lead to bleeding via the following mechanisms:
  a. *Deficient thrombin generation on the proper cellular surface due to*
     i. Deficiencies of factor VIII or IX (hemophilia A and B) required for propagation phase of thrombin generation
     ii. Deficiencies of factors in the final common pathway of thrombin generation (factors II, V, and X)
     iii. Deficiency of factor XI (required for "overdrive" of coagulation)
  b. *Defect in fibrin polymerization*
     i. Deficiencies or abnormalities of fibrinogen
     ii. Deficiency of factor XIII (required for cross linking fibrin)
  c. *Defect in primary hemostasis*
     i. Von Willebrand disease
     ii. Platelet dysfunction (adhesion or aggregation defect)
     iii. Thrombocytopenia
  d. *Abnormal fibrinolysis*
- Defects in the anticoagulant system lead to thrombosis via the following mechanisms:
  a. *Unopposed or excess generation of thrombin*
     i. Antithrombin deficiency
     ii. Prothrombin G20210A mutation
  b. *Insufficient inactivation of procoagulant proteins*
     i. Activated protein C resistance including that due to the factor V Leiden mutation
     ii. Deficiencies of protein C or protein S

SUGGESTED READINGS

1. Amirlak I, Amirlak B. Haemolytic uraemic syndrome: An overview. *Nephrology (Carlton)*. 2006;11:213–218.
2. Folsom AR, Aleksic N, Wang L, et al. Protein C, antithrombin, and venous thromboembolism incidence: A prospective population-based study. *Arterioscler Thromb Vasc Biol*. 2002;22:1018–1022.
3. Levy GG, Nichols WC, Lian EC, et al. Mutations in a member of the ADAMTS gene family cause thrombotic thrombocytopenic purpura. *Nature*. 2001;413:488–494.
4. Levy JH, Hursting MJ. Heparin-induced thrombocytopenia, a prothrombotic disease. *Hematol Oncol Clin North Am*. 2007;21: 65–88.
5. McCrae KR, ed. *Thrombocytopenia*. 1st ed. New York: Taylor and Francis; 2006.
6. Moake JL, Rudy CK, Troll JH, et al. Unusually large plasma factor VIII: von Willebrand factor multimers in chronic relapsing thrombotic thrombocytopenic purpura. *N Engl J Med*. 1982;307: 1432–1435.
7. Nichols WL, Hultin MB, James AH, et al. von Willebrand disease (VWD): Evidence-based diagnosis and management guidelines, the National Heart, Lung, and Blood Institute (NHLBI) Expert Panel report (USA). *Haemophilia*. 2008;14:171–232.
8. Peyvandi F, Cattaneo M, Inbal A, et al. Rare bleeding disorders. *Haemophilia*. 2008;14(suppl 3):202–210.

9. Roberts HR, Hoffman M, Monroe DM. A cell-based model of thrombin generation. *Semin Thromb Hemost*. 2006;32(suppl 1): 32–38.

10. Schafer AI, Levine MN, Konkle BA, et al. Thrombotic disorders: Diagnosis and treatment, hematology. *Am Soc Hematol Educ Program*. 2003;520–539.

11. Seri M, Pecci A, Di Bari F, et al. MYH9-related disease: May-Hegglin anomaly, Sebastian syndrome, Fechtner syndrome, and Epstein syndrome are not distinct entities but represent a variable expression of a single illness. *Medicine (Baltimore)*. 2003; 82:203–215.

16

Molecular Basis of Lymphoid and Myeloid Diseases

Joseph R. Biggs ▪ Dong-Er Zhang

DEVELOPMENT OF THE BLOOD AND LYMPHOID ORGANS

Hematopoietic Stem Cells

All hematopoietic cells are derived from hematopoietic stem cells (HSCs) that are capable of both self-renewal and differentiation into all blood cell lineages. In mammals, HSCs are produced at sequential sites beginning with the yolk sac and followed by an area surrounding the dorsal aorta called the aorta-gonad mesonephros (AGM) region, the fetal liver, and finally the bone marrow. Yolk sac hematopoiesis is termed primitive because it produces mainly red blood cells and is transient and rapidly replaced by definitive hematopoiesis. Definitive hematopoiesis involves the colonization of the fetal liver, thymus, spleen, and bone marrow by HSCs that have migrated from earlier sites of formation. The AGM has long been viewed as the principal site of HSC production, but recent studies have suggested that the yolk sac may also contribute to the adult hematopoietic system. It is not yet clear whether HSCs from the fetal liver circulate to the adult bone marrow and are the source of adult hematopoiesis, or if the fetal liver and bone marrow are seeded with HSCs at the same time during development. All types of mature hematopoietic cells arise from differentiation of the HSCs. Figure 16.1 illustrates normal hematopoietic development.

HSCs, like all stem cells, depend on their microenvironment (or niche) for normal self-renewal and differentiation. The adult bone marrow is the most widely studied HSC niche, and experimental evidence to date suggests that HSCs may associate with either osteoblasts or with vascular cells, as illustrated in Figure 16.2. It has been proposed that the precise site of association may regulate HSC function. This regulation is commonly thought to be mediated by cell-cell interactions between the HSC and the osteoblast or vascular cell, and by chemokines secreted by components of the niche.

Hematopoietic Differentiation and the Role of Transcription Factors

All types of mature blood cells are produced by lineage-restricted differentiation of HSCs. This process is believed to be regulated by a relatively small group of transcription factors, some required for HSC formation and others for differentiation. In the most commonly presented model of hematopoietic differentiation (Figure 16.1), long-term or quiescent HSCs are mobilized from the niche to become proliferating short-term HSCs. The short-term HSCs then differentiate into common myeloid progenitors (CMPs) or common lymphoid progenitors (CLPs). The CMPs give rise to megakaryocyte/erythroid precursors (MEPs) and granulocyte/macrophage precursors (GMPs); the MEPs differentiate into red blood cells and megakaryocytes; and GMPs produce mast cells, eosinophils, neutrophils, and monocyte/macrophages. B- and T-lymphocytes differentiate from the CLPs. There is some evidence for alternative pathways, such as the possibility that MEPs do not originate from CMPs but from an earlier precursor. Each stage of this process is controlled by specific transcription factors. The identity of these factors (as with HSC transcription factors) has been determined largely through the study of conventional or conditional gene knock-outs in mice and other model organisms. The fact that most of these factors show lineage- and stage-restricted expression also provides information about their function.

Transcription factors essential for the formation of HCSs include SCL/tal-1 and its partner LMO2, as well as Runx1 and its partner CBFβ. The histone methyltransferase MLL, which is necessary to maintain HOX

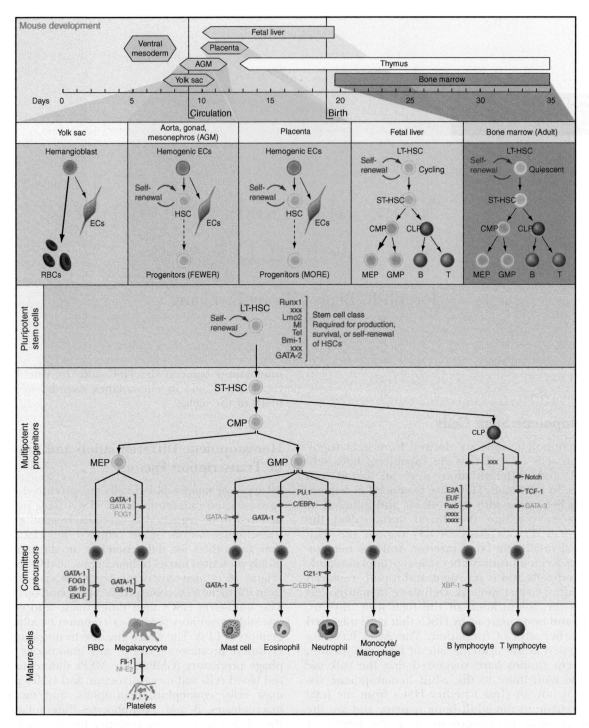

Figure 16.1 Hematopoietic development. The upper panel shows stages of hematopoiesis in the mouse. Hematopoietic stem cells (HSCs) are derived from the ventral mesoderm, and sequential sites of hematopoiesis include the yolk sac, the aorta-gonad-mesonephros (AGM) region, the fetal liver, placenta, and bone marrow. The types of cells produced at each site are illustrated in the middle panel. The main function of primitive hematopoiesis, which occurs in the yolk sac, is to produce red blood cells. The relative contribution of HSCs produced in the AGM region and the placenta to the final pool of adult HSCs remains unknown. Definitive hematopoiesis involves the colonization of the fetal liver, thymus, spleen, and bone marrow. In definitive hematopoiesis, long-term HSCs produce short-term HSCs, which in turn give rise to common myeloid progenitors (CMPs) and common lymphoid progenitors (CLPs). CMPs produce megakaryocyte/erythroid progenitors (MEPs) and granulocyte/macrophage progenitors (GMPs). CLPs produce B and T lymphocytes. The lower panel shows transcription factors that regulate hematopoiesis in mammals. The stages at which hematopoietic development is blocked in the absence of a given factor, as determined through gene knockout, are indicated by red loops. The factors in red are associated with oncogenesis; those in black have not yet been found mutated in hematologic malignancies. Among the genes required for HSC production, survival, or self-renewal are *MLL, Runx1, TEL/EV6, SCL/tal1,* and *LMO2*. These genes account in toto for the majority of known leukemia-associated translocations in patients. Reproduced with permission from Orkin SH and Zon LI, *SnapShot: Hematopoiesis Cell* 2008;132:172.e1.

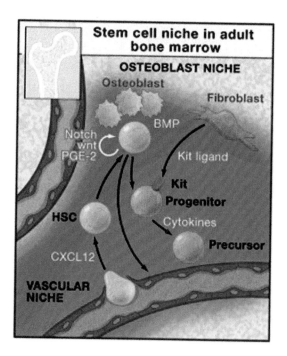

Figure 16.2 Stem cell niche in adult bone marrow. HSCs are found in the osteoblast niche adjacent to osteoblasts that are under the regulation of bone morphogenetic protein (BMP). Pathways involving Notch, wnt, and PGE-2 stimulate HSC self-renewal. HSCs are also found adjacent to blood vessels (the vascular niche). The chemokine CXCL12 regulates the migration of HSCs from the circulation to the bone marrow. The osteoblast and vascular niches in vivo lie in close proximity or may be interdigitated. The marrow space also contains stromal cells that support hematopoiesis including the production of cytokines, such as c-Kit ligand, that stimulate stem cells and progenitors. Other cytokines, including interleukins, thrombopoietin, and erythropoietin, also influence progenitor function and survival. Reproduced with permission from Orkin SH and Zon LI, *Cell* 2008;132:631–644.

gene expression, also has a vital role in hematopoiesis. In the absence of SCL/tal-1 and LMO2, failure of both primitive and definitive hematopoiesis is observed. In the absence of Runx1 or MLL, HSCs do not appear in the AGM region of the mouse embryo. A striking observation is that this set of transcription factors controlling HSC development account for the majority of known leukemia-associated translocations in patients. These translocations either deregulate the expression of the locus or generate chimeric fusion proteins.

A second set of transcription factors is required for differentiation of HSCs into specific types of mature blood cells, and the transcription factors involved in the development of HSCs also have roles in later hematopoietic development. Like the factors that control HSC development, these lineage-specific factors have been identified largely through the study of gene knockout models. As examples, loss of the factor GATA-1 or its cofactor FOG results in failure of erythroid and megakaryocytic differentiation, while mice deficient in the transcription factor

C/EBPα lack GMPs and granulocytes. Figure 16.1 illustrates the point of action of many transcription factors involved in hematopoiesis.

Hematopoietic transcription factor levels can be controlled by both transcriptional and post-transcriptional mechanisms. Recent studies on microRNAs (miRNAs) suggest that they provide an additional mechanism for controlling hematopoietic transcription factor levels. MiRNAs bind to the 3′-untranslated region of mRNAs and suppress translation, and several have been shown to affect the levels of transcription factors in hematopoietic cells.

Hematopoietic Differentiation and the Role of Signal Transduction

During the processes of proliferation and differentiation, cells respond to external signals such as growth factors or cell-cell contacts. Growth factors act by binding to a specific cell surface receptor and activating intracellular cascades which stimulate or suppress downstream transcription factors. Many of the receptors that regulate normal hematopoiesis are receptor tyrosine kinases (RTKs), such as cFMS (receptor for macrophage colony-stimulating factor/colony-stimulating factor-1), FMS-related tyrosine kinase (FLT3, receptor for FLT3-ligand), c-KIT (receptor for stem cell factor), and platelet-derived growth factor receptor (PDGFR). Ligand binding activates the tyrosine kinase activity of these RTKs, which then phosphorylate tyrosine residues on associated proteins, thereby triggering cascades in which intracellular kinases are sequentially phosphorylated and activated, until finally the signal is transmitted to nuclear transcription factors. Hematopoietic RTKs usually activate several such cascades, including the Ras/Raf/ERK pathway, the PIK3/Akt pathway, and the JAK/STAT pathway. The activation of these pathways usually favors cell proliferation and survival. Figure 16.3 illustrates several of the pathways commonly activated in response to receptor tyrosine kinases.

Many leukemias are associated with mutations which cause constitutive activation of RTKs (the receptor tyrosine kinase is continuously active, even in the absence of ligand). This results in a continuous signal to the cell favoring growth and survival. Two observations about mutations in leukemia led to the proposal of the two-hit model of leukemogenesis. First, many leukemia patients possess two types of mutations, one affecting a hematopoietic transcription factor, such as *RUNX1/AML1*, and the other affecting a receptor tyrosine kinase or signal-transduction molecule, such as *FLT3*. Second, studies in model systems have shown that many of the leukemia-associated mutations found in patients are unable to induce leukemia by themselves, but can induce leukemia in combination with other mutations. Therefore, the two-hit model proposes that induction of leukemia requires the presence of two types of mutations: (i) a class I mutation in a receptor or signal-transduction molecule which confers a

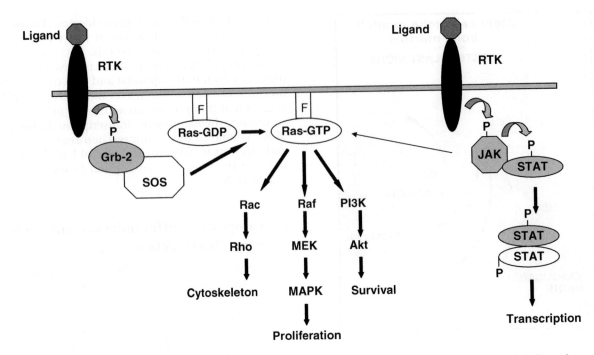

Figure 16.3   Signal transduction pathways involved in leukemia. The drawing on the left shows signaling of a receptor tyrosine kinase (RTK) through Ras. Ligand binding causes phosphorylation of Grb-2 by the RTK and formation of a Grb-2/SOS complex. Interaction of Grb-2/SOS with farnesylated (F) Ras-GDP causes conversion to active Ras-GTP, which in turn phosphorylates Rac, Raf, and PI3K leading to stimulation of their respective pathways. The drawing on the right shows the JAK/STAT pathway. Ligand binding causes RTK phosphorylation of JAK, which may then activate the Ras pathway and phosphorylate STATs. The STATs form homodimers or heterodimers with other STATs and translocate to the nucleus where they activate transcription of specific target genes.

proliferative or survival advantage, and (ii) a class II mutation in a hematopoietic transcription factor which impairs differentiation. This model is almost certainly true for many leukemias, but it is not yet clear if all or most leukemias require mutation of more than one gene.

Spleen

The spleen is a lymphoid organ that also serves as a blood filter. The arteries of the spleen are ensheathed by lymphocytes, which form the white pulp. The white pulp is further subdivided into a T-cell domain and a B-cell domain. The spleen, along with the lymph nodes, is a major repository for lymphocytes and a major site of adaptive immune response to foreign antigens. The remaining internal portion of the spleen is composed of red pulp, which is designed to filter foreign matter from the bloodstream, including damaged blood cells.

Thymus

Mature mammalian T-cells originate in the bone marrow or fetal liver as pluripotent precursors, and these cells then migrate to the thymus, where they proliferate extensively and differentiate into the various mature T-cell lineages. The sole function of the thymus is to serve as the site of T-cell differentiation. Beginning at puberty, the thymus involutes and

shrinks, until it eventually consists of groups of epithelial cells depleted of lymphocytes.

Lymph Nodes

Lymph nodes are small glands located in many parts of the body, mainly in the neck, under the arms, and in the groin. Lymph vessels drain fluid from tissue, which then enters the lymph nodes via afferent lymphatic vessels. Lymph nodes are composed of structures known as lobules, subdivided into structures called follicles. Immature B-cells originating in the bone marrow home to follicles where they interact with follicular dendritic cells (FDCs). FDCs trap antigen-antibody complexes that may be collected from lymph carried into the follicle. If a B-cell encounters its antigen displayed on an FDC, it is stimulated to proliferate, and these proliferating B-cells form distinctive germinal centers, which are referred to as secondary follicles. A large number of B-cells undergo apoptosis during this process of proliferation and differentiation. In contrast to B-cells, T-cells migrate to the paracortex and interfollicular cortex and survey dendritic cells. The dendritic cells that interact with T-cells form a separate class from those that interact with B-cells; they collect and process antigens in tissue and then migrate to the lymph nodes. The interfollicular cortex and DCU serve as corridors for the movement of B- and T-cells. Several afferent lymphatic vessels enter the lymph node, but each delivers a stream of lymph to a specific lobule. After passing over the lobules, all the lymph streams exit the lymph node through a single efferent vessel. Thus, individual lobules are exposed to different sets of antigens and cells collected from a specific drainage area by an individual afferent vessel. The constant flow of lymph containing cells and antigens collected from tissue allows the lymph nodes, like the spleen, to serve as a major site of interaction between foreign antigens and lymphocytes.

MYELOID DISORDERS

Anemia

Anemia is a condition in which the blood contains a lower than normal number of red blood cells (RBCs), or RBCs that do not contain enough hemoglobin. Anemia may be caused either by lower than normal production of RBCs or higher than normal rates of RBC destruction. Nongenetic causes of anemia include blood loss, iron deficiency, lack of folic acid (vitamin B12), or chronic disease, all of which can impede the production of RBCs. Higher than normal rates of RBC destruction can be caused by inherited disorders such as sickle cell anemia and thalassemia, and certain enzyme deficiencies. Hemolytic anemia occurs when the immune system mistakenly attacks RBCs. Anemia may also be caused by myelodysplastic syndromes, defined as one or more secondary blood cytopenias (cell loss) caused by bone marrow dysfunction.

Neutropenia

Neutropenia may occur as chronic idiopathic neutropenia or severe congenital neutropenia. Chronic idiopathic neutropenia is defined as any unexplained reduction in neutrophil count to below average. Chronic idiopathic neutropenia (CIN) is believed to result from impaired bone marrow granulopoiesis, but the precise molecular mechanism remains unknown. It has been demonstrated that the bone marrow of CIN patients contains activated T-cells producing IFNγ and Fas-ligand, as well as increased local production of TNFα and TGFβ1 and decreased production of IL-10. These changes may lead to increased apoptosis of neutrophil precursors.

Severe congenital neutropenia (SCN) is characterized by life-long neutropenia with an absolute neutrophil count under 0.5×10^9/liter, recurrent bacterial infections, and arrest of neutrophil maturation at the promyelocyte stage. Approximately 60% of patients with SCN carry mutations in the neutrophil elastase (ELA2) gene. These patients fall into the categories of dominant inheritance of the disease or spontaneous acquisition of the disease. Mutations in the ELA2 gene are also present in patients with cyclic hematopoiesis, in which the number of neutrophils and other blood cells oscillates in weekly phases. Neutrophil elastase is a protease found in the granules of mature neutrophils. The ELA2 mutations found in patients with SCN or cyclic hematopoiesis induce the unfolded

protein response (UPR) and apoptosis. Protein folding occurs in the lumen of the endoplasmic reticulum; misfolded proteins trigger the UPR, which leads to attenuation of translation, expression of ER-resident chaperones, and ER-associated degradation pathways. If this adaptive response is overwhelmed, apoptosis is induced. Patients with SCN also display a deficiency of the transcription factor LEF-1, leading in turn to reduced levels of the LEF-1 targets C/EBPα, cyclin D1, c-myc, and survivin. The LEF-1 deficiency (not coupled to a mutation in the gene) is present in SCN patients with either mutation of ELA2 or the gene for HS-1-associated protein X (HAX-1). This suggests that LEF-1 deficiency may synergize with the ELA2 or HAX-1 mutations to promote neutropenia. In contrast to other SCN patients, those who acquired the disease through recessive inheritance lack mutations in the ELA2 gene but carry mutations in the HAX-1 gene. This form of neutropenia was first described by Rolf Kostmann and is also known as Kostmann's disease. HAX-1 is a mitochondria-targeted protein, containing Bcl-2 homology domains, and is critical for maintaining the inner mitochondrial membrane potential. Loss of HAX-1 function causes increased apoptosis in myeloid cells. It therefore appears that mutations in either ELA2 or HAX-1 contribute to neutropenia by causing enhanced levels of apoptosis in myeloid precursor cells.

Myelodysplastic Syndromes

Myelodysplastic syndromes (MDS) are diagnosed at a rate of 3.6/100,000 people in the United States. The MDS occur primarily in older patients (>60 years), but occasionally in younger patients. Anemia, bleeding, easy bruising, and fatigue are common, and splenomegaly or hepatosplenomegaly may occasionally be present. The MDS are characterized by abnormal bone marrow and blood cell morphology. The bone marrow is usually hypercellular, but approximately 15% of patients have hypoplastic bone marrow. Circulating granulocytes are often severely reduced and hypogranular or hypergranular. Early, abnormal myeloid progenitors are identified in the marrow in varying percentages, depending on the type of MDS. Abnormally small megakaryocytes may be seen in the marrow and hypogranular or giant platelets in the blood. The MDS are classified according to cellular morphology, etiology, and clinical features.

Knowledge of the molecular defects underlying the refractory anemias is currently limited. Recurrent deletions of 5q31, 7q22, and 20q12 in MDS suggest that loss of unidentified tumor suppressor genes within these regions contributes to development of MDS. Recently mutations in the RUNX1 gene have been associated with development of some types of MDS. This suggests that loss of RUNX1 transcription factor activity, which normally regulates hematopoietic stem cell development and differentiation of some hematopoietic lineages, may contribute to the development of some types of refractory anemias.

Myelodysplastic/Myeloproliferative Diseases

Myelodysplastic/myeloproliferative diseases have features of both myelodysplastic syndromes and myeloproliferative disorders. A greater than normal number of stem cells develop into one or more types of more mature cells, and the blood cell number increases; but there is also some degree of failure to mature properly. The three main types of myelodysplastic/myeloproliferative diseases are chronic myelomonocytic leukemia (CMML), juvenile myelomonocytic leukemia (JMML), and atypical chronic myelomonocytic leukemia (aCML). A myelodysplastic/myeloproliferative disease that does not match any of the previous types is referred to as myelodysplastic/myeloproliferative disease, unclassifiable (MDS/MPD-UC).

Chronic Myelomonocytic Leukemia

Chronic myelomonocytic leukemia (CMML) is characterized by the overproduction of myelocytes and monocytes, as well as immature blasts. Gradually, these cells replace other cell types, such as red cells and platelets in the bone marrow, leading to anemia or easy bleeding. The CMML bone marrow may exhibit hypercellularity (75% of cases), a blast count of less than 20%, granulocytic and monocytic proliferation, micromegakaryocytes or megakaryocytes with lobated nuclei (80% of cases), and fibrosis (30% of cases).

Juvenile Myelomonocytic Leukemia

Juvenile myelomonocytic leukemia (JMML) accounts for 2% of all childhood leukemias. The three required criteria for a diagnosis of JMML are no Philadelphia chromosome (BCR/ABL fusion gene), peripheral blood monocytosis greater than 1×10^9/liter, and fewer than 20% blasts in the blood and bone marrow. The presence of two or more of the following minor criteria is also required: fetal hemoglobin increased for age, immature granulocytes in the peripheral blood, a white blood cell count greater than 1×10^9/liter, a clonal chromosomal abnormality, and granulocyte-macrophage colony-stimulating factor (GM-CSF) hypersensitivity of myeloid progenitors.

A distinctive characteristic of JMML leukemic cells is their spontaneous proliferation in vitro due to their hypersensitivity to GM-CSF. This hypersensitivity has been attributed to altered Ras pathway signaling as a result of mutually exclusive mutations affecting one of the pathway regulatory molecules including the genes for RAS, PTPN11, and NF1. Thirty percent of JMML patients display PTPN11 mutations, while 15–20% display either RAS mutations or NF1 mutations. Ras is a GTP-dependent protein (G-protein) localized at the inner side of the cell membrane, and transduces signal from growth factor receptors to downstream effectors. Protein-tyrosine phosphatase, nonreceptor-type, 11 (PTPN11) encodes the SHP-2 protein, which transmits signals from growth factor receptors to Ras. Neurofibromatosis, type 1 (NF1) is a tumor suppressor gene that inactivates Ras through

acceleration of Ras-associated GTP hydrolysis. Activating mutations in RAS or PTPN11 or inactivating mutations in NF1 in JMML cells all result in enhancement of signaling through the Ras pathway and increased stimulus to proliferate (Figure 16.3). This finding has stimulated interest in molecules that inhibit the Ras pathway as possible therapeutic agents.

Chronic Myeloid Leukemia

Chronic myeloid leukemia (CML) is characterized by less than 10% blasts and promyelocytes in peripheral blood and bone marrow. The transition from chronic phase to the accelerated phase and later blastic phase may occur gradually over a period of 1 year or more, or it may appear suddenly (blast crisis). Examination of bone marrow shows a shift of the myeloid cells to more immature forms that increase in number as the disease progresses. The bone marrow is hypercellular, but the spectrum of mature and immature granulocytes is similar to that in normal marrow. The percentage of lymphocytes is reduced in both blood and marrow, and the ratio of erythroid to myeloid cells is usually reduced.

The leukemic cells of almost all CML patients contain a distinctive cytogenetic abnormality, the Philadelphia chromosome. The Philadelphia chromosome is formed by a reciprocal translocation between the long arms of chromosomes 9 and 22, and results in the fusion of the ABL gene on chromosome 9 to the BCR gene on chromosome 22. The resulting fusion gene, BCR-ABL, produces a fusion protein containing the oligomerization and serine/threonine kinase domains of BCR at the amino terminus and most of the ABL protein at the carboxy-terminus. ABL is a nonreceptor tyrosine kinase, and its activity is normally tightly regulated in cells. The fusion of BCR sequences constitutively activates the ABL tyrosine kinase, transforming ABL into an oncogene. BCR-ABL, with the aid of mediator proteins, associates with Ras and stimulates its activation; through stimulation of the Ras-Raf pathway, BCR-ABL increases growth factor-independent cell growth (Figure 16.4). BCR-ABL is also associated with the Janus kinase and signal transducer and activator of the transcription (JAK-STAT) pathway. BCR-ABL activates the phosphatidylinositol-3-kinase (PI3K) pathway, suppressing programmed cell death or apoptosis. BCR-ABL is associated with cytoskeletal proteins leading to a decrease in cell adhesion, and activates pathways that lead to an increase in cell migration.

Knowledge of the role of BCR-ABL in the development of CML led to the discovery of imatinib, a small molecule ABL kinase inhibitor, a highly effective therapy for early phase CML. However, some patients develop resistance to imatinib, usually caused by point mutations in the kinase domain of BCR-ABL that reduce sensitivity to imatinib. To overcome this problem, a second generation of BCR-ABL inhibitors are under development, as well as inhibitors which target oncogenic signaling pathways downstream of BCR-ABL.

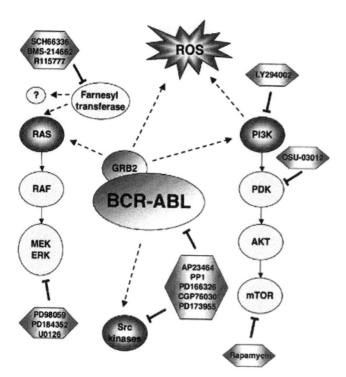

Figure 16.4 Targeting signaling pathways of BCR-ABL.
The BCR-ABL onco protein chronically activates many different downstream signaling pathways to confer malignant transformation in hematopoietic cells. For example, efficient activation of PI3K, Ras, and reactive oxygen species (ROS) requires autophosphorylation on Tyr177, a Grb-2 binding site in BCR-ABL. Also, activation of Src family tyrosine kinases have been implicated in the BCR-ABL-related disease process. A selection of some inhibitors and pathways discussed in the text is illustrated. Reproduced with permission from Walz C and Sattler M, *Critical Reviews in Oncology/Hematology* 2006; 57:145–164.

Chronic Neutrophilic Leukemia

Chronic neutrophilic leukemia (CNL) is a rare disorder characterized by peripheral blood neutrophilia (greater than 25×10^9/liter) and hepatosplenomegaly. The bone marrow is hypercellular, and there is no significant dysplasia in any cell lineage. Cytogenetic studies are normal in 90% of patients.

Approximately 20% of CNL patients are positive for the *JAK2* V617F mutation. JAK2 is a member of the Janus family of tyrosine kinases (JAK1, JAK2, JAK3, and TYK2), which are cytoplasmic kinases that mediate signaling downstream of cytokine receptors (Figure 16.3).

Polycythemia Vera

In polycythemia vera (PV) too many red blood cells are made in the bone marrow, and the blood becomes thickened with red blood cells. The extra red cells may collect in the spleen, causing it to swell, or may cause bleeding problems and clots.

Between 65% and 97% (depending on the study) of polycythemia vera patients have the *JAK2* V617F

mutation. As mentioned in the previous section, JAK2 is one of a family of cytoplasmic tyrosine kinases which mediate signaling by growth factor receptors. The V617F mutation changes JAK2 amino acid 617 from valine to phenylalanine, and creates a constitutively active form of JAK2. *JAK2* V617F renders cells hypersensitive to the growth-stimulating effects of the erythroid growth factor erythropoietin and other growth factors. Constitutive activation of the JAK/STAT, PI3K, ERK, and Akt signal-transduction pathways is also observed in the presence of *JAK2* V617F, all of which may promote cell proliferation.

The role of *JAK2* V617F in myeloproliferative disorders has stimulated interest in JAK inhibitors as possible therapeutic agents, although no current JAK inhibitors are candidates because of their lack of specificity. Current effort is focused on the development of specific JAK2 inhibitors for use in clinical trials.

Essential Thrombocythemia

Essential thrombocythemia (ET) causes an abnormal increase in the number of platelets in the blood and bone marrow. This may inhibit blood flow and lead to problems such as stroke or heart attack. Like PV patients, a high percentage (approximately 50%) of essential thrombocythemia patients carry the *JAK2* V617F mutation. This raises the question of how a single mutation can give rise to several different diseases. Among the theories suggested are transformation of different types of hematopoietic stem or progenitor cells, different genetic backgrounds, or the effects of additional somatic mutations. Transformation of different types of hematopoietic cells is considered unlikely by some because of the belief that the *JAK2* V617F mutation occurs at the level of the hematopoietic stem cell in all cases. It is considered more likely that the activity of *JAK2* V617F is modified by gene dosage, *JAK2* polymorphisms, or cooperating mutations in JAK2-interacting proteins.

Chronic Idiopathic Myelofibrosis

Chronic idiopathic myelofibrosis (CIMF) is characterized by the production of too few red blood cells and too many white cells and platelets. An important constant is the production of too many megakaryocytes, which results in overproduction of platelets and cytokine release in the bone marrow. The cytokines stimulate the development of fibrous tissue in the marrow. Megakaryocytes can become so abnormal that platelet production is decreased in some patients.

Approximately 50% of patients with myelofibrosis carry the *JAK2* V617F mutation. As discussed previously, the *JAK2* V617F mutation is also found in patients with PV or ET. Which of these diseases develops as a result of the mutation is thought to depend on genetic background or the presence of secondary mutations. The three diseases are also related to some degree. About 10–15% of cases of myelofibrosis begin as either PV or ET.

Chronic Eosinophilic Leukemia/ Hypereosinophilic Syndrome

In chronic eosinophilic leukemia (CEL), a clonal proliferation of eosinophilic precursors results in persistently increased numbers of eosinophils in the blood, bone marrow, and peripheral tissues.

CEL may be divided into three classes based on the presence or absence of mutations in the *PDGFRA* (platelet-derived growth factor receptor A) or *PDGFRB* genes. The mutations all take the form of fusion genes, and the most widely studied is the *FIP1L1–PDGFRA* fusion. This fusion causes constitutive activation of the PDGFRA tyrosine kinase activity by disruption of the PDGFRA autoinhibitory juxtamembrane motif. The active FIP1F1–PDGFRA kinase stimulates the JAK/STAT, PI3K, ERK, and Akt signal-transduction pathways leading to increased survival and proliferation. Other *PDGFRA* fusion mutations are associated with CEL, including *KIF5B-PDGFRA*, *BCR-PDGFRA*, and *CDK5RAP2–PDGFRA*. However, in all cases the *PDGFRA* breakpoint occurs in the region of the juxtamembrane motif.

Systemic Mastocytosis

Systemic mastocytosis (SM) is a rare disease in which too many mast cells are found in skin, bones, joints, lymph nodes, liver, spleen, and the gastrointestinal tract. The *FIPILI-PDGFRA* fusion causes a disease with symptoms of both chronic eosinophilic leukemia and systemic mastocytosis (CEL-SM). Remaining cases of SM may be divided into those associated with a mutation in the *c-KIT* gene and those without this mutation. The oncogene *c-KIT* encodes a receptor tyrosine kinase. c-KIT and its ligand stem cell factor (SCF) are required for the growth and survival of normal mast cells. SCF ligation to c-Kit activates the Ras/Raf/Erk cascade, the PI3K/Akt pathway, the shp/rac/JNK/ c-jun pathway, and the NFκB pathway. Activation of the PI3K/Akt and NFκB pathways have been shown to be necessary for mast cell proliferation. Systemic mastocytosis patients carry mutations which cause ligand-independent activation of c-Kit.

Stem Cell Leukemia-Lymphoma Syndrome

Stem cell leukemia-lymphoma syndrome (SCLL) is characterized by concurrent lymphoma, myeloid proliferation/eosinophilia often evolving to AML, and cytogenetic abnormalities involving 8p11. Both myeloid and lymphoid cells exhibit one of a number of 8p11 translocations, all involving the fibroblast growth factor receptor-1 (*FGFR1*) gene. These translocations cause constitutive activation of *FGFR1* by fusing a dimerization domain from the fusion partner to *FGFR1*, a mechanism similar to that observed with the *PDGFRB* fusions associated with CEL-UMPD. *FGFR1* is a receptor tyrosine kinase which when activated stimulates the Ras/Raf/Erk and PI3K/Akt signal transduction cascades.

Acute Myeloid Leukemia

Normal myeloid stem cells eventually develop into granulocytes, macrophage/monocytes, and megakaryocytes. In acute myeloid leukemia (AML), myeloid stem cells usually develop into a type of immature white blood cell called myeloblasts which are abnormal and do not differentiate. For many years, the different categories of AML were described by the French-American-British (FAB) classification scheme. The eight FAB subtypes are M0 (undifferentiated AML); M1 (myeloblastic, without maturation); M2 (myeloblastic, with maturation); M3 (promyelocytic), or acute promyelocytic leukemia (APL); M4 (myelomonocytic); M4eo (myelomonocytic together with bone marrow eosinophilia); M5 monoblastic leukemia (M5a) or monocytic leukemia (M5b); M6 (erythrocytic), or erythroleukemia; M7 (megakaryoblastic).

Beginning in 1997, the *World Health Organization* (WHO) developed a new classification scheme for acute myeloid leukemias that attempts to incorporate morphology, cytogenetics, and molecular genetics. The WHO scheme also reduced the required blast percentage in the blood or bone marrow for a diagnosis of AML from 30% to 20%. The category of AML with characteristic genetic abnormalities is associated with high rates of remission and favorable prognosis.

AML with t(8;21)(q22;q22)

One of the most common genetic abnormalities in AML is (8;21)(q22;q22), which accounts for 5–12% of all cases of AML. t(8;21) generates the fusion gene *AML1-ETO*, which fuses sequences coding the amino-terminal portion of the transcription factor *RUNX1* (formerly *AML1*) to almost the entire coding region of *RUNX1T1* (formerly *ETO* or *MTG8*). The resulting fusion protein contains the DNA-binding domain (runt domain) of RUNX1 fused to the RUNX1T1 co-repressor protein. Numerous studies in model systems have demonstrated that expression of the AML1-ETO protein alone is insufficient to induce leukemia but can induce leukemia in cooperation with other mutations. Expression of AML1-ETO does lead to some inhibition of myeloid, lymphoid, and erythroid differentiation, as well as promotion of stem cell self-renewal. This is thought to predispose hematopoietic stem cells to leukemia development. It was formerly believed that AML1-ETO changed gene expression patterns by dominant-negative suppression of RUNX1 target genes. However, subsequent gene expression studies found the AML1-ETO activated as many genes as it repressed, suggesting that AML1-ETO promotes leukemogenesis by complex effects on gene expression. Model studies suggest that the carboxy-terminus of the ETO/RUNX1T1 protein actually suppresses leukemia development, since mutations or deletions in this region allow AML1-ETO to promote leukemia development without the need for additional mutations. Presumably when full-length AML1-ETO is expressed, such mutations are needed to overcome the antileukemogenic effects of its carboxy-terminal sequences.

AML with inv(16)(p13q22) or t(16;16)(p13q22)

AML with inv(16)(p13q22) or t(16;16)(p13q22) comprises 10–12% of all cases of AML, and is predominant in younger patients. This type of AML was formerly classified as FAB type M4.

Both inv(16)(p13q22) and t(16;16)(p13q22) result in the fusion of the core binding factor-b (CBFb) gene located at 16q22 to the smooth muscle myosin heavy chain (*MYH11*) gene at 16p13. CBFb has no DNA-binding domain, but forms a heterodimer with the AML1/RUNX1 transcription factor and stabilizes AML1/RUNX1 binding to DNA. Since the RUNX1 and CBFb proteins function as a heterodimeric transcription factor, the leukemic fusion protein AML1-ETO and the CBFb/MYH11 fusion protein are predicted to disrupt expression of a similar set of target genes. CBFb/MYH11 binds to RUNX1 with a much higher affinity than CBFb, and two mechanisms have been proposed by which CBFb/MYH11 may disrupt normal RUNX1/CBFb activity. CBFb/MYH11 may sequester RUNX1 in the cytoplasm through the interaction of the MYH11 region with the actin cytoskeleton, or the MYH11 sequences may recruit co-repressors when bound with RUNX1 to promoters in the nucleus. It is not yet clear if CBFb/MYH11 utilizes one or both mechanisms. As was observed with AML1-ETO, expression of CBFb/MYH11 in model systems was not sufficient for leukemogenesis unless secondary mutations are introduced.

Acute Promyelocytic Leukemia—AML with t(15;17)(q22q12)

Acute promyelocytic leukemia (APL) comprises 5–8% of all cases of AML and is found as typical APL or microgranular APL. Common features of typical APL include promyelocytes with kidney-shaped or bilobed nuclei and cytoplasm densely packed with large granules. Features of microgranular APL include bilobed nuclei and, scarce or absent granules. APL was formerly classified as FAB type M3.

In over 98% of cases, the retinoic acid receptor alpha (*RARa*) gene at 17q12 is fused to the *PML* gene at 15q22. In rare cases, *RARa* is fused to another gene, including *PLZF*, *NuMa*, *NPM*, or *STAT5b*. Retinoid signaling is transmitted by two families of nuclear receptors, retinoic acid receptor (RAR) and retinoid X receptor (RXR), which form RAR/RXR heterodimers. In the absence of ligand, the RAR/RXR heterodimer binds to target gene promoters and represses transcription. When a ligand (such as retinoic acid) binds to the complex, it induces a conformational change which transforms the heterodimer into a transcriptional activator. The PML-RARA fusion protein created by t(15;17)(q22q12) binds to RAR/RXR target genes and acts as a potent transcriptional repressor which is not activated by physiological concentrations of ligand. This is due to the fact that all the oncogenic fusion partners of RARA provide a dimerization domain, which results in a dimerized fusion protein with two corepressor binding sites instead of the one found in the RAR/RXR

complex. However, recent studies suggest that the PML-RARA fusion protein must have other oncogenic properties, since enforced corepressor binding onto RARA does not initiate APL in model systems. Recent models suggest PML-RARA leukemogenesis combines enhanced corepressor recruitment and relaxed target specificity to both enhance repression of some genes and target genes not normally bound by RAR/RXR. This disruption of normal gene expression is thought to affect two pathways: myeloid progenitor cell self-renewal and promyelocyte differentiation.

APL is highly sensitive to treatment with all-*trans* retinoic acid (ATRA), which overcomes the enhanced repression by PML-RARA and induces differentiation of leukemic cells. Although effective, treatment with ATRA alone will cause progressive resistance to the drug, resulting in relapse in 3–6 months. To overcome this problem, treatment of APL now employs a combination of ATRA and other agents, such as the proapoptotic arsenic compound arsenic trioxide (ATO). ATRA and ATO are believed to synergistically enhance differentiation signaling pathways in leukemic cells.

AML with 11q23 Abnormalities

AML with 11q23 abnormalities are associated with aberration of the *MLL* gene and comprise 5–6% of all cases of AML. Two groups of patients show a high frequency of this type of AML: infants and adults with therapy-related AML, usually occurring after treatment with topoisomerase inhibitors. The latter is classified separately and will be discussed later. Common morphologic features include monoblasts and promyelocytes predominant in the bone marrow and showing strong positive nonspecific esterase reactions. AML due to 11q23 abnormalities can be associated with acute myelomonocytic, monoblastic, and monocytic leukemias (FAB M4, M5a, and M5b classifications) and more rarely with leukemias with or without maturation (FAB M2 and M1).

The *MLL* gene encodes a DNA-binding protein that methylates histone H3 lysine 4 (H3K4). *MLL* knockout studies indicate that MLL is necessary for proper regulation of Hox gene expression. Hox genes are a family of transcription factors that regulate many aspects of tissue development. The precise mechanism by which MLL regulates gene expression has not yet been determined. All *MLL* translocations contain the first 8–13 exons of MLL and a variable number of exons from a fusion partner gene. At least 52 *MLL* fusion partner genes have been described, and these fusion partners have diverse functions. Some are nuclear proteins involved in control of transcription and chromatin remodeling; others are cytoplasmic proteins which interact with the cytoskeleton. All MLL fusion proteins have lost the domain necessary for H3K4 methylation. It is believed that leukemogenesis mediated by MLL fusion proteins involves disruption of normal gene expression patterns regulating stem cell differentiation and self-renewal. In some cases the MLL fusion is believed to reactivate the self-renewal program in committed myeloid progenitors. The protein domains contributed by the *MLL* fusion partners are believed to contribute to leukemogenesis through their effects on transcription, chromatin remodeling, and protein-protein interactions.

AML Associated with FLT3 Mutation

Activating mutations in the Fms-like tyrosine kinase (*FLT3*) gene are present in 20–30% of all cases of *de novo* AML. Although *FLT3* mutations can be associated with all the major leukemic translocations (*AML1-ETO*, *CBFb-MYH11*, *PML-RARα*, *MLL* fusions), the majority of cases of AML with *FLT3* mutations are cytogenetically normal. Other common clinical features are leukocytosis and monocytic differentiation.

Two major types of *FLT3* mutations are found in AML patients. An internal tandem duplication (ITD) of the region of the gene encoding the juxtamembrane domain is found in 25–35% of adult and 12% of childhood AML. The second type of mutation is a missense mutation in the activation loop of the tyrosine kinase domain, commonly affecting codon D835. Both types of mutations result in constitutive phosphorylation of the FLT3 receptor in the absence of ligand and activation of downstream signaling pathways including the PI3K/AKT pathway and the Ras/Raf/ERK pathway.

Studies using model systems have shown that mutated *FLT3* alone is not enough to cause the development of AML. In this respect, FLT3 is similar to fusion proteins such as AML1-ETO or CBFb-MYH11. It is hypothesized that all these mutated proteins require the presence of secondary mutations for AML development.

Therapy-Related AML and MDS

This class includes both AML and MDS that arise after chemotherapy or radiation therapy. These diseases are classified according to the mutagenic agents used for treatment, but it can be difficult to attribute a secondary AML to a specific agent because treatment often involves multiple mutagenic agents.

Alkylating Agent-Related AML Alkylating agent-related AML usually occurs 5–6 years after exposure to the agent. Typically, this condition is first observed as an MDS with bone marrow failure. Some cases evolve into AML, which may correspond to acute myeloid leukemia with maturation (FAB class M2), acute monocytic leukemia (M5b), AMML (M4), erythroleukemia (M6a), or acute megakaryoblastic leukemia (M7).

Cytogenetic abnormalities are observed in more than 90% of cases of therapy-related AML/MDS. Complex abnormalities are the most common finding, often including chromosomes 5 and 7. Recent studies have shown that many patients with therapy-related MDS or AML carry point mutations in *p53* (24% of cases), *RUNX1* (16% of cases), or various other oncogenes. An association between *p53* point mutations and chromosome 5 aberrations, and between *RUNX1* mutations and chromosome 7 aberrations has been observed, suggesting that these sets of mutations may cooperate in the development of therapy-related AML/MDS.

Topoisomerase II Inhibitor-Related AML Topoisomerase II inhibitor-related AML may develop in patients treated with the topoisomerase II inhibitors etoposide, teniposide, doxorubicin, or 4-epi-doxorubicin. Development of AML is observed approximately two years after treatment, and is most commonly diagnosed as acute monoblastic or myelomonocytic leukemia.

AML resulting from treatment with topoisomerase poisons such as etoposide are predominantly associated with translocations of the *MLL* gene at 11q23. Of leukemias that are associated with *MLL*, 5–10% are therapy-related. Translocations involving other genes associated with leukemogenesis, such as *RUNX1/AML1*, *CBFβ*, and *PML-RARA*, have also been observed.

LYMPHOCYTE DISORDERS

Disorders of lymphocytes include deficiency of lymphocytes (lymphopenia) and overproliferation of lymphocytes. Overproliferation of lymphocytes is due to either reactive proliferation of lymphocytes (lymphocytosis) or to neoplastic problems.

Lymphopenia

Lymphopenia is defined by less than 1500 lymphocytes/microliter of blood in adults and less than 3000 lymphocytes/microliter of blood in children. Lymphopenia is relatively rare compared to other leukopenias involving granulocytic cells mentioned previously. Some lymphopenias are due to genetic abnormalities, which are categorized as congenital immunodeficiencies. Most lymphopenias are due to viral infection, chemotherapy, radiation, undernutrition, immunosuppressant drug reaction, and autoimmune diseases.

Lymphocytosis

Lymphocytosis can be divided into relative lymphocytosis and absolute lymphocytosis. 20–49% of human white blood cells are lymphocytes. When the percentage exceeds 40%, it is recognized as relative lymphocytosis. When the total lymphocyte count in blood is more than 4000/microliter in adults, 7000/microliter in older children, and 9000/microliter in infants, the patient is diagnosed with absolute lymphocytosis.

The best known lymphocytosis is infectious mononucleosis. This disease is due to an infection of Epstein-Barr Virus (EBV). EBV infection at an early age will not show any specific symptoms. However, infection in adolescents and young adults can cause more severe problems (kissing disease), such as fever, sore throat, lymphadenopathy, splenomegaly, hepatomegaly, and increased atypical lymphocytes in blood. EBV is a member of the herpesvirus family and infects B-lymphocytes. In a minority of infected B-cells, EBV infection occurs in the lytic form, which induces cell lysis and virus release. In a majority of cells, EBV infection is nonproductive and the virus is maintained in latent form. The cells with latent viruses are activated and undergo proliferation, and also produce specific antibodies against the virus. The massive expansion of monoclonal or oligoclonal cytotoxic CD8+ T-cells presented as atypical lymphocytes in peripheral blood is the major feature of infectious mononucleosis. Such strong humoral and cellular responses to EBV eventually highly restrict EBV infection.

Neoplastic Problems of Lymphocytes

Lymphocytic leukemia and lymphoma are the two major groups of lymphoid neoplasms. Leukemia and lymphoma do not have a very clear distinction, and the use of the two terms can be confusing. In general, the term lymphocytic leukemia is used for neoplasms involving the general area of the bone marrow and the presence of a large number of tumor cells in the peripheral blood; lymphomas show uncontrolled growth of a tissue mass of lymphoid cells. However, quite often, especially at the late stage of lymphoma, tumor cells originating from the lymphoma mass may spread to peripheral blood and produce a phenotype similar to leukemia.

The second important issue is the classification of lymphocytic neoplasms. According to the WHO classification, lymphocytic neoplasms are divided into five major categories: (i) precursor B-cell neoplasms, (ii) peripheral B-cell neoplasms, (iii) precursor T-cell neoplasms, (iv) peripheral T-cell neoplasms, and (v) Hodgkin's lymphoma.

It is important to mention that all lymphoid neoplasms develop from a single transformed lymphoid cell. Furthermore, the transformation happens after the rearrangement of antigen receptor genes, including T-cell receptors and immunoglobulin heavy and light chains. Therefore, antigen receptor patterns are generally used to distinguish monoclonal neoplasms from polyclonal reactive lymphadenopathy.

Although there are many different lymphocytic malignancies, the majority of adult lymphoid neoplasms are one of four diseases: follicular lymphoma, large B-cell lymphoma, chronic lymphocytic leukemia/small lymphocytic lymphoma, and multiple myeloma; and the majority of childhood lymphoid neoplasms are one of two diseases: acute lymphoblastic leukemia/lymphoma and Burkitt lymphoma.

Acute Lymphoblast Leukemia/Lymphoma

Acute lymphoblast leukemia/lymphoma (ALL) is most common between the age of 2 and 5 years although it affects both adults and children. The majority of ALL is pre-B-cell leukemia. Pre-T-cell leukemia is often reported in adolescent males. Morphologically, it is difficult to separate T and B lineage ALL. Furthermore, patients with T and B ALL also present similar symptoms. Therefore, flow cytometry studies to identify the expression of specific cell surface markers are generally used to distinguish the lineage and differentiation of ALL.

Various chromosomal locus translocations are associated with the development of ALL, such as those involving the MLL gene (chromosome 11q23), the TCRβ enhancer (chromosome 7q34), the TCRα/δ enhancer (chromosome 14q11), the E2A gene (chromosome 19p13), and the PAX5 gene (chromosome 9p13), t(12;21)(p13;q22), and t(9;22)(q34;q11).

An additional translocation, t(12;21), is identified in approximately 25% of childhood pre-B-cell ALL. The critical fusion protein generated from this translocation is ETV6-RUNX1 (also known as TEL-AML1), which contains 336 amino acids from the N-terminal region of ETV6 and almost the entire RUNX1 protein. Quite frequently, another allele of ETV6 is also lost in t(12;21) ALL patient samples. This finding suggests that TEL is a potential tumor suppressor gene. ETV6-RUNX1 can form dimers via the ETV6 helix-loop-helix domain and contains the RUNX1 DNA binding domain. Therefore, it is believed that ETV6-RUNX1 affects the expression of RUNX1 target genes to promote leukemia development. Interestingly, the ETV6-RUNX1 fusion gene has been identified in neonatal blood spots of children who developed leukemia between 2 and 5 years of age, suggesting that t(12;21) is not sufficient for leukemogenesis without additional malignant promoting factors.

The Philadelphia chromosome caused by t(9;22) is the most frequently identified chromosomal translocation in adult ALL. The Philadelphia chromosome encodes the fusion protein BCR-ABL. The constitutive activation of the ABL tyrosine kinase and the interaction of this fusion protein with various signaling regulators and proto-oncogene products as indicated in Figure 16.4 also promote B-ALL development.

Through use of high-resolution single-nucleotide polymorphism arrays and genomic DNA sequencing to study 242 B-ALL patient samples, the PAX5 gene has been identified as the most frequent target of somatic mutation. Approximately 32% of samples present either deletion or point mutation of the PAX5 gene and result in decreased expression or partial loss of its function. PAX5 is also known as B-cell specific activating protein (BSAP), which plays a crucial role during B lineage commitment and differentiation.

The NOTCH signaling pathway plays important roles during hematopoiesis, especially in T-cell lineage development. The interaction of cell surface NOTCH receptors and their ligands of the Delta-Serrate-Lag2 family induces two-step proteolytic cleavage of the NOTCH protein and generates the intercellular domain of NOTCH (ICN) fragment. ICN translocates to the nucleus and activates target gene expression via interaction with the DNA binding transcription factor CSL, displacement of transcription repressors, and recruitment of transcription activators to the DNA binding complexes (Figure 16.5). A NOTCH activating mutation involving somatic alteration of the NOTCH1 gene has been identified in over 50% of T-ALL patients. Furthermore, the FBW7 ubiquitin E3 ligase responsible to the degradation of ICN is also mutated in T-ALL patient samples and cell lines, which increases the cellular concentration of ICN and

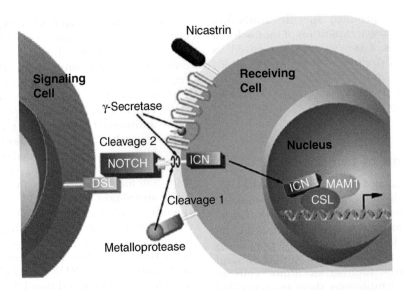

Figure 16.5   NOTCH signaling. Interaction of NOTCH and delta serrate ligand (DSL) stimulates proteolytic cleavage of NOTCH by metalloproteases and γ-secretase. This leads to the release of the intracellular ICN domain, which translocates to the nucleus where it interacts with the DNA binding protein CSL, displaces corepressors and recruits co-activators (MAM1), thereby converting CSL from a repressor to an activator of gene expression. Reproduced with permission from Armstrong SA and Look AT, *J. Clin. Oncol.* 2005;23:6306.

further enhances NOTCH signaling. The best known NOTCH target gene related to cancer development is the MYC oncogene.

Chronic Lymphocytic Leukemia/Small Lymphocytic Lymphoma

Chronic lymphocytic leukemia (CLL) is characterized by the presence of over 5000/microliter of mature-appearing lymphocytes in peripheral blood and a specific range of immunophenotypes. Small lymphocytic lymphoma (SLL) refers to a small percentage of cases in which the tumor cells have a similar immunophenotype to CLL but are restricted to lymph nodes without blood and bone marrow involvement. Due to currently unclear genetic factors, CLL is rare in Asian populations but is the most common form of leukemia in North America and Europe.

The cytogenetic abnormalities detected by fluorescence *in situ* hybridization (FISH) in CLL are mainly chromosome trisomies and deletions. Analysis of 325 CLL patient samples identified trisomy 12 (18%) and deletions on chromosomes 13q (55%), 11q (16%), and 17p (7%). Importantly, chromosome 11q and chromosome 17p deletions are related to poor prognosis and an advanced stage of this disease. The well-known tumor suppressor gene *p53* is located in the deleted region of chromosome 17. ATM kinase that regulates *p53* activity is located in the deleted region of chromosome 11. Since *p53* is a critical inhibitor of cell cycle progression and most chemotherapy drugs target *p53*-dependent pathways, it is valuable to evaluate cytogenetic conditions before treating CLL patients.

Death-associated protein kinase 1 (DAPK1) is a pro-apoptotic protein. Genetic and epigenetic studies have revealed that reduced DAPK1 expression is associated with both familial and sporadic CLL development. The

expression of two additional pro-apoptotic proteins, BAX and BCL-XS, is also downregulated in CLL. Furthermore, several antiapoptosis proteins, such as BCL-XL, BAG1, and MCL1, show increased expression in CML. These results suggest that CLL cells have a decreased apoptosis rate, which may contribute to the accumulation of relatively mature B-cells in patients.

Follicular Lymphoma

The neoplastic cells of follicular lymphoma are derived from germinal center B-cells or cells differentiated toward germinal center B-cells, but unlike CLL cells, they do not express CD5 on their surface. Furthermore, they present in either a pure follicular pattern or mixed with follicular and diffused areas. The occurrence of this disease is also affected by genetic background. It is one of the most common lymphocytic neoplasms in North America and Europe, but less common in Asia. Follicular lymphoma is a disease of late life with a peak of detection between 60 and 70 years of age.

BCL2 is highly expressed in neoplastic cells in over 90% of follicular lymphoma patients. Therefore, BCL2 immunostaining is used to distinguish normal follicles from follicular lymphoma. This high expression of BCL2 is due to a specific chromosomal translocation [t(14;18)(q32;q21)] that generates a fusion between the immunoglobulin heavy chain enhancer on chromosome 14 and the *BCL2* gene on chromosome 18. This was one of the earliest chromosomal translocations related to cancer development to be discovered. The translocation breakpoint on chromosome 14 is at the functional diversity region-joining region (D-J) joint, indicating that mistaken recombination involving the recombination enzymes is the molecular mechanism of generating this translocation.

BCL2 is a strong antiapoptotic factor. Normally, most B-cells should be terminated via apoptosis if they are not challenged by specific antigens. With the overexpression of BCL2, follicular lymphoma cells are able to overcome normal apoptotic signals and avoid termination. Therefore, the prolonged life span of follicular cells due to this defect in apoptotic elimination contributes to the development of follicular lymphoma. However, additional cytogenetic lesions besides t(14;18)(q32;q21) are generally observed in most follicular lymphoma cells, which include trisomies, monosomies, deletions, amplifications, and chromosome translocations. These observations suggest that additional mutations besides the overexpression of BCL2 are required for the development of follicular lymphoma.

Diffuse Large B-Cell Lymphoma

The name diffuse large B-cell lymphoma (DLBCL) is based on the morphology and behavior of this group of malignant cells. They typically express B-cell markers but lack terminal deoxytransferase and are distinguished from mantle cell lymphoma by the lack of cyclin D1 overexpression. DLBCL cells are large and diffusely invade the lymph nodes and extranodal areas. However, this is a highly heterogeneous disease. DLBCL is generally identified in older patients with a median age over 60 and with almost equal distribution between male and female. This is the most common lymphocytic malignancy in adults.

BCL6 is a zinc finger transcription factor and is encoded by a gene located on chromosome 3q27. During normal B-cell development, BCL6 is specifically expressed in germinal center B-cells and plays a critical role during B-cell differentiation and in the formation of the germinal center. Importantly, *BCL6*-involved chromosomal translocations are the most commonly detectable genetic abnormalities in DLBCL, occurring in 35–40% of cases. One major chromosomal translocation leads to immunoglobulin heavy chain regulatory element directed BCL6 expression. The general feature of these translocations is the constitutive expression of *BCL6*. A significant target of BCL6 related to disease development is *p53*. BCL6 inhibits *p53* expression by directly binding to the *p53* regulatory element and initiating the formation of a histone deacetylase complex which modifies local chromatin structure to generate an inactive condition and represses *p53* transcription. Decreased *p53* leads to a reduced rate of apoptosis in response to DNA damage, resulting in the proliferation of malignant clones.

BCL2 is another deregulated gene in DLBCL. *BCL2* is overexpressed in B-cells with t(14;18)(q32;q21), a hallmark of follicular lymphoma. This chromosomal translocation is also observed in 15% of DLBCL, which may come from the transformation of follicular lymphoma to DLBCL. However, abnormally high levels of *BCL2* expression occur in about 50% of DLBCL, indicating the involvement of other mechanisms to induce *BCL2* expression in this disease.

Burkitt Lymphoma

In 1958, Denis Burkitt reported a special type of jaw tumor in African children, and these tumors were later named highly malignant Burkitt lymphoma. Burkitt lymphoma cells are generally monomorphic medium-sized cells (bigger than ALL cells and smaller than DLBCL cells) and with round nuclei and multiple nucleoli. These cells are extremely hyperproliferative and are also highly apoptotic. Close to 100% of Burkitt lymphoma cells are positive for the proliferation marker Ki-67.

According to the WHO, Burkitt lymphoma can be further divided into three categories: endemic, sporadic (nonendemic), and immunodeficiency-associated. The common feature of Burkitt lymphoma is the chromosomal translocation-induced overexpression of the *c-Myc* proto-oncogene. The most common form of translocation is t(8;14)(q24;q32), which leads to immunoglobulin heavy chain regulatory element directed expression of the *c-Myc* gene. In rare cases, the immunoglobulin κ or λ chain locus (instead of the heavy chain locus) is involved in the translocation [t(2;8)(p12;q24) or t(8;22)(q24;q11)], each of which leads to the overexpression of *c-Myc*. Interestingly, *c-Myc* was the first gene known to be involved in a chromosome translocation-associated neoplasm via the study of t(8;14)(q24;q32) in Burkitt lymphoma. Since *c-Myc* is also overexpressed in other forms of leukemia and lymphoma, it is believed that other genetic lesions also play critical roles in the development of Burkitt lymphoma.

Endemic Burkitt lymphoma is found mainly among children living in the malaria belt of equatorial Africa. The common sites of endemic Burkitt lymphoma are kidney and jaw. Furthermore, most endemic Burkitt lymphomas are also positive for EBV. Studies suggest that EBV infections occur long before the translocation of *c-Myc*. Several EBV proteins enhance cell proliferation and inhibit apoptosis, which can provide premalignant activation conditions for further lymphoma development. Furthermore, the high geographic correlation of endemic Burkitt lymphoma and malaria infection also raises the possible involvement of malaria in the development of Burkitt lymphoma. B-cell proliferation is activated during malaria infection.

Sporadic Burkitt lymphoma has no geographic preference. Furthermore, it also has fewer age restrictions and is detected in adults. Besides *c-Myc* related chromosome translocations, sporadic Burkitt lymphoma patients are generally EBV negative. The lymph nodes and terminal ileum are the common sites of this type of lymphoma.

Multiple Myeloma

Multiple myeloma is the most important and common plasma cell neoplasm. Plasma cells are mature immunoglobulin-producing cells. Plasma cell neoplasms are a group of neoplastic diseases of terminally differentiated monoclonal immunoglobulin-producing B-cells. They are generally referred to as myeloma. The monoclonal immunoglobulin produced by these cells is considered the M factor of myeloma. In normal plasma cells, the production of immunoglobulin heavy chain and light chain is well balanced. Under neoplastic conditions, normal balance may not be maintained, resulting in the overproduction of either heavy chain or light chain. The free light chains are known as Bence Jones proteins. Multiple myeloma is generally preceded by a premalignant condition called monoclonal gammopathy of undetermined significance (MGUS). MGUS is quite common in older people. About 20% of patients with MGUS will develop myeloma, generally multiple myeloma.

Multiple myeloma is a disease with multiple masses of neoplastic plasma cells in the skeletal system, which is generally associated with pain, bone fracture, and renal failure.

Like other lymphoid malignancies, multiple myeloma is related to the overexpression of various regulators of cell proliferation and survival due to chromosomal translocations which place these genes under the control of immunoglobulin regulator elements, mainly the immunoglobulin heavy chain locus on chromosome 14. The five most common chromosomal translocations involving the immunoglobulin heavy chain locus are those involving the *cyclin D1* gene on chromosome 11q13 (16%), the *cyclin D3* gene on chromosome 6p21 (3%), the *MAF* gene on chromosome 16q23, the *MAFB* gene on chromosome 20q12, and the *MMSET* and fibroblastic growth factor receptor 3 (*FGFR3*) on chromosome 4p16 (15%). The well-established MAF targets integrin 7 and cyclin D2 are important in communication with the cellular microenvironment and in regulating cell cycle progression, respectively. MMSET is a histone methyl transferase and is likely involved in the regulation of chromatin structure and protein-protein interactions to regulate gene expression. FGFR3 is a receptor tyrosine kinase and its activation directly promotes cell proliferation and survival.

Hodgkin's Lymphoma

Thomas Hodgkin, in 1832, was the first to give a detailed report about the macroscopic pathology of the disease currently named Hodgkin's lymphoma. In contrast to non-Hodgkin's lymphoma, Hodgkin's lymphoma starts from a single lymph node or a chain of nodes and spreads in an orderly way from one node to another. Further microscopic analysis revealed that Hodgkin's lymphoma presents a very unique type of malignant cells, called Reed-Sternberg cells (Figure 16.6). The classical morphology of these cells is large size (20–50 micrometers), relative abundance, amphophilic and homogeneous cytoplasm, and two mirror-image nuclei (owl eyes) with one eosinophilic nucleolus in each nucleus. Reed-Sternberg cells occupy only a small portion of the tumor mass. The majority of the cells in the tumors are reactive lymphocytes, macrophages, plasma cells, and eosinophils, which are attracted to the surrounding malignant Reed-Sternberg cells by their secreted cytokines. The WHO classification has divided Hodgkin's lymphoma into several subcategories.

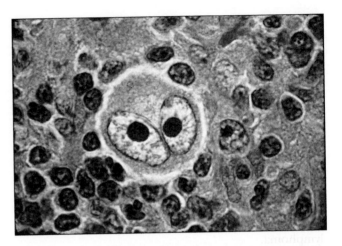

Figure 16.6   Reed-Sternberg cell. The figure shows the typical characteristics of the Reed-Sternberg cell: large size (20–50 micrometers), amphophilic and homogeneous cytoplasm, and two mirror-image nuclei (owl eyes) with one eosinophilic nucleolus in each nucleus. Reed-Sternberg cells only occupy a small portion of the tumor mass. Reproduced with permission from a website maintained by the Department of Pathology, Stanford University (http://hematopathology.stanford.edu/).

Hodgkin's lymphoma affects people of relatively young age. It is one of the most common forms of cancer in young adults, with an average age at diagnosis of 32 years. Nowadays, highly developed radiation therapy and chemotherapy treatments have made Hodgkin's lymphoma a curable cancer. However, about 20% of patients still die from this disease. Furthermore, successfully treated patients have a higher risk of dying from late toxicities, such as secondary malignancies and cardiovascular diseases.

KEY CONCEPTS

- All types of differentiated blood cells originate from hematopoietic stem cells.
- Growth factor receptor signaling and expression of specific hematopoietic transcription factors regulate hematopoietic cell differentiation.
- Mutations that cause hyperactive growth factor receptor signaling often contribute to myeloid or lymphoid cancers.
- Mutational inactivation of transcription factors that promote differentiation often contributes to myeloid or lymphoid cancer.
- Cancer-causing mutations can be spontaneous or inherited.

SUGGESTED READINGS

1. Dzierzak E, Speck NA. Of lineage and legacy: The development of mammalian hematopoietic stem cells. *Nature Immunol.* 2008;9:129–136.
2. Fabbri M, Garzon R, Andreeff M, et al. MicroRNAs and noncoding RNAs in hematological malignancies: Molecular, clinical and therapeutic implications. *Leukemia.* 2008;22:1095–1105.

3. Gilliland DG, Griffin JD. The roles of FLT3 in hematopoiesis and leukemia. *Blood.* 2002;100:1532–1542.

4. Iwasaki H, Akashi K. Myeloid lineage commitment from the hematopoietic stem cell. *Immunity.* 2007;26:726–740.

5. Kiel MJ, Morrison SJ. Uncertainty in the niches that maintain haematopoietic stem cells. *Nature Rev Immunol.* 2008;8:290–301.

6. Kim SI, Bresnick EH. Transcriptional control of erythropoiesis: Emerging mechanisms and principles. *Oncogene.* 2007;26: 6777–6794.

7. Laird DJ, von Andrian UH, Wagers AJ. Stem cell trafficking in tissue development, growth, and disease. *Cell.* 2008;132:612–630.

8. Ma L, Sun B, Hood L, et al. Molecular profiling of stem cells. *Clin Chim Acta.* 2007;378:24–32.

9. Matsumura I, Mizuki M, Kanakura Y. Roles for deregulated receptor tyrosine kinases and their downstream signaling molecules in hematologic malignancies. *Cancer Sci.* 2008;99:479–485.

10. Nutt SL, Kee BL. The transcriptional regulation of B-cell lineage commitment. *Immunity.* 2007;26:715–725.

11. Okuda T, van Deursen J, Hiebert SW, et al. AML1, the target of multiple chromosomal translocations in human leukemia, is essential for normal fetal liver hematopoiesis. *Cell.* 1996;84:321–330.

12. Orkin SH. Priming the hematopoietic pump. *Immunity.* 2003; 19:633–634.

13. Orkin SH, Zon LI. Hematopoiesis: An evolving paradigm for stem cell biology. *Cell.* 2008;132:631–644.

14. Rothenberg EV. Negotiation of the T lineage fate decision by transcription-factor interplay and microenvironmental signals. *Immunity.* 2007;26:690–702.

15. Rothenberg EV, Moore JE, Yui MA. Launching the T-cell-lineage developmental programme. *Nature Rev Immunol.* 2008;8:9–21.

16. Rosenbauer F, Tenen DG. Transcription factors in myeloid development: Balancing differentiation with transformation. *Nature Rev Immunol.* 2007;7:105–117.

17. Samokhvalov IM, Samokhvalova NI, Nishikawa S. Cell tracing shows the contribution of the yolk sac to adult haematopoiesis. *Nature.* 2007;446:1056–1061.

18. Steelman LS, Abrams SL, Whelan J, et al. Contributions of the Raf/MEK/ERK, PI3K/PTEN/Akt/mTOR and Jak/STAT pathways to leukemia. *Leukemia.* 2008;22:686–707.

19. Wang Q, Stacy T, Binder M, et al. Disruption of the Cbfa2 gene causes necrosis and hemorrhaging in the central nervous system and blocks definitive hematopoiesis. *Proc Natl Acad Sci USA.* 1996;93:3444–3449.

20. Yu BD, Hess JL, Horning SE, et al. Altered Hox expression and segmental identity in Mll-mutant mice. *Nature.* 1995;378:505–508.

17

Molecular Basis of Diseases of Immunity

David O. Beenhouwer

INTRODUCTION

The immune system protects against infection with microorganisms. This system is remarkable in both its complexity and its effectiveness. Without a functioning immune system, humans cannot survive past the first few months of life. However, until the most recent century, infections were by far the most common cause of death among humans keeping the average life expectancy to about 25–30 years. The current life expectancy in developing countries reflects progress in control of infectious diseases including improved hygiene, vaccines, and antimicrobial drugs.

This chapter covers major syndromes of immune dysfunction. These include diseases of deficient immunity, hyperactive immunity (hypersensitivity), and dysregulated immunity (autoimmune diseases). To understand the pathophysiology of these syndromes, one must have an understanding of how the normal immune system functions. This chapter begins with a brief summary of the important cells and molecules of the immune system.

NORMAL IMMUNE SYSTEM

The immune system is made up of several types of cells that carry out specialized functions. Through a remarkable array of specialized cell surface molecules (receptors), immune cells recognize, respond, and adapt to their environment and to foreign invaders. Immune system responses can be thought of as rapid and preprogrammed (innate) or slower but more flexible (adaptive).

Cells

The white blood cells that make up the immune system originate in the bone marrow. These cells then circulate to the peripheral tissues in the bloodstream. They also travel in the lymphatic system, which is a network of ducts connecting lymph nodes throughout the body.

All blood cells derive from hematopoietic stem cells, which first differentiate into either lymphoid or myeloid progenitor cells. Lymphoid progenitors give rise to lymphocytes, natural killer (NK) cells, and dendritic cells. Myeloid progenitors give rise to granulocytes (including neutrophils, eosinophils, mast cells and basophils), monocyte/macrophages, and dendritic cells.

B-Cells

B-cells produce antibodies or immunoglobulins. They carry receptors on their cell surface (B-cell receptor, BCR). The BCR is a membrane-bound version of the antibody molecule the B-cell will secrete when activated. The BCR is associated with other molecules capable of transducing signals to the cell. Antibodies recognize and bind to portions of soluble molecules called antigens. Once activated, B-cells may differentiate into plasma cells, which are specialized for producing massive quantities of antibody.

T-Cells

There are two major classes of T-cells: (i) T helper cells and (ii) cytotoxic T-cells. T helper cells express CD4 on the surface membrane and activate other cells such as B-cells and macrophages. Cytotoxic T-cells (CTLs) express CD8 on their surface membrane and recognize and kill infected cells. T-cells also express an antigen receptor on their cell surface (T-cell receptor or TCR). These antigen receptors are somewhat different from antibodies in that they recognize antigen only when it is bound to a specific cell surface molecule called major histocompatibility complex (MHC).

NK Cells

Part of both innate and adaptive immune responses, NK cells recognize and destroy abnormal cells, such as cancer cells and infected cells. NK cells express receptors for antibodies (Fc receptors) and are capable of recognizing and destroying antibody-coated cells by a process known as antibody-dependent cell-mediated cytotoxicity (ADCC). When stimulated, NK cells release perforin, which forms pores in the cell membrane. Granzymes are also released, which enter through these pores and can then trigger apoptosis in the target cell.

Dendritic Cells

When immature, dendritic cells reside in many host tissues, particularly the skin and mucosa, where they avidly ingest pathogens in a process called phagocytosis. Upon pathogen uptake, these cells mature and travel to lymphoid tissues including the lymph nodes and the spleen. Inside the dendritic cell, pathogens are digested by various enzymes into fragments. Upon maturation, these cells then act as antigen-presenting cells (APCs), displaying pathogen fragments on their cell surface for lymphocytes to recognize. These cells are the primary link between the innate and adaptive immune systems.

Macrophages

Macrophages are phagocytic cells that engulf pathogens and cellular debris and play a major role in the innate immune response. Pathogens are taken up into specialized intracellular vesicles called phagosomes, where they are destroyed. Macrophages can also present antigens to lymphocytes. These cells are distributed widely in many different organs and tissues where they may further differentiate and acquire specialized functions.

Granulocytes

Granulocytes are cells that have densely staining granules in their cytoplasm. The granulocytes include neutrophils, eosinophils, mast cells, and basophils. Neutrophils are the most abundant white blood cell and play an essential role in innate protection against bacterial infection. They are phagocytic cells that contain a vast array of microbicidal molecules. Eosinophils carry Fc receptors for a particular form of antibody (IgE). Upon binding multicellular organisms (parasites) coated with IgE, eosinophils discharge their toxic granules, which contain both destructive enzymes and vasoactive substances. Mast cells trigger local inflammatory response to antigen by releasing granules containing histamine and other vasoactive substances. The function of basophils is not well defined.

Molecules

Complement

The complement system consists of a cascade of serum proteins that are involved in both innate and adaptive immunity. There are three pathways through which complement can be activated: classical, alternative, and lectin. C3b can bind covalently to pathogens, where it acts as an opsonin, targeting pathogens for uptake and destruction by phagocytes bearing complement receptors. C5a is a potent mediator of vascular permeability and can also recruit neutrophils and monocytes, in a process known as chemotaxis, to areas of inflammation. The membrane attack complex (MAC) is formed by C5–9 and can form lytic membrane pores in cells or pathogens without cell walls, such as gram-negative bacteria. Initiation of the classical pathway involves C1q binding to antibody molecules complexed with antigen, either on a pathogen surface or in the form of soluble antibody-antigen complexes. In a similar fashion, initiation of the lectin binding pathway involves mannose binding lectin (MBL) binding to exposed densely spaced mannose residues on pathogen surfaces. On mammalian cells, most mannose residues are masked by other sugars, especially sialic acid. The alternative pathway is initiated by ongoing spontaneous C3 hydrolysis. C3b binds factor B, which then initiates the alternative pathway. Host cells are protected from destruction by this pathway (and others) by several complement regulatory proteins residing on the cell surface.

Antibodies

Antibodies (immunoglobulins) are a family of glycoproteins that are produced by B-cells and plasma cells. Antibody molecules are composed of two identical light (L) chain polypeptide chains and two identical heavy (H) polypeptide chains linked together by disulfide bonds. The majority of the molecule consists of one of five protein sequences known as the constant region. However, the amino terminal end is made up of an apparently infinite variety of sequences known as the variable region. The variable (V) region constitutes the antigen binding site. This region is separated from the rest of the molecule by a flexible hinge. The carboxyl terminal portion of the antibody is called the Fc region and can bind to Fc receptors, activate complement, and mediate antibody half-life.

The V region of the heavy chain is composed of three segments: V_H, D_H, J_H. The light chain V region is composed of two segments: V_L and J_L. The extraordinary diversity of the V region is generated during B-cell development when gene segments irreversibly rearrange in a process called VDJ or VJ recombination. In humans, there are about 6000 different VDJ combinations and 320 different VJ combinations generating a possible 2 million different antibody V regions with different specificities. Upon stimulation with T-cells, B-cells can induce a process called somatic hypermutation, which introduces random mutations in V region genes. This process leads to fine-tuning of the antigen binding site, thereby increasing antigen affinity.

Antibody effector function is determined by the heavy chain or isotype (IgM, IgG, IgA, or IgE). Isotypes have different properties including the ability to mediate phagocytosis (IgG), ADCC (IgG and IgE), and complement activation (IgG and IgM). IgA is particularly resistant to proteolysis and is important in host defense at mucosal surfaces. IgE binds to high affinity Fc receptors (FcεRI) on mast cells and eosinophils inducing degranulation. A circulating Fc receptor known as the neonatal Fc receptor (FcRn) binds IgG and contributes to its remarkably long half-life (up to 3 weeks).

T-Cell Receptors

The TCR has a similar structure to the antibody Fab region and consists of two transmembrane glycoprotein chains, TCRα and TCRβ. The generation of diversity of the TCR is very similar to that for antibodies. In contrast to antibodies, TCRs bind to peptides in complex with major histocompatibility complex (MHC) displayed on the surface of APCs. CD4$^+$ T-cells interact with antigen bound to MHC class II, and CD8$^+$ T-cells interact with antigen presented in the context of MHC class I.

Major Histocompatibility Complex

All nucleated cells express major histocompatibility complex (MHC) class I. Cells infected internally (for instance, in viral infection) will process pathogen peptides and present them on the surface in complex with MHC class I. There they can be recognized by CD8$^+$ T-cells, which are specialized to kill cells recognized in this fashion via apoptosis. MHC class II is expressed by APCs, including dendritic cells, macrophages, and B-cells. CD4$^+$ T-cells recognize peptides processed by APCs and displayed on their surface complexed with MHC class II. MHC class I and II genes are also known as human leukocyte antigen (HLA) genes. There are three class I α-chain genes: HLA-A, -B, and -C. There are three pairs of MHC class II α-chain and β-chain genes: HLA-DR, HLA-DP, and HLA-DQ. Each MHC protein binds a different range of peptides. HLA genes are highly polymorphic. This leads to population-based immunologic diversity. As described in following sections, several autoimmune diseases are linked to specific HLA alleles.

Cytokines

Cytokines are proteins secreted by cells that interact with specific receptors on other cells, thereby affecting function. Macrophages produce a wide array of these molecules, including proinflammatory cytokines, like interleukin (IL)- 1β, IL-6, and tumor necrosis factor (TNF), the Th1 cytokine IL-12, and the anti-inflammatory cytokine IL-10. T-cells also produce several cytokines, including those that characterize Th1 (IL-2 and IFN-γ) and Th2 (IL-4, IL-5, and IL-13) responses.

Chemokines

Chemokines are cytokines that act specifically to attract cells bearing specific receptors. The CC chemokines have two adjacent cysteines near the amino terminus, while the CXC chemokines have another amino acid separating the two cysteines. CC chemokines bind to CC receptors (CCRs) and CXC chemokines bind to CXC receptors (CXCRs). Some important chemokines include CXCL-8 (IL-8), which attracts neutrophils bearing CXCR1 and 2; CCL-2 (MCP-1), CCL-3 (MIP-1α), and CCL-5 (RANTES), which attract monocytes and macrophages; and CCL-11 (eotaxin-1), which attracts eosinophils bearing CCR3.

Toll-Like Receptors

Microorganisms produce several molecules that can be immediately recognized as foreign by their general repetitive structural qualities not shared by host molecules, known as pathogen-associated molecular patterns (PAMPs). For example, viruses often express double-stranded RNA (dsRNA), bacteria express unmethylated repeats of the dinucleotide CpG, gram-negative bacteria express lipopolysaccharide (LPS), and some bacteria possess flagella. The innate immune system recognizes these PAMPs through pattern recognition receptors (PRRs). There are several PRRs expressed in the human immune system, including MBL and toll-like receptors (TLRs). Humans make 10 TLRs, which are expressed on macrophages, dendritic cells, and other cells, and recognize various PAMPs. Some PAMPs recognized by specific TLRs include dsRNA (TLR-3), LPS (TLR-4), flagellin (TLR-5), and CpG DNA (TLR-9). Ligation of TLRs triggers a rapid preprogrammed response (innate immune response). At the same time, the more slowly reactive adaptive immune response is initiated, often in a particular fashion depending on which TLRs were activated. Therefore, while TLRs play a primary role in the innate response, they are also responsible for initiating and shaping adaptive immune responses.

Immune Responses

Innate Immune Responses

The immune system must be capable of responding immediately to a foreign invader. This response, mediated by soluble and cell-attached PRRs, is called an innate immune response. Important cells of the innate immune response include macrophages, NK cells, and granulocytes. Soluble molecules, such as complement, also play important roles. Recognition of a foreign invader leads to release of cytokines, in particular TNF, as well as chemokines. This leads to infiltration of the infection site with neutrophils, macrophages, and dendritic cells as well as lymphocytes. Most breaches in host defenses are controlled by the innate immune response.

Adaptive Immune Responses

The adaptive immune response primarily involves lymphocytes with their wide diversity of antigen receptors. The adaptive immune response continues to be adjusted during infection and is set in motion following activation of the innate response. APCs, particularly dendritic cells, ingest pathogens and associated antigens and travel from the site of infection to lymph nodes where they encounter dense collections of T-cells and B-cells. APCs process antigens and present them complexed with MHC class II. T-cells that bind these peptide-MHC complexes are activated to proliferate and produce cytokines that further direct immune responses. This process leads to clonal amplification of pathogen-specific T-cells. Specialized APCs, called follicular dendritic cells (FDCs), do not process antigens into fragments, but carry larger fragments and even entire pathogens on their cell surface, where they can be presented to B-cells, which are then clonally amplified as well. B-cells can also process antigen and present it to T-cells in the context of MHC II. T-cells recognizing this antigen can then provide signals to B-cells leading to maturation of the antibody response including isotype switching and somatic hypermutation.

An important caveat of lymphocyte activation is that it requires two signals: (i) antigen binding, which signals through the antigen receptor, and (ii) a second signal provided by another cell. For T-cells, this second signal is provided by APCs. For B-cells, the second signal is usually provided by activated T-cells. The second signal is delivered through the interaction of certain cell surface molecules, such as CD40–CD40 ligand and CD28–B7, known as co-stimulatory molecules. If lymphocytes receive signaling only through the antigen receptor without co-stimulation, the lymphocyte either becomes anergic (immunologically inactive) or dies via apoptosis. This process ensures that the powerful adaptive immune system is only activated by foreign pathogens initiating an inflammatory response, and not to inert or host antigens.

Another important feature of the adaptive immune response is the development of antigenic memory or recall. The initial adaptive response to a pathogen, known as the primary response, usually takes 7–10 days to reach full effectiveness. In a process that is still rather unclear, during the primary (and subsequent) response, some clonally amplified lymphocytes, known as memory cells, are produced that remain circulating in the host for many years. Upon re-exposure to the same pathogen, these cells are then rapidly stimulated, leading to quick and efficient eradication. This response is also known as a secondary response and requires only a few days to reach maximum intensity

and efficacy. For most responses, memory does fade over time. This is why many vaccines require booster immunizations to remain effective.

In the past, there has been a tendency to think about the adaptive immune response as two separate entities—humoral immunity (antibody) and cell-mediated immunity (T-cell and APC). It is now clear that effective responses to most infections require both types of responses and that this dichotomy is largely a didactic one.

MAJOR SYNDROMES

Now that the many important cells and molecules involved in immunity have been introduced, we will turn to a discussion of pathological states of the immune system. One such state is the hypersensitivity reaction, in which environmental antigens induce an immune response leading to destructive inflammation. Another pathological state is immunodeficiency, in which specific defects in the immune system render it unable to control infection. Finally, autoimmunity arises when the immune system mistakes host tissue as foreign and causes damage.

Hypersensitivity Reactions

In certain cases, harmless environmental antigens may induce an adaptive immune response, and upon re-exposure to the same antigen, an inflammatory state ensues. These antigens are referred to as allergens, and the hypersensitivity responses to them are also known as allergic reactions. Allergic reactions may range from mildly itchy skin to significant illnesses such as asthma and to life-threatening situations such as anaphylactic shock. Gell and Coombs divided hypersensitivity reactions into four main types for didactic purposes (Table 17.1), although distinguishing these practically may be difficult, as antibody-mediated and cell-mediated immune responses may overlap or occur simultaneously.

Type I or Immediate Hypersensitivity

Immediate hypersensitivity is classically mediated by IgE and may occur as a local (such as asthma) or systemic (anaphylaxis) reaction. The initial phase occurring

| Table 17.1 | Hypersensitivity Reactions |

Type	Antigen	Mediator	Effector Mechanism	Examples
I Immediate	Soluble antigen, allergen	IgE	FcεRI activation of mast cells	Asthma, anaphylaxis
II Antibody-mediated	Cell-associated antigen	IgM, IgG	FcγR bearing cells, complement	Drug reaction
III Immune-complex	Soluble antigen	IgG	FcγR bearing cells, complement	Serum sickness, Arthus reaction
IV Delayed-type hypersensitivity	Soluble antigen, cell-associated antigen	T-cells	Macrophage activation, direct cytotoxicity	Tuberculin reaction, contact dermatitis

within minutes following exposure to allergen is characterized by vasodilation, vascular leakage, smooth muscle spasm, and glandular secretions. A late phase reaction, lasting several days may then occur and is characterized by infiltration of tissues with eosinophils and CD4$^+$ T-cells, as well as tissue destruction.

During the sensitization phase of a type I response, allergen is presented to T-cells by APCs, and signals are generated that cause differentiation of naïve CD4$^+$ T-cells into Th2 cells. IL-4 and IL-13 secreted by Th2 cells stimulate B-cells to produce IgE. Allergen-specific IgE then binds to high-affinity FcεRI on mast cells. During the effector phase, the allergen binds to cytophilic IgE, thus cross-linking FcεRs and leading to activation of mast cells. This is followed by immediate degranulation with release of preformed mediators such as histamine and proteolytic enzymes, which cause smooth muscle contraction, increase vascular permeability, and break down tissue matrix proteins. At the same time, there is induction of chemokines, cytokines, leukotrienes, and prostaglandins. IL-5 promotes eosinophil production, activation, and chemotaxis. IL-4 and IL-13 promote and amplify Th2 responses. TNF, platelet activating factor (PAF), and macrophage inflammatory protein (MIP-1) induce efflux of effector leukocytes from the bloodstream into tissues. Leukotriene B4 is chemotactic for eosinophils, neutrophils, and monocytes. Leukotrienes C4 and D4 are potent (>1000-fold more active than histamine) mediators of vascular permeability and smooth muscle contraction. Prostaglandin D2 causes increased mucous secretion.

Susceptibility to Type I hypersensitivity reactions, termed atopy, is familial. However, the genetic basis for this predisposition has not been clearly established and is probably polygenic. Most allergens that trigger a type I reaction are low-molecular weight, highly soluble proteins that enter the body via the mucosa of the respiratory and digestive tracts.

Type II or Antibody-Mediated Hypersensitivity

Type II hypersensitivity is mediated by IgM or IgG targeting membrane-associated antigens. A sensitization phase leads to production of antibodies that recognize substances or metabolites that accumulate in cellular membrane structures. In the effector phase, target cells become coated with antibodies, a process termed opsonization, which leads to cellular destruction by three mechanisms: (i) phagocytosis, (ii) complement-dependent cytotoxicity (CDC), and (iii) ADCC. First, IgG or IgM antibodies coating target cells can bind to Fc receptors present on cells such as macrophages and neutrophils and mediate phagocytosis. IgG or IgM antibodies can also activate complement via the classical pathway. This leads to deposition of C3b, which can mediate phagocytosis. Complement activation also leads to production of the MAC, which forms pores in the cellular membrane resulting in cytolysis (CDC). Finally, IgG antibodies can bind FcγRIII on NK cells and macrophages, thus mediating release of granzymes and perforin and resulting in cell death by apoptosis (ADCC).

The most common causes of type II reactions are medications including penicillins, cephalosporins, hydrochlorothiazide, and methyldopa, which become associated with red blood cells or platelets leading to anemia and thrombocytopenia. The mechanisms involved in type II hypersensitivity also play a role in cellular destruction by autoantibodies.

Type III or Immune Complex Reaction

In this reaction, antibodies bind antigen to form immune complexes, which become deposited in tissues where they elicit inflammation. It is important to note that the formation of immune complexes occurs during many infections and is an effective method of antigen clearance. However, under certain circumstances these complexes sometimes escape clearance by the reticuloendothelial system and are deposited in tissues including the kidney, joints, and blood vessels. The immune complexes fix complement and bind to leukocytes via Fc receptors. Activation of the complement cascade leads to production of vasoactive mediators C5a and C3a, allowing immune complexes to deposit in vessel walls and tissues. The result is tissue edema and deposition of immune complexes in the vessel walls and surrounding tissue. At the same time, chemotactic factors are produced leading to PMN and monocyte recruitment, which upon further activation cause tissue destruction.

Examples of Type III reactions include serum sickness and the Arthus reaction. Serum sickness was originally described in patients suffering from diphtheria who were treated with immune horse serum (antiserum). Currently, the most common causes of serum sickness-like illness are antibiotics (such as cephalosporins and sulfonamides) and blood products. The Arthus reaction is a local immune complex-mediated vasculitis, usually observed as edema and necrosis in the skin occurring several hours following antigen exposure. As with Type II reactions, the mechanisms involved in Type III hypersensitivity also play a role in certain autoimmune diseases.

Type IV or Delayed-Type Hypersensitivity

While the first three types of hypersensitivity are mediated primarily by antibodies, delayed-type hypersensitivity (DTH) is mediated by T-cells. During the sensitization phase, naïve CD4$^+$ T-cells are exposed to antigens with an induction of an adaptive Th response. In the effector phase, antigen is carried by APCs to lymph nodes where it is presented to memory T-cells, which become activated and then travel back to the site of antigen deposition where they may be stimulated to secrete IFN-γ, thereby initiating a Th1 response and tissue inflammation mediated primarily by macrophages.

The classic example of DTH is the tuberculin skin reaction, which occurs 24–48 hours following intradermal injection of a purified protein derivative (PPD) prepared from *Mycobacterium tuberculosis*. Contact dermatitis from poison ivy represents another common example of DTH.

Immunologic Deficiencies

The first inherited immunodeficiency was described by Bruton over 50 years ago. Many of these diseases have devastating pathological consequences for afflicted individuals. An understanding of the underlying mechanisms responsible for these disease processes has been invaluable in our current understanding of immune system function. There are currently over 200 described inherited immunodeficiencies. A systematic review of these deficiencies is well beyond the scope of this chapter. A few of the more common and important ones will be discussed. In addition, there are acquired immunodeficiencies, which were primarily iatrogenic until the arrival of the Acquired Immune Deficiency Syndrome (AIDS) epidemic in the late 1970s.

Primary Immunodeficiencies

These are inherited disorders that primarily affect immune system cell function. Most are rare with prevalences less than 1 in 50,000 individuals. As would be expected, the less severe tend to be more common. Certain types of infections associate with different classes of immunodeficiency (Table 17.2). Many primary immunodeficiencies lead to significant infections early in life. Deficiencies in multiple lymphocyte lineages tend to be devastating. Breastfeeding neonates with antibody deficiencies usually remain well for a time after birth due to passive transfer of maternal immunoglobulin in breast milk during the first 6–9 months of life. Then the affected individual develops upper and lower respiratory tract infections including sinusitis, otitis, and pneumonia. Most of these infections are caused by bacteria such as *Streptococcus pneumoniae* and *Haemophilus influenzae* that have polysaccharide capsules, making them resistant to phagocytosis without effective antibody opsonization. If untreated, these recurrent infections cause significant destruction and scarring of lung tissue, resulting in bronchiectasis. Fortunately, treatment with pooled immunoglobulin is effective in treating antibody deficiencies. More recently, it has been recognized that deficiency in one molecule may predispose individuals to only one or a few types of infections. We are only beginning to recognize these syndromes, and it is likely several more will be discovered.

X-Linked Agammaglobulinemia (Bruton's Disease) This condition presents with recurrent respiratory tract infections, lack of tonsilar tissue, and a normal white blood cell count with normal lymphocyte percentage. When flow cytometry for surface membrane immunoglobulin (BCR) is performed in these patients, few if any B-cells are found. Serum immunoglobulins are nonexistent. The defect is in Bruton's tyrosine kinase (*Btk*), which is essential for B-cell development and survival. B-cells develop in the bone marrow from pluripotent hematopoietic stem cells, first becoming pro-B cells, then pre-B-cells when the immunoglobulin heavy and light chain V region gene segments undergo rearrangement. During the next stage, the immature B-cell, the BCR is expressed on the cell surface (sIgM), and these cells leave the bone marrow to undergo further maturation steps. These cells eventually develop into mature B-cells and then plasma cells. At the pre-B-cell stage, Btk signaling appears to be essential for directing light chain gene rearrangement events. Thus, defective Btk leads to arrest of B-cell differentiation at the pre-B-cell stage.

Hyper-IgM Syndrome Hyper-IgM syndrome, which is characterized by the presence of normal or elevated serum levels of IgM and low IgG and IgA, may be caused by one of at least 10 gene defects. The most common (and the most clinically severe) of these is an X-linked deficiency in CD40 ligand (CD40L). This receptor is transiently expressed on activated T-cells and interacts with CD40 molecules constitutively expressed on the surface of B-cells and APCs. T-cell interaction with B-cells via

---

> **Table 17.2**   Infections in Immunodeficiencies

Host Defense Affected	Clinical Example	History	Relevant Pathogens
T-cells	AIDS	Disseminated infections Opportunistic infections Persistent viral infections	*Pneumocystis jiroveci* *Cryptococcus neoformans* Herpes viruses
B-cells	X-linked agammaglobulinemia	Recurrent respiratory infections Chronic diarrhea Aseptic meningitis	*Streptococcus pneumoniae* *Haemophilus influenzae* *Giardia lamblia* Enteroviruses
Phagocytes	Chronic granulomatous disease	Gingivitis Aphthous ulcers Recurrent pyogenic infections	*Staphylococcus aureus* *Burkholderia cepacia* *Serratia marcescens* *Aspergillus* spp.
Complement	Late complement component (C5, C6, C7, C8, or C9) deficiency	Recurrent bacteremia Recurrent meningitis	*Neisseria* spp.

CD40–CD40L interaction leads to immunoglobulin class switching and differentiation into plasma cells. In the absence of this signal, B-cells only produce IgM. Other T-cell/B-cell interactions can induce class switching, so some small amounts of IgG and IgA may be seen. However, the impaired production of IgG and IgA in CD40L deficiency leads to susceptibility to recurrent bacterial infections, particularly those involving the respiratory tract caused by encapsulated organisms. Importantly, T-cells also interact with APCs via CD40–CD40L. Thus, CD40L deficiency leads to significant impairment in both antibody production and cell-mediated immunity, resulting in a clinical presentation that may include both infections with encapsulated bacteria and intracellular organisms.

Complement Deficiency Defects have been described for most of the components of complement. There is a propensity for individuals with C3 deficiency and with classical component defects (C1, C2, and C4) to develop systemic lupus erythematosus (SLE). While this is not well understood, classical pathway clearance of apoptotic cells and immune complexes may be important. C2 deficiency is relatively common, occurring in 1 in 20,000 individuals. These individuals have an increased propensity to develop infections, SLE, and myocardial infarctions. Defects in MBP are relatively common, and there is an association with infections in children. Finally, defects in the terminal components (C5–9) lead to a remarkable susceptibility (~5000-fold risk) to recurrent infections caused by pathogenic *Neisseria* spp., especially *N. meningitidis*. Interestingly, mortality due to these infections is 10-fold lower than nondeficient individuals.

Severe Combined Immunodeficiency (SCID) The severe combined immunodeficiency disorder is characterized by the absence of T-cell differentiation, resulting in severely reduced or absent T-cells. At least 10 autosomal recessive or X-linked genetic defects have been described, which result in several possible phenotypes based on B and NK cell development. The most common cause of SCID is an X-linked deficiency in the common γ chain shared by receptors for several cytokines including IL-2 and IL-4. This form is characterized by complete absence of T-cells and NK cells. When untreated, SCID results in death within the first year of life from overwhelming opportunistic infections. Improved survival has been reported in infants receiving stem cell transplants. Gene therapy has also been used successfully in several cases.

Common Variable Immunodeficiency (CVID) Common variable immunodeficiency (CVID) is a heterogeneous syndrome, probably comprising several distinct diseases, characterized by recurrent infections and low antibody levels (IgG and IgA and/or IgM). The syndrome is usually diagnosed in the fourth decade of life and with a typical delay of >15 years from first symptoms to diagnosis. The estimated prevalence of CVID is about 1 in 25,000 individuals, making it the most prevalent immunodeficiency requiring medical attention. The main clinical features are recurrent infections of the respiratory tract, chronic diarrhea, autoimmune disease, and malignancy. While most gene defects are unknown, 10–15% are caused by a defect in the transmembrane activator and calcium modulator and cyclophillin interactor (*TACI*) gene. TACI is a member of the TNF receptor family and is expressed on B-cells and activated T-cells. Ligands for TACI include B-cell activating factor of the TNF family (BAFF) and a proliferation inducing ligand (APRIL). Both ligands bind other receptors on B-cells and T-cells. However, in response to ligation by APRIL, TACI mediates isotype switching to IgG and IgA. Similarly, BAFF mediates IgA switching. The penetrance of CVID phenotype in families carrying a mutant TACI gene is quite variable. Recently, a defect in BAFF receptor has also been identified as a cause of CVID.

Chronic Granulomatous Disease (CGD) Phagocytes produce reactive oxygen species (including superoxide and hydrogen peroxide) to kill ingested pathogens. NADPH oxidase is a five-subunit enzyme that catalyzes the production of superoxide from oxygen for this purpose. Over 400 genetic defects in NADPH have been described that result in chronic granulomatous disease (CGD), which is characterized by recurrent, indolent bacterial and fungal infections caused by catalase positive organisms including *Staphylococcus aureus* and *Aspergillus* spp. Most mutations occur in the gp91*phox* gene located on the X chromosome. There are also autosomal recessive forms of the disease primarily involving p47*phox*. The lack of microbicidal oxygen species due to NADPH deficiency leads to a severe defect in intracellular killing of phagocytosed organisms. However, organisms that lack catalase produce peroxide as a byproduct of oxidative metabolism, which accumulates and leads to effective intracellular killing of these microbes. Tissue pathology is characterized by granulomas and lipid-filled histiocytes in liver, spleen, lymph nodes, and gut. The disease is treated with antibiotic prophylaxis, and chronic administration of interferon-γ may also be beneficial.

Inherited Susceptibility to Herpes Encephalitis (UNC-93B and TLR-3 Deficiencies) The paradigm of a single gene lesion conferring vulnerability to multiple infections has been challenged by several recent discoveries. While over 80% of young adults are infected with herpes simplex virus type 1 (HSV-1), development of herpes simplex encephalitis (HSE) is quite rare (1/250,000 patient years). Jean Laurent Casanova and colleagues hypothesized that susceptibility to developing HSE was inherited as a monogenic trait. There was evidence that impaired interferon responses might predispose to HSE. Therefore, they screened otherwise healthy children with a history of HSE and found two unrelated children whose leukocytes had defective interferon production in response to HSV-1 antigens. These children had no evidence of increased susceptibility to infections. However, it was determined that they had impaired responses to agonists of TLR-7, TLR-8, TLR-9, and possibly TLR-3. These responses were similar to those recently reported by Bruce Beutler's group in mice lacking UNC-93B, an endoplasmic reticulum protein involved in activation by TLR-3, TLR-7, and TLR-9. The children identified by Casanova completely lacked mRNA encoding UNC-93B. Each was homozygous for a different mutation in the UNC-93B gene: one had a 4 base deletion and one had a point mutation leading to alternative splicing. Previously, children with a deficiency in interleukin 1 receptor-associated kinase-4 (IRAK-4) had been described. These patients fail to signal through TLR-7, TLR-8, or TLR-9, and show a propensity to developing certain bacterial infections but not HSE. Therefore, Casanova and colleagues proposed that the HSE susceptibility of the UNC-93B deficient children was due to impaired TLR-3 induction of interferon. One year later the same group identified two more otherwise healthy children with HSE and a single point mutation in the *tlr3* gene. TLR-3 is highly expressed in the central nervous system. Together, these data strongly suggest an important role for TLR-3 in protection against HSE and indicate that a single genetic defect may predispose to infection (or severe manifestations of infection) by a specific pathogen. The possibility remains that the TLR-3 pathway may also be important in protection against encephalitis cause by other viruses.

Inherited Susceptibility to Tuberculosis Several recently described genetic defects predispose individuals to severe disease caused by *Mycobacterium tuberculosis* and other less virulent mycobacterial species. The proteins encoded by these genes all play a role in Th1 responses and include IL-12, receptors for interferon-γ and IL-12, and STAT1, which is involved in signaling via the interferon-γ receptor.

Acquired Immunodeficiencies

Exposure to a variety of factors such as infectious agents, immunosuppressive drugs, and environmental conditions are much more prevalent causes of immunodeficiency than the genetic defects described above. Malnutrition is the most common cause of immunodeficiency worldwide. Metabolic diseases, such as diabetes, hepatic cirrhosis, and chronic kidney disease also lead to immunosuppression. Many viral and bacterial infections can result in transient immunosuppression. Infection with the human immunodeficiency virus (HIV) can lead to a chronic and severe state of immunosuppression known as the Acquired Immune Deficiency Syndrome (AIDS).

Acquired Immune Deficiency Syndrome It is currently estimated that over 40 million people are infected with HIV, with the majority in sub-Saharan Africa, and South and Southeast Asia. HIV is a retrovirus that contains 9 genes: *gag, pol, env, tat, rev, nef, vif, vpr, vpu*. The *gag* gene product is split by the HIV protease into 5 structural proteins. The *pol* gene product is split into three enzymes: integrase, reverse transcriptase, and protease. The *env* gene product is cleaved to produce two envelope proteins, gp120 and gp41, which together constitute gp160. The other 5 genes encode regulatory

proteins. The virus is shed into body fluids and the bloodstream. HIV infection is most typically spread via sexual contact, but transmission via contaminated hypodermic needles and blood products can also occur as can transmission from mother to infant.

HIV enters CD4$^+$ T-cells through interaction of gp160 on the viral envelope and CD4 along with a co-receptor, either CCR5 or CXCR4, on the cell surface. Individuals expressing a *ccr5* mutant gene are protected from HIV infection. Once inside the cell, the virus uncoats and the reverse transcriptase, which is complexed to the RNA genome of the virus, transcribes viral RNA into double-stranded DNA. This is then transported to the cell nucleus where, with the help of virally encoded integrase, it is inserted into the cellular DNA. Thus, the virus establishes lifelong infection of the cell. Replication of the viral genome occurs along with cell replication.

Acute HIV infection is characterized by high viremia, immune activation, and CD4$^+$ T-cell lymphopenia. The acute phase of infection lasts several weeks. Patients often develop nonspecific symptoms of fever, rash, headaches, and myalgias. Occasionally, opportunistic infections may occur in this period. The acute phase is followed by a period of clinical latency generally characterized by absence of significant symptoms. During this period, viral replication continues in the lymphoid tissue, resulting in lymphadenopathy, and there is increased susceptibility to certain infections such as tuberculosis. The latent period lasts several years. Higher viral loads predict shorter clinical latency. While initial host immune responses control viral infection somewhat, they inevitably fail. CD4$^+$ T-cells continue to decline, and eventually the host becomes susceptible to opportunistic infections, including pneumocystis pneumonia, cryptococcal meningitis, disseminated cytomegalovirus (CMV), and mycobacterial infections. If untreated, death occurs on average about 10 years after infection.

While the virus targets CD4$^+$ T-cells, abnormalities in all parts of the immune system occur during HIV infection. There is profound disruption of lymphoid tissue architecture, resulting in an inability to mount responses against new antigens and severely impaired memory responses. CD4$^+$ T-cells that are not killed directly are dysregulated. They have decreased IL-2 production and IL-2 receptor expression, resulting in diminished capacities to proliferate and differentiate. CD28 expression is also reduced. T-cell receptor Vβ repertoire can be significantly reduced in advanced HIV infection. CD8$^+$ T-cells also have decreased IL-2 receptor and CD28 expression. CTL activity is reduced as is production of cytokines and chemokines, including those that block HIV replication. B-cells are hyperactivated by gp41 in HIV infection. gp120 acts as a superantigen for B-cells carrying the V$_H$3 variable region. This leads to a gap in the B-cell antibody repertoire. B-cell dysregulation explains the propensity to develop infections with encapsulated bacteria. B-cells also express proinflammatory cytokines such as TNF and IL-6, which enhance HIV replication.

Macrophages express CD4 and other HIV co-receptors and become infected with HIV. They serve as important reservoirs for HIV. Infection of macrophages also leads to functional abnormalities, including decreased IL-12 secretion, increased IL-10 production, decreased antigen uptake, and impaired chemotaxis. HIV infection of brain tissue macrophages or microglial cells plays a major role in HIV encephalopathy. Other immune cells shown to be dysregulated in HIV infection include NK cells and neutrophils.

Treatment with multiple drugs that target HIV enzymes, known as combined antiretroviral therapy (cART), is effective in reducing viremia and restoring normal CD4$^+$ T-cell counts. cART has drastically reduced mortality rates in HIV infection. After 2–3 weeks, patients treated with cART may develop a severe inflammatory response to pre-existing opportunistic infections known as the immune reconstitution inflammatory syndrome (IRIS). No treatment regimen has been successful in eradicating infection.

Autoimmune Diseases

T-lymphocytes and B-lymphocytes have incredible diversity in antigen recognition. They also carry a potent armamentarium, capable of destroying cells and causing damaging inflammation. While some pathogen-associated antigens are quite distinct from host molecules, there is often no fundamental difference between host antigens and those of pathogens. Therefore, autoreactive lymphocytes will most certainly arise. The host needs to eliminate these self-reactive lymphocytes or suffer self-destruction. Macfarlane Burnet originally formulated the clonal deletion theory, proposing that all self-reactive lymphocytes (forbidden clones) are destroyed during development of the immune system. In fact, some autoreactive T-cells and B-cells exist in many individuals that do not develop autoimmune disease. Autoimmunity occurs when T-cells or B-cells are activated and cause tissue destruction in the absence of ongoing infection.

Tolerance is the process that neutralizes these autoreactive T-cells and B-cells. Autoimmunity is the failure of tolerance mechanisms. B-cells may produce antibodies that recognize cell surface proteins, and these may directly cause disease by (i) initiating destruction of host cells or (ii) mimicking receptor ligand, and causing hyperactivation. Antibodies that bind intracellular antigens, such as many of those formed in SLE, are generally believed to be secondary to the autoimmune process itself. Autoreactive B-cells may be deleted in the bone marrow or the lymph nodes and spleen. B-cells must receive a second signal (co-stimulation) following B-cell receptor ligation by antigen. Most often this second signal is delivered by T-cells. Without T-cell help, antigen binding B-cells will die.

Autoreactive T-cells are removed by two separate processes: central and peripheral tolerance. During maturation, T-cells leave the bone marrow and travel to the thymus, where they encounter endogenous peptides complexed with MHC. If the receptors on a given T-cell

bind these complexes with significant affinity, the T-cell is directed to die by apoptosis in a process called negative selection. Not all self-antigens are presented in the thymus, and some autoreactive T-cells escape to the periphery. Similar to B-cells, presentation of antigens to T-cells in the absence of co-stimulation leads to deletion. Activated T-cells express Fas on their cell surface. If they encounter Fas ligand, they will undergo apoptosis. Some tissues such as the eye constitutively express Fas ligand. When activated, T-cells expressing Fas (CD95) enter the anterior chamber of the eye, they encounter Fas ligand and undergo apoptosis without causing tissue damage. Another molecule involved in peripheral tolerance is cytotoxic T-lymphocyte-associated protein (CTLA-4 or CD152), which binds to B7-1 (CD80) on T-cells and B7-2 (CD86) on B-cells with higher affinity than the co-stimulatory molecule CD28. Thus, CTLA-4 can prevent B-cell and T-cell activation. CTLA-4 polymorphisms are associated with an increased predisposition to autoimmune diseases, including SLE, autoimmune thyroiditis, and type 1 diabetes.

Recently, a new subset of T-cells has been described called regulatory T-cells (Tregs). These cells play an important role in controlling the magnitude and the quality of immune responses. Tregs express CD4 and CD25 (alpha chain of the IL-2 receptor) on their surface. They also produce a transcription factor, forkhead box P3 (FoxP3), which is critical for the differentiation of thymic T-cells into Tregs. FoxP3 interacts with other transcription factors to control expression of ~700 gene products, and leads to repression of IL-2 production and activation of CTLA-4 and CD25 expression. Tregs mediate immunosuppression by several mechanisms. First, they may compete with specific naïve T-cells for antigen presented by APCs. Second, they may downregulate APCs via CTLA-4-dependent mechanisms. Finally, they may interact with effector T-cells to either kill them or inactivate them with immunosuppressive cytokines such as IL-10. The role of abnormal Tregs in autoimmunity is still being established. Defects in certain genes specifically associated with Tregs may lead to autoimmunity. In addition to CTLA-4, polymorphisms in IL-2 and CD25 are associated with susceptibility to autoimmune diseases. Mutations in *foxP3* lead to an immunodeficiency syndrome, IPEX (immune dysregulation, polyendocrinopathy, enteropathy, X-linked), associated with autoimmune diseases in endocrine organs.

Tolerance can be broken by several mechanisms. Infections are thought to be the main exogenous cause of autoimmunity. Infections can damage barriers, leading to release of sequestered antigens. Superantigens, which stimulate polyclonal T-cell activation without the need for co-stimulation, are produced by certain microbes. Also, infection can produce significant inflammation with production of inflammatory cytokines and other co-stimulatory molecules, which may activate autoreactive lymphocytes (bystander activation). Another trigger of autoimmunity is a seemingly appropriate immune response to microbial antigens that mimic host antigens. For example, in the demyelinating disease Guillain-Barré syndrome, antibody cross-reactivity has been demonstrated between human gangliosides and *Campylobacter jejuni* lipopolysaccharide. In fact, some microbial antigens may have evolved to resemble host antigens in order to evade the host immune response. Drugs, such as procainamide, are another significant exogenous trigger of autoimmunity.

Several autoimmune diseases have been described. These may be either systemic (such as SLE) or organ-specific (for instance, type 1 diabetes). Some of the more common autoimmune diseases are considered in the following sections.

Systemic Lupus Erythematosus

Systemic lupus erythematosus (SLE) is a diverse systemic autoimmune syndrome with a significant range of symptoms and disease severity. The etiologies of SLE have not been clearly established. An antecedent viral illness often precedes the onset of SLE, and there is a temporal association with EBV infection and SLE. MHC genes (such as HLA-A1, B8, and DR3) are linked to SLE. Deficiencies in the early complement component cascade (C1q, C2, or C4) are strongly associated with SLE. Several other genes may also be linked to SLE. The syndrome probably represents several distinct diseases that result in similar manifestations. There is a major propensity for females to develop this disease (9:1 versus males), implicating a role for female hormones (or male hormones as protective). However, the basis for this gender preference has not been established.

The hallmark of SLE is the development of autoantibodies, particularly those that bind double-stranded DNA (anti-dsDNA). Many other autoantibodies have been described in SLE and are associated with certain clinical manifestations. Antibodies to dsDNA, Sm antigen, and C1q are correlated with kidney disease. Antibodies to Ro and La antigens are associated with fetal heart problems, while antibodies to phospholipids are associated with thrombotic events and fetal loss in pregnancy. Tissue damage by autoantibodies has been most well studied with anti-dsDNA causing kidney damage. There are two possible mechanisms involved. First, anti-dsDNA may bind to fragments of DNA released by apoptotic cells (nucleosomes) in the bloodstream, forming immune complexes. These complexes are deposited in the glomerular basement membrane in the kidney and cause disease by Type III reactions. Second, anti-dsDNA may cross-react with another antigen expressed on kidney cells. One possible candidate for this antigen is α-actinin, which crosslinks actin and is important in maintaining the function of renal podocytes.

Type 1 Diabetes Mellitus

Pancreatic β-cells produce insulin, which plays an essential role in glucose metabolism. In type 1 diabetes, cell-mediated destruction of β-cells occurs, leading to complete insulin deficiency. Susceptibility to development of type 1 diabetes is linked to HLA-DR3/4 and DQ8 genes and is also associated with polymorphisms in the gene for CTLA-4. Potential exogenous triggers include viral infections (such as congenital rubella), chemicals (such as nitrosamines), and foods (including early exposure to cow's milk proteins). Interaction with environmental triggers in an individual with genetic predisposition leads to infiltration of the pancreatic islets with CD4$^+$ and CD8$^+$ T-cells, B-cells, and macrophages, and the production of auto-antibodies. Autoantibodies to β-cell antigens can precede the onset of type 1 diabetes by years. However, there is no direct evidence that these antibodies play a role in pathogenesis. Th1-activated autoreactive CD4$^+$ T-cells together with cytotoxic CD8$^+$ T-cells induce apoptosis of β-cells mediated via Fas-Fas ligand interaction and release of cytotoxic molecules. Clinical expression of disease occurs only after >90% of the β-cells are destroyed.

Multiple Sclerosis

Multiple sclerosis (MS) is a demyelinating disease presumed to be of autoimmune etiology. The disease presents with neurologic deficits and generally has a relapsing course followed by a progressive phase. The disease is more common in females (2:1 versus men), and there appears to be a genetic predisposition. As with SLE and type 1 diabetes, there is an association with certain MHC, including HLA-DR2 and DQ6. However, the inciting factor is believed to be environmental. The disease occurs more commonly in temperate climates, and relapses are often preceded by viral respiratory tract infections. The hallmark of MS is the central nervous system inflammatory plaque containing a perivascular infiltration of myelin-laden macrophages and T-cells (both CD4$^+$ and CD8$^+$). These plaques tend to form in the white matter and involve both the myelin sheath and oligodendrocytes. There is often widespread axonal damage. Myelin-reactive T-cells can be isolated from individuals with and without MS. However, in the former, these T-cells are activated with a Th1 phenotype, whereas in the latter, these cells are naïve. There is significantly increased antibody production in the central nervous system of MS patients. B-cells recovered from cerebrospinal fluid of MS patients also display an activated phenotype.

Celiac Disease

Celiac disease occurs in genetically predisposed individuals following exposure to gluten. The disease results in diarrhea and malabsorption. It is unique among autoimmune diseases, as the environmental precipitant is known. Strict avoidance of the antigen, which is found in wheat, leads to remittance of symptoms in most subjects. Gliadin is the alcohol soluble fraction of gluten and is the primary antigen leading to an inflammatory reaction in the small intestine, characterized by chronic inflammatory infiltrate and villous atrophy. An enzyme, tissue transglutaminase, deamidates gliadin peptides in the lamina propria, increasing their immunogenicity. Gliadin-reactive T-cells recognizing antigen in the context of HLA-DQ2 or DQ8 lead to an inflammatory Th1 phenotype. While these HLA types are necessary for the development of celiac disease, they are not sufficient, and other genetic factors and environmental exposures also play a role in developing disease. IgA antibodies directed against gliadin, tissue transglutaminase, and connective tissue (such as antiendomysial and antireticulin) are found in individuals with celiac disease. These patients are also susceptible to developing several types of cancers, most notably adenocarcinoma of the small intestine and enteropathy-associated T-cell lymphoma.

KEY CONCEPTS

- Upon reexposure to environmental antigens called allergens, an inflammatory state can ensue, known as a hypersensitivity or allergic reaction. Four types of hypersensitivity states are classically recognized. Type I or immediate hypersensitivity is mediated by IgE and may occur as a local (e.g., asthma) or systemic (anaphylaxis) reaction. Type II hypersensitivity is mediated by IgM or IgG targeting cell membrane-associated antigens. The most common cause of type II reactions is medications, such as penicillin, which become associated with red blood cells or platelets leading to anemia and thrombocytopenia. Type III hypersensitivity occurs when antibodies bind antigen to form immune complexes that become deposited in tissues where they elicit inflammation. Examples of Type III reactions include serum sickness and the Arthus reaction. Type IV or delayed-type hypersensitivity (DTH) is mediated by T cells. Examples of DTH include the tuberculin skin reaction and contact dermatitis from poison ivy.

- Individuals with antibody deficiencies often remain well for a time after birth due to passive transfer of immunoglobulin from breast milk during the first 6 to 9 months of life. Then they develop upper and lower respiratory tract infections including sinusitis, otitis, and pneumonia, primarily caused by encapsulated bacteria such as *S. pneumoniae*. If left untreated, these recurrent infections cause significant destruction and scarring of lung tissue, resulting in bronchiectasis.

- X-linked agammaglobulinemia is caused by a defect in Bruton's tyrosine kinase (Btk), which leads to arrest of B cell differentiation at the Pre-B cell stage. The syndrome is manifested by a complete lack of serum immunoglobulins and a lack of tonsilar tissue.

- The most common form of hyper-IgM syndrome is caused by an X-linked deficiency in CD40L, which is expressed on the surface of T cells. In the absence of B cell-T cell interaction via CD40-CD40L, there is no immunoglobulin class switching and B cells only produce IgM. In CD40L deficiency, there is significant impairment in both antibody production and cell-mediated immunity, resulting in a clinical presentation that may include infections with encapsulated bacteria and intracellular organisms.

- Defects in the terminal components of complement (C5-9) lead to a remarkable susceptibility to recurrent infections caused by pathogenic *Neisseria* spp.

- In chronic granulomatous disease (CGD), phagocytes are unable to produce reactive oxygen species to kill ingested pathogens due to genetic defects in the five subunit enzyme NADPH oxidase. The disease is characterized by recurrent, indolent infections caused by catalase positive organisms including *S. aureus* and *Aspergillus* spp.

- HIV is a retrovirus that enters $CD4^+$ T cells through interaction of gp160 on the viral envelope and CD4 along with a coreceptor, either CCR5 or CXCR4, on the cell surface. HIV infection leads to a progressive decline in $CD4^+$ T cells, leading to opportunistic infections including pneumocystis pneumonia, cryptococcal meningitis, disseminated CMV, and mycobacterial infections.

- The hallmark of SLE is the development of autoantibodies, particularly those that bind double-stranded DNA (anti-dsDNA). Deficiencies in the early complement component cascade (C1q, C2, or C4) are strongly associated with SLE.

- In type 1 diabetes, cell-mediated destruction of pancreatic β-cells occurs, leading to complete insulin deficiency.

- Celiac disease occurs in genetically predisposed individuals following exposure to wheat gluten. Th1-activated $CD4^+$ T cells recognizing the alcohol soluble fraction of gluten (gliadin) leads to an inflammatory reaction in the small intestine characterized by chronic inflammatory infiltrate and villous atrophy, resulting in diarrhea and malabsorption. Strict avoidance of the antigen leads to remittance of symptoms in most subjects.

SUGGESTED READINGS

1. Abbas A, Lichtman A, Pillai S. *Cellular and Molecular Immunology.* New York: WB Saunders; 2007.
2. Bhaskaran K, Hamouda O, Sannes M, et al. Changes in the risk of death after HIV seroconversion compared with mortality in the general population. *JAMA.* 2008;300:51–59.
3. Bruton OC. Agammaglobulinemia. *Pediatrics.* 1952;9:722–727.
4. Casanova JL, Abel L. Inborn errors of immunity to infection: The rule rather than the exception. *J Exp Med.* 2005;202:197–201.
5. Casrouge A, Zhang SY, Eidenschenk C, et al. Herpes simplex virus encephalitis in human UNC-93B deficiency. *Science.* 2006; 314:308–312.
6. Castigli E, Geha RS. Molecular basis of common variable immunodeficiency. *J Allergy Clin Immunol.* 2006;117:740–746.
7. Durandy A, Peron S, Fischer A. Hyper-IgM syndromes. *Curr Opin Rheumatol.* 2006;18:369–376.
8. Fortin A, Abel L, Casanova JL, et al. Host genetics of mycobacterial diseases in mice and men: Forward genetic studies of BCG-osis and tuberculosis. *Annu Rev Genomics Hum Genet.* 2007;8:163–192.
9. Frohman EM, Racke MK, Raine CS. Multiple sclerosis—The plaque and its pathogenesis. *N Engl J Med.* 2006;354:942–955.
10. Gell PGH, Coombs RRA. *Clinical Aspects of Immunology.* Oxford: Blackwell; 1968.
11. Goldsby R, Kindt T, Osborne B. *Kuby Immunology.* New York: WE Freeman; 2006.
12. Heyworth PG, Cross AR, Curnutte JT. Chronic granulomatous disease. *Curr Opin Immunol.* 2003;15:578–584.

13. Jonsson G, Truedsson L, Sturfelt G, et al. Hereditary C2 deficiency in Sweden: Frequent occurrence of invasive infection, atherosclerosis, rheumatic disease. *Medicine (Baltimore)*. 2005;84:23–34.

14. Murphy KM, Travers P, Walport M. *Janeway's Immunobiology.* New York: Garland; 2007.

15. Oksenhendler E, Gerard L, Fieschi C, et al. Infections in 252 patients with common variable immunodeficiency. *Clin Infect Dis.* 2008;46:1547–1554.

16. Parham P. *The Immune System.* New York: Garland; 2005.

17. Rahman A, Isenberg DA. Systemic lupus erythematosus. *N Engl J Med.* 2008;358:929–939.

18. Rajan TV. The Gell-Coombs classification of hypersensitivity reactions: A re-interpretation. *Trends Immunol.* 2003;24:376–379.

19. Sakaguchi S, Yamaguchi T, Nomura T, et al. Regulatory T cells and immune tolerance. *Cell.* 2008;133:775–787.

20. van der Vliet HJ, Nieuwenhuis EE. IPEX as a result of mutations in FOXP3. *Clin Dev Immunol.* 2007;2007:89017.

18

Molecular Basis of Pulmonary Disease

Carol F. Farver . Dani S. Zander

INTRODUCTION

In recent years, understanding of the molecular pathology of lung diseases has expanded significantly, enabling development of new therapies, diagnostic procedures, and prognostic markers. This chapter highlights key advances in this area, presenting this information in the context of the clinical and pathologic features of the various neoplastic and non-neoplastic disease entities discussed.

NEOPLASTIC LUNG AND PLEURAL DISEASES

Lung cancer is an important cause of morbidity and mortality throughout the world. Over 200,000 people are diagnosed with lung cancer and over 150,000 die annually due to the disease in the United States. Changes in incidence parallel changes in the prevalence of tobacco smoking, the most important risk factor for the disease, and are showing a decline in men and stabilization in women. The *World Health Organization* (WHO) classification scheme is the most widely used system for histologic classification of these neoplasms (Table 18.1). The most common types of malignant epithelial tumors can be grouped into the categories of nonsmall cell lung cancers (NSCLCs) and small cell carcinomas (SCLCs). This grouping and the specific histologic type of the neoplasm have relevance for prognosis and treatment planning. NSCLCs include adenocarcinomas (ACs), squamous cell carcinomas (SqCCs), large cell carcinomas, adenosquamous carcinomas, and sarcomatoid carcinomas. SCLCs include cases of pure and combined small cell carcinoma.

Lung cancer often declares itself by causing a cough, shortness of breath, chest pain or tightness, and/or hemoptysis (coughing up blood). Pneumonia can develop due to airway obstruction by the neoplasm. Constitutional symptoms of fever, weight loss, and malaise can occur. Symptoms related to local invasion of the chest wall, nerves, superior vena cava, esophagus, or heart can call attention to the existence of the neoplasm, or the disease can present itself in metastatic form. Some tumors elaborate hormones and produce endocrine syndromes including Cushing syndrome, syndrome of inappropriate antidiuretic hormone, hypercalcemia, carcinoid syndrome, gynecomastia, and others. Hypercoagulability commonly accompanies lung cancers, leading to venous thrombosis and less frequently nonbacterial thrombotic endocarditis or disseminated intravascular coagulation. Other associated findings can include anemia, granulocytosis, eosinophilia, clubbing of the fingers, myasthenic syndromes, dermatomyositis/polymyositis, and transverse myelitis.

Pathologic diagnosis and histologic classification of lung cancer occur via examination of tissue or cytology samples. Fiberoptic bronchoscopy is commonly performed to collect these samples. Cytologic evaluation of sputum samples can provide a diagnosis in some cases, particularly for centrally located neoplasms such as SqCC and SCLC. Peripheral neoplasms frequently are biopsied by percutaneous fine needle aspiration or core needle biopsy performed under radiologic guidance. If a pleural effusion is present, cytologic examination of the pleural fluid can be diagnostic. Pleural biopsy, mediastinoscopy with biopsy, and wedge biopsy of the lung can also be performed, depending upon the specific clinical and radiologic findings. Biopsy of a metastatic lesion, if present, can establish a pathologic diagnosis and determine the stage of the tumor.

For patients with lung cancer, survival is closely related to tumor stage. The *American Joint Commission*

Table 18.1   World Health Organization histological classification of tumors of the lung

Malignant epithelial tumors
Squamous cell carcinoma
  Papillary
  Clear cell
  Small cell
  Basaloid
Small cell carcinoma
  Combined small cell carcinoma
Adenocarcinoma
  Adenocarcinoma, mixed subtype
  Acinar adenocarcinoma
  Papillary adenocarcinoma
  Bronchioloalveolar carcinoma
    Nonmucinous
    Mucinous
    Mixed or indeterminate
  Solid adenocarcinoma with mucin production
    Fetal adenocarcinoma
    Mucinous ("colloid") carcinoma
    Mucinous cystadenocarcinoma
    Signet ring adenocarcinoma
    Clear cell adenocarcinoma
Large cell carcinoma
  Large cell neuroendocrine carcinoma
    Combined large cell neuroendocrine carcinoma
  Basaloid carcinoma
  Lymphoepithelioma-like carcinoma
  Clear cell carcinoma
  Large cell carcinoma with rhabdoid phenotype
Adenosquamous carcinoma
Sarcomatoid carcinoma
  Pleomorphic carcinoma
  Spindle cell carcinoma
  Giant cell carcinoma
  Carcinosarcoma
  Pulmonary blastoma
Carcinoid tumour
  Typical carcinoid
  Atypical carcinoid
Salivary gland tumors
  Mucoepidermoid carcinoma
  Adenoid cystic carcinoma
  Epithelial-myoepithelial carcinoma
Preinvasive lesions
  Squamous carcinoma *in situ*
  Atypical adenomatous hyperplasia
  Diffuse idiopathic pulmonary neuroendocrine
    hyperplasia

Mesenchymal tumours
Epithelioid haemangioendothelioma
Angiosarcoma
Pleuropulmonary blastoma
Chondroma
Congenital peribronchial myofibroblastic
  tumour
Diffuse pulmonary lymphangiomatosis
Inflammatory myofibroblastic tumour
Lymphangioleiomyomatosis
Synovial sarcoma
  Monophasic
  Biphasic
Pulmonary artery sarcoma
Pulmonary vein sarcoma
Benign epithelial tumours
Papillomas
  Squamous cell papilloma
    Exophytic
    Inverted
  Glandular papilloma
  Mixed squamous cell and glandular papilloma
Adenomas
  Alveolar adenoma
  Papillary adenoma
  Adenomas of salivary-gland type
    Mucous gland adenoma
    Pleomorphic adenoma
    Others
  Mucinous cystadenoma
Lymphoproliferative tumours
Marginal zone B-cell lymphoma of the MALT
  type
Diffuse large B-cell lymphoma
Lymphomatoid granulomatosis
Langerhans cell histiocytosis
Miscellaneous tumours
Hamartoma
Sclerosing hemangioma
Clear cell tumour
Germ cell tumours
  Teratoma, mature
  Immature
  Other germ cell tumours
Intrapulmonary thymoma
Melanoma
Metastatic tumours

Reprinted with permission from Travis WD, Brambilla E, Müller-Hermelink HK, et al., eds. World Health Organization Classification of Tumours. Pathology and Genetics of Tumours of the Lung, Pleura, Thymus and Heart. Lyon: IARCPress, 2004.

on Cancer TNM staging system is used widely for staging NSCLCs. Stage is determined based on a combination of patient characteristics such as size and extent of the neoplasm (T) status of nodal metastasis (N), and distant metastasis (M). For SCLCs, disease is classified as limited (restricted to one hemithorax) or extensive. The overall 5-year survival for patients with lung cancer is relatively poor (13.4% for men and 17.9% for women), in part related to the advanced stage at which many lung cancers are diagnosed. Patients diagnosed with localized disease fare substantially better than those with regional or distant disease. The 5-year survival rates are 49.0%, 15.3%, and 2.8%, and the 10-year survival rates are 37.8%,

10.3%, and 1.6% for patients with localized disease, disease with regional involvement, and disease with distant spread, respectively. Treatment decisions depend upon stage, histology, and comorbid conditions. Surgical resection is preferred for treatment of localized NSCLCs, provided there is no medical contraindication. For these patients, lobectomy or more extensive resection (depending upon tumor extent) usually is performed, with intraoperative mediastinal lymph node sampling or dissection to assess pathologic stage. Some patients also benefit from chemotherapy and radiotherapy. Advanced NSCLCs and most SCLCs are treated primarily with chemotherapy and radiotherapy.

Common Molecular Genetic Changes in Lung Cancer

Multiple, complex, stepwise genetic, and epigenetic changes involving allelic losses, chromosomal instability and imbalance, mutations in tumor suppressor genes (TSGs) and dominant oncogenes, epigenetic gene silencing through promoter hypermethylation, and aberrant expression of genes participating in control of cell proliferation and apoptosis play roles in the genesis of lung cancer. Oncogenes involved in the pathogenesis of lung cancer include *MYC*, *K-RAS* (predominantly ACs), *Cyclin D1*, *BCL2*, and *ERBB* family genes such as *EGFR* (epidermal growth factor receptor; predominantly ACs) and *HER2/neu* (predominantly ACs). TSG abnormalities involving *TP53*, *RB*, *p16^{INK4a}*, and new candidate TSGs on the short arm of chromosome 3 (*DUTT1, FHIT, RASFF1A, FUS-1, BAP-1*) are common. New therapeutic agents have been developed from our expanding knowledge about the molecular pathogenesis of lung cancer (Figure 18.1). These targeted agents include EGFR pathway inhibitors, VEGF/VEGFR pathway inhibitors, Ras/Raf/MEK pathway inhibitors, PI3K/Akt/PTEN pathway inhibitors, tumor suppressor gene therapies, proteasome inhibitors, HDAC inhibitors, telomerase inhibitors, and others.

In cancers, deletions and amplifications of chromosomal regions harboring TSGs and oncogenes often occur. Allelic losses of 3p have been reported in 96% of the lung cancers studied and in 78% of the precursor lesions, with larger segments of allelic loss noted in more SCLCs and SqCCs than in ACs and preneoplastic/preinvasive lesions. 8p21-23 deletions are frequent and early events in pulmonary carcinogenesis, and other common regions of deletion include 4q, 9p, 10p, 10q, chromosome 18, and chromosome 21. Deletions at 17p13 (*TP53*), 13q14 (*RB*), 9p21 (*p16^{INK4a}*), 8p21-p23, and several regions of 3p are more frequent in SqCCs than ACs. Areas of frequent amplification in SqCCs and ACs include 5p, chromosome 7, 8q, 11q13, 19q, and 20q. However, ACs display higher frequencies of deletion of chromosome 6, 8p, 9q, 15q, and chromosome 16 than SqCCs, and possess small regions of amplification on chromosomes 12 and 14 not seen in SqCCs. Other studies using cell lines have shown deletions of 2q and 13q in AC but amplifications in SqCC and deletion of 17p and amplification of 3q in both types of tumors.

Inactivation of recessive oncogenes is believed to occur through point mutation of the first allele followed by chromosomal deletion, translocation, or other alteration (such as hypermethylation of the gene promoter region) of the second allele. Inactivating mutations in the TSG *TP53*, which encodes the p53 protein, are the most frequent mutations in lung cancers, occurring in up to 50% of NSCLCs and over

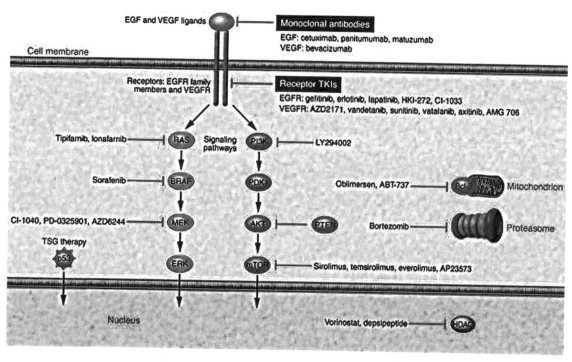

Figure 18.1 Targeted therapies are focused upon key oncogenic pathways in lung cancer. These agents are designed to interfere with lung cancer cell proliferation, inhibition of apoptosis, angiogenesis, and invasion. EGFR = epidermal growth factor receptor; VEGFR = vascular endothelial growth factor receptor; TKIs = receptor tyrosine kinase inhibitors; TSG = tumor suppressor gene; PR = proteasome; PDK1 = pyruvate dehydrogenase kinase isoenzyme 1; PTEN = phosphatase and tensin homolog. Reprinted with permission from Sun, S., Schiller, J.H., Spinola, M., and Minna, J.D. (2007) New molecularly targeted therapies for lung cancer. *J Clin Invest* 117:2740–2750.

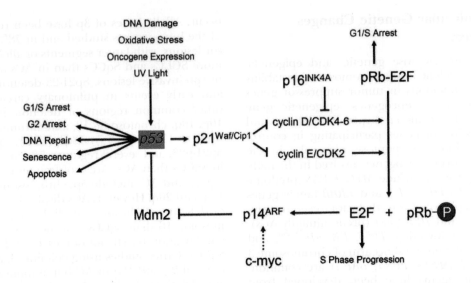

Figure 18.2   The *p53* and retinoblastoma (Rb) pathways. A phosphate residue on Rb protein is indicated with a purple P. UV = ultraviolet. Reprinted with kind permission of Springer Science+Business Media, from Stelter, A.A. and Xie, J. (2008) *Molecular oncogenesis of lung cancer.* In: *Molecular Pathology of Lung Diseases* (Zander, D.S., Popper, H.H., Jagirdar, J., Haque, A.K., Cagle, P.T., and Barrios, R., eds.), pp. 169–175, Springer, New York, NY.

70% of SCLCs, and are largely attributable to direct DNA damage from cigarette smoke carcinogens. *p53* protein is a transcription factor and a key regulator of cell cycle progression. DNA damage, oncogene expression, or other stimuli trigger *p53*-dependent responses including initiating cell cycle arrest, apoptosis, differentiation, and DNA repair. Loss of *p53* function can lead to inappropriate progression through the dysregulated cell cycle checkpoints and allow the inappropriate survival of genetically damaged cells.

The p16^{INK4a}-cyclin D1-CDK4-Rb pathway (Figure 18.2) is another important tumor suppressor pathway that plays a central role in controlling the G1 to S phase transition of the cell cycle and often is disrupted in lung cancers. This pathway interfaces with the *p53* pathway through p14^{ARF} and p21$^{Waf/Cip1}$. Thirty to 70% of NSCLCs contain mutations of p16^{INK4a} leading to p16^{INK4a} inactivation. Rb expression is lost in almost 90% of SCLCs and smaller numbers of NSCLCs, and mutational mechanisms usually responsible include deletion, nonsense mutations, and splicing abnormalities that lead to truncated Rb protein. p16^{INK4a} leads to hypophosphorylation of the Rb protein, causing arrest of cells in the G1 phase. The active, hypophosphorylated form of Rb regulates the transcription factors E2F1, E2F2, and E2F3, which are essential for progression through the G1/S phase transition. Loss of p16^{INK4a} protein or increased complexes of cyclin D-CDK4/6 or cyclin E-CDK2 leads to hyperphosphorylation of Rb, evasion of cell cycle arrest, and progression into S phase. NSCLCs lacking detectable alterations in p16^{INK4a} and Rb may have abnormalities of cyclin D1 and CDK4, causing inactivation of the Rb pathway. Epigenetic gene silencing (hypermethylation of the 5′ CpG island) of TSGs also occurs frequently during pulmonary carcinogenesis. Methylation rates of

p16^{INK4a} and *APC* in current or former smokers significantly exceed the rates in never-smokers and methylation rates of *APC*, *CDH13*, and *RARbeta* are significantly higher in ACs than in SqCCs.

Proto-oncogene activation and growth factor signaling are important in pulmonary carcinogenesis. The tyrosine kinase EGFR, K-RAS, and HER2/neu are important in ACs. A related pathway, the phosphoinositide 3-kinase (PI3K)/Akt/mammalian target of rapamycin (mTOR) pathway, often is dysregulated in pulmonary carcinogenesis and may mediate the effects of several tyrosine kinase receptors on proliferation and survival in NSCLC and SCLC, including EGFR, c-Met, c-Kit, and IGF-IR. The mTOR inhibitor rapamycin and its analogues are under evaluation for their efficacy in patients with lung cancer. Point mutations of RAS family proto-oncogenes (most often at *K-RAS* codons 12, 13, or 61) are detected in 20 to 30% of lung ACs and 15 to 50% of all NSCLCs, but agents that block RAS signaling have not shown significant activity as single-agent therapies in untreated NSCLC or relapsed SCLC. MYC family genes (*MYC, MYCN,* and *MYCL*) with roles in cell cycle regulation, proliferation, and DNA synthesis are more frequently activated in SCLCs than in NSCLCs. Vascular endothelial growth factor (VEGF) is a homodimeric glycoprotein that is overexpressed in many lung cancers and mediates multiple aspects of angiogenesis via three receptors, VEGFR-1, VEGF2, and VEGF3. Monoclonal antibodies to VEGF (bevacizumab) and tyrosine kinase inhibitors to VEGFRs have shown some favorable results in subsets of patients with NSCLC, but some reports suggest an increased rate of severe pulmonary hemorrhage in patients with SqCC.

MicroRNAs are nonprotein-coding, endogenous, small RNAs that regulate gene expression by translational repression, mRNA cleavage, and mRNA decay initiated by miRNA-guided rapid deadenylation. Some microRNAs such as *let-7* may play roles in carcinogenesis by functioning as oncogenes or tumor suppressors, negatively regulating TSGs and/or genes that control cell differentiation or apoptosis. A microRNA signature, which includes 40 to 45 microRNAs that are aberrantly expressed, has been discovered for lung cancer. The roles of microRNAs in prediction of clinical outcome and as therapeutic agents are being explored.

Adenocarcinoma and Its Precursors

Clinical and Pathologic Features

AC is defined in the WHO classification scheme as "a malignant epithelial tumour with glandular differentiation or mucin production, showing acinar, papillary, bronchioloalveolar or solid with mucin growth patterns or a mixture of these patterns." It is currently the most frequent histologic type of lung cancer in parts of the world, and is also the most common type of lung cancer in never-smokers and in women. Most ACs arise in the periphery of the lung and may invade the pleura and chest wall. Radiographs may show one or multiple nodules, ground-glass opacities, or mixed solid and ground-glass lesions. Grossly, they appear as gray-white or glistening nodules or masses, sometimes with necrosis or cavitation, or they can appear as a zone of consolidation mimicking pneumonia (usually bronchioloalveolar carcinoma) (Figure 18.3). Histologically, ACs can display a variety of patterns alone or in combination (Figure 18.4A). These include acinar, papillary, bronchioloalveolar, and solid arrangements. Less common subtypes include fetal AC, mucinous (or colloid) AC, mucinous cystadenocarcinoma, signet ring AC, and clear cell AC. ACs range from very well-differentiated tumors with extensive gland formation and little cytoatypia, to poorly differentiated, solid tumors that require a mucin stain to be categorized as ACs. Invasiveness is indicated by infiltrative neoplastic glands with an associated fibroblastic (desmoplastic) response or by cells in the lumens of blood vessels or lymphatics.

Atypical adenomatous hyperplasia (AAH) is believed to be a precursor lesion for peripheral pulmonary ACs, and is defined in the WHO scheme as "a localized proliferation of mild to moderately atypical cells lining involved alveoli and, sometimes, respiratory bronchioles, resulting in focal lesions in peripheral alveolated lung, usually less than 5 mm in diameter and generally in the absence of underlying interstitial inflammation and fibrosis" (Figure 18.5). AAH lies on a histologic continuum with bronchioloalveolar carcinoma (BAC), an *in situ* (noninvasive) form of AC in which the neoplastic cells grow along alveolar septa (lepidic growth) without invasion of stroma or vasculature (Figure 18.4B). BACs generally exceed 1 cm in diameter and display greater cytoatypia than AAH. AAH is found in approximately 3% of patients

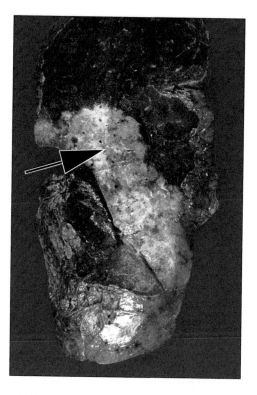

Figure 18.3 Bronchioloalveolar carcinoma. The tan tumor (arrow) replaces a large portion of the normal lung parenchyma.

without lung cancer at autopsy and 16 to 35% of lung resection specimens with AC. Both BAC and AAH are thought to arise from epithelial cells located at the junction between the terminal bronchiole and alveolus.

Molecular Pathogenesis

Primary lung ACs show recurrent copy-number alterations including many known proto-oncogenes and TSGs. Amplifications are found on 14q, 12q (*MDM2, CDK4*), 12p (*KRAS*), 8q (*MYC*), 7p (*EGFR*), 19q (*CCNE1*), 17q (*ERBB2*), 11q (*CCND1*), 5p (*TERT*), and others. Common deletions occur on 9p (*CDKN2A/CDKN2B*), 5q, 7q, 10q (*PTEN*), and other loci. Amplification of chromosome 14q13.3 occurs in 12% of samples. This region includes *NKX2-1*, which encodes a lineage specific transcription factor (thyroid transcription factor-1 [TTF-1]) that activates transcription of target genes including the surfactant proteins. Immunohistochemical staining for TTF-1 can be clinically helpful for determining a primary origin for the tumor in the lung (Figure 18.4D). Nuclear TTF-1 expression also has been associated with improved prognosis in subsets of patients with AC. *EGFR* and *K-RAS* mutations are mutually exclusive mutational events that occur in separate subsets of pulmonary ACs. EGFR is a receptor tyrosine kinase whose activation by ligand binding leads to activation of cell

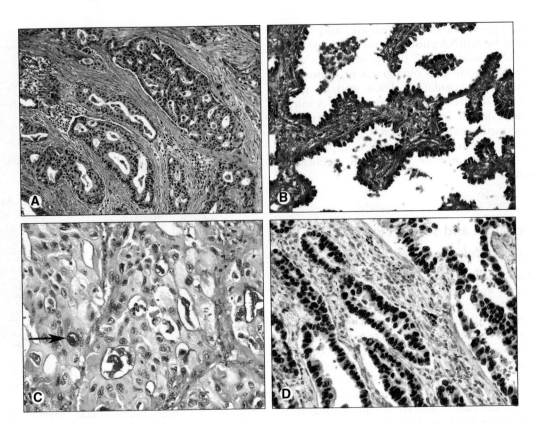

Figure 18.4   Adenocarcinoma. (A) Acinar pattern. The tumor consists of abnormal glands lined by columnar tumor cells with abundant cytoplasm and mildly pleomorphic nuclei. The desmoplastic stromal background indicates invasion. **(B)** Bronchioloalveolar pattern. Columnar tumor cells with a hobnail appearance and enlarged, hyperchromatic nuclei line thickened alveolar septa. The cells remain on the surface of the alveolar septa, and do not invade the lung tissue (an *in situ* lesion). **(C)** Mucicarmine stain. Intracytoplasmic (arrow) and luminal mucin stains dark pink; mucin production indicates glandular differentiation. **(D)** Immunohistochemical stain for thyroid transcription factor-1 (TTF-1). The brown-stained nuclei are positive for TTF-1. TTF-1 is expressed in the majority of pulmonary adenocarcinomas and small cell carcinomas, as well as in follicular cells in the thyroid.

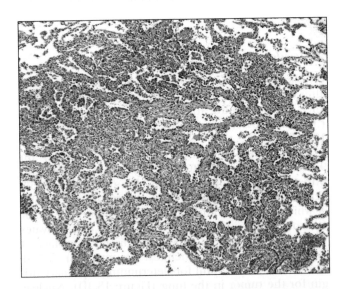

Figure 18.5   Atypical adenomatous hyperplasia. This lesion, which has been defined as a precursor lesion for peripheral pulmonary adenocarcinomas, consists of a well-circumscribed nodule measuring several millimeters in diameter, in which alveolar septa are lined by mildly moderate atypical cells.

signaling pathways such as Ras/mitogen-activated protein kinase (MAPK) and phosphatidylinositol-3-kinase, in turn propagating signals for proliferation, blocking of apoptosis, differentiation, motility, invasion, and adhesion. Tumor-acquired mutations in the tyrosine kinase domain of *EGFR* are found in approximately 5 to 10% of NSCLCs in the United States, and are associated with AC histology, never-smoker status, East Asian ethnicity, and female gender. The mutations are most often in-frame deletions in exon 19, single missense mutations in exon 21, or in-frame duplications/insertions in exon 20. The presence of an *EGFR* mutation is associated with a significant, favorable response to EGFR tyrosine kinase inhibitor therapy, and EGFR amplification and protein overexpression have been reported to correlate with survival after EGFR tyrosine kinase inhibitor therapy. K-Ras is a member of the Ras family of proteins, which serve as signal transducers between cell membrane-based growth factor signaling and the MAPK pathways; mutations in *K-RAS* are associated with smoking, male gender, and poorly differentiated tumors. *HER2*, a member of the EGFR family of receptor tyrosine kinases, is uncommonly mutated in NSCLC, primarily in ACs

(2.8%), never-smokers (3.2%), people of Asian ethnicity (3.9%), and women (3.6%). Epigenetic differences also exist between *EGFR-* and *K-RAS*-mediated carcinogenesis. *EGFR* mutation is reportedly less common in tumors with $p16^{INK4a}$ and *CDH13* methylation than in those without, whereas *K-RAS* mutation is significantly higher in $p16^{INK4a}$ methylated cases than in unmethylated cases.

Squamous Cell Carcinoma and Its Precursors

Clinical and Pathologic Features

SqCC is defined in the WHO classification scheme as "a malignant epithelial tumour showing keratinization and/or intercellular bridges, that arises from bronchial epithelium." This type of NSCLC is closely linked to cigarette smoking and typically arises in a mainstem, lobar, or segmental bronchus, producing a central mass on radiologic studies. Many of these neoplasms have an endobronchial component that can cause airway obstruction, leading to postobstructive pneumonia, atelectasis, or bronchiectasis. Symptoms of these associated processes can prompt evaluation of the patient and lead to discovery of the neoplasm. Gross examination characteristically reveals a tan or gray mass arising from a large bronchus, sometimes with a visible endobronchial component (Figure 18.6). Necrosis and cavitation are common. Key microscopic features indicating squamous differentiation include keratinization, sometimes with formation of keratin pearls and intercellular

bridges (Figure 18.7). These features vary in extent from very extensive in well-differentiated cases to minimal in poorly differentiated cases. Likewise, cytoatypia and mitoses can range from minimal to marked and frequent. Invasive nests and sheets of tumor infiltrate through tissues, stimulating a fibroblastic response, or are seen inside vascular or lymphatic spaces.

Basal cells in the bronchial epithelium are believed to represent the progenitor cells for invasive SqCC, and the progression to SqCC is believed to begin with basal cell hyperplasia, followed by squamous metaplasia, squamous dysplasia, carcinoma *in situ*, and invasive SqCC (Figure 18.8). Regression of the precursor lesions for invasive SqCC can occur, particularly the earlier lesions. Like invasive SqCC, these precursor lesions arise in major bronchi, and often are visible adjacent to the invasive tumor. However, the precursor lesions are not invasive as they do not extend through the basement membrane of the bronchial epithelium. Grossly, they may be inconspicuous or appear as flat, tan, or red bronchial mucosal discolorations, or tan wart-like excrescences. Microscopically, they demonstrate a range of squamous changes (Figure 18.9). As the severity of dysplasia increases, the epithelium thickens, maturation is increasingly impaired, and mitoses appear at more superficial levels of the epithelium (lower third in mild or moderate dysplasia, lower two-thirds in severe dysplasia, or throughout the epithelium in carcinoma *in situ*). The basilar zone expands, there is reduced flattening of the superficial squamous cells, and cell size, pleomorphism, and

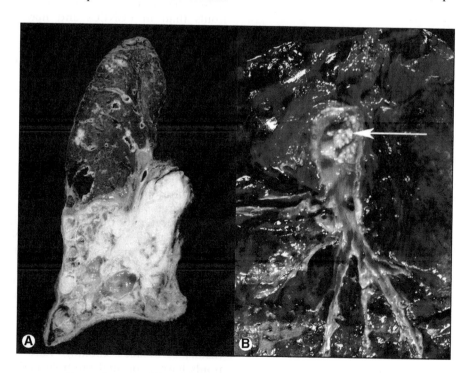

Figure 18.6   Squamous cell carcinoma. (A) Invasive squamous cell carcinoma with postobstructive pneumonia and abscesses. This tan tumor lies in the central (perihilar) area of the lung. Distal to the tumor, the lung has extensive cystic changes reflecting abscesses and bronchiectasis, as well as a background of tan consolidation representing pneumonia. **(B)** Squamous cell carcinoma with a warty endobronchial component (arrow).

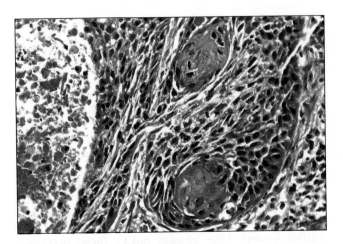

Figure 18.7 Invasive squamous cell carcinoma. This tumor consists of cells with hyperchromatic, pleomorphic nuclei, and eosinophilic cytoplasm. Two keratin pearls are present (center) and a portion of the tumor is necrotic (left).

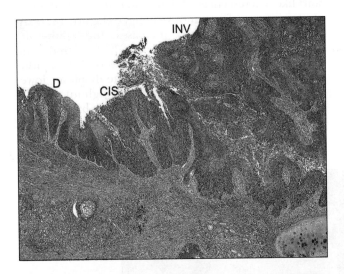

Figure 18.8 Dysplasia, squamous cell carcinoma *in situ*, and invasive squamous cell carcinoma. The dysplastic squamous epithelium (D) demonstrates increased thickness of the basal layer with mild squamous atypia. The atypia is full thickness in the area of carcinoma in situ (CIS), which is contiguous with invasive tumor (INV).

anisocytosis usually increase. There is coarsening of the chromatin and nucleoli, nuclear angulations and folding may appear. In carcinoma *in situ*, there is minimal or no maturation from the base to the superficial aspect of the epithelium and the atypical nuclear features and mitoses are seen throughout the entire thickness of the epithelium.

Molecular Pathogenesis

Some investigators have evaluated SqCCs and precursor lesions for loss of heterozygosity (LOH) at multiple chromosomal regions frequently deleted in lung cancer and found multiple, sequentially occurring allele-specific molecular changes in separate, apparently clonally independent foci, early in the pathogenesis of SqCCs of the lung, suggesting a field cancerization effect. Clones of cells with allelic loss were noted in 31% of cases of histologically normal epithelium and 42% of specimens with hyperplasia or metaplasia, allelic losses at 3p and 9p occurred frequently and early in the evolution of the process, and *TP53* allelic loss was found in many histologically advanced lesions (dysplasia and CIS). The sequential molecular events leading to invasive SqCC are illustrated in Figure 18.9.

Neuroendocrine Neoplasms and Their Precursors

Clinical and Pathologic Features

Pulmonary neuroendocrine (NE) neoplasms include the high-grade neoplasms small cell carcinoma (SCLC) and large cell neuroendocrine carcinoma (LCNEC), the low-grade neoplasm typical carcinoid, and atypical carcinoid, which occupies in an intermediate position in the spectrum of biologic aggressiveness. 5-year and 10-year survival rates for typical carcinoid are 87% and 87%, 56%, and 35% for atypical carcinoid, 27% and 9% for LCNEC, and 9% and 5% for SCLC, respectively. These tumors display NE architectural features including organoid nesting, trabecular arrangements, rosette formation, and palisading. These patterns are more prominent in carcinoids than in LCNECs and may not be visible in some SCLCs. Typical carcinoids show less than two mitoses per $2 \, mm^2$ (10 HPF) and lack necrosis (Figure 18.10A), whereas atypical carcinoids have 2 to 10 mitoses per $2 \, mm^2$ (10 HPF) and minimal necrosis, which is often punctate. SCLC manifests itself as small undifferentiated tumor cells with scant cytoplasm and finely granular chromatin, and absent or inconspicuous nucleoli (Figure 18.11). Nuclear molding is characteristic for SCLC, necrosis is common, and the mitotic rate frequently exceeds 60 mitoses per $2 \, mm^2$. Combined SCLCs include an SCLC component and one or more histologic types of NSCLC. LCNECs consist of large tumor cells with varying degrees of NE architectural patterns, necrosis, a high mitotic rate (median of 70 per $2 \, mm^2$), and NE differentiation reflected in immunohistochemical staining for one or more NE markers (chromogranin A, synaptophysin, CD56, and/or CD57), or the presence of neurosecretory granules on ultrastructural examination.

Patients with carcinoids are typically younger and less likely to smoke than those with SCLCs and LCNECs, most of whom have a current or previous history of tobacco smoking. Rare patients with carcinoids have associated multiple endocrine neoplasia 1 (MEN1) syndrome or diffuse idiopathic pulmonary neuroendocrine cell hyperplasia (DIPNECH). DIPNECH is a diffuse proliferation of single cells, NE bodies, and linear proliferations of pulmonary NE cells

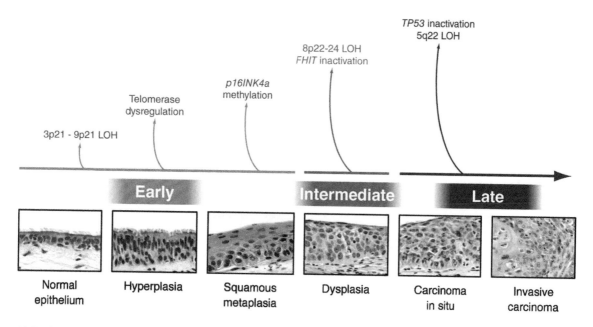

Figure 18.9   Histologic and molecular changes in the development of pulmonary squamous cell carcinoma. These changes occur in a stepwise fashion, beginning in histologically normal epithelium. LOH = loss of heterozygosity. Reprinted with kind permission from Wistuba, I.I. and Gazdar, A.F. (2006) Lung cancer preneoplasia. *Annu Rev Pathol* **1**, 331–348.

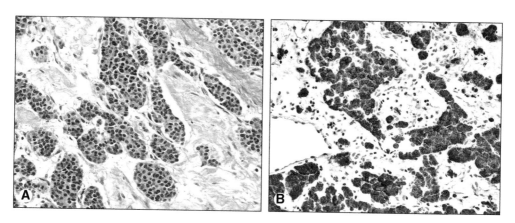

Figure 18.10   Typical carcinoid. (A) This tumor consists of nests of uniform tumor cells with round or ovoid nuclei with little cytoatypia, fine chromatin, and a moderate amount of cytoplasm. No necrosis or mitoses are observed. The stromal background is hyalinized. **(B)** An immunohistochemical stain for synaptophysin is positive (brown-stained cytoplasm), reflecting neuroendocrine differentiation.

that may reside in the bronchial or bronchiolar epithelia. DIPNECH may be accompanied by carcinoids and tumorlets, and the NE proliferations may represent precursors for carcinoids. However, precursor lesions for SCLC and LCNEC have not been identified.

Molecular Pathogenesis

Molecular markers expressed by all categories of pulmonary NE tumors include chromogranin A, synaptophysin (Figure 18.10B), and N-CAM (CD56). Overexpression of gastrin-releasing peptide, calcitonin, other peptide hormones, the insulinoma-associated 1 (*INSM1*) promoter and the human achaete-scute

homolog-1 (*hASH1*) gene have also been reported. Thyroid transcription factor–1 (TTF–1) is expressed by 80 to 90% of SCLCs, 30 to 50% of LCNECs and zero to 70% of carcinoids.

SCLCs are aneuploid tumors with high frequencies of deletions on chromosomes 3p (includes *ROBO1/ DUTT1, FHIT, RASSF1, β-catenin, Fus1, SEMA3B, SEMA3F, VHL,* and *RARβ*), 4q (includes the pro-apoptotic gene *MAPK10*), 5q, 10q (includes the pro-apoptotic gene *TNFRSF6*), 13q (*Rb*), and 17p (*TP53*), and gains on 3q, 5p, 6p, 8q, 17q, 19q, and 20q. More than 90% of SCLCs and SqCCs demonstrate large segments of allelic loss on chromosome 3p, in areas encompassing multiple candidate TSGs. *TP53* mutations are found in

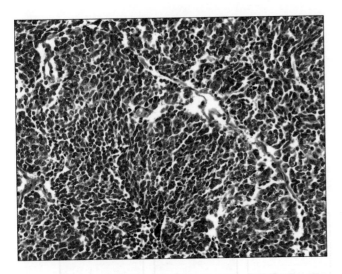

Figure 18.11   Small cell carcinoma. Small cell carcinomas typically display sheets of small tumor cells with scant cytoplasm and nuclei demonstrating fine chromatin. Numerous mitoses and apoptotic cells are characteristic.

80 to 90% of SCLCs, many LCNECs, fewer atypical carcinoids, and virtually no typical carcinoids. Dysregulation of *p53* produces downstream effects on Bcl-2 and Bax. Anti-apoptotic Bcl-2 predominates over pro-apoptotic Bax in the high-grade NE carcinomas, but the reverse is true for carcinoids. Alterations compromising the p16^{INK4a}/cyclin D1/Rb pathway of G1 arrest and leading to evasion of cell cycle arrest are frequent in high-grade pulmonary NE carcinomas (92%), but less common in atypical carcinoids (59%) and typical carcinoids. Mutations in the *RB1* gene exist in many SCLCs, with associated loss of function of the gene product. Overexpression of *MDM2* (a transcriptional target of *p53*) or p14^{ARF} loss in SCLC may also lead to evasion of cell cycle arrest through the *p53* and Rb pathway (Figure 18.2). Epigenetic inactivation of TSGs also occurs in NE tumors. *RASSF1A* is a potential TSG that undergoes epigenetic inactivation in virtually all SCLCs and many NSCLCs through hypermethylation of its promoter region. However, NE tumors have lower frequencies of methylation of *p16*, *APC*, and *CDH13* (H-cadherin) than NSCLCs. Promoter methylation of *CASP8*, which encodes the apoptosis-inducing cysteine protease caspase 8, was also found in 35% of SCLCs, 18% of carcinoids, and no NSCLCs, suggesting that *CASP8* may function as a TSG in NE lung tumors.

Genes frequently amplified in SCLCs include the oncogenes *MYC* (8q24), *MYCN* (2p24), and *MYCL1* (1p34), and the anti-apoptotic genes *TNFRSF4* (1p36), *DAD1* (14q11), *BCL2L1* (20q11), and *BCL2L2* (14q11). The Myc proteins are transcription factors that are important in cell cycle regulation, proliferation, and DNA synthesis, and can induce p14^{ARF}, leading to apoptosis through p53 if cellular conditions do not favor proliferation. Some typical and atypical carcinoids possess mutations of the *multiple endocrine*

neoplasia 1 (*MEN1*) gene on chromosome 11q13 or LOH at this locus, whereas these abnormalities occur with lower frequencies in SCLCs and LCNECs, supporting separate pathways of tumorigenesis. *MEN1* encodes for the nuclear protein menin, which is believed to play several roles in tumorigenesis by linking transcription factor function to histone-modification pathways, in part through interacting with the activator-protein-1-family transcription factor JunD.

Telomeres play an important role in the protection of chromosomes against degradation, and telomerases, the enzymes that synthesize telomeric DNA strands, counterbalance losses of DNA during cell divisions. High telomerase activity is noted in over 80% of SCLCs and LCNECs versus 14% or fewer typical carcinoids. Human telomerase mRNA component (hTERC) and human telomerase reverse transcriptase (hTERT) mRNA are expressed in most typical carcinoids, and virtually all atypical carcinoids, LCNECs and SCLCs.

Mesenchymal Neoplasms

Mesenchymal neoplasms (Table 18.1) are much less common in the lung than are epithelial neoplasms, and information about molecular pathogenesis is available for several of them. Pulmonary inflammatory myofibroblastic tumor (IMT) occurs primarily in children and younger adults, and is composed of myofibroblastic cells, collagen, and inflammatory cells (Figure 18.12A). Many IMTs demonstrate clonal abnormalities with rearrangements of chromosome 2p23 and the *anaplastic lymphoma kinase* (*ALK*) gene. The rearrangements involve fusion of tropomyosin (TPM) N-terminal coiled-coil domains to the ALK C-terminal kinase domain, producing two ALK fusion genes, *TPM4-ALK* and *TPM3-ALK*, which encode oncoproteins with constitutive kinase activity. Though typically a soft tissue malignancy, synovial sarcoma uncommonly arises in the pleura or lung and often takes an aggressive course. Like their soft tissue counterparts, more than 90% of pulmonary and pleural synovial sarcomas demonstrate a chromosomal translocation t(X;18)(*SYT-SSX*). Pulmonary hamartomas are benign neoplasms consisting of mixtures of cartilage, fat, connective tissue, and smooth muscle (Figure 18.12B), which present as so-called coin lesions on chest radiographs. Most of these tumors show abnormalities of chromosomal bands 6p21, 12q14-15, or other regions, corresponding to mutations of high-mobility group (HMG) proteins, a family of nonhistone chromatin-associated proteins that serve an important role in regulating chromatin architecture and gene expression.

Pleural Malignant Mesothelioma
Clinical and Pathologic Features

Malignant mesothelioma (MM) is an uncommon, aggressive tumor arising from mesothelial cells on serosal surfaces, primarily the pleura and peritoneum.

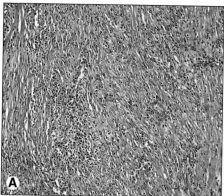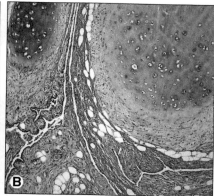

Figure 18.12   Mesenchymal neoplasms. (A) Inflammatory myofibroblastic tumor. The tumor consists of a proliferation of cytologically bland spindle cells in a background of collagen, with abundant lymphocytes and plasma cells. **(B)** Hamartoma. A hamartoma typically includes the components of mature cartilage, adipose tissue, and myxoid or fibrous tissue, as shown here.

Exposure to asbestos fibers known as amphiboles (crocidolite and amosite) is the most important risk factor for MM, and the neoplasm characteristically presents itself multiple decades after exposure. Smaller numbers of cases have been linked to radiation, a non-asbestos fiber known as erionite, and other processes associated with pleural scarring, and a role for simian virus 40 (SV40) in the genesis of this tumor remains controversial. Pleural MM usually arises in males over the age of 60. Hemorrhagic pleural effusion associated with shortness of breath, chest wall pain, weight loss, and malaise are common presenting features. From the time of diagnosis, the median survival is 12 months. Treatment may include surgery, chemotherapy, radiotherapy, immunotherapy, or other treatments, often in combination. Gross pathologic features of MM include firm tan pleural nodules that coalesce to fill the pleural cavity, form a thick rind around the lung, extend along the interlobar fissures, and invade the adjacent lung, diaphragm, and chest wall (Figure 18.13A). Further spread can occur into the pericardial cavity and around other mediastinal structures, and lymph node and distant metastases can also develop.

Histologically, MM can display a wide variety of patterns (Figure 18.13B, C). Epithelioid mesothelioma consists of round, ovoid, or polygonal cells with eosinophilic cytoplasm, and round nuclei with little cytoatypia. Tumor cells form sheets, tubulopapillary structures, or gland-like arrangements, and some neoplasms can have a myxoid appearance due to production of large amounts of hyaluronate. Sarcomatoid mesothelioma is composed of malignant-appearing spindle cells occasionally accompanied by mature sarcomatous components (osteosarcoma, chondrosarcoma, others). Desmoplastic mesothelioma consists of variably atypical spindle cells in a dense collagenous matrix and can be a diagnostic challenge due to its bland appearance and resemblance to organizing pleuritis. In biphasic mesotheliomas, epithelioid and sarcomatoid elements each comprise at least 10% of the tumor.

Immunohistochemistry is extremely helpful for the diagnosis of MM and for distinction from other malignancies. Immunoreactivity with cytokeratin 5/6, calretinin (Figure 18.13D), HBME-1, D2-40, and other antibodies supports mesothelial differentiation. Since AC is an important entity in the differential diagnosis of epithelioid mesothelioma, a panel approach using calretinin and cytokeratin 5/6, with other antibodies reactive with ACs (CEA, MOC-31, Ber-EP4, leu M1, B72.3, and others) often is employed. Electron microscopy can demonstrate long thin microvilli in many MMs with an epithelioid component, assisting in diagnosis. Pan-cytokeratin staining is helpful for supporting a diagnosis of sarcomatoid or desmoplastic MM, as opposed to sarcoma, since most (but not all) sarcomas will not stain for pan-cytokeratin. Other mesenchymal markers can also be selected for assisting in the differentiation of MM from histologically similar sarcomas.

Molecular Pathogenesis

In most patients, exposure to asbestos fibers is believed to trigger the pathobiological changes leading to MM. Currently, it is believed that asbestos may act as an initiator (genetically) and promoter (epigenetically) in the development of MMs. In MM, a variety of genetic abnormalities have been reported including deletions of 1p21-22, 3p21, 4p, 4q, 6q, 9p21, 13q13-14, 14q and proximal 15q, monosomy 22, and gains of 1q, 5p, 7p, 8q22-24, and 15q22-25. The most common genetic abnormality in MM is a deletion in 9p21, the locus of the TSG *CDKN2A* encoding the tumor suppressors $p16^{INK4a}$ and $p14^{ARF}$, which participate in the *p53* and Rb pathways and inhibit cell cycle progression (Figure 18.2). Also, SV40 large T antigen, present in some MMs, inactivates the TSG products Rb and *p53*, raising the possibility that asbestos and SV40 could act as cocarcinogens in MM and suggesting that perturbations of Rb-dependent and *p53*-dependent growth-regulatory pathways may be involved in the pathogenesis of MM. Other common findings include

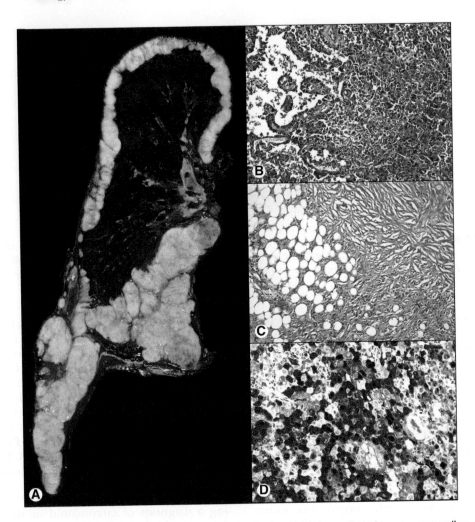

Figure 18.13 Malignant mesothelioma. (A) The tan/white tumor involves the entire pleura, surrounding and compressing the underlying parenchyma. **(B)** Malignant mesothelioma, epithelioid. This neoplasm consists of sheets of polygonal cells with pleomorphic nuclei and also forms some papillary structures (left). **(C)** Malignant mesothelioma, desmoplastic. Abundant dense collagen is characteristic of this tumor, and is shown in the upper right. Tumor cells are spindle shaped and relatively cytologically bland. The tumor infiltrates adipose tissue, which is helpful in confirming that the tumor is a mesothelioma, as opposed to organizing pleuritis. **(D)** Immunohistochemical stain for calretinin. The tumor demonstrates cytoplasmic and nuclear staining (brown) for calretinin, which is expressed by many epithelioid malignant mesotheliomas.

inactivating mutations with allelic loss in the TSG neurofibromin 2 (*NF2*), and inactivation of *CDKN2A/p14^{ARF}* and *GPC3* (another TSG) by promoter methylation. Loss of *CDKN2A/p14^{ARF}* also results in MDM2-mediated inactivation of *p53*. However, unlike many other epithelial tumors, mutations in the *TP53*, *RB*, and *RAS* genes are rare in MMs.

Evidence also suggests that the Wnt signal transduction pathway and the phosphatidylinositol 3-kinase (PI3-K/AKT) pathway are frequently activated in MMs. Wnt pathway activation leads to cytoplasmic accumulation of β-catenin and its translocation to the nucleus, apparently through involvement of upstream components of the pathway such as the disheveled proteins. Interactions with TCF/LEF transcription factors promote expression of multiple genes including C-*MYC* and *Cyclin D*. The Wilms' tumor gene (*WT1*) also is expressed in most MMs, but its role in the pathogenesis of MM is unclear. Finally, EGFR signaling

in MMs recently has been investigated and it appears that common EGFR mutations conferring sensitivity to gefitinib are not prevalent in human MM.

OBSTRUCTIVE LUNG DISEASES

Obstructive lung diseases are characterized by a reduction in airflow due to airway narrowing. This airflow reduction occurs, in general, by two basic mechanisms: (1) inflammation and injury of the airway resulting in obstruction by mucous and cellular debris within and around the airway lumen, and (2) destruction of the elastin fibers of the alveolar walls causing loss of elastic recoil and subsequent premature collapse of the airway during the expiratory phase of respiration. The major obstructive lung diseases are asthma, emphysema, chronic bronchitis, and bronchiectasis.

Asthma

Clinical and Pathologic Features

Asthma is a chronic inflammatory disease of the airways that affects more than 150 million people worldwide. In the United States, asthma affects approximately 8 to 10% of the population and is the leading cause of hospitalization among children less than 15 years of age. Clinically, the disease is defined as a generalized obstruction of airflow with a reversibility that can occur spontaneously or with therapy. It is characterized by recurrent wheezing, cough, or shortness of breath resulting from airway hyperactivity and mucous hypersecretion. The hyperresponsiveness is a result of acute bronchospasm and can be elicited for diagnostic purposes using histamine or methacholine challenges. The key feature of these symptoms is that they are variable—worse at night or in the early morning, and in some people worse after exercise. It has previously been assumed that these symptoms are separated by intervals of normal physiology. However, evidence is now accumulating that asthma can cause progressive lung impairment due to chronic morphologic changes in the airways.

The pathologic changes to the airways in asthma consist of a thickened basement membrane with epithelial desquamation, goblet cell hyperplasia, and subepithelial elastin deposition. In the wall of the airway, smooth muscle hypertrophy and submucosal gland hyperplasia are also present (Figure 18.14). In acute asthma exacerbations, a transmural chronic inflammatory infiltrate with variable amounts of eosinophilia may be present, resulting in epithelial injury and desquamation that can become quite pronounced. One sees clumps of degenerating epithelial cells mixed with mucin in the lumen airway. These aggregates of degenerating cells are referred to as Creola bodies and can be seen in expectorated mucin from these patients. Also present in these sputum samples are Charcot-Leyden crystals, rhomboid-shaped structures that represent breakdown products from eosinophil cytoplasmic granules (Figure 18.15). The changes seen in the walls of these airways represent long-term airway remodeling caused by prolonged inflammation. This remodeling may play a role in the pathophysiology of asthma. The amount of airway remodeling is highly variable from patient to patient, but remodeling has been found even in patients with mild asthma. Currently, the effect of the treatment on this chronic pathology is unclear.

Molecular Pathogenesis

The pathogenesis of asthma is complex and most likely involves both genetic and environmental components. Most experts now see it as a disease in which an insult initiates a series of events in a genetically susceptible host. No single gene accounts for the familial component of this disease. Genetic analysis of these patients reveals a prevalence of specific HLA alleles, polymorphisms of *FceRiB*, *IL-4*, and *CD14*. Asthma can be classified using a number of different schema. Most commonly it

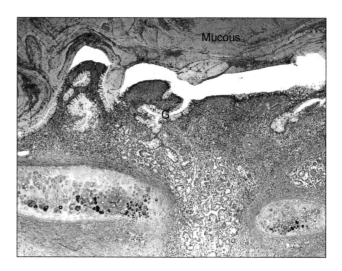

Figure 18.14   Asthma. The bronchial wall from a patient with asthma shows marked inflammation with eosinophils, mucosal goblet cell hyperplasia (G), and an increase in the smooth muscle.

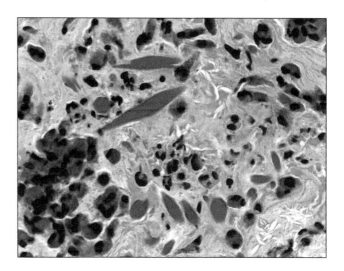

Figure 18.15   Asthma. Charcot-Leyden crystals are rhomboid-shaped structures within a mucous plug from an airway of an asthmatic patient. In addition, there are abundant eosinophils. These crystals are made of breakdown products of eosinophils, including major basic protein.

is divided into two categories: atopic (allergic) and nonatopic (nonallergic). Atopic asthma results from an allergic sensitization usually early in life and has its onset in early childhood. Nonatopic asthma is late-onset and, though the immunopathology has not been as well studied, probably has similar mechanisms to atopic asthma. Although this nosology is convenient for purposes of understanding the mechanisms of the disease, most patients manifest a combination of these two categories with overlapping symptoms.

The pathogenetic mechanisms of both types encompass a variety of cells and their products.

These include airway epithelium, smooth muscle cells, fibroblasts, mast cells, eosinophils, and T-cells. The asthma response includes two phases: an early response comprising an acute bronchospastic event within 15 to 30 minutes after exposure, and a late response that peaks approximately 4 to 6 hours and that can have prolonged effects. To understand this complex response, it is best to divide it into three components: (1) a type 1 hypersensitivity response, (2) acute and chronic inflammation, and (3) bronchial hyperactivity.

Type 1 Hypersensitivity In general, human asthma is associated with a predominance of type 2 helper cells with a CD4+ phenotype. These Th2-type cells result from the uptake and processing of viral, allergen, and environmental triggers that initiate the episode. The processing includes the presentation of these triggers by the airway dendritic cells to naive T cells (Th0), resulting in their differentiation into populations of Th1 and Th2. The Th2 differentiation is a result of IL-10 release by the dendritic cells and the Th2 cells then further propagate the inflammatory reaction in two ways. First, they release a variety of cytokines such as IL-4, IL-5, and IL-13 that mediate a wide variety of responses. IL-4 and IL-13 stimulate B-cells and plasma cells to produce IgE that in turn stimulates mast cell maturation and the release of multiple mediators, including histamine and leukotrienes. Second, these Th2 cells secrete IL-5 that, together with IL-4, also stimulates mast cells to secrete histamine, tryptase, chymase, and the cysteinyl leukotrienes causing the bronchoconstrictor response that occurs rapidly after the exposure to the allergen. IL-5 from these lymphocytes also recruits eosinophils to the airways and stimulates the release of the contents of their granules, including eosinophil cationic protein (ECP), major basic protein (MBP), eosinophil peroxidase, and eosinophil-derived neurotoxin. These compounds not only induce the bronchial wall hyperactivity but also are responsible for the increased vascular permeability that produces the transmural edema in the airways.

Th0 cells can also differentiate into Th1 cells as a result of IL-12 also produced by dendritic cells. These Th1 cells produce interferon-γ (IFN-γ), IL-2, and lymphotoxin, which play a role in macrophage activation in delayed-type hypersensitivity reactions as is seen in such diseases as rheumatoid arthritis and tuberculosis. These Th1 cells are predominantly responsible for defense against intracellular organisms and are more prominent in normal airways and in airways of patients with emphysema than in asthmatics. However, in severe forms of asthma, Th1 cells are recruited and have the capacity to secrete tumor necrosis factor α (TNF-α) and IFN-γ, which may lead to the tissue-damaging immune response one sees in these airways.

Acute and Chronic Inflammation The role of acute and chronic inflammatory cells, including eosinophils, mast cells, macrophages, and lymphocytes, in asthma is evident in the abundance of these cells in airways, sputum, and bronchoalveolar samples from patients with this disease. The number of eosinophils in the airways correlates with the severity of asthma and the amount of bronchial hyperresponsiveness. Proteins released by these cells including ECP, MCP, and eosinophil-derived neurotoxin cause at least some of the epithelial damage seen in the active form of asthma. Neutrophils are prominent in the more acute exacerbations of asthma and probably are recruited to these airways by IL-8, a potent neutrophil chemoattractant released by airway epithelial cells. These cells also release proteases, reactive oxygen species (ROS), and other proinflammatory mediators that, in addition to the epithelial damage, also contribute to the airway destruction and remodeling that occur in the more chronic forms of this disease. The susceptibility of the epithelium in asthma to this oxidant injury may be increased due to decreased antioxidants such as superoxide dismutase in these lungs. Finally, mast cells are activated to release an abundance of mediators through the binding of IgE to FcεRI, high-affinity receptors on their surface. Allergens bind to IgE molecules and induce a cross-linking of these molecules leading to activation of the mast cell and release of a number of mediators, most notably histamine, tryptase, and various leukotrienes, including leukotriene D_4 (LTD_4), and interact with the smooth muscle to induce contraction and the acute bronchospastic response.

Bronchial Hyperactivity The cornerstone of asthma is the hyperactive response of the airway smooth muscle. The mechanism by which this occurs combines neural pathways and inflammatory pathways. As stated, the inflammatory component of this response comes predominantly from the mast cells. The major neural pathway involved is the nonadrenergic noncholinergic (NANC) system. Although cholinergic pathways are responsible for maintaining the airway smooth muscle tone, it is the NANC system that releases bronchoactive tachykinins (substance P and neurokinin A) that bind to NK2 receptors on the smooth muscle and cause the constriction that characterizes the acute asthmatic response.

In addition to these acute mechanisms, the airway also undergoes structural alterations to its formed elements. In the mucosa, these changes include goblet cell hyperplasia and basement membrane thickening. Within the submucosa and airway wall, increased deposition of collagen and elastic fibers results in fibrosis and elastosis and both the smooth muscle cells and the submucosal glands undergo hypertrophy and hyperplasia. These irreversible changes are a consequence of chronic inflammatory insults on the airways through mechanisms that include release of fibrosing mediators such as transforming growth factor-β (TGFβ) and mitogenic mediators such as epidermal and fibroblast growth factors (EGF, FGF). The exact mechanisms by which this occurs are not clearly defined, but the similarity of these factors with those involved in branching morphogenesis of the developing lung has led to a focus on the effect of inflammation on the interaction of the epithelium with the underlying mesenchymal cells.

Chronic Obstructive Pulmonary Disease (COPD)—Emphysema

Clinical and Pathologic Features

The term chronic obstructive pulmonary disease (COPD) applies to emphysema, chronic bronchitis, and bronchiectasis, those diseases in which airflow limitation is usually progressive, but, unlike asthma, not fully reversible. The prevalence of COPD worldwide is estimated at 9 to 10% in adults over the age of 40. Though there are different forms of COPD with different etiologies, the clinical manifestations of the most common forms of the disease are the same. These include a progressive decline in lung function, usually measured as decreased forced expiratory flow in 1 second (FEV1), a chronic cough, and dyspnea. Emphysema and chronic bronchitis are the most common diseases of COPD and are the result of cigarette smoking. As such, they usually exist together in most smokers. Chronic bronchitis is defined clinically as a persistent cough with sputum production for at least three months in at least two consecutive years without any other identifiable cause. Patients with chronic bronchitis typically have copious sputum with a prominent cough, more commonly get infections, and typically experience hypercapnia and severe hypoxemia. Emphysema is the destruction and permanent enlargement of the air spaces distal to the terminal bronchioles without obvious fibrosis. These patients have only a slight cough, while the overinflation of the lungs is severe.

The pathologic features of COPD are best understood if one considers the whole of COPD as a spectrum of pathology that consists of emphysematous tissue destruction, airway inflammation, remodeling, and obstruction. The lungs of patients with COPD usually contain all of these features, but in varying proportions. The pathologic features of chronic bronchitis include mucosal pathology that consists of epithelial inflammation, injury, and regenerative epithelial changes of squamous and goblet cell metaplasia. In addition, the submucosa shows changes of remodeling with smooth muscle hypertrophy and submucosal gland hyperplasia. These changes are responsible for the copious secretions characteristic of this clinical disease, although studies have reported no consistent relationship between these pathologic features of the large airways and the airflow obstruction.

The pathology definition of emphysema is an abnormal, permanent enlargement of the airspaces distal to the terminal bronchioles accompanied by destruction of the alveolar walls without fibrosis. The four major pathologic patterns of emphysema are defined by the location of this destruction. These include centriacinar, panacinar, paraseptal, and irregular

emphysema. The first two of these are responsible for the overwhelming majority of the clinical disease. Centriacinar emphysema (sometimes referred to as centrilobular) represents 95% of the cases and is a result of destruction of alveoli at the proximal and central areas of the pulmonary acinus, including the respiratory bronchioles (Figure 18.16B). It predominantly affects the upper lobes (Figure 18.16A). Panacinar emphysema, usually associated with α1-antitrypsin (αAT) deficiency, results in a destruction of the entire pulmonary acinus from the proximal respiratory bronchioles to the distal area of the acinus, and affects predominantly the lower lobes (Figure 18.17). The remaining two types of emphysema, paraseptal and irregular, rarely are associated with clinical disease. In paraseptal emphysema, the damage is to the distal acinus, the area that abuts the pleura at the margins of the lobules. Damage in this area may cause spontaneous pneumothoraces, typically in young, thin men. Irregular emphysema is tissue destruction and alveolar enlargement that occurs adjacent to scarring, secondary to the enhanced inflammation in the area. Though this is a common finding in a scarred lung, it is of little if any clinical significance to the patient.

Though the emphysema in these lungs plays the dominant role in causing the obstruction, small airway pathology is also present. Respiratory bronchiolitis refers to the inflammatory changes found in the distal airways of smokers. These consist of pigmented macrophages filling the lumen and the peribronchiolar airspaces and mild chronic inflammation and fibrosis around the bronchioles. The pigment in these macrophages represents the inhaled particulate matter of the cigarette smoke that has been phagocytized by these cells. The macrophages in turn release proteases, which destroy the elastic fibers in the surrounding area resulting in the loss of elastic recoil and the obstructive symptoms.

Molecular Pathogenesis

In general, COPD is a result of inflammation of the large airways that produces the airway remodeling characteristic of chronic bronchitis as well as inflammation of the smaller airways that results in the destruction of the adjacent tissue and consequent emphysema. The predominant inflammatory cells involved in this process are the alveolar macrophages, neutrophils, and lymphocytes. The main theories of the pathogenesis of COPD support the interaction of airway inflammation with two main systems in the lung: the protease–antiprotease system and the oxidant–antioxidant system. These systems help to protect the lung from the many irritants that enter the lung via the large pulmonary surface area that interfaces with the environment.

In the protease–antiprotease system, proteases are produced by a number of cells, including epithelial cells and inflammatory cells that degrade the underlying lung matrix. The most important proteases in the lung are the neutrophil elastases, part of the serine protease family, and the metalloproteinases (MMPs)

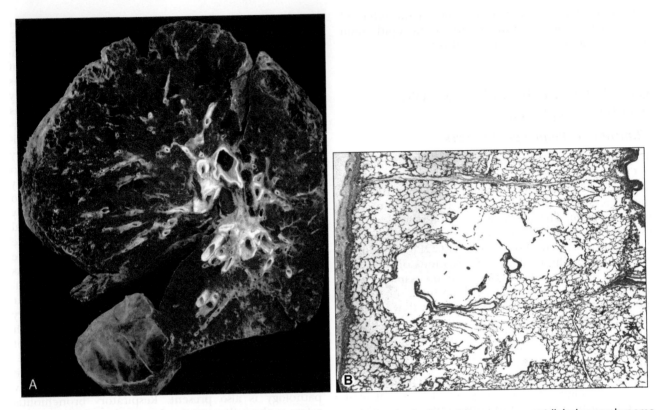

Figure 18.16 Centrilobular emphysema. (A) This sagittal cut section of a lung contains severe centrilobular emphysema with significant tissue destruction in the upper lobe and bulla forming in the upper and lower lobes. **(B)** Tissue destruction in the central area of the pulmonary lobule is demonstrated in this lung with a mild centrilobular emphysema. The pattern of tissue destruction is in the area surrounding the small airway where pigmented macrophages release proteases in response to the cigarette smoke.

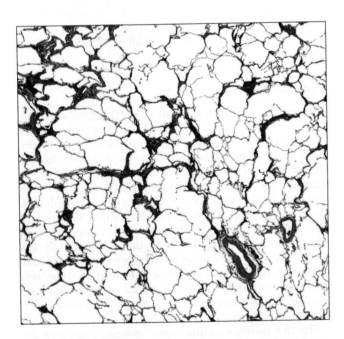

Figure 18.17 Panacinar emphysema. Tissue destruction in panacinar emphysema occurs throughout the lobule producing a diffuse loss of alveolar walls unlike that of centrilobular emphysema (see Figure 18.16) with more irregular holes in the tissue.

produced predominantly by macrophages. These proteases can be secreted in response to invasion by environmental irritants, most notably infectious agents such as bacteria. In this setting, their role is to enzymatically degrade the organism. However, proteases can also be secreted by both inflammatory and epithelial cells in a normal lung to repair and maintain the underlying lung matrix proteins. To protect the lung from unwanted destruction by these enzymes, the liver secretes antiproteases that circulate in the bloodstream to the lung and inhibit the action of the proteases. In addition, macrophages that secrete MMPs also secrete tissue inhibitors of metalloproteinases (TIMPs). A delicate balance of proteases and antiproteases is needed to maintain the integrity of the lung structure. An imbalance that results in a relative excess of proteases (either by overproduction of proteases or underproduction of their inhibitors) leads to tissue destruction and the formation of emphysema.

This imbalance occurs in different ways in the two major types of emphysema, centriacinar and panacinar. In centriacinar emphysema, caused primarily by cigarette smoking, there is an overproduction of proteases primarily due to the stimulatory effect of chemicals within the smoke on the neutrophils and macrophages. Though the exact mechanism is not completely understood, most studies support that nicotine from the cigarette smoke acts as a chemoattractant and ROS (also contained in the smoke) stimulates an increased release of neutrophil elastases and MMPs from activated macrophages, leading to the destruction of the elastin in the alveolar spaces. This inflammatory cell activation may come about through the activation of the transcription factor NF-κB that leads to TNF-α production. In addition, the elastin peptides themselves may attract additional inflammatory cells to further increase the protease secretion and exacerbate the matrix destruction.

Unlike centriacinar emphysema, panacinar emphysema most commonly is caused by a genetic deficiency of antiproteases, usually due to αAT deficiency, a condition that affects approximately 75,000 people in the United States. αAT deficiency is due to a defect in the gene that encodes the protein α1-antitrypsin, a glycoprotein produced by hepatocytes and the main inhibitor of neutrophil elastase. The affected gene is the SERPINA1 gene (formerly known as P1), located on the long arm of chromosome 14 (14q31-32.3). The genetic mutations that occur have been categorized into four groups: base substitution, in-frame deletions, frame-shift mutations, and exon deletions. These mutations usually result in misfolding, polymerization, and retention of the aberrant protein within the hepatocytes, leading to decreased circulating levels. αAT deficiency is an autosomal codominant disease with over 100 allelic variants, of which the M alleles (M1-M6) are the most common. These alleles produce normal serum levels of a less active protein. Individuals who manifest the lung disease are usually homozygous for the alleles Z or S (ZZ and SS phenotype) or heterozygous for the 2 M alleles (MZ or SZ phenotype). An αAT concentration in plasma of less than 40% of normal confers a risk for emphysema. In individuals with the ZZ genotype, the activity of αAT is approximately one fifth of normal.

The second system in the lung involved in the pathogenesis of emphysema is the oxidant–antioxidant system. As in the protease system, the lung is protected from oxidative stress in the form of ROS by antioxidants produced by cells in the lung. ROS in the lung include oxygen ions, free radicals, and peroxides. The major antioxidants in the airways are enzymes including catalase, superoxide dismutase (SOD), glutathione peroxidase, glutathione S-transferase, xanthine oxidase, and thioredoxin, as well as nonenzymatic antioxidants including glutathione, ascorbate, urate, and bilirubin. The balance of oxidants and antioxidants in the lung prevents damage by ROS. However, cigarette smoke increases the production of ROS by neutrophils, eosinophils, macrophages, and epithelial cells. Evidence that damage to the lung epithelium and matrix is a direct result of ROS includes the presence of exhaled H_2O_2 and 8-isoprostane, decreased plasma antioxidants, and increased plasma and tissue levels of oxidized proteins, including various lipid peroxidation products. In addition to this direct effect, ROS may also induce a pro-inflammatory response that recruits more inflammatory cells to the lung. In animal models, cigarette smoke induces

the expression of proinflammatory cytokines such as IL-6, IL-8, TNF-α, and IL-1 from macrophages, epithelial cells, and fibroblasts, perhaps through activation of the transcription factor NF-κB. Finally, there is some evidence that cigarette smoke further disturbs the oxidant–antioxidant balance in the lung by depleting antioxidants such as ascorbate and glutathione.

Bronchiectasis

Clinical and Pathologic Features

Bronchiectasis represents the permanent remodeling and dilatation of the large airways of the lung most commonly due to chronic inflammation and recurrent pneumonia. These infections usually occur because airway secretions and entrapped organisms cannot be effectively cleared. This pathology dictates the clinical features of the disease, which include chronic cough with copious secretions and a history of recurrent pneumonia. The five major causes of bronchiectasis are infection, obstruction, impaired mucociliary defenses, impaired systemic immune defenses, and congenital. These may produce either a localized or diffuse form of the disease. Localized bronchiectasis is usually due to obstruction of airways by mass lesions or scars from previous injury or infection. Diffuse bronchiectasis can result from defects in systemic immune defenses in which either innate or adaptive immunity may be impaired. Diseases due to the former include chronic granulomatous disease (CGD) and diseases due to the latter include agammaglobulinemia/hypogammaglobulinemia and severe combined immune deficiencies. Defects in the mucociliary defense mechanism, responsible for physically clearing organisms from the lung, may also cause diffuse bronchiectasis. These include ciliary dyskinesias that result in cilia with aberrant ultrastructure and cystic fibrosis (CF). Congenital forms of bronchiectasis are rare, but do exist. The most common include Mounier-Kuhns's syndrome and Williams-Campbell syndrome, the former causing enlargement of the trachea and major bronchi due to loss of bronchial cartilage and the latter causing diffuse bronchiectasis of the major airways probably due to a genetic defect in the connective tissue.

The pathology of bronchiectasis is most dramatically seen at the gross level. One can see dilated airways containing copious amounts of infected secretions and mucous plugs localized either to a segment of the lung or diffusely involving the entire lung as in CF (Figure 18.18). Microscopic features include chronic inflammatory changes similar to those of chronic bronchitis but with ulceration of the mucosa and submucosa leading to destruction of the smooth muscle and elastic in the airway wall and the characteristic dilatation and fibrosis. These enlarged airways contain mucous plugs comprising mucin and abundant degenerating inflammatory cells, a result of infections that establish themselves in these airways following the loss of the mucociliary defense

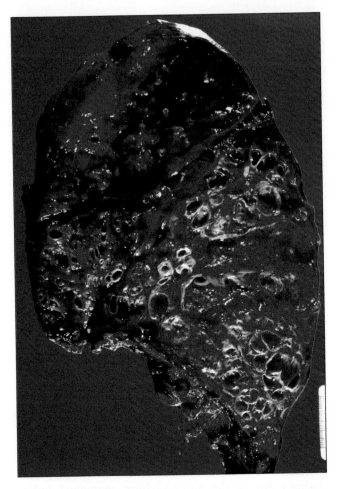

Figure 18.18   Cystic fibrosis. This sagittal cut section of a lung from a patient with cystic fibrosis demonstrates a diffuse bronchiectasis illustrated by enlarged, cyst-like airways. This is the typical pathology for cystic fibrosis. The remainder of the lung contains some red areas of congestion.

mechanism. Bacteria may be found in these plugs, most notably *P. aeruginosa*.

Molecular Pathogenesis

The pathogenetic mechanism of bronchiectasis is complex and depends upon the underlying etiology. In general, the initial damage to the bronchial epithelium is due to aberrant mucin (cystic fibrosis), dysfunctional cilia (ciliary dyskinesias), and ineffective immune surveillance (defects in innate and antibody-mediated immunity), leading to a cycle of tissue injury, repair and remodeling that ultimately destroys the normal airway. The initial event in this cycle usually involves dysfunction of the mucociliary mechanism that inhibits the expulsion from the lungs of organisms and other foreign substances that invade the airways. This may be due to defects in the cilia or the mucin. Ciliary defects are found in primary ciliary dyskinesia, a genetically heterogeneous disorder, usually inherited as an autosomal recessive trait that produces immotile cilia with clinical manifestations in the lungs,

sinuses, middle ear, male fertility, and organ lateralization. Over 250 proteins make up the axoneme of the cilia, but mutations in two genes, *DNAI1* and *DNAH5*, which encode for proteins in the outer dynein arms, most frequently cause this disorder. In CF the main defect affects the mucin. In patients with this autosomal recessive condition, there is a low volume of airway surface liquid (ASL) causing sticky mucin that inhibits normal ciliary motion and effective mucociliary clearance of organisms. This is due to a defect in the *cystic fibrosis transmembrane conductance regulator* (*CFTR*) gene located on chromosome 7, which encodes a cAMP-activated channel that regulates the flow of chloride ions in and out of cells and intracellular vacuoles, helping to maintain the osmolality of the mucin. This protein is present predominantly on the apical membrane of the airway epithelial cells though it is also involved in subapical, intracellular trafficking and recycling during the course of its maturation within these cells. This genetic disease manifests in multiple other organs that depend upon chloride ion transport to maintain normal secretions, including the pancreas, intestine, liver, reproductive organs, and sweat glands.

The genetic mutations in CF influence the CFTR trafficking in the distal compartments of the protein secretary pathway, and various genetic mutations produce different clinical phenotypes of the disease. Over 1600 mutations of the *CFTR* gene have been found. However, only four of these mutations occur at a frequency of greater than 1%. These mutations are grouped into five classes according to their functional deficit:

Group I: CFTR is not synthesized
Group II: CFTR is inadequately processed
Group III: CFTR is not regulated
Group IV: CFTR shows abnormal conductance
Group V: CFTR has partially defective production or processing

Approximately 70% of CF patients are in Group II and have the same mutation, F508Δ CFTR, a deletion of phenylalanine at codon 508. In these patients, most of the CFTR protein is misfolded and undergoes premature degradation within the endoplasmic reticulum, though a small amount of the CFTR protein is present on the apical membrane and does function normally. CF patients may have a combination of genetic mutations from any of the five groups. However, those patients with the most severe disease involving both the lungs and pancreas usually carry at least two mutations from Group I, II, or III.

Systemic immune deficiencies cause bronchiectasis through the establishment of persistent infection and inflammation. There are four major categories of immune deficiencies. The first category consists of a number of genetic diseases that cause either agammaglobulinemia or hypogammaglobulinemia. These include X-linked agammaglobulinemia (XLA) and common variable immunodeficiency (CVI). XLA is caused by a mutation of the *Bruton's tyrosine kinase* (*BTK*) gene that results in the virtual absence of all immunoglobulin isotypes and of circulating B lymphocytes. In CVI there is a marked reduction in IgG and IgA and/or IgM, associated with defective antibody response to protein and polysaccharide antigens. As expected, both of these diseases increase susceptibility to infections from encapsulated bacteria. The second category of immune deficiency is hyper-IgE syndrome, a disease with markedly elevated serum IgE levels that is characterized by recurrent staphylococcal infections. The third category is chronic granulomatous disease (CGD), a genetically heterogeneous group of disorders that have a defective phagocytic respiratory burst and superoxide production, inhibiting the ability to kill *Staphylococcus* spp. and fungi such as *Aspergillus* spp. Finally, severe combined immune deficiency (SCID) comprises a group of disorders with abnormal T-cell development and B-cell and/or natural killer cell maturation and function, predisposing these patients to *Pneumocystis jiroveci* and viral infections.

INTERSTITIAL LUNG DISEASES

Idiopathic Interstitial Pneumonias—Usual Interstitial Pneumonia

Clinical and Pathology Features

The idiopathic interstitial pneumonias (IIPs) comprise a group of diffuse infiltrative pulmonary diseases with a similar clinical presentation characterized by dyspnea, restrictive physiology, and bilateral interstitial infiltrates on chest radiography. Pathologically, these diseases have characteristic patterns of tissue injury with chronic inflammation and varying amounts of fibrosis. By recognizing these patterns, a pathologist can classify each of these entities and predict prognosis. However, the pathologist cannot establish the etiology, since these pathologic patterns can be seen in multiple clinical settings.

The pathologic classification of these diseases, originally defined by Liebow and Carrington in 1969, has undergone important revisions over the past 35 years with the latest revision by the *American Thoracic Society/European Respiratory Society* in 2003. The best-known and most prevalent entity of the IIPs is idiopathic pulmonary fibrosis (IPF), which is known pathologically as usual interstitial pneumonia (UIP). UIP is a histologic pattern characterized by patchy areas of chronic lymphocytic inflammation with organizing and collagenous-type fibrosis. These patients usually present with gradually increasing shortness of breath and a nonproductive cough after having had symptoms for many months or even years. Imaging studies usually reveal bilateral, basilar disease with a reticular pattern. Therapy begins with corticosteroids, advancing to more cytotoxic drugs such as methotrexate and cytoxan, but most current therapies are not effective in stopping the progression of the disease. The current estimates are that 20/100,000 males and 113/100,000 females have the disease, most of whom progress to respiratory failure and death within 5 years.

The pathology is characterized by a leading edge of chronic inflammation with fibroblastic foci that begin in different areas of the lung at different times. These processes produce a variegated pattern of fibrosis, usually referred to as a temporally heterogeneous pattern of injury. Because it occurs predominantly in the periphery of the lung involving the subpleura and interlobular septae, the gross picture is one of more advanced peripheral and basilar disease (Figure 18.19A). The progression from inflammation to fibrosis includes interstitial widening, epithelial injury and sloughing, fibroblastic infiltration, and organizing fibrosis within the characteristic fibroblastic foci (Figure 18.20B). Deposition of collagen by fibroblasts occurs in the latter stages of repair. The presence of the abundant collagen produces stiff lungs that are unable to clear the airway secretions, leading to recurrent inflammation of the bronchiolar epithelium with eventual fibrosis and breakdown of the airway structure. This remodeling produces mucous-filled ectatic spaces giving rise to the gross picture of honeycomb spaces, which is seen in the advanced pathology (Figure 18.19B).

Molecular Pathogenesis

Theories of the pathogenesis of idiopathic pulmonary fibrosis have evolved over the past decade. Early theories favored a primary inflammatory process, whereas current theories favor the concept that the fibrosis of the lung proceeds independently of inflammatory events and develops from aberrant epithelial and epithelial-mesenchymal responses to injury to the alveolar epithelial cells (AECs). The AECs consist of two populations: the type 1 pneumocytes and the type 2 pneumocytes. In normal lungs, type 1 pneumocytes line 95% of the alveolar wall and type 2 pneumocytes line the remaining 5%. However, in lung injury, the type 1 cells, which are exquisitely fragile, undergo cell death and the type 2 pneumocytes serve as progenitor cells to regenerate the alveolar epithelium. Though some studies have suggested that repopulation of the type 2 cells depends on circulating stem cells, this concept remains to be fully proven. According to current concepts, the injury and/or apoptosis of the AECs initiates a cascade of cellular events that produce the scarring in these lungs. Studies of AECs in lungs from patients with IPF have shown ultrastructural evidence of cell injury and apoptosis as well as expression of proapoptotic proteins. Further, inhibition of this apoptosis by blocking a variety of proapoptotic mechanisms such the Fas-Fas ligand pathway, angiotensin and TNF-α production, and caspase activation can stop the progression of this fibrosis.

The result of the AEC injury is the migration, proliferation, and activation of the fibroblasts and myofibroblasts that lead to the formation of the characteristic fibroblastic foci of the UIP pathology and the deposition and accumulation of collagen and elastic fibers in the alveoli (Figure 18.20). This unique pathology may be a result of the increased production of profibrotic factors such as transforming growth factor-α

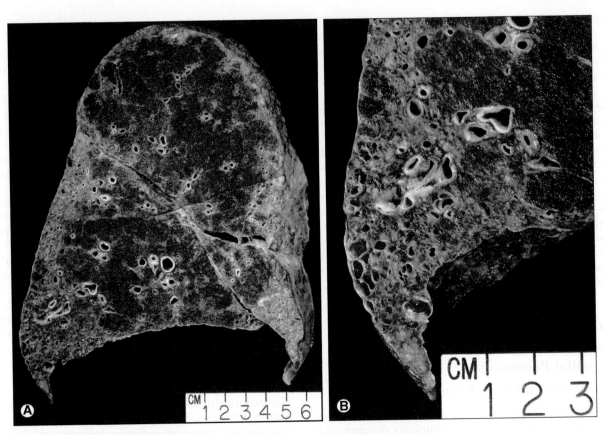

Figure 18.19   Usual interstitial pneumonia. (A) A sagittal cut of a lung involved by usual interstitial pneumonia reveals the peripheral and basilar predominance of the dense, white fibrosis. **(B)** A higher power view of the left lower lobe highlights the remodeled honeycomb spaces in the area of the lung with the end-stage disease.

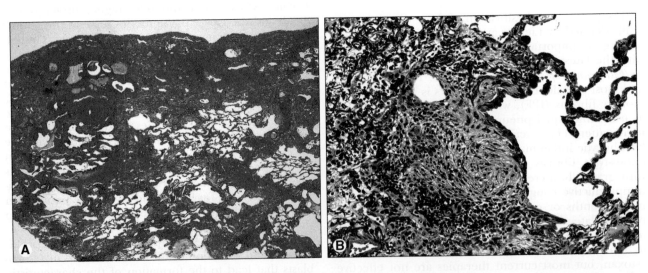

Figure 18.20   Usual interstitial pneumonia. (A) The microscopic features of UIP lungs are characterized by inflammation and fibrosis that demonstrate the temporally heterogeneous pattern of pathologic injury with normal, inflamed, and fibrotic areas of the lung all seen at a single lower power view. **(B)** The leading edge of inflammation is represented by deposition of new collagen in fibroblastic foci. These consist of fibroblasts surrounded by collagen containing mucopolysaccharides highlighted in blue by this connective tissue stain.

(TGFα) and TGFβ, fibroblastic growth factor-2, insulin-like growth factor-1, and platelet-derived growth factors. An alternative pathway might involve overproduction of inhibitors of matrix degradation such as TIMPs (tissue inhibitors of matrix production). In support of the former mechanism, fibroblasts isolated from the lungs of IPF patients exhibit a profibrotic secretory phenotype.

Multiple factors, such as environmental particulates, drug or chemical exposures, and viruses may trigger the initial epithelial injury but genetics also plays a role. Approximately 2 to 20% of patients with IPF have a family history of the disease with an inheritance pattern of autosomal dominance with variable penetrance. Genetic mutations have been implicated in this familial form of IPF. One large kindred has been reported with a mutation in the gene encoding surfactant protein C (*SFTPC*). The severity of lung disease associated with these mutations is highly variable.

Lymphangioleiomyomatosis

Clinical and Pathologic Features

Lymphangioleiomyomatosis (LAM) is a rare systemic disease of women, usually in their reproductive years (average age of 35 years), that is characterized by a proliferation of abnormal smooth muscle cells giving rise to cysts in the lungs, abnormalities in the lymphatics, and abdominal tumors, most notably in the kidneys. In addition to sporadic cases (denoted as S-LAM), LAM also affects 30% of women with tuberous sclerosis (denoted as TSC-LAM), a genetic disorder with variable penetrance associated with seizures, brain tumors, and cognitive impairment. Global estimates indicate that TSC-LAM may be as much as five- to

10-fold more prevalent than S-LAM though at least some suggest that TSC-LAM may have a milder clinical course than S-LAM. Clinically, LAM patients usually present with increasing shortness of breath on exertion, obstructive symptoms, spontaneous pneumothoraces, and chylous effusions, or with abdominal masses consisting of either angiomyolipomas and/or lymphangiomyomas. Chest imaging studies characteristically reveal hyperinflation with flattened diaphragms and thin-walled cystic changes. Mortality at 10 years from the onset of symptoms is 10 to 20%.

LAM appears as small, thin-walled cysts (0.5–5.0 cm) randomly throughout both lungs (Figure 18.21A). Microscopically, LAM lungs contain a diffuse infiltration of smooth muscle cells, predominantly around lymphatics, veins, and venules (Figure 18.21B). Most notably, one finds smooth muscle cells in the subpleural with hemosiderin-laden macrophages in the adjacent field, and the macrophages are also seen on bronchoalveolar lavage specimens from these patients. The hemosiderin pigment in these lungs is thought to be secondary to microhemorrhages from the obstruction of the veins. The smooth muscle cells in LAM react to antibodies to HMB-45, a premelanosomal protein. Other melanosome-like structures are also found in LAM cells, suggesting that these cells have characteristics of both smooth muscle and melanosomes.

Molecular Pathogenesis

The lesional cells in LAM are smooth muscle-like with both spindled and epithelioid morphology. These cells are the same in both S-LAM and TSC-LAM and are a clonal population although they lack other features of malignancy. Molecular studies reveal that the

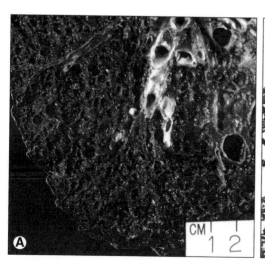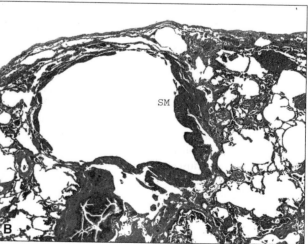

Figure 18.21   Lymphangioleiomyomatosis. (A) The sagittal cut section of an upper lobe from an explanted lung from a patient with LAM demonstrates cystic features of the red/brown lung parenchyma that are characteristic of this disease. **(B)** The microscopic view of the LAM lung reveals cysts lined by spindled smooth muscle cells **(SM)**. Scattered macrophages surrounding these cysts contain brown hemosiderin pigment.

abnormal LAM cell proliferation is caused by mutations in one of two genes linked to tuberous sclerosis, tuberous sclerosis complex 1 or 2 (*TSC1* or *TSC2*). These two genes control cell growth and differentiation through the Akt/mammalian target of rapamycin (mTOR) signaling pathway. In this pathway, a growth factor receptor (such as insulin or PDGF receptors) becomes phosphorylated when an appropriate ligand binds, resulting in activation of downstream effectors and ultimately Akt. The gene products of *TSC1* and *TSC2* are hamartin and tuberin, which act as dimers to maintain Rheb (a member of the Ras family) in a GDP-loaded state via statins, acting as a break to the Akt/mTOR pathway, thereby retarding protein synthesis and cell growth. In LAM cells, loss-of-function mutations in these two genes removes this inhibition, leading to enhanced Rheb activation, mTOR activation (with raptor), and subsequent phosphorylation of downstream molecules, which result in uncontrolled cell growth, angiogenesis, and damage to the lung tissue (Figure 18.22).

The abnormal proliferation of LAM cells is thought to damage the lung through overproduction of MMPs that degrade the connective tissue of the lung architecture, destroy the alveolar integrity, and result in cyst formation with air trapping. These destructive capabilities of the LAM cells are enhanced by their secretion of the angiogenic factor VEGF-C, which is thought to cause the proliferation of lymphatic channels throughout the lung.

PULMONARY VASCULAR DISEASES

Pulmonary Hypertension

Clinical and Pathologic Features

Pulmonary hypertension consists of a group of distinct diseases whose pathology is characterized by abnormal destruction, repair, remodeling, and proliferation of all compartments of the pulmonary vascular tree, including arteries, arterioles, capillaries, and veins. The classification of these diseases has undergone a number of revisions. The most recent revision groups these diseases based on both their pathologic and clinical characteristics. There are five major disease categories in the current classification system: (1) pulmonary arterial hypertension (PAH); (2) pulmonary hypertension with left heart disease; (3) pulmonary hypertension associated with lung disease and/or hypoxemia; (4) pulmonary hypertension due to chronic thrombotic and/or embolic disease; and (5) miscellaneous causes, including sarcoidosis, pulmonary Langerhans cell histiocytosis X, and lymphangioleiomyomatosis. The clinical course of most patients with pulmonary hypertension begins with exertional dyspnea, and progresses through chest pain, syncope, increased mean pulmonary artery pressures, and eventually, right heart failure. The rate of this clinical progression varies among patients, from a few months to many years. Treatment of these diseases focuses on blocking the mediators involved in the pathogenesis of the diseases. However, current therapies rarely prevent progression of the disease and lung transplantation provides the only hope for long-term survival.

The major group of this classification, PAH, can be divided into familial PAH, idiopathic PAH, PAH associated with other conditions (such as connective tissue diseases, HIV, and congenital heart disease), and PAH secondary to drugs and toxins (such as anorexigens, cocaine, and amphetamines). In these diseases, the primary pathology is localized predominantly in the small pulmonary arteries and arterioles. However, two other diseases in this group, pulmonary veno-occlusive disease and pulmonary capillary hemangiomatosis, involve predominantly other components of the pulmonary vasculature, the veins and the capillaries, respectively. The pathologic changes seen in the pulmonary vessels of these patients primarily reflect injury to and repair of the endothelium. Early pathologic changes include medial hypertrophy and intimal fibrosis that narrows and obliterates the vessel lumen. These are followed by remodeling and revascularization, producing a proliferation of abnormal endothelial-lined spaces. These structures are known as plexogenic lesions and are the pathognomonic feature of PAH (Figure 18.23). In the most severe pathologic lesions, these abnormal vascular structures become dilated or angiomatoid-like and may develop features of a necrotizing vasculitis with transmural inflammation and fibrinoid necrosis.

Molecular Pathogenesis

Though the exact pathogenetic mechanism of PAH remains unknown, research over the past 10 years has begun to offer some clues. The familial form of PAH, with a 2:1 female-to-male prevalence, has an autosomal dominance inheritance pattern with low penetrance. The genetic basis for this has been found to be germline mutations in the gene encoding the bone morphogenetic protein receptor type 2 (*BMPR2*). These mutations account for approximately 60 to 70% of familial PAH and 10 to 25% of patients with sporadic PAH. Approximately 140 *BMPR2* mutations have been identified in familial PAH, each resulting in a loss of receptor function, either through alteration in transcription of the gene through missense, nonsense, or frameshift alterations in the codon or by RNA splicing mistakes.

The mechanism by which a single mutation to the *BMPR2* gene induces vascular smooth muscle proliferation and decreased apoptosis is not completely understood, but it most likely involves defects in the BMPR2 signaling pathway. BMPR2 is a receptor for a family of cytokines (BMPs) that are members of the TGFβ superfamily of proteins that play a role in the growth and regulation of many cells, including those of the pulmonary vasculature. In the vascular smooth muscle cells of the lung, TGFβ signaling causes a proliferation of smooth muscle in pulmonary arterioles and BMPR2 signaling causes an inhibition of the proliferation of these cells, favoring an apoptotic environment. The BMPR2 signaling occurs

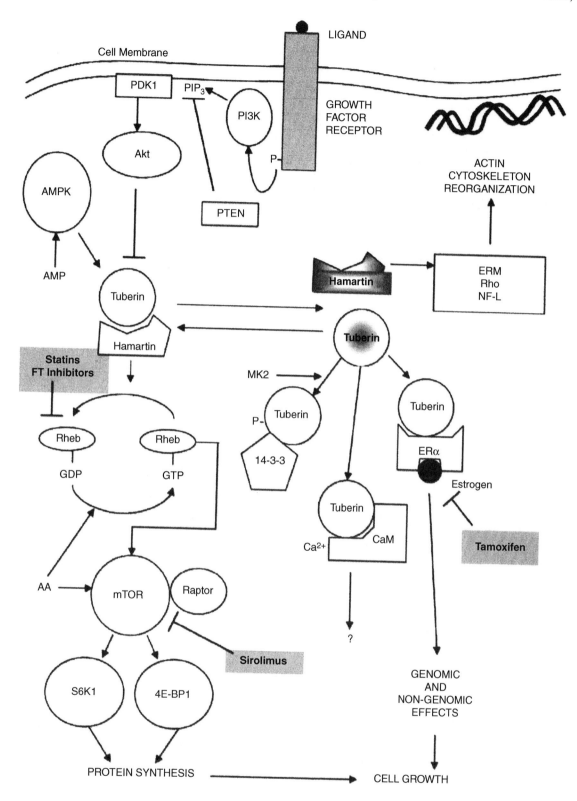

Figure 18.22 Signal transduction pathways involving the *TSC1* and *TSC2* gene products, hamartin and tuberin. *Arrowheads* indicate activating or facilitating influences; *flat-headed lines* indicate inhibitory influences. The harmartin-tuberin dimer maintains Rheb in a GDP-loaded state, thereby preventing activation of mTOR, which requires activated Rheb-GTP. Growth and energy signals tend to inhibit this function of the hamartin-tuberin complex, permitting mTOR activation. The sites of action of several drugs with therapeutic potential in LAM are indicated in gray-shaded boxes. AA, amino acids; FT, farnesyltransferase. Reprinted with permission from Juvet, S.C., McCormack, F.C., Kwiatkowski, D.J., and Downey, G.P. (2007) Molecular pathogenesis of lymphangioleiomyomatosis: Lessons learned from orphans. *Am J Respir Cell Mol Biol* 36:398–408.

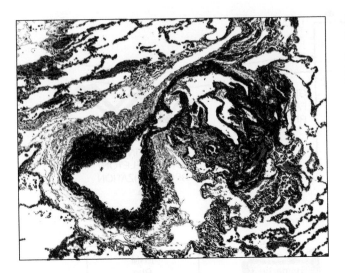

Figure 18.23   Pulmonary hypertension. A plexogenic lesion in a lung from a patient with idiopathic pulmonary hypertension reveals slit-like spaces (upper right corner) emerging from a pulmonary artery. These remodeling vascular spaces represent the irreversible damage done to these vessels in this disease.

through an activation of a receptor complex (BMPR1 and BMPR2) that leads to phosphorylation and activation of a number cytoplasmic mediators, most notably the Smad proteins (homologs to the drosophila protein: Mothers against decapentaplegic). These Smad proteins, especially the Smad 1, Smad 5, and Smad 8 complex with Smad 4, translocate to the nucleus where they target gene transcription that induces an antiproliferative effect in the cell. In familial PAH, the *BMRPR2* gene mutation may lead to insufficient protein product and subsequent decreased protein function, in this case decreased BMPR2 receptor function, decreased Smad protein activation, and decreased antiproliferative effects in the vascular smooth muscle cells. The imbalance between the proproliferative effects of the TGFβs and the antiproliferative effects of the BMPs results in the formation of the vascular lesions of PAH (Figure 18.24).

Despite these advances, questions regarding the pathogenesis of PAH remain. Most notably, why do only 10 to 20% of patients with the mutation develop clinical disease? Some speculate that genes confer susceptibility but a second hit is required to develop the clinical disease, such as modifier genes or environmental triggers, perhaps drugs or viral infections. In addition, though *BMPR2* mutations have been found in both the familial and the idiopathic form of PAH, they are present in only 30% of all PAH patients, suggesting that further research is needed to uncover additional etiologic agents.

DEVELOPMENTAL ABNORMALITIES

Surfactant Dysfunction Disorders

Clinical and Pathologic Features

Surfactant dysfunction disorders represent a heterogeneous group of inherited disorders of surfactant metabolism, found predominantly in infants and children.

Pulmonary surfactant includes both phospholipids and surfactant proteins, designated surfactant proteins A, B, C, and D (SP-A, SP-B, SP-C, SP-D), synthesized and secreted by type II cells beginning in the canalicular stage of lung development. Damage to type II cells during this time period can lead to acquired surfactant deficiencies. However, more commonly these deficiencies are the result of genetic defects of the surfactant proteins themselves.

The major diseases are caused by genetic defects in the surfactant proteins B (*SFTPB*, chromosome 2p12-p11.2); C (*SFTPC*, chromosome 8p21); and adenosine triphosphate (APT)-binding cassette transporter subfamily A member 3 (*ABCA3*, chromosome 16p13.3). Defects in *SFTPB* and *ABCA3* have an autosomal recessive inheritance pattern and defects in *SFTPC* have an autosomal dominant pattern. SP-B deficiency is the most common. It presents at birth with a rapidly progressive respiratory failure and chest imaging studies showing diffuse ground glass infiltrates. The gross pathology in these lungs consists of heavy, red, and congested parenchyma with microscopic features that range from a PAP-like pattern to a chronic pneumonitis of infancy (CPI) pattern. In SP-B deficiency, the PAP pattern predominates with a histologic picture of cuboidal alveolar epithelium and eosinophilic PAS-positive material within the alveolar spaces that appears with disease progression. In the late stages of the disease, the alveolar wall thickens with a chronic inflammatory infiltrate and fibroblasts. This alveolar proteinosis-type pattern of injury can be confirmed with immunohistochemical studies that establish the absence of SP-B within this surfactant-like material. Diseases due to *ABCA3* or *SFTPC* deficiency may present within a week of birth or years later; the former has a poor prognosis, but the latter has a more variable prognosis with some patients surviving into adulthood. Indeed, SP-C mutations also have been recognized in some families as a cause of interstitial pneumonia and pulmonary fibrosis in adults. The pathology of SP-C deficiency has more CPI features and less proteinosis. In contrast, *ABCA* deficiency can have either PAP or CPI features, with the former present early in the disease and the latter present in more chronically affected lungs.

Molecular Pathogenesis

The SP-B gene (*SPTPB*) is approximately 10 kb in length and is located on chromosome 2. There are over 30 recessive loss-of-function mutations associated with the *SPTPB* gene. However, the most common mutation is a GAA substitution for C in codon 121, found in about 70% of the cases. The lack of SP-B leads to an abnormal proportion of phosphatidylglycerol and an accumulation of a pro-SP-C peptide, leading to the alveolar proteinosis-like pathology.

SP-C protein deficiency is due to a defect in the *SFTPC* gene localized to human chromosome 8. There are approximately 35 dominantly expressed mutations in *SFTPC* that result in acute and chronic lung disease. Approximately 55% of them arise spontaneously and

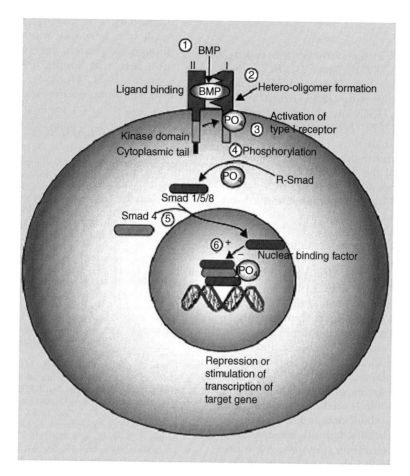

Figure 18.24   Bone-morphogenetic-protein signaling pathway. 1 and 2: BMPR1 and BMPR2 are present on most cell surfaces as homodimers or hetero-oligomers. With ligand (bone morphogenetic proteins; BMP) binding, a complex of ligand, two type 1 receptors, and two type II receptors is formed. 3: After ligand, two type I receptors and two type II receptors phosphorylate the type I receptor in its juxtamembrane domain. 4: The activated type I receptor then phosphorylates a receptor-regulated Smad (R-Smad); thus the type I receptors determine the specificity of the signal. 5: Once activated by phosphorylation, the R-Smads interact with the common mediator Smad 4 to form hetero-oligomers that are translocated to the nucleus. 6: In the nucleus, the Smad complex interacts with transcription factors and binds to DNA to induce or suppress transcription of target genes. Reprinted with permission from Runo, J.R. and Loyd, J.E. (2003) Primary pulmonary hypertension. *Lancet* 316:1533–1544.

the remainder are inherited. The most common mutation is a threonine substitution for isoleucine in codon 73 (I73T), found in 25% of the cases, including both sporadic and inherited disease. This mutation leads to a misfolding of the SP-C protein, which inhibits its progression through the intracellular secretory pathway, usually within the Golgi apparatus or the endoplasmic reticulum. The absence of SP-C within the alveolar space causes severe lung disease in mouse models. Infants with documented mutated proSP-C protein, the larger primary translation product from which SP-C is proteolytically cleaved, can have respiratory distress syndrome (RDS) or CPI. In older individuals, pathologic patterns observed in the lungs with these mutations include nonspecific interstitial pneumonitis (NSIP) and UIP. In this affected adult population, the pathology and age of disease presentation varies even within familial cohorts, suggesting the

involvement of a second hit, perhaps an environmental factor.

The ABCA3 protein is a member of the family of ATP-dependent transporters, which includes the cystic fibrosis transmembrane conductance regulator (CFTR), and is expressed in epithelial cells. Mutation of this gene results in severe respiratory failure that is refractory to surfactant replacement. The cellular basis for the lack of surfactant in patients with this genetic mutation is not known. The presence of abnormal lamellar bodies within the type II cells by ultrastructural analysis suggests a disruption in the normal surfactant synthesis and packaging in this disease. There is some evidence that this gene contains promoters that share elements consistent with their activation by the transcription factors TTF-1 and Foxa7 and deletions in either or both of these genes may play a role in this disease.

KEY CONCEPTS

- Oncogenes involved in the pathogenesis of lung cancer include *MYC, K-RAS, Cyclin D1, BCL2,* and *ERBB* family genes such as *EGFR* and *HER2/neu.* TSG abnormalities involving *TP53, RB, p16^INK4a,* and new candidate TSGs on the short arm of chromosome 3 (*DUTT1, FHIT, RASFF1A, FUS-1, BAP-1*) are common in lung cancers. The most common genetic abnormality in MM is a deletion in 9p21, the locus of the TSG *CDKN2A* encoding the tumor suppressors p16^{INK4a} and p14^{ARF}, which participate in the *p53* and Rb pathways and inhibit cell cycle progression.
- Targeted therapeutic agents in use and under investigation for treatment of lung cancers include EGFR pathway inhibitors, VEGF/VEGFR pathway inhibitors, Ras/Raf/MEK pathway inhibitors, PI3K/Akt/PTEN pathway inhibitors, tumor suppressor gene therapies, proteasome inhibitors, HDAC inhibitors, telomerase inhibitors, and others.
- Precursor lesions have been defined for ACs, SqCCs, and carcinoids, and manifest histologic and molecular alterations that overlap with the corresponding neoplasms.
- The two main categories of asthma are atopic (allergic) and nonatopic (nonallergic), the former with an onset in early childhood and the latter with late onset during adulthood. Both are associated with a Type 1 hypersensitivity response with a predominance of Type 2 helper cells with CD4+ phenotype.
- Emphysema, the loss of elastic fibers in the lung, is the result of an imbalance of elastases, which destroy elastic fibers and antielastases, which inhibit this destruction. Smoking causes an excess of elastase production and a centrilobular pattern of emphysema. A decrease in the circulating level of one antielastase (αAT) is the result of a genetic mutation on the *SERPINA1* gene and results in a panacinar pattern of emphysema.
- Usual interstitial pneumonitis (UIP) is the pathologic pattern most associated with the clinical disease idiopathic pulmonary fibrosis (IPF). Though the cause of this disease is unknown, it is most likely the result of injury to the alveolar epithelial cells (AEC) that activates fibroblasts and myofibroblasts to deposit collagen and elastic fibers in the form of fibroblastic foci in these lungs.
- Lymphangioleiomyomatosis is characterized by abnormal smooth muscle cells in the lung, a result of genetic mutations in either the *TSC1* or *TSC2* gene causing a loss of inhibition in the mTOR pathway and uncontrolled cell growth.
- Pulmonary arterial hypertension (PAH) is the result of a germline mutation in the bone morphogenetic protein receptor type 3 (*BMPR2*) in most forms of familial PAH and some forms of sporadic PAH. However, a minority of patients with the mutation develop clinical disease.

SUGGESTED READINGS

1. American Thoracic Society/European Respiratory Society. International multidisciplinary consensus classification of the idiopathic interstitial pneumonias: The Joint Statement of the American Thoracic Society (ATS), and the European Respiratory Society (ERS) was adopted by the ATS Board of Directors, June 2001, and by the ERS Executive Committee, June 2001. *Am J Respir Crit Care Med.* 2002;165:277–304.
2. Austin ED, Loyd JE. Genetics and mediators in pulmonary arterial hypertension. *Clin Chest Med.* 2007;28:43–57.
3. Barnes PJ. Immunology of asthma and chronic obstructive pulmonary disease. *Nature Rev Immunology.* 2008;8:183–192.
4. Beasley MB, Lantuejoul S, Abbondanzo S, et al. The P16/cyclin D1/Rb pathway in neuroendocrine tumors of the lung. *Hum Pathol.* 2003;34:136–142.
5. Bullard JE, Wert SE, Whitsett JA, et al. ABCA3 mutations associated with pediatric interstitial lung disease. *Am J Respir Crit Care Med.* 2005;172:1026–1031.
6. Churg A, Cosio M, Wright JL. Mechanisms of cigarette smoke-induced COPD: Insights from animal models. *Am J Physiol Lung Cell Mol Physiol.* 2008;294:L612–L631.
7. Gutierrez M, Giaccone G. Antiangiogenic therapy in nonsmall cell lung cancer. *Curr Opin Oncol.* 2008;20:176–182.
8. Hamvas A, Cole FS, Nogee L. Genetic disorders of surfactant proteins. *Neonatology.* 2007;91:311–371.
9. Juvet SC, McCormack FX, Kwiatkowski DJ, et al. Molecular pathogenesis of lymphangioleiomyomatosis: Lessons learned from orphans. *Am J Respir Cell Mol Biol.* 2007;36(4):398–408.
10. Kohnlein T, Welte T. Alpha-1-antitrypsin deficiency: Pathogenesis, clinical presentation, diagnosis, and treatment. *Am J of Med.* 2008;121:3–9.
11. Lawson WE, Loyd JE. The genetic approach in pulmonary fibrosis. *Proc Am Thorac Soc.* 2006;3:345–349.
12. Lee AY, Raz DJ, He B, et al. Update on the molecular biology of malignant mesothelioma. *Cancer.* 2007;109:1454–1461.
13. Righi L, Volante M, Rapa I, et al. Neuroendocrine tumours of the lung. A review of relevant pathological and molecular data. *Virchows Arch.* 2007;451(suppl 1):S51–S59.
14. Shigematsu H, Gazdar AF. Somatic mutations of epidermal growth factor receptor signaling pathway in lung cancers. *Int J Cancer.* 2006;118:257–262.
15. Sun S, Schiller JH, Spinola M, et al. New molecularly targeted therapies for lung cancer. *J Clin Invest.* 2007;117:2740–2750.
16. Taraseviciene-Stewart L, Voelkel NF. Molecular pathogenesis of emphysema. *J Clin Invest.* 2008;118:394–402.
17. Travis WD, Brambilla E, Müller-Hermelink HK, et al., eds. *World Health Organization Classification of Tumours. Pathology and Genetics of Tumours of the Lung, Pleura, Thymus and Heart.* Lyon: IARC Press; 2004.
18. Weir BA, Woo MS, Getz G, et al. Characterizing the cancer genome in lung adenocarcinoma. *Nature.* 2007;450:893–898.
19. Wistuba II. Genetics of preneoplasia: Lessons from lung cancer. *Curr Mol Med.* 2007;7:3–14.
20. Zander DS, Popper HH, Jagirdar J, et al., eds. *Molecular Pathology of Lung Diseases.* New York, NY: Springer; 2008.

19

Molecular Basis of Diseases of the Gastrointestinal Tract

Antonia R. Sepulveda . Dara L. Aisner

GASTRIC CANCER

Gastric carcinoma is the fourth most frequent cancer worldwide, with highest rates in Asia and countries such as Eastern Europe and areas of Central and South America, while it is less frequent in Western countries. Overall, gastric cancer represents the second most common cause of death from cancer (approximately 700,000/year). About two-thirds of the cases occur in developing countries and 42% in China alone. In the United States, 21,259 new cases of stomach cancer were estimated in 2007. Most gastric cancers develop as sporadic cancers, without a well-defined hereditary predisposition. A small proportion of gastric cancers arise as a consequence of a hereditary predisposition caused by specific inherited germ line mutations in critical cancer-related genes.

Nonhereditary Gastric Cancer

Risk Factors

Most nonhereditary gastric cancers arise in a background of chronic gastritis, which is most commonly caused by *H. pylori* infection of the stomach. There may be clustering of gastric cancer within some affected families. Relatives of patients with gastric carcinoma have an increased risk for gastric carcinoma of about 3-fold. Patients with both a positive family history and infection with a CagA-positive *H. pylori* strain were reported to have a greater than 8-fold risk of gastric carcinoma as compared to others without these risk factors.

The pathogenesis of gastric cancer is multifactorial, resulting from the interactions of host genetic susceptibility factors, environmental exogenous factors that have carcinogenic activity such as dietary elements and smoking, and the complex damaging effects of chronic gastritis. The diets that have been implicated in increased risk of gastric cancer are predominantly salted, smoked, pickled, and preserved foods (rich in salt, nitrite, and preformed *N*-nitroso compounds), and diets with reduced vegetables and vitamin intake.

Chronic infection of the stomach by *H. pylori* is the most common form of chronic gastritis. Therefore, *H. pylori* is the most significant known risk factor for the development of gastric cancer. *H. pylori* was classified as a human carcinogen based on a strong epidemiological association of *H. pylori* gastritis and gastric cancer. Overall, the risk of gastric cancer in patients with *H. pylori* gastritis is 6-fold higher than that of the population without *H. pylori* gastritis. The risk of gastric cancer increases exponentially with increasing grade of gastric atrophy and intestinal metaplasia, the risk of gastric cancer being about 90 times higher in patients with severe multifocal atrophic gastritis affecting the antrum and corpus of the stomach than in individuals with normal noninfected stomachs. However, even nonatrophic gastritis, as compared to healthy non-*H. pylori* infected individuals, raises the gastric cancer risk to approximately 2-fold.

Additional evidence supporting the association of *H. pylori* and gastric cancer includes (i) prospective studies demonstrating gastric cancer development in 2.9% of infected patients over a period of about 8 years, and in 8.4% of patients with extensive atrophic gastritis and intestinal metaplasia during a 10-year surveillance; (ii) development of animal models that develop gastric cancer associated with *H. pylori* infection, including Mongolian gerbils and mice; and (iii) eradication of *H. pylori* infection in patients with early gastric cancer resulted in the decreased appearance of new cancers. *H. pylori* eradication reduced the incidence of gastric cancer in patients without atrophy and intestinal metaplasia at baseline, suggesting that eradication may contribute to preventing progression to gastric cancer.

Stepwise Progression of H. pylori Gastritis to Gastric Carcinoma: Histologic Changes of the Gastric Mucosa

The chronicity associated with *H. pylori* gastritis is critical to the carcinogenic potential of the infection. *H. pylori* infection is generally acquired during childhood and persists throughout life unless the patient undergoes eradication treatment. Gastric cancer develops several decades after acquisition of the infection, associated with the progression of mucosal damage with development of specific histological alterations. *H. pylori* infection of the stomach activates both humoral and cellular inflammatory responses within the gastric mucosa. Despite a continuous inflammatory response, *H. pylori* organisms are able to evade the host immune mechanisms and persist in the mucosa, causing chronic gastritis. Histologically, the progression of *H. pylori*-associated chronic gastritis to gastric cancer is characterized by a stepwise acquisition of mucosal changes, starting with chronic gastritis, progressive damage of gastric glands resulting in mucosal atrophy, replacement of normal gastric glands by intestinal metaplasia, and development of dysplasia and carcinoma in some patients (Figure 19.1).

Stomach cancers are classified according to the World Health Organization classification based on their grade of differentiation into well, moderately, and poorly differentiated adenocarcinomas. In addition, gastric adenocarcinomas can be categorized into intestinal and diffuse types (Lauren's classification), based on the morphologic features on H&E stained tumor sections. Gastric cancers arising on the inflammatory background of *H. pylori*-associated chronic gastritis are most commonly intestinal type adenocarcinoma, which are predominantly well to moderately differentiated adenocarcinomas, but diffuse type tumors, which are poorly differentiated or are signet ring cell carcinomas, also occur and may develop in patients with gastric ulcers and chronic active gastritis (Figure 19.1).

Progression to gastric cancer is higher in patients with extensive forms of atrophic gastritis with intestinal metaplasia involving large areas of the stomach, including the gastric body and fundus. This pattern of gastritis has been described as pangastritis or multifocal atrophic gastritis. Extensive gastritis involving the gastric body and fundus results in hypochlorhydria, allowing for bacterial overgrowth and increased carcinogenic activity in the stomach through the conversion of nitrites to carcinogenic nitroso-*N* compounds. *H. pylori*-associated pangastritis is frequently seen in the relatives of gastric cancer patients, which may contribute to gastric cancer clustering in some families.

Molecular Mechanisms Underlying Gastric Epithelial Neoplasia Associated with H. pylori Infection

How *H. pylori* gastritis promotes gastric carcinogenesis involves an interplay of mechanisms that include (i) a long-standing inflammation in the mucosa, with increased oxidative damage of gastric epithelium, (ii) a number of alterations of epithelial and inflammatory cells induced by *H. pylori* organisms and by released bacterial products, (iii) inefficient host response to the induced damage, and (iv) mechanisms of response related to host genetic susceptibility, which may mediate the variable levels of damage in different individuals.

H. pylori bacterial products and factors released by activated or injured epithelial and inflammatory cells both contribute to persistent chronic inflammatory response involving the activation of innate and acquired immune responses in the infected gastric mucosa. This chronic and continuously active mucosal inflammatory infiltrate is in part responsible for potential damage to the epithelium, through the release of oxygen radicals, and the production of chemokines that may alter the normal regulation of molecular signaling in epithelial cells. *H. pylori* organisms and released bacterial products, including the *H. pylori* virulence factors (CagA and VacA), which may directly alter the gastric epithelial cells as well as inflammatory cells, leading to alterations of signaling pathways, gene transcription, and genomic modifications. These changes lead to modifications in cell behavior, such as increased apoptosis and proliferation, as well as increased rates of mutagenesis. Once the neoplastic program is activated, then gastric cancer may progress through mechanisms similar to gastric cancers that do not arise in a background of *H. pylori* infection.

A number of genetic susceptibility factors that increase the risk of gastric cancer development in *H. pylori*-infected patients have been identified. Interleukin-1 (IL-1) gene polymorphisms, in IL-1beta and IL-1RN (receptor antagonist), have been shown to increase the risk of gastric cancer and gastric atrophy in the presence of *H. pylori*. Individuals with the IL-1B-31*C or IL-1B-511*T, and the IL-1RN*2/*2 genotypes are at increased risk of hypochlorhydria, gastric atrophy, and gastric cancer in response to *H. pylori* infection. A gene polymorphism that may affect the function of OGG1, a protein involved in the repair of mutations induced by oxidative stress, was reported frequently in patients with intestinal metaplasia and gastric cancer, suggesting that deficient OGG1 function may contribute to increased mutagenesis during gastric carcinogenesis.

Mechanisms and Spectrum of Epigenetic Changes, Mutagenesis, and Gene Expression Changes of the Gastric Epithelium Induced by H. pylori Infection

Epigenetic modification and mutagenesis precede tumor development and accompany neoplastic progression during gastric carcinogenesis. The combined effects of epigenetic modifications, mutagenesis, and functional gene expression changes in gastric epithelial cells in response to inflammatory mediators and bacterial virulence factors result in abnormal gene expression and function in the various stages of progression to gastric cancer (gastritis, intestinal metaplasia, dysplasia, and cancer). The effects of *H. pylori* organisms on gastric epithelial cells are likely to occur primarily during the phase

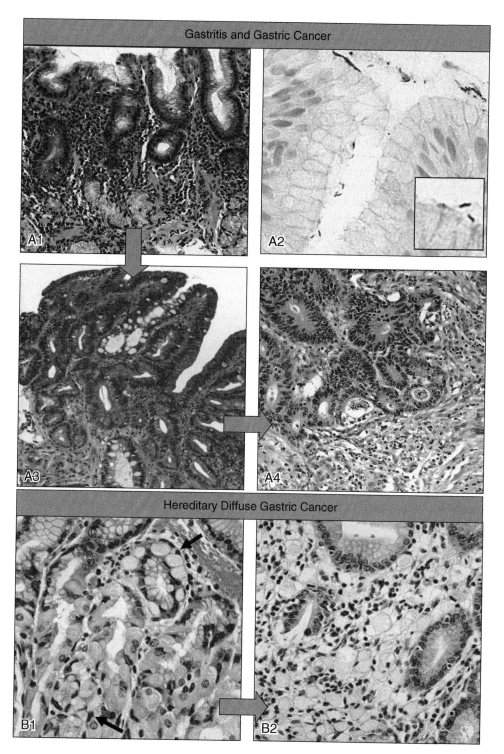

Figure 19.1   Gastric carcinogenesis: Stepwise progression of *H. pylori*-associated gastric cancer (panels A1–A4), and hereditary diffuse gastric cancer (panels B1 and B2). Panel A1. Chronic active gastritis involving the mucosa of the gastric antrum (H&E stain, original magnification 10×). **Panel A2.** Immunohistochemical stain highlights *H. pylori* organisms with typical S and comma shapes, seen at higher magnification in the inset. *H. pylori* organisms typically appear attached or adjacent to the gastric surface and foveolar epithelium (original magnification 40×). **Panel A3.** Gastric mucosa with intestinal metaplasia and low-grade dysplasia/adenoma (H&E stain, original magnification 10×). **Panel A4.** Gastric carcinoma of intestinal type (moderately differentiated adenocarcinoma) (H&E stain, original magnification 10×). **Panels B1 and B2** (Courtesy of Dr. Adrian Gologan, Jewish General Hospital, McGill University). Gastric mucosa of patient with hereditary diffuse gastric cancer (HDGC) with *in situ* signet ring cell carcinoma (arrow) (B1) and invasive signet ring cell carcinoma expanding the lamina propria between the gastric glands (E) (H&E stain, original magnification 20×).

of gastritis and intestinal metaplasia, while additional molecular events associated with neoplastic progression from dysplasia to invasive cancer may be independent of *H. pylori*. However, the background inflammatory milieu associated with ongoing chronic infection may influence the mechanisms of neoplastic progression.

During gastric carcinogenesis a number of genes are regulated by CpG methylation of the promoter regions at CpG sites, with potential promoter inactivation. For example, some investigators have suggested that there are five different classes of methylation behaviors in chronic gastritis, intestinal metaplasia, gastric adenoma, and gastric cancer: (i) genes methylated in gastric cancer only (*GSTP1* and *RASSF1A*); (ii) genes showing low methylation frequency in chronic gastritis, intestinal metaplasia, and gastric adenoma, but significantly higher methylation frequency in gastric cancer (*COX-2*, *MLH1*, and *p16*); (iii) low and similar methylation frequency in all gastric lesions (*MGMT*); (iv) genes with high and similar methylation frequency in all gastric lesions (*APC* and *E-cadherin*); and (v) genes showing an increasing methylation frequency along the oncogenic progression (*DAP-kinase*, *p14*, *THBS1*, and *TIMP3*).

The mechanisms that regulate CpG methylation and gene silencing during *H. pylori*-associated gastritis and resulting gastric mucosal lesions are not currently known. Recent studies showed that proinflammatory interleukin-1-beta polymorphisms were associated with CpG island methylation of target genes, and CpG methylation of the *E-cadherin* promoter was induced in cells treated with IL-1β. These data suggest that components of the inflammatory cascade induced by *H. pylori* may contribute to orchestration of the epigenetic response in *H. pylori*-associated carcinogenesis.

Mutations are likely to accumulate during *H. pylori* chronic gastritis because of increased damaging factors in the mucosa and also because of overall deficiency of some DNA repair functions. DNA damage during *H. pylori* gastritis is caused primarily by reactive oxygen species (ROS), and reactive nitrogen species (RNS). Additionally, when mucosal atrophy develops, the resulting reduced acid levels may allow the overgrowth of other bacteria and activation of environmental carcinogens with mutagenic activity.

ROS are generated by inflammatory cells, as well as by gastric epithelial cells after activation by *H. pylori* bacterial products and cellular released cytokines. The increased ROS levels are associated with increased expression of inducible nitric oxide synthase (iNOS) and increased production of nitric oxide (NO). Increased cyclooxygenase (COX2) has also been reported in *H. pylori*-associated gastritis and may contribute to increased mutagenesis through oxidative stress. Further, with reduced levels of oxygen radical scavengers, such as glutathione and glutathione-S-transferase, relatively higher levels of oxygen radicals may accumulate in the mucosa of *H. pylori*-infected patients. DNA 8-hydroxydeoxyguanosine (8OHdG) can be used as a marker for oxidative DNA damage. The gastric mucosa with *H. pylori* gastritis and preneoplastic lesions (intestinal metaplasia and atrophy) contain increased levels of DNA

8-hydroxydeoxyguanosine (8OHdG) and the levels of 8OHdG in the gastric mucosa significantly decrease after eradication of *H. pylori*, supporting the role of active infection in the accumulation of mutations. Mutations associated with oxidative damage include point mutations in genes involved in gastric carcinogenesis such as *p53*.

Mutations accumulating during *H. pylori* gastritis may be enhanced because of a relatively deficient DNA repair system, with persistence of ROS-induced mutations and uncorrected DNA sequence replication errors that are transmitted to future epithelial cell generations. Several DNA repair systems are required for correction of DNA damage occurring during *H. pylori* gastritis: (i) the DNA mismatch repair system, which repairs DNA replication-associated sequence errors, and (ii) several other proteins that primarily repair DNA lesions induced by oxidative and nitrosative stress, including MGMT and OGG1 glycosylase. Several studies have reported a role of relative DNA mismatch repair deficiency in the mutation accumulation during *H. pylori* infection. In experimental conditions, when gastric epithelial cells are co-cultured with *H. pylori* organisms, the levels of DNA mismatch repair proteins, including MSH2 and MLH1, are greatly reduced, and both point mutations and frameshift/microsatellite-type mutations accumulate in the *H. pylori*-exposed cells. *In vivo* microsatellite instability was reported in 13% cases of chronic gastritis, 20% of intestinal metaplasias, 25% of dysplasias, and 38% of gastric cancers, indicating a stepwise accumulation of microsatellite instability (MSI) during gastric carcinogenesis. Microsatellite instability has been detected in intestinal metaplasia from patients with gastric cancer indicating that MSI can occur in preneoplastic gastric mucosa.

Deficient function of other genes involved in the repair of oxidative stress-induced mutations may also contribute to mutagenesis during *H. pylori* gastritis. Repair of 8-OHdG is accomplished by DNA repair proteins including a polymorphic glycosylase (OGG1). This protein may be less efficient in carriers of a gene polymorphism that was reported frequently in patients with intestinal metaplasia and gastric cancer. O6-methylguanine-DNA methyltransferase (MGMT) function includes the repair of O6-alkylG DNA adducts. In the absence of functional MGMT, these adducts mispair with T during DNA replication, resulting in G-to-A mutations. MGMT-promoter methylation has been reported in a subset of cases of *H. pylori* gastritis and in various stages of gastric carcinogenesis, suggesting a possible role for this DNA repair protein in gastric carcinogenesis.

Mutational, Epigenetic Gene Expression and MicroRNA Patterns of Gastric Intestinal Metaplasia, Dysplasia/Adenoma, and Cancer

Intestinal metaplasia, dysplasia, and carcinoma represent cell populations with a clonal origin that manifest epigenetic and genetic alterations incurred by the non-neoplastic epithelium, as well as additional events that occur during neoplastic progression. Figure 19.2

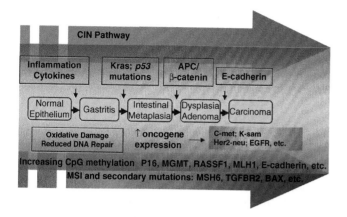

Figure 19.2   Stepwise progression of molecular events during gastric carcinogenesis.

represents the main molecular events that characterize gastric carcinogenesis.

A high level of MSI is associated with loss of expression and promoter hypermethylation of *MLH1* in gastric adenomas and cancer. MSI was reported in 17–35% of gastric adenomas, and in 17–59% of gastric carcinomas. Gastric tumors with MSI-H may also harbor frameshift mutations in the coding regions of cancer-related genes, such as *BAX, IGFRII, TGFβRII, MSH3,* and *MSH6.* In MSI-H adenomas, frameshift mutations of *TGFβRII* were detected in 38–88% of the cases, *BAX* in 13%, *MSH3* in 13%, and *E2F-4* in 50% of the cases.

Mutations of *p53* and *APC* genes have been reported in intestinal metaplasia and gastric dysplasia. *p53* mutations at exons 5 to 8 resulting in G:C to A:T transitions are detected in gastric carcinogenesis. *APC* mutations, including stop-codon and frameshift mutations, were reported in 46% of gastric adenomas and 5q allelic loss in 33% of informative cases of gastric adenoma and in 45% of carcinomas. K-*ras* mutations at codon 12 were reported in 14% of biopsies with atrophic gastritis and in less than 10% of adenomas, dysplasia, and carcinomas.

In gastric carcinogenesis CpG island methylation occurs in genes such as *MLH1, p14, p15, p16, E-cadherin, RUNX3,* thrombospondin-1 (*THBS1*), tissue inhibitor of metalloproteinase 3 (*TIMP-3*), COX-2, and *MGMT.* A number of genes were re-expressed in gastric cancer cell lines after treatment with the demethylating agent 5-aza-2′-deoxycytidine, identifying putative candidate genes involved in gastric carcinogens through epigenetic silencing.

Genome-wide gene expression analysis with microarrays has yielded a huge amount of information on gene expression of the lesions associated with gastric carcinogenesis, with identification of specific profiles that characterize gastritis, intestinal metaplasia, intestinal versus diffuse-type gastric adenocarcinoma, and different gastric cancer prognostic groups. One study reported a signature of diffuse-type cancers which exhibited altered expression of genes related to cell-matrix interaction and extracellular-matrix components, whereas intestinal-type cancers had a pattern of enhancement of cell growth. Another study reported several combinations of genes that could discriminate between normal and

tumor samples, and intestinal metaplasia cases were characterized by a gene expression signature resembling that of adenocarcinoma, supporting the notion that intestinal metaplasia tissue might progress to cancer.

MicroRNAs (miRNAs) are small noncoding RNAs that have been shown to regulate gene expression and may be aberrantly expressed in cancer. There are only a few studies addressing miRNA alterations in gastric carcinogenesis. Available data indicates that most gastric cancers showed overexpression of *miR-21*, while *miR-218–2* was consistently downregulated.

Familial Gastric Cancer

Genetic and Molecular Basis of Familial and Hereditary Gastric Cancer

Familial gastric cancer may be inherited as an autosomal dominant disease, occurring as the main tumor in hereditary diffuse gastric cancer (HDGC) or as one of the types of tumors in a number of cancer-predisposing syndromes, but it may also occur as family clustering of gastric cancer for which the etiology is likely multifactorial. Familial gastric cancer represents less than 10% of all stomach cancers. HDGC associated with germline mutations in the *E-cadherin* gene account for 30% to 40% of known cases of hereditary diffuse gastric cancer. In up to 70% of the cases of familial gastric cancer, the underlying genetic defect is unknown. Gastric carcinomas represent one of the types of tumors occurring in the following hereditary cancer syndromes: (i) germline mutations in the *E-cadherin* gene (CDH1) underlie some but not all HDGC families of note; (ii) Li-Fraumeni syndrome, associated with germline *p53* mutations; (iii) hereditary nonpolyposis colon cancer (HNPCC), with most HNPCC-associated gastric cancers representing the intestinal type; (iv) gastric cancer may also occur in Peutz-Jeghers syndrome (PJS), where hamartomatous polyps in the stomach occur in approximately 24% of patients, but the overall risk of gastric cancer is small; (v) patients with familial adenomatous polyposis (FAP) frequently develop fundic gland polyps in the stomach, and may develop gastric adenomatous polyps in about 10% of individuals with FAP, but the risk of gastric cancer is small.

Hereditary Diffuse Gastric Cancer: Genetic Basis

The *CDH1* gene, which encodes the protein E-cadherin, is the only gene known to be associated with HDGC. Mutations in other genes may account for susceptibility to HDGC, but the evidence is limited. Gastric cancer occurs in 5.7% of families with the *BRCA2* 6174delT mutation, but the type of gastric cancer in these families was not characterized.

The human *CDH1* gene consists of 16 exons that span 100 kb. An excess of 30 germline pathologic mutations has been reported in families with HDGC. The identified mutations are scattered throughout the gene and are truncating mutations, caused by frameshift mutations, exon/intron splice site mutations, point mutations, and missense mutations.

Hereditary Diffuse Gastric Cancer: Molecular Mechanisms, Clinical and Pathologic Features

Natural History and Pathologic Features The average age of diagnosis of HDGC is 38 years, ranging from 14–69 years, with most cases occurring before the age of 40 years. The lifetime risk of gastric cancer by age 80 years is 67% for men and 83% for women.

Histologically the adenocarcinomas in patients with HDGC are characteristically poorly differentiated carcinomas with signet ring morphology (signet ring cell carcinomas). Tumor foci are initially confined mostly in the superficial zone of the gastric mucosa and appear to arise in the lower proliferative zone of the gastric foveolae. Tumor foci may be multifocal. *In situ* signet ring cell carcinoma, characterized by disorderly oriented signet ring cells within glands or foveolar epithelium, may be observed (Figure 19.1). Grossly, the tumors extend through the gastric layers and gastric wall with the development of so-called linitis plastica. These tumors do not arise in a background of intestinal metaplasia and gastric atrophy or dysplastic preneoplastic lesions and are not involved in the progression to cancer. E-cadherin and β-catenin immunohistochemistry shows reduced to absent expression in both the *in situ* and invasive areas of the tumor.

Molecular Mechanisms and Pathologic Correlates In hereditary diffuse gastric cancer with known germline mutations, E-cadherin loss of function caused by pathologic mutations underlies the development of cancer. However, up to 70% of cases of familial gastric cancer of diffuse type do not have a well-defined genetic defect. In HGDC cases, loss of E-cadherin expression is also associated with a transcriptional downregulation of the wild-type *CDH1* allele by promoter hypermethylation.

Loss of E-cadherin expression is seen in most diffuse gastric cancers, including sporadic-type gastric carcinomas of diffuse type. The E-cadherin/catenin complex is important to suppress invasion and metastasis and cell proliferation. Somatic mutation of E-cadherin is associated with increased activation of epidermal growth factor receptor (EGFR) followed by enhanced recruitment of the downstream signal transduction, and activation of *Ras*. The activation of EGFR by E-cadherin mutants in the extracellular domain explains the enhanced motility of cancer cells in the presence of an extracellular mutation of E-cadherin.

Hereditary Diffuse Gastric Cancer: Genetic Testing and Clinical Management Criteria for consideration of *CDH1* molecular genetic testing in individuals with diffuse gastric cancer have been recommended. The criteria are as follows: (i) two or more cases of gastric cancer in a family, with at least one diffuse gastric cancer diagnosed before 50 years of age; (ii) three or more cases of gastric cancer in a family, diagnosed at any age, with at least one documented case of diffuse gastric cancer; (iii) an individual diagnosed with diffuse gastric cancer before 45 years of age; (iv) an individual diagnosed with both diffuse gastric cancer and lobular breast cancer (but no other criteria met); (v) one family member diagnosed with diffuse gastric cancer and another with lobular breast cancer (but no other criteria met); (vi) one family member diagnosed with diffuse gastric cancer and another with signet ring colon cancer (but no other criteria met).

COLORECTAL CANCER

Worldwide colorectal cancer ranks fourth in frequency in men and third in women, with approximately one million cases annually. Men and women are similarly affected. In the United States there are approximately 150,000 new cases of colorectal cancer per year. Most colorectal cancers do not develop in association with a hereditary cancer syndrome and are known as sporadic colon cancer. Several hereditary colon cancer syndromes have been characterized (Table 19.1). The most frequent is Lynch syndrome or HNPCC, representing 3–4% of colorectal cancers. Familial adenomatous polyposis (FAP) represents about 1%, and the remaining cancer syndromes (Table 19.1) are responsible for less than 1% of colorectal cancers.

Sporadic Colon Cancer

Colon cancers generally arise from precursor lesions of the colonic epithelium described histologically as dysplasia or adenoma, with progression to high-grade dysplasia (previously referred to as carcinoma *in situ*) and invasive adenocarcinomas (Figure 19.3). Adenomas are classified into tubular, tubulovillous, or villous adenomas. The epithelium that constitutes colonic adenomas displays cytologic features of dysplasia. In the initial steps, adenomas consist of epithelium with low-grade dysplasia, which may progress to high-grade dysplasia and invasive adenocarcinomas (Figure 19.3). Recently, another type of neoplastic lesions described as serrated adenomas has been described in the colon (Figure 19.4).

The molecular pathways of colon cancer development include a stepwise acquisition of mutations, epigenetic changes, and alterations of gene expression, resulting in uncontrolled cell division and manifestation of invasive neoplastic behavior. The molecular changes underlying colonic neoplasia correlate relatively well with histopathological variants. However, more than one pathway can lead to the development of adenocarcinomas with similar morphology. The main molecular pathways of colorectal cancer development characterized to date include (i) the chromosomal instability pathway (CIN), (ii) the microsatellite instability pathway (MSI), and (iii) the CpG island methylator pathway (CIMP) (Figure 19.5).

Most colon cancers that develop through the CIN or tumor suppressor pathway develop frequent cytogenetic abnormalities and allelic losses (Figure 19.5). The molecular mechanisms underlying CIN are poorly understood. CIN appears to result from deregulation of the DNA replication checkpoints and mitotic-spindle checkpoints. Mutation of the mitotic checkpoint regulators *BUB1* and *BUBR1*, and amplification of *STK15* are seen in a

Table 19.1   Gastrointestinal Hereditary Cancer Syndromes and Other Polyposis Syndromes

Syndrome	Inheritance	Key Clinical Features	Gene	Gene Product Function
Familial adenomatous polyposis and variants:	Autosomal dominant	>100 colonic adenomas, near 100% lifetime risk for colorectal carcinoma	APC	Growth inhibitory: β-catenin sequestration; targeting of β-catenin for destruction
- Gardner		FAP plus CHRPE, osteomas, desmoid tumors		
- Turcot*		FAP plus medulloblastoma, glioblastoma		
- Attenuated FAP		>15 but <100 colonic adenomas		
MYH-associated polyposis	Autosomal recessive	FAP-like presentation; no APC mutation identifiable	MYH	DNA damage repair
Hereditary nonpolyposis colorectal cancer (HNPCC; Lynch syndrome) and variants:	Autosomal dominant	70–85% lifetime risk for colorectal carcinoma; predisposition for extracolonic malignancy including endometrial	MSH2 MLH1 MSH6 PMS2 MLH3	DNA mismatch repair
- Muir-Torre		HNPCC plus sebaceous tumors and extracolonic malignancies		
Peutz-Jeghers	Autosomal dominant	Hamartomatous gastrointestinal polyps; predisposition for multiple extracolonic malignancies	STK11	Serine/threonine kinase
Cowden and **BRR syndrome	Autosomal dominant	Hamartomatous gastrointestinal polyps	PTEN	Protein phosphatase
Juvenile polyposis	Autosomal dominant	Hamartomatous gastrointestinal polyposis; increased risk for colorectal carcinoma	SMAD4 BMPR1A ENG	TGF beta signaling

CHRPE: Congenital Hypertrophy of the Retinal Pigment Epithelium.
*Two-thirds of Turcot syndrome occurs in patients with APC gene mutation and one third in DNA mismatch repair gene mutation.
** BRR: Bannayan-Ruvalcaba-Riley syndrome

subset of CIN-colon cancers. Among these colorectal cancers, the dominant genomic abnormality is inactivation of tumor suppressor genes, such as *APC* (chromosome 5q), *p53* (chromosome 17p), *DCC* (deleted in colon cancer), *SMAD2*, and *SMAD4* (chromosome 18q). *APC* gene mutations occur early in colonic neoplasia. *APC* mutations are detected in aberrant crypt foci (ACF), the lesion that precedes the development of adenomas, and have been detected in 50% of sporadic adenomas and 80% of sporadic colon cancers (Figure 19.5). *DCC* gene loss occurs late in neoplastic progression, with frequent deletion in carcinomas (73%) and in high-grade adenomas (47%). The tumor suppressor gene *p53* is also involved in the later steps of colon carcinogenesis. The *p53* protein has DNA binding activity, contains a transcription activation domain, and regulates target genes that mediate cell cycle arrest, apoptosis, and DNA repair. In the sporadic colonic carcinogenesis pathway, *p53* gene mutations occur in adenomas with high-grade dysplasia (50%) and carcinomas (75%) (Figure 19.5). The most common mutations are missense point mutations of one allele followed by deletion of the second allele, resulting in LOH of the *p53* gene locus on chromosome 17p.

K-*ras* mutations have been observed in the first stages of colonic neoplasia in aberrant crypt foci. Sporadic, FAP-associated, and CpG island methylator colorectal neoplasms have the highest rates of activating K-*ras* mutations (50–80%), while K-*ras* mutations are rare in both sporadic MSI-H and HNPCC-associated cancers. In sporadic colorectal carcinogenesis, K-*ras* mutation mainly occurs during the formation of ACF (Figure 19.5). Of note, in FAP, somatic mutation of *APC* predominantly occurs during ACF formation, followed by K-*ras* mutation. Inactivation of *APC* results in activation of Wnt/beta-catenin signaling, which in turn can induce chromosomal instability in colon cancer.

The MSI pathway is seen in approximately 15% of sporadic colorectal cancers. Tumors arising through the MSI pathway are characterized by an underlying deficiency of DNA mismatch repair proteins. The main DNA mismatch repair proteins are MLH1, MSH2, MSH6, and PMS2. Loss of function of one of the main DNA mismatch repair proteins results in high levels of mutagenesis, a molecular phenotype described as MSI. Through the use of a recommended panel of five microsatellite loci, tumors can be classified by their levels of microsatellite instability. Tumors with high-level MSI (MSI-H) reveal loss of expression of main DNA mismatch repair proteins in tumor cells. In sporadic carcinomas MSI is usually caused by loss of expression of MLH1, secondary to *MLH1* promoter hypermethylation, while in HNPCC the underlying defect is caused by inherited mutations of one of the DNA mismatch repair genes. The deficiency of DNA mismatch repair may facilitate the accumulation of

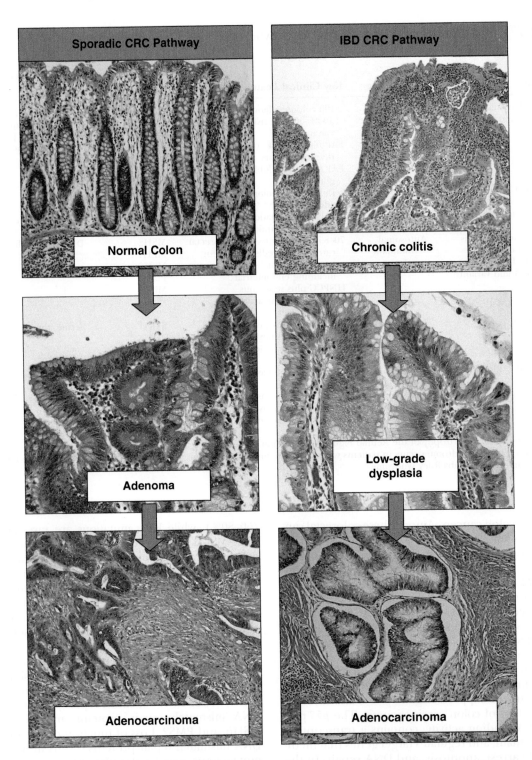

Figure 19.3   Stepwise progression of colorectal neoplasia: progression of neoplasia in sporadic colorectal cancer and in inflammatory bowel disease-associated colitis. In the sporadic colorectal cancer pathway, adenomas characterized in the early stages by low-grade epithelial dysplasia precede the development of high-grade dysplasia, which may then progress to invasive adenocarcinoma. In IBD-associated neoplasia, the background colonic mucosa reveals variable degrees of chronic colitis, and eventually foci of low-grade dysplasia develop, which in turn may progress to high-grade dysplasia and invasive adenocarcinoma. The morphologic features of the neoplastic lesions significantly overlap between sporadic colorectal cancer and IBD-associated neoplasia, but the inflammatory environment that characterizes chronic colitis dictates a number of different molecular mechanisms of neoplastic development and progression.

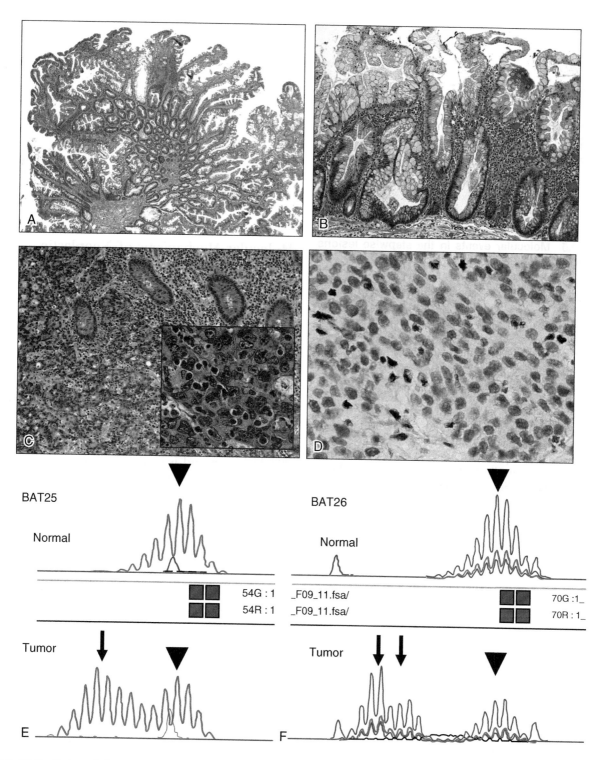

Figure 19.4   Colorectal cancer pathways: histopathology and molecular correlates. Panels A and B: The serrated pathway. Adenocarcinomas that develop through the serrated pathway arise from serrated polyps that include traditional serrated adenomas. (A) and sessile serrated adenomas (B) (H&E stain, original magnification 5× and 10×, for panels A and B, respectively). Traditional serrated adenomas show both serrated architecture and dysplasia similar to that seen in adenomatous mucosa, whereas sessile serrated adenomas reveal architectural abnormalities but no evidence of classic dysplasia. Panels C and D: The microsatellite instability pathway. Poorly differentiated colonic adenocarcinoma with prominent intratumoral lymphocytes, best seen in the inset (C) (H&E stain, original magnification 10×). By immunohistochemistry, the tumor cells are negative for MSH2, while the surrounding lymphocytes and stromal cells show preserved expression of MSH2 protein in the non-neoplastic cell nuclei (D) immunohistochemistry, original magnification 10×). (E) The presence of microsatellite instability in the tumor DNA is demonstrated by microsatellite instability at the microsatellite markers *BAT25* and *BAT26* characterized by the appearance of new PCR amplification peaks of smaller size (tailed arrows) as compared to non-neoplastic DNA from the same patient (arrow tip).

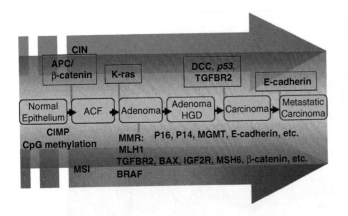

Figure 19.5   Molecular events in the stepwise lesions of sporadic colorectal carcinogenesis.

secondary mutations in cancer-related genes. In both sporadic and familial MSI-H colon cancers, mutational inactivation of the *TGFβRII* gene by MSI occurs in 80% of tested cancers and is frequently a late event, occurring at the transition from adenoma to cancer. Studies comparing HNPCC cancers and sporadic MSI colon cancers show that HNPCC tumors are more frequently characterized by aberrant nuclear beta-catenin. Aberrant *p53* expression, 5q loss of heterozygosity, and K-*ras* mutation is infrequent in both MSI-cancer groups. Despite sharing a similar underlying pathway (DNA mismatch repair deficiency), there are differences between sporadic CRC with MSI and HNPCC-associated tumors. Sporadic MSI-H cancers are more frequently poorly differentiated, mucinous, and proximally located than HNPCC tumors. In sporadic MSI-H cancers, contiguous adenomas are frequently serrated lesions, while traditional adenomas are usually seen in HNPCC. Lymphocytic infiltration is common in both types of tumors (Figure 19.4).

The CpG island methylator phenotype (CIMP) pathway is characterized by widespread CpG methylation in neoplasms. Epigenetic regulation is important in cancer development and progression. The presence of extensive CpG methylation in the CIMP pathway may result in inactivation of important genes involved in tumorigenesis. For example, the DNA repair protein MGMT may be inactivated by promoter hypermethylation in sporadic colorectal cancer (39–42%), and hypermethylation can be detected in 49% of adenomas. Recent studies indicate that the CpG island methylator phenotype underlies microsatellite instability in most cases of sporadic colorectal cancers associated with *MLH1* hypermethylation, and is tightly associated with *BRAF* mutation in sporadic colorectal cancer. However, only about one-third of CIMP-positive tumors are MSI-H. Thus, CIMP appears to be independent of microsatellite status. Several studies have confirmed that CIMP-positive tumors are associated with *BRAF* mutations, wild-type *TP53*, inactive WNT/β-catenin, and low level of genomic instability of the CIN type.

CIMP phenotype is frequent in lesions of the serrated pathway. *BRAF* mutations are also associated with

serrated colorectal lesions, while K-*ras* mutations are associated with hyperplastic polyps and adenomas with tubulovillous morphology. Sessile serrated adenomas characteristically have frequent *BRAF* mutation (78%), but have rare K-*ras* mutations (11%). In contrast, hyperplastic polyps and tubulovillous adenomas show frequent K-*ras* mutation (70% and 55%, respectively), whereas *BRAF* mutation is rare in these lesions (20% and 0%, respectively). *MLH1* promoter methylation is frequent in serrated polyps from patients with cancers showing MSI but not in the lesions from patients with MSS cancers, suggesting that serrated adenomas may give rise to sporadic colorectal carcinomas with MSI.

Molecular Mechanisms of Neoplastic Progression in Inflammatory Bowel Disease

Natural History of Neoplasia in Inflammatory Bowel Disease

Patients with inflammatory bowel disease have an increased risk of dysplasia and colorectal cancer, associated with long-standing chronic colitis. Colorectal adenocarcinomas in IBD develop from foci of low-grade dysplasia, which may progress to high-grade dysplasia and ultimately invasive carcinoma (Figure 19.3). The risk of colon cancer for patients with IBD was reported to increase by 0.5–1.0% every year after 8–10 years of diagnosis. CRC in IBD has an incidence 20-fold higher and is detected on average in patients 20 years younger than colorectal cancer in the non-IBD population.

Molecular Mechanisms of CRC Development and Progression in IBD-Associated Colitis

In inflammatory bowel disease-associated colitis, there is a continued inflammatory environment of the mucosa (Figure 19.3) with damage of the colonic epithelium associated with increased cell proliferation and deregulation of apoptosis. Driving factors in the inflammation-associated cancer pathway include the increased oxidative damage with associated mutagenesis caused by the heightened inflammatory infiltration of the mucosa, resulting in a stepwise progression of neoplastic lesions. The molecular players of IBD-associated carcinogenesis are similar to those seen in sporadic colorectal carcinogenesis, but the timing of occurrence of molecular events is different (Figure 19.6).

During chronic colitis, there is activation of NF-κB in the epithelium. NF-κB activates the expression of COX2; several proinflammatory cytokines including IL-1, TNFα, IL-12p40, and IL-23p19; antiapoptotic factor inhibitor of apoptosis protein (IAP); and B-cell leukemia/lymphoma (Bcl-xL) (Figure 19.5). Prostaglandins and cytokines such as IL-6 are released in the inflammatory milieu and activate intracellular serine-threonine kinase Akt signaling, with inhibition of proapoptotic factors *p53*, BAD, and FoxO1, and increased cell survival.

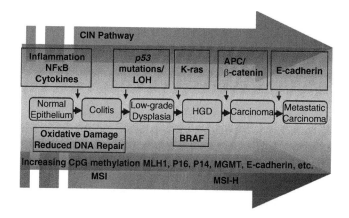

Figure 19.6 Molecular events in the stepwise lesions of inflammatory bowel disease-associated neoplasia.

In colitis-associated cancers, genetic instability includes CIN and MSI, similar to sporadic colon cancers, occurring in 85% and 15% of the cases, respectively. In UC-associated colon cancers, *APC* mutations or LOH of the *APC* locus are seen in 14% to 33% of cancers. *APC* mutations occur late in UC-neoplasia, in the transition from high-grade dysplasia to carcinoma. This is in contrast with sporadic colon carcinogenesis, where the *APC* gene is mutated in the majority (80%) of tumors and occurs early in the neoplastic process. This significant difference in the timing of genetic events in IBD-associated versus sporadic colon carcinogenesis is likely due to the inflammatory environment in IBD. Since other molecular mechanisms, such as the activation of NF-κB in colonic epithelial cells by inflammatory-cell released cytokines, promote cell proliferation and inhibit apoptosis, APC inactivation may not be necessary to drive the early steps of colitis-associated carcinogenesis.

The tumor suppressor gene *p53* is mutated at a high rate in both sporadic and UC colon cancers, but in contrast to sporadic cancers, in UC *p53* mutations occur earlier in the colitis-associated carcinogenesis pathway. Mutations in *p53* were detected in 19% of biopsies without dysplasia, with the frequency increasing with higher grades of dysplasia. This is in contrast with sporadic CRC in which *p53* mutations and LOH are associated with the progression from high-grade adenoma to cancer. Early loss of function of *p53* in UC-associated carcinogenesis contributes to the rapid progression to colon cancer observed in these patients. In the inflamed mucosa, early mutation of *p53* is associated with inflammation-related oxidative damage. The reduced DNA repair, also caused by the inflammatory environment, together with the increased mutation burden and the reduced ability to remove cells with significant mutations, results in clonal selection, expansion of neoplastic cells, and tumor development.

In MSI-positive UC colon cancers, the cause of MSI-H is loss of MLH1 protein expression associated with *MLH1* by promoter hypermethylation. However, MSI in the non-neoplastic mucosa is significantly more frequent in the colitic mucosa as compared to the background mucosa of sporadic colon cancers. This

may be explained by reduced expression and function of DNA repair enzymes induced by oxidative stress free radicals.

Several genes have been reported to be hypermethylated in dysplasia and/or cancer in UC, including genes that have been reported as targets of CpG island methylation in sporadic colorectal cancer, such as the *MLH1* promoter, and the *p16INK4a* promoter regions. Some investigators have reported an increased methylation level in high-grade dysplasia (HGD) of patients with UC versus controls without UC, for estrogen receptor (*ESR1*), *MYOD*, *p16* exon 1, and *CSPG2*, and hypermethylation of three of these genes had similar methylation in the colitic mucosa of patients with HGD. The *p16* methylation levels averaged 2% in the mucosa of controls, 3% in UC patients who had mucosa without dysplasia, 8% in the normal-appearing epithelium of patients with HGD/CA, and 9% in the dysplastic epithelium (HGD/CA). Additionally, methylation was present not just in the neoplastic mucosa but also in the non-neoplastic-appearing epithelium from UC patients with HGD/cancer, suggesting that the increased levels of methylation are widespread in the inflammation-afflicted colon and occur early in the process of tumorigenesis, preceding the histological appearance of dysplasia.

Methylation of the hyperplastic polyposis gene 1 (*HPP1*) was observed in 50% of UC adenocarcinomas and in 40% of dysplasias. In contrast, no non-neoplastic UC mucosa showed *HPP1* methylation. Methylation of the *CDH1* promoter was detected in 93% of the patients with dysplastic biopsy samples, in contrast to only 6% of the patients without dysplasia and by immunohistochemistry areas of dysplasia displayed reduced E-cadherin expression levels. In IBD, promoter hypermethylation of the gene for DNA repair protein O6-methylguanine-DNA methyltransferase (*MGMT*) was detected in 16.7% adenocarcinomas and in 3.7% of mucosal samples with mild inflammation. Notably, *MGMT* is more frequently methylated in sporadic adenomas and carcinomas than in IBD. Extensive methylation characteristic of the CIMP phenotype was observed in 17% of the UC-related cancers, while global DNA methylation measured with a LINE-1 assay was seen in 58% of UC-associated cancers.

Activation of the Raf/MEK/ERK (MAPK) kinase pathway, either through *ras* or *BRAF* mutation, was detected in 27% of all UC-related cancers. Nondysplastic UC mucosa of patients with UC-cancer did not show *BRAF* mutations, indicating that *BRAF* mutations are not an initiating event in UC-related carcinogenesis, but are associated with mismatch-repair deficiency through *MLH1* promoter hypermethylation in advanced lesions. Conversely, K-*ras* mutations may occur in dysplasia, but also in villous regeneration and active colitis. K-*ras* mutations are inversely correlated with *BRAF* mutations in UC cancers.

IBD-Associated Neoplasia: Diagnosis and Clinical Management

The most significant predictor of the risk of malignancy in IBD is the presence of dysplasia on colonic biopsies. Colonoscopy with biopsies to rule out dysplasia is currently used during follow-up of patients with IBD. The current pathological diagnosis of dysplasia relies on a classification based on morphological criteria established in the early 1980s, and includes the following categories: (i) negative for dysplasia; (ii) indefinite/indeterminate for dysplasia; and (iii) positive for low-grade dysplasia (LGD), high-grade dysplasia (HGD), or invasive cancer.

Hereditary Nonpolyposis Colorectal Cancer

Hereditary Nonpolyposis Colorectal Cancer: Genetic and Molecular Basis

HNPCC is an autosomal dominant cancer predisposition syndrome, characterized by early onset of CRC and tumors from other organs, including endometrium, ovary, urothelium, stomach, brain, and sebaceous glands. HNPCC with underlying DNA mismatch repair mutations and characteristic MSI-positive tumors represent the HNPCC/Lynch syndrome, while a group of patients with clinical features of HNPCC but lacking DNA mismatch repair gene mutations represent a separate group of tumors. HNPCC/Lynch syndrome is estimated to represent 4–6% of all colorectal cancer cases. At the molecular level, HNPCC/Lynch syndrome patients inherit germline mutations in one of the DNA mismatch repair genes, leading to defects in the corresponding DNA mismatch repair proteins. Mutations are found most commonly in the *MLH1* and *MSH2* and less frequently in the *MSH6* and *PMS2* genes.

HNPCC/Lynch syndrome is subdivided into Lynch syndrome I, characterized by colon cancer susceptibility, and Lynch syndrome II, which shows all the features of Lynch syndrome I, but patients are also at increased risk for carcinoma of the endometrium, ovary, and other sites. The spectrum of tumors that occur in patients with HNPCC is referred to as HNPCC cancers and includes carcinomas of the colon and rectum, small bowel, stomach, biliary tract, pancreas, endometrium, ovary, urinary bladder, ureter, and renal pelvis. Sebaceous gland adenomas and keratoacanthomas are part of the Muir-Torre syndrome, and tumors of the brain, usually glioblastomas, are seen in the Turcot syndrome (Table 19.1).

Known germline mutations of DNA MMR genes in HNPCC affect the coding region of *MLH1* (approximately 40%), *MSH2* (approximately 40%), *MSH6* (approximately 10%), and *PMS2* (approximately 5%).

Hereditary Nonpolyposis Colorectal Cancer: Natural History, Clinical and Pathologic Features, and Molecular Mechanisms

The average age of presentation of colorectal cancer in HNPCC patients is 45 years of age. Tumors are often multiple or associated with other synchronous or metachronous neoplasms of the HNPCC cancer spectrum. In addition to colorectal cancers, patients may present with tumors of the Muir-Torre syndrome or with the Turcot syndrome.

In HNPCC/Lynch syndrome patients, the lifetime risk of colorectal cancer is up to 80%, and it is up to 60% for endometrial carcinoma. Colorectal cancers are located in the proximal colon in two-thirds of cases, and have detectable microsatellite instability (MSI) in more than 90% of the cases.

A few histopathologic features are characteristically associated with MSI-H colorectal adenocarcinomas that may suggest the possibility of HNPCC (Figure 19.4). These features are not specific to HNPCC cancers, in that they are also seen frequently in sporadic MSI-H tumors and in some tumors that are MSS. Three major histopathologic groups of MSI-H cancers can be recognized: (i) poorly differentiated adenocarcinomas, also described as medullary-type cancers; (ii) mucinous adenocarcinomas and carcinomas with signet ring cell features; and (iii) well-to-moderately differentiated adenocarcinomas. The presence of prominent intratumoral-infiltrating lymphocytes (TILs) is the most predictive finding of MSI-H status (Figure 19.4). TILs may be particularly numerous in poorly differentiated cancers, but also occur in the other morphologic types of HNPCC-associated cancers. The intratumoral-infiltrating lymphocytes are CD3-positive T-cells and most are CD8-positive cytotoxic T-lymphocytes. Peritumoral lymphocytic inflammation and lymphoid aggregates forming a Crohn's-like reaction are also frequent in MSI-H carcinomas.

Patients with HNPCC develop adenomas more frequently and at an earlier age than noncarriers of DNA mismatch repair gene mutations. In HNPCC, patients develop one or few colonic adenomas of traditional type (tubular adenomas and tubulovillous adenomas). The progression from adenoma to invasive carcinoma occurs rapidly, in many patients being less than 3 years, contrasting to a mean of 15 years in patients without HNPCC. Detection of colonic adenomas in HNPCC mutation carriers occurs at a mean age of 42–43 years (range 24–62 years). Compared to sporadic adenomas, HNPCC adenomas are more frequently proximal in the colon, and more frequently show high-grade dysplasia. In HNPCC a significant association was found between MSI-H and high-grade dysplasia in adenomas, with associated loss of either MLH1 or MSH2. Based on these findings it was recommended that immunohistochemical staining/MSI testing of large adenomas with high-grade dysplasia in young patients (younger than 50 years) may be performed to help identify patients with suspected HNPCC.

The deficient or absent DNA mismatch repair protein(s) in HNPCC/Lynch syndrome is dictated by the gene that carries a germline mutation. Epigenetic silencing through CpG methylation of *MLH1*, which is the underlying mechanism of MSI in sporadic colon cancer, is rare in HNPCC tumors. In addition, the patients with unambiguous germline mutation in DNA mismatch repair genes do not appear to carry *BRAF*-activating mutations in their tumors. As in sporadic MSI carcinomas, loss of DNA mismatch repair may lead to accumulation of mutations in cancer-related genes, such as the *TGFβRII* gene. HNPCC tumors show frequent aberrant nuclear beta-catenin, but aberrant *p53* expression, 5q loss of heterozygosity, and K-*ras* mutation are uncommon, similar to sporadic MSI cancers.

Hereditary Nonpolyposis Colorectal Cancer: Molecular Diagnosis, Clinical Management, and Genetic Counseling

To identify patients with HNPCC, criteria known as the Amsterdam criteria were established in the early 1990s, with later revisions. Because the Amsterdam criteria do not include some patients with known germline MMR gene mutations, another set of guidelines, known as the Bethesda guidelines, was established to help decide whether or not a patient should undergo further molecular testing to rule out HNPCC. Through use of the Amsterdam II criteria, patients are diagnosed with HNPCC when the following criteria are present: (i) the family includes three or more relatives with an HNPCC-associated cancer, (ii) one affected patient is a first-degree relative of the other two, (iii) two or more successive generations are affected, (iv) cancer in one or more affected relatives is diagnosed before the age of 50 years, (v) familial adenomatous polyposis is excluded in any cases of colorectal cancer, and (vi) tumors are verified by pathological examination. Alternatively, one of the following criteria (Modified Amsterdam) needs to be met: (i) very small families, which cannot be further expanded, can be considered to have HNPCC with only two colorectal cancers in first-degree relatives if at least two generations have the cancer and at least one case of colorectal cancer was diagnosed by the age of 55 years; (ii) in families with two first-degree relatives affected by colorectal cancer, the presence of a third relative with an unusual early onset neoplasm or endometrial cancer is sufficient; (iii) if an individual is diagnosed before the age of 40 years and does not have a family history that fulfills the preceding criteria (Amsterdam II and Modified Amsterdam criteria), that individual is still considered as having HNPCC.

If an individual has a family history that is suggestive of HNPCC but does not fulfill the Amsterdam, modified Amsterdam, or young age at onset criteria, that individual is considered to be HNPCC variant, or familial colorectal cancer X. A large group of patients representing 60% of all cases who meet Amsterdam I or Amsterdam II criteria for HNPCC do not have characteristic features of MMR deficiency. Compared to the MSI-HNPCC patients, the MSS HNPCC patient's age at diagnosis is 6 years higher on average, and most colorectal cancers appear on the left side of the colon. The underlying genetic defect for these tumors is not yet known.

The most recently revised Bethesda criteria recommend testing of patients to rule out HNPCC if there is one of the following criteria: (i) patient is diagnosed with colorectal cancer before the age of 50 years; (ii) presence of synchronous or metachronous colorectal or other HNPCC-related tumors (stomach, urinary bladder, ureter and renal pelvis, biliary tract, brain [glioblastoma], sebaceous gland adenomas, keratoacanthomas, and small bowel), regardless of age; (iii) colorectal cancers with a high-microsatellite instability morphology (presence of tumor-infiltrating lymphocytes, Crohn's-like lymphocytic reaction, mucinous or signet ring cell differentiation, or medullary growth pattern) that was diagnosed before the age of 60 years; (iv) colorectal cancer patient with one or more first-degree relatives with colorectal cancer or other HNPCC-related tumors, and one of the cancers must have been diagnosed before the age of 50 years; and (v) colorectal cancer patient with two or more relatives with colorectal cancer or other HNPCC-related tumors, regardless of age. Given recent knowledge of the molecular changes underlying MSI-sporadic as compared to HNPCC-associated cancers, algorithms have been established to determine whether a patient has sporadic MSI-positive cancer or HNPCC/Lynch syndrome (Figure 19.7).

When a patient with HNPCC cancer is identified by the Amsterdam criteria or by the revised Bethesda criteria, the next step is to evaluate MSI with the MSI test and/or immunohistochemical (IHC) analysis of tumors for MSH2 and MLH1, MSH6, and PMS2 DNA mismatch repair proteins.

The MSI test is based on the evaluation of instability in small DNA segments that consist of repetitive nucleotides of generally 100 to 200 base pairs in length, called microsatellite regions or short tandem repeats (STRs). The most used set of microsatellite markers was recommended by an NCI consensus group and consists of two mononucleotide repeat markers (*BAT25* and *BAT26*) and three dinucleotide repeat markers (*D2S123*, *D5S346*, and *D17S250*). The results of the MSI test using the NCI panel of five microsatellite markers are reported as MSI-High (MSI-H), MSI-Low (MSI-L), and microsatellite stable (MSS). MSI-H tumors show MSI in at least two of the five markers, MSI-Low (MSI-L) tumors show MSI in only one marker, and no instability is detected in any of the five markers in MSS tumors. Figure 19.4 illustrates a colorectal cancer with loss of expression of MSH2 in the tumor cells and associated MSI-H identified by microsatellite instability at both the *BAT25* and *BAT26* markers (Figure 19.4).

If the tumor tissue reveals MSI-H and/or there is loss of expression of one of the DNA repair proteins by immunohistochemistry, germline testing should be performed for the gene encoding the deficient protein, after appropriate genetic counseling of the patient. If tissue testing is not feasible, or if there is sufficient clinical evidence of HNPCC, it is acceptable to proceed directly to germline analysis of the *MSH2* and/or *MLH1* genes. The likelihood of finding a germline mutation in the *MLH1/MSH2* genes of patients with colorectal cancer tumors that are not MSI-H is low.

If no loss of expression of *MSH2* or *MLH1* is seen in MSI-H tumors or if the tumor is MSI-L or MSS but there is suspicion of HNPCC, evaluation of other MMR genes, in particular *MSH6* and *PMS2*, should be performed, first by immunohistochemical stains,

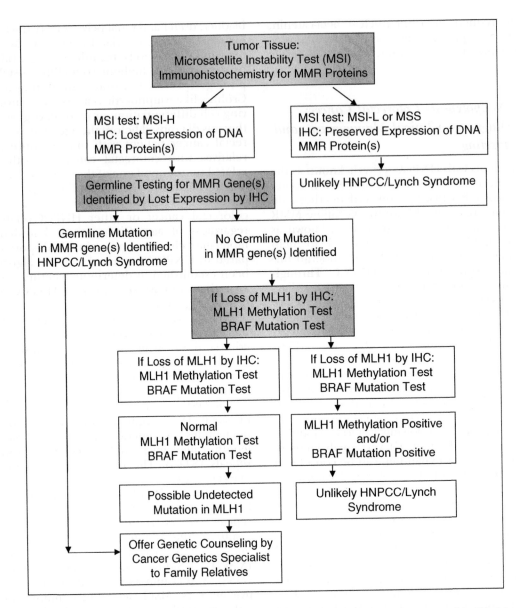

Figure 19.7 **Algorithm for molecular testing of HNPCC/Lynch syndrome colorectal cancer.** Modified from Lindor NM, Petersen GM, Hadley DW, et al. Recommendations for the care of individuals with an inherited predisposition to Lynch syndrome: A systematic review. Reprinted with permission from *JAMA*. 2006;296:1507–1517.

followed by germline mutational analyses. Identification of a germline mutation in index cancer patients is important because it confirms a diagnosis of HNPCC, and the identified mutation may be used to screen at-risk relatives who may be mutation carriers. If the tumor tissue revealed loss of expression of MLH1 by immunohistochemistry, but no mutation in the DNA mismatch repair genes underlying HNPCC/Lynch syndrome are found, two other tests, the *MLH1* methylation and *BRAF* mutation tests, may help discriminate between a sporadic MSI-tumor and HNPCC tumor with undetected *MLH1* mutation (Figure 19.7). Through use of quantitative methylation analyses, HNPCC patients showed no or low level of *MLH1* promoter methylation, in contrast to high levels of methylation (greater than a cutoff

value of 18% methylation) in sporadic MSI cancers. In addition, none of the patients with unambiguous germline mutation in DNA mismatch repair genes demonstrated *BRAF* mutation. Therefore, adding *BRAF* mutation and *MLH1* methylation tests in the algorithm for testing of MSI-H colon tumors with loss of expression of MLH1 protein in the tumor can help determine whether a tumor is likely to be a sporadic or an HNPCC-associated tumor (Figure 19.7).

In families that meet strict clinical criteria for HNPCC, germline mutations in *MSH2* and *MLH1* have been found in 45–70% of the families, and germline mutations in these two genes account for 95% of HNPCC cases with an identified mutation. The reported data show that despite extensive testing there is still a significant number of families without an identified germline mutation that accounts for HNPCC. The germline mutations that occur in *MSH2* and *MLH1* are widely distributed throughout the two genes. Two hundred fifty-nine pathogenic mutations and 45 polymorphisms have been reported in *MLH1*, and 191 pathogenic mutations and 55 polymorphisms have been identified in *MSH2*.

Individuals diagnosed as carriers of DNA mismatch repair gene mutations seen in the HNPCC/ Lynch syndrome are recommended to have colonoscopic surveillance staring at early age. Colonoscopy is recommended every 1 to 2 years starting at ages 20 to 25 years (age 30 years for those with *MSH6* mutations), or 10 years younger than the youngest age of the person diagnosed in the family. Although there is limited evidence regarding efficacy, the following are also recommended annually: (i) endometrial sampling and transvaginal ultrasound of the uterus and ovaries (ages 30–35 years); (ii) urinalysis with cytology (ages 25–35 years); (iii) history, examination, review of systems, education, and genetic counseling regarding Lynch syndrome (age 21 years). For individuals who will undergo surgical resection of a colon cancer, subtotal colectomy is favored. Further, evidence supports the efficacy of prophylactic hysterectomy and oophorectomy.

Familial Adenomatous Polyposis (FAP) and Variants

Familial Adenomatous Polyposis: Genetic Basis

FAP is a cancer predisposition syndrome characterized by numerous adenomatous colorectal polyps, with virtually universal progression to colorectal carcinoma at an early age. It accounts for less than 1% of all colorectal carcinoma cases in the United States and affects 1 in 8000–10,000 individuals. The majority of cases of FAP are caused by germline mutations in the *APC* gene on chromosome 5q, and in its inherited form, FAP is transmitted in an autosomal dominant fashion, although up to a third of cases may present as *de novo* germline mutations.

Familial Adenomatous Polyposis: Natural History and Clinical, Molecular, and Pathologic Features

Patients affected by FAP have a nearly 100% lifetime risk for the development of colorectal carcinoma in the absence of aggressive treatment, which often includes prophylactic colectomy. In addition, patients have a 90% lifetime risk for the development of upper gastrointestinal tract polyps, with a 50% risk of developing advanced duodenal polyposis by the age of 70 years.

Clinically, the common manifestation of FAP and its variants is the presence of numerous (sometimes in excess of 1000) adenomatous polyps distributed throughout the colon and rectum. The characteristic gross findings, combined with the histologic findings

and predictable progression to colorectal carcinoma, have historically made FAP a robust model system to better understand carcinogenesis development and progression. There are several variants of FAP, including Gardner Syndrome, Turcot Syndrome, and Attenuated FAP (AFAP), summarized in Table 19.1. Gardner syndrome is characterized by multiple extra-colonic manifestations including osteomas, desmoid tumors, dental abnormalities, ophthalmologic abnormalities including congenital hypertrophy of retinal pigment epithelium (CHRPE), and cutaneous cysts. Some suggest that some degree of these extra-intestinal manifestations may be identified with close scrutiny in typical FAP patients. Turcot syndrome is the association of the colorectal polyposis and brain tumors, most commonly medulloblastoma. Attenuated FAP demonstrates a reduction in the number of colonic polyps, usually falling short of the 100 polyps necessary for a diagnosis of FAP, but with sufficient colonic polyposis (frequently over 15) to raise suspicion for an underlying polyposis syndrome.

One of the most striking features of FAP and its variants (except AFAP) is the presence of hundreds, perhaps even thousands, of polyps throughout the colon and rectum, leading to a carpet appearance of the colorectal mucosa (Figure 19.8). The polyps are frequently sessile, and appear as early as late childhood/early adolescence, requiring that endoscopic screening in familial cases begin early in life. The histologic features of polyps in FAP are essentially indistinguishable from those seen in sporadic adenomas; however, it is common to identify lesions at various stages in the dysplasia-adenoma-carcinoma sequence, further underscoring the multistep nature of carcinogenesis (Figure 19.8).

The majority of cases of FAP and its variants are attributed to germline mutation in the *APC* gene, located at the 5q21–22 chromosome locus. The protein product of the *APC* gene serves as a key mediator in the Wnt pathway of signal transduction for cellular growth and proliferation (Figure 19.8). Wnt binding to cellular receptors initiates a series of downstream signals which result in increased transcription of cell growth and proliferation-associated genes, through the effect of the protein β-catenin. β-catenin is a transcriptional regulator which must be localized to the nucleus in order to impart its effect on transcription. In the absence of a Wnt-mediated growth signal, APC serves as part of a complex which destabilizes β-catenin through phosphorylation, thereby targeting it for destruction by the proteasome, thus preventing its nuclear localization and transcriptional effects (Figure 19.8A). In the presence of Wnt-mediated growth signal, the APC-β-catenin protein complex is disrupted, and phosphorylation cannot occur, resulting in increased stability and nuclear localization of β-catenin (Figure 19.8B). As is common in many examples of neoplasia, the disease state is one which mimics the activated state, and mutations in APC commonly cause a disruption of the protein complex which destabilizes β-catenin, resulting in constitutive activation of the Wnt pathway (Figure 19.8C).

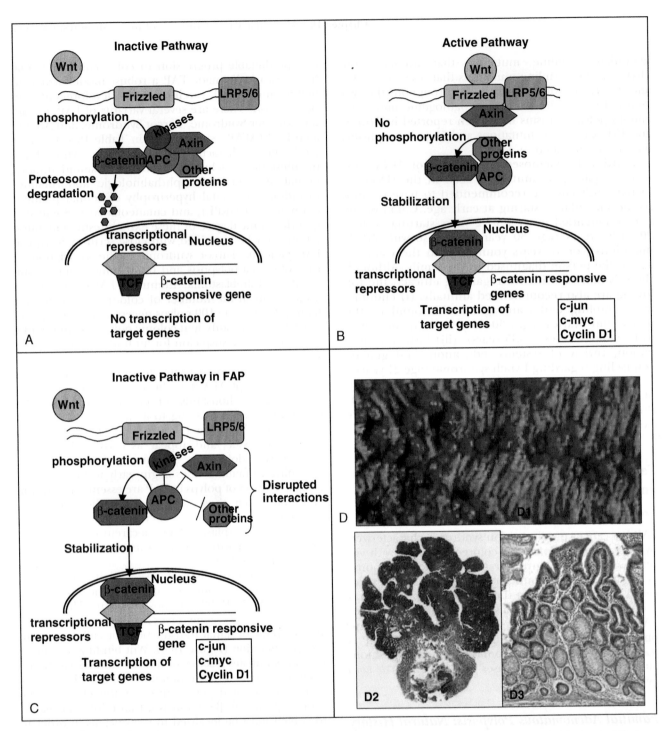

Figure 19.8 The Wnt pathway and familial adenomatous polyposis. (A) Signaling of the Wnt pathway is mediated through the Frizzled family of receptors, and a co-receptor LRP5 or 6. In the absence of ligand, the pathway is inactive through the negative regulation of the downstream effector β-catenin. When present in sufficient quantity in the nucleus, β-catenin stimulates transcription of target genes. Lack of signaling through the Frizzled receptor results in sequestration of β-catenin in a multiprotein complex including APC, Axin, and other proteins, which exert a negative effect both through the cytoplasmic sequestration of β-catenin and by targeting it for destruction by the proteasome through phosphorylation. **(B)** In the presence of Wnt ligand, the Frizzled receptor and LRP5/6 form a complex which results in the recruitment and sequestration of Axin at the cell surface, thereby inhibiting the kinase activity of the APC complex. This results in the stabilization and nuclear translocation of β-catenin, thus resulting in transcription of target genes. **(C)** In the setting of an *APC* gene mutation, the multiprotein APC/Axin/β-catenin complex is disrupted, most commonly due to truncation of the APC protein in domains responsible for protein-protein interaction. This results in stabilization and nuclear translocation of β-catenin, with the net effect of constitutively active transcription of growth-promoting genes otherwise under tight regulation. **(D)** Typical appearance of polyps in FAP, which are indistinguishable from spontaneous nonsyndromic polyps, but are numerous and may eventually cover most of the surface of the colon with a carpet appearance **(D1**; From the files of the Department of Pathology, Hospital of the University of Pennsylvania). Tubular adenoma **(D2)** and adenomatous colonic mucosa **(D3)** at an early stage before the development of larger adenomas and adenocarcinomas (H&E stain, original magnification 5×).

Familial Adenomatous Polyposis and Related Syndromes: Molecular Diagnosis, Clinical Management, and Genetic Counseling

As the majority of cases of FAP have been attributed to mutation in *APC*, there has been extensive investigation of the function of its protein product. In addition to its role in FAP, the understanding of the role of *APC* bears special relevance, as over 70% of nonsyndromic colorectal carcinomas are found to have somatic mutation of *APC*. *APC* codes for a 312 kDa protein that is expressed in a wide range of tissues and is thought to participate in several cellular functions including Wnt-mediated signaling, cell adhesion, cell migration, and chromosomal segregation. Within the APC protein, there are multiple domains which are responsible for these varied functions, including an oligomerization domain; an armadillo domain region, which is thought to be involved in binding of APC to proteins related to cell morphology and motility; β-catenin binding domain, axin

binding domain; and a microtubule binding domain (Figure 19.9). The majority of germline mutations associated with FAP are either frameshift or nonsense mutations which lead to a truncated protein product, thereby disrupting the interaction with β-catenin, leading to its stabilization and subsequent downstream signaling. Two hotspots have been identified for germline mutations in *APC*, at codons 1061 and 1309, and account for 17% and 11% of all germline APC mutations, respectively (Figure 19.9). The region between codons 1286 and 1513 is termed the mutation cluster region (MCR) to reflect the observation that this region encompasses many of the identified *APC* mutations.

Both within and beyond the MCR, there is some degree of association between the location of the germline *APC* mutation and the clinical phenotype. Mutations within the MCR are associated with a profuse polyposis, with the development of over 5000 colorectal polyps. Attenuated polyposis is seen in the settings of mutations in the 5′-end of the *APC* gene

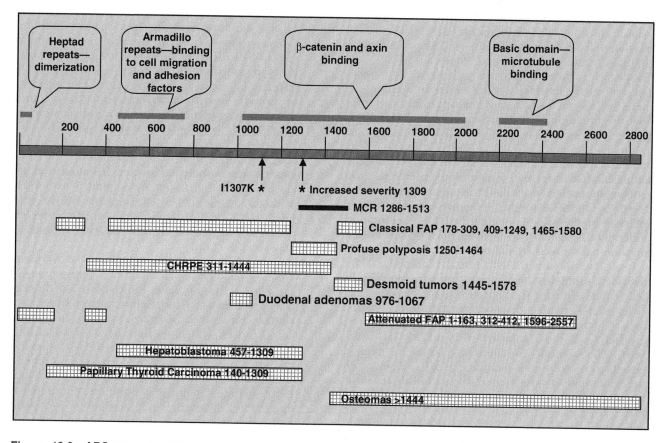

Figure 19.9 APC gene mutation and phenotype correlation in FAP. The figure illustrates the *APC* gene functional domains and mutation-phenotype correlations. The APC protein product consists of 2843 amino acids with multiple functional domains, including a dimerization domain, microtubule domain, and binding sites for β-catenin and axin (pictured as gray boxes). Mutations within certain codon ranges may correlate to a clinical phenotype as depicted in the yellow boxes. Germline mutations generally fall within the entire spectrum of depicted mutational sites, while somatic mutations tend to cluster within the mutation cluster region (MCR). Of note, a profuse polyposis phenotype is seen in patients in whom a germline mutation between codons 1250 and 1464 is seen, with the clinical presentation of >5000 polyps. Mutation at codon 1309 is associated with profuse polyposis and earlier onset of disease. I1307K represents a mutation common in the Ashkenazi Jewish population.

(exons 4 and 5), the alternatively spliced form of exon 9, or the 3'-distal end of the gene. An intermediate phenotype is observed with mutations between codons 157 and 1249 and 1465 and 1595. Similarly, genotype-phenotype correlations for extra-intestinal manifestations have been identified. Desmoid tumors and upper gastrointestinal lesions are clinically the most pressing extra-intestinal manifestations, as there is significant morbidity and mortality associated with them. The association of FAP with desmoid tumor formation has been correlated to mutations downstream of codon 1400. An unequivocal correlation between upper gastrointestinal tumors and specific mutations has not been established, although there are data to suggest that some specific mutations are associated with a higher rate of these lesions, including mutations beyond codon 1395, those beyond codon 934, and within codons 564 to 1465. CHRPE is associated with mutations falling between codons 311 and 1444 (Figure 19.9).

Well-established methodologies are in place for screening patients in whom a diagnosis of FAP is a consideration. Recommendations for indications and approach have been suggested by the American Gastroenterological Association, and the primary indications include clinically high suspicion for FAP (>100 colorectal adenomas), first-degree relatives of FAP patients, >20 cumulative colorectal adenomas (suspected AFAP), and first-degree relatives of patients with AFAP. In *de novo* cases or in cases in which a family mutation is unknown, there are several high-throughput approaches to conduct the mutational analysis. Sequencing of the entire coding region is the gold standard for diagnosis, although other methods such as protein truncation tests and mutation scanning approaches can be used. In some settings, initial targeting of common mutational hotspots is the preferred method for initial screening, followed by more extensive sequence analysis if the common mutations are not identified. Once a mutation is identified, targeted genetic testing can be carried out for potentially affected family members.

In a subset of patients, a discrete *APC* mutation is not identified using first-line approaches to diagnosis. In some cases, adjusting the approach to screen individual alleles of *APC* yields a diagnostic mutation or, in some cases, large exonic or entire gene deletions. However, in other cases, these approaches do not identify a discrete *APC* mutation. An alternative target for genetic testing has recently been identified as *MUTYH*, in which biallelic mutations are detected in a significant minority of *APC* mutation-negative cases of polyposis, and is correlated to cases of attenuated FAP (AFAP). There have been a number of approaches for identifying the underlying molecular changes in *APC* mutation-negative cases not resolved by the evaluation of individual alleles or *MUTYH* gene, including examination of epigenetic regulation of the *APC* gene, evaluation of genes encoding other proteins involved in the β-catenin pathway such as Axin, evaluation of allelic mRNA ratios, and evaluation of somatic *APC* mosaicism. Germline hypermethylation of the *APC* gene has been shown not to be a significant cause of FAP in *APC* mutation-negative cases. One study has shown unbalanced *APC* allelic mRNA

expression in cases in which no other discrete mutation was identified. Interestingly, in the tumor specimens from these cases, there was loss of the remaining wild-type allele, indicating that the reduced dosage of *APC* and unbalanced *APC* allelic mRNA expression may create a functional haploinsufficiency that engenders the same predisposition to colorectal carcinogenesis. The mechanism of unbalanced allelic mRNA expression is unclear. Mutations in *AXIN2* have also been rarely identified in *APC* mutation-negative FAP.

Surveillance of those affected by FAP begins early in life with annual sigmoidoscopy or colonoscopy, beginning at age 10 to 12, followed by prophylactic colectomy, usually by the time the patient reaches his or her early 20s. Additionally, regular endoscopy with full visualization of the stomach, duodenum, and periampullary region has been recommended; however, the optimal timing of these screening evaluations is not well established and is generally managed based on the severity of upper gastrointestinal disease burden.

KEY CONCEPTS

- Gastric carcinoma is the fourth most frequent cancer worldwide and the second most common cause of death from cancer. Most gastric cancers develop as sporadic cancers, without a well-defined hereditary predisposition. The majority of sporadic gastric cancers arises in a background of chronic gastritis, which is most commonly caused by *H. pylori* infection of the stomach.

- The pathogenesis of gastric cancer is multifactorial, resulting from the interactions of host genetic susceptibility factors, environmental exogenous factors that have carcinogenic activity such as dietary elements (salted, smoked, pickled, and preserved foods) and smoking, and the complex damaging effects of chronic gastritis related to *H. pylori* infection.

- Colorectal cancer is one of the most frequent types of cancers worldwide with approximately 1 million cases annually. Most colorectal cancers do not develop in association with a hereditary cancer syndrome. However, several hereditary colon cancer syndromes have been characterized, including Lynch syndrome (HNPCC), familial adenomatous polyposis (FAP), and several others.

- The molecular pathways of colon cancer development include a stepwise acquisition of mutations, epigenetic changes, and alterations of gene expression, resulting in uncontrolled cell division, and manifestation of invasive neoplastic behavior. The major molecular pathways of colorectal cancer development include (1) the chromosomal instability pathway (CIN), (2) the microsatellite instability pathway (MSI), and (3) the CpG island methylator pathway (CIMP).

- HNPCC is an autosomal dominant cancer predisposition syndrome, characterized by early onset CRC (accounting for 4–6% of CRC) and tumors in other organs (including endometrium, ovary, urothelium, stomach, brain, and sebaceous glands). HNPCC patients inherit germline mutations in one of the DNA mismatch repair genes (most commonly in the *MLH1* and *MSH2*), leading to defects in the corresponding DNA mismatch repair.

- FAP is a cancer predisposition syndrome characterized by numerous adenomatous colorectal polyps, with uniform progression to colorectal carcinoma at an early age, accounting for more than 1% of all CRC cases in the United States. The majority of FAP cases are related to autosomal dominant inheritance of germline mutations in the *APC* gene on chromosome 5q. Clinically, the common manifestation of FAP is the presence of numerous (sometimes >1000) adenomatous polyps distributed throughout the colon and rectum.

SUGGESTED READINGS

1. Boland CR, Thibodeau SN, Hamilton SR, et al. A National Cancer Institute Workshop on Microsatellite Instability for cancer detection and familial predisposition: Development of international criteria for the determination of microsatellite instability in colorectal cancer. *Cancer Res.* 1998;58:5248–5257.
2. Brenner H, Arndt V, Sturmer T, et al. Individual and joint contribution of family history and *Helicobacter pylori* infection to the risk of gastric carcinoma. *Cancer.* 2000;88:274–279.
3. Correa P. *Helicobacter pylori* and gastric carcinogenesis. *Am J Surg Pathol.* 1995;19:S37–S43.
4. Correa P, Houghton J. Carcinogenesis of *Helicobacter pylori.* *Gastroenterology.* 2007;133:659–672.
5. Goel A, Arnold CN, Niedzwiecki D, et al. Characterization of sporadic colon cancer by patterns of genomic instability. *Cancer Res.* 2003;63:1608–1614.
6. Grady WM. Genetic testing for high-risk colon cancer patients. *Gastroenterology.* 2003;124:1574–1594.
7. Kang GH, Lee S, Cho NY, et al. DNA methylation profiles of gastric carcinoma characterized by quantitative DNA methylation analysis. *Lab Invest.* 2008;161–170.
8. Kang GH, Lee S, Kim JS, et al. Profile of aberrant CpG island methylation along the multistep pathway of gastric carcinogenesis. *Lab Invest.* 2003;83:635–641.
9. Lengauer C, Kinzler KW, Vogelstein B. Genetic instability in colorectal cancers. *Nature.* 1997;386:623–627.
10. Lindblom A. Different mechanisms in the tumorigenesis of proximal and distal colon cancers. *Curr Opin Oncol.* 2001;13:63–69.
11. Nielsen M, Hes FJ, Nagengast FM, et al. Germline mutations in APC and MUTYH are responsible for the majority of families with attenuated familial adenomatous polyposis. *Clin Genet.* 2007;71:427–433.
12. Parsonnet J, Friedman GD, Vandersteen DP, et al. *Helicobacter pylori* and the risk of gastric carcinoma. *N Eng J Med.* 1991;325:1127–1131.
13. Rustgi AK. The genetics of hereditary colon cancer. *Genes Dev.* 2007;21:2525–2538.
14. Strickler JG, Zheng J, Shu Q, et al. p53 mutations and microsatellite instability in sporadic gastric cancer: When guardians fail. *Cancer Res.* 1994;54:4750–4755.
15. Tomlinson I, Ilyas M, Johnson V, et al. A comparison of the genetic pathways involved in the pathogenesis of three types of colorectal cancer. *J Pathol.* 1998;184:148–152.
16. Toyota M, Ahuja N, Suzuki H, et al. Aberrant methylation in gastric cancer associated with the CpG island methylator phenotype. *Cancer Res.* 1999;59:5438–5442.
17. Toyota M, Ahuja N, Ohe-Toyota M, et al. CpG island methylator phenotype in colorectal cancer. *Proc Natl Acad Sci U S A.* 1999;96:8681–8686.
18. Vasen HF, Watson P, Mecklin JP, et al. New clinical criteria for hereditary nonpolyposis colorectal cancer (HNPCC, Lynch syndrome) proposed by the International Collaborative group on HNPCC. *Gastroenterology.* 1999;116:1453–1456.
19. Vogelstein B, Fearon ER, Hamilton SR, et al. Genetic alterations during colorectal-tumor development. *N Engl J Med.* 1988;319:525–532.
20. Wu MS, Lee CW, Shun CT, et al. Distinct clinicopathologic and genetic profiles in sporadic gastric cancer with different mutator phenotypes. *Genes Chromosomes Cancer.* 2000;27:403–411.

20

Molecular Basis of Liver Disease

Satdarshan P. Singh Monga ▪ Jaideep Behari

This chapter is dedicated to Dr. Pramod Behari, a pioneering neurosurgeon and wonderful father, and to Dr. Gurdarshan Singh Monga, a loving father and a deeply caring and highly respected family physician.

INTRODUCTION

Liver diseases are a cause of global morbidity and mortality. While the predominance of a specific liver disease varies with geographical location, the breadth of hepatic diseases affecting the underdeveloped, developing, and developed countries is phenomenal, with diseases ranging from infectious diseases of the liver to neoplasia and obesity-related illnesses. This chapter will apprise readers of the progress made toward unraveling the molecular aberrations of several liver diseases that has led to a better understanding of the disease biology with the hope of eventually yielding improved diagnostic, prognostic, and therapeutic tools.

MOLECULAR BASIS OF LIVER DEVELOPMENT

Embryonic liver development is characterized by timely and precise regulatory signals that enable hepatic competence of the foregut endoderm, hepatic specification, and induction followed by hepatic morphogenesis. Clearly, the preceding events are governed by molecular signals that are highly temporal, cell specific, and tightly regulated (Figure 20.1). Liver in mouse begins to arise from the definitive gut endoderm at the embryonic day 8.5 (E8.5) or the 7–8 somite stage. At this time the Foxa family of transcription factors specifies the endoderm to express hepatic genes in the process of hepatic competence. Next, fibroblast growth factor 1 (FGF1) and FGF2, which originate from the cardiac mesoderm, initiate the expression of liver-specific genes in the endoderm. FGF8, which

is important for the morphogenetic outgrowth of the liver, is also expressed during this stage. The hepatic bud next migrates into the septum transversum mesenchyme under the direction of bone morphogenic protein 4 (BMP4) signaling, which is essential for hepatogenesis.

This stage is followed by the phase of embryonic liver growth characterized by the expansion and proliferation of the resident cells within the hepatic bud. Several transcription factors including Hex, Gata6, and Prox1 are the earliest known mediators of this phase (Figure 20.1). Once the hepatic program is in full swing, the liver growth continues and is now labeled as the stage of hepatic morphogenesis. The epithelial cells at this stage are now considered the hepatoblasts, or the bipotential progenitors, which means that they are capable of giving rise to both major lineages of the liver, the hepatocytes, and the biliary epithelial cells. Hepatoblasts will be undergoing expansion while maintaining their dedifferentiated state during this stage. While distinct from the traditionally known stem cell renewal, this event marks the expansion of a lineage-restricted progenitor population. Several key players at this stage include the HGF/c-Met, β-catenin, TGFβ, embryonic liver fodrin (Elf), FGF8, FGF10, Foxm1b, and Hlx, which are regulating the proliferation and survival of resident cells as well as regulating their survival. Additional factors at this stage have also been identified although the mechanisms are less clear. These include components of NFκB, c-jun, XBP1, K-ras, and others. At this time the general architecture of the liver is beginning to be established, including the formation of sinusoids and the development of hepatic vasculature.

The final stage is characterized by the differentiation of hepatoblasts to mature functional cell types: the hepatocytes and the biliary epithelial cells. For maturation into the hepatocytes, a huge change in cell morphology is observed from earlier stages to E17, when the resident epithelial cells acquire a cuboidal morphology with definitive cell polarity and clear cytoplasm after losing their blast characteristics such as the high nuclear

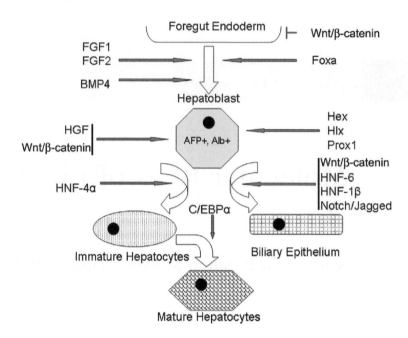

Figure 20.1   Summary of molecular signaling during liver development in mouse. Abbreviations: FGF—Fibroblast growth factor; BMP—Bone morphogenic protein; AFP—α-fetoprotein; Alb—Albumin; HGF—Hepatocyte growth factor; HNF—Hepatocyte nuclear factor; C/EBPα—CCAAT enhancer binding protein-alpha.

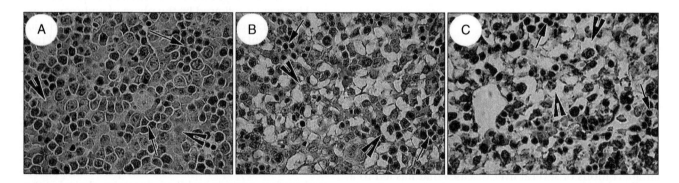

Figure 20.2   H&E of developing mouse livers. (A) Several hematopoietic cells (arrow) are seen interspersed among hepatoblasts (arrowhead), which display large nuclei and scanty cytoplasm, in an E14 liver section. **(B)** Fewer hematopoietic cells (arrow) are observed among the hepatocytes (arrowhead) which begin to show cuboidal morphology and large clear cytoplasm as well as begin to display polarity, in an E17 liver section. **(C)** Similar cuboidal morphology of hepatocytes (arrowhead) is seen in an E19 mouse liver.

to cytoplasmic ratio (Figure 20.2). At the center of the hepatoblast to hepatocyte differentiation process are the liver-enriched transcription factors such as hepatocyte nuclear factor (HNF4α) transcription factors and CCAAT enhancing binding protein-α (C/EBPα). HNF4α is essential for differentiation toward a hepatocyte phenotype, as well as formation of the parenchyma. The liver-enriched transcription factors enable functioning of the fetal hepatocytes by directing the expression of various genes that are classically associated with hepatocyte functions at this stage, including the cytochrome P450s, metabolic, and synthetic enzymes. The hepatocytes at this stage clearly show glycogen accumulation and have been shown to exhibit many functions of adult hepatocyte including xenobiotic metabolism. Again several signaling pathways have

been shown to play a role in regulating hepatocyte maturation by regulating the expression of liver-enriched transcription factors. Some of these pathways include HGF, EGF, FGFs, Wnt/β-catenin signaling, and others.

While the signals involved in lineage commitment of hepatoblasts to biliary epithelium are not fully understood, there is evidence of the involvement of HNF6 (One-cut-1; OC-1), HNF1β, OC-2, TGFβ, and activin. HNF6 promotes hepatocyte over biliary development by activating HNF1β, which in turn attenuates early biliary cell commitment. It has been shown that HNF6 and OC-2 double knockouts display cells with both biliary and hepatocyte programs activated, suggesting their role in segregating the two lineages. They do so by regulating the TGFβ/activin signaling gradient in developing livers by controlling expression of TGFβ, TGFβ receptor type II

(TBRII), activin B, α2-macroglobulin (TGFβ-antagonist), and follistatin (activin antagonist). At E12.5, TGFβ activity is high in the vicinity of portal vein, where hepatoblasts differentiate into biliary cells. While lineage specification begins at E12.5, bile duct differentiation becomes apparent at E15.5, where biliary cells are organized around branches of portal veins and express cytokeratins, show basal lamina (laminin-positive) toward the portal side of the ductal plate, and lack HNF4α. HNF6 is required for bile duct morphogenesis at this stage. The Notch signaling pathway is also known to play a role in the development of biliary epithelia. Various results support the fact that Jagged/Notch interactions may activate a cascade of events involving HNF6, HNF1β, and *Pkhd1* gene encoding the ciliary protein polyductin or fibrocystin. Additional roles of Foxm1b and Wnt/β-catenin signaling have also been reported in biliary differentiation, although the mechanism is less clear.

This somewhat simplistic outline of embryonic liver development does not take into account the complexity involved in the expression of these growth and transcription factors. Liver development is clearly not a linear process. Rather, there is a significant overlap between gene expression patterns that blur the lines between one stage of liver development and the next. Additionally, activation of one gene may initiate a feedback mechanism that regulates cross-talk between different cell populations. Finally, timing remains critical since certain pathways can act at different stages to inhibit or stimulate certain stages or processes of the hepatic development. A classic example is the Wnt/β-catenin pathway, which needs to be initially repressed to induce the hepatic program in the foregut endoderm, but immediately following that stage, it becomes indispensable. Also, the same pathway plays a role in hepatoblast expansion and survival, but a later stage is indispensable for its maturation into hepatocyte, at the same time playing an early role in biliary differentiation.

MOLECULAR BASIS OF LIVER REGENERATION

The liver is a unique organ with an innate ability to regenerate, and rightfully so, since it is the gatekeeper to a variety of absorbed materials (both nutrients and toxins) through the intestines. Thus, with an overwhelming ongoing assault on a daily basis, the liver must regenerate as and when necessary in order to continue its functions of synthesis, detoxification, and metabolism. This capacity of regeneration is ascertained by activation of multiple signaling pathways as evidenced in many animal models and studies where surgical loss of liver mass triggers the process of regeneration. This ensures proliferation and expansion of all cell types of the liver to enable restoration of lost hepatic mass. It is imperative to identify the various pathways that form the basis of initiation, continuation, and termination of the regeneration process to understand the dysregulation that is often seen in aberrant growth in benign and malignant liver tumors. Simultaneously, understanding such a complex mechanism would be critical for regenerative medicine, stem cell

transdifferentiation, hepatic tissue engineering, and cell-based therapies. The list of factors independently shown to play important roles in hepatocyte proliferation during regeneration is exhaustive and will be discussed concisely henceforth.

Partial hepatectomy triggers a sequence of events that proceed in an orderly fashion to restore the lost mass within 7 days in rats, 14 days in mice, and 8–15 days in humans. Following these periods, the liver lobules become larger, and the thickness of hepatocyte plates is doubled as compared to the prehepatectomy livers. However, over several weeks there is gradual lobular and cellular reorganization, leading to an unremarkable and undistinguishable liver histology from a normal liver.

Within minutes after hepatectomy, there are specific changes in the gene expression as well as at post-translational levels, which complement each other and lead to a well-orchestrated event of regeneration, during which time the hepatocyte functions are maintained for the functioning of the animal. More than 95% of hepatocytes will undergo cell proliferation during the process of regeneration. The earliest known signals involve both the growth factors and cytokines. While it is difficult to lay out the exact chronology of events, it is important to say that many events are concomitant, ensuring proliferation and maintaining liver functions at the same time. The earliest events observed include activation of uPA enabling activation of plasminogen to plasmin, which induces matrix remodeling that leads to, among other events, activation of HGF from the bound hepatic matrix. HGF is a known hepatocyte mitogen that acts through its receptor c-Met, a tyrosine kinase, and is a master effector of hepatocyte proliferation and survival both *in vitro* and *in vivo*. Similarly EGF, which is continually present in the portal circulation, is also assisting in hepatocyte proliferation. Other factors activated at this time include the Wnt/β-catenin pathway, Notch/Jagged pathway, norepinephrine, serotonin, and TGFα. These factors are working in autocrine and paracrine fashion, and cell sources of various factors include the hepatocytes, Kupffer cells, stellate cells, and sinusoidal endothelial cells. Concurrently, the TNFα and IL-6 are being released from the Kupffer cells and have been shown to be important in normal liver regeneration through genetic studies. These pathways are known to act through NFκB and Stat3 activation. Bile acids have also been shown to play a vital role in normal liver regeneration through activation of transcription factors such as FoxM1b and c-myc, necessary for cell cycle transition as well, and decreased hepatocyte proliferation after partial hepatectomy was observed in animals depleted for bile acids with the use of cholestyramine or in the *FXR*-null mice. The eventual goal of these changes is to initiate cell cycle in hepatocytes with a successful G1 to S phase transition dependent on the key cyclins such as A, D, and E to ensure DNA synthesis and mitosis. Additional growth factors such as FGF, PDGF, and insulin are important in regeneration and might be playing an important role in providing homeostatic support to the regenerating liver.

In liver regeneration, while the hepatocytes are the main cell type undergoing proliferation, division of all cells native to the liver is observed at specific times after partial hepatectomy. The peak hepatocyte proliferation in rats is observed at 24 hours after hepatectomy followed by a peak proliferation of the biliary epithelial cells at 48 hours, Kupffer and stellate cells at 72 hours, and sinusoidal endothelial cells at 96 hours.

How does the liver know when to stop? This is complex and incompletely understood. While the liver mass is restored after 14 days in mice and 7 days in rats, it may exceed its original mass, when transient apoptosis occurs and the overall liver mass matches the prehepatectomy liver mass. Based on the mitoinhibitory action of TGFβ on the hepatocytes, it has been suggested to be the terminator of regeneration. However, changes in TGFβ (produced by stellate cells) expression mimics that of many proproliferation factors during regeneration and begins at 2–3 hours after hepatectomy in rats and remains elevated until 72 hours. However, the receptors for TGFβ are downregulated during regeneration, and hence the hepatocytes are resistant to excess TGFβ presence. In addition, during regeneration, TGFβ protein is lost first periportally and then gradually toward the central vein. Just behind the leading edge of loss of TGFβ is the wave of hepatocyte mitosis, suggesting that somehow TGFβ is balancing the act of quiescence and proliferation even during regeneration to perhaps keep the growth regulated and to maintain a certain number of hepatocytes in a nonproliferative and differentiated state, to continue their functions necessary for the animal's survival. However, whether TGFβ is the final cytokine that enables termination of the regenerative process has not been shown convincingly and still remains an open-ended question.

As can be appreciated, multiple cytokines and growth factors are activated in response to partial-hepatectomy where two-thirds of the liver is surgically resected. Liver regeneration is guided by a significant signaling redundancy, which is paramount to inducing the much-needed cell proliferation within the regenerating liver. In addition, when all else fails, additional cell types such as the oval cells can be called upon to restore hepatocytes. These cells are the facultative hepatic progenitors or transient amplifying progenitor cells that originate from the biliary compartment and appear in the periportal areas when the hepatocytes are unable to proliferate. These cells become hepatocytes and rescue the regenerative process.

ADULT LIVER STEM CELLS IN LIVER HEALTH AND DISEASE

Despite the liver's capacity to regenerate and the presence of redundant signaling enabling regeneration on most occasions, there is sometimes a need for stem cell activation in the liver. This term is often used to depict appearance and expansion of the hepatic progenitors in the liver. The basal presence of these facultative stem cells or oval cells or transiently amplifying hepatic progenitors in a normal liver remains debated.

Classically, the activation of adult stem cells is observed as atypical ductular proliferation, which can go on to differentiate into polygonal or intermediate hepatocytes and finally mature into hepatocytes. The oval cell response is typically dictated by the kind of injury (biliary versus hepatocytic versus mixed), which in turn determines the proportion of ductular response to intermediate hepatocytes observed.

In rodents, various models have been optimized to induce activation of stem cells. The basic premise behind these models is the presence of an injury to the hepatocytes after disabling the proliferative capacity of the hepatocytes. This is classically attained in rats by acetylaminofluorene (AAF), which crosslinks DNA in hepatocytes, followed by two-thirds partial hepatectomy, and leads to appearance of oval cells. In mice, the models are more complicated since AAF does not work well. Alternatives used are the administration of diet containing 3,5-diethoxycarbonyl-1,4-dihydrocollidine (DDC) diet. However, DDC primarily causes a biliary injury that also leads to periportal hepatocyte injury, and this model has been recently characterized as a mouse model of PBC and PSC. CDE diet has been used successfully in rats and mice to induce oval cell activation. It should also be emphasized that remarkable heterogeneity was identified between the oval cells observed in response to specific protocols and at least partly could be explained by the differences in the kind of injury that directs oval cell activation toward hepatocyte or biliary differentiation, for the maintenance of liver function, while on these protocols.

Various markers have been applied to detect these oval cells. By histology, these cells are smaller than the hepatocytes and possess high nuclear to cytoplasmic ratio and typically are seen in the periportal region. These cells are concomitantly positive for biliary (CK19, CK7, A6, OV6), hepatocyte (HepPar-1, albumin), and fetal hepatocyte markers (α-fetoprotein). However, at any given time, only a subset of these cells are positive for all markers. This reflects the different stages of differentiation occupied by the cells or the basic heterogeneity of these cell populations. Reactive ducts are also positive for neural cell adhesion molecules (NCAM) or vascular cell adhesion molecules (VCAM). Additional surface markers for the oval cells have been identified in rodent studies, which gives an advantage of cell sorting. The six unique markers include CD133, claudin-7, cadherin 22, mucin-1, Ros1 (oncogene v-ros), and γ-aminobutyrate, type A receptor π (Gabrp).

Using these models, several pathways have been shown to play an important role in the appearance, expansion, and differentiation of stem cells. These factors include HGF/c-met, interleukin-6, TGFα, TGFβ, EGF, Wnt/β-catenin pathway, PPARs, IGF, and others. These signaling pathways function in an autocrine or paracrine manner to induce oval cell activation.

Progenitor cell activation has been seen in patients after various forms of hepatic injury. An acute ductular reaction is observed in the setting of submassive necrosis due to hepatitis, drugs, alcohol, or cholestatic disease, which is subtle during the early stages. Noteworthy hepatic progenitor activation has been observed in alcoholic and nonalcoholic fatty liver disease (ALD and NAFLD) patients as well. These scenarios are associated with increased lipid peroxidation, generation of reactive oxygen species, and additional features of elevated oxidative stress, which is a known inhibitor of hepatocyte proliferation. A positive correlation between the oval cell response and the stage of hepatic fibrosis and fatty liver disease has been identified. In viral hepatitis also, oval cell activation is observed. This is classically seen at the periportal site and is usually proportional to the extent of inflammatory infiltrate. These scenarios also bring into perspective the progenitor or oval cell origin of a subset of hepatocellular cancers (HCC). There are established advantages of stem cells being a target of transformation. It is feasible that the oval cell activation in the pathologies as discussed previously, while providing a distinct advantage of maintenance of hepatic function, might also serve as a basis of neoplastic transformation in the right microenvironmental milieu. This Jekyll and Hyde hypothesis is supported by the observation that several early HCC lesions possess hepatic progenitor signatures at genetic and protein levels. However, conclusive studies to this end are still missing.

MOLECULAR BASIS OF HEPATOCYTE DEATH

Hepatocyte death is a common hallmark of much hepatic pathology. This is often seen as diffuse or zonal hepatocyte death and dropout. In many cases the hepatocyte death occurs due to death receptor activation, which leads to hepatocyte apoptosis and ensuing liver injury. These mechanisms of liver injury have been identified in hepatitis, inflammatory hepatitis, alcoholic liver disease, ischemia reperfusion injury, and cholestatic liver disease.

Fas Activation-Induced Liver Injury

This mode of death is known to be associated with liver diseases such as viral hepatitis, inflammatory hepatitis, Wilson's disease, cholestasis, and alcoholic liver disease. FasL, present on inflammatory cells or Fas-activating agonistic antibodies such as Jo-2 injection (in experimental models), leads to Fas activation resulting in massive hemorrhagic liver injury with extensive hepatocyte apoptosis and necrosis (Figure 20.3). Most mice die within 4–6 hrs after Jo-2 injection. Upon activation, homotrimeric association of Fas receptors

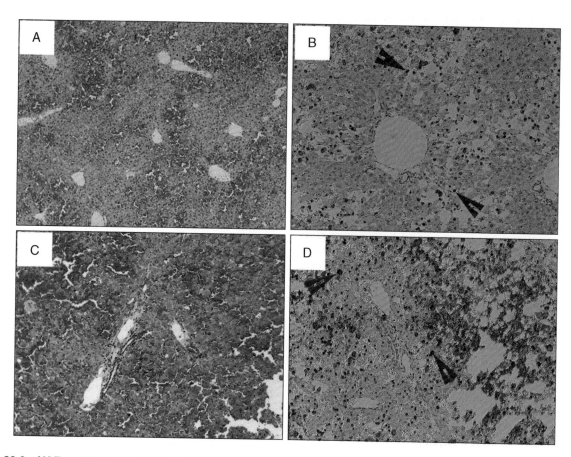

Figure 20.3   H&E and TUNEL immunohistochemistry exhibiting apoptotic cell death after Fas- and TNFα-mediated liver injury. (A) H&E shows massive cell death 6 hours after Jo-2 antibody administration. **(B)** Several TUNEL-positive apoptotic nuclei (arrowheads) are evident in the same liver. **(C)** H&E shows massive cell death 7 hours after D-galactosamine/LPS administration in mice. **(D)** Several TUNEL-positive apoptotic nuclei (arrowhead) are evident in the same liver.

occurs, which recruits Fas-associated death domain protein (FADD) adapter molecule, and initiator caspase-8. This complex, or the death-inducing signaling complex (DISC), requires mitochondrial involvement and cytochrome c release, which is inhibited by antideath Bcl-2-family proteins (Bcl-2 and Bcl-x_L). Cytochrome c stimulates the activation of caspase-9 and then caspase-3, inducing cell death. Interestingly, pro-death Bcl-2 family proteins Bid, Bax, and Bak are also needed for hepatocyte death induced by Fas. In fact, Bid is cleaved by caspase-8 after Fas activation, and cleaved Bid translocates to mitochondria. This in turn activates Bax or Bak on the mitochondria to stimulate the release of apoptotic factors such as cytochrome c. At the same time Bid also induces mitochondrial release of Smac/DIABLO, which inactivates the inhibitors of apoptosis (IAP). The role of IAPs is at the level of caspase-3 activation, which occurs in two steps. The first step is caspase-8-induced severance of larger subunit of caspase-3. The second step is the removal of the prodomain by its autocatalytic activity, which is essential for caspase-3 activation and inhibited by IAPs. While caspase-8 can directly activate caspase-3, bypassing the mitochondrial involvement, the execution of the entire pathway ensures the process of death in the hepatocytes. Thus, overall it has been suggested that relative expression and activity levels of caspase-8, Bid, and other modulators of this pathway, such as inhibitors of apoptosis (IAPs) and Smac/DIABLO, would finally determine whether Fas-induced hepatocyte apoptosis will or will not utilize mitochondria to induce cell death.

A recent discovery unveils another important regulatory step in the Fas-mediated cell death. Under normal circumstances, the Fas receptor was shown to be sequestered with c-Met, the HGF receptor, in hepatocytes. This makes the Fas receptor unavailable to the Fas-ligand. Upon HGF stimulation, this complex was destabilized and hepatocytes became more sensitive to Fas-agonistic antibody. On the contrary, transgenic mice overexpressing extracellular domain of c-Met stably sequestered Fas receptor and hence were resistant to anti-Fas-induced liver injury. Recently, lack of this Fas antagonism by Met was identified in fatty liver disease. While this explains how HGF could be prodeath at high doses or in combination with other death signals, a paradoxical effect of HGF on promoting cell survival is also observed, albeit at lower doses. This effect is usually also observed with other growth-promoting factors such as TGFα and are mediated by elevated expression of Bcl-x_L that inhibits mitochondrial release of cytochrome c and inhibits Bid-induced release of Smac/DIABLO.

TNFα-Induced Liver Injury

This mode of hepatocyte death is commonly observed in ischemia-perfusion liver injury and alcoholic liver disease. In mice, TNFα is induced by bacterial toxin administration such as lipopolysaccharide (25–50 μg/kg LPS). Since it was identified that LPS alone initiates NFκB-mediated protective mechanisms, an inhibitor of transcription (D-Galactosamine) or translation (Cycloheximide) is used before the LPS injection, for successful execution of cell death. This induced massive hepatocyte death due to apoptosis, as seen by TUNEL immunohistochemistry (Figure 20.3). TNFα binds to the TNFα-R1 on hepatocytes to induce receptor trimerization and DISC formation. DISC is composed of TNFα-R1 and TNFR-associated death domain (TRADD), which can recruit FADD and caspase-8 via an unknown mechanism, to further activate caspase-3.

One of the important effects of TNFα stimulation is the unique and concomitant activation of the NF-κB pathway as a protective mechanism. How this occurs is not fully understood, but is relevant as an ongoing protective mode for maintaining hepatic homeostasis.

MOLECULAR BASIS OF NONALCOHOLIC FATTY LIVER DISEASE

Accumulation of triglycerides in the liver in the absence of significant alcohol intake is called nonalcoholic fatty liver disease (NAFLD). The term NAFLD encompasses a spectrum of liver disease ranging from simple steatosis to steatosis with inflammation called nonalcoholic steatohepatitis (NASH). The latter condition can progress to fibrosis, cirrhosis, and hepatocellular cancer.

While NAFLD has been associated with many drugs, genetic defects in metabolism, and abnormalities in nutritional states, it is most commonly associated with the metabolic syndrome. The metabolic syndrome is a group of related clinical features linked to visceral obesity, including insulin resistance (IR), dyslipidemia, and hypertension. NAFLD is strongly associated with and considered the hepatic manifestation of the metabolic syndrome. The prevalence of the metabolic syndrome, and therefore also of NAFLD, has been increasing in parallel with the obesity and diabetes epidemic, and NAFLD is now the most common cause of abnormal liver enzyme elevation in the United States.

While simple steatosis has a benign course, NASH represents the progressive form of the disease. NASH is characterized by necroinflammatory activity, hepatocellular injury, and progressive fibrosis. The pathogenesis of NAFLD/NASH is incompletely understood, although significant strides have been made in recent years to unravel its underlying molecular processes (Figure 20.4). A two-hit hypothesis has been proposed to explain the pathogenesis of NAFLD and the progression of simple steatosis to NASH that occurs in a subset of patients with steatotic livers. The first hit consists of triglyceride deposition in hepatocytes. A second hit consisting of another cellular event then leads to inflammation and hepatocyte injury and the subsequent manifestations of the disease.

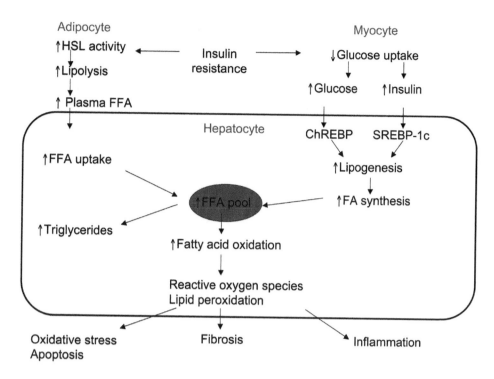

Figure 20.4   Molecular basis of nonalcoholic steatohepatitis. Insulin resistance results in decreased insulin-induced inhibition of hormone-sensitive lipase (HSL) activity in adipocytes resulting in high rates of lipolysis and increased plasma free fatty acid (FFA) levels. High plasma FFA levels increase FFA flux to the liver. In myocytes, insulin resistance results in decreased glucose uptake and high plasma glucose and insulin levels. In the liver, hyperinsulinemia activates SREBP-1c, leading to increased transcription of lipogenic genes. High plasma glucose levels simultaneously activate ChREBP, which activates glycolysis and lipogenic genes. SREBP-1c and ChREBP act synergistically to convert excess glucose to fatty acids. Increased FFA flux and increased lipogenesis increases the total hepatic FFA pool. The increased hepatic FFA pool can either be converted to triglycerides for storage or transport out of the liver as very low density lipoprotein or undergo oxidation in mitochondria, peroxisomes, or endoplasmic reticulum. Increased reactive oxygen species generated during fatty acid oxidation causes lipid peroxidation, which subsequently increases oxidative stress, inflammation, and fibrosis.

Factors Leading to the Development of Hepatic Steatosis: The First Hit

The accumulation of triglycerides in the liver is the essential characteristic of NAFLD/NASH and can be observed in patients as well as experimental models (Figure 20.5). It results from aberration in metabolic processes in the liver, as well as extrahepatic tissue such as skeletal muscle and adipose tissue that is mediated by insulin resistance (IR). Normally, insulin acts on the insulin receptor of myocytes to tyrosine phosphorylate insulin receptor substrate (IRS). IRS, in turn, activates phosphatidyl inositol 3-kinase and protein kinase B, and results in translocation of the glucose transporter to the plasma membrane with resultant rapid uptake of glucose from the blood into the myocyte. The net effect is decreased blood glucose and therefore decreased insulin secretion from the pancreas. In the setting of obesity, fat-laden myocytes become resistant to the signaling effects of insulin. This inability of skeletal muscle to take up glucose from the circulation on insulin stimulation leads to elevated blood glucose levels and increased insulin secretion from the pancreas, with metabolic consequences in the liver as described in the following text.

Besides skeletal muscle, IR also causes increased blood glucose via decreased insulin action in the liver. Normally, the liver plays an important role in maintaining blood glucose levels regardless of the nutritional state. In the fasting state, gluconeogenesis in the liver results in hepatic glucose production that maintains blood glucose level. However, in the fed state when blood glucose levels are elevated, glucose is converted to pyruvate via glycolysis, which then enters the Krebs cycle to form citrate and ultimately acetyl-CoA, which is utilized for fatty acid biosynthesis. This process is normally activated by insulin, which is also a potent inhibitor of hepatic glucose production. However, in the setting of IR, insulin is unable to suppress hepatic glucose production, which, along with decreased myocyte glucose uptake, leads to high blood glucose levels and high circulating insulin levels. The net effect of these changes is increased hepatic lipogenesis and triglyceride accumulation mediated by the synergistic action of two transcription factors: carbohydrate response element binding protein (ChREBP) and sterol regulatory element-binding protein-1c (SREBP-1c). There is evidence from both animal experiments and human studies to support the conclusion that there is increased hepatic fatty acid synthesis in the setting of IR.

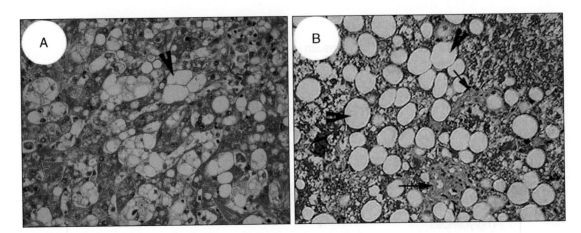

Figure 20.5   Fatty liver disease along with hepatic fibrosis. (A) Patient with NAFLD shows significant macrovesicular (arrowhead) and microvesicular steatosis (blue arrowhead) in a representative H&E stain. **(B)** Masson trichrome staining of liver from mouse on methionine and choline-deficient diet for 4 weeks reveals hepatic fibrosis (arrow) along with macrovesicular steatosis (arrowheads).

ChREBP is a member of the basic helix-loop-helix-leucine zipper (b-HLH-Zip) transcription factor that after activation by glucose translocates from the cytosol into the nucleus and binds to the E-box motif in the promoter of liver-type pyruvate kinase (L-PK), an important glycolytic enzyme. L-PK catalyzes the formation of pyruvate, which enters the Krebs cycle and generates citrate and ultimately acetyl-CoA, which is then utilized for the *de novo* synthesis of fatty acids. Studies with ChREBP knockout mice also revealed that ChREBP also independently stimulates the transcription of fatty acid synthetic enzymes. Therefore, in the setting of IR and hyperglycemia, ChREBP mediates the conversion of glucose to fatty acids by upregulating both glycolysis as well as lipogenesis.

Besides glucose, insulin can also regulate *de novo* hepatic fatty acid synthesis via a second transcription factor, SREBP-1c, also a member of the b-HLH-Zip family of transcription factors. There are three SREBP isoforms, but the SREBP-1c isoform is the major isoform in the liver that can activate expression of fatty acid biosynthetic genes and stimulate lipogenesis. The important role of SREBP-1c in development of hepatic steatosis was established by the demonstration that transgenic mice overexpressing SREBP-1c have increased lipogenesis and develop fatty liver. Furthermore, in the *ob/ob* mouse model of genetic obesity and insulin resistance, disruption of the *Srebp-1* gene caused a reduction in hepatic triglyceride accumulation. Since SREBP-1c is activated by insulin, it would be expected to be inactive in the setting of IR. Surprisingly, the protein is activated by insulin even when there is IR, resulting in increased fatty acid biosynthesis in the liver.

A third process that favors hepatic steatosis results from IR in the adipose tissue. In the fat-laden adipocytes in obese individuals, IR causes defective insulin-mediated inhibition of hormone-sensitive lipase and therefore increased lipolysis and increased free fatty acid (FFA) release into the circulation. Increased FFA uptake into the liver from the circulation also favors the development of hepatic steatosis and inflammation.

Besides SREBP-1c and ChREBP, a transcription factor belonging to the nuclear hormone receptor family, peroxisome-proliferator activated receptor-γ (PPAR-γ) may also contribute to hepatic steatosis. While normal hepatic expression of PPAR-γ is low, increased expression of the transcription factor has been noted in animal models of steatosis. Furthermore, in the *ob/ob* mouse model of insulin resistance, deletion of PPAR-γ from the liver markedly decreases hepatic steatosis, suggesting an important role for PPAR-γ in the development of NAFLD.

Once hepatic steatosis is established, outcomes are variable. In some patients with NAFLD there is no further damage to the liver and prognosis in terms of liver-related mortality is good. However, in some patients there is hepatic necrosis, inflammation, hepatocyte apoptosis, and fibrosis. The reason some patients develop NASH is not understood, but several mechanisms have been postulated, as described in the next section.

Progression of Steatosis to NASH: The Second Hit

Absence of noninvasive biomarkers to differentiate NASH from simple steatosis and lack of animal models that completely recapitulates the pathophysiology of human NASH have limited research into the processes that promote this condition. However, several important insights have been gained in recent years on the pathogenesis of NASH and include oxidative stress, lipid peroxidation, mitochondrial dysfunction, inflammatory cytokines and adipokines, and activation of pathways of cell death.

Oxidative Stress and Lipid Peroxidation

Reactive oxygen species (ROS) refers to several short-lived, pro-oxidant chemicals including hydroxyl radical, singlet oxygen molecules, hydrogen peroxide, and superoxide anions. These pro-oxidant molecules lead to oxidative damage to macromolecules in the cell when they overwhelm the protective antioxidant mechanisms of the cell.

Functional inactivation of essential cellular biomolecules causes either cell death or production of inflammatory mediators via redox-sensitive transcription factors such as Nrf-1 and NFκB. Studies have shown increased serum markers of oxidative stress, and markers of oxidative damage for DNA, and proteins besides lipid peroxides in NASH. Additionally, antioxidant factors such as glutathione S-transferase and catalase are reduced. Polyunsaturated fatty acids (PUFAs) in the cell can undergo peroxidation by ROS, resulting in the formation of malondialdehyde (MDA) and trans-4-hydroxy-2-nonenal (4–HNE) byproducts.

Lipid peroxides contribute to liver injury by increasing production of TNFα, increasing the influx of inflammatory cells, impairing protein and DNA synthesis, and depleting levels of protective cellular antioxidants like glutathione.

Mitochondrial Dysfunction

Several lines of evidence suggest that hepatic mitochondrial dysfunction has a role in the pathogenesis of NASH. Ultrastructural studies have shown the presence of hepatocyte megamitochondria with paracrystalline inclusions in patients with NASH. Functional differences include impaired ability to synthesize ATP after a fructose challenge, which causes transient liver ATP depletion and lower expression levels of mitochondrial DNA-encoded proteins and lower activity of complexes of the mitochondrial respiratory chain in patients with NASH.

Role of Signaling Pathways of Inflammation, Proinflammatory Cytokines, and Adipokines

Inflammation in NASH results from cross-talk between hepatocytes and nonparenchymal cells (activated Kupffer cell, stellate cells, and sinusoidal endothelial cells) mediated by soluble factors, as well as proinflammatory molecules released from visceral adipose tissue. The NFκB pathway has been extensively studied for its role in steatohepatitis and is upregulated in NASH patients and in animal models of the disease. However, its role in pathogenesis of NASH is complex, and while activation of NFκB signaling in liver cells induces inflammation and steatosis, its inactivation also causes steatohepatitis and liver cancer in a mouse model. Deletion of the JNK1 isoform of c-Jun N-terminal kinase (JNK), a mediator of TNF-induced apoptosis, is protective in a mouse model of steatohepatitis, suggesting that JNK signaling is important in the pathogenesis of NASH.

Proinflammatory cytokines TNFα and IL-6 are increased in NASH patients. In animal models of steatohepatitis, decreasing TNFα expression causes decreased hepatic steatosis, cell injury, and inflammation, suggesting a role of this cytokine in the pathogenesis of NASH. However, targeted disruption of TNFα or its receptor TNFR1 does not protect against lipid peroxidation and

hepatocyte injury in diet-induced steatohepatitis. Since inactivation of NFκB, of which TNFα is an effector, does protect hepatocytes in this setting, TNFα may represent one of many inflammatory mediators of steatohepatitis.

Adipose tissue, particularly visceral fat, plays an important role in the pathogenesis of IR and development of NASH. Adipocytes are hormonally active and produce several adipokines that affect insulin action and metabolic processes in the liver. Among the important adipokines are leptin and adiponectin. Leptin levels are increased in the plasma of patients with NAFLD. However, leptin has an antisteatotic role and is protective. Therefore, it has been suggested that NAFLD may be associated with leptin resistance and a defective response to leptin.

Adiponectin levels, on the other hand, are low in patients with NASH, and lower levels are associated with more severe liver injury. Adiponectin has been shown to increase insulin sensitivity, regulate FFA metabolism, and inhibit gluconeogenesis. Adiponectin also has strong anti-inflammatory effects that are mediated by suppressing TNFα levels and activity.

Cell Death

Patients with NASH have increased hepatocyte FasL expression that can trigger apoptosis. When FasL interacts with Fas, it eventually induces apoptotic cascade, leading to cell death and hepatic injury.

MOLECULAR BASIS OF ALCOHOLIC LIVER DISEASE

The histological spectrum of alcohol-induced liver disease spans simple steatosis, to steatohepatitis (characterized by inflammatory cell infiltration, hepatocyte ballooning, apoptosis, and necrosis in addition to fat accumulation), fibrosis, and cirrhosis. Steatosis is mainly macrovesicular and most prominent in the centrilobular region. Early stages of alcohol-induced liver injury are reversible, and prevention of steatosis in experimental models has been shown to prevent development of inflammation and fibrosis.

Pathways of Alcohol Metabolism in the Liver

There are three enzymatic pathways of ethanol metabolism in the liver, and all three contribute to ethanol-induced liver injury. First, oxidation of ethanol occurs through the cytosolic alcohol dehydrogenase (ADH) isoenzymes that produce acetaldehyde, which is then converted into acetate, and these reactions result in the reduction of nicotinamide adenine dinucleotide (NAD) to NADH, its reduced form. Excessive ADH-mediated hepatic NADH generation has multiple important metabolic effects due to inability of hepatocytes to maintain redox homeostasis. This redox imbalance results in inhibition of the Krebs cycle and fatty acid oxidation, which promotes hepatic steatosis. However, with chronic ethanol consumption, the redox state is normalized. Instead, alterations in regulation of lipid metabolism genes appear to be important in this setting. Second, alcohol is metabolized by the microsomal ethanol oxidizing system (MEOS), of which cytochrome P450 2E1 (Cyp2E1) is the key enzyme, although other cytochrome P450 enzymes such as Cyp1A2 and Cyp3A4 also play a role. Cyp2E1 is induced by chronic ethanol consumption, and its induction is responsible for the metabolic tolerance to alcohol in alcohol-dependent individuals. Induction of Cyp2E1 plays a role in alcohol-induced liver injury through the production of reactive oxygen species that can overwhelm the cellular defense systems leading to mitochondrial injury, and further exacerbation of oxidative stress. Finally, ethanol is also metabolized via a nonoxidative pathway by fatty acid ethyl ester (FAEE) synthase. The product of this reaction, fatty acid ethyl esters (FAEEs), accumulates in the plasma, lysosomal, and mitochondrial membranes and can interfere with their functioning and with cellular signal-transduction pathways.

Changes in Expression of Genes Involved in Lipid Metabolism on Chronic Alcohol Exposure

Recent studies suggest that the development of alcoholic and nonalcoholic steatohepatitis (NASH) share many common pathogenic mechanisms. Insight into the molecular mechanisms of alcoholic-induced steatohepatitis has been obtained through the investigation of pathways and regulatory molecules involved in lipid homeostasis that have previously been shown to play a role in development of NASH. Chronic ethanol consumption affects both oxidation of fatty acids and increases *de novo* lipogenesis, thereby causing accumulation of fat in the liver. Peroxisome proliferators-activated receptor-α (PPARα), a member of the nuclear hormone receptor superfamily, is the master regulator of genes involved in free fatty acid transport and oxidation. PPARα forms a dimer with retinoid X receptor (RXR) and binds to the peroxisome proliferators response element (PPRE) in the promoter regions of its target genes. PPARα targets include mitochondrial and peroxisomal fatty acid oxidation pathway genes, apolipoprotein genes, and the membrane transporter, carnitine palmitoyl-transferase I (CPT I), which allows long-chain acyl-CoAs to enter mitochondria to initiate β-oxidation. PPARα also regulates CPT I activity and therefore fatty acid oxidation indirectly via regulation of the enzyme malonyl-CoA decarboxylase, which degrades malonyl-CoA, an allosteric regulator of CPT I. Both *in vitro* in primary hepatocyte cultures and hepatoma cells, and *in vivo*, ethanol has been shown to decrease β-oxidation of fatty acids by interfering with PPARα DNA binding and transcription-activation. Decreased oxidation of fatty acids by ethanol leads to accumulation of fat in the liver.

The second mechanism by which ethanol causes steatosis is by increasing the rate of fat synthesis in the liver by increasing expression levels of lipogenic enzymes that are regulated by the transcription factor SREBP. There are three isoforms of SREBP, called SREBP1a,

SREBP1c, and SREBP2. The isoform SREBP1a is mainly expressed in cultured cells, SREBP1c regulates fatty acid synthesis in the liver, and SREBP2 regulates cholesterol synthesis. SREBPs are synthesized as precursor proteins and are attached to the endoplasmic reticulum and nuclear membranes. When the protein is activated, there is proteolytic cleavage of the NH2-terminal fragment (called the mature form of the protein), which then enters the nucleus and binds to target genes via the sterol response elements (SREs). Recent studies have shown that ethanol-fed mice have increased SREBP1 expression; increase in the amount of mature SREBP1 in the liver; and corresponding upregulation of expression of SREBP-target genes, such as fatty acid synthase, stearoyl-coA desaturase, and ATP citrate lyase, in the fatty acid biosynthetic pathway.

The third critical metabolic regulatory molecule in the liver affected by ethanol is AMP-activated protein kinase (AMPK), which is a metabolic switch regulating pathways of hepatic triglyceride and cholesterol synthesis. AMPK can regulate SREBP1 expression at the transcriptional and post-transcriptional levels. Ethanol has been shown to inhibit hepatic AMPK activity, and treatment with AMPK activators partially blocks ethanol-mediated increase in expression of SREBP-dependent genes *in vitro*. Furthermore, adiponectin, a hormone derived from adipocytes that is an activator of AMPK, has been shown to alleviate both alcoholic and nonalcoholic fatty liver disease in mice.

Inflammatory Cytokines and Role of Kupffer Cells

Besides altering metabolic pathways, chronic ethanol feeding results in altered expression of several inflammatory mediators, including reactive oxygen species, cytokines, and chemokines. An important source of inflammatory mediators from alcohol consumption is hepatic Kupffer cells which produce TNFα, an important mediator of the inflammatory response in mammals. Administration of TNFα to mice causes development of fatty liver, activation of SREBP, and increased fatty acid synthesis. Other studies have shown lipopolysaccharide-mediated increased reactive oxygen species production by Kupffer cells with chronic alcohol feeding and normalization of this response by adiponectin.

Activation of NFκB by alcohol has been shown in experimental models as well as in livers of patients with alcoholic hepatitis. NFκB activation and its target gene expression play an important role in inflammatory response to bacterial endotoxin and in the activation of the innate immune system in response to necrotic cells. In isolated Kupffer cells, chronic ethanol feeding has been shown to increase NFκB binding to the TNFα promoter on LPS stimulation.

MOLECULAR BASIS OF HEPATIC FIBROSIS AND CIRRHOSIS

Chronic liver injury is often associated with a wound healing process in the liver that is commonly referred to as liver fibrosis, which in the advanced stage is termed cirrhosis. Liver cirrhosis can range in its presentation from being asymptomatic to liver failure. Hepatic fibrosis sets in many pathological scenarios including but not limited to alcoholic liver disease, NAFLD, viral hepatitis, autoimmune disorders, Wilson's disease, cholestatic liver disease, and others. The process of fibrosis entails excessive deposition of extracellular matrix in the liver, especially collagen (Figure 20.5). This compromises the hepatic architecture leading eventually to cirrhosis and hepatic failure.

The fundamental cell type involved in the process of hepatic fibrosis is the hepatic stellate cell (HSC). Chronic insult to the liver leads to the activation of hepatic stellate cells. The process of HSC activation entails increased DNA synthesis and proliferation, activation of profibrotic target genes, and increased cell contractility. The end result of this process is hepatic fibrosis and eventually cirrhosis. It should be mentioned that the overall impact of the fibrosis is not only due to excessive deposition of extracellular matrix and loss of functional hepatocyte compartment, but also indirect effects on circulation secondary to constriction of sinusoids. How do HSCs undergo activation? A multitude of insults converge onto this cell type, which upon activation bring about the fibrotic phenotype owing to downstream genetic changes. Broadly speaking, the two major insults, including ASH and NASH, both seem to be mediating HSC activation through increased oxidative stress, albeit via unique mechanisms. Alcohol, which is the leading cause of hepatic fibrosis in Western countries, has been shown to elevate oxidative stress after being metabolized by CYP2E1 to generate reactive oxygen species that are known to directly interact with HSC. In addition, alcohol is also metabolized to acetaldehyde, which in turn has been shown to be fibrogenic. Lastly, Kupffer cells, an additional nonparenchymal cell in the liver, are also involved in the generation of both acetaldehyde and alcohol-induced lipid peroxidation products in alcoholic liver disease. It is worth noting that activation of HSC can very well be secondary to stimuli produced by other cell types present during the injury, including the hepatocytes, Kupffer cells, sinusoidal endothelial cells, and the circulating inflammatory cells.

What signaling mechanisms induce the HSC activation and in turn induce the target gene expression in HSC? Several relevant pathways have been identified, and the most prominent ones include the PDGF axis and TGFβ/Smad signaling pathway. PDGF is a potent mitogen for HSC, and its cognate receptor PDGFR expression goes up during HSC activation. PDGF activation has also been shown to stimulate PI3 kinase, which also induces HSC proliferation. In addition, recruitment of Ras to the PDGF receptor also leads to ERK activation via sequential activation of Raf-1, MAPK-1/2, ERK-1, and ERK-2. Nuclear ERK can regulate target genes responsible for proliferation of HSCs. Additionally JNK activation has been shown to positively regulate HSC proliferation. TGFβ is a potent profibrogenic cytokine and is known to be produced by variety of cells including HSC, hepatocytes, and others. TGFβ

signaling eventually leads to target gene expression, and several of these targets are the hallmark signatures of HSC activation. TGFβ has been shown to stimulate synthesis of collagens, decorin, elastin, and others. Increased expression of KLF6, a transcription factor and a tumor suppressor, which acts as a chaperone for collagen, has also been reported to be increased during HSC activation. Some relevant targets of KLF6 include collagen 1α, TGF-β1 and its receptors, and urokinase plasminogen activator (uPA), which activates latent TGF-β1. Also, connective tissue growth factor (CTGF), which is regulated by TGFβ, is also upregulated during HSC activation, as well as in chronic viral hepatitis, bile duct ligation, and other related scenarios. PPAR-γ ligands have been shown to inhibit HSC proliferation and activation in cell cultures. Sustained basal expression of PPAR-γ in HSC has been suggested to maintain their quiescence. Quiescent HSCs are typically characterized by the presence of perinuclear lipid droplets containing vitamin A or retinoid. These droplets are lost following activation of HSC, when the retinoid is released as retinol. While the causal relationship of this event to HSC activation is unclear, increasing focus is shifting on retinoic acid receptors in the nuclei including RAR and RXR. It is important to point out that a decrease in RXR and PPAR-γ receptor is associated with HSC activation, and the converse has been reported in HSC quiescence.

Once HSC activation occurs, several relevant genes have been shown to be upregulated. The most pertinent include the extracellular matrix genes such as type I collagen (α1 and α2), type III collagen, laminin, and fibronectin, and proteoglycans such as decorin, hyaluronan, heparin sulfate, and chondroitin sulfate. In addition, several genes that are essential for matrix remodeling such as MMP-2, MMP-9, and TIMP-1 have also been shown to be elevated. Lastly, genes such as ICAM-1 and α-SMA are upregulated as well. In addition, several studies have reported a genome-wide analysis and either strengthened the existing findings or identified additional aberrations, which would, in the years to come, be exploited for understanding the biology and in turn devising novel therapies for hepatic fibrosis and eventually cirrhosis.

Cirrhosis is often defined as the advanced stage of hepatic fibrosis, which is accompanied by development of regenerative nodules amidst the fibrous bands and follows a chronic liver injury. This pathology leads to vascular distortions, impaired parenchymal flow leading to portal hypertension, and end-stage liver disease. Thus, the molecular pathology of cirrhosis is a continuum of the hepatic fibrosis. An interesting point to remember is that initially during the process of hepatic fibrosis, increased fibrogenesis is being counterbalanced by factors negatively regulating ECM deposition. However, as chronic insult to the liver continues, the process of fibrogenesis, observed as continued activation of myofibroblasts derived from HSC and perivascular fibroblasts, exceeds fibrinolysis. It remains a conundrum to be able to successfully predict the risk of developing cirrhosis in patients with same hepatic pathology. The most promising advances that are being reported include the identification of genetic polymorphisms that are able to predict the risk of progression of hepatic fibrosis. Studies reporting single nucleotide polymorphisms (SNPs) as predictors of progression of fibrosis are beginning to trickle in. Recently, cytokine SNPs were identified in successfully predicting disease progression. Similarly SNPs in the DDX5 gene successfully predicted fibrosis progression in Hepatitis C patients as well.

MOLECULAR BASIS OF HEPATIC TUMORS

Primary liver tumors presenting as hepatic masses are classified based broadly as benign or malignant. The common benign tumors in the liver include hemangiomas, focal nodular hyperplasia (FNH), and liver cell adenoma or hepatic adenoma (HA). Malignant tumors are usually classified based on the lineage of transformed cells and various clinicopathological characteristics. Broadly, the tumors originating from hepatic progenitors or hepatoblasts are referred to as hepatoblastoma (HB) and the tumors originating from more mature hepatocytes are referred to as hepatocellular cancer (HCC).

Benign Liver Tumors

Hemangioma

Cavernous hemangioma tops the list of the benign tumors of the liver. The tumor arises from the endothelial cells that line the blood vessels within the liver and is thought to entail ectasia rather than hypertrophic or hyperplastic events. The tumor is composed of large vascular channels lined by endothelial cells and collagen lining and separated by connective tissue septa. These endothelial cells have been shown to possess immunocytochemical properties of vascular rather than sinusoidal endothelial cells. The tumors derive their blood supply from the hepatic artery. The pathogenesis of this tumor is poorly understood.

Focal Nodular Hyperplasia

Focal nodular hyperplasia (FNH) occurs at a higher frequency in females and occurs between the ages of 20 and 50. The tumor is usually multinodular, composed of a few hepatocyte-thick plates, and is thought to occur as a hyperplastic response to either a preexisting developmental arterial malformation or to increased blood flow, thus leading to cellular hyperplasia. There is well-known association of FNH with vascular disorders such as Rendu-Osler-Weber syndrome or hereditary telangiectasia. The molecular basis of FNH remains largely obscure. Recently, transcriptome analysis of FNH identified activation of the Wnt/β-catenin pathway without any mutations in the β-catenin gene. While the significance of these findings is unclear, especially since these tumors are thought to be a result of vascular disturbances, these observations might be a result of alternate mechanisms of β-catenin activation such as growth factor-dependent activation. Also, levels of genes encoding for angiopoietin 1 and 2 are altered in FNH, with levels of the ANGPT1/ANGPT2 ratio being greater in all FNH cases examined.

Hepatic Adenoma

Hepatic adenomas (HA) are benign liver tumors that occur in greater frequency in females and are observed as benign proliferation of hepatocytes in an otherwise normal liver. These monoclonal tumors have been classically associated with the use of estrogen-containing oral contraceptives or androgen-containing steroid anabolic drugs. Such tumors are usually observed as solitary masses and are asymptomatic. Additionally, glycogen storage diseases, especially Type I (von Gierke) and Type III, are also known risk factors for the development of hepatic adenoma. However, in these circumstances, HAs oftentimes occur as multiple lesions with a greater propensity to undergo malignant transformation. Macroscopically, HAs present as solitary, yellowish masses due to lipid accumulation and give a pseudo-encapsulated appearance because of the compression of adjacent hepatic tissue and can range from 0.5–15 cm in diameter. Some of the histological features include areas of fatty deposits and hemorrhage, cord-like or plate-like arrangements of larger hepatocytes containing excessive glycogen and fat, sinusoidal dilatation (the result of the effects of arterial pressure, as these tumors lack a portal venous supply), absent bile ductules, and presence of few and nonfunctioning Kupffer cells (Figure 20.7). The extensive hypervascularity and lack of a true capsule make this tumor prone to hemorrhage. In view of these risks, surgical resection is often recommended.

Significant subsets of hepatic adenomas display inactivating mutations in *HNF1α* or *TCF1* gene. The tumors in this scenario display marked steatosis and excess glycogen accumulation. These observations were explained by the role of HNF1α in regulating liver fatty acid-binding protein (L-FABP), which plays a role in fatty acid trafficking in hepatocytes, and glucose-6-phosphate increase, respectively. HA with *HNF1α* inactivation displays an extremely low risk of malignant transformation. In another subset of HAs, Wnt/β-catenin activation is observed secondary to mutations in the *CTNNB1* gene that eventually affect the degradation of the β-catenin protein. These adenomas can range from 15% to 46%, but the numbers might be closer to the lower percentage when only *CTNNB1* mutations as a mechanism of β-catenin activation are taken into account. These tumors show frequent cytological abnormalities and pseudo-glandular formation and have been shown to occur at abnormally higher frequency in males. Most importantly, these HAs have been shown to possess a higher propensity for malignant transformation. There is also yet another group of HAs that display acute phase inflammation histologically. Such tumors show sinusoidal dilatation, inflammatory infiltrates, and vessel dystrophy.

Malignant Liver Tumors

Hepatoblastoma

Hepatoblastoma (HB) is a rare primary tumor of the liver (incidence is 1 in 1,000,000 births), but is the most frequent liver tumor in children under the age of 3 years. This tumor is classified based on cellular composition and differentiation. The epithelial group of tumors comprises fetal, embryonal, mixed (embryonal and fetal), macrotrabecular, and small cell undifferentiated subtypes (Figure 20.6). Most HB are sporadic but have been reported as a component of the Beckwith-Wiedemann syndrome (BWS), where its incidence is higher than the general population or in the familial adenomatous polyposis (FAP), which occurs due to germline mutation in the adenomatous polyposis coli (*APC*) gene.

The most relevant molecular aberration in the HB is anomalous activation of Wnt/β-catenin signaling, also referred to as the canonical Wnt signaling pathway. Sequence analysis of the β-catenin gene (*CTNNB1*) has revealed missense mutations or interstitial deletions in a significant subset in up to 90% of HBs. These events lead to nuclear and/or cytoplasmic accumulation of β-catenin in HB as identified by immunohistochemistry and coincide with upregulation of several targets of the Wnt pathway, such as cyclin-D1 (Figure 20.6). While immunohistochemistry for β-catenin has been attempted to classify HB for prognosis as well as histological subtypes, a more mechanistic insight will be necessary to dissect out the innocent from the inciting β-catenin redistribution.

Hepatocellular Cancer

Hepatocellular cancer (HCC) is the most common primary tumor of the liver, accounting for 85% of all primary malignant tumors. It is the fifth most common malignancy worldwide and third most common cause of death related to cancers. Common risk factors of HCC include hepatitis, chronic alcohol abuse, toxins such as aflatoxins, and nonalcoholic fatty liver disease (NAFLD). This disease, which used to be a common malignancy in underdeveloped or developing countries, is now on the rise in developed countries. The increasing incidence of HCC in developed countries is attributed to the increasing incidence of hepatitis C and hepatitis B, as well as of NAFLD. HCC afflicts three times more men than women and overall incidence increases with age, especially in the Western world, although trends are now changing. In fact peak incidence of HCC in a recent study was between the ages of 45 and 60 years.

Much of the HCC occurs in the background of cirrhosis. In fact most of the chronic liver insults eventually lead to cirrhosis. During this process, liver function is ascertained by the presence of regenerating nodules, which are a function of the regenerative capacity of surviving hepatocytes. However, some of these nodules evolve into low-grade and high-grade dysplastic nodules, which then lead to HCC. Similarly, a minor subset of HAs proceeds to evolve into HCC (Figure 20.7). Histologically, HCC may be well differentiated to poorly differentiated, depending on degree of nuclear atypia, anaplasia, and nucleolar

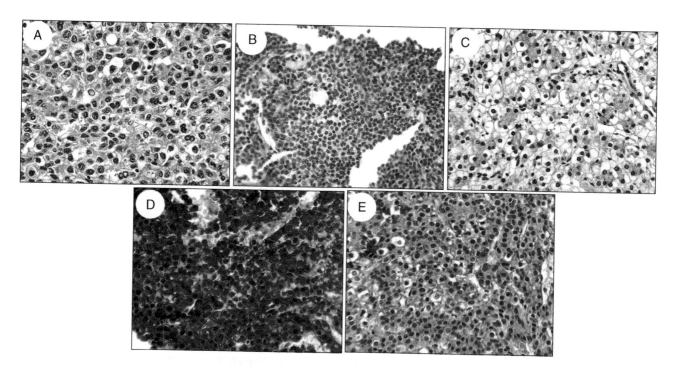

Figure 20.6   Histology and immunohistochemistry for β-catenin in pediatric hepatoblastomas. (A) H&E displaying embryonal pattern of hepatoblastoma. **(B)** H&E displaying small cell embryonal hepatoblastoma. **(C)** H&E displaying fetal hepatoblastoma. **(D)** Nuclear and cytoplasmic localization of β-catenin in an embryonal hepatoblastoma. **(E)** Nuclear localization of β-catenin along with some membranous staining in a fetal hepatoblastoma.

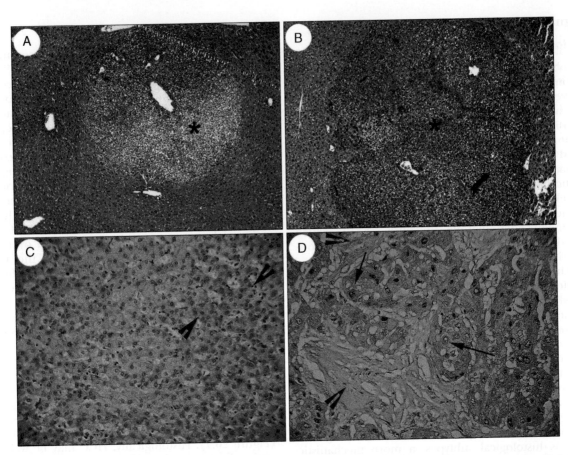

Figure 20.7 Histology of hepatic adenoma and HCC in mouse model of chemical carcinogenesis and HCC in patients. (A) A hepatic adenoma (*) is evident on H&E in mice 6 months after diethylnitrosamine (DEN) injection. **(B)** HCC (*) is evident in mouse liver 9 months after DEN injection. **(C)** H&E show abnormal trabecular pattern (arrowheads) in HCC patient. **(D)** H&E showing fibrolamellar variant of HCC in a patient with lamellar fibrosis (arrowhead) pattern surrounding large tumor cells (arrows).

prominence, and the tumor cells usually arrange in a trabecular pattern composed of uneven layers of hepatocytes, with occasional mitosis (Figure 20.7). Thus, overall, there is a progression of the preneoplastic events into neoplasia, and clearly, understanding the molecular basis of this evolution will have strong clinical implications.

The possibility of a cancer stem cell origin of some of the HCC is also being entertained. Cancer stem cells are formed by mutations in the normal existing stem cells or a progenitor cell within a tissue. It has also been suggested that perfectly mature cells such as hepatocytes can dedifferentiate into progenitor-like cells and also be targets of mutations.

During the process of hepatocarcinogenesis, multiple genetic alterations have been identified ranging from point mutations in individual genes to gain or loss of chromosome arms. Several candidate genes include *c-myc* (8q), *Cyclin-A2* (4q), *Cyclin-D1* (11q), *Rb1* (13q), *AXIN1* (16p), *p53* (17p), *IGFR-II/M6PR* (6q), *p16* (9p), *E-Cadherin* (16q), *SOCS* (16p), and *PTEN* (10q). The most frequently mutated genes in HCC include *p53*, *PIK3CA*, and *CTNNB1* (β-catenin gene). Various signaling pathways that become aberrantly active in HCC include the Wnt/β-catenin,

EGFR, TGFα, VEGFR, IGF, and HGF/Met pathways. Similarly, after any of the preceding factors (excluding Wnt signaling) activates the receptor tyrosine kinases, the signal is transduced via Ras/MAPK, PI3kinase, or Jak/Stat pathways. Success of Sorafenib in HCC is attributable to identification of such molecular aberrations in HCC.

Various agents that are implicated in HCC pathogenesis have shown some preference for the pathways they inflict. Aflatoxins are metabolized by cytochrome P450 to form an active metabolite, which leads to DNA adduct formation. Highest *p53* somatic mutations in fact are noted in HCCs that occur secondary to aflatoxin exposure. A high rate of *p53* mutations is also observed in hemochromatosis-related HCC. The three leading causes of HCC in the Western world (HCV, HBV, and alcoholic liver disease) mostly employ common molecular and genetic pathways for tumorigenesis. These include the *Rb1*, *p53*, and *Wnt* pathways.

Fibrolamellar HCC One of the uncommon variants of HCC that deserves a mention is the fibrolamellar HCC (FL-HCC). This usually occurs in younger patients (5–35 years) and in a noncirrhotic hepatic background. The histology is characterized by a lamellar pattern of fibrosis surrounding larger tumor cells that contain abundant granular cytoplasm and prominent nucleoli (Figure 20.7). While initially thought to have better prognosis, FL-HCC appears to be equally aggressive with a 45% 5-year survival. In fact, the improved prognosis appears to be due to the absence of cirrhosis in FL-HCC as compared to existing cirrhosis in the conventional HCC cases. The molecular basis of this variant of HCC remains largely obscure. Recently, increased EGFR protein expression has also been identified in significant numbers of FL-HCC, and this increase appeared to be secondary to gains of chromosome 7.

KEY CONCEPTS

- There exists a great deal of cellular and molecular complexity during liver development. This process is not linear and demonstrates a significant overlap between various developmental stages. Cell-cell interactions are at the heart of the normal development as are the signaling mechanisms. The complexity, however, can be appreciated by many examples that reveal temporal activation of one pathway, which in turn initiates a feedback mechanism to not only regulate itself but also cross-talks with other pathways and this can happen in the same or between different cell populations. It must be highlighted that timing remains critical for normal liver development, since certain pathways can be repressive, stimulatory, or bystanders for some cells albeit at different stages of the hepatic development.

- The significance of understanding the molecular basis of hepatic development is beneficial for the fields of hepatic tissue engineering, stem cell differentiation and hepatic regenerative medicine, especially due to scarcity of organs for orthotopic liver transplantation requiring studies on alternate sources of hepatocytes. For successful, persistent, and reproducible derivation of functioning hepatocytes from sources such as ES cells, mesenchymal stem cells, hematopoietic stem cells or stem cells derived from skin, adipose tissue, and placenta it will be critical to identify precise signaling pathways that need to be temporally turned "on" or "off." Many of the molecular signaling pathways that are associated with normal liver development have also been shown to play key roles in hepatocarcinogenesis. Thus studying regulation of liver development might disclose "physiological ways" to tame the "oncogenic" pathways through identification of novel cross talks and negative regulators of such pathways.

- Multiple cytokines and growth factors are activated in response to partial hepatectomy. Liver regeneration is guided by a significant signaling redundancy, which is paramount to inducing the much needed cell proliferation within the regenerating liver. In addition, when all else fails, additional cell types such as the oval cells or hepatic progenitors from various sources can be called upon to restore. Overall, the redundancy in liver regenerative processes, at the level of signaling, and at the level of cells, ensures the health of this indispensable Promethean organ.

- Based on the observation of the presence of progenitors or transiently amplifying cells in liver pathologies such as hepatitis, fibrosis, and other forms of chronic hepatic injuries, which often predisposes the liver to neoplastic transformation, the cancer stem cells as a source of a subset of HCC is often suggested. It is feasible that the oval cell activation observed in various hepatic pathologies, while providing a distinct advantage of maintenance of hepatic function, might also serve as a basis of neoplastic transformation in the right microenvironmental milieu. This Jekyll and Hyde hypothesis needs further investigation.

- While simple steatosis or fatty liver as a component of the non-alcoholic fatty liver disease (NAFLD) represents a benign form of the disease, non-alcoholic steatohepatitis (NASH) represents the progressive form and is characterized by necroinflammatory activity, hepatocellular injury, and progressive fibrosis. A "two-hit" hypothesis has been proposed to explain the pathogenesis of NAFLD and the progression of simple steatosis to NASH that occurs in a subset of patients. The "first-hit" consists of triglyceride deposition in hepatocytes. A "second-hit" consisting of another cellular event then leads to inflammation and hepatocyte injury and the subsequent manifestations of the disease.

- Hepatocellular cancer typically occurs in the background of cirrhosis. In fact, most chronic liver insults eventually lead to cirrhosis. During this process, liver function is ascertained by the presence of regenerating nodules, which are a function of the regenerative capacity of surviving hepatocytes. However, some of these nodules evolve into low- and high-grade dysplastic nodules, which then lead to HCC, and a clear understanding of the molecular basis of this "evolution" will have strong clinical implications. Aberrations in many receptor tyrosine kinases and other pathways such as the Wnt/β-catenin signaling have been reported in HCC and agents targeting these pathways are at various stages of development for chemoprevention and chemotherapy.

SUGGESTED READINGS

1. Choudhury J, Sanyal AJ. Insulin resistance in NASH. *Front Biosci.* 2005;10:1520–1533.
2. Clotman F, Jacquemin P, Plumb-Rudewiez N, et al. Control of liver cell fate decision by a gradient of TGF beta signaling modulated by Onecut transcription factors. *Genes Dev.* 2005;19:1849–1854.
3. Diehl AM, Li ZP, Lin HZ, et al. Cytokines and the pathogenesis of non-alcoholic steatohepatitis. *Gut.* 2005;54:303–306.
4. Dorrell C, Grompe M. Liver repair by intra- and extrahepatic progenitors. *Stem Cell Rev.* 2005;1:61–64.

5. Duncan SA. Mechanisms controlling early development of the liver. *Mech Dev.* 2003;120:19–33.
6. El-Serag HB. Hepatocellular carcinoma: Recent trends in the United States. *Gastroenterology.* 2004;127:S27–34.
7. Fausto N. Liver regeneration and repair: Hepatocytes, progenitor cells, and stem cells. *Hepatology.* 2004;39:1477–1487.
8. Friedman SL. Mechanisms of hepatic fibrogenesis. *Gastroenterology.* 2008;134:1655–1669.
9. Jelnes P, Santoni-Rugiu E, Rasmussen M, et al. Remarkable heterogeneity displayed by oval cells in rat and mouse models of stem cell-mediated liver regeneration. *Hepatology.* 2007;45:1462–1470.
10. Jung J, Zheng M, Goldfarb M, et al. Initiation of mammalian liver development from endoderm by fibroblast growth factors. *Science.* 1999;284:1998–2003.
11. Lee CS, Friedman JR, Fulmer JT, et al. The initiation of liver development is dependent on Foxa transcription factors. *Nature.* 2005;435:944–947.
12. Lee JS, Heo J, Libbrecht L, et al. A novel prognostic subtype of human hepatocellular carcinoma derived from hepatic progenitor cells. *Nat Med.* 2006;12:410–416.
13. Llovet JM, Di Bisceglie AM, Bruix J, et al. Design and endpoints of clinical trials in hepatocellular carcinoma. *J Natl Cancer Inst.* 2008;100:698–711.
14. Michalopoulos GK. Liver regeneration. *J Cell Physiol.* 2007;213:286–300.
15. Odom DT, Zizlsperger N, Gordon DB, et al. Control of pancreas and liver gene expression by HNF transcription factors. *Science.* 2004;303:1378–1381.
16. Reddy JK, Rao MS. Lipid metabolism and liver inflammation. II. Fatty liver disease and fatty acid oxidation. *Am J Physiol Gastrointest Liver Physiol.* 2006;290:G852–858.
17. Schuppan D, Afdhal NH. Liver cirrhosis. *Lancet.* 2008;371:838–851.
18. Sell S, Leffert HL. Liver cancer stem cells. *J Clin Oncol.* 2008;26:2800–2805.
19. Stoick-Cooper CL, Moon RT, Weidinger G. Advances in signaling in vertebrate regeneration as a prelude to regenerative medicine. *Genes Dev.* 2007;21:1292–1315.
20. Villanueva A, Newell P, Chiang DY, et al. Genomics and signaling pathways in hepatocellular carcinoma. *Semin Liver Dis.* 2007;27:55–76.
21. Yovchev MI, Grozdanov PN, Joseph B, et al. Novel hepatic progenitor cell surface markers in the adult rat liver. *Hepatology.* 2007;45:139–149.
22. Zaret KS. Regulatory phases of early liver development: Paradigms of organogenesis. *Nat Rev Genet.* 2002;3:499–512.
23. Zucman-Rossi J, Jeannot E, Nhieu JT, et al. Genotype-phenotype correlation in hepatocellular adenoma: New classification and relationship with HCC. *Hepatology.* 2006;43:515–524.

21

Molecular Basis of Diseases of the Exocrine Pancreas

Matthias Sendler . Julia Mayerle . Markus M. Lerch

ACUTE PANCREATITIS

Acute pancreatitis presents clinically as a sudden inflammatory disorder of the pancreas, and is caused by premature intracellular activation of pancreatic proteases leading to (i) self-destruction of acinar cells and (ii) autodigestion of the organ. Necrotic cell debris resulting from this process produces a systemic inflammatory reaction, which may lead to multiorgan failure in due course. The incidence of acute pancreatitis differs regionally from 20 to 120 cases per 100,000 population. Acute pancreatitis varies considerably in severity and can be categorized into two forms of the disease. The majority of cases (85%) present with a mild form of disease, classified as edematous pancreatitis, with absent or only transient extrapancreatic organ failure. In the remaining minority of cases (15%), pancreatitis follows a severe course accompanied by sustained multiorgan failure. This latter form of pancreatitis is commonly referred to as severe necrotizing pancreatitis. Severe necrotizing pancreatitis is associated with high mortality (10–20%) and may lead to long-term complications such as the formation of pancreatic pseudocysts or impairment of exocrine and endocrine function of the pancreatic gland.

In many patients (approximately 50%) the underlying cause of acute pancreatitis is the migration of a gallstone resulting in obstruction of the pancreatic duct at the papilla of Vater. Development of acute pancreatitis is frequently triggered by alcohol abuse. In 25–40% of acute pancreatitis patients, increased or excess alcohol consumption is regarded as the cause of the disease. Removal of the underlying disease-causing agent results in complete regeneration of the pancreas and preserved exocrine and endocrine function in the majority of cases. Recurrent attacks of the disease can result from chronic alcohol abuse, repeated gallstone passage, genetic predispositions, sphincter dysfunction, metabolic disorders, or pancreatic duct strictures. All of these factors can contribute to the development of

chronic pancreatitis as well. In the remaining cases (10–20%), no apparent clinical cause or etiology of the disease can be identified. These cases are referred to as idiopathic pancreatitis. It became clear during the past decade that previously unknown genetic factors play a major role in these cases. The initial cellular mechanism causing acute pancreatitis is probably independent of the underlying etiology of the disease.

Early Events in Acute Pancreatitis and the Role of Protease Activation

Acute pancreatitis is an inflammatory disorder whose pathogenesis is not well understood. The pancreas is known as the enzyme factory of the human organism, producing and secreting large amounts of potentially hazardous digestive enzymes, many of which are synthesized as pro-enzymes known as zymogens. Under physiological conditions the pancreatic digestive enzymes are secreted in response to hormonal stimulation. Activation of the pro-enzymes (or zymogens) requires hydrolytic cleavage of their activation peptide by protease enzymes. After entering the small intestine, the pancreatic zymogen trypsinogen is first activated to trypsin by the intestinal protease enterokinase (or enteropeptidase). Trypsin then proteolytically processes other pancreatic enzymes to their active forms. Under physiological conditions pancreatic proteases remain inactive during synthesis, intracellular transport, secretion from acinar cells, and transit through the pancreatic duct. They are activated only upon reaching the lumen and brush border of the small intestine (Figure 21.1).

More than a century ago, Hans Chiari proposed that the underlying pathophysiological mechanism for the development of pancreatitis was autodigestion of the exocrine pancreatic tissue by proteolytic enzymes. Today, this theory is well accepted. Nevertheless, this theory suggests that disease results from the premature

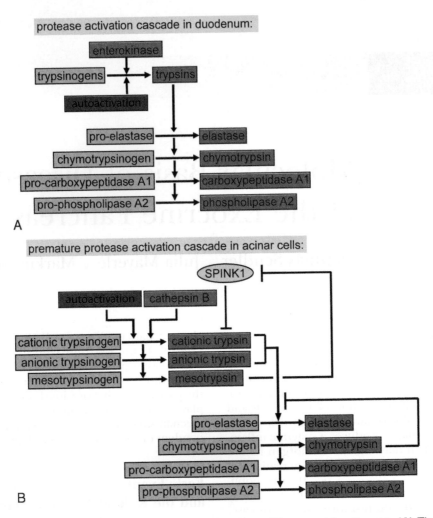

Figure 21.1 Activation of pancreatic proteases under normal conditions and in disease. (A) The protease activation cascade in the duodenum. Enterokinase activates trypsinogen by proteolytic cleavage of the trypsin activation peptide (TAP), and autoactivation contributes to this process. Trypsin then activates other digestive pro-enzymes in a cascade-like fashion. **(B)** Within acinar cells, premature activation of trypsinogen by cathepsin B is involved in the setting of the proteolytic cascade. Intracellular trypsin may also activate other digestive pro-enzymes, in spite of the presence of the trypsin inhibitor SPINK-1 and the trypsin-degrading enzyme chymotrypsin C.

intracellular activation of zymogens and that this occurs in the absence of enterokinase. Furthermore, this theory suggests that protease inactivation takes place despite several physiological defense mechanisms, including the synthesis of endogenous protease inhibitors and the storage of proteases in a membrane-confined compartment of zymogen granules.

Much of our current knowledge regarding the onset of pancreatitis was not gained from studies involving the human pancreas or patients with pancreatitis, but from animal or isolated cell models.

Data from various animal models suggest that after the initial insult a variety of pathophysiological factors determine disease onset. These include (i) a block in secretion, (ii) the co-localization of zymogens with lysosomal enzymes, (iii) the activation of trypsinogen and other zymogens, and (iv) acinar cell injury. *In vitro* and *in vivo* studies have demonstrated the importance of premature zymogen activation in the pathogenesis of pancreatitis since the inception of the hypothesis by

Chiari. The activation of trypsinogen and other pancreatic zymogens can be demonstrated in pancreatic homogenates from experimental animals and this zymogen activation appears to be an early event. Trypsin activity is detected as early as 10 minutes after supramaximal stimulation with the cholecystokinin-analogue cerulein in rats and increases over time. The activation of trypsinogen requires the hydrolytic cleavage of a 7–10 amino acid propeptide called trypsin activation peptide (TAP) on the N-terminus of trypsinogen. Measuring an increase of immunoactive TAP after cerulein-induced pancreatitis in rats showed trypsinogen activation in the secretory compartment of acinar cells. Furthermore, TAP was detected in the serum and urine of patients with pancreatitis, and the amount of TAP appears to correlate with the severity of the disease. The activity of trypsin and elastase increases in the early phase in acute experimental pancreatitis. In addition to the activation peptide of trypsinogen (TAP), the activation peptide of carboxypeptidase A1 (PCA1) can be identified in serum at an early stage of pancreatitis. Premature activation of these pro-enzymes leads to the development of necrosis and to autodigestion of pancreatic tissues. Recent studies defining the localization of early activation of pro-enzymes suggest that these processes, which lead to pancreatitis and pancreatic tissue necrosis, originate in acinar cells (Figure 21.1).

In conclusion, premature intracellular activation of zymogens to active proteases in the secretory compartment of acinar cells results in acinar necrosis and contributes to the onset of pancreatitis. As a direct result of cellular injury, the acinar cells release chemokines and cytokines which initiate the later events in pancreatitis, including recruitment of inflammatory cells into the tissue. Trypsin seems to be the key enzyme in the process of activating other digestive pro-enzymes prematurely, and one of the crucial questions in understanding the pathophysiology of acute pancreatitis is to identify the mechanism which prematurely activates trypsinogen inside acinar cells. However, it must be noted that the term trypsin, as defined by the cleavage of specific synthetic or protein substrates, comprises a group of enzymes whose individual role in the initial activation cascade may differ considerably.

The Mechanism of Zymogen Activation

One hypothesis for the initiation of the premature activation of trypsinogen suggests that during the early stage of acute pancreatitis, pancreatic digestive zymogens become co-localized with lysosomal hydrolases. Recent data show that the lysosomal cysteine proteinase cathepsin B may play an important role for the activation of trypsinogen. Many years ago *in vitro* data demonstrated activation of trypsinogen by cathepsin B. Most lysosomal hydrolases are synthesized as inactive pro-enzymes, but in contrast to digestive zymogens, they are activated by post-translational processing in the cell. During protein sorting in the Golgi system, lysosomal hydrolases are sorted into prelysosomes, whereas zymogens are packaged into condensing vacuoles. The sorting of lysosomal hydrolases depends on a mannose-6-phosphate-dependent pathway, which leads to a separation of lysosomal hydrolases from other secretory proteins and to the formation of prelysosomal vacuoles. However, this sorting is incomplete. Under physiological conditions, a significant fraction of hydrolases enter the secretory pathway. It has been suggested that these mis-sorted hydrolases play a role in the regulation of zymogen secretion. In acute pancreatitis the separation of digestive zymogens and lysosomal hydrolases is impaired. This leads to further co-localization of lysosomal hydrolases and zymogens within cytoplasmic vacuoles of acinar cells. This co-localization has also been shown in electron microscopy, as well as in subcellular fractions isolated by density gradient centrifugation. The redistribution of cathepsin B from the lysosome-enriched fraction was noted within 15 minutes of the start of pancreatitis induction, and trypsinogen activation was observed in parallel. There are two main theories trying to explain

the co-localization of cysteine and serine proteases: (i) fusion of lysosomes and zymogen granules or (ii) incorrect sorting of zymogens and hydrolases in the process of vacuole maturation. Wortmannin, a phosphoinositide-3-kinase inhibitor, prevents the intracellular mis-sorting of hydrolases and zymogens, and subsequently prevents trypsinogen activation to trypsin during acute pancreatitis.

Further experiments focused on cathepsin B as the main enzyme driving the intracellular activation of trypsinogen. Cathepsin B is the most abundant lysosomal hydrolase in acinar cells. Pretreatment of rat pancreatic acini with E64d, a cell-permeable cathepsin B inhibitor, leads to complete inhibition of cathepsin B and completely abrogates trypsinogen activation. Final evidence that cathepsin B is involved in activation of trypsinogen during cerulein-induced experimental pancreatitis comes from experiments in cathepsin B knockout mice. In these animals, after induction of experimental pancreatitis, trypsin activity was reduced to less than 20% compared to wild-type animals and the severity of the disease was markedly ameliorated. These data showed unequivocally the importance of cathepsin B for the pathogenesis of acute pancreatitis (Figure 21.1).

The cathepsin B theory implies one further critical point—that trypsinogen is expressed and stored in the presence of different potent intrapancreatic trypsin inhibitors. To activate trypsinogen, cathepsin B needs to override these defensive mechanisms to initiate the premature intracellular activation cascade. Recently, it has become clear that cathepsin B can not only activate cationic and anionic trypsinogen, but also mesotrypsinogen. Mesotrypsin, the third trypsin isoform expressed in the human pancreas, is resistant against trypsin inhibitors like SPINK-1 or soybean trypsin inhibitor (SBTI). Moreover, mesotrypsin is able to degrade trypsin inhibitors. Under physiological conditions, mesotrypsin is activated in the duodenum by enterokinase, where it degrades exogenous trypsin inhibitors to ensure normal tryptic digestion. Mesotrypsin rapidly inactivates trypsin inhibitors like SPINK-1 by proteolytic cleavage *in vitro*. Therefore, activation of trypsins by cathepsins might not only trigger a proteolytical cascade, but also involve the removal of trypsin inhibitors such as SPINK-1 via the activation of mesotrypsin.

Taken together, these experimental observations represent compelling evidence that cathepsin B can contribute to premature intracellular zymogen activation and the initiation of acute pancreatitis not only through co-localization with trypsinogen, but also through activation of mesotrypsin, rendering endogenous pancreatic protease inhibitors inactive.

The Degradation of Active Trypsin

During the early phase of pancreatitis, trypsinogen and other zymogens are rapidly activated, while later in the disease course their activity declines to physiological levels, suggesting degradation of the active enzymes. This phenomenon has been termed autolysis or autodegradation. Since this process self-limits autoactivation of trypsinogen, it is regarded as a safety mechanism to counteract premature zymogen activation.

One theory to possibly explain how uncontrolled trypsinogen activation can be antagonized is based on the existence of a serine protease that is capable of trypsin degradation. In 1988 Heinrich Rinderknecht discovered an enzyme which rapidly degrades active cationic and anionic trypsin and named this protease enzyme Y. Recent *in vitro* data suggest that the autodegradation of trypsin is a very slow process and that most of trypsin degradation is not mediated by trypsin itself but by another enzyme. Chymotrypsin C has the capability to proteolytically cleave cationic trypsin at Leu81–Glu82 in the Ca^{2+} binding loop. This leads to rapid autodegradation and catalytic inactivation of trypsin by additional cleavage at the Arg122–Val123. However, chymotrypsin C has also the ability to induce trypsin-mediated trypsinogen autoactivation by proteolytic cleavage at the Phe18–Asp19 position of cationic trypsinogen. The balance between autoactivation and autodegradation of cationic trypsin mediated by chymotrypsin C is regulated via the Ca^{2+} concentration. In the presence of 1 mM Ca^{2+}, degradation of trypsin is blocked and autoactivation of trypsinogen is induced. Under physiological conditions in the duodenum, high Ca^{2+} concentrations facilitate the activation of trypsinogen to promote digestion. In the absence of high Ca^{2+} concentrations, chymotrypsin C degrades active trypsin and protects against premature activation of trypsin (Figure 21.2).

Calcium Signaling

The second messenger calcium plays an important role in multiple different intracellular processes such as metabolism, cellular secretion, cell differentiation, and cell growth. Under physiological conditions, pancreatic acinar cells maintain a Ca^{2+} gradient across the plasma membrane with low intracellular concentration and high extracellular concentration of calcium. In response to hormonal stimulation, Ca^{2+} is released from intracellular stores to regulate signal-secretion coupling. In pancreatic acinar cells, acetylcholine (ACh) and cholecystokinin (CCK) regulate the secretion of digestive enzymes via the generation of repetitive local cytosolic Ca^{2+} signals. In response to secretagogue stimulation with ACh or CCK, Ca^{2+} is initially released from intracellular stores near the apical pole of acinar cells. This induces the fusion of zymogen granules with the apical plasma membrane and activation of Ca^{2+}-dependent Cl^- channels in the apical membrane. The pattern of intracellular calcium signal in response to secretagogue stimulation is dependent on the neurotransmitter or hormone concentration. ACh at physiological concentrations elicits repetitive Ca^{2+} spikes and oscillation of Ca^{2+} concentrations, but these oscillations are restricted to the secretory pole of the cell. High concentrations of cholecystokinin lead to short-lasting spikes followed by longer Ca^{2+} transients that spread to the entire cell. Each oscillation is associated with a burst of exocytotic activity and the release of zymogen into the duct lumen. In contrast, supramaximal stimulation of acinar cells induces a completely different pattern of Ca^{2+} signals. Instead of oscillatory activity observed with physiological doses of cholecystokinin, there is a much larger rise followed by a sustained elevation associated to a block of enzyme secretion and premature intracellular protease activation. Ca^{2+} is released from the endoplasmic reticulum (ER) in response to stimulation. The ER is located in the basolateral part of the acinar cell with extensions in the apical part enriched with zymogen granules. While the entire ER contains Ca^{2+}, release of Ca^{2+} in response to cholecystokinin or ACh occurs only at the apical pole due to the higher density of Ca^{2+} release channels at the apical pole of the ER. Two types of

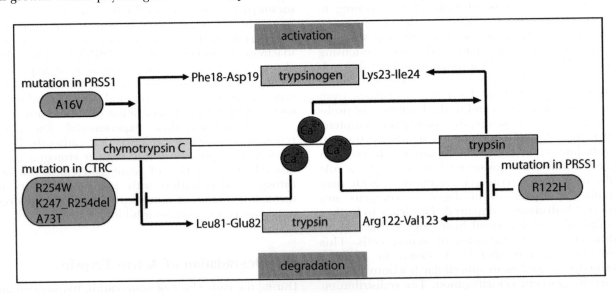

Figure 21.2   Chymotrypsin C has different functions in the processing of trypsin or trypsinogen. The major role in pancreatitis is degrading active trypsin. This function is disturbed by mutations within the *CTRC* gene or by high levels of Ca^{2+}. The activation of trypsinogen to trypsin can also be mediated by chymotrypsin C. Trypsin itself has the capability to autoactivate or to self-degrade. The R122H mutation results in the decreased autolysis of active trypsin.

Ca^{2+} channels are expressed in the ER of acinar cells, inositol-triphosphate-receptors (IP_3) and ryanodine receptors. Both of these channels are required for apical Ca^{2+} peaks. ACh activates phospholipase C (PLC) and initiates the Ca^{2+} release via the intracellular messenger IP_3, whereas cholecystokinin does not activate PLC but increases the intracellular concentration of nicotinic acid adenine dinucleotide phosphate (NAADP) in a dose-dependent manner. The higher density of Ca^{2+} channels in the apical part of the ER explains the initiation of Ca^{2+} signals in the granule part of the cytoplasm. The apical, zymogen granule-enriched part of the acinar cell is surrounded by a barrier of mitochondria which absorb released calcium and prevent higher Ca^{2+} concentrations from expanding beyond the apical part of acinar cells. The spatially limited release of Ca^{2+} at the apical pole also prevents an unregulated chain reaction across gap junctions, which would affect neighboring cells. The mitochondrial Ca^{2+} uptake leads to increased metabolism and generation of ATP. ATP is required for the reuptake of Ca^{2+} in the ER via the sarcoplasmic endoplasmic reticulum calcium ATPase (SERCA), and for exocytosis across the apical membrane. Thus, Ca^{2+} homeostasis plays a crucial role for maintaining signal-secretion coupling in pancreatic acinar cells (Figure 21.3).

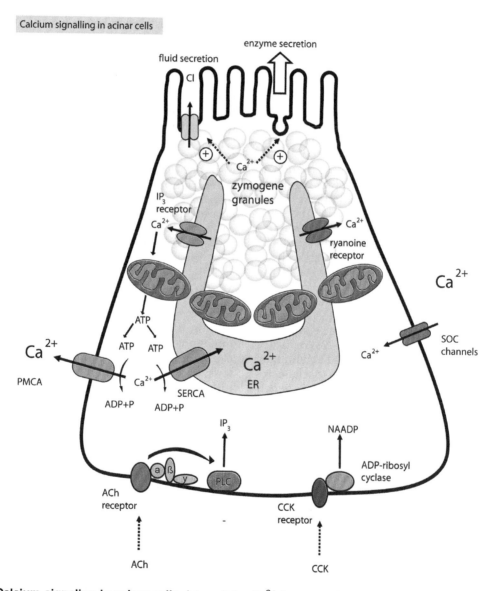

Figure 21.3   Calcium signaling in acinar cells. Intracellular Ca^{2+} homeostasis regulates the secretion of zymogens and fluid. Acetylcholine (ACh) regulates via the second messenger inositol-3-phosphate the apical Ca^{2+} influx from intracellular stores (ER). Cholecystokinin leads to the production of nicotinic acid adenine dinucleotide phosphate (NAADP), which interacts with the ryanodine receptor in the ER membrane and also regulates the apical Ca^{2+} influx. The plasma membrane calcium ATPase (PMCA) or the sarcoplasmic endoreticulum calcium ATPase (SERCA) regulates cytoplasmic Ca^{2+} decrease. A mitochondrial barrier inhibits a global Ca^{2+} increase by absorbing free Ca^{2+} from the apical zymogen-enriched part of the acinar cell. Interference in the calcium homeostasis leads to a global Ca^{2+} increase in the cytoplasm, which results in premature activation of zymogens.

Elevated Ca^{2+} concentrations in the extracellular compartment or within acinar cells is known to be a risk factor for the development of acute pancreatitis. Disturbances in the Ca^{2+} homeostasis of pancreatic acinar cells occur early in the secretagogue-induced model of pancreatitis. An attenuation of Ca^{2+} elevation in acinar cells results from exposure to the cytosolic Ca^{2+} chelator BAPTA-AM, which also prevents zymogen activation, proving that Ca^{2+} is essential for zymogen activation. The sustained elevation that follows the initial intracellular Ca^{2+} spike induced by supramaximal concentrations of cerulein is also attenuated in the complete absence of intracellular calcium and appears to depend on extracellular Ca^{2+}. In the absence of extracellular Ca^{2+}, the activation of trypsinogen induced by supramaximal doses of cerulein is also attenuated, suggesting that the initial and transient rise in Ca^{2+} caused by the release of calcium from the internal stores is not sufficient to permit trypsinogen activation. In contrast, interference with high calcium plateaus by the natural Ca^{2+} antagonist magnesium or a Ca^{2+} chelator *in vivo* abolishes trypsinogen activation as well as pancreatitis.

Acute pancreatitis is characterized by the pathologic activation of zymogens within pancreatic acinar cells. The process requires a rise in cytosolic Ca^{2+} from undefined intracellular stores. Zymogen activation is thereby mediated by ryanodine receptor-regulated Ca^{2+} release, and early zymogen activation takes place in a supranuclear compartment that overlaps in distribution with the ryanodine receptor. Furthermore, *in vivo* inhibition of the ryanodine receptor results in a loss of zymogen activation. Therefore, Ca^{2+} release from the ryanodine receptor mediates zymogen activation but not enzyme secretion.

Recent reports have shown that metabolites of alcohol metabolism can have a pathological effect on acinar cell Ca^{2+} homeostasis, suggesting a possible pathogenic mechanism in alcoholic pancreatitis. Nonoxidative metabolites like fatty acid ethyl esters (FAEE) and fatty acids (FA) can cause Ca^{2+}-dependent acinar cell necrosis. It was previously demonstrated that FAEEs are generated in acinar cells incubated with clinically relevant concentrations of ethanol. FAEEs activate the IP_3 receptor after which Ca^{2+} is released from the ER. In contrast, FAs do not activate Ca^{2+} channels but decrease ATP levels in the cytoplasm, which results in impaired Ca^{2+} reuptake into the ER. Subsequently, increased intracellular Ca^{2+} levels contribute to premature zymogen activation.

Ca^{2+} not only is an important second messenger, but has a direct effect on the activation, activity, and degradation mechanisms of trypsin. Experiments with purified human anionic and cationic trypsinogen in the absence of Ca^{2+} show markedly reduced autoactivation compared to high Ca^{2+} concentrations. Moreover, in cytoplasmic vacuoles developing upon supramaximal cholecystokinin stimulation in acinar cells, Ca^{2+} concentrations decrease rapidly to levels which are much lower than optimal for trypsinogen autoactivation. This mechanism could represent a protective mechanism of the endosomes to prevent damage from premature trypsinogen activation.

CHRONIC AND HEREDITARY PANCREATITIS

Chronic pancreatitis is clinically defined as recurrent bouts of a sterile inflammatory disease characterized by persistent and often progressive and irreversible morphological changes, typically causing pain and permanent impairment of pancreatic function. Chronic pancreatitis histologically represents a transformation of focal necrosis into perilobular and intralobular fibrosis of the parenchyma, pancreatic duct obstruction by pancreatic stones and tissue calcification, and the development of pseudocysts. In the course of the disease, progressive loss of endocrine and exocrine function can be monitored. It should be noted that the clinical distinction between acute and chronic pancreatitis is becoming pathophysiologically ever more blurred, and similar or identical onset mechanisms may play a role. These mechanisms include premature and intracellular activation of digestive proteases. Much of our present knowledge about this process has been generated since a genetic basis for pancreatitis was first reported in 1996. Hereditary pancreatitis represents a genetic disorder associated with mutations in the cationic trypsinogen gene and presents with a disease penetrance of up to 80%. Patients with hereditary pancreatitis suffer from recurrent episodes of pancreatitis which do progress in the majority to chronic pancreatitis. The disease usually begins in early childhood, but onset can vary from infancy to the sixth decade of life.

Mutations Within the *PRSS1* Gene

Hereditary pancreatitis is associated with genetic mutations in the cationic trypsinogen gene, suggesting that mutations in a digestive protease (such as trypsin) can cause the disease. Hereditary pancreatitis typically follows an autosomal dominant pattern of inheritance with an 80% disease penetrance. The gene coding for cationic trypsinogen (*PRSS1*) is approximately 3.6 kb in size, is located on chromosome 7, and contains 5 exons. The precursor of cationic trypsinogen is a 247 amino acid protein, the first 15 amino acids of which represent the signal sequence; the next 8 amino acids, the activation peptide; and the remaining 224 amino acids form the backbone and catalytic center of the digestive enzyme. Presently, there are two known mutations within the same codon in the *PRSS1* gene: (i) His-122-trypsinogen shows increased autoactivation, and (ii) Cys-122 trypsinogen has reduced activity. The first of these mutations was discovered in 1996, exactly one century after Chiari proposed his theory on autodigestion of the pancreas as a pathogenic mechanism of pancreatitis. Whitcomb et al. reported a mutation in exon 3 of the cationic trypsinogen gene (*PRSS1*) on chromosome 7 (7q36) that was strongly associated with hereditary chronic pancreatitis. This single point mutation (CGC to CAC) causes an arginine to histidine (R to H) substitution at position 22 of the cationic trypsinogen gene (R122H). The amino acid exchange is located in the hydrolysis site of trypsin and can prevent the autodegradation of active trypsin (Figure 21.2). Once trypsin has been activated intracellularly, the R122H mutation interferes with the elimination of active trypsin by autodegradation. This conclusion was derived from *in vitro* data using recombinant R122H mutated trypsinogen. Using the same *E. coli* based system, Sahin-Toth and coworkers showed that the R122H mutation leads to an increase in trypsinogen autoactivation. Therefore, the R122H mutation represents a dual gain of function mutation which facilitates intracellular trypsin activity and results in a higher stability of R122H-trypsin.

Shortly after the identification of the R122H mutation a second mutation was reported in kindreds with hereditary pancreatitis. The R122C mutation is a single amino acid exchange affecting the same codon as the R122H mutation. In contrast to the R122H mutation, the R122C mutation causes a decreased trypsinogen autoactivation, and biochemical studies demonstrated a 60–70% reduced activation in the Cys-122 trypsinogen mutant induced by either enterokinase or cathepsin B activation. The amino acid exchange of the R122C mutation leads to altered cysteine-disulfide bonds and consequently a misfolded protein structure with reduced catalytic activity. In recent years several other attempts have been made to elucidate the pathophysiological role of trypsinogen in the onset of pancreatitis, but the issue is still a matter of intense research.

Since the initial discovery, several other mutations (24 to date) in the trypsinogen gene have been reported, but the R122H mutation is still the most common. In addition to the R122H mutation, five other mutations in different regions of the *PRSS1* gene associated with hereditary pancreatitis have been biochemically characterized: A16V, D22G, K23R, E79K, and N29I. It has been suggested that these mutations may have different structural effects on the activation and activity of trypsinogen.

The nucleotide substitution from A-T in exon 2 of the *PRSS1* gene (AAC to ATC) in codon 29 of cationic trypsinogen results in an asparagine to isoleucine amino acid change (N29I). This amino acid substitution affects the protein structure of trypsinogen, and seems to stabilize the enzyme. The N29I mutation does not affect the autoactivation of trypsinogen but impairs enzyme degradation *in vivo*. The N29I mutation causes a slightly milder course of hereditary pancreatitis compared to the R122H mutation, the onset of disease occurs somewhat later, and the need for in-hospital treatment is lower.

The A16V mutation results in an amino acid exchange from alanine to valine in the signal peptide of trypsinogen. This mutation is rare and in contrast to R122H and N29I mutations, which are burdened with an 80% penetrance for chronic pancreatitis, develops only in one out of seven carriers of the A16V mutation. Two other mutations were found in the signal peptide of cationic trypsinogen, D22G and K23R. Both mutations result in an increase of autoactivation of trypsinogen but are resistant to cathepsin B activation. Furthermore, in contrast to wild-type trypsinogen,

expression of active trypsin and mutated trypsinogens (D22G, K23R) reduced cell viability of AR4–2J cells. This suggests that mutations in the activation peptide of trypsinogen play an important role in premature protease activation, but the biochemical mechanisms remain unsolved.

The EUROPAC-1 study compared genotype to phenotype characteristics of hereditary pancreatitis patients. This study confirmed the importance of *PRSS1* mutations associated with chronic pancreatitis. In a multilevel proportional hazard model employing data obtained from the European Registry of Hereditary Pancreatitis, 112 families in 14 countries (418 affected individuals) were collected: 58 (52%) families carried the R122H mutation, 24 (21%) carried the N29I mutation, and 5 (4%) carried the A16V mutation, while 2 families had rare mutations, and 21 (19%) had no known *PRSS1* mutation. The median time to the start of symptoms for the R122H mutation is 10 years of age (8 to 12 years of age, 95% confidence interval), 14 years of age for the N29I mutation (11 to 18 years of age, 95% confidence interval), and 14.5 years of age for mutation-negative patients (10 to 21 years of age, 95% confidence interval; $P = 0.032$). The cumulative risk at 50 years of age for exocrine pancreas failure was 37.2% (28.5–45.8%, 95% confidence interval), 47.6% for endocrine failure (37.1–58.1%, 95% confidence interval), and 17.5% for pancreatic resection for pain (12.2–22.7%, 95% confidence interval). Time to resection was significantly reduced for females ($P < 0.001$) and those with the N29I mutation ($P = 0.014$). Pancreatic cancer was diagnosed in 26 (6%) of 418 affected patients (Table 21.1).

The remaining *Prss1* mutations are very rare and are for the most part detected only in a single family or a single patient. Mutations like P36R, K92N, or G83E were each found in only one patient with idiopathic chronic pancreatitis. The biochemical consequences of these mutations do not result in increased stability or autoactivation of trypsinogen compared to the R122H or N29I mutations. This observation causes difficulties explaining the underlying pathogenic mechanism for these rare mutations.

Mutations Within the *PRSS2* Gene

The fact that mutations in the *PRSS1* gene encoding for cationic trypsinogen are associated with hereditary pancreatitis could suggest that genetic alterations of the anionic trypsinogen gene (*PRSS2*) could also be associated with chronic pancreatitis. The E79K mutation (related to a G to A mutation at codon 237) reduces autoactivation of cationic trypsinogen by 80–90% but leads to a 2-fold increase in the activation of anionic trypsinogen, suggesting a potential role of *PRSS2*. A direct link between the development of chronic pancreatitis and mutations in the *PRSS2* gene was not established until 2006. Genetic analysis of the *PRSS2* gene in 2466 chronic pancreatitis patients and 6459 healthy individuals revealed an increased rate of a rare mutation in the anionic trypsinogen gene in control subjects. A variant of codon 191 (G191R) was present in

Table 21.1   Most Common Mutations Associated with Pancreatitis

Gene	Mutation	Comments	Frequency
PRSS1	R122H	increased autoactivation and decreased autolysis of cationic trypsin	most common mutation (>500)
	R122C	decreased autoactivation of trypsinogen, decreased autolysis of trypsin, also decreased trypsin activity	5 affected carriers
	N29I	increased autoactivation of trypsinogen	second most common mutation (>160)
	A16V	increased autoactivation of trypsinogen	25 affected carriers
	D22G	increased autoactivation of trypsinogen	rare, 2 carriers
	K23R	increased autoactivation of trypsinogen	rare, 2 carriers
	E79K	increased activation of anionic trypsinogen, decreased autoactivation of trypsinogen	8 affected carriers
SPINK1	N34S	no functional defect reported	
	R65Q	60% loss of protein expression	
	G48E	nearly complete loss of protein expression	
	D50E	nearly complete loss of protein expression	
	Y54H	nearly complete loss of protein expression	
	R67C	nearly complete loss of protein expression	
	L14P and L14R	rapid intracellular degradation of SPINK1	
CTRC	R254W	decreased activity of chymotrypsin C	
	K247_R254del	loss of function of chymotrypsin C	
	A37T	decreased trypsin degradation of chymotrypsin C	
CFTR	various >1000	decreased fluid secretion from acinar cells	

220/6459 (3.4%) control subjects, but in only 32/2466 (1.3%) affected individuals. Biochemical analysis of the recombinantly expressed G191R protein showed a complete loss of tryptic function after enterokinase or trypsin activation, as well as a rapid autolytic proteolysis of the mutant. This was the first report of a loss of trypsin function (in *PRSS2*) that has a protective effect on the onset of pancreatitis.

Mutations in the Chymotrypsin C Gene

Because trypsin degradation is thought to represent a protective mechanism against pancreatitis, Sahin-Toth and coworkers hypothesized that loss-of-function variants in trypsin-degrading enzymes would increase the risk of pancreatitis. Since they knew that chymotrypsin can degrade all human trypsins and trypsinogen isoforms with high specificity, they sequenced the chymotrypsin C gene (*CTRC*) in a German cohort suffering from idiopathic and hereditary pancreatitis. They detected two variants in the *CTRC* gene in association with hereditary and idiopathic chronic pancreatitis. Mutation of codon 760 (C to T) resulted in an R254W variant form of the protein that occurred with a frequency of 2.1% (19/901) in affected individuals compared to 0.6% (18/2804) in healthy controls. In addition, a deletion mutation (c738_761del24) resulting in a K247–R254del variant protein was found in 1.2% of affected individuals compared to 0.1% in controls. In a confirmative cohort of different ethnic background, the authors detected a third mutation in affected patients with a frequency of 5.6% (4/71) compared to 0% (0/84) in control individuals. This mutation leads to an amino acid exchange at position 73 (A73T) resulting from a G to A mutation at codon 217. The assumed pathogenic mechanism of *CTRC*

mutations is based on lowered enzyme activity in the R254W variant and a total loss of function in the deletion mutation (K247–R254del) and the A73T variant (Figure 21.2). Thus, chymotrypsin C is an enzyme that can counteract the disease-causing effect of trypsin, and loss of function mutations impair its protective role in pancreatitis by letting prematurely activated trypsin escape its degradation.

Mutations in Serine Protease Inhibitor Kazal-Type 1

Shortly after the identification of mutations in the trypsinogen gene in hereditary pancreatitis, another important observation was made by Witt et al. This group found that mutations in the *SPINK-1* gene (the pancreatic secretory trypsin inhibitor or PSTI, OMIM 167790) can be associated with idiopathic chronic pancreatitis in children. *SPINK-1* mutations can be frequently detected in cohorts of patients who do not have a family history but also have none of the classical risk factors for chronic pancreatitis. The most common mutation is found in exon 2 of the *SPINK-1* gene (AAT to AGT), which leads to an asparagine to serine amino acid change (N34S). Homozygote and heterozygote N34S mutations were detected in 10–20% of patients with pancreatitis compared to 1–2% of healthy controls, suggesting that *SPINK-1* is a disease-modifying factor. Structural modeling of SPINK-1 predicted that the N34S region near the lysine41 residue functions as the trypsin-binding pocket of SPINK-1 and that the N34S mutation changes the structure of the trypsin-binding pocket of SPINK-1, resulting in decreased inhibitory capacity of SPINK-1. In contrast to the computer-modeled prediction, *in vitro* experiments using recombinant N34S *SPINK-1* and wild-type

SPINK-1 demonstrated identical trypsin inhibitory activities. Recently, two novel *SPINK-1* variants affecting the secretory signal peptide have been reported. Seven missense mutations occurring within the mature peptide of PSTI associated with chronic pancreatitis were analyzed for their expression levels. The N34S and the P55S mutation results neither in a change of PSTI activity nor in a change of expression. The R65Q mutation involves substitution of a positively charged amino acid by a noncharged amino acid, causing ~60% reduction of protein expression. G48E, D50E, Y54H, and R67C mutations, all of which occur in strictly conserved amino acid residues, cause nearly complete loss of PSTI expression. As the authors had excluded the possibility that the reduced protein expression may have resulted from reduced transcription of unstable mRNA, they concluded that these missense mutations probably cause intracellular retention of their respective mutant proteins. In addition, two novel mutants have been described recently. A disease-associated codon 41 T to G alteration was found in two European families with an autosomal dominant inheritance pattern, whereas a codon 36 G to C variant was identified as a frequent alteration in subjects of African descent. L14R and L14P mutations resulted in rapid intracellular degradation of the protein and thereby abolished SPINK-1 secretion, whereas the L12F variant showed no such effect. The discovery of *SPINK-1* mutations in humans provides additional support for a role of active trypsin in the development of pancreatitis. SPINK-1 is believed to form the first line of defense in inhibiting prematurely activated trypsinogen in the pancreas.

CFTR Mutations: A New Cause of Chronic Pancreatitis

Cystic fibrosis is an autosomal-recessive disorder with an estimated incidence of 1 in 2500 individuals and is characterized by pancreatic exocrine insufficiency and chronic pulmonary disease. The extent of pancreatic involvement varies between a complete loss of exocrine and endocrine function, to nearly unimpaired pancreatic function. In 1996, Ravnik-Glavac et al. were the first to report mutations in the cystic fibrosis gene in patients with hereditary chronic pancreatitis. Analysis of larger cohorts revealed recurrent episodes of pancreatitis in 1–2% of patients with cystic fibrosis and normal exocrine function, and rarely in patients with exocrine insufficiency as well. Compared to an unaffected population, 17–26% of patients who suffer from idiopathic pancreatitis carry mutations in *CFTR*. Chronic pancreatitis now represents, in addition to chronic lung disease and infertility due to vas deferens aplasia, a third disease entity associated with mutations in the *CFTR* gene. It is important to note that pancreatic exocrine insufficiency in patients with cystic fibrosis is a different disease entity and not to be confused with chronic pancreatitis in the presence of *CFTR* mutations. CFTR is a chloride channel, regulated by $3',5'$-cAMP and phosphorylation and is essential for the control of epithelial ion transport. The

level of executable protein function determines the type and the severity of the disease phenotype. *CFTR* knockout mice show a more severe form of experimental pancreatitis induced by supramaximal cerulein stimulation compared with wild-type animals. The underlying hypothesis of the *CFTR*-related pancreatic injury is a disrupted fluid secretion, which leads to impaired secretion of pancreatic digestive enzymes in response to stimulation. Today more than 1000 mutations within the *CFTR* gene are known, and several of them have been reported in direct association with chronic pancreatitis. For healthy subjects who are heterozygous carriers of *CFTR* mutations, the risk of developing pancreatitis is about 2–4-fold.

KEY CONCEPTS

- Pancreatitis has long been considered an auto-digestive disorder in which the pancreas is destroyed by its own digestive proteases.
- Under physiological conditions, pancreatic proteases are synthesized as inactive precursor zymogen and stored by acinar cells in zymogen granules. Independent of a pathological stimulus that triggers the disease, the pathophysiological events that eventually lead to tissue destruction begin within the acinar cells and involve premature intracellular activation of pancreatic zymogens. This results in acinar cell necrosis followed by a systemic inflammatory response.
- Genetic studies on patients suffering from hereditary pancreatitis suggest that the serine protease trypsin, which is activated by the lysosomal cysteine proteinase cathepsin B, contributes significantly to the development of acute pancreatitis.
- The inhibition of active trypsin by PSTI, as well as the degradation of active trypsin mediated by chymotrypsin C in a calcium-dependent manner, is a protective mechanism of the pancreas against pancreatitis.
- The intracellular rise of Ca^{2+} concentration is a prerequisite for the activation of proteases at the apical pol of acinar cells. The ryanodine receptor plays an important role for the intracellular Ca^{2+} release.
- The R122H mutation within the PRSS1 gene (cationic trypsinogen) is the most frequent mutation associated with chronic pancreatitis and is inherited in an autosomal dominant pattern with an 80% penetrance. This gain of function mutation leads to increased autoactivation of trypsinogen. Since the initial discovery of this mutation, several other mutations within trypsinogen have been reported to be associated with chronic pancreatitis.
- In contrast to mutations within the PRSS1 gene, mutations within the gene encoding for anionic trypsinogen (PRSS2) are protective against idiopathic chronic pancreatitis. The G191R mutation results in a complete loss of tryptic function and has a higher frequency in healthy individuals compared to affected subjects.

- The degradation of active trypsin is thought to be a protective mechanism against pancreatitis. Different mutations within the chymotrypsin C gene, a trypsin degrading enzyme that leads to a loss of function of trypsin, are associated with the development of pancreatitis.
- Mutations within the pancreatic secretory trypsin inhibitor (SPINK-1) results in a increased risk to develop pancreatitis. The most common SPINK-1 mutation is N34S; a change in the trypsin binding pocket might result in a decreased inhibitory capacity of SPINK-1.
- Different mutations within the CFTR gene confer an increased risk for the development of chronic pancreatitis in patients with sufficient exocrine pancreatic function.

SUGGESTED READINGS

1. Chen JM, Ferec C. Molecular basis of hereditary pancreatitis. *Eur J Hum Genet.* 2000;8(7):473–479.
2. Cohn JA, Friedman KJ, Noone PG, et al. Relation between mutations of the cystic fibrosis gene and idiopathic pancreatitis. *N Engl J Med.* 1998;339(10): 653–658.
3. Dimagno MJ, Lee SH, Hao Y, et al. A proinflammatory, antiapoptotic phenotype underlies the susceptibility to acute pancreatitis in cystic fibrosis transmembrane regulator (-/-) mice. *Gastroenterology.* 2005;129(2):665–681.
4. Halangk W, Lerch MM, Brandt-Nedelev B, et al. Role of cathepsin B in intracellular trypsinogen activation and the onset of acute pancreatitis. *J Clin Invest.* 2000;106(6):773–781.
5. Halangk W, Lerch MM. Early events in acute pancreatitis. *Clin Lab Med.* 2005;25(1):1–15.
6. Hirota M, Ohmuraya M, Baba H. The role of trypsin, trypsin inhibitor, and trypsin receptor in the onset and aggravation of pancreatitis. *J Gastroenterol.* 2006;41(9):832–836.
7. Howes N, Lerch MM, Greenhalf W, et al. Clinical and genetic characteristics of hereditary pancreatitis in Europe. *Clin Gastroenterol Hepatol.* 2004;2(3): 252–261.
8. Kruger B, Albrecht E, Lerch MM. The role of intracellular calcium signaling in premature protease activation and the onset of pancreatitis. *Am J Pathol.* 2000;157(1):43–50.
9. Lerch MM, Gorelick FS. Early trypsinogen activation in acute pancreatitis. *Med Clin North Am.* 2000;84(3):549–563.
10. Lerch MM, Halangk W, Kruger B. The role of cysteine proteases in intracellular pancreatic serine protease activation. *Adv Exp Med Biol.* 2000;477:403–411.
11. Pandol SJ, Saluja AK, Imrie CW, et al. Acute pancreatitis: Bench to the bedside. *Gastroenterology.* 2007;132(3):1127–1151.
12. Petersen OH. Ca²⁺ signalling and Ca²⁺-activated ion channels in exocrine acinar cells. *Cell Calcium.* 2005;38(3–4):171–200.
13. Petersen OH. Ca²⁺-induced pancreatic cell death: Roles of the endoplasmic reticulum, zymogen granules, lysosomes and endosomes. *J Gastroenterol Hepatol.* 2008;23(suppl 1):S31–S36.
14. Rosendahl J, Witt H, Szmola R, et al. Chymotrypsin C (CTRC) variants that diminish activity or secretion are associated with chronic pancreatitis. *Nature Genetics* 2008;40:78–82.
15. Szmola R, Kukor Z, Sahin-Toth M. Human mesotrypsin is a unique digestive protease specialized for the degradation of trypsin inhibitors. *J Biol Chem.* 2003;278(49):48580–48589.
16. Szmola R, Sahin-Toth M. Chymotrypsin C (caldecrin) promotes degradation of human cationic trypsin: Identity with Rinderknecht's enzyme Y. *Proc Natl Acad Sci U S A.* 2007;104(27):11227–11232.
17. Whitcomb DC, Gorry MC, Preston RA, et al. Hereditary pancreatitis is caused by a mutation in the cationic trypsinogen gene. *Nat Genet.* 1996;14(2): 141–145.
18. Witt H, Luck W, Hennies HC, et al. Mutations in the gene encoding the serine protease inhibitor, Kazal type 1 are associated with chronic pancreatitis. *Nat Genet.* 2000;25(2):213–216.
19. Witt H. Chronic pancreatitis and cystic fibrosis. *Gut.* 2003;52(suppl 2):ii31–ii41.
20. Witt H, Sahin-Toth M, Landt O, et al. A degradation-sensitive anionic trypsinogen (PRSS2) variant protects against chronic pancreatitis. *Nat Genet.* 2006;38(6):668–673.

22

Molecular Basis of Diseases of the Endocrine System

Alan Lap-Yin Pang . Wai-Yee Chan

INTRODUCTION

This chapter will limit its attention to several well-established hormonal systems. There are a large number of genes whose mutations are now known to be the cause of or be associated with endocrine disorders, and we will direct our attention to the more common ones.

THE PITUITARY GLAND

The pituitary gland is located at the base of the brain. The pituitary gland functions as a relay between the hypothalamus and target organs by producing, storing, and releasing hormones that affect different target organs in the regulation of basic physiological functions (such as growth, stress response, reproduction, metabolism, and lactation). The anterior pituitary, the largest part of the gland, is composed of distinct types of hormone-secreting cells: growth hormone (GH) by the somatotrophs, thyroid-stimulating hormone (TSH) by the thyrotrophs, adrenocorticotropin (ACTH) by the corticotrophs, follicle-stimulating hormone (FSH) and luteinizing hormone (LH) by the gonadotrophs, prolactin (PR) by the lactotrophs, and melanocyte-stimulating hormone (MSH) by the melanotrophs. The posterior pituitary stores and releases hormones (such as antidiuretic hormone and oxytocin) produced by the hypothalamus.

The function of the pituitary gland is regulated by the hypothalamus. Secretion of anterior pituitary hormones (except prolactin) is stimulated or suppressed by specific hypothalamic releasing or inhibiting factors, respectively. The release of these factors is regulated by feedback mechanisms from hormones produced by the target organs as exemplified by the hypothalamus-pituitary-thyroid axis shown in Figure 22.1. The same feedback mechanism also acts on the pituitary to fine-tune the production of pituitary hormones. At the molecular level, secretion of a specific pituitary hormone is triggered by binding of the respective hypothalamic releasing hormone to the corresponding membrane receptors on specific hormone-producing cells in the anterior pituitary. The pituitary hormone is released into the bloodstream, and its binding to the cell surface receptors in target organs triggers hormone secretion to carry out the relevant physiological functions. Therefore, malfunctioning of the pituitary gland can be caused directly by intrinsic deficits of the gland or indirectly by disturbance in any part of the hypothalamic-pituitary-target organ axis.

Pituitary disorders are caused by either hyposecretion (hypopituitarism) or hypersecretion. Hypopituitarism is the deficiency of a single or multiple pituitary hormones. Target organs usually become atrophic and function abnormally due to the loss of their stimulating factors from the pituitary. Hypopituitarism can be acquired by physical means, e.g., traumatic brain injury, tumors that destroy the pituitary or physically interfere with hormone secretion, vascular lesions, and radiation therapy to the head or neck. Congenital hypopituitarism is caused mainly by abnormal pituitary development. Mutations in genes encoding transcription factors controlling Rathke's pouch formation, cellular proliferation, and differentiation of cell lineages have been identified. These mutants generally result in combined pituitary hormone deficiency (CPHD), which is characterized by pituitary malformation and a concomitant or sequential loss of multiple anterior pituitary hormones. Isolated pituitary hormone deficiency (IPHD) is caused by mutations in transcription factors controlling the differentiation of a particular anterior pituitary cell lineage or mutations affecting the expression or function of individual hormones or their receptors. Pituitary hypersecretion disorders are usually caused by ACTH (Cushing's syndrome) and GH (gigantism or acromegaly).

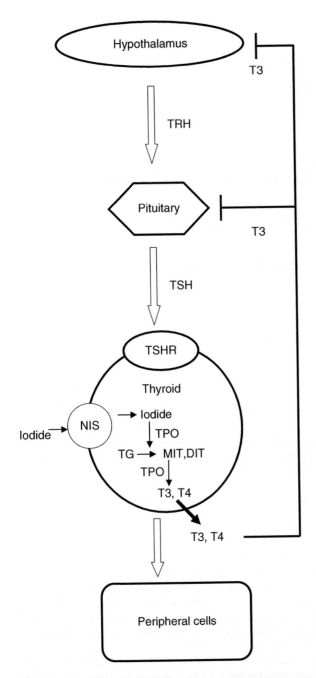

Figure 22.1 The hypothalamus-pituitary-thyroid axis. Hypothalamic thyrotropin-releasing hormone (TRH) stimulates pituitary thyroid-stimulating hormone (TSH) secretion. TSH binds to TSH receptors (TSHR) in the thyroid gland (target organ) to trigger thyroid hormone (T3 and T4) secretion. Thyroid hormones are released into the bloodstream and elicit their physiological functions in peripheral cells through receptor-mediated mechanism. Meanwhile, thyroid hormones inhibit further hypothalamic TRH and pituitary TSH secretion through negative feedback. A steady circulating level of thyroid hormones is thus achieved. The example shown reflects the principal mode of regulation of the hypothalamus-pituitary-target organ axis. ⇓ indicates stimulation, ⊢ indicates inhibition. NIS: sodium iodide symporter; TPO: thyroid peroxidase; TG: thyroglobulin; MIT: mono-iodotyrosyl; DIT: di-iodotyrosyl.

Genes That are Involved in Combined Pituitary Hormone Deficiency

Homeobox Expressed in ES Cells 1

HESX1 is a paired-like homeobox transcriptional repressor essential to the early determination and differentiation of the pituitary. It is one of the earliest markers of the pituitary primordium. Both autosomal recessive and autosomal dominant inheritance have been suggested in HESX1 deficiency.

Thirteen mutations of *HESX1* have been reported. Five mutations are found in homozygous individuals; and eight mutations are found in heterozygotes. Heterozygotes mostly demonstrate incomplete penetrance and milder phenotypes compared to homozygotes. The mutated proteins display diminished DNA binding capacity or impaired recruitment of co-repressor molecules or impaired interaction with PROP1, which is believed to disrupt the normal timing of PROP1-dependent pituitary program.

LIM Homeobox Protein 3 and 4

LHX3 and *LHX4* are LIM-homeodomain transcription factors important to early pituitary development and maintenance of mature anterior pituitary cells. Inheritance of LHX3 deficiency syndrome follows an autosomal recessive pattern. A hypoplastic pituitary is commonly observed. An enlarged anterior pituitary or microadenoma may be found in some patients. Except for gene deletion, the mutant proteins display compromised DNA binding ability and variable degree of loss of gene transactivation capacity.

The inheritance pattern of LHX4 deficiency syndrome is autosomal dominant. All reported mutations are associated with CPHD and/or IPHD, and affected patients show anterior pituitary hypoplasia. The LHX4 mutant proteins show reduced or complete loss of DNA binding and transactivation properties on target gene promoters.

Prophet of Pit-1

PROP1 encodes a paired-like transcription factor restricted to the developing anterior pituitary. PROP1 is required for the determination of somatotrophs, lactotrophs, and thyrotrophs, and differentiation of gonadotrophs. *PROP1* mutations are the leading causes of CPHD, accounting for 30–50% of familial cases. Nevertheless, the presentation of deficiencies (disease onset and severity) is variable. The majority of patients have a hypoplastic or normal anterior pituitary.

Most *PROP1* mutations are found in homozygous and compound heterozygous patients. However, three of the mutations (IVS2–2A>T, c.301_302delAG, and p.R120C) are identified in heterozygotes even the inheritance of the disorder follows an autosomal recessive mode. *PROP1* mutations mostly affect the DNA-binding homeodomain, which leads to reduced or abolished DNA binding and/or gene transactivation activity of the transcription factor.

POU Domain, Class 1, Transcription Factor 1

POU1F1, also known as PIT-1, encodes a POU domain protein essential to the terminal differentiation and expansion of somatotrophs, lactotrophs, and thyrotrophs. It functions as a transcription factor that regulates the transcription of itself and other pituitary hormones and their receptors, including GH, PRL, TSH beta subunit (TSHβ), TSH receptor (TSHR), and growth hormone releasing hormone receptor (GHRHR).

Mutations of POU1F1 have been shown to be responsible for GH, PRL, and TSH deficiencies. Both autosomal recessive and autosomal dominant inheritance have been suggested, and both normal and hypoplastic anterior pituitaries have been observed. Compound heterozygous patients appear to have a more severe disease phenotype. The POU1F1 mutations are mainly found in the POU-specific and POU-homeodomains, which are important interfaces for high-affinity DNA binding on GH and PRL genes and protein-protein interaction with other transcription factors. Consequently, DNA-binding and/or gene transactivation capacity is impaired. In some cases the mutant proteins may also act as dominant inhibitors of gene transcription.

Zinc Finger Protein Gli2

GLI family of transcription factors is implicated as the mediators of Sonic hedgehog (Shh) signals in vertebrates. In humans, Shh signaling is associated with the forebrain defect holoprosencephaly. Heterozygous mutations have been identified in 7 of 390 holoprosencephaly patients who also displayed malformed anterior pituitary and pan-hypopituitarism. All mutations are shown or predicted to lead to premature termination of the GLI2 protein.

Growth Hormone

GH1 is a 191 amino acid (~22 kDa) single-chain polypeptide hormone synthesized and secreted by somatotrophs in the anterior pituitary. It is the major player in many physiological processes related to growth and metabolism. For instance, GH1 displays lactogenic, diabetogenic, lipolytic, and protein anabolic effects, and promotes salt and water retention. The biological effect of GH1 is mainly mediated by insulin-like growth factor 1 (IGF1). IGF1 is generated predominantly in the liver, and acts through its own receptors to stimulate cell proliferation and maturation in cartilage, bone, muscle, and other tissues.

GH deficiency leads to growth retardation in children and GH insufficiency in adults. Isolated growth hormone deficiency (IGHD) refers to the conditions associated with childhood growth failure owing to the absence of GH action, and not associated with other pituitary hormone deficiencies or organic lesions.

Based on the presence or absence of GH secretion, clinical characteristics, and inheritance patterns, familial IGHD is classified into four types: IGHD types 1a and 1b, type 2, and type 3. IGHD type 1 is characterized by its autosomal recessive mode of transmission.

IGHD type 1a is the most severe form of IGHD, with affected individuals developing severe growth retardation by 6 months of age. Anti-GH antibodies often develop in patients undergoing GH replacement therapy. The phenotype of IGHD type 1b is milder but more variable than type 1a. Patients with IGHD type 1a show no detectable serum GH, whereas low but detectable levels of GH are observed in type 1b patients, and they usually respond well to GH treatment. IGHD type 2 is inherited in an autosomal dominant mode. Affected patients do not always show marked growth retardation. Disease phenotypes vary according to the type of the GH1 mutation. IGHD type 3 is an X-linked recessively inherited disorder with a highly variable disease phenotype. The precise genetic lesion that causes IGHD type 3 is still unknown.

GHRH-GH-IGF1 Axis

The GHRH-GH-IGF1 axis plays a central role in the regulation of somatic growth. The hormones and receptors along the axis represent the hot spots for GHD.

Growth Hormone Releasing Hormone Receptor (GHRHR)

GHRHR is a G-protein-coupled receptor (GPCR) expressed specifically in somatotrophs. It is essential to the synthesis and release of GH in response to GHRH, and the expansion of somatotrophs during the final stage of pituitary development. Biallelic mutation in GHRHR is a frequent cause of IGHD type 1b. The mode of inheritance of GHRHR deficiency follows an autosomal recessive pattern. However, heterozygous mutant alleles have also been reported. The mutant receptors are unable to bind to their ligands or elicit cyclic AMP (cAMP) responses after GHRH treatment. In vitro studies show that mutated GHRHR display dominant negative effect on wild-type GHRHR functions. Patients with homozygous mutations tend to display a more severe GHD, but the overall disease phenotype is variable.

Growth Hormone

Human GH1 resides on chromosome 17q24.2. This chromosomal region contains a cluster of GH-like genes (namely GH1, CSHP, CSH1, GH2, and CSH2) arising from gene duplication. The highly homologous nature of the intergenic and coding regions makes the GH locus susceptible to gene deletions arising from homologous recombination.

Both autosomal dominant and autosomal recessive inheritance have been suggested for IGHD type 1a. The severity of disease phenotype due to GH1 mutation varies, with heterozygous mutations giving a milder phenotype than homozygotes. The disorder typically results from deletion or inactivating mutation of GH1. A large number of these mutations were found in heterozygous patients. The most frequent GH1 mutations involve the donor splice site of intron 3, which lead to the skipping of exon 3. The exon skipping event leads to an internal deletion of amino acid residues 32–71 and the production of a smaller (17.5 kDa) GH. This smaller GH inhibits normal GH secretion in a dominant negative manner. Five missense mutations lead to the production of biologically inactive GH that can counteract GH action, impair GHR binding and signaling, or prevent productive GHR dimerization.

GH Receptor

GHR encodes a single transmembrane domain glycoprotein belonging to the cytokine receptor superfamily. In mature form, GHR encodes 620 amino acid residues, with the first 246 residues constituting the extracellular domain for GH binding and receptor dimerization, the middle 24 residues representing the transmembrane domain, and the last 350 residues forming the intracellular domain for GH signaling. GHR is predominantly expressed in liver. The binding of GH triggers receptor dimerization and activation of the downstream signaling process leading to IGF-1 production. The absence of GHR activity causes an autosomal recessive disorder called Laron's syndrome. Affected patients display clinical features similar to IGHD but are characterized by low level of IGF-I and increased level of GH. A vast majority of GHR mutations impact the extracellular domain of GHR. The mutated GHRs display reduced affinity to GH. In other cases, GHR mutations lead to defective homodimer formation and thus defective GH binding. The intracellular domain of GHR is lost in certain splice site mutants: the truncated receptor is unable to anchor to the cell membrane. Although they can bind GH, these GHR mutants fail to activate downstream signaling processes; and they can deplete the binding sites from functional GHRs in a dominant negative manner.

Insulin-Like Growth Factor 1

IGF1, originally called somatomedin C, is a 70-amino acid polypeptide hormone. IGF1 is the major mediator of prenatal and postnatal growth. It is produced primarily in liver and serves as an endocrine (as well as paracrine and autocrine) hormone mediating the action of GH in peripheral tissues such as muscle, cartilage, bone, kidney, nerves, skin, lungs, and the liver itself. In the circulation IGF1 exists in a complex with IGF binding protein 3 (IGFBP3) and an acid-labile subunit (ALS). IGF1 deficiency follows an autosomal recessive mode of transmission. Currently, only four IGF1 mutations have been described, which leads to low or no IGF1 expression or impaired affinity to IGF1 receptors.

GH Hypersecretion

Excessive production of GH causes IGF1 overproduction which leads to gigantism in children and acromegaly in adults. Almost all GH hypersecretion syndromes are caused by benign pituitary GH-secreting adenomas, either as isolated disorders or associated with other genetic conditions such as multiple endocrine neoplasia (MEN1), McCune Albright's syndrome, neurofibromatosis, Carney's complex, or pituitary adenoma predisposition caused by inactivating mutation of the aryl hydrocarbon receptor-interacting protein (AIP) gene. Activating mutation of the gene encoding G-protein subunit $G_{s\alpha}$ and inactivating mutations of GHR, which lead to overactivation of GHRHR and impaired negative feedback on GH production, respectively, are identified in GH-secreting pituitary tumors. In rare cases, GH hypersecretion can be attributed to a hypothalamic tumor or carcinoid that ectopically secretes GHRH.

THE THYROID GLAND

The thyroid, one of the largest endocrine glands, is composed of two types of secretory elements: the follicular and parafollicular cells. The former are responsible for thyroid hormone production, and the latter elaborate calcitonin. The factors that control thyroidal embryogenesis and account for the majority of cases of thyroid agenesis or dysgenesis are presently poorly understood.

Although the thyroid is capable of some independent function, thyroid hormone production is regulated by the hypothalamic-pituitary-thyroid axis (Figure 22.1). Hypothalamic tripeptide thyrotropin-releasing hormone (TRH) stimulates pituitary TSH production. TSH in turn stimulates the thyroid follicular cells to secrete prohormone thyroxine (T4) and its active form triiodothyronine (T3). Since thyroid hormone has a profound influence on metabolic processes, it plays a key role in normal growth and development.

Hypothyroidism

Impaired activity of any component of the hypothalamic-pituitary-thyroid axis will result in decreased thyroid function. Hypothyroidism leads to a slowing of gross metabolic rates. The common causes are faulty embryogenesis, inflammation (most often autoimmune thyroiditis: Hashimoto's thyroiditis), or a reduction in functioning thyroid tissue due to prior surgery, radiation damage, and cancer.

Congenital hypothyroidism is the most common congenital endocrine disorder with an incidence of about 1 in 3500 births. The majority of cases (85%) are nongoitrous and are due to faulty embryogenesis (athyreosis, agenesis, or dysembryogenesis). A small number are goitrous, the result of enzyme defects in hormone biosynthesis (dyshormonogenesis).

Childhood onset (juvenile) hypothyroidism is characterized by thyroid enlargement, slow growth, delayed neuromuscular development, delayed tooth eruption, slow speech, dry skin, hoarse voice, coarse facial features, and retarded bone age. Pubertal development may be early or late depending on age of onset and severity of the hormonal deficiency. The majority of cases of juvenile hypothyroidism are due to chronic lymphocytic (Hashimoto's) thyroiditis.

Hypothyroidism in the adult is most often seen in middle age. Females are 8 times more likely to develop this disease than males. It is most common in individuals with a family history of thyroid disease.

The genes responsible for the development of thyroid follicular cells include thyroid transcription factor 1 (*TIF1*, also known as *TITF1*, *NKX21*, or *T/EBP*), thyroid transcription factor 2 (*TTF2*, also known as *TITF2*, *FOXE1*, or *FKHL15*), paired box transcription factor 8 (*PAX8*), *TSH*, and *TSHR*. Mutation of these genes accounts for about 5% of hypothyroidism patients. *TITF1*, *TITF2*, and *PAX8* mutations give rise to non-syndromic congenital hypothyroidism, while mutations of *TSH* and *TSHR* cause syndromic congenital hypothyroidism. The presence of congenital goiter (which accounts for the remaining 15% of the cases) has been linked to hereditary defects in the enzymatic cascade of thyroid hormone synthesis. Mutations in sodium iodide symporter (*NIS*), pendrin (*PDS*, also known as *SLC264A*), thyroid peroxidase (*TPO*), thyroid oxidase 2 (*THOX2*, also known as dual oxidase DUOX2, *LNOX2*), and thyroglobulin (*TG*) affect organification of iodide and have been linked to congenital hypothyroidism.

Etiologically, three types of congenital hypothyroidism can be distinguished: (i) central congenital hypothyroidism including hypothalamic (tertiary) and pituitary (secondary) hypothyroidism, (ii) primary hypothyroidism (impairment of the thyroid gland itself), and (iii) peripheral hypothyroidism (resistance to thyroid hormone).

Central Congenital Hypothyroidism

Central hypothyroidism may be caused by pituitary or hypothalamic diseases leading to deficiency of TSH or TRH. TRH or TRH receptor mutation is very rare. The majority of cases of central congenital hypothyroidism are caused by *TSH* mutations.

Thyroid-Stimulating Hormone TSH is a heterodimeric glycohormone sharing a common α chain with FSH, LH, and choriogonadotropin (CG). Transmission of congenital isolated TSH deficiency follows an autosomal recessive pattern. Patients with homozygous and compound heterozygous mutations in *TSHβ* have been identified in different ethnic populations. Seven different mutations have been identified. Nonsense mutations lead to a truncated TSHβ subunit, while other mutations may cause a change in conformation. Thus, mutated TSHβ may be immunologically active but biologically inactive, depending on how significant the mutation is on the structure of the protein. There is also variability in the clinical presentation of patients carrying different mutations.

TSH Receptor-Inactivating Mutations The effects of TSH on thyroid follicular cells are mediated by TSHR, a cell surface receptor which, together with the receptor for FSH (FSHR) and the receptor for LH and CG (LHR), forms a subfamily of the GPCR family. This subfamily of glycohormone receptors is characterized by the presence of a large amino-terminal extracellular domain (ECD) involved in hormone-binding specificity. The signal of hormone binding is transduced by a transmembrane domain (TMD) containing 7 α-helices. The intracytoplasmic C-terminal tail (ICD) of the receptor is coupled to $G_{s\alpha}$. Binding of hormone to the receptor causes activation of $G_{s\alpha}$ and the adenylyl cyclase cascade resulting in augmented synthesis of cAMP. The receptor is also coupled to G_q and activates the inositol phosphate cascade.

A subgroup of patients with familial congenital hypothyroidism due to TSH unresponsiveness has homozygous or compound heterozygous loss-of-function mutations in *TSHR*. These mutations are distributed throughout the receptor with no obvious mutation hot spot. Monoallelic heterozygous inactivating mutations have been reported. In these cases, the other TSHR allele is probably inactivated by unrecognized mutations in the intronic, 5'-noncoding region, or 3'-noncoding region, because there is no evidence to support a dominant negative effect of an inactivating *TSHR* mutation.

Depending on where the mutation is located, the effects can be defective receptor synthesis due to premature truncation, accelerated mRNA or protein decay, abnormal protein structure, abnormal trafficking, subnormal expression on cell membrane, ineffective ligand binding, and altered signaling. All mutant TSHRs are associated with defective cAMP response to TSH stimulation. There is good correlation between the phenotype of resistant patients observed *in vivo* with receptor function measured *in vitro*. Patients with mutated TSHR with residual activities are also less severely affected than those with activity void mutated TSHR.

Thyroid Dyshormonogenesis

Iodide is accrued from the blood into thyroidal cells through the sodium/iodide symporter (NIS). It is attached to the tyrosyl residue of TG through the action of thyroid peroxidase (TPO) in the presence of hydrogen peroxide (H_2O_2). Two iodinated TG couple with di-iodotyrosyl (DIT) to form T4; one DIT and one TG couple with mono-iodotyrosyl (MIT) to form T3. The coupling reaction is also catalyzed by TPO. TG molecules that contain T4 and T3 upon complete hydrolysis in lysosomes yield the iodothyroxines, T4 and T3. Disorders resulting in congenital primary hypothyroidism have been identified in all major steps in thyroid hormonogenesis.

Sodium Iodide Symporter Homozygous or compound heterozygous mutation of *NIS* causes iodide transport defect and an uncommon form of congenital hypothyroidism characterized by a variable degree of hypothyroidism and goiter. It also results in low or no radioiodide uptake by the thyroid and other NIS-expressing organs and a low saliva-to-plasma iodide ratio.

Pendrin Pendred's syndrome is an autosomal recessive disorder first described in 1896 and characterized by the triad of deafness, goiter, and a partial organification defect. While many patients with Pendred's syndrome are euthyroid, others have subclinical or overt hypothyroidism.

The causative gene, *PDS* (also known as *SLC26A4*), belongs to the solute carrier family 26A. It encodes a 780-amino acid protein, pendrin, which is predominantly expressed in thyroid, inner ear, and kidney. More than 100 different mutations of *PDS* have been described in patients with classical Pendred's syndrome. The majority of *PDS* mutations are missense mutations. Individuals with Pendred's syndrome from consanguineous families are usually homozygous for *PDS* mutations, while sporadic cases typically harbor compound heterozygous mutations.

In thyroid follicular cells, pendrin is inserted into the apical membrane and acts as an iodide transporter for apical iodide efflux. Abnormality of pendrin affects iodide transport and may lead to iodide organification defects. However, this is often mild or absent, leading to the hypothesis that an as-yet-undefined mechanism may compensate for the lack of pendrin.

Thyroid Peroxidase Thyroid peroxidase (TPO) is a key enzyme in thyroid hormone biosynthesis. It catalyzes both iodination and coupling of iodotyrosine residues in TG. Human *TPO* is located on chromosome 2p25, with 17 exons. The mRNA is about 3 kb encoding a thyroid-specific glycosylated hemoprotein of 110 kDa bound at the apical membrane of thyrocytes. The majority of patients with congenital hypothyroidism have defects in the synthesis or iodination of TG that are attributable to TPO deficiency. Subnormal or absence of TPO activities results in thyroid iodide organification giving rise to goitrous congenital hypothyroidism. Total iodide organification defect (TIOD) is inherited in an autosomal recessive mode.

TPO mutations are one of the most frequent causes of thyroid dyshormonogenesis. Most cases of congenital hypothyroidism are caused by homozygous or compound heterozygous mutations of TPO. TIOD is often associated with inactivating mutations in TPO. Mutant TPO alleles with residual activities are associated with milder thyroid hormone insufficiency or partial iodide organification defects (PIOD). Mutations of *TPO* have also been identified in follicular thyroid carcinoma and adenoma. Most of the inactivating mutations of *TPO* are found in exons 8 and 9 which constitute part of the catalytic site of the enzyme. Mutations affecting protein folding will disturb the structure of the enzyme. The fact that TPO is glycosylated suggests that any mutation affecting a potential glycosylation site (N-X-S/T), such as the missense mutation at nucleotide 391 which causes the replacement of Ser131 by Pro and disturbs the glycosylation site at Asn129, may alter TPO activity. Mutations affecting the membrane spanning region will disturb the insertion of the enzyme into the plasma membrane or alter the normal cellular distribution or trafficking of the enzyme.

Thyroid Oxidase 2 THOX2 generates H_2O_2 which is used by TPO in the oxidation of iodine prior to iodination of the tyrosyl residues of TG. There are two highly homologous *THOX* genes, namely *THOX1* and *THOX2*, which are 16kb apart on chromosome 15q15.3. *THOX2* is preferentially expressed in thyroid whereas *THOX1* is preferentially expressed in airway epithelium and skin. Only THOX2 is involved in H_2O_2 generation in the thyroid. The *THOX2* mRNA encodes a protein of 1548 amino acids, 26 of which comprise a signal peptide that is cleaved off to produce the mature protein. The mature protein is glycosylated. Defects in THOX2 result in a lack or shortage of H_2O_2 and consequently hypothyroidism.

All *THOX2* mutations are identified as congenital hypothyroidism in childhood. The exception is the p.R1110Q mutation which is found in a homozygous adult woman whose thyroid function remains almost normal until the age of 44. Homozygous and compound heterozygous inactivating mutations in *THOX2* lead to complete disruption of thyroid hormone synthesis and are associated with severe and permanent congenital hypothyroidism. Compound heterozygotes with one or both mutant alleles having residual activity have permanent but mild (subclinical) hypothyroidism, while heterozygotes are associated with milder, transient hypothyroidism. It is unusual that a genetic defect results in a transient phenotype at a younger age. This can be due to a changing requirement of thyroid hormone with age.

Thyroglobulin TG is the most abundant protein in the thyroid. It serves as the storage of iodine. Some of its tyrosine residues are iodinated by the action of TPO, and the iodinated tyrosines are coupled to form T3 and T4. TG is produced by thyroid follicular cells and secreted into the follicular lumen. Mutations of gene encoding factors responsible for the development and growth of thyroid follicular cells, such as *TTF1*, *TTF2*, *PAX8*, and *TSHR*, cause dysgenetic congenital hypothyroidism. A rarer cause of congenial goitrous hypothyroidism is mutation of *TG* resulting in structural defects of the protein.

TG encodes a protein of 2768 amino acids of which the N-terminal 19 amino acids constitute a signal peptide. Mutations have been found in exons and introns. Most mutations alter the folding, stability, or intracellular trafficking of the mature protein. Interestingly, some nonsense mutations (e.g., p.R277X) cause the production of a smaller peptide which is sufficient for the synthesis of T_4 in the N-terminal domain. This results in a milder phenotype.

TG deficiency is transmitted in an autosomal recessive mode. Affected individuals are either compound heterozygous or homozygous. The patient phenotypes range from mild to severe goitrous hypothyroidism. No clear phenotype-genotype correlation has been observed.

Resistance to Thyroid Hormone

Resistance to thyroid hormone is characterized by reduced clinical and biochemical manifestations of thyroid hormone action relative to the circulating hormone levels. Patients have high circulating free thyroid hormones, reduced target tissue responsiveness, and an inappropriately normal or elevated TSH.

The thyroid hormones, T3 and T4, exert their target organ effects via nuclear thyroid hormone receptors (TR). After entering target cells, T4 undergoes deiodination to T3, which enters the nucleus and binds to TR. Upon T3 binding, the conformation of TR changes, causing the release of co-repressors and the recruitment of co-activators resulting in the activation or repression of target gene transcription.

Thyroid Hormone Receptor

There are two TR genes, *TRα* and *TRβ*, located on chromosomes 17 and 3, respectively. Both genes produce two isoforms, *TRα1* and *TRα2* by alternative splicing and *TRβ1* and *TRβ2* by utilization of different transcription start sites. TR binds as a monomer, or with greater affinity as homodimers or heterodimers, to thyroid hormone response elements (TREs) in target genes to regulate gene expression. The TR protein has a central DNA binding domain; the carboxy-terminal of TR contains the T3-ligand binding domain, the transactivation, and the dimerization domains. Unlike the other TRs, TRα2 does not bind thyroid hormone. TRs are expressed at different levels in different tissues and at different stages of development. The biological effects of different TRs are compensatory to some degree. However, some thyroid hormone effects are TR isoform specific.

Thyroid hormone resistance is associated with heterozygous autosomal dominant mutations in the *TRβ*. No *TRα* mutation has been reported, and 10% of families with thyroid hormone hyposensitivity do not have *TRβ* mutation. The vast majority of the mutations cluster within three CpG-rich hot spots, namely amino acids 234–282 (hinge domain), amino acids 310–353, and amino acids 429–460 (ligand-binding domain). Many of the mutations are found in different families, and the most common of all *TRβ* mutations is p.R338W. Most mutated receptors display reduced T3 binding or abnormal interaction with one of the cofactors involved in thyroid hormone action. Somatic *TRβ* mutations have been identified in TSH-producing pituitary adenomas.

Dimerization of TRs results in dominant negative effects of mutant TRβs. Besides impairing gene transactivation, the mutant receptors interfere with the function of the wild-type receptor. On the other hand, individuals expressing a single wild-type *TRβ* allele are normal if they lost the other allele due to deletion. Mutations in the hinge domain of TRβ cause impairment of transactivation function and thus resistance to thyroid hormone. Differences in the degree of hormonal resistance in different tissues are due to the absolute and relative levels of TRβ and TRα.

Thyroid Hormone Cell Transporter

Thyroid hormone action requires transport of the hormone from extracellular compartments across the cell membrane, which is mediated by the protein encoded by monocarboxylate transporter gene 8 (*MCT8*). Clinical features of patients with *MCT8* mutations are those of Allan-Herndon-Dudley syndrome (AHDS). These patients have an unusual combination of a high concentration of active thyroid hormone and a low level of inactive metabolite (reverse T3). The severity of the disease is variable among families.

Human *MCT8* (also known as *SLC16A2*) is an X-linked gene. It belongs to the family of genes named *SLC16*, the products of which catalyze the proton-linked transport of monocarboxylates, such as lactate, pyruvate, and ketone bodies. *MCT8* encodes two putative proteins of 613 and 539 amino acids, respectively, by translation from two in-frame start sites. However, it is unknown whether there are two human MCT8 proteins expressed *in vivo*. MCT8 is expressed in a number of tissues, in particular liver and heart. In the pituitary, it is expressed in folliculostellate cells rather than TSH-producing cells.

A variety of missense, nonsense, insertion and deletion mutations of *MCT8* have been reported. In a female patient with full-blown AHDS clinical features, *MCT8* was disrupted at the X-breakpoint of a *de novo* translocation t(X:9)(q13.2;p24). The majority of the mutations results in complete loss of MCT8 transporter function, which can be due to reduced protein expression, impaired trafficking to plasma membrane, or reduced substrate affinity. Missense *MCT8* mutations (such as p.S194F, p.V235M, p.R271H, p.L434W, and p.L568P) show residual activities. A missense mutation affecting the putative translational start codon (c.1 A>T, p.M1L) is not pathogenic, implying that the methionine at amino acid position 75 may serve as the translational start of a functional MCT8 protein.

FAMILIAL NONAUTOIMMUNE HYPERTHYROIDISM

Patients with a family history of thyrotoxicosis with early disease onset typically have thyromegaly, absence of typical signs of autoimmune hyperthyroidism, and recurrence after medical treatment. A number of germline constitutive-activating mutations of *TSHR* have been identified in patients with nonautoimmune hyperthyroidism. The inheritance of the trait follows an autosomal dominant pattern. The majority of these mutations are located in exon 10 of *TSHR*, encoding the transmembrane and intracellular domains. So far only two amino acid residues with 3 different mutations (p.R183K, p.S281N, and p.S281I) which occur outside exon 10 and in exons encoding the extracellular domain have been identified. Mutation hot spots include p.Ala623 and p.Phe631. Aside from germline activating mutations, a number of sporadic mutations have also been identified in patients with nonautoimmune hyperthyroidism. Germline *TSHR* mutations usually present later in life with milder clinical manifestation, whereas sporadic *TSHR* mutations cause severe thyrotoxicosis and goiter with neonatal or infancy onset. Biological activities of the mutant TSHRs differ in terms of adenylyl cyclase and inositol phosphate pathway activation. Most mutations cause only constitutive activation of the cAMP pathway, whereas a few mutants activate both cAMP and phospholipase C cascades.

The expressivity of the activating mutations may be affected by environmental factors such as iodine uptake. Even among cases with the same mutation of TSHR, phenotypic expression can be quite variable with respect to the age of onset and severity of hyperthyroidism and thyroid growth.

Somatic activating mutations of TSHR have been identified as the cause of solitary toxic adenomas and multinodular goiter and thyroid carcinomas, though less frequently. Near 80% of toxic thyroid adenomas (hot nodules) exhibit TSHR gain-of-function mutations.

THE PARATHYROID GLAND

Histologically, the parathyroid glands comprise two types of cells: chief cells and oxyphil cells; each type may appear in clusters or occasionally mixed. Chief cells have a pale staining cytoplasm with centrally placed nuclei. The cytoplasm contains elongated mitochondria and two sets of granules: secretory granules that take a silver stain and others that stain for glycogen. The oxyphil cells are fewer in number, larger than the chief cells, and their cytoplasm has more mitochondria. The oxyphil cells do not appear until after puberty and their function is unclear.

Parathyroid hormone (PTH) is secreted by the chief cells of the parathyroid glands. It is synthesized as a mature peptide of 84 amino acids with a signal peptide of 25 amino acids. The hormone has two main targets: bones and kidneys. In bone, PTH activates lysosomal enzymes in osteoclasts to mobilize calcium; in kidneys it promotes calcium resorption and phosphate and bicarbonate excretion by activating the cAMP signaling pathway. The synthesis and secretion of PTH is regulated by serum ionized calcium concentration.

Calcium Homeostasis

The level of calcium in the blood is kept within a very narrow range between 2.2 and 2.55 mM and is present in three forms: (i) bound, (ii) complexed, and (iii) free. About 30% is bound to protein, mainly albumin; 10% is chelated, mostly to citrate; and the remaining 60% is free or ionized. In humans, the regulation of the ionized fraction of serum calcium depends on the interaction of PTH and the active form of vitamin D-1,25-dihydroxycholecalciferol [$1,25(OH)_2D$].

Vitamin D is a prohormone which requires hydroxylation for its conversion to the active form. Vitamin D from food (ergocalciferol, D_2) is absorbed in the proximal gastrointestinal tract and is also produced in the skin by exposure to ultraviolet light (cholecalciferol, D_3). It is carried by a vitamin D binding protein to the liver where it is hydroxylated to form 25-hydroxy vitamin D. The latter equilibrates with a large stable pool, and the serum concentration of 25-hydroxy vitamin D is the best clinical measure of nutritional vitamin D status. The 25-hydroxy vitamin D is then further hydroxylated in kidneys to $1,25(OH)_2D$ under the influence of PTH to become the physiologically active hormone.

Decreasing serum ionized calcium, acting via a transmembrane G-protein-coupled calcium-sensing receptor, stimulates PTH secretion. PTH in turn acts on osteoclasts to release calcium from bones and inhibits calcium excretion from kidneys. PTH, hypocalcemia, and hypophosphatemia furthermore augment the formation of $1,25(OH)_2D$ which stimulates absorption of calcium from the gut and aids PTH in mobilizing calcium from bone, while at the same time inhibiting 24-hydroxylation. The reverse occurs when calcium levels rise, and PTH secretion is inhibited. Whereas 24-hydroxylation is induced, 21-hydroxylation is inhibited, resulting in a shift to the inactive metabolite of vitamin D. Movement of calcium from bone and gut into the extracellular fluid compartment ceases, and so does renal reabsorption, thereby restoring serum calcium to its normal physiological concentration. Any disturbance in this regulation may result in hypocalcemia or hypercalcemia. Symptoms and signs may vary from mild, asymptomatic incidental findings to serious life-threatening disorders depending on the duration, severity, and rapidity of onset. Hypocalcemia is the result either of increased loss of calcium from the circulation or insufficient entry of calcium into the circulation, and hypercalcemia is the reverse. Since PTH is the main player in this scenario, its influence is central.

Hypoparathyroidism

Hypoparathyroidism is characterized by levels of PTH insufficient to maintain normal serum calcium concentration. The clinical manifestations of hypoparathyroidism are those of hypocalcemia.

The most common cause of hypoparathyroidism is injury to the parathyroid glands during head and neck surgery or due to autoimmunity (either isolated or as a polyglandular syndrome). Familial forms of congenital and acquired hypoparathyroidism are relatively rare. Both familial isolated hypoparathyroidism and familial hypoparathyroidism accompanied by abnormalities in multiple organ systems have been described.

Familial Isolated Hypoparathyroidism

Several forms of familial isolated hypoparathyroidism with autosomal dominant, autosomal recessive, and X-linked recessive inheritance have been described. Mutations affecting the signal peptide of pre-pro-PTH have been reported in families with hypoparathyroidism following an autosomal recessive or autosomal dominant inheritance pattern. The mutant pre-pro-PTH cannot be cleaved, resulting in the degradation of mutant peptide in endoplasmic reticulum.

Glial cells missing (GCM) belongs to a small family of key regulators of parathyroid gland development.

There are two forms: *GCMA* and *GCMB*. The mouse homolog of *GCMB*, Gcm2, is expressed exclusively in parathyroid glands. Five different homozygous mutations of *GCMB* have been identified in patients with isolated hypoparathyroidism. These mutations either disrupt the DNA binding domain or remove the transactivation domain of GCMB, leading to inactivation of the protein. Aside from mutations of *PTH* and *GCMB*, X-linked recessive hypoparathyroidism due to mutation of a gene localized to chromosome Xq26-q27 has been reported in two multigenerational kindreds. The candidate gene has not been cloned yet.

Familial Hypoparathyroidism as Part of a Complex Congenital Defect

Hypoparathyroidism may be part of a polyglandular autoimmune disorder or associated with multiple organ abnormalities. Because of the involvement of multiple organs in these disorders, it is likely the abnormalities are not intrinsic to the parathyroid gland but consequential to other abnormalities. Hypoparathyroidism occurs in HAM syndrome (Hypoparathyroidism, Addison's Disease, Mucocutaneous Candidiasis, also known as Polyglandular Failure Type 1 syndrome or Autoimmune Polyendocrinopathy-Candidiasis-Ectodermal Dystrophy (APECED). This syndrome is caused by mutations of *AIRE1* (autoimmune regulator type 1). Cells destined to become parathyroid glands emerge from the third and fourth pharyngeal pouches. Failure of development of the derivatives of these two pouches results in absence or hypoplasia of the parathyroids and thymus and gives rise to DiGeorge syndrome (DGS). Most cases of DGS are sporadic, but autosomal dominant inheritance of DGS has been observed. The vast majority of cases display imbalanced translocation or microdeletion of 22q11.2. So far, there is no consensus on the gene(s) whose mutation(s) is(are) responsible for DGS. The broad spectrum of abnormalities in DGS patients and the large genomic region in the deletion suggest the possibility of involvement of multiple genes. Similarly, abnormality of *PTH* is excluded in familial Hypoparathyroidism, sensorineural Deafness, and Renal dysplasia (HDR syndrome), which is inherited in an autosomal dominant mode. Deletion analyses of HDR patients indicate the absence of *GATA3* is responsible for HDR. GATA3 is a member of the GATA-binding family of transcription factors, and is expressed in the developing parathyroid gland, inner ear, kidney, thymus and central nervous system. Thus, *GATA3* mutations will affect parathyroid development. These mutations affect DNA-binding, mRNA and protein stability, or reduced protein synthesis of GATA3. The pathogenic mechanism responsible for HDR syndrome is believed to be GATA3 haploinsufficiency. Other diseases with hypoparathyroidism but with no genetic abnormality of *PTH* include several mitochondrial disorders, Kenney-Caffey and Sanjad-Sakati syndrome, Barakat syndrome, and Blomstrand's disease.

Hyperparathyroidism

Primary hyperparathyroidism is a genetically heterogeneous disorder, and usually occurs as a sporadic disorder. It may occur at any age, but is most common in the 60+ age group and more common in females. It results when PTH is secreted in excess due to a benign parathyroid adenoma, parathyroid hyperplasia, multiple endocrine neoplasia (MEN1 or MEN2A), or cancer. An elevated serum calcium together with low serum phosphorus is suggestive of hyperparathyroidism and becomes diagnostic when supported by increased urinary calcium excretion and elevated PTH levels.

Familial Isolated Hyperparathyroidism

Familial isolated hyperparathyroidism (FIHP) is a rare disorder characterized by single or multiple glandular parathyroid lesions without other syndromes or tumors. The mode of transmission is autosomal dominant. Most FIHP kindreds have unknown genetic background. Studies of 36 kindreds revealed that a few families are associated with multiple endocrine neoplasia gene 1 (*MEN1*), with inactivating calcium-sensing receptor (*CaSR*) mutations, or DNA markers near the Hyperparathyroidism-Jaw Tumor (HPT-JT) syndrome locus at 1q25-q32.

Multiple Endocrine Neoplasia Type 1

Multiple Endocrine Neoplasia Type 1 (MEN1) is an autosomal dominant disorder with a high degree of penetrance. Over 95% of patients develop clinical manifestations by the fifth decade. The most common feature of MEN1 is parathyroid tumors which occur in about 95% of patients. Tumors of the pancreas and anterior pituitary also occur in a significant, albeit smaller percentage of patients. MEN1 is caused by germline mutations of *MEN1* which is located on chromosome 11q13. The gene encodes MEN1N, a tumor suppressor and a nuclear protein that participates in transcriptional regulations, genome stability, cell division, and proliferation. MEN1 tumors frequently have loss of heterozygosity (LOH) of the *MEN1* locus. Somatic abnormalities of *MEN1* have been reported in MEN1 and non-MEN1 endocrine tumors.

There are a total of 565 different *MEN1* mutations among the 1336 mutations reported, with 459 being germline and 167 being somatic (61 mutations occur both as germline and somatic), scattered throughout *MEN1*. The majority of these mutations are predicted to give rise to truncated or destabilized proteins. Four mutations, namely, c.249_252delGTCT (deletion at codons 83–84), c.1546–1547insC (insertion at codon 516), c.1378C>T (R460X), and c.628_631delACAG (deletion at codons 210–211), represent potential mutation hot spots. No phenotype-genotype correlations have been observed in MEN1. In fact, similar *MEN1* mutations have been observed in FIHP and MEN1 patients. The reason for these altered phenotypes is not known.

Multiple Endocrine Neoplasia Type 2

Multiple Endocrine Neoplasia Type 2 (MEN2) describes the association of medullary thyroid carcinoma, pheochromocytomas, and parathyroid tumors. There are three clinical variants of MEN2, the most common of which is MEN2A, which is also known as Sipple's syndrome. MEN2A is inherited in an autosomal dominant manner. All MEN2A patients develop medullary thyroid carcinoma (MTC), 50% develop bilateral or unilateral pheochromocytoma, and 20–30% develop primary hyperparathyroidism. Primary hyperparathyroidism can occur by the age of 70 in up to 70% of patients. MEN2A is caused by activating point mutations of the *RET* proto-oncogene. *RET* encodes a tyrosine kinase receptor with cadherin-like and cysteine-rich extracellular domains, and a tyrosine kinase intracellular domain. In 95% of patients, MEN2A is associated with mutations of the cysteine-rich extracellular domain, and missense mutation in codon 634 (Cys to Arg) accounts for 85% of MEN2A mutations.

The second clinical variant is MEN2B. These patients exhibit a relative paucity of hyperparathyroidism and development of marfanoid habitus, mucosal neuromas, and intestinal ganglioneuromatosis. Of these patients, 95% present with mutation in codon 918 (Met to Thr) of the intracellular tyrosine kinase domain of *RET* proto-oncogene. The third clinical variant of MEN2 is MTC-only. This variant is also associated with missense mutation of RET in the cysteine-rich extracellular domain, and most mutations are in codon 618. The precise mechanism of the genotype-phenotype relationship in the three clinical variants is unknown.

Hyperparathyroidism-Jaw Tumor Syndrome

Hyperparathyroidism-jaw tumor syndrome (HPT-JT syndrome) is an autosomal dominant disorder characterized by the development of parathyroid adenomas and carcinomas and fibro-osseous jaw tumors. The causative gene, *HRPT2*, encodes a ubiquitously expressed 531-amino acid protein called parafibromin. *HRPT2* is a tumor-suppressor gene. Its inactivation is directly involved in the predisposition to HPT-TJ syndrome and in the development of some sporadic parathyroid tumors. The function of the gene has not been elucidated. To date, 13 different heterozygous inactivating mutations that predict truncation of parafibromin have been reported.

Calcium-Sensing Receptor and Related Disorders

Extracellular ionized calcium concentration regulates the synthesis and secretion of PTH as well as parathyroid cell proliferation. The action of calcium is mediated by the calcium-sensing receptor (CaSR). Human CaSR is a member of family CII of the GPCR superfamily. The protein has a large extracellular domain with 612 amino acids, a transmembrane domain of 250 amino acids containing 7 transmembrane helices, and an intracellular domain of 216 amino acids. The receptor is heavily glycosylated, which is important for normal cell membrane expression. CaSR forms homodimers via intramolecular disulfide linkages within the extracellular domain. The extracellular domain is responsible for calcium binding. However, the stoichiometry between CaSR and calcium ions is unknown. The transmembrane domain is responsible for transducing the signal of calcium binding, while the intracellular domain interacts with $G_{\alpha i}$ or $G_{\alpha o}$ and activates different signal transducing pathways, depending on the cell line. The critical pathway(s) through which CaSR mediates its biological effects has not been defined.

CaSR is expressed in parathyroid cells, kidneys, bones, along the gastrointestinal tract, and in other tissues that are not directly involved in calcium homeostasis. However, the highest cell surface expression of CaSR is found in parathyroid cells, C-cells of the thyroid, and kidneys. Parathyroid cells are capable of recognizing small perturbation in serum calcium and respond by altering the secretion of PTH. Mutations of *CaSR* can result in loss-of-function or gain-of-function of the receptor. Heterozygous mutations of *CaSR* are the cause of a growing number of disorders of calcium metabolism, which typically manifest as asymptomatic hypercalcemia or hypocalcemia, with relative or absolute hypercalciuria or hypocalciuria. On the other hand, homozygous or compound heterozygous inactivating mutations produce a severe and sometimes lethal disease if left untreated.

Disorders Due to Loss-of-Function Mutations of CaSR

There are two hypercalcemic disorders caused by inactivating mutations in *CaSR*: familial hypocalciuric hypercalcemia and neonatal severe primary hyperparathyroidism.

Familial Hypocalciuric Hypercalcemia A disorder that must be recognized as distinct from primary hyperparathyroidism is familial hypocalciuric hypercalcemia (FHH), also known as familial benign hypercalcemia (FBH). The hypercalcemia is usually mild and asymptomatic, and is associated with a reduction rather than an increase in urinary calcium excretion. In the majority of cases the cause appears to be an autosomal dominant loss-of-function mutation in *CaSR*. The mutation in FHH reduces the receptor sensitivity to calcium. In the parathyroid glands, this defect means that a higher than normal serum calcium concentration is required to reduce PTH release. In the kidney, this defect leads to an increase in tubular calcium and magnesium resorption. The net effect is hypercalcemia, hypocalciuria, and frequently hypermagnesemia, as well as normal or slightly higher serum PTH concentration.

About two-thirds of the FHH kindreds studied have unique heterozygous mutations. These mutations are described in the *CaSR* mutation database (http://www.casrdb.mcgill.ca). Some of these mutations cause a more severe hypercalcemia by inhibiting the wild-type CaSR in mutant-wild-type heterodimers. There are ~30% of FHH patients without an identifiable mutation. Linkage analyses have linked FHH in some of these patients to the long and short arms of chromosome 19.

Neonatal Severe Primary Hyperparathyroidism   Neonatal severe primary hyperparathyroidism (NSHPT) represents the most severe expression of FHH. Symptoms manifest very early in life with severe hypercalcemia, bone demineralization, and failure to thrive. Most NSHPT patients have either homozygous or compound heterozygous *CaSR* inactivating mutations. There are also three cases of NSHPT in which *de novo* mutation was found in the extracellular domain of CaSR. The patients have only one copy of *CaSR* mutated, and no *CaSR* mutations are found in the parents. There are also asymptomatic patients with homozygous *CaSR* inactivating mutations. Thus, the severity of symptoms may be affected by factors other than mutant gene dosage.

Disorders Due to Gain-of-Function Mutation of the CaSR

There are two hypocalcemic disorders associated with gain of CaSR function due to activating mutations of the receptor. These mutated receptors are more sensitive to extracellular calcium required for PTH secretion and cause reduced renal calcium resorption. The two disorders are Autosomal Dominant Hypocalcemia (ADH) and Bartter syndrome type V.

Autosomal Dominant Hypocalcemia   ADH is a familial form of isolated hypoparathyroidism characterized by hypocalcemia, hyperphosphatemia, and normal to hypoparathyroidism. Inheritance of the disorder follows an autosomal dominant mode. The patients are generally asymptomatic. A significant fraction of cases of idiopathic hypoparathyroidism may in fact be ADH.

More than 80% of the reported ADH kindreds have CaSR mutations. There are 44 activating mutations of CaSR reported that produce a gain of CaSR function when expressed in *in vitro* systems. The majority of the ADH mutations are missense mutations within the extracellular domain and transmembrane domain of CaSR. The mechanism of CaSR activation by these mutations is not known. Interestingly, almost every ADH family has its own unique missense heterozygous CaSR mutation. Most ADH patients are heterozygous. The only deletion-activating mutation occurs in a homozygous patient in an ADH family. However, there is no apparent difference in the severity of the phenotype between heterozygous and homozygous patients.

Bartter Syndrome Type V   In addition to hypocalcemia, patients with Bartter syndrome type V have hypercalciuria, hypomagnesemia, potassium wasting, hypokalemia, metabolic alkalosis, elevated renin and aldosterone levels, and low blood pressure. Four different activating mutations of CaSR (p.K29E, p.L125P, p.C131W, and p.A843E) have been identified in patients with Bartter syndrome type V. Functional analyses show that these CaSR mutations result in a

more severe receptor activation when compared to the other activating mutations described.

THE ADRENAL GLAND

The adrenal glands are two crescent-shaped structures located at the superior pole of each kidney. Each gland is composed of two separate endocrine tissues, the inner medulla and the outer cortex, which have different embryonic origins. The adrenal medulla is responsible for the production of the catecholamines, epinephrine and norepinephrine, which are involved in fight-or-flight responses. The adrenal cortex produces a number of steroid hormones that regulate diverse physiological functions. Despite the difference in origin and physiology, the hormones produced by the medulla and cortex often act in a concerted manner.

The adrenal cortex is composed of three distinct cell layers. The outermost layer (zona glomerulosa) secretes mineralocorticoids, the most important of which is aldosterone. Aldosterone primarily affects the distal renal tubular sodium-potassium exchange mechanism to regulate circulatory volume. The synthesis and release of aldosterone is mainly influenced by the renin-angiotensin system in response to variations in renal blood flow.

The middle layer of the adrenal cortex (zona fasciculata) secretes glucocorticoids, the most important of which is cortisol. The major biological function of cortisol is to maintain blood glucose level and blood pressure and to modulate stress and inflammatory responses. Secretion of cortisol is regulated by the hypothalamus-pituitary-adrenal axis. The hypothalamic corticotropin-releasing hormone (CRH) stimulates ACTH secretion by the pituitary corticotrophs through a ligand-receptor-mediated mechanism. ACTH triggers cortisol production in adrenocortical cells by binding to melanocortin receptor 2 (MC2R, which is the ACTH receptor). A series of enzymatic reactions is initiated to mediate the uptake of cholesterol and the biosynthesis of cortisol. The circulating cortisol feeds back to the pituitary and hypothalamus to suppress further ACTH secretion.

The innermost layer of the adrenal cortex (zona reticularis) produces adrenal androgens. The production of adrenal androgens is also regulated by ACTH. In the fetal adrenal cortex (during weeks 7–12 of gestation), androgens are secreted and regulate the differentiation of male external genitalia. However, in adults the contribution of androgens from adrenal glands is quantitatively insignificant.

Congenital Primary Adrenal Insufficiency

Adrenal insufficiency is generally referred to an inadequate production of cortisol by the adrenals. Primary adrenal insufficiency is caused by defects of the adrenals themselves. Congenital primary adrenal insufficiency is caused by inactivating mutations of genes responsible for normal adrenal development or cortisol production. Depending on the severity, a deficiency of cortisol alone or with other adrenal steroids would occur.

Adrenal Hypoplasia Congenita

Adrenal hypoplasia congenita (AHC) is the underdevelopment of adrenal glands that typically results in adrenal insufficiency during early infancy. Two forms of AHC are known: (i) an autosomal recessive miniature adult form and (ii) an X-linked cytomegalic form. Patients affected by the latter display low or absent glucocorticoids, mineralocorticoids, and androgens and do not respond to ACTH stimulation. The adrenals are atrophic and structurally disorganized, with no normal zona formation in the cortex. Impaired sexual development, owing to hypogonadotropic hypogonadism, manifests in affected males who survive childhood.

The cytomegalic form of AHC is an X-linked disorder due to deletion or inactivating mutation of *NR0B1*, which encodes DAX1 (dosage sensitive sex reversal-AHC critical region on the X chromosome gene 1). As a result, females are unaffected carriers and males are affected by *NR0B1* mutation. DAX1 encodes an orphan nuclear receptor that interacts with SF1 (Steroidogenic Factor 1; encoded by *NR5A1*). Binding of DAX1 inhibits SF1-mediated transactivation of genes involved in the development of the hypothalamus-pituitary-adrenal-gonad axis and biosynthesis of steroid hormones. More than 100 mutations of *NR0B1* have been described. The mutant DAX1 proteins cannot repress SF1 transactivation activity. Phenotypic heterogeneity, in terms of severity and onset of disorder, is observed in patients with different or even the same *DAX1* mutations, suggesting that other genetic, epigenetic, or environmental factors may be involved in AHC. Alternatively, the presence of residual activity in *DAX1* mutants (positional effect of mutations) may determine the extent of phenotypic variation. The expression of a novel *NR0B1* splice variant in the adrenal glands is also suspected to influence AHC phenotype.

Congenital Adrenal Hyperplasia

Congenital adrenal hyperplasia (CAH) is an inborn metabolic error in adrenal steroid biosynthesis. Cholesterol is transported from the cytoplasm into mitochondria in adrenocortical cells by steroidogenic acute regulatory protein (StAR), and is first converted to pregnenolone, the common precursor of all adrenal steroids, by rate-limiting cholesterol side-chain cleavage enzyme P450scc (encoded by *CYP11A1*). Adrenal steroids are produced under the action of various cytochrome P450s, with some of them participating in more than one steroidogenic pathway. Depending on the enzymatic defect, an impaired production of glucocorticoids (and mineralocorticoids or androgens) can occur starting *in utero*. The feedback to the pituitary is undermined such that ACTH production becomes excessive. Consequently, the adrenals are overstimulated and hyperplastic. The overproduced steroid precursors are shunted to the androgen biosynthetic pathway, leading to androgen overproduction. Genital ambiguity in newborn females (owing to excessive fetal exposure *in utero*) and precocious pseudo-puberty in both sexes are commonly observed in CAH patients. In severe cases, salt-wasting CAH occurs with life-threatening vomiting and dehydration within the first few weeks of life.

21-Hydroxylase Deficiency The most frequent CAH variant is 21-hydroxylase deficiency (21-OHD), caused by inactivating mutations or deletion of *CYP21A2*, and accounts for ~95% of classical CAH. In many cases the term *CAH* refers to 21-OHD. The failure of conversion of progesterone and 17-hydroxyprogesterone leads to their accumulation and shunting to the androgen biosynthetic pathway. Based on the severity of enzyme defect, three clinical forms of 21-OHD CAH are recognized: the most severe classical salt-wasting form (due to aldosterone insufficiency and androgen excess), the moderately severe classical simple-virilizing form (due to androgen excess), and the least severe nonclassical form.

11β-Hydroxylase Deficiency The second most common cause of CAH is 11β-hydroxylase deficiency (11β-OHD). The enzyme, 11β-hydroxylase (encoded by *CYP11B1*), catalyzes the conversion of deoxycorticosterone and 11-deoxycortisol to corticosterone and cortisol, respectively. Similar to 21-OHD, the enzymatic defect leads to virilizing CAH due to accumulation of immediate steroid precursors. The accumulation of deoxycorticosterone also causes salt retention and hypertension.

A variant form of CAH related to 11β-OHD is caused by chimeric gene formation due to unequal crossing over. *CYP11B2*, which is >95% identical to and is located ~40 kb upstream of *CYP11B1*, encodes an aldosterone synthase that is exclusively expressed in zona glomerulosa to catalyze the conversion of deoxycorticosterone to corticosterone. The high degree of homology renders the two isoforms susceptible to crossing over during DNA replication, with regulatory sequence from one fused to the coding region of the other. Use of *CYP11B2* promoter leads to no production of chimeric CYP11B2/B1 protein in zona fasciculata because the promoter is inactive in this region of the adrenal. Thus, carriers of this mutation display 11β-OHD. Conversely, the *CYP11B1* promoter-driven expression of *CYP11B1/B2* chimera, which retains aldosterone synthase activity, leads to overproduction of mineralocorticoids and results in an autosomal dominant disorder called glucocorticoid-suppressible hyperaldosteronism (or familial hyperaldosterone type 1). A rare case of salt-wasting CAH with 11-OHD due to homozygous internal deletion of the *CYP11B2/B1* chimera has also been reported.

Other Less Common Steroidogenic Enzyme Deficiency in CAH The key enzyme in steroid hormone biosynthesis is 3β-hydroxysteroid dehydrogenase (3β-HSD). It catalyzes the conversion of progesterone, 17-OH progesterone, and androstenedione from their respective Δ^5 steroid precursors. As a result, inactivation of 3β-HSD leads to incomplete genital development and impaired aldosterone synthesis due to a deficiency of all classes of adrenal steroids. In humans, *HSD3B2* is the isoform expressed specifically in the adrenals and gonads.

CYP17A1 encodes an enzyme with dual (17α-hydroxylase and 17,20-lyase) activities, mediating the biosynthesis of cortisol and sex steroid precursors. Inactivating mutations of the gene frequently produce combined enzyme deficiencies, displaying hypertension, hypokalemia, and sexual infantilism. Isolated 17,20-lyase deficiency also occurs when lyase activity is selectively impaired by mutations of amino acid residues (such as p.R347 or p.R358) that are crucial to the interaction between CYP17A1 and its redox partners (P450 oxidoreductase and cytochrome b5) in the P450-electron donor complex.

Although uncommon, Lipoid Congenital Adrenal Hyperplasia (LCAH) represents the most severe form of CAH. Life-threatening mineralocorticoid and glucocorticoid deficiencies are common in infants and children. Male infants are undervirilized, and puberty is delayed in both sexes. The adrenals and gonads are enlarged and filled with lipid globules. The majority of cases of LCAH is derived from defective cholesterol transport into the mitochondria by inactivating mutation of *StAR*, and to a lower extent, defective pregnenolone conversion from cholesterol by inactivating mutation of P450scc. The lack of substrate leads to adrenal steroid deficiency and absence of feedback for ACTH suppression. The elevated level of ACTH excessively stimulates cholesterol uptake and adrenal cell growth, resulting in adrenal hyperplasia.

ACTH Resistance Syndromes

ACTH resistance syndromes represent a group of disorders that lead to unresponsiveness to ACTH in the production of glucocorticoid by adrenal cortex.

Familial Glucocorticoid Deficiency Familial Glucocorticoid Deficiency (FGD) is a rare autosomal recessive disease characterized by cortisol deficiency, but without mineralocorticoid deficiency and pituitary structural defects. It is usually diagnosed during the neonatal period or early childhood. Specifically, FGD patients show an extremely high plasma ACTH level, which indicates a resistance to ACTH action. They almost always develop hyperpigmentation. Adrenarche is absent in children with FGD.

FGD is caused by defects of melanocortin receptor 2 (MC2R). Based on gene mutations, FGD is subdivided into two types. FGD type 1 is caused by inactivating mutations of MC2R. FGD type 2 is caused by inactivating mutations of the MC2R accessory protein (MRAP) which assists the trafficking and expression of MC2R at the cell surface. Genetic defects of MC2R and MRAP, however, account for 45% of FGD cases, suggesting that more genes may be involved in the etiology of FGD.

Triple A Syndrome Triple A syndrome is a complex disorder characterized by adrenal failure, alacrima, and achalasia. Not all patients exhibit adrenal failure, but of those with adrenal failure ~80% will have isolated glucocorticoid deficiency. Mutations of the causative gene, AAAS, are identified in patients with the syndrome. However, defects in AAAS do not account for all cases of Triple A syndrome, and the expression pattern of the gene correlates poorly to pathology.

Secondary Adrenal Insufficiency

Secondary adrenal insufficiency can be attributed to a lack of ACTH. The adrenal glands become atrophic and cortisol, but not aldosterone, production is extremely low or undetectable. For acquired cases, ACTH deficiency may result from pituitary tumor or other physical lesions that prevent ACTH secretion. Meanwhile, congenital cases are caused by disorders of the hypothalamus-pituitary axial components that control ACTH production.

The hypothalamic corticotropin-releasing hormone (CRH) binds to CRH receptor (CRHR) on pituitary corticotrophs to stimulate ACTH production through enzymatic cleavage of its precursor propiomelanocortin (POMC) under the action of prohormone convertase 1 (PC-1). POMC deficiency syndrome, owing to recessive inactivating mutations of *POMC*, have been described. However, mutations of *TPIT*, which encodes a T-box transcription factor important to POMC expression and terminal differentiation of pituitary POMC-expressing cells, cause adrenal insufficiency more frequently.

Generalized Glucocorticoid Resistance/Insensitivity

Generalized glucocorticoid resistance is a rare genetic condition characterized by generalized partial target tissue insensitivity to glucocorticoid. Patients show increased levels of ACTH and cortisol, resistance of the hypothalamus-pituitary-adrenal axis to dexamethasone suppression, but no clinical sign of hypercortisolism. Production of mineralocorticoids and androgens is increased as a result of excessive ACTH. The disorder is caused by inactivation of the glucocorticoid receptor-α isoform (GRα), which functions as a ligand-dependent transcription factor. Almost all known mutations are heterozygous. In some cases, the mutated receptors display a dominant negative effect on wild-type receptors.

Hypercortisolism (Cushing's Syndrome)

Excessive hormone production by the adrenal cortex may result from abnormal pituitary ACTH stimulation, pituitary tumors, ectopic ACTH produced by a neoplasm, or pathology within the adrenals themselves.

Cushing's syndrome is caused by excessive circulating cortisol, which affects 10–15 in every million people at age 20–50. Cushing's syndrome is primarily caused by ACTH-secreting pituitary adenoma (also known as Cushing's disease) which accounts for ~70% of all cases of ACTH-dependent Cushing's syndrome. Other causes include adrenal hyperplasia or neoplasia (the second most common genetic cause) and ectopic ACTH production by other tumors. All defects result in adrenal gland overgrowth and cortisol overproduction. The elevated level of cortisol promotes protein catabolism, muscle wasting, skin thinning, and conversion to fat without weight gain, giving the characteristic appearance such as moon faces, buffalo hump, and central obesity. Otherwise, the cause of Cushing's syndrome is mostly iatrogenic, resulting from prolonged steroid ingestion (e.g., from glucocorticoid medication).

ACTH-independent Cushing's syndrome constitutes the remaining subgroup of the disorder. It is well known to be caused by adrenocortical tumors or inherent endocrine tumor forming-diseases such as MEN1 and primary pigmented nodular adrenocortical disease. An activating mutation of *MC2R* (p.F278C) is associated with the disorder and adrenal hyperactivity. Other endocrine factors such as LH/hCG emerge to play a role in disease etiology.

PUBERTY

Puberty refers to the physical changes by which a child becomes an adult capable of reproduction. The maturation of the reproductive system involves almost the entire endocrine system, and occurs in a phasic manner. During fetal development, neuroendocrine cells appear in the rostral forebrain, whence they migrate to an area in the hypothalamus to become the gonadotropin-releasing hormone (GnRH) pulse generator. During infancy, gonadotropin secretion is inhibited centrally by a sensitive negative feedback control, which keeps the reproductive system quiescent.

The onset of puberty is contingent upon maturation of the central pulse generator and disinhibition and reactivation of the hypothalamic-pituitary-gonadal axis. Temporal and developmental changes in GnRH pulse frequency have differential effects on FSH and LH. Slow GnRH pulse frequencies preferentially stimulate FSH synthesis and release, whereas higher frequencies preferentially stimulate LH production. The pulsatile LH secretion promotes growth and maturation of the gonads (ovaries and testes), sex hormone production, and development of secondary sex characteristics. As puberty progresses, there is an increase in the amplitude and frequency of LH pulses, leading eventually to adult steroid production and gametogenesis. The sex hormones stimulate growth as well as function and transformation of the brain, bones, muscle, and skin in addition to the reproductive organs. Growth accelerates at the onset of puberty and stops at its completion.

Before puberty, the body differences between the sexes are mainly confined to the genitalia. During puberty, major differences of size, form, shape, composition, and function develop. The time of onset and subsequent course of hormonal and physical changes during puberty are highly variable and influenced by genetic makeup, nutrition, and general health. Breast budding in girls and testicular enlargement in boys are usually the first signs of puberty. Detailed studies of the progression of the physical changes with puberty (Tanner) demonstrate peak growth velocity in girls early in puberty before menarche, whereas in boys it occurs later (in the second half of puberty). The time interval between breast budding and menarche is on average 2½ years, and the average age at menarche is between 12½ and 13 years. During the first 2 years, menstrual cycles are often irregular and nonovulatory.

In addition to the striking changes brought about by activation of the pituitary-gonadal axis, secretion of androgens by the adrenal gland also increases. However, gonadal and adrenal androgen production (adrenarche) is not causally related temporally, and discordance in the two is seen in a number of situations. Adrenarche occurs in response to rising adrenal androgen secretion, usually precedes the pubertal rise in gonadotropins and sex steroids by about 2 years, and is associated with differentiation and growth of the adrenal zona reticularis. The major adrenal androgens are androstenedione and epiandrosterone. They are weak androgens, which in the female, directly or by peripheral conversion, account for about 50% of serum testosterone and are largely responsible for female sexual hair development. Unlike the major gonadal sex steroids, adrenal androgens do not play an important role in the adolescent growth spurt.

Any disturbance in the integrity of the brain, the hypothalamic-pituitary axis, or the peripheral endocrine system may interfere with normal puberty. Absence of one or more trophic or pituitary hormones interferes with hormonal effectiveness or end organ response, which may be on a genetic or acquired basis. Lack of signs of puberty in a female 13 years of age or a male 14 years of age becomes a concern and deserves evaluation.

Delayed Puberty

The most common cause of failure of pubertal development in the female is gonadal dysgenesis, a disorder of the X chromosome.

For boys, the most common cause of delay in the onset of puberty is constitutional. It is often diagnosed in retrospect if boys initiate puberty spontaneously before the age of 18.

Hypogonadotropic Hypogonadism

Isolated hypogonadotropic hypogonadism (IHH) describes the condition which is the consequence of defects in the pulsatile release of gonadotropins or the deficiency of gonadotropin action. A number of conditions are associated with the Mendelian forms, including developmental defects of the hypothalamus or abnormal pituitary gonadotropin secretion. Structural defects of LH and FSH affect the action of gonadotropins. Mutations in three genes (*KAL1*, *GNRHR*, and *FGF1*) account for most of the known cases of IHH.

X-linked Kallmann's Syndrome and KAL1 When associated with anosmia or hyposmia, IHH is referred to as Kallmann's syndrome. IHH is caused by GnRH deficiency due to failure of embryonic migration of GnRH-synthesizing neurons, while anosmia is related to hypoplasia or aplasia of the olfactory bulbs. The phenotype associated with Kallmann's syndrome mutations varies significantly, indicating the potential influence of modifying genes and other factors. There are two forms of Kallmann's syndrome: an X-linked form caused by mutations of *KAL1* and an autosomal-dominant form caused by mutations of *KAL2*.

The X-linked *KAL1* encodes an approximately 100 kD extracellular-matrix glycoprotein, anosmin-1, which shares homology with neural cell adhesion molecule. Anosmin-1 has been suggested to provide a scaffold to direct neuronal migration of both GnRH and olfactory neurons to their proper embryonic destination. Mutations in *KAL1* account for approximately 10–15% of anosmic male patients in sporadic cases but may comprise 30–60% of familial cases. *KAL1* mutations have not been described in normosmic male patients or in female patients. The majority of the mutations result in the formation of truncated proteins. Some mutations affect the formation of disulfide bond in the N-terminal cysteine-rich region (WAP domain), whereas other mutations affect the structure of the four fibronectin type III (FNIII) domains of anosmin-1. The WAP and the first FNIII domain are believed to be essential for the binding of anosmin-1 with heparin sulfate and its other ligands.

Autosomal Dominant Kallmann's Syndrome and KAL2/FGFR1 The autosomal dominant form of Kallmann's syndrome is caused by inactivating mutation of *KAL2/FGFR1* which encodes fibroblast growth factor receptor 1 (FGFR1). FGFR1 is a single spanning transmembrane receptor expressing in GnRH neurons. The protein has three immunoglobulin-like domains, a heparin-binding domain, and two tyrosine kinase domains.

Mutations in *KAL2/FGFR1* occur in approximately 7–10% of patients (male and female) with autosomal dominant Kallmann's syndrome. Although most *KAL2/FGFR1* mutations have been reported in Kallmann's patients, mutations of this gene in normosmic IHH patients have also been reported. Mutations of *KAL2/FGFR2* show reduced penetrance and variable expressivity.

Missense mutations tend to cluster in the first immunoglobulin domain of FGFR1, implicating the importance of this domain in FGFR1 function. Nonsense mutations cluster at the C-terminal tyrosine kinase domain and result in truncated receptors lacking the autophosphorylated tyrosine residues, potentially impeding the signaling activity of the receptor.

Normosmic IHH and Gonadotropin-Releasing Hormone Receptor (GnRHR) About 50% of familial cases of normosmic IHH are associated with loss-of-function

GnRHR mutations. *GnRHR* encodes a protein of 327 amino acids and is a member of the GPCR family. GnRHR is expressed on the cell surface of pituitary gonadotrophs. The ligand for GnRHR is gonadotropin-releasing hormone-1 (GnRH), a decapeptide which is derived from a 92-amino acid pre-pro-protein. It is released in a pulsatile manner in the pre-optic area of the hypothalamus and delivered to the anterior pituitary gland. There, it binds and activates GnRHR, resulting in the synthesis and release of gonadotropins (LH and FSH). No GnRH mutation causing IHH has been reported.

Inactivating mutations of *GnRHR* are transmitted in an autosomal recessive mode. Most patients are compound heterozygotes. The mutations either interfere with GnRH binding or affect signaling activity of the receptor. *GnRHR* mutations are present in 1–4.6% of all IHH patients, while it is 6–11% in autosomal recessive IHH families. In most cases the phenotype correlates with the functional alterations of the GnRHR *in vitro*. Patients with complete inactivating GnRHR variants on both alleles present with severe hypogonadism, while patients homozygous for a partially inactivating GnRHR variant present with partial hypogonadism.

More recently, another GPCR, GPR54, has been found to be associated with autosomal recessive IHH. Inactivating mutations of *GRP54* have been described in homozygous and compound heterozygous IHH patients. Impaired signaling capacity of the mutated receptor has been observed. GPR54 is the receptor for kisspeptins, which are potent stimulators of LH and FSH secretion. Kisspeptin-induced GPR54 signaling is also believed to be a major control point for GnRH release. No kisspeptin gene mutation has been described. Phenotypic expression of *GPR54* inactivating mutation does not differ from that of *GnRHR* mutation.

Isolated Gonadotropin Deficiency LH and FSH are the main regulators of gonadal steroid secretion, pubertal maturation, and fertility. A small number of hypogonadotropic hypogonadism patients have been found to carry loss-of-function mutations of the hormone-specific β subunit of LH and FSH. Both LH and FSH are heterodimers with an α subunit which is shared among the glycohormones. The β subunit of each hormone confers specificity. *LHβ* is located on chromosome 19q13.32, and *FSHβ* is on 11p13. Loss-of-function mutation of the β subunit renders the hormone inactive, giving rise to hypogonadotropic hypogonadism. No mutation of the common α subunit and CGβ subunit is known.

Inactivating mutations of FSHβ interfere with dimerization with the α subunit and render the hormone inactive. Females with homozygous inactivating *FSHβ* mutations have sexual infantilism and infertility because of a lack of follicular maturation, primary amenorrhea, low estrogen production, undetectable serum FSH, and increased LH. Male homozygotes are all normally masculinized, with normal to delayed puberty, but azoospermic.

Missense and splicing mutations of LHβ have been described. The missense mutations either interfere with the formation of active heterodimers or affect the interaction with LH receptor (LHR). The splicing mutation produces a truncated LHβ subunit and abrogates LH secretion. Male homozygotes are normally masculinized at birth but lack postnatal sexual differentiation and hypogonadal with delayed pubertal development, absence of mature Leydig cells, spermatogenic arrest, and absence of circulating LH and testosterone. The only woman with the homozygous splicing mutation has normal pubertal development, secondary amenorrhea, and infertility. All heterozygotes are fertile and have normal basal gonadotropin and sex-steroid levels.

Hypergonadotropic Hypogonadism

Hypergonadotropic hypogonadism and infertility/subfertility in both sexes are the result of gonadotropin resistance caused by inactivating mutations in receptors of the two gonadotropins, LH and FSH. Both LHR and FSHR are members of GPCR family. The extracellular domains (ECDs) of the receptors convey ligand specificity and are different among the receptors, while the TMD is highly homologous among species and among the glycohormone receptors.

Leydig Cell Hypoplasia (LCH)—Inactivating Mutations of LHR LHR is shared between LH and CG, thus the name LH/CG receptor. CG exerts its effect during early embryogenesis to induce Leydig cell maturation. LH also promotes steroidogenesis by Leydig cells, especially around the period of puberty. Inactivating mutations of LHR are recessive in nature. In males with homozygous or compound heterozygous inactivating mutations, a loss of LHR function causes resistance to LH stimulation, resulting in failure of testicular Leydig cell differentiation. This gives rise to Leydig Cell Hypoplasia (LCH), also called Leydig Cell Agenesis. A number of features distinguish LCH from other forms of male pseudohermaphroditism. LCH patients are genetic males with a 46 XY karyotype. The hormonal profile of them shows elevated serum LH level, normal to elevated FSH level, and low testosterone level, which is unresponsive to CG stimulation. Clinical presentation of LCH is variable, ranging from hypergonadotropic hypogonadism with microphallus and hypoplastic male external genitalia to a form of male pseudohermaphroditism with female external genitalia. In between are patients with variable degree of masculinization of the external genitalia. LCH patients show no development of either male or female secondary sexual characteristics at puberty. In females, LHR inactivation causes hypergonadotropic hypogonadism and primary amenorrhea with subnormal follicular development and ovulation, and infertility.

Currently, 23 inactivating mutations, distributed throughout LHR, have been identified in LCH patients, including single base substitutions (missense and nonsense mutations) and in-frame and out-of-frame loss-of-function insertion mutations. The majority of the inactivating mutations are found in homozygous patients. There are a number of LCH kindreds in which LHR mutations have not been identified, indicating that inactivating mutations are very heterogeneous or that LCH is caused by mutation of LHR as well as other gene(s).

Effect of the different mutations on LHR activity is variable. Depending on the location of the mutation in the receptor protein, it can cause diminished hormone binding, reduced surface expression, abnormal trafficking, or reduced coupling efficiency, all of which result in reduction or abolition of signal transduction triggered by hormone binding. Clinical presentation of LCH patients can be correlated with the amount of residual activity of the mutated receptor. Mutated LHRs of patients with the most severe phenotype, i.e., male pseudohermaphroditism, have zero or minimal signal transduction activity. On the other hand, patients with male hypogonadism have mutated LHRs with reduced, but not abolished, signal transduction.

Ovarian Dysgenesis—Inactivating Mutation of FSHR FSHR mediates the action of FSH. In females, FSH function is essential for ovarian follicular maturation. In males, FSH regulates Sertoli cell proliferation before and at puberty, and participates in the regulation of spermatogenesis. Loss-of-function mutations of FSHR are thus expected to be found in connection with hypergonadotropic hypogonadism associated with retarded follicular maturation and anovulatory infertility in women, and small testicles and impaired spermatogenesis in normally masculinized males.

FSHR is located on chromosome 2p, next to LHR. Inactivating mutations of FSHR are all missense mutations found in the ECD and transmembrane domain (TMD). The inheritance of these mutations follows an autosomal recessive mode. The most frequently detected mutation is p.A198V. Female patients homozygous with this mutation present with ovarian dysgenesis that includes hypergonadotropic hypogonadism, primary or early onset secondary amenorrhea, variable pubertal development, hypoplastic ovaries with impaired follicle growth, high gonadotropin and low estrogen levels. Compound heterozygous female patients with totally or partially inactivating FSHR mutations have less prominent phenotypes. The male phenotype of inactivating FSHR mutations is less assuring. The five male subjects with homozygous p.A198V mutation are normally masculinized, with moderately or slightly decreased testicular volume, normal plasma testosterone, normal to elevated LH but high FSH levels, and variable spermatogenic failure. In some cases fertility is maintained, suggesting that FSH action is not compulsory for spermatogenesis.

There is good correlation between the phenotype and the degree of receptor inactivation, as well as the site of mutation and its functional consequences. All mutations in the ECD cause a defect in ligand binding and targeting of FSHR to the cell membrane. In the TMD, the mutations have minimal effect on ligand binding but impair signal transduction to various extents.

Precocious Puberty

Pubertal development is considered precocious if it occurs before 7 or 8 years of age in girls or 9 years in boys. It is more common in girls and is mostly idiopathic. In contrast, precocious puberty in boys is due to some underlying pathology in half of the cases.

Precocious pubertal development that results from premature activation of the hypothalamic-pituitary-gonadal axis, and hence gonadotropin dependent, is termed central precocious puberty (CPP). Precocious sexual development that is not physiological, not gonadotropin dependent, but the result of abnormal sex hormone secretion, is termed sexual precocity or gonadotropin-independent precocious puberty (also called pseudo-puberty or PPP). CPP is not uncommon in otherwise normal healthy girls.

Clinically, the earliest sign of puberty in girls is breast enlargement (thelarche) in response to rising estrogen secretion, which usually precedes but may accompany or follow the growth of pubic hair. The earliest sign of puberty in boys is an increase in the size of testes and in testosterone production, followed by penile and pubic hair growth.

Gonadotropin-Dependent Precocious Puberty

A recent study of 156 children with idiopathic CPP indicates 27.5% of it to be familial. Segregation analysis reveals autosomal dominant transmission with incomplete sex-dependent penetrance. Studies of a girl with CPP show a heterozygous single base substitution leading to the replacement of Arg386 by Pro (c.G1157C) in the carboxy-terminal tail of GPR54. This mutation leads to prolonged activation of intracellular signaling pathways in response to kisspeptin. The kisspeptin-GPR54 signaling complex has been proposed as a gatekeeper of pubertal activation of GnRH neurons and the reproductive axis. Whether mutations affecting the kisspeptin-GPR54 duet are the genetic causes of gonadotropin-dependent precocious puberty need further study.

Gonadotropin-Independent Precocious Puberty

Familial Male-Limited Precocious Puberty (FMPP)—Activating Mutations of LHR Gain-of-function mutations, resulting in constitutive activation of LHR, cause luteinizing hormone releasing hormone (LHRH)-independent isosexual precocious puberty in boys. Constitutive activity of LHR leads to stimulation of testicular Leydig cells in the fetal and prepubertal period in the absence of the hormone, resulting in autonomous production of testosterone and pubertal development at a very young age. This autosomal dominant condition is termed familial male-limited precocious puberty (FMPP) or testotoxicosis. Signs of puberty usually appear by 2–3 years of age in boys with FMPP. These patients have pubertal to adult levels of testosterone, while the basal and LHRH-stimulated levels of gonadotropins are appropriate for age, i.e., prepubertal. There is also lack of a pubertal pattern of LH pulsatility.

Activating mutations of *LHR* have no apparent effect on female carriers.

All activating mutations of the *LHR* identified in FMPP patients are single base substitutions. Sixteen activating mutations in exon 11 of *LHR* have been identified among over 120 kindreds with FMPP. These mutations affect 13 amino acids. Half of the mutations are located in transmembrane helix VI (transmembrane VI), which represent 67.5% of all mutations identified in FMPP patients. The most frequently mutated amino acid is Asp578, with >55% of all activating mutations affecting this amino acid. The most common mutation is the c.T1733G transition, which results in the replacement of Asp578 by Gly. This mutation represents >51% of all mutations identified in FMPP patients so far. With the exception of the replacement of Ile542 by Leu in transmembrane V, all mutations result in the substitution of amino acids which are conserved among the glycohormone receptors.

LHR is prone to mutation. Among the kindreds of FMPP with confirmed molecular diagnoses, over 25% are caused by new mutations. Rare mutations occur more often in patients of non-Caucasian ethnic background. Detection of *LHR* mutations is unsuccessful in about 18% of patients diagnosed to have FMPP.

All FMPP mutations reside in the TMD. Agonist affinity of the mutated receptors is largely unchanged, while cell surface expression is either the same or reduced when compared to the wild-type receptor. All FMPP mutations have been shown to confer constitutive activity to the mutated LHR by *in vitro* expression studies. There is no consensus on the phenotype-genotype correlation in FMPP. However, mutations that give the highest basal level of cAMP in *in vitro* assays are associated with an earlier age of pubertal development.

Besides the germline mutations found in FMPP patients, a somatic activating mutation of *LHR* (p.D578H) has also been identified in tumor tissue of a number of patients with testicular neoplasia. Even though a couple of FMPP patients developed testicular neoplasia, the p.D578H mutation has never been found as a germline mutation in any FMPP patient. Its presence is confined to patients with testicular neoplasia.

Activating Mutation of FSHR *FSHR* is not prone to mutation. So far, there is only one gain-of-function *FSHR* mutation identified in a single case. The patient had been previously hypophysectomized but maintained normal spermatogenesis in spite of undetectable gonadotropins. He has a heterozygous activating mutation p.D578G in the third intracytoplasmic loop of FSHR. *In vitro* expression of the mutated receptor shows elevated basal activity of the receptor.

Ovarian Hyperstimulation Syndrome Ovarian hyperstimulation syndrome (OHSS) is an iatrogenic complication of ovulation-induction therapy. In its most severe form, this syndrome involves massive ovarian enlargement and the formation of multiple ovarian cysts and can be fatal. OHSS can arise spontaneously during pregnancy owing to a broadening of the specificity of

FSHR for hCG at high concentrations of the hormone. So far, 6 missense mutations of FSHR (p.S128Y, p.T449I, p.T449A, p.I545T, p.D567N, and p.D567G) have been identified in OHSS patients. The transmission of these mutations follows an autosomal dominant mode. These mutations relax the ligand specificity of the receptor in such a way that it also binds and becomes activated by hCG. Response of the receptor to TSH was also found for mutations that are located in the TMD. It is possible the promiscuous stimulation of follicles during the first trimester of pregnancy by hCG results in excessive follicular recruitment observed in this disorder.

Acknowledgment

This work was supported in part by the Intramural Research Program of the NIH, Eunice Kennedy Shriver National Institute of Child Health and Human Development.

KEY CONCEPTS

- The pituitary acts as a relay between the hypothalamus and target organs (the hypothalamus-pituitary-target organ axis) in the regulation of normal physiology by hormone secretion. The production of pituitary hormones, and their releasing factors from the hypothalamus, is subject to negative feedback regulation from the target organs. The action of pituitary hormones and their release are mediated through the ligand-receptor binding mechanism. Inactivating mutations of the genes encoding pituitary hormones or their receptors lead to pituitary hormone deficiency phenotypes. Activating mutations of hormone receptors, or the loss of negative feedback regulation on pituitary hormone secretion, can lead to hypersecretion of pituitary hormones and/or overstimulation of target organs.

- The major hormone-secreting cells in the pituitary are located at the anterior pituitary. Genes that are essential to the development of pituitary hormone-secreting cells have been identified, including *HESX1*, *LHX3/4*, *PROP1*, *POU1F1*, and *GLI2*. Loss-of-function mutations of these genes lead to abnormal pituitary development and are the major causes of congenital hypopituitarism.

- The majority of cases of congenital hypothyroidism is nongoitrous and is due to faulty embryogenesis (athyreosis, agenesis, or dysembryogenesis). A small number are goitrous, the result of enzyme defects in hormone biosynthesis (dyshormonogenesis). Etiologically, three types of congenital hypothyroidism can be distinguished: central congenital hypothyroidism including hypothalamic (tertiary) and pituitary (secondary), primary (impairment of the thyroid gland itself), and peripheral (resistance to thyroid hormone). The majority of cases of central congenital hypothyroidism are caused by *TSH* mutations. A subgroup of patients with familial congenital hypothyroidism is the result of loss-of-function mutations in *TSHR*. Constitutive activating

mutations of *TSHR* give rise to familial nonautoimmune hyperthyroidism. Disorders resulting in congenital primary hypothyroidism have been identified in all major steps in thyroid hormonogenesis.

- In humans, the regulation of the ionized fraction of serum calcium depends on the interaction of PTH and the active form of vitamin D-1,25-dihydroxycholecalciferol [1,25(OH)$_2$D]. Familial isolated hypoparathyroidism can be caused by mutations of *PTH* and *Glial cells missing* (*GCM*). On the other hand, primary hyperparathyroidism is genetically heterogeneous. Mutations of the Calcium-Sensing Receptor (*CaSR*) can result in loss-of-function or gain-of-function of the receptor. Disorders due to loss-of-function mutations of *CaSR* include familial hypocalciuric hypercalcemia and neonatal severe primary hyperparathyroidism, whereas disorders due to gain-of-function mutations of the CaSR include Autosomal Dominant Hypocalcemia (ADH) and Bartter syndrome type V.

- The adrenal cortex of adrenal glands produces mineralocorticoids and glucocorticoids for homeostasis, and adrenal androgens for fetal external genitalia development. The cause of congenital primary adrenal insufficiency can be attributed to defective development of the adrenal glands (Adrenal Hypoplasia Congenita; AHC) or inborn errors in adrenal steroid biosynthesis (Congenital Adrenal Hyperplasia; CAH). The cytomegalic form of AHC is an X-linked disorder caused by inactivating mutations of *NR0B1* (*DAX1*). CAH is caused most frequently by inactivating mutations of *CYP21A2* and *CYP11B1*, which encode enzymes that regulate the biosynthesis of mineralocorticoids and glucocorticoids. Excessive adrenal hormone production results in hypercortisolism, which is caused mostly by ACTH-secreting pituitary adenoma (Cushing's disease).

- Mutations in three genes (*KAL1*, *GNRHR*, and *FGF1*) account for most of the known cases of Isolated Hypogonadotropic Hypogonadism (IHH). A small number of hypogonadotropic hypogonadism patients have been found to carry loss-of-function mutations of the hormone-specific β subunit of LH and FSH. On the other hand, hypergonadotropic hypogonadism and infertility/subfertility in both sexes are the result of gonadotropin resistance caused by inactivating mutations in receptors of LH and FSH; inactivating mutations of the *LHR* cause Leydig Cell Hypoplasia (LCH); those of FSHR cause Ovarian Dysgenesis. Activating mutations of LHR are responsible for the development of familial male-limited precocious puberty (FMPP).

SUGGESTED READINGS

1. Bhagavath B, Layman LC. The genetics of hypogonadotropic hypogonadism. *Semin Reprod Med.* 2007;25:272–286.
2. Chan LF, Clark AJ, Metherell LA. Familial glucocorticoid deficiency: Advances in the molecular understanding of ACTH action. *Horm Res.* 2008;69:75–82.

3. Charmandari E, Kino T, Ichijo T, et al. Generalized glucocorticoid resistance: Clinical aspects, molecular mechanisms, and implications of a rare genetic disorder. *J Clin Endocrinol Metab.* 2008;93:1563–1572.

4. Cohen LE, Radovick S. Molecular basis of combined pituitary hormone deficiencies. *Endocr Rev.* 2002;23:431–442.

5. D'Souza-Li L. The calcium-sensing receptor and related diseases. *Arq Bras Endocrinol Metabol.* 2006;50:628–639.

6. Ferraz-de-Souza B, Achermann JC. Disorders of adrenal development. *Endocr Dev.* 2008;13:19–32.

7. Ferris RL, Simental AA, Jr. Molecular biology of primary hyperparathyroidism. *Otolaryngol Clin North Am.* 2004;37:819–831.

8. Huhtaniemi I, Alevizaki M. Gonadotrophin resistance. *Best Pract Res Clin Endocrinol Metab.* 2006;20:561–576.

9. Layman LC. Hypogonadotropic hypogonadism. *Endocrinol Metab Clin North Am.* 2007;36:283–296.

10. Mehta A, Dattani MT. Developmental disorders of the hypothalamus and pituitary gland associated with congenital hypopituitarism. *Best Pract Res Clin Endocrinol Metab.* 2008;22:191–206.

11. Miyai K. Congenital thyrotropin deficiency—from discovery to molecular biology, postgenome and preventive medicine. *Endocr J.* 2007;54:191–203.

12. Mullis PE. Genetics of growth hormone deficiency. *Endocrinol Metab Clin North Am.* 2007;36:17–36.

13. Nimkarn S, New MI. Prenatal diagnosis and treatment of congenital adrenal hyperplasia owing to 21-hydroxylase deficiency. *Nat Clin Pract Endocrinol Metab.* 2007;3:405–413.

14. Park SM, Chatterjee VK. Genetics of congenital hypothyroidism. *J Med Genet.* 2005;42:379–389.

15. Refetoff S. Resistance to thyroid hormone: One of several defects causing reduced sensitivity to thyroid hormone. *Nat Clin Pract Endocrinol Metab.* 2008;4:1.

16. Refetoff S, Dumitrescu AM. Syndromes of reduced sensitivity to thyroid hormone: Genetic defects in hormone receptors, cell transporters and deiodination. *Best Pract Res Clin Endocrinol Metab.* 2007;21:277–305.

17. Rivolta CM, Targovnik HM. Molecular advances in thyroglobulin disorders. *Clin Chim Acta.* 2006;374:8–24.

18. Thakker RV. Genetics of endocrine and metabolic disorders: Parathyroid. *Rev Endocr Metab Disord.* 2004;5:37–51.

19. Toogood AA, Stewart PM. Hypopituitarism: Clinical features, diagnosis, and management. *Endocrinol Metab Clin North Am.* 2008;37:235–261.

20. Walenkamp MJ, Wit JM. Genetic disorders in the growth hormone—insulin-like growth factor-I axis. *Horm Res.* 2006;66:221–230.

23

Molecular Basis of Gynecologic Diseases

Samuel C. Mok . Kwong-kwok Wong . Karen Lu .
Karl Munger . Zoltan Nagymanyoki

INTRODUCTION

Gynecologic diseases in general are diseases involving the female reproductive tract. These diseases include benign and malignant tumors, pregnancy-related diseases, infection, and endocrine diseases. Among them, malignant tumor is the most common cause of death. In recent years, the etiology of some of these diseases has been revealed. For example, human papillomavirus (HPV) infection has been shown to be one of the major etiological factors associated with cervical cancer. Inactivation of tumor suppressor gene *BRCA1* has been implicated in hereditary ovarian cancer. In spite of these findings, the molecular bases of most of the diseases remain largely unknown. In this chapter, we will focus on discussing benign and malignant tumors of female reproductive organs as well as pregnancy-related diseases, which have relatively well-understood molecular bases.

BENIGN AND MALIGNANT TUMORS OF THE FEMALE REPRODUCTIVE TRACT

Cervix

Infections of the genital mucosa with human papillomaviruses represent the most common virus-associated sexually transmitted disease, and at age 50 approximately 80% of all females will have acquired a genital HPV infection sometime during their life. At present, approximately 630 million individuals worldwide have a genital HPV infection, with an incidence of approximately 30 million new infections per year. Currently in the United States there are in excess of 20 million people with genital HPV infections, with an estimated annual incidence of 6.2 million new infections. Genital HPV infections are particularly prevalent in sexually active younger individuals. Most of these infections

are transient and may not cause any overt clinical disease or symptoms. Nonetheless, the total annual cost of clinical care for genital HPV infections exceeds $3 billion in the United States alone.

HPVs Associated with Cervical Lesions and Cancer

Human papillomaviruses are members of the Papillomaviridae family. They have a tropism for squamous epithelial cells and cause the formation of generally benign hyperplastic lesions that are commonly referred to as papillomas or warts. Papillomaviruses contain closed circular double-stranded DNA genomes of approximately 8,000 base pairs that are packaged into ~55 nm non-enveloped icosahedral particles. Their genomes consist of three regions; the early (E) region encompasses up to 8 open reading frames (ORFs) designated E followed by a numeral, with the lowest number designating the longest ORF, and the late (L) region encodes the major and minor capsid proteins, L1 and L2, respectively. Only one of the two DNA strands is actively transcribed, and early and late ORFs are encoded on the same DNA strand. A third region, referred to as the long control region (LCR), the upstream regulatory region (URR), or the noncoding region (NCR), does not have significant coding capacity and contains various regulatory DNA sequences that control viral genome replication and transcription (Figure 23.1A).

In excess of 100 HPV types have been described. HPVs are classified as genotypes based on their nucleotide sequences. A new HPV type is defined when the entire genome has been cloned and the sequence of the L1 open reading frame (ORF), the most conserved ORF among papillomaviruses, is less than 90% identical to a known HPV type. HPVs with higher sequence identity are referred to as subtypes (90% to 98% identity) or variants (>98% identity).

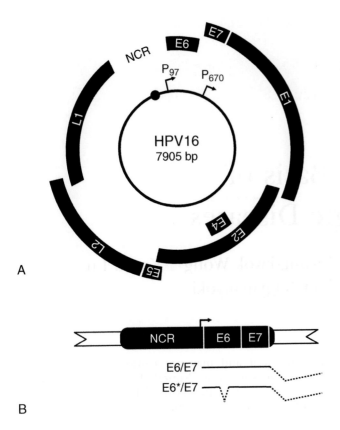

Figure 23.1   The HPV genome. (A) Schematic representation of the HPV16 genome. The double-stranded circular DNA genome is represented by the central circle. Early (E) and late (L) genes are all transcribed from one of the two DNA strands in each of the three possible reading frames. The noncoding region (NCR) does not have extensive coding potential but contains the viral origin of replication (designated by the black circle) as well as the major early promoter, P_{97}, designated by an arrow. The differentiation-specific late promoter, P_{670}, is contained within the E7 ORF. See text for details. **(B)** Structure of the minimal HPV16 genome fragment that is consistently retained after integration into a host chromosome. The major E6/E7 transcripts are shown underneath. See text for details.

Approximately 30 HPV types infect mucosal epithelia, and these viruses are further classified as low risk or high risk depending on the propensity for malignant progression of the lesions that they cause. Low-risk HPVs, such as HPV6 and HPV11, cause genital warts, whereas high-risk HPVs, such as HPV16 and HPV18, cause intraepithelial neoplasia that can progress to frank carcinoma. Harald zur Hausen's group discovered the association of HPVs with anogenital tract lesions and isolated HPVs from genital warts. Using these sequences as hybridization probes under low stringency conditions, they succeeded in detecting HPV sequences in cervical carcinomas. The most abundant high-risk HPVs are HPV16 and HPV18, which are detected in approximately 50% and 20% of all cervical carcinomas, respectively. HPV18 appears to be frequently associated with adenocarcinomas, whereas HPV16 is mostly detected in squamous cell carcinomas. The following sections are focused on a review of mucosal high-risk HPVs and their contributions to cervical lesions and cancers.

HPV Infection and Life Cycle

The HPV life cycle is closely linked to the differentiation program of the infected squamous epithelial host cell. HPVs infect basal cells, a single layer of actively cycling cells in the squamous epithelium. Basal cells are not readily accessible for viral infection, as they are protected by several layers of differentiated cells that have withdrawn for the cell division cycle. These cell layers are essential for the mechanical stability of the skin and shield the proliferating basal cells from environmental genotoxic insults. HPVs can gain access to basal cells through microabrasions caused by mechanical trauma. Basal-like cells at the cervical squamocolumnar transformation zone in the cervix, however, are particularly accessible and vulnerable to HPV infection. It has been postulated that within the cervical transformation zone, reserve cells, which can give rise to squamous or columnar epithelia, may be physiologically relevant targets for HPV infection. The mechanisms of viral entry remain relatively poorly understood but are thought to involve initial binding to heparin sulfate on the cell surface followed by receptor binding and viral uptake, although there is controversy regarding the identity of the virus receptor.

Following infection, HPV genomes are maintained at a low copy number in the nuclei of infected cells and can persist in basal epithelial cells for decades. The productive phase of the viral life cycle, which includes HPV genome amplification, production of capsid proteins, and packaging of newly synthesized genomes, however, occurs exclusively in the terminally differentiated layers of the infected tissue. HPVs are nonlytic, and infectious viral particles are sloughed off with the terminally differentiated, denucleated scales where they remain infectious over extended periods of time.

HPVs encode two proteins, E1 and E2, which directly contribute to viral genome replication. The E1 origin-binding protein is the only virally encoded enzyme and has intrinsic ATPase and helicase activities. E1 forms a complex with the E2 protein, the major HPV-encoded transcriptional regulatory protein. E2 binds with high affinity to specific DNA sequences $ACCN_6GGT$ in the viral regulatory region, whereas E1 binds to the AT-rich replication origin sequences with relatively low affinity. The origin sequence is often flanked by E2 binding sites resulting in high-affinity binding of the E1/E2 complex to the origin of replication.

With the exception of the E1/E2 origin-binding complex, HPVs do not encode enzymes that are necessary for viral genome replication and co-opt the host DNA synthesis machinery. Since high-copy-number HPV genome replication and viral progeny synthesis are confined to terminally differentiated cells, which are growth arrested and thus intrinsically incompetent for DNA replication, a major challenge for the viral life cycle is to maintain and/or re-establish a replication-competent milieu in these cells.

The HPV E7 protein contributes to induction and/or maintenance of S-phase competence in differentiating keratinocytes through several mechanisms. Perhaps most importantly, HPV E7 proteins bind to the retinoblastoma tumor suppressor protein pRB and the related p107 and p130 pocket proteins. These proteins have been implicated in regulating G1/S phase transcription through members of the E2F family of transcription factors. The G1 specific pRB/E2F complex is a transcriptional repressor that inhibits S-phase entry. In normal cells, pRB is phosphorylated by cyclin/cdk complexes in late G1, the pRB/E2F complex dissociates, and DNA-bound E2Fs act as transcriptional activators. The pRB/E2F complex re-forms when pRB is dephosphorylated at the end of mitosis. This regulatory loop is subverted by HPV E7 proteins, which can associate with pRB and abrogate the inhibitory activity of pRB/E2F complexes. E7 proteins encoded by low-risk HPV associate with pRB with lower affinity than high-risk HPV E7 proteins. Additionally, high-risk HPV E7 proteins induce proteasome-mediated degradation of pRB. Moreover, E7 proteins abrogate the action of cyclin-dependent kinase inhibitors (CKIs) p21^{CIP1} and p27^{KIP1}, which regulate cell cycle withdrawal during epithelial cell differentiation, thereby uncoupling epithelial cell differentiation and cell cycle withdrawal. This leads to the formation of hyperplastic lesions, warts, and is necessary for production of progeny virus.

Detection of HPV-Associated Lesions

Papanicolaou tests (also known as Pap tests) are named after their inventor, Georgios Papanicolaou, and serve to detect HPV-associated lesions in the cervix. Upon implementation, this relatively inexpensive test dramatically reduced the incidence and mortality rates of cervical cancer. In the United States the current recommendation is for women to have a Pap test performed at least once every 3 years, and in 2003 approximately 65.6 million Pap tests were performed. The current test involves collecting exfoliated epithelial cells from the outer opening of the cervix and either directly smearing the cells on a slide, or immediately preserving and storing the cells in fixative liquid medium followed by automated processing into a monolayer. In either format, cells are stained and examined for cytological abnormalities. Liquid-based monolayer cytology appears to have a reduced rate of false-positivity presumably due to standardized specimen preparation and immediate fixation of the sampled cells. While nuclear features are currently used for diagnoses, attempts are under way to identify new biomarkers for high-risk HPV-associated cervical lesions. The most promising biomarker is p16^{INK4A}, an inhibitor of cdk4/cdk6 cyclin D complexes, although the molecular basis of p16^{INK4A} overexpression in cervical cancer is currently unknown.

The *American Cancer Society* (ACS) and the *American College of Obstetricians and Gynecologists* (ACOG) also recommend that women over the age of 30 be tested for the presence of HPV DNA. The only currently FDA-approved HPV testing method is based on nucleic acid hybridization and can distinguish between absence and presence of the most frequent low-risk or high-risk HPV types but does not allow identification of individual HPV types. A number of different PCR-based HPV typing methods have been developed for research purposes, but these are currently not FDA approved. While some studies have suggested that HPV testing may be more effective in identifying cervical lesions than Pap smears, this issue clearly requires additional study.

Diagnosis and Treatment

Genital warts are a frequent manifestation of low-risk mucosal HPV infections and are diagnosed based on appearance. Such lesions have a very low propensity for malignant progression, and they often regress spontaneously. However, patients generally insist on their removal. No HPV-specific antivirals currently exist for such applications. Standard therapeutic modalities include surgical excision, laser therapy, cryotherapy, topical administration of various caustic chemicals, or immunomodulating agents. In rare cases, low-risk HPV infections of the genital tract can also cause serious disease. One example is the giant condyloma of Buschke-Lowenstein that is caused by infection with low-risk HPVs. In such patients, the immune system is unable to control and/or clear the infection. Although these are slow-growing lesions, they are highly destructive to adjacent normal tissue and eventually can form local and distant metastases.

Cytological abnormalities detected by Pap tests are classified according to the Bethesda system as atypical squamous cells (ASC) or squamous intraepithelial lesions (SIL). ASC are further classified as Atypical Squamous Cells of Undetermined Significance (ASCUS) or Atypical Squamous Cells, cannot exclude High-grade squamous intraepithelial lesions (ASC-H), whereas SILs are designated low-grade (LSIL) or high-grade (HSIL). LSILs are followed up by additional Pap tests, whereas HSILs require analysis by colposcopy. The procedure involves application of an acetic acid-based solution to the cervix whereupon lesions appear as white masses upon evaluation with a colposcope. When lesions are detected, a biopsy is performed and the tissue is examined histologically. Lesions are classified as cervical intraepithelial neoplasias (CIN), carcinoma in situ, or invasive cervical carcinoma. Treatments for CIN include cryotherapy, laser ablation, or loop electrosurgical excision, whereas carcinomas are treated by surgery and/or chemotherapy.

Prevention and Vaccines

Condoms reduce, but do not negate, the risk of infections with HPVs. In addition, preclinical studies in a mouse model suggest that the polysaccharide carrageen greatly inhibits HPV transmission, whereas the spermicidal compound nonoxynol-9 appears to increase HPV transmission, but these studies await confirmation by clinical studies in humans.

The first generation prophylactic HPV vaccines consist of recombinant HPV L1 proteins that self-assemble into virus-like particles (VLPs). Gardasil® was developed by Merck and has been FDA approved for use in girls and young women of 9 to 26 years of age. It is a quadrivalent formulation that contains VLPs of the most prevalent low-risk (HPV6 and HPV11) and high-risk HPVs (HPV16 and HPV18). It is administered as three doses over the course of 6 months and promises to be highly efficacious in providing type-specific protection from new infections with these HPV types. As such, this vaccine has the potential to reduce the burden of cervical carcinoma and genital warts by up to 70%. Since these prophylactic vaccines lead to the development of humoral immune responses, they are not predicted to affect potential HPV infections at the time of vaccination. Given that cervical cancer generally develops decades after the initial infection, it has been estimated that incidence and mortality rates of cervical cancer will not decrease for 25 to 40 years. Moreover, it is not clear whether other nonvaccine high-risk HPV types will become more prevalent as HPV16 and HPV18 are removed from the biological pool. Hence, recommendations for cervical cytology screening (Pap smears) remain unchanged for vaccinated individuals. The minor capsid protein L2 contains linear, cross-neutralizing epitopes that may afford more general protection from HPV infections and appears to be an excellent candidate for the development of second-generation vaccines.

Contributions of HPV Oncoproteins to Induction of Genomic Instability

Cervical cancers generally develop years or decades after the initial infection, and these tumors have suffered a multitude of genomic aberrations. The acquisition of some of these genomic alterations appears to define certain stages of disease progression. The action of the high-risk HPV E6 and E7 oncoproteins on telomerase activity and the *p53* and pRB tumor suppressors are sufficient to lead to extended uncontrolled proliferation and cellular immortalization, but acquisition of additional host genome mutations is necessary for malignant progression (Figure 23.2). A defining biological activity of high-risk HPV E6/E7 proteins is their ability to subvert genomic integrity. Hence, high-risk HPV E6/E7 oncoproteins not only contribute to initiation but also play a key role in malignant progression.

There are two principal mechanisms that lead to genomic instability. Subversion of cell cycle checkpoint mechanisms and DNA repair pathways can lead to perpetuation of mutations induced by environmental triggers such as UV irradiation or exposure to chemical compounds that cause DNA damage. Alternatively, genomic instability can be triggered by active mechanisms that cause genomic destabilization, which have been collectively referred to as a mutator phenotype. Expression of HPV E6/E7 oncoproteins causes genomic instability by both of these mechanisms.

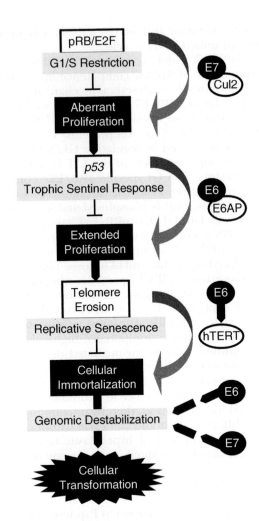

Figure 23.2 **Schematic depiction of some of the major biochemical and biological activities of high-risk HPV oncogenes and how they may cooperate in the development of cervical disease and cancer.**

HPV16 E7 has activities of a mitotic mutator and its expression in primary human epithelial cells causes several types of mitotic abnormalities. These include induction of supernumerary centrosomes, lagging chromosomes, and anaphase bridges. Of these, induction of supernumerary centrosomes has been studied in greatest detail. Centrosome-associated multipolar mitoses are a histopathological hallmark of high-risk HPV-associated cervical lesions. HPV16 E7 induces centrosome abnormalities through multiple, cooperating pathways. HPV16 E7 expression causes aberrant activation of cdk2 through several mechanisms, including E2F-mediated transcriptional activation of expression of the cdk2 catalytic subunits cyclins E and A as a consequence of pRB/p107/p130 degradation and inactivation of the cdk2 inhibitors $p21^{CIP1}$ and $p27^{KIP1}$. Induction of supernumerary centrosomes by HPV16 E7 is strictly dependent on cdk2 activity. In addition, HPV16 E7 associates with the centrosomal regulatory protein gamma-tubulin and inhibits its loading on centrosomes. As a consequence, the process of centrosome duplication is uncoupled from the cell division cycle, resulting in the synthesis of multiple daughter centrioles from a single maternal centriole template. Whereas expression of HPV16 in primary cells effectively induces centrosome overduplication, E6 co-expression is necessary for induction of multipolar mitoses.

HPV16 E7 expression also causes a higher incidence of DNA double strand breaks, which can lead to breakage fusion bridge cycles and chromosomal translocations. Specific recurrent chromosomal translocations are well-documented drivers of hematological malignancies and, more recently, similar translocations have also been documented to contribute to the genesis of human solid tumors. Chromosomal translocations are regularly detected in cervical cancer specimens, and it will be interesting to determine whether some of these directly contribute to cervical carcinogenesis. The mechanistic basis of the formation of mitoses with lagging chromosomal material still awaits full investigation.

Fanconi anemia (FA) patients frequently develop squamous cell carcinomas, and it has been reported that oral cancers arising in FA patients are more frequently HPV-positive than in the general population. The FA pathway, which is normally activated by DNA cross-linking and stalled replication forks, is triggered in response to HPV16 E7 expression, and HPV16 E7 induces DNA double strand breaks more efficiently in FA patient-derived cell lines. Hence, the increased incidence of HPV-associated carcinomas in FA patients may be related to more potent induction of genomic instability by the HPV E7 protein.

The ability of the HPV16 E6 protein to contribute to genomic destabilization is to a large part based on its ability to inactivate the *p53* tumor suppressor. As a consequence, the DNA damage-induced G1/S checkpoint is nonfunctional, and HPV16 E6-expressing cells also exhibit mitotic checkpoint defects. HPV16 E6-mediated *p53* degradation also causes subversion of a postmitotic checkpoint that is specifically triggered in cells when they re-enter a G1-like state after a failed mitosis. Such cells have a tetraploid rather than the normal diploid set of chromosomes and contain two centrosomes rather than one. Cells with *p53* defects will disregard this checkpoint, re-enter S-phase, and may eventually undergo tetrapolar mitosis, which can lead to generation of aneuploid progeny. Consistent with this model, HPV16 E6-expressing cells show marked nuclear abnormalities.

In addition to passive mechanisms of cell cycle checkpoint subversion, HPV16 E6 may also contribute to genomic destabilization through active mutator mechanisms. HPV16 E6 has been shown to associate with single-strand DNA break repair protein XRCC1 and induce degradation of O6-methylguanine-DNA methyltransferase, which is also involved in single-strand DNA break repair. It has also been reported that HPV16 E6 expression decreases the fidelity of DNA end joining. Moreover, while HPV16 E6 expression does not induce centrosome overduplication, it greatly increases the incidence of multipolar mitoses in cells that contain supernumerary centrosomes due to HPV16 E7 expression.

Uterine Corpus

The uterine corpus represents the second common site for malignancy of the female reproductive tract. These neoplasms can be divided into epithelial, mesenchymal tumors, and trophoblastic tumors.

Epithelial Tumors

Uterine cancer is the most common gynecologic cancer in the United States. In 2008, the American Cancer Society estimates that there will be 40,100 new cases of uterine cancer, as compared to 21,650 new cases of ovarian cancer and 11,070 new cases of cervical cancer. Most uterine cancers are adenocarcinomas and develop from the endometrium, the inner lining of the uterus. They are therefore referred to as endometrial cancers. Risk factors for the development of endometrial cancer include obesity, unopposed estrogen use, polycystic ovary syndrome, insulin resistance and diabetes, and estrogen-secreting ovarian tumors. Given the spectrum of risk factors, the development of endometrial cancer has been strongly associated with an excess of systemic estrogen and a lack of progesterone. This hyperestrogenic state is presumed to result in endometrial hyperproliferation and endometrial cancer. Since the early 1990s, there has been increased research into understanding the molecular pathogenesis of endometrial cancer.

Endometrial cancer can be broadly divided into Type I and Type II categories, based on risk factors, natural history, and molecular features. Women with Type I endometrial cancers have the classic risk factors as stated previously, have tumors with low-grade endometrioid histology, frequently have concurrent complex atypical hyperplasia, and typically present with early-stage disease. Molecular features important in Type I endometrial cancers include presence of ER and PR receptors, microsatellite instability (MSI), *PTEN* mutations, *KRAS* mutations, and *Beta catenin* mutations. Women with Type II endometrial cancers are typically older, nonobese, have tumors with serous and clear cell histologies as well as high-grade endometrioid histology, and may present with more advanced stage disease. In general, Type II tumors have a poorer clinical prognosis, in part due to their propensity to metastasize even with minimal myometrial invasion. The molecular alterations commonly seen in Type II cancers include *p53* mutations and chromosomal aneuploidy. While not all tumors fall neatly into these categories, the classification is helpful to broadly define endometrial cancers. A number of investigators are working toward developing more specific and rational targeted therapies. This section will focus on defining the molecular changes that have been described for endometrial hyperplasia, as well as for Type I and Type II endometrial cancers, with an emphasis on therapeutic relevance.

Endometrial Hyperplasia Endometrial hyperplasia is a proliferation of the endometrial glands without evidence of frank invasion. According to the *World Health*

Organization (WHO), there are four categories of endometrial hyperplasia: (i) simple hyperplasia, (ii) simple hyperplasia with atypia, (iii) complex hyperplasia, and (iv) complex hyperplasia with atypia. Complex hyperplasia with atypia, or complex atypical hyperplasia (CAH), is considered a precursor lesion to endometrial cancer. Simple hyperplasia is characterized by proliferating glands of irregular size, separated by abundant stroma. Cytologically, there is no atypia. Simple hyperplasia with atypia is characterized by mild architectural changes in the proliferating glands and cytologic atypia, but is rarely found in clinical practice. Complex hyperplasia is characterized by more densely crowded and irregular glands, although intervening stroma is present. Cytologically, there is no atypia. CAH is characterized by densely crowded and irregular glands with intervening stroma, as well as cytologic atypia based on enlarged nuclei, irregular nuclear membranes, and loss of polarity of the cells. From a clinical standpoint, only CAH is associated with a substantial risk of developing endometrial cancer. One study estimated that the risk of progression to endometrial cancer is 1% for simple hyperplasia, 3% for complex hyperplasia, and 28.6% for CAH. In another study, complex hyperplasia without atypia had a 2% risk of progression, and CAH had a 51.8% risk of progression. Furthermore, a nested case-control study of 138 patients with endometrial hyperplasia found that simple hyperplasia and complex hyperplasia without atypia had minimal (relative risk of 2 and 2.8, respectively) risk of progression to cancer. However, CAH had a substantial (relative risk of 14) risk of progression. Recently a *Gynecologic Oncology Group* study (GOG 167) demonstrated that in 289 cases of community-diagnosed CAH, 123 (43%) cases had a concurrent Grade I endometrial cancer. Of the 123 cases of endometrial cancer, 38 (31%) had some degree of invasion into the myometrium. In addition, there was substantial discrepancy between the community-diagnosed CAH and subsequent pathologic review by gynecologic pathologists, with both under- and overdiagnosis. Finally, even when a panel of expert gynecologic pathologists reviewed cases of CAH, there was only modest reproducibility with a kappa value of 0.28. The current clinical management of CAH is simple hysterectomy and bilateral salpingo-oophorectomy. In young women who desire future fertility, progestins (including megestrol acetate) have been used to successfully reverse CAH. An initial dilatation and curettage with careful pathologic evaluation is necessary prior to initiating treatment. In addition, an MRI to rule out an invasive process can also be helpful. Close surveillance with every-3-month endometrial sampling is also recommended.

Risk factors for CAH and Type I endometrial cancers are the same, including obesity, diabetes and insulin resistance, unopposed estrogen use, and polycystic ovarian syndrome. Certain, but not all, molecular features are shared. *PTEN* mutations, *KRAS* mutations, and MSI have all been identified in complex endometrial hyperplasias and likely represent early molecular alterations in the pathogenesis of endometrioid endometrial cancer. A novel estrogen-regulated gene, *EIG121*, shows similar increase in expression in CAH and grade I endometrial cancers. *IGFI-R* is also increased and activated in both CAH and grade I endometrial cancers.

Endometrioid Endometrial Cancer Endometrioid endometrial cancer accounts for approximately 70–80% of newly diagnosed cases of endometrial cancer, and these cancers are considered Type I. Risk factors for the development of endometrioid endometrial cancer are associated in general with an excess of estrogen and a lack of progesterone. The most common genetic changes in endometrioid endometrial cancers include mutations in *PTEN*, MSI, mutations in *K-RAS*, and *beta-catenin*. Somatic mutations in the *PTEN* tumor suppressor gene are the most common genetic defect in endometrioid endometrial cancers and have been reported to occur in approximately 40–50% of cancer cases. *PTEN* is a tumor suppressor gene located on chromosome 10. It encodes a lipid phosphatase that acts to negatively regulate AKT. The importance of *PTEN* inactivation in endometrial carcinogenesis has been also demonstrated in a *PTEN* heterozygous mouse model. In this model, 100% of the heterozygote mice will develop endometrial hyperplasia by 26 weeks of age, and approximately 20% will develop endometrial carcinoma. In a study using this mouse model, phosphorylation of AKT followed by activation of ERα was demonstrated in the mouse endometrium. Reduction of endometrial ERα levels and activity reduced the development of endometrial hyperplasia and cancer. In humans, germline *PTEN* mutations are the underlying genetic defect in individuals with Cowden's syndrome, which is a hereditary syndrome characterized by skin and gastrointestinal hamartomas and increased risk of breast and thyroid cancers. Women with Cowden's syndrome also are at increased risk for endometrial cancer.

In addition to *PTEN* mutations, mutation and inactivation of other components of the PIK3CA/AKT/mTOR pathway have been described. In one study, mutations in the oncogene phosphatidylinositol-3-kinase (*PIK3CA*) were found in 39% of endometrial carcinomas, but only 7% of CAH cases. Another investigation reported a 36% rate of mutations in *PIK3CA* and found that there was a high frequency of tumors with both *PIK3CA* and *PTEN* mutations. Our group described loss of *TSC2* and *LKB1* expression in 13% and 21% of endometrial cancers, with the subsequent activation of mTOR. A heterozygous *LKB1* mouse model has been described recently that develops highly invasive endometrial adenocarcinomas. In humans, germline *LKB1* mutations are responsible for Peutz-Jeghers syndrome. It has not previously been reported that women with Peutz-Jeghers syndrome are at increased risk for endometrial cancer. Clearly, somatic abnormalities in the components of the PTEN/AKT/TSC2/MTOR pathway have been identified in a substantial number of endometrial cancers. Currently, clinical trials are either under way or recently completed examining mTOR inhibitors, including CCI-779 and RAD-001. Further investigation of whether specific alterations in the pathway correlates to response to therapy will be necessary.

Approximately 30% of sporadic endometrioid endometrial cancers demonstrate MSI. MSI identifies tumors that are prone to DNA replication repair errors. Microsatellites are well-defined short segments of repetitive DNA (example: CACACA) scattered throughout the genome. Tumors that demonstrate gain or loss of these repeat elements at specific microsatellite loci, when compared to normal tissue, are considered to have MSI. MSI occurs in approximately 30–40% of all endometrioid endometrial cancers, but rarely in Type II endometrial cancers. In addition, MSI also occurs in approximately 20% of all colon cancers.

The mechanism of MSI is due to either a somatic hypermethylation or silencing of the *MLH1* promoter or an inherited defect in one of the mismatch repair genes (*MSH1, MSH2, MSH6, PMS2*). An inherited defect in one of the mismatch repair genes is the cause of Lynch syndrome, a hereditary cancer predisposition syndrome. Individuals with Lynch syndrome are at increased risk for colon and endometrial cancer, as well as cancers of the stomach, ovary, small bowel, and ureters. Women with Lynch syndrome have a 40–60% lifetime risk of endometrial cancer, which equals or exceeds their risk of colon cancer. While the risk of endometrial cancer is high in these women, overall Lynch syndrome accounts only for approximately 2–3% of all endometrial cancers. Endometrial cancers that develop in women with Lynch syndrome almost uniformly demonstrate MSI.

Somatic hypermethylation of *MLH1* and MSI occurs in approximately 30% of all endometrioid endometrial cancers and is an early event in the pathogenesis. MSI has been identified in CAH lesions. It is presumed that MSI specifically targets tumor suppressor genes, resulting in the development of cancer. Reported target genes include *FAS, BAX, IGF*, the insulin-like growth factor 2 receptor (*IGFIIR*), and transforming growth factor β receptor type 2 (*TGFβRII*). Studies focusing on the clinical significance of MSI have been mixed. While some investigators have found an association of MSI with a more aggressive clinical course, a more recent study examining only endometrioid endometrial cancers found no difference in clinical outcome between tumors demonstrating MSI and those without.

The microsatellite instability assay, as well as immunohistochemistry (IHC) for MSH1, MSH2, MSH6, PMS2, can be very useful as a screening tool to identify women with endometrial cancer as having Lynch syndrome. While collecting and interpreting a family history is helpful, these molecular tools can be useful in targeting certain populations that may be at higher risk for Lynch syndrome, such as women under the age of 50.

KRAS mutations have been identified in 20–30% of endometrioid endometrial cancers. There is a higher frequency of *KRAS* mutations in cancers that demonstrate MSI. Mutations in *beta-catenin* have been seen in approximately 20–30% of endometrioid endometrial cancers. These mutations occur in exon 3 and result in stabilization of the beta catenin protein as well as nuclear accumulation. Nuclear beta-catenin plays an important role in transcriptional activation. One study demonstrated that *PTEN*, MSI, and *KRAS* mutations frequently co-exist. However, beta catenin alterations usually are not seen in these other abnormalities. *Beta catenin* mutations have been identified in CAH, suggesting that this is an early step in the pathogenesis of endometrial cancer.

Type II Endometrial Cancers Type II endometrial cancers have a more aggressive clinical course and include poorly differentiated endometrioid tumors as well as papillary serous and clear cell endometrial cancers. Patients with Stage I papillary serous endometrial cancer have a 5-year overall survival of 74%, significantly lower than the 90% 5-year overall survival for women with endometrioid endometrial cancer. The average age of diagnosis of patients with papillary serous endometrial cancers is 68 years, and risk factors typically associated with Type I endometrial cancers are not present. In addition, the genetic alterations seen in Type I endometrial cancers, as described previously, are not frequently found in papillary serous endometrial cancers. Microarray studies examining Type 1 versus Type 2 cancers have shown a distinct set of genes that are up- and downregulated in Type 1 versus Type 2 cancers. *TP53* gene mutations are common in uterine papillary serous carcinomas. Her-2/neu overexpression by immunohistochemistry (IHC) was observed in 18% of uterine papillary serous carcinomas. Of note, fewer tumors that demonstrated Her-2/neu overexpression by IHC showed *Her-2/neu* gene amplification. Several studies have reported that Her-2/neu overexpression is associated with a poor overall survival for UPSC. Interestingly, one study reported that overexpression of Her-2/neu in uterine papillary carcinoma occurs more frequently in African American women.

Mesenchymal Tumors Endometrial mesenchymal tumors are derived from the mesenchyme of the corpus composed of cells resembling those of proliferative phase endometrial stroma. Among them, uterine fibroids are the most common mesenchymal tumors of the female reproductive tract, which represent benign, smooth muscle tumors of the uterus. Recent studies have shown that the lifetime risk of fibroids in a woman over the age of 45 years is >60%, with incidence higher in African Americans than in Caucasians. The course of fibroids remains largely unknown and the molecular basis of this disease remains to be determined. Recent molecular and cytogenetic studies have revealed genetic heterogeneity in various histological types of uterine fibroids. Chromosome 7q22 deletions are common in uterine leiomyoma, the most common type of uterine fibroids. Several candidate genes, including *ORC5L* and *LHFPL3*, have been identified, but their roles in the pathogenesis of the disease have not been elucidated. Loss of a portion of chromosome 1 is common in the cellular form of uterine leiomyomata.

Endometrial stromal sarcoma is the most common malignant mesenchymal tumor of the uterus. They often arise from endometriosis. Genetic studies suggested that abnormalities of chromosomes 1, 7, and 11 may play a role in tumor initiation or progression in uterine sarcomas. Fusion of two zinc fingers (*JAZF1* and *JJAZ1*) by translocation t(7;17) has also been described.

Ovary and Fallopian Tube

Multiple benign and malignant diseases have been identified in the ovary and the fallopian tube. The most common ones are polycystic ovary syndrome (PCOS), and benign and malignant tumors of the ovary. While molecular studies in all these diseases have revealed changes in multiple genes, the etiology of most of these diseases remains largely unknown.

Polycystic Ovary Syndrome

Polycystic ovary syndrome (PCOS) is a common heterogeneous endocrine disorder associated with amenorrhea, hyperandrogenism, hirsutism, insulin resistance, obesity, and a 5–10-fold increased risk of type 2 diabetes mellitus. It is a leading cause of female infertility. The inherited basis of this disease was established by epidemiologic studies demonstrating an increased prevalence of PCOS, hyperandrogenemia, insulin resistance, and altered insulin secretion in relatives of women diagnosed with PCOS. From these pathways, several genes have been studied, including genes involved in steroid hormone biosynthesis and metabolism (*StAR, CYP11, CYP17, CYP19 HSD17B1-3, HSD3B1-2*), gonadotropin and gonadal hormone actions (*ACTR1, ACTR2A-B, FS, INHA, INHBA-B, INHC, SHBG, LHCGR, FSHR, MADH4, AR*), obesity and energy regulation (*MC4R, OB, OBR, POMC, UCP2-3*), insulin secretion and action (*IGF1, IGF1R, IGFBPI1-3, INS VNTR, IR, INSL, IRS1-2, PPARG*), and others. PCOS appears to be associated with the absence of the four-repeat-units allele in a polymorphic region of the *CYP11A* gene, which encodes cytochrome P450scc. Alteration of serine phosphorylation also seems to be involved in the post-translational regulation of 17,20-lyase activity (*CYP17*). About 50 genes have been demonstrated to have association with PCOS. Linkage and association studies identified a hotspot of candidate genes on chromosome 19p13.3, but to date, no genes are universally accepted as important in the pathogenesis of PCOS. Further confirmatory and functional studies are needed to identify key genes in the pathogenesis of PCOS.

Benign Ovarian Cysts

Benign ovarian cysts are a very common condition in premenopausal women. During normal ovulation each month, the follicle (or cyst) created by the ovaries bursts harmlessly. For unknown reasons, this normal physiologic process may sometimes go wrong. The follicle may continue to swell with fluid without releasing its egg, or hormones secreting tissue (corpus luteum) that prepare for pregnancy may persist even though the egg has not been fertilized. Subsequently, a cyst (or fluid-filled sac), which may be as small as a grape or as large as a tennis ball, is formed. Mutation of the *FOXL2* gene has been found in a patient with a large ovarian cyst. Such a cyst is usually benign and disappears within a couple of months. However, if such a cyst persists after several months, it may become a benign semisolid cyst. The most common semisolid cyst is a dermoid cyst, so-called because it is made up of skin-like tissue and can be removed by laparoscopic surgery.

Borderline Tumors

Borderline tumors account for 15–20% of epithelial ovarian tumors, and was first recognized by Howard Taylor in 1929. He described a group of women of reproductive age with large ovarian tumors whose course was rather indolent. In the early 1970s, the *Federation of Gynecologists and Obstetricians* defined these so-called semi-malignant tumors as borderline ovarian tumors (BOTs). Later on, at the 2003 WHO workshop, the term low malignant potential (LMP) became an accepted synonym for BOTs.

Borderline tumors with different cell types (serous, mucinous, endometrioid, clear cell, transitional, and mixed epithelial cells) have been reported. However, the serous and mucinous are the most common types. Mutational analyses have identified several gene mutations (Table 23.1) in borderline tumors. Both *BRAF* and *KRAS* mutations are very common in borderline serous tumors. However, *BRAF* mutation has not been found in borderline mucinous tumors. Moreover, the frequency of *KRAS* mutation is higher in borderline mucinous tumor than in serous tumors. On the other hand, *CTNNB1* and *PTEN* mutations have been found in borderline endometrioid tumors. The difference in mutation spectrum indicates different pathogenic pathways for these histological subtypes.

Whether serous BOT (SBOT) will progress to invasive carcinoma is still controversial. In a recent review of the clinical outcome of 276 patients with SBOTs, approximately 7% of the patients had recurrent disease as invasive low-grade serous carcinoma. In another report, when patients with SBOTs were followed up for a longer period, over 30% of the patients had recurrent disease as low-grade carcinoma over a period of 3 to 25 years. Studying genetic changes in different types of ovarian tumors provides insight into the pathogenic pathways for ovarian cancer. Mutations in *KRAS* have been found in 63% of mucinous BOTs and 75% of invasive mucinous ovarian cancers. These data suggest that *KRAS* mutations are involved in the development of mucinous BOTs and support the notion that mucinous BOTs may represent a phase of development along the pathologic continuum between benign and malignant mucinous tumors.

Table 23.1 Common Somatic Mutations in Human Sporadic Ovarian Tumors

	Gene Name	Number of Samples Screened	Number of Positive Samples	Percent Mutated
Ovarian adenoma				
	KRAS	152	15	9%
	BRAF	59	3	5%
	BRCA1	13	0	0%
	BRCA2	9	0	0%
	PIK3CA	8	0	0%
Borderline tumors				
Serous	BRAF	191	80	41%
	KRAS	141	37	26%
	ERBB2	21	2	9%
	PIK3CA	23	1	4%
	CDKN2A	15	0	0%
Mucinous	KRAS	5	4	80%
	CDKN2A	10	3	30%
	SMAD4	5	1	20%
	BRAF	28	0	0%
	BRCA1	5	0	0%
Endometrioid	CTNNB1	9	8	88%
	PTEN	8	1	12%
	KRAS	8	0	0%
Carcinomas				
Serous	KRAS	447	38	8%
	CDKN2A	227	14	6%
	BRCA1	246	11	4%
	BRAF	250	6	2%
	PIK3CA	230	5	2%
Mucinous	KRAS	143	62	43%
	CDKN2A	60	12	20%
	PTEN	36	6	16%
	PIK3CA	30	2	6%
	AATK	2	2	100%
Clear cell	KRAS	119	10	8%
	CDKN2A	23	5	21%
	PIK3CA	20	5	25%
	BRCA2	5	2	40%
	BRAF	53	1	1%
Endometrioid	CTNNB1	240	65	27%
	PTEN	92	23	25%
	BRAF	107	22	20%
	KRAS	176	12	6%
	PIK3CA	67	10	14%

However, the *KRAS* mutation rate in serous BOTs is significantly higher than that in serous invasive cancers. These data suggest that serous BOTs and invasive carcinomas may have different pathogenic pathways, and only a small percentage of serous BOTs may progress to invasive cancers. A mucinous cystadenoma would give rise to a mucinous borderline ovarian tumor (BOT), a subset of which may progress to invasive low-grade and perhaps high-grade mucinous carcinomas. A serous cystadenoma would give rise to a serous BOT, in which benign and borderline features are rare, in contrast to their frequent presence in mucinous BOTs.

Malignant Tumors

Ovarian cancer is a general term that represents a diversity of cancers that are believed to originate in the ovary. Over 20 microscopically distinct types can be identified, which can be classified into three major groups, epithelial cancers, germ cell tumors, and specialized stromal cell cancers. These three groups correspond to the three distinct cell types of different functions in the normal ovary: (i) the epithelial covering may give rise to the epithelial ovarian cancers, (ii) the germ cells may give rise to the germ cell tumors, and (iii) the steroid-producing cells may give rise to the specialized stromal cell cancers.

Epithelial Ovarian Tumors The majority of malignant ovarian tumors in adult women are epithelial ovarian cancer. Based on the histology of the tumor cells, they are classified into different categories: serous, mucinous, endometrioid, clear cell, transitional, squamous, mixed, and undifferentiated. From the Sanger Center's Catalogues of Somatic Mutations in Cancer (COSMIC) database, we have extracted the most frequently identified mutations for each histological subtype (Table 23.1). In addition to these common mutations, hundreds of rare mutations have also been identified (http://www.sanger.ac.uk/genetics/CGP/cosmic/).

Besides genetic analysis, other high-throughput methods have been exploited to understand the pathogenic pathways in ovarian cancer. It is hypothesized that most ovarian tumors develop from ovarian inclusion cysts arising from the OSE. In most cases of serous cancer, the majority of serous BOTs develop directly from ovarian inclusion cysts, without the cystadenoma stage. Alternatively, the epithelial lining of the cyst develops into a mucinous or serous cystadenoma by following one of two distinct pathogenic pathways. Our recent expression profiling analysis of serous BOTs and serous low- and high-grade carcinomas suggested that serous BOTs and low-grade carcinomas may represent developmental stages along a continuum of disease development and progression. Thus, we hypothesize that a majority of high-grade serous carcinomas derive *de novo* from ovarian cysts or ovarian endosalpingiosis. This hypothesis is consistent with the two-tier system recently proposed for the grading of serous carcinomas.

Loss-of-heterozygosity (LOH) studies have been widely used to identify minimally deleted chromosomal regions where tumor suppressor genes may reside. We previously identified several common loss regions by using 105 microsatellite markers to perform detailed deletion mapping on chromosomes 1, 3, 6, 7, 9, 11, 17, and X in BOTs, invasive ovarian cancers, and serous surface carcinoma of the ovary. Except at the androgen receptor (AR) locus on the X chromosome, BOTs showed significantly lower LOH rates (0–18%) than invasive tumors at all loci screened, suggesting that the LOH rate at autosomes is less important in the development of BOTs than more advanced tumors. Based on these and other results, the importance of LOH at the AR locus in BOTs and invasive cancers remains undetermined. However, several studies have found differences in LOH rates on other chromosomes. LOH at the *p73* locus on 1p36 was found in both high-grade and low-grade ovarian and surface serous carcinomas, but not in borderline ovarian tumors. In one study, LOH rates at 3p25, 6q25.1–26, and 7q31.3 were significantly higher in high-grade serous carcinomas than in low-grade serous carcinomas, mucinous carcinomas, and borderline tumors. In another study, LOH rates at a 9-cM region on 6q23-24 were significantly higher in surface serous carcinomas than in serous ovarian tumors. In addition, multiple minimally deleted regions have been identified on chromosomes 11 and 17. LOH rates at a 4-cM region on chromosome 11p15.1 and an 11-cM region on chromosome 11p15.5 were found only in serous invasive tumors, and the LOH rates in 11p15.1 and 11p15.5 were significantly higher in high-grade than low-grade serous tumors. Similarly, significantly higher LOH rates were identified at the *TP53* locus on 17p13.1 and the *NF1* locus on 17q11.1 in high-grade serous carcinomas than in low-grade and borderline serous tumors and all mucinous tumors. LOH at the region between *THRA1* and D17S1327, including the *BRCA1* locus on 17q21, was found exclusively in high-grade serous tumors. In general, the fact that LOH rates at multiple chromosomal sites were significantly higher in serous than in mucinous tumor subtypes

suggests that serous and mucinous tumors may have different pathogenic pathways. Since tumor suppressor genes are implicated to be located in chromosomal regions demonstrating LOH, further analysis of the genes located in the minimally deleted regions may provide insights into genes that may be important for the pathogenesis of serous and mucinous ovarian tumors. Studying genetic changes in endometriosis, endometrioid, and clear cell ovarian cancer also provides insights into pathogenic pathways in the development of these tumor types. Endometriosis is highly associated with endometrioid and clear cell carcinomas (28% and 49%, respectively), in contrast to its very low frequency of association with serous and mucinous carcinomas of the ovary (3% and 4%, respectively). This fact furnishes strong evidence that endometriosis is a precancerous lesion for both endometrioid and clear cell carcinomas. One study showed that endometrioid but not serous or mucinous epithelial ovarian tumors had frequent *PTEN/MMAC* mutations. In addition, the loss of PTEN immunoreactivity has been reported in a significantly higher percentage of clear cell and endometrioid ovarian cancers than cancers of other histologic types. Some investigators demonstrated that the expression of oncogenic *KRAS* or conditional *PTEN* deletion within the OSE induces preneoplastic ovarian lesions with an endometrioid glandular morphology. Furthermore, the combination of the two mutations in the ovary led to the induction of invasive and widely metastatic endometrioid ovarian adenocarcinomas. These data further suggest that tumors with different histologic subtypes may arise through distinct developmental pathways.

Germ Cell Cancer The ovarian germ cell tumors commonly occur in young women and can be very aggressive. Fortunately, chemotherapy is usually effective in preventing it from recurring. The pathogenesis of these tumors is not well understood. Several subtypes exist, which are dysgerminoma, yolk sac tumor, embryonal carcinoma, polyembryoma, choriocarcinoma, immature teratoma, and mixed GCTs. A recent study has identified *c-KIT* mutation and the presence of Y-chromosome material in dysgerminoma. Furthermore, DNA copy number changes were detected by comparative genomic hybridization in dysgerminomas. The most common changes in DGs were gains from chromosome arms 1p (33%), 6p (33%), 12p (67%), 12q (75%), 15q (42%), 20q (50%), 21q (67%), and 22q (58%). However, overexpression or mutation of *p53* was not observed in ovarian germ cell tumors.

Stromal Cancer The specialized stromal cell tumors (granulose cell tumors, thecal cell tumors, and Sertoli-Leydig cell tumors) are uncommon. One interesting characteristic of these tumors is that they can produce hormones. Granulosa and thecal cell tumors are frequently mixed and can produce estrogen, which will result in premature sexual development and short stature if the tumors are developed in young girls. Sertoli-Leydig cell tumors produce male hormones, which will cause defeminization. Subsequently, male pattern baldness, deep voice, excessive hair growth, and enlargement of the clitoris will develop in patients with Sertoli-Leydig cell tumors. Overexpression of the *BCL2* gene in Sertoli-Leydig cell tumor of the ovary has been reported. Fortunately, these specialized stromal cell cancers are usually not aggressive cancers and involve only one ovary.

Tumors of the Fallopian Tube The fallopian tubes are the passageways that connect the ovaries and the uterus. Fallopian tube cancer is very rare. It accounts for less than 1% of all cancers of the female reproductive organs. Only 1,500 to 2,000 cases have been reported worldwide, primarily in postmenopausal women. There is some evidence that women who inherit a mutation in the *BRCA1* gene, a gene already linked to breast and ovarian cancer, seem to have an increased risk of developing fallopian tube cancer. In one recent analysis of several hundred women who were carriers of the *BRCA1* gene mutation, the incidence of fallopian tube cancer was increased more than 100-fold. Likewise, a substantial proportion of women with the diagnosis of fallopian tube cancer test positively for either the *BRCA1* or *BRCA2* gene mutation. Recent recommendations suggest that any woman with a diagnosis of fallopian tube cancer be tested for the *BRCA* mutations. Based on the pathological study of fallopian tubes from *BRCA1* carrier, another hypothesis is being proposed that some serous ovarian tumors may be developed from the secretory fimbrial cells of the tube.

Vagina and Vulva

Vaginal intraepithelial neoplasia and squamous cell carcinoma are the most common tumor types affecting the vagina. Persistent infection with high-risk HPV (such as HPV16 and HPV18) is a major etiological factor for both diseases. Overexpression of *p53* and Ki67 has been identified in 19% and 75% of vaginal cancer cases, respectively.

Similar to vaginal malignancies, squamous cell carcinoma is the most common form of neoplasia in the vulva. Precursor lesions include vulvar intraepithelial neoplasia (VIN), the simplex (differentiated) type of VIN, lichen sclerosis, and chromic granulomatous. While vaginal carcinomas and VIN have been shown to be associated with HPV, usually of type 16 and 11, the simplex form of VIN and other precursor lesions are not HPV-associated. Genetic studies identified both *TP53* and *PTEN* mutations in VIN, as well as in carcinomas, suggesting that these molecular changes are early events in the pathogenesis of vulvar cancers. Other genetic changes including chromosome losses of 3p, 4p, 5q, 8p, 22q, Xp, 10q, and chromosome gains of 3q, 8p, and 11q have been reported. In addition, high frequencies of allelic imbalances have been found in both HPV-positive and HPV-negative lesions at multiple chromosomal loci located on 1q, 2q, 3p, 5q, 8p, 8q, 10p, 10q, 11p, 11q, 15q, 17p, 18q, 21q, and 22q.

Other less common epithelial malignancies of the vulva include Paget disease and Bartholin gland carcinoma. Nonepithelial vulvar malignancies include malignant melanomas, which accounts for 2–10% of vulvar malignancies. The molecular basis of most of these diseases remains largely unknown.

DISORDERS RELATED TO PREGNANCY

Conception, maintenance of the pregnancy, and delivering a baby are the most important functions of the female reproductive tract. Proper implantation of the embryo is a crucial step in a healthy pregnancy. Failure of this process can compromise the life of the fetus and the mother. During implantation, trophoblastic cells basically invade into the endometrium in a controlled fashion and transform the spiral arteries to supply the growing fetus with oxygen and nutrition. Failure of this process results in fetal hypoxia and growth retardation, which might lead to complications like abortion, preeclampsia, and preterm delivery. On the other hand, uncontrolled invasion can lead to deep implantation (placenta accreta, increta, percreta), or in gestational trophoblastic diseases, it can result in persistent disease, metastases, and development of neoplasia.

The trophoblastic invasion is regulated by at least three factors: (i) the endometrial environment (including adhesion molecules and vascularization at the implantation site), (ii) the maternal immune system, and (iii) the invasive and proliferative potentials of the trophoblastic cells. Complications following normal conceptions are mainly due to the failure of the maternal endometrium or immune system, whereas complications following genetically abnormal conceptions are most often the result of abnormal trophoblastic cell function. Preeclampsia and gestational trophoblastic diseases represent serious conditions that can lead to maternal death. This fact has put these diseases into the focus of clinical and molecular researchers. The molecules discussed here may also play a role in other implantation-related diseases, such as unsuccessful IVF cycles, recurrent spontaneous abortions, preterm deliveries, and others.

In gestational trophoblastic diseases (GTD), the problem is the overly aggressive behavior of the trophoblastic cells. The trophoblastic cells in GTD are hyperplastic and have much more aggressive behavior than in normal pregnancy. Trophoblastic cells might invade into the myometrium and persist after evacuation, which, without adequate chemotherapy, can compromise the life of the mother. Trophoblastic hyperplasia in GTD also can result in true gestational trophoblastic neoplasia (GTN).

In preeclampsia, hypertension and vascular dysregulation are the indirect results of the inadequate placentation due to different combinations of the previously mentioned factors. To date, the primary mechanisms that account for failure of trophoblastic cells to invade and transform the spiral arteries are still unknown. However, the mechanisms and factors leading to the classic symptoms after fetal stress seem to be identified, and clinical trials have been started to control these factors to prevent preeclampsia.

Gestational Trophoblastic Diseases

Gestational trophoblastic disease is a broad term covering all pregnancy-related disorders in which the trophoblastic cells demonstrate abnormal differentiation and hyperproliferation. Molar pregnancies are the most common gestational trophoblastic diseases, which are characterized by focal (partial mole) or extensive (complete mole) trophoblastic cell proliferation on the surface of the chorionic villi. Gestational trophoblastic neoplasms, like gestational choriocarcinoma, can arise from molar pregnancies (5–20%) and on very rare occasions from normal pregnancies (0.01%).

Partial molar pregnancy is a result of the union of an ovum and two sperms. Therefore, the genetic content of the fetus is imbalanced with overrepresentation of paternal genetic material. Since the maternal chromosomes are present, the embryo can develop to a certain point. The complete mole is the result of a union between an empty ovum and generally two sperms. Rarely, the conception happens with one sperm, and the genetic content of the sperm doubles in the ovum. In this case, the maternal chromosomes are not present. Therefore, no embryo develops. Some investigators have described familial recurrent hydatidiform molar pregnancies where balanced biparental genomic contribution could be found. In these cases, gene mutation is suspected on the long arm of chromosome 19, but to date the responsible gene has not been identified.

Epigenetic Changes in Trophoblastic Diseases

Due to unbalanced conception, gene expression in molar pregnancy differs greatly from that of a normal pregnancy. The gene expression differences can partly be explained by the lack or excess of chromosome sets, but more importantly epigenetic parental imprinting has a substantial role in the modification of the gene expression profile. Maternally imprinted genes are normally expressed from one set of paternal chromosomes, while in molar pregnancies these genes are expressed from two or more sets of paternal chromosomes. On the other hand, paternally imprinted genes are normally expressed only from the maternal allele, but in complete moles these genes cannot be expressed at all due to the lack of maternal chromosomes in complete gestational trophoblastic diseases. Complete molar pregnancy, as a unique uniparental tissue, is in the focus of epigenetic studies of placental differentiation. Several imprinted genes ($P57^{kip2}$, IGF-2, HASH, HYMAI, P19, LIT-1) have been investigated in complete moles, and paternally imprinted $p57^{kip2}$ is now used as a diagnostic tool to differentiate complete mole from partial mole and hydropic pregnancy by detecting maternal gene expression.

Immunology and Trophoblastic Diseases

As partial and complete moles are foreign tissues in the maternal uterus, vigorous immune response would be expected from the mother's side. On the other hand, trophoblastic cells express several immunosuppressive factors to attenuate the maternal immune response to protect the semiallogeneic fetus.

Trophoblastic cells, unlike other human cells, lack class I major histocompatibility complexes (MHC). Instead, they express atypical MHCs like human leukocyte antigen (HLA) G, E, and F. The atypical HLAs do not present antigens to the immune cells, but they are still able to inactivate the natural killer (NK) cells. Thus, these atypical HLAs protect the semiallogeneic fetus from the immune cells. The expression of immunosuppressive factors shows significant differences between normal and molar trophoblastic cells.

Table 23.2 lists the molecules which may play a role in the development of systematic and local gestational immunosuppression. In gestational trophoblastic diseases,

Table 23.2   Immunosuppressive Molecules Secreted by Trophoblastic Cells

	Increases the Expression	Ligand	Immune Cell or System Affected	Function
HLA molecules (C, G, E, F)	IL-10	KIR	NK	Inhibition
Soluble HLA G	IL-10	KIR	Tc cell	Apoptosis
Fas L	Th2 cytokines and hCG	Fas	Tc cell	Apoptosis
Cytokines (TGFβ, IL-10, IL-4, IL-6)	Unknown	Cytokine receptors	Th cells, Tc cells, NK, trophoblastic cells	Suppression HLA G, E, F overexpression
Indoleamine 2,3-dioxygenase	γ-INF	L-tryptophan	T-cell, complement system	Inhibition of T-cell maturation by tryptophan depletion, complement inhibition
DAF, MCP, CD59	Unknown	Multiple	Complement system	MAC inhibition
PI-9	Unknown	Granzyme B	Tc cell, NK cell killing	Granzyme B inhibitor
hCG	Unknown	hCG receptor	Trophoblastic immune cells	Fas L overexpression hCG → Progesterone→PIBF

IL—Interleukin; HLA—Human leukocyte antigen; Fas L—Fas ligand; DAF—Decay accelerating factor; MCP—Membrane cofactor protein; INF—Interferon; KIR—Killer inhibitor receptor; NK—Natural killer; Tc—Cytotoxic T cells; Th—Helper T-cells; MAC—Membrane attack complex; PIBF—Progesterone-induced blocking factor.

the hyperproliferation of the trophoblastic results in an even higher level of immunosuppressive factors in the mother. Furthermore, some of the immunosuppressive factors in GTD and GTN are elevated not only because of the higher number of hyperplastic trophoblastic cells, but also due to overexpression of these factors by the trophoblastic cells. For example, soluble and membrane-bound HLA-G was shown to be overexpressed in molar pregnancies and gestational trophoblastic neoplasms. Besides, molar villous fluid effectively suppressed the cytotoxic T-cells *in vitro*. One study found that the soluble IL-2 receptor level was significantly higher in the serum of mothers who subsequently developed persistent gestational trophoblastic disease.

Oncogene and Tumor Suppressor Gene Alterations in Trophoblastic Diseases

As molar pregnancies and especially gestational trophoblastic neoplasms demonstrate excessive trophoblastic cell proliferation, extensive studies were undertaken to characterize the expression of tumor suppressor genes and oncogenes. One study demonstrated that both complete mole and choriocarcinoma were characterized by overexpression of *p53*, *p21*, and *Rb* tumor suppressor genes and *c-myc*, *c-erbB-2*, and *bcl-2* oncogenes, while partial mole and normal placenta generally do not strongly express these molecules. *p53* and *Rb* molecules are mainly found in the nuclei of cytotrophoblastic cells, while *p21* could be seen in syncytiotrophoblast cells. *DOC-2/hDab2* tumor suppressor gene expression, on the other hand, was significantly stronger in normal placenta and partial mole compared to complete mole and choriocarcinoma.

In one study, investigators compared hydatidiform mole gene expression profile to normal placenta gene expressions by microarray method and identified 508 differentially expressed gene. Most of these genes are members of MAP-RAS kinase, Wnt, and Jak-STAT5 pathways. Some other genes, like versican, might have a role in cell migration or drug interactions. These findings provide important insight into the development of the GTN. However, to date it is not clear which gene has the primary role to develop gestational trophoblastic disease or neoplasia.

Detection of Trophoblastic Diseases

Gestational trophoblastic neoplasms are mostly known as chemosensitive malignancies. Human choriogonadotropin (hCG) follow-up is used in every case to diagnose persistent disease or GTN before applying chemotherapy. Several studies have tried to identify a reliable molecular or histologic marker which could predict persistent disease at the time of evacuation, but to date none of them seems to be sensitive and specific enough to change the current prognostic scoring and hCG follow-up in clinical practice. However, some promising results may help to diagnose GTN and predict prognosis in the future. A recent investigation demonstrated that hyperglycosylated hCG (hCG-H)

appears to reliably identify active trophoblastic malignancy, but unfortunately hCG-H alone is unable to predict persistence at the time of primary evacuation. Serum IL-2 and soluble IL-2 receptor levels at the time of evacuation have been demonstrated to have association with the clinical outcome and persistence. IL-1, EGFR, c-erbB-3, and antiapoptotic MCL-1 staining on trophoblastic cells also significantly correlated with the development of persistent postmolar gestational trophoblastic neoplasia. Further studies of potential molecules are needed to find a reliable marker or marker panel to predict persistent gestational trophoblastic diseases.

Molecular Basis of Preeclampsia

To date, the etiology of the disease is not known. In preeclampsia, the failure of the trophoblastic implantation indirectly leads to serious consequences. The implantation is sufficient to supply the embryo to a certain point, but the capacity of the not completely transformed spiral arteries is limited. Therefore, the fetus and the placenta suffer from hypoxic injury. Preeclampsia is characterized by endothelial dysfunction, which produces altered quantities of vasoactive mediators resulting in general vasoconstriction.

Recent studies demonstrated that soluble fms-like tyrosine kinase-1 (sFlt-1 or sVEGFR-1) might have a crucial role in developing hypertension and proteinuria in pregnant women. sFlt-1 is a tyrosine kinase protein which disables proteins that cause blood vessel growth. sFlt-1 prevents the effect of VEGF and placental growth factor (PLGF) on vascular structures. In women developing preeclampsia, the serum level of sFlt-1 is significantly elevated, and the levels of VEGF and PLGF are markedly decreased compared to normotensive women. The sFlt-1 level starts to rise 5 weeks before the onset of preeclampsia. In animal studies, exogenous application of sFlt-1, similar to anti-VEGF bevacizumab, results in hypertension and proteinuria in pregnant rats.

Together with sFlt-1, endoglin serum level correlates with the progression and severity of preeclampsia. These molecules might also serve as diagnostic tools for preeclampsia. Studies have shown that sFlt-1 and endoglin are useful tools to differentiate preeclampsia from other types of hypertension in pregnancy. These findings also hold the possibility of developing a new medical therapy for preeclampsia targeting angiogenic factors.

KEY CONCEPTS

- High-risk HPVs contribute to the genesis of almost all human cervical carcinomas and have also been associated with a number of other anogenital malignancies including vulvar, anal, and penile carcinomas.
- HPV-associated cervical cancers are quite unique in that they represent the only human solid tumor

for which the initiating carcinogenic agents have been identified at a molecular level. The fact that the high-risk HPV E6 and E7 oncoproteins contribute to initiation as well as progression and that they are necessary for the maintenance of the transformed phenotype of cervical cancer cells suggests that these proteins and/or the processes that they regulate should provide targets for intervention.

- Uterine cancer is the most common gynecologic cancer in the United States. Most uterine cancers are adenocarcinomas and develop from the endometrium, the inner lining of the uterus.
- Risk factors for the development of endometrial cancer include obesity, unopposed estrogen use, polycystic ovarian syndrome, insulin resistance and diabetes, and estrogen-secreting ovarian tumors.
- Endometrial cancer can be broadly divided into Type I and Type II categories, based on risk factors, natural history, and molecular features.
- The majority of malignant ovarian tumors in adult women are epithelial ovarian cancer, which can be classified into serous, mucinous, endometrioid, clear cell, transitional, squamous, mixed, and undifferentiated. They all have different pathogenetic pathways.
- *BRCA1* and *BRCA2* are the key genes involved in the development of familial ovarian cancer.
- Based on genetic analysis, it has been hypothesized that low-grade and high-grade serous ovarian cancers are developed through a two-tier system.
- Gestational trophoblastic disease is a broad term covering all pregnancy-related disorders, in which the trophoblastic cells demonstrate abnormal differentiation and hyperproliferation. Molar pregnancies are the most common gestational trophoblastic diseases, which are characterized by focal (partial mole) or extensive (complete mole) trophoblastic cell proliferation on the surface of the chorionic villi.
- Epigenetic parental imprinting has a substantial role in the modification of the expression profile in gestational trophoblastic diseases.

SUGGESTED READINGS

1. Barr E, Tamms G. Quadrivalent human papillomavirus vaccine. *Clin Infect Dis.* 2007;45:609–617.
2. Bell DA, Longacre TA, Prat J, et al. Serous borderline (low malignant potential, atypical proliferative) ovarian tumors: Workshop perspectives. *Hum Pathol.* 2004;35:934–948.
3. Bonome T, Lee JY, Park DC, et al. Expression profiling of serous low malignant potential, low-grade, and high-grade tumors of the ovary. *Cancer Res.* 2005;65:10602–10612.
4. Brinton LA, Berman ML, Mortel R, et al. Reproductive, menstrual, and medical risk factors for endometrial cancer: Results from a case-control study. *Am J Obstet Gynecol.* 1992;167:1317–1325.
5. Feltmate CM, Batorfi J, Fulop V, et al. Human chorionic gonadotropin follow-up in patients with molar pregnancy: A time for reevaluation. *Obstet Gynecol.* 2003;101:732–736.
6. Genest DR, Dorfman DM, Castrillon DH. Ploidy and imprinting in hydatidiform moles. Complementary use of flow cytometry and immunohistochemistry of the imprinted gene product p57KIP2 to assist molar classification. *J Reprod Med.* 2002; 47:342–346.
7. Hashiguchi Y, Tsuda H, Inoue T, et al. PTEN expression in clear cell adenocarcinoma of the ovary. *Gynecol Oncol.* 2006;101:71–75.
8. Jarboe E, Folkins A, Nucci MR, et al. Serous carcinogenesis in the fallopian tube: A descriptive classification. *Int J Gynecol Pathol.* 2008;27:1–9.
9. Kato HD, Terao Y, Ogawa M, et al. Growth-associated gene expression profiles by microarray analysis of trophoblast of molar pregnancies and normal villi. *Int J Gynecol Pathol.* 2002; 21:255–260.
10. Lee C, Laimins LA. The differentiation-dependent life cycle of human papillomaviruses in keratinocytes. In: Garcea RL, DiMaio D, eds. *The Papillomaviruses.* New York: Springer; 2007: 45–68.
11. Levine RL, Cargile CB, Blazes MS, et al. PTEN mutations and microsatellite instability in complex atypical hyperplasia, a precursor lesion to uterine endometrioid carcinoma. *Cancer Res.* 1998;58:3254–3258.
12. Lu KH, Wu W, Dave B, et al. Loss of tuberous sclerosis complex-2 function and activation of mammalian target of rapamycin signaling in endometrial carcinoma. *Clin Cancer Res.* 2008;14: 2543–2550.
13. MacDonald ND, Salvesen HB, Ryan A, et al. Frequency and prognostic impact of microsatellite instability in a large population-based study of endometrial carcinomas. *Cancer Res.* 2000;60:1750–1752.
14. Munger K, Baldwin A, Edwards KM, et al. Mechanisms of human papillomavirus-induced oncogenesis. *J Virol.* 2004;78:11451–11460.
15. Risinger JI, Hayes K, Maxwell GL, et al. PTEN mutation in endometrial cancers is associated with favorable clinical and pathologic characteristics. *Clin Cancer Res.* 1998;4:3005–3010.
16. Risinger JI, Maxwell GL, Chandramouli GV, et al. Microarray analysis reveals distinct gene expression profiles among different histologic types of endometrial cancer. *Cancer Res.* 2003;63:6–11.
17. Schiffman M, Castle PE, Jeronimo J, et al. Human papillomavirus and cervical cancer. *Lancet.* 2007;370:890–907.
18. Shih Ie M, Kurman RJ. Ovarian tumorigenesis: A proposed model based on morphological and molecular genetic analysis. *Am J Pathol.* 2004;164:1511–1518.
19. Trimble CL, Kauderer J, Zaino R, et al. Concurrent endometrial carcinoma in women with a biopsy diagnosis of atypical endometrial hyperplasia: A Gynecologic Oncology Group study. *Cancer.* 2006;106:812–819.

24

Molecular Pathogenesis of Diseases of the Kidney

Amy K. Mottl . Carla Nester

INTRODUCTION

The ability to define the pathological basis of kidney diseases has advanced substantially since the completion of the Human Genome Project. Moreover, advances in technology have been greater in the past decade than has ever been seen in modern medicine. These advances combine to open the door to a richer understanding of both normal and pathological processes involved in kidney function and consequent diseases.

Kidney disease is defined as either decreased ability to filter products of metabolism (such as creatinine) or the loss of protein in the urine (proteinuria). The ability of the kidney to filter metabolic substances into the urine is referred to as the glomerular filtration rate (GFR). With worsening proteinuria or deterioration in GFR, there is a greater risk of end-stage kidney disease (ESKD) requiring dialysis or transplantation, as well as increased risk of cardiovascular disease. The molecular pathogenesis of kidney diseases with Mendelian modes of inheritance are most well understood. Three such entities will be reviewed including focal segmental glomerulosclerosis (FSGS), Fabry disease, and polycystic kidney diseases. Additionally, Bartter's and Gitelman's diseases will be discussed as examples of renal tubular disorders, which can result in electrolyte and acid/base disturbances. First, we give a brief description of the structure and important physiologic mechanisms in the normal functioning kidney.

NORMAL KIDNEY FUNCTION

The kidney is composed of millions of nephrons, the individual functional units responsible for urine production. The glomerulus is the filtration component of the nephron that allows toxic substances accumulating in the body to pass from the blood to the tubules of the nephron from which they are excreted in the urine. The filtration barrier is comprised of fenestrated vascular endothelial cells, the glomerular basement membrane, and interdigitating epithelial cells (podocytes) connected by slit diaphragms (Figure 24.1). Water, ions, and small solutes are able to permeate the filtration barrier, but larger proteins, such as albumin, cannot pass through. This filtrate is collected in Bowman's capsule and is then modified as it moves through subsequent tubules of the nephron. Tubular cells possess ion channels that are important to maintaining proper sodium, potassium, and acid/base balance, as well as to the nephron's ability to concentrate or dilute urine.

FOCAL SEGMENTAL GLOMERULOSCLEROSIS

Focal segmental glomerulosclerosis (FSGS) is considered a disease, but in actuality is a histopathologic finding with numerous etiologies, each of which can vary dramatically in clinical course. FSGS is diagnosed with kidney biopsy. Light microscopy is characterized by scarring (sclerosis) of select portions (segments) of glomeruli. This should be differentiated from global sclerosis in which the entire glomerulus is scarred, and which can be a normal finding with age. Generally, not all glomeruli are involved, accounting for the focal portion of its name. FSGS leads to proteinuria, frequently resulting in decreased GFR and ESKD requiring dialysis or transplantation.

Clinical Presentation of Focal Segmental Glomerulosclerosis

The clinical presentation of FSGS is extremely variable. Onset can occur from the neonatal period to the seventh decade. The disease course can be self-limited, slowly progressive, or rapidly deteriorate to ESKD.

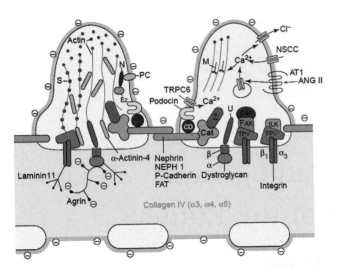

Figure 24.1 Glomerular filtration barrier demonstrating two podocyte foot processes bridged by the slit membrane, the glomerular basement membrane (GBM), and the porous capillary endothelium. Reproduced with permission from *Trends Mol Med.* 2005;11(12):527–530.

Patients present with proteinuria, which may be discovered on routine urinalysis, or they can be symptomatic with nephrotic syndrome. This is characterized by nephrotic range proteinuria (>3 grams/24 hours), edema, hypoalbuminemia, and hyperlipidemia. FSGS is the most common cause of acquired chronic kidney disease in children. The vast majority of FSGS is idiopathic. Idiopathic FSGS affects African Americans more than Caucasians, and is more likely to present in young adulthood. In the United States, idiopathic FSGS is responsible for approximately 4% of ESKD.

There are five histologic variants of FSGS, depending on the glomerular location and nature of the sclerotic lesion. The relevance of the histologic variant to etiology, clinical presentation, and response to treatment has been investigated. The collapsing variant has a predilection for African Americans, is resistant to treatment, and typically follows an aggressive course with a 3-year renal survival rate of 33%. This pathologic entity is found frequently in FSGS due to toxic exposures such as heroin or pamidronate. The tip variant is more common in Caucasians, is more likely to abate with immunosuppression, and has a better prognosis, with a 3-year renal survival rate of 76%. Perihilar FSGS most often is found in secondary forms of FSGS due to hyperfiltration states such as obesity and hypertension. The cellular variant is the rarest and is thought to possibly represent an earlier stage of FSGS. The final variant form of FSGS is termed FSGS not-otherwise-specified, and may represent a later stage of FSGS.

Familial FSGS is classified according to its Mendelian inheritance pattern and early (childhood) versus late (adult) onset. These forms of FSGS are more consistent in their clinical presentation and multiple genes have been discovered as the cause. Congenital nephrotic syndrome of the Finnish type (CNF) was first reported in Finland, explaining the derivation of its name. However, CNF since has been identified in multiple geographic populations. It is characterized by massive proteinuria, often 20–30 grams/24 hour *in utero,* and is inherited in an autosomal recessive pattern. Without bilateral nephrectomy and transplantation, complications from nephrotic syndrome result in an exceedingly high mortality rate within the first few months of life. It is caused by mutation of the gene for nephrin (*NPHS1*), a critical protein of the slit diaphragm component of the filtration barrier. A substantial portion of children who receive kidney transplants develop recurrence of disease, thought likely due to autoantibodies to nephrin or the glomerular basement membrane (GBM). Carriers of the mutation may have proteinuria, but typically do not develop chronic renal insufficiency.

A less severe form of early-onset autosomal recessive FSGS is caused by mutation of the podocin gene (*NPHS2*). Age of onset ranges from a few months to 5 years. This form of FSGS is resistant to steroid treatment and progresses rapidly to ESKD. Posttransplant FSGS is relatively uncommon. *NPHS2* mutations can also be found in sporadic FSGS, and have been identified mostly in children, but there are also reports in adults. Autosomal dominant (AD), adult-onset FSGS has been identified in families with mutations in either the alpha-actinin 4 (*ACTN4*) or transient receptor potential 6 (*TRPC6*) genes. Disease onset can range from adolescence to the fifth decade and there is variable progression to ESKD.

Pathogenesis of Focal Segmental Glomerulosclerosis

In the normal kidney, numerous branches emanate from the podocyte body, terminating on distinct capillary vessels where they are referred to as foot processes. Disruption, or effacement, of the podocyte foot processes occurs invariably in FSGS. The underlying mechanism of foot process effacement is not completely understood and may be due to multiple mechanisms. Alterations in the cell membrane, the structural integrity of the slit diaphragm, and/or actin cytoskeleton and signaling among these components can result in podocyte damage. The slit diaphragm is integral to maintaining cell polarity, regulation of the cell cycle, and intracellular signaling of external conditions. The intricate actin cytoskeleton of the podocyte maintains the normal architecture of the foot processes, and is capable of responding to a changing environmental milieu. Destabilization of the actin cytoskeleton leads to foot process detachment and effacement.

Podocyte damage initiates a vicious cycle of cytokine production, proteinuria, and hyperfiltration. Specifically, TGFβ, SMAD, VEGF, and angiotensin have been implicated in this process. These events lead to up regulation of the inflammatory response with recruitment of T-cells and macrophages. Cell polarity and maintenance of the cell cycle is lost, resulting in apoptosis. As a terminally differentiated cell, the ability of the podocyte to regenerate is limited, resulting in podocytopenia and exposure of the GBM. Collagen matrix deposition ensues, resulting in sclerosis and ultimately, obliteration of the capillary lumen.

FSGS is most often sporadic with no identifiable cause. However, many cases of FSGS are secondary to specific etiologic conditions. Environmental insults resulting in hyperfiltration can lead to FSGS, including hypertension, obesity, and unilateral renal agenesis. Viral infection with human immunodeficiency virus (HIV), parvovirus, and cytomegalovirus have been associated with some cases of FSGS. Other cases have been related to toxic exposures such as heroin, pamidronate, and interferon-α. FSGS can demonstrate familial aggregation with Mendelian inheritance patterns. This has led to the discovery of multiple genes which, when mutated, lead to disruption of the filtration barrier and subsequent proteinuria and sclerosis.

Some forms of FSGS may be the result of a circulating permeability factor, which damages the basement membrane and disrupts the adherence of the foot process. This is best exemplified in recurrent FSGS in those who have undergone kidney transplantation. Posttransplant FSGS can sometimes be treated successfully with plasmapheresis, which removes proteins, and presumably this circulating permeability factor, from the blood.

Genetics of Focal Segmental Glomerulosclerosis

Mutations in genes coding for important proteins in the slit diaphragm or the actin cytoskeleton can lead to monogenic forms of FSGS, and some predict responsiveness to treatment. To date, there are at least seven genes in which certain mutations can lead to FSGS. For illustrative purposes, this text will cover four such genes, including *NPHS1*, *NPHS2*, *ACTN4*, and *TRPC6* (Figure 24.1).

Nephrin is a slit diaphragm protein, which not only forms one of the physical barriers to solute passage, but is also important to intracellular signaling and apoptosis. Certain mutations in nephrin result in CNF, the most severe clinical form of FSGS. Its mode of inheritance is autosomal recessive. Two mutations account for the vast majority of CNF, Fin major and Fin minor. Fin major is a frameshift mutation caused by deletion of nucleotides 121 and 122. Fin minor is a nonsense mutation resulting in a premature stop codon at amino acid 1109.

Podocin is a protein found at the base of the foot process that functions as both a structural protein in the slit diaphragm and also a signaling molecule. It acts with lipid rafts in the apical region of the cell membrane and recruits nephrin and *CD2AP*, another slit diaphragm-associated protein. Some podocin mutations result in autosomal recessive transmission of steroid-resistant nephrotic syndrome. Recently, several missense mutations have been associated with sporadic FSGS.

ACTN4 mutations can result in autosomal dominant adult-onset FSGS. *ACTN4* anchors the actin cytoskeleton to the cell membrane, and certain mutants have been found to increase the affinity of binding. However, the podocyte's ability to anchor to the GBM appears to be decreased with *ACTN4* mutations, as demonstrated in animal models.

TRPC6 is a cation selective channel of the podocyte, localized near the slit diaphragm. It results in AD adult-onset FSGS with variable progression to ESKD depending on the specific mutation. *TRPC6* regulates calcium entry into the cell, with certain mutations resulting in increased permeability, especially in the presence of angiotensin II. The effect of increased intracellular calcium is unknown. However, some investigators have suggested that the actin cytoskeleton may be pathologically altered.

Treatment of Focal Segmental Glomerulosclerosis

The mainstay of FSGS treatment is suppression of proteinuria with inhibitors of the renin-angiotensin system and immunosuppression with corticosteroids and cyclosporine. Genetic forms of FSGS typically do not respond to immunosuppressive treatment, especially in children. Sporadic FSGS is more likely to abate than familial FSGS. However, response is still quite poor. Ongoing efforts aim to identify pathogenetic mechanisms predicting response to immunosuppression. However, to date none has been consistently identified.

FABRY DISEASE

Fabry Disease is an X-linked lysosomal storage disease that affects multiple organ systems, including the kidney. It is a rare disorder, with early-onset disease affecting approximately one in 40,000 males and late-onset disease affecting as many as one in 3,100 males. Fabry Disease results from deficiency of α-galactosidase (α-gal), a ubiquitous enzyme crucial to the metabolism of instrinsic cellular components, glycosphingolipids. These molecules accumulate intracellularly and result in disruption of normal physiology with irreversible damage. Clinical manifestations begin in childhood. The average life expectancy is approximately 55 years for men and 70 years for women.

Clinical Manifestations of Fabry Disease

The age of onset of classical Fabry Disease is in early childhood, typically by the age of 5 to 6 years in boys and 9 to 10 years in girls. The most common symptom during this period is acroparesthesias, tingling and burning in the hands and feet, which can become extremely disabling. Gastrointestinal symptoms are common and can include abdominal pain, nausea, vomiting, and diarrhea. Hypohydrosis, decreased sweating, is also typical and can result in exercise intolerance. During adolescence, approximately 40% of Fabry patients develop angiokeratomas (Figure 24.2a). These are painless,

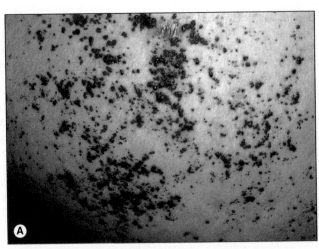
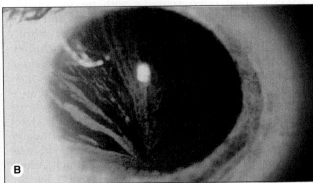

Figure 24.2   Typical finding of (A) angiokeratoma and (B) corneal verticillata in Fabry Disease. Reproduced with permission from (a) *Lancet.* 2008;372:1427–1435; and (b) *Survey of Ophthalmol.* 2008;53(4):416–423.

nonpruritic red or purple nodules, typically over the central abdomen and groin region. Sensorineural hearing loss can occur at any age, but the prevalence increases with age. Corneal verticillata (Figure 24.2b), or corneal whorls, are linear opacities emanating from a single point in the cornea but do not interfere with visual acuity. These are seen in most male and many female patients and are diagnostic of Fabry disease.

Prior to the initiation of dialysis, the most common cause of death in males with Fabry Disease was ESKD. By early adulthood, many males and some females manifest with proteinuria and by age 50, most men will have decreased GFR. The mean time from decreased GFR to ESKD is 4 years ($\pm$3 years). There is a correlation between the age of onset of kidney disease and the residual α-gal enzyme activity, with those who have undetectable α-gal activity presenting with decreased GFR in their twenties versus forties for those with ≥1% α-gal activity. Some *GLA* mutations result in renal variant Fabry, whereby no symptoms manifest until adulthood when they present with kidney disease.

There are multiple cardiac and cerebrovascular manifestations of Fabry disease. The most common is left ventricular hypertrophy (LVH) and subsequent diastolic dysfunction, which can be severe. Premature atherosclerotic disease is a frequent cause of morbidity and mortality, with a first myocardial infarction or stroke occurring in men in their forties and women in their fifties. Conduction abnormalities are frequent, with shortened PR interval and resting bradycardia being the most common. Ventricular arrhythmias can sometimes be the first presenting sign of Fabry disease, and is the most common cause of cardiac death. Similar to renal variant Fabry, there are *GLA* mutations that result in cardiac variants of the disease in which the first clinical manifestation is cardiac disease.

Pathogenesis of Fabry Disease

Glycosphingolipids are constituent components of the plasma cell membrane. They typically undergo degradation after being endocytosed into the lysosome.

With the absence of the intralysosomal degrading enzyme α-galactosidase A, these neutral glycosphingolipids, mostly globotriaosylceramide (GB3), progressively accumulate. Accumulation has been found to occur not only within the lysosomes, but also within the nucleus, cytoplasm, and cell membranes. Using immunohistochemical techniques, GB3 staining occurs most strongly in the kidney and heart. Virtually all cardiac tissue types stain positive, including endocardium, myocardium, endothelial cells of the coronary arteries and intracardiac nerves, and vasa nervorum. Staining within the kidney is more variable and mostly involves the glomerular and tubular epithelial cells. Accumulation of GB3 within the brain is almost entirely within the vasculature and in the skin typically is isolated to dermal blood vessels and epidermal fibroblasts.

The mechanism by which GB3 accumulation results in the clinical pathology of Fabry Disease is unknown. It does result in accelerated atherosclerotic disease, possibly due to the induction of reactive oxygen species, apoptosis, and interference with the fibrinolytic system. Glycosphingolipid accumulation may also interfere with intracellular trafficking between endosomal organelles and the endoplasmic reticulum. Signal transduction between cells, specifically via endothelial nitric oxide synthase (eNOS), may be disrupted by the accumulation of GB3 in the cell membrane.

Genetics of Fabry Disease

Fabry Disease results from mutation of the α-gal (*GLA*) gene located on the X chromosome. Since males possess only a single X chromosome, all those that carry a mutation in *GLA* manifest symptoms. In contrast, heterozygous females have a more variable course. The heterogeneity of Fabry Disease in females is due to X chromosome mosaicism. During embryonic development in the female, one X chromosome is randomly inactivated in each cell so that the transcriptional dosage of the X chromosome is the same in both genders. Different cell lines of the same cell type, within the same organ, may have a different activated X chromosome, and this is termed mosaicism. If there is a skewing of cell lines with the mutated X chromosome, a female can manifest with more severe Fabry Disease, similar to that of a male. The effect of skewed X-inactivation is best evidenced by monozygotic twins, wherein one twin is composed primarily of cells expressing the mutant gene and manifests with classic Fabry Disease versus the other twin who may remain completely asymptomatic.

Clinical manifestations are also heterogeneous among males, depending on the genetic mutation and its effect on enzyme level and function. Over 300 mutations spanning all seven exons have been identified. More than 70% are point mutations, mostly missense or nonsense, and 25% represent small deletions. Mutations that result in a complete loss of enzyme function, such as nonsense mutations or rearrangements, traditionally yield phenotypes consistent with classic Fabry Disease. Male patients with these mutations will have no enzyme activity. Missense mutations, resulting in amino acid substitutions, may give rise to later onset Fabry symptoms such as the renal or cardiac variants. Specifically, conservative substitutions generally result in a later onset and milder disease course than those resulting in nonconservative substitutions. Moreover, the location of the mutation can also account for phenotypic variability of Fabry, depending on whether it is the active site of the enzyme that is affected or protein folding or enzyme stability.

Although Fabry Disease is a simple Mendelian disorder, the complexity of the disease process is revealed by the lack of genotype-phenotype correlation. Different families with the same genetic mutation, and even different family members within the same family, may manifest with different disease characteristics such as the age of onset, organ systems involved, and severity of disease. The basis for this lack of genotype–phenotype correlation is unknown, but it is likely that modifying genes and epigenetic factors play a large role.

Diagnosis of Fabry Disease

The nonspecific nature of symptoms in Fabry Disease often results in delayed diagnosis. The mean time from the onset of symptoms to correct diagnosis has been reported to be 15 to 20 years. The findings of angiokeratoma and corneal verticillata are virtually diagnostic. Once Fabry Disease is suspected, the diagnosis is fairly simple in men who can be diagnosed by measuring peripheral leukocyte or plasma α-gal activity. Levels less than 20% of normal are diagnostic and less than 35% of normal are suggestive of Fabry disease. However, enzyme activity levels in women are highly variable and rarely useful, as they can be within the normal range even in highly symptomatic women.

In women with suspected Fabry Disease, genetic testing for a known causative mutation should be performed. This is done by gene sequencing and will only be diagnostic if a known causative mutation is identified. Due to genetic heterogeneity, the occasion may arise that a previously unidentified mutation exists in a family. Hence, sequencing results from a

questionably affected female would be difficult to interpret. If there are affected male relatives, a higher yield approach would be to sequence the male and then test for the specific mutation in the female. Prenatal diagnosis also occurs by sequencing for the known familial mutation.

Treatment of Fabry Disease

Intravenous enzyme replacement therapy (ERT) with α-galactosidase became available in the United States in 2003 and is the first specific therapy for Fabry Disease. The enzyme preparation utilized in ERT for Fabry Disease is called agalsidase. Antibody formation to agalsidase is common among treated patients, especially in males with <1% α-gal activity. The significance of these antibodies remains unknown, but *in vitro* studies suggest they may be inhibitory. ERT is generally well tolerated, although infusion-related reactions such as fever, chills, and rigors are frequently reported, especially during the first several treatments. There are no parameters to follow for assessing treatment efficacy. Levels of GB3 are variable and generally do not correlate with symptomatic improvement.

Agalsidase has been demonstrated to slow the progression of GFR decline and cardiac hypertrophy, decrease pain crises, and improve tissue clearance of GB3. Outcomes are best in those without significant kidney disease (GFR ≥ 55 mL/min/1.73 m^2), although there are a few reports of decreased kidney and cardiovascular events even in advanced Fabry Disease. In addition, improvement in hearing, sweating, and gastrointestinal symptoms have been reported in patients treated with therapy. Treatment is recommended for all adolescent boys or older. Treatment recommendations for children and women are less well defined, but those who are symptomatic should likely receive therapy. Clearly, early diagnosis and treatment are important to assessing the maximal beneficial effects of ERT in Fabry Disease.

POLYCYSTIC KIDNEY DISEASE

Though there are multiple diseases of the kidney that may present with the pathological finding of cyst formation, classically, this terminology is used to refer to two specific disorders that in most cases have a genetic basis for the pathology: (1) autosomal dominant polycystic kidney disease and (2) autosomal recessive polycystic kidney disease. Differences in the patterns of inheritance, clinical presentation, and appearance of the kidneys reflect the differences in the underlying pathogenesis of these disorders.

Autosomal Dominant Polycystic Kidney Disease

Autosomal dominant polycystic kidney disease (ADPKD) is the most commonly inherited renal disease, with a prevalence of one in 400 to one in 1000 individuals. There are estimated to be 600,000 cases in the United

States and over 12 million cases worldwide. Although cyst formation begins *in utero*, disease onset has a bimodal pattern, with most patients presenting in the third to fifth decades of life and a clinically significant subset of patients presenting in infancy and childhood. Progression to renal failure occurs in nearly 50% of patients with ADPKD by age 60. ADPKD accounts for approximately 2.5–3.0% of cases of ESRD in the United States.

Clinical Presentation and Diagnosis of Autosomal Dominant Polycystic Kidney Disease

ADPKD cyst formation begins *in utero*, but signs of the disease may not be detected for several decades. Over time, multiple renal cysts that arise from distal nephron segments result in renal enlargement and progressive loss of kidney function. Although bilateral renal involvement is reported most frequently, approximately 15% of cases have asymmetric renal involvement—a finding that is more common in children. The typical adult presentation includes decreased renal concentrating ability (polyuria), hematuria, and flank pain resulting from cyst expansion. Hypertension is noted in 70–80% of affected individuals, and typically predates the development of renal insufficiency. Hypertension likely results from activation of the renin-angiotensin-aldosterone system (RAAS). When this occurs, cyst compression of the surrounding parenchyma causes local ischemia.

In addition to kidney cysts, patients with ADPKD may develop cystic lesions of the liver, pancreas, and occasionally the lung. Berry aneurysms within the brain usually involve the circle of Willis and have been reported in 10–15% of cases with a tendency to cluster in families. Coronary artery aneurysms, cardiac valve anomalies, ovarian cysts, inguinal or ventral hernias, colonic diverticula, and subcutaneous cysts also have been reported. Intracranial aneurysms are a severe presentation and may result in headache or even stroke. On the other end of the spectrum, nearly 50% of individuals with ADPKD will remain undiagnosed based on the silent nature of cyst formation. ADPKD can present as an incidental finding or as surveillance of apparently unaffected family members of known cases. ADPKD may be discovered as a result of a workup for abdominal mass or hypertension.

In patients with a positive family history, the diagnosis of ADPKD is established by radiologic evidence of bilateral, fluid-filled renal cysts. Because of cost and safety, ultrasonography is most commonly used as the imaging modality. Typical findings include large kidneys and extensive cysts scattered throughout both kidneys. The Ravine ultrasonographic criteria use patient age and family history to suggest the diagnosis of ADPKD (Table 24.1). Important issues related to the diagnosis of ADPKD include the presence or absence of a family history of the disease, the numbers and types of renal cysts, and the age of the patient. The presence of liver or pancreatic cysts confirms a diagnosis of ADPKD.

Table 24.1 Ravine Ultrasonographic Criteria for Diagnosing Autosomal Dominant Polycystic Kidney Disease

Age	Positive Family History	Negative Family History
<30 years	2 cysts bilaterally or unilaterally	5 cysts bilaterally
30–60 years	4 cysts bilaterally	5 cysts bilaterally
>60 years	8 cysts bilaterally	8 cysts bilaterally

Based on: Evaluation of Ultrasonographic Diagnostic Criteria for Autosomal Dominant Polycystic Kidney Disease 1, *Lancet.* 1994;343:824–827.

Genetic testing is available for both *PKD1* and *PKD2*. Current genotype testing identifies approximately 70% of the known pathogenic mutations, and hence diagnosis is more often made on radiologic grounds.

Genetics and Pathogenesis of Autosomal Dominant Polycystic Kidney Disease

As the name indicates, ADPKD typically is inherited in an autosomal dominant fashion and results from the inheritance of mutations in one of two genes: *PKD1* or *PKD2*. Eighty-five percent of families with ADPKD have a mutation in the *PKD1* gene and another 15% in *PKD2*. Family history is absent in nearly 10% of cases, suggesting that there are also a significant number of sporadic mutations leading to ADPKD.

PKD1 is located on the short arm of chromosome 16 (16p13.3 region) and its gene product is Polycystin-1 (PC1). Polycystin-1 is expressed predominantly in the distal convoluted tubule and collecting ducts of the kidney. PC1 is a primary cilium transmembrane glycoprotein that plays a role in regulating tubular and vascular development in the kidney as well as other organs (liver, brain, heart, and pancreas). These cilia detect environmental signals and increase the flow of calcium through a cation channel formed in the plasma membrane by Polycystin-2 (PC2), the gene product of *PKD2*. Normal calcium flux results in inhibition of cell proliferation. When calcium entry is impaired by a mutation in one of these proteins, proliferative cell pathways predominate, leading to renal cyst formation. From this standpoint, the cysts of ADPKD are a benign renal tubular neoplasm that expands by increasing the mass of proliferating epithelial cells that surround a cavity filled with fluid.

The pathogenesis of ADPKD is attributed primarily to a two-hit mechanism, in which an inherited germline mutation is compounded by a second somatic mutation. In a cell that has one allele carrying the genetic mutation, a second hit to the remaining normal allele triggers a sequence of events that leads to the proliferation of the tubule cell as described earlier. Once a cyst reaches 2 mm in diameter, it separates from the parent nephron and will function as an autonomous fluid-filled sac/tumor. Cysts expand and destroy normally functioning renal tissue, and this provokes the complications of polycystic kidney disease.

PC2 is a member of the transient receptor potential channel (TRPC) superfamily of nonselective cation channels. As described earlier, PC1 and PC2 function as a regulated ion channel complex. Though mutations in *PKD1* are associated with earlier disease onset and a more severe disease course than mutations in *PKD2*, a mutation of either polycystin protein can disrupt this normal function of the cilium and result in a similar pathology. PC2 also functions as a Ca^{2+} release channel in the endoplasmic reticulum and has been localized to the apical primary cilia of epithelial cells, where the PC1–PC2 complex is thought to participate in flow-induced mechanosensation.

Treatment of Autosomal Dominant Polycystic Kidney Disease

There are currently no treatments that have been shown in randomized trials to slow the formation of cysts or slow disease progression in patients with ADPKD. Therefore, basic treatment measures aimed at ameliorating the effects of associated findings (such as hypertension) take a primary role. Blood pressure should be controlled to less than 130/80 in adults and to the normal range for sex- and age-matched children. A retrospective analysis has shown that an Angiotensin-converting-enzyme inhibitor (ACEI) or an Angiotensin II receptor blocker (ARB) is associated with preservation of renal function in patients with ADPKD. Salt restriction plays a role in hypertension control (as well as experimentally in lowering vasopressin as a potential growth factor for cysts).

The role of the RAAS system in ADPKD progression is the focus of the ongoing *HALT Progression of Polycystic Kidney Disease* (*HALT PKD*) study. This randomized, double-blind, placebo-controlled study is assessing whether pharmacologic interruption of the RAAS affects disease progression or the rate of decline in renal function in ADPKD.

Although there is currently no therapy that would replace the function of the abnormal protein, there are experimental therapies being considered based on a best science approach. For instance, it is known that cyclic AMP (cAMP) increases the proliferation of epithelial cells in cyst walls and increases the rate of fluid secretion into cysts. Therefore stimuli that result in increased cAMP production should be avoided on a theoretical basis. This includes caffeine and theophylline, as well as beta-adrenergic agonists.

Arginine vasopressin (AVP) is a potent activator of renal adenyl cyclase, another protein that plays a role in epithelial cell proliferation. From a nonpharmacologic standpoint, suppression of arginine vasopressin release by high water intake appears to limit cyst formation. The arginine vasopressin V2 receptor antagonists such as tolvaptan have been shown to inhibit the development of polycystic kidney disease in animal models. Additionally, the results of phase 2 and phase 2–3 clinical trials suggest that tolvaptan is safe and well tolerated in ADPKD. A phase 3, placebo-controlled, double-blind study will determine whether tolvaptan is effective in slowing down the progression of cystic kidney disease. Although data from randomized controlled trials are lacking, on theoretical grounds it may be reasonable to encourage generous water intake to ensure that endogenous AVP is suppressed and while we await the results of the V2 receptor antagonist trials. When a therapy such as this is considered, the risk of hyponatremia must always be considered.

The mammalian target of rapamycin (mTOR) is a protein kinase and a central regulator of cell growth and proliferation. Building on the observations that ADPKD is associated with dysregulated cell proliferation, animal studies have tested sirolimus, an mTOR inhibitor, and have noted reduced cyst formation. Human trials are now being considered.

Autosomal Recessive Polycystic Kidney Disease

Autosomal Recessive Polycystic Kidney Disease (ARPKD) is a severe form of cystic disease that affects primarily the kidneys and biliary tract. ARPKD occurs in approximately one in 20,000 live births and can involve a wide spectrum of phenotypes, depending on the type of mutation and age of presentation. In affected fetuses the principal pathological findings are enlarged, echogenic, cystic kidneys. Oligohydramnios occurs due to insufficient fetal urine production. Nearly half of affected neonates die shortly after birth, as a result of the pulmonary hypoplasia that results when urine production is decreased. Among neonatal survivors, morbidity and mortality results from severe systemic hypertension, renal insufficiency, and portal hypertension due to portal-tract hyperplasia and fibrosis.

Clinical Presentation and Diagnosis of Autosomal Recessive Polycystic Kidney Disease

Clinical manifestations of ARPKD are quite variable but usually include a significant impairment in renal function. In severe prenatal cases, involvement of the kidneys leads to massively enlarged echogenic kidneys, oligohydramnios, and pulmonary hypoplasia. Approximately 30% of these neonates die as a result of pulmonary hypoplasia from the oligohydramnios sequence. Respiratory distress and pneumothorax often worsen the clinical picture. In the severe perinatal form of ARPKD, the kidneys are markedly enlarged because of the cumulative effect of dilatation of all the collecting ducts. Most infants with ARPKD have an elevated serum creatinine, oliguria, and hyponatremia during the first days of life. Infants and children typically develop systemic hypertension, and approximately 60% of cases progress to ESRD by 20 years of age.

Although most infants presenting in the perinatal period ultimately require renal transplantation, the age at transplantation is very variable and occasionally can be delayed until adulthood.

Some children present with liver-predominant symptoms, but all will develop variable degrees of congenital hepatic fibrosis and subsequent portal hypertension. Patients may manifest with hepatic cysts and biliary dilatation often complicated by acute bacterial cholangitis. Portal hypertension and its associated complications tend to become progressively more severe with age but also can occur as the predominant clinical feature initially. The rates of progression of hepatic and renal disease can vary, even among patients carrying the same *PKHD1* mutation, and are independent of each other.

Pathogenesis of Autosomal Recessive Polycystic Kidney Disease

To date, all typical cases of ARPKD are linked to a single locus on chromosome 6p12, *PKHD1*. *PKHD1* is a large gene containing 66 coding exons, which encode several alternatively spliced isoforms. The gene codes for the protein fibrocystin, a hepatocyte growth factor receptor-like protein that functions on the primary cilia of the renal collecting duct and biliary epithelial cells. The most severe form of ARPKD involves two protein-truncating mutations, whereas milder forms of the disease typically have one or more missense mutations. Dysfunction of fibrocystin leads to abnormal ciliary signaling, which is normally required for regulation of proliferation and differentiation of renal and biliary epithelial cells. The exact function of the numerous isoforms has not been defined. The widely varying clinical spectrum of ARPKD may depend, in part, on the nature and number of splice variants that are disrupted by specific *PKHD1* mutations.

Both the polycystins and fibrocystin are essential for maintenance of the differentiated, polarized, predominantly reabsorptive tubular epithelial phenotype, and for the normal maintenance of low rates of proliferation. Functional disruption of one of these proteins produces many cellular biochemical abnormalities with excessive cell proliferation and excessive fluid secretion predominating.

To date, greater than 300 different *PKHD1* mutations are known. Of these, 60% are missense mutations and 40% are predicted to truncate the protein. Approximately one third of *PKHD1* mutations are unique to a single family. Some ancestral mutations are common in particular populations. In fact, the T36M mutation of northern European origin accounts for approximately 17% of mutant alleles. The average reported mutation detection rate is approximately 80%.

Treatment of Autosomal Recessive Polycystic Kidney Disease

Similar to the treatment of ADPKD, blood pressure control with ACEI and ARBs is likely to be quite useful in the treatment of ARPKD. As this renal disease occurs much more frequently in childhood, nutritional issues are likely to require more precise management. Finally, as with ADPKD, there are no proven therapies that will reduce cyst size or stop cyst formation and in general therapies should be considered experimental at this time. Studies in the *PKHD1*-deficient rat demonstrate that V2R antagonists retard renal cystic disease progression, making V2 receptor antagonists an interesting therapeutic option. In addition, small studies have shown that pharmacologic targeting of the EGF receptor pathway rescues the renal and biliary lesions in animal models of ARPKD. These observations suggest that pharmacologic interruption of cAMP-related pathways and EGF receptor-related axis may provide promising therapeutic strategies for patients with ARPKD.

A significant issue for the pharmacological management of ARPKD is the age of the affected patient and the lack of either controlled or adequate kinetic data to support the use of the experimental drugs being tried in ADPKD in this patient population. Nonetheless, each of the drugs with theoretical promise in ADPKD may be considered in this patient population in the future.

DISORDERS OF RENAL TUBULAR FUNCTION

There are many disorders that affect the individual tubular segments of the nephron. The following two examples, although not common, are classic in their presentation and involve the handling of sodium chloride by the kidney—one of the most important functions of the kidney. Bartter's and Gitelman's Syndromes are a group of autosomal recessive disorders characterized by an abnormality in one of the transporters involved in sodium chloride reabsorption in the nephron. Since the main determinant of intravascular volume (and water reabsorption) is salt reabsorption, the resultant renal salt wasting leads to hypotension and a host of other signs and symptoms for affected patients. There is considerable overlap in the presentations of these similar diseases.

Bartter's Syndrome

The estimated prevalence of Bartter's Syndrome is one per million people. Classic Bartter's Syndrome involves an abnormality that limits the ability of the nephron to reabsorb sodium chloride in the Loop of Henle—an important segment of the nephron. The Loop of Henle is the point of the nephron where pharmacological salt wasting is intentionally encouraged by the use of the so-called loop diuretics (i.e., furosemide) in order to cause a diuresis in individuals who have edema and require the loss of water. The key associated biochemical findings in Bartter's Syndrome include hypokalemia and metabolic alkalosis. Additionally, salt wasting and hypotension can trigger hyperreninemia and hyperaldosteronemia in this patient population.

Clinical Presentation and Diagnosis of Bartter's Syndrome

Classic Bartter's Syndrome presents in early life. The initial presentation can be in the perinatal period with polyhydramnios (the *in utero* equivalent of polyuria) and preterm delivery or in the first few years of life with polyuria, polydipsia, failure to thrive, and frequent episodes of dehydration. The diagnosis of Bartter's Syndrome is by exclusion. Surreptitious vomiting and diuretic use are the two other major causes of unexplained hypokalemia and metabolic alkalosis in a normotensive or hypotensive patient. Vomiting can be ruled out by detecting a low urine chloride. The use of diuretics is either obvious from the history or can be detected chemically by a urine assay for diuretics.

Distinguishing Bartter's Syndrome patients from Gitelman's Syndrome patients is difficult given the marked variability of clinical phenotypes and the overlapping ages at presentation. The genetic diagnosis of this disorder is possible. However, genetic diagnosis is not widely used clinically, primarily because of its availability in only a few research labs, the heterogeneity of the mutations encountered, and the high cost of mutational testing.

Pathogenesis of Bartter's Syndrome

Bartter's Syndrome involves defects in key transport proteins in the thick ascending limb of the Loop of Henle. On the apical membrane, the Na-2Cl-K co-transporter (NKCC2), encoded by *SLC12A1*, transports sodium and potassium down an electrochemical gradient. For enhancement of the efficiency of this transporter, potassium is recirculated across the apical membrane through the renal outer medullary potassium (ROMK) channel. The apical recycling of potassium establishes a transepithelial voltage gradient

that drives the paracellular reabsorption of calcium and magnesium. On the basolateral membrane, sodium is actively transported from the cell via the Na-K-ATPase, whereas chloride exits through at least two chloride channels, predominantly ClC-Kb (encoded by *CLCNKB*), and to a lesser degree, ClC-Ka (encoded by *CLCNKA*). The subunit protein, barttin (encoded by *BSDN*), regulates the function of these chloride channels. Mutations in *SLC12A1*, *ROMK1*, *CLCNKB*, and *BSDN* cause Bartter syndrome types I, II, III, and IV, respectively (Figure 24.3a). Within the various causes of Bartter's Syndrome, there appears to be no direct correlation between the clinical phenotype and the underlying genotypic abnormality.

Treatment of Bartter's Syndrome

The mainstay of treatment for Bartter's Syndrome is replacement of fluid, sodium chloride, and potassium chloride. Chronic hypokalemia can be addressed further with either amiloride or aldosterone antagonists. Circulating prostaglandins, which often are elevated in Bartter syndromes I and II, and which are integral to maintaining the electrolyte abnormalities in this syndrome, can exacerbate the polyuria and cause fever as well as gastrointestinal side effects. The cyclooxygenase inhibitors indomethacin, ibuprofen, and COX-2 inhibitors may be effective therapeutic additions in this setting. Some patients, particularly those with Bartter III, have hypomagnesemia that necessitates oral magnesium replacement therapy.

Gitelman's Syndrome

The estimated prevalence of Gitelman's Syndrome is one per 400,000 individuals. Gitelman's Syndrome is also an autosomal recessive disorder of renal tubular function that limits the ability of the nephron to

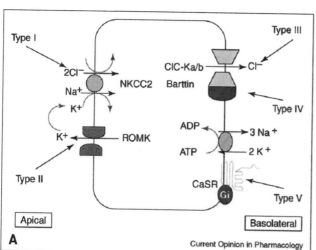

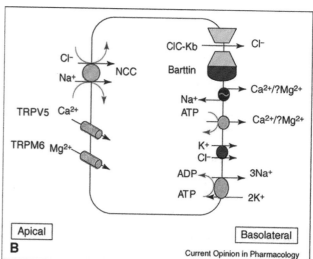

Figure 24.3   Cellular defects in Gitelman's Syndrome and Bartter's Syndrome. (A) A composite DCT cell showing the transport proteins relevant to Gitelman's syndrome. **(B)** A TAL cell showing the transport proteins relevant to the five types of Bartter's syndrome currently defined (see text for details). Reproduced with permission from *Curr Opin Pharmacol.* 2006;6:208–213.

reabsorb sodium chloride. However, in this particular disorder it is an abnormality of the distal tubule. Gitelman's Syndrome tends to be a more benign condition compared to Bartter's Syndrome, and often is not diagnosed until late in childhood or even into adulthood.

Clinical Presentation and Diagnosis of Gitelman's Syndrome

Gitelman's Syndrome is diagnosed in childhood or adulthood often because of weakness, tetany, or joint pain. It is not uncommon for the diagnosis in asymptomatic patients to be prompted by an incidental finding of hypokalemia on routine laboratory studies. Patients with Gitelman's Syndrome have milder renal salt wasting and often have normal or only slightly low blood pressures. Hypokalemia and metabolic alkalosis are important to the diagnosis and as mentioned earlier, hypomagnesemia and hypocalciuria are relatively common.

Similar to Bartter's Syndrome, diagnosis of Gitelman's Syndrome is by exclusion. Surreptitious vomiting and diuretic use are again in the differential and they can be ruled out by history, urine chloride, and a urine assay for diuretics. As mentioned previously, distinguishing Bartter's Syndrome patients from Gitelman's Syndrome patients is difficult and there may be little clinical utility in doing so. When this cannot be done, the designation of salt wasting nephropathy may be preferred. The genetic diagnosis of this disorder is possible. However, genetic diagnosis is not used clinically because of its limited availability, the heterogeneity of the mutations, and the high cost of mutational testing.

Pathogenesis of Gitelman's Syndrome

The vast majority of patients with Gitelman's Syndrome have heterogeneous defects in *SLC12A3*, the gene that encodes the sodium chloride cotransporter NCCT located in the distal tubule (Figure 24.3b). The distal tubule is the point of the nephron where pharmacological salt wasting is intentionally encouraged by the use of the thiazide diuretics (i.e., hydrochlorothiazide) in order to cause a diuresis in individuals who have edema and require the loss of water. There are a small number of patients with the Gitelman phenotype who have been shown to have mutations in *CLCNKB*, a chloride channel that is also thought to play a role in Bartter's Syndrome. Hypocalciuria and hypomagnesuria are classical features of Gitelman Syndrome, but the pathogenic mechanisms for this finding are not known.

Treatment of Gitelman's Syndrome

As with Bartter's Syndrome, the mainstay of treatment is the replacement of fluids, sodium chloride, and potassium chloride. These patients are also more likely to require magnesium replacement. Chronic hypokalemia can be addressed further with either amiloride or aldosterone antagonists. Circulating prostaglandins are not elevated in Gitelman's Syndrome and therefore inhibitors are of little benefit in Gitelman's Syndrome.

KEY CONCEPTS

- The pathogenesis of focal segmental glomerulosclerosis involves effacement of podocyte foot processes. This can be due to derangements in structural components and intracellular signaling of the slit diaphragm, podocyte cell membrane, or actin cytoskeleton. Genes important to both familial and sporadic FSGS include nephrin (*NPHS1*), podocin (*NPHS2*) *ACTN4*, and *TRPC6*. Treatment response to FSGS is not favorable, and a substantial proportion of patients progress to ESKD.
- Fabry Disease is an X-linked lysosomal storage disease due to mutation in the gene for the α-galactosidase enzyme that is responsible for catabolism of glycosphingolipids, important components of cell membranes. The classical form of Fabry Disease begins in early childhood with acroparesthesias and angiokeratomas, and progresses to early-onset kidney and cardiovascular disease in adulthood. X-inactivation and mosaicism in women and severity of genetic mutation, modifying genes, and epigenetics in all patients account for the wide variability in disease manifestations.
- Renal cysts of tubular epithelial cell origin are a hallmark of autosomal dominant polycystic kidney disease (ADPKD) and autosomal recessive polycystic kidney disease (ARPKD). Causative genetic loci are *PKD1* and *PKD2*, which code for the polycystins, and *PKHD1*, which codes for fibrocystin in ADPKD and ARPKD, respectively. Due to the size and complexity of these genes, diagnosis is still made predominantly using radiologic findings and family history.
- Bartter's and Gitelman's Syndromes are salt wasting diseases of the kidney that lead to acid-base and electrolyte disturbances. They can be caused by mutation in a number of genes that encode proteins of the thick ascending limb in Bartter's Syndrome or in the distal tubule in Gitelman's Syndrome.

SUGGESTED READINGS

1. Banikazemi M, Bultas J, Waldek S, et al. For the Fabry Disease Clinical Trial Study Group. Agalsidase beta therapy for advanced Fabry disease: A randomized trial. *Annals of Int Med.* 2007; 146 (2):77–86.
2. Bergmann C, Senderek J, Sedlacek B, et al. Spectrum of mutations in the gene for autosomal recessive polycystic kidney disease (ARPKD/PKHD1). *J Am Soc Nephrol.* 2003;14:76–89.
3. D'Agati VD. The spectrum of focal segmental glomerulosclerosis: New insights. *Curr Opin Nephrol Hypertens.* 2008;17:271–281.
4. Eng CM, Germain DP, Banikazemi M, et al. Fabry disease: Guidelines for the evaluation and management of multi-organ system involvement. *Genet Medicine.* 2006;8(9):539–548.
5. Igarashi P, Somlo S. Genetics and pathogenesis of polycystic kidney disease. *J Am Soc Nephrol.* 2002;13:2384–2398.
6. Kriz W. TRPC6—A new podocyte gene involved in focal segmental glomerulosclerosis. *Trends Mol Med.* 2005;11(12):527–530.

7. Kurtz I. Molecular pathogenesis of Bartter's and Gitelman's ?syndromes. *Kidney Int.* 1998;54(4):1396–1410.

8. Naesens M, Steels P, Verberckmoes R, et al. Bartter's and Gitelman's syndromes. From gene to clinic. *Nephron Physiology.* 2004;96:65–78.

9. Pastores GM, Lien YH. Biochemical and molecular genetic basis of Fabry disease. *J Am Soc Nephrol.* 2002;13:S130–S133.

10. Ravine D, Sheffield L, Danks DM, et al. Evaluation of ultrasonographic diagnostic criteria for autosomal dominant polycystic kidney disease 1. *Lancet.* 1994;343:824–827.

11. Reidy K, Kaskel FJ. Pathophysiology of focal segmental glomerulosclerosis. *Pediatr Nephrol.* 2007;22:350–354.

12. Rossetti S, Kubly VJ, Consugar MB, et al. Incompletely penetrant PKD1 alleles suggest a role for gene dosage in cyst initiation in polycystic kidney disease. *Kidney Int.* 2009;75(8):848–855.

13. Ruf RF, Lichtenberger A, Karle SM, et al. Patients with mutations in NPHS2 (podocin) do not respond to standard steroid treatment of nephrotic syndrome. *J Am Soc Nephrol.* 2004;15:722–732.

14. Schaefer E, Mehta A, Gal A. Genotype and phenotype in Fabry disease. Analysis of the Fabry Outcome Survey. *Acta Paediatrica.* 2005;94(suppl 447):87–92.

15. Schrier RW, McFann KK, Johnson AM. Epidemiological study of kidney survival in autosomal dominant polycystic kidney disease. *Kidney Int.* 2003;63:678–685.

16. Seyberth HW. An improved terminology and classification of Bartter's-like syndromes. *Nat Clin Pract Nephrol.* 2008;4(10):560–567.

17. Tryggvason K, Pattraka J, Wartiovaara J. Hereditary proteinuria syndromes. *N Eng J Med.* 2006;354:1387–1401.

18. Unwin RJ, Capasso G. Bartter's and Gitelman's syndromes: Their relationship to the actions of loop and thiazide diuretics. *Curr Opin Pharmacol.* 2006;6:208–213.

19. Zarate YA, Hopkin RJ. Lysosomal storage disease 3: Fabry's disease. *Lancet.* 2008;372:1427–1435.

25

Molecular Pathogenesis of Prostate Cancer: Somatic, Epigenetic, and Genetic Alterations

Carlise R. Bethel . Angelo M. De Marzo . William G. Nelson

INTRODUCTION

Prostate cancer is one of the most common malignancies in Western men, and risk factors include age, race, inherited genes, and environmental factors such as diet. Although the precise molecular mechanisms underlying carcinogenesis and progression are currently unknown, histological studies indicate a multistage developmental progression. The existence of premalignant lesions has been demonstrated in the prostate based on shared histological and molecular features with adenocarcinoma, as well as prevalence and severity. Ample evidence supports the hypothesis that prostate cancer can progress from intraepithelial neoplasia to invasive carcinoma, and ultimately metastasis and androgen-independent lethal disease. Prostatic intraepithelial neoplasia (PIN) is a term currently used to describe the closest precursor lesion to carcinoma. Histologically, the transition from normal epithelium to PIN involves nuclear atypia, epithelial cell crowding, and some component of basal cell loss. Autopsy studies suggest that PIN lesions precede the appearance of carcinoma. High-grade PIN lesions are often spatially associated with carcinoma and may exhibit molecular alterations similar to those found in prostate tumor cells.

Glandular atrophy is frequently observed in aging male prostates. Histologically, luminal spaces appear dilated with flattened epithelial linings. Atrophic lesions in the prostate consist of a number of histological variants, and some of these have also been proposed to be potential precursors to prostate cancer based in part on their frequent occurrence in proximity to carcinoma. Proliferative inflammatory atrophy (PIA), a term used to describe a range of these morphologies, includes simple atrophy and post-atrophic hyperplasia. PIA lesions are highly proliferative and often associated with inflammation. Epithelial cells within PIA foci express both luminal and basal markers. The lesions likely result from cellular injury initiated by inflammation and/or carcinogen insult, and at times show evidence of transition to high-grade PIN and, rarely, to carcinoma. Given the intermediate phenotype of these cells and the characteristic expression of stress-associated genes, PIA lesions are considered to be the morphological manifestation of prostate epithelial damage and regeneration. Some of the genetic and epigenetic alterations observed in PIN and adenocarcinoma have also been found in PIA lesions, albeit to a lesser degree. These findings suggest a potential link between a subset of atrophic lesions and adenocarcinoma.

We have proposed a multistep progression model that leads from normal epithelium to focal prostatic atrophy to PIN and then to carcinoma. In this model, ongoing injury to the prostate epithelium, as a result of inflammation and/or carcinogen exposures, results in cell damage and cell death. Cell regeneration ensues, and this is manifest morphologically as prostatic atrophy. These atrophic cells may then undergo somatic genome alterations during self-renewal, including methylation of the *GSTP1* promoter, telomere shortening, activation of MYC, leading to neoplastic transformation.

Prostate cancer arises from the accumulation of genetic and epigenetic alterations. Cytogenetic analyses have demonstrated the prevalence of chromosomal aberrations associated with prostate tumorigenesis, and many genes that map to deleted or amplified regions have been investigated for their roles in disease progression. In recent years, genome-wide profiling of prostate tumors has led to the identification of a number of biomarkers and pathways that are altered in prostate cancer, some of which may be potential molecular targets for therapy. Despite the advancing knowledge of genes altered in prostate adenocarcinoma, the precise molecular pathways, their combinatorial relation, and the ordering of events utilized in the development of preneoplastic lesions and adenocarcinoma are still being refined. To elucidate mechanisms and to begin to test new therapies and preventative strategies for human prostate cancer, a number of groups have been developing novel animal models of prostate cancer.

These models will help to define many aspects of the molecular and cellular pathogenesis of prostate cancer, such as the role of inflammation, angiogenesis, and stromal-epithelial interactions. In light of previous reviews that extensively address our current knowledge of molecular alterations in prostate tumor progression, the primary objectives of this chapter will be to highlight recent advances in our understanding of the molecular pathology of prostatic adenocarcinoma and the latest developments in mouse prostate cancer models.

HEREDITARY COMPONENT OF PROSTATE CANCER RISK

Prostate cancer is known to have a hereditary component, and a large effort has been devoted to uncovering familial prostate cancer genes. A limited set of germline polymorphisms and mutations have been associated with increased prostate cancer risk. Based on linkage analysis, the inherited susceptibility locus Hereditary prostate cancer 1 (*HPC1*), which encodes Ribonuclease L (RNASEL), was mapped to chromosomal region 1. Germline mutations in the Macrophage Scavenger Receptor 1 (*MSR1*) locus have also been linked to prostate cancer risk. The *MSR1* gene is located at chromosomal region 8p22 and is expressed in infiltrating macrophages. Many other loci have been implicated as well, and further characterization of these genes in the etiology of prostate cancer is warranted. Most recently several groups have used Genome Wide Association Studies (GWAS) to show a number of SNPs at novel loci related to prostate cancer risk, several of which map to chromosome 8q24.

SOMATIC ALTERATIONS IN GENE EXPRESSION

Acquired somatic gene alterations in prostate cancer cells include cytosine methylation alterations within CpG dinucleotides, point mutations, deletions, amplifications, and telomere shortening. Inactivation of classical tumor suppressors, for example, *TP53* and

Retinoblastoma (*RB*), has been found in prostate tumors and cancer cell lines. However, these alterations are far more common in advanced hormone-refractory and/or metastatic cancers. As with other epithelial cancers, cytogenetic studies using fluorescence *in situ* hybridization (FISH) and comparative genomic hybridization (CGH) have identified chromosomal regions frequently gained and lost in prostate cancer. The most common chromosomal abnormalities are losses at 8p, 10q, 13q, and 16q, and gains at 7p, 7q, 8q, and Xq.

MYC

MYC protein functions as a nuclear transcription factor that impacts a wide range of cellular processes including cell cycle progression, metabolism, ribosome biogenesis, and protein synthesis. The *MYC* oncogene maps to chromosome 8q24, and this region is amplified in a number of human tumors. The complexity of MYC-regulated transcriptional networks has been under intense study since the late 1990s, yet the precise role of the *MYC* oncogene during neoplastic transformation and its direct molecular targets in prostate tumorigenesis remain largely unknown. Initial reports of increased *MYC* gene expression in PIN lesions and ~30% of tumors, combined with the correlation of 8q24 amplification with high Gleason grade and metastatic carcinoma, suggested that alterations in MYC were associated with advanced disease. This was in contrast to increased levels of *MYC* mRNA detected in most prostate cancers, including low-grade cases. However, recent evidence from our group suggests that MYC upregulation at the protein level is an early and common event in primary prostate cancer cases. Using an improved antibody for immunohistochemical analyses on tissue microarrays, nuclear MYC staining was increased in the luminal epithelial cells of PIN, PIA, and carcinoma lesions compared to benign tissues. Interestingly, FISH analysis revealed a positive correlation between gain of 8q24 and Gleason grade, but not overall MYC protein levels. These findings, in conjunction with the fact that human MYC overexpression is able to initiate prostate cancer in transgenic mice, suggest a key role for MYC upregulation in the initiation of prostate tumorigenesis that is likely independent of gene amplification. Indeed, further investigations of the mechanisms that underlie MYC overexpression in prostate cancer progression are warranted.

NKX3.1

NKX3.1 is one of several candidate tumor suppressor genes located on chromosome 8p. The *NKX3.1* gene encodes a homeodomain transcription factor that is the earliest known marker of prostate epithelium during embryogenesis. Its expression persists in the epithelial cells of the adult gland and is required for maintenance of ductal morphology and the regulation of cell proliferation. Human *NKX3.1* maps within the minimal deletion interval of chromosomal region 8p21. Loss of heterozygosity at 8p21 has been observed in

63% of high-grade PIN foci and up to ~70% of prostate tumors, although this percentage may be significantly less according to recent findings. Methylation of CpG dinucleotides upstream of the *NKX3.1* transcriptional start site and the existence of germline variants within the homeodomain have been reported. However, the order of occurrence relative to 8p21 loss and prostate cancer initiation is unknown. In loss of function analyses, *NKX3.1* homozygous mutant mice do not develop invasive carcinoma. However, epithelial hyperplasia and PIN lesions are observed with age in these mice. These phenotypes are also observed in heterozygous mice, albeit to a lesser extent. When taken together, human and mouse studies support the notion that haploinsufficiency of *NKX3.1* plays a role in prostate cancer development.

Several studies have analyzed NKX3.1 expression by immunohistochemistry in human prostate cancer specimens. In studies of PIN and carcinoma, both decreased intensity and loss of NKX3.1 protein staining compared to benign tissue have been reported. Decreased expression correlated with high Gleason score, advanced tumor stage, the presence of metastatic disease, and hormone-refractory disease. Recently, our immunohistochemical analyses with a novel NKX3.1 antibody revealed a dramatic decrease in the level of NKX3.1 in PIA lesions. As NKX3.1 is thought to regulate cell proliferation, loss of expression in atrophic epithelial cells may contribute to increased proliferative capacity. This may serve to amplify any genetic changes that may occur in epithelial cells within these regenerative lesions. Consistent with previous reports, NKX3.1 staining intensity was also significantly diminished in PIN and carcinoma lesions. Diminished NKX3.1 protein expression correlated with 8p loss in high-grade tumors, but not PIN or PIA. Intriguingly, the levels of *NKX3.1* mRNA and protein were found to be discordant in 7 of 11 carcinoma cases, confirming previous reports and suggesting that multiple mechanisms underlie sporadic loss of NKX3.1 for tumor progression.

PTEN

The Phosphatase and tensin homologue (*PTEN*) gene is a well-characterized tumor suppressor that maps to chromosomal region 10q23 and acts as a negative regulator of the phosphatidylinositol 3-kinase/ AKT (PI3K/AKT) signaling, used for cell survival. PTEN inhibits growth factor signals sent through PI3 kinase by dephosphorylating the PI3K product, phosphatidylinositol 3,4,5-trisphosphate (PIP3). Loss of PTEN expression results in the downstream activation (phosphorylation) of AKT, an inhibitor of apoptosis and promoter of cell proliferation. Aberrant PTEN expression has been implicated in numerous cancers, including metastatic prostate cancers, which at times exhibit homozygous deletion of *PTEN*. However, the majority of primary prostate cancer cases with genetic alterations in *PTEN* harbor loss of heterozygosity at the *PTEN* locus without mutations in the remaining allele. That PTEN is a haploinsufficient tumor suppressor in the prostate

is evident by prostate-specific disruption of *PTEN* in mice, which results in PIN lesions, followed by invasive carcinoma and/or metastasis with age. Cooperativity exists between *NKX3.1* and *PTEN* in prostate tumorigenesis, as compound *NKX3.1* and *PTEN* heterozygous mice rapidly develop invasive carcinoma and androgen-independent disease.

Androgen Receptor

The role of AR is central to prostate pathobiology, as the prostate is dependent on androgens for normal growth and maintenance. AR is a nuclear steroid hormone receptor with high expression in the luminal epithelial cells and little expression in basal epithelial cells. In the absence of ligand, AR is inactive and bound to heat shock chaperone proteins. Upon binding of the active form of testosterone (dihydrotestosterone, DHT), AR is released from heat shock proteins and translocates to the nucleus where it physically associates with cofactors to regulate target gene transcription. AR activity is essential for the development of prostate cancer, and AR expression is evident in high-grade PIN and most adenocarcinoma lesions. Androgen ablation, through the use of antiandrogens, castration, or gonadotropin superagonists, is the mainline therapy for advanced metastatic carcinoma. While the majority of patients respond to this treatment, it eventually fails and the tumors become androgen independent. As tumors progress from androgen dependence to a hormone refractory state, AR can be amplified at the Xq12 region or mutated to respond to a range of ligands for androgen-independent activation and tumor growth. In recent years, AR has been reported to engage in crosstalk with other mitogenic signaling pathways such as PI3K/AKT and MAPK as a means of enhancing androgen-independent tumor progression.

TMPRSS2-ETS Gene Fusions

Ets-related gene-1 (ERG) is an oncogenic transcription factor upregulated in the majority of primary prostate cancer cases. Evidence of recurrent chromosomal rearrangements in prostate cancer was first reported by Tomlins et al. Through use of a bioinformatic approach to analyze DNA microarray studies, gene fusions between the promoter/enhancer region of the androgen-responsive *TMPRSS2* gene and members of the *ETS* family, including *ERG* and *ETV1*, were found in ~90% of prostate tumors with known overexpression of *ERG*. This discovery has fueled efforts to characterize the functional implications of aberrant chromosomal fusions on tumor progression and clinical outcome. Presently, there are conflicting data regarding the presence of *TMPRSS-Erg* gene fusions and clinical outcome. While initial studies suggested these fusions were not present in PIN lesions, it has become clear that some PIN lesions indeed harbor fusions, and this suggests that gene fusion may lead to neoplastic transformation itself and not specifically to the invasive phenotype. Additionally, overexpression of *ERG* in the transgenic mouse prostates results in basal cell loss and PIN lesions, although this latter finding is controversial. These studies demonstrate that upregulation of *ERG* may contribute to transformation of prostate epithelial cells but is insufficient to initiate prostate cancer.

p27

The cyclin-dependent kinase inhibitor p27^{kip1} is a candidate tumor suppressor encoded by the *CDKN1B* gene. To prevent cell cycle progression, p27^{kip1} binds to and inhibits cyclin E/CDK2 and cyclin A/CDK2 complexes. Although mutations are rare, loss of p27^{kip1} expression results in hyperplasia and malignancy in many organs, including the prostate. In normal and benign prostate tissues, p27^{kip1} is expressed highly in most luminal epithelial cells, and much more variably in basal cells. However, p27^{kip1} expression is decreased in most high-grade PIN and carcinoma lesions. Several studies have shown decreased p27^{kip1} to be positively correlated with increased proliferation, PSA relapse, high tumor grade, and advanced stage. The molecular mechanism by which p27 protein is decreased in prostate cancer has not been clarified, although post-transcriptional regulation of p27^{kip1} is evident in carcinoma cases which express high levels of *p27^{kip1}* mRNA but lack protein expression.

Additional support for p27^{kip1} as a tumor suppressor comes from loss-of-function analyses in mice. Prostates from p27^{kip1}-deficient animals exhibit hyperplasia and increased size. Given the cooperativity observed between *NKX3.1* and *PTEN* in tumorigenesis, the relationship between *NKX3.1* and *p27* loss has also been investigated. Both genes have been shown to exhibit haploinsufficiency for tumor suppression. Anterior prostates of 36-week-old double mutant mice displayed epithelial hyperplasia and dysplasia that were more severe than those observed in the single *p27* mutant mice. However, combined loss of p27^{kip1}, *PTEN*, and *NKX3.1* in bigenic and trigenic mouse models results in prostate hyperplasia followed by tumor development. These findings suggest that prostate tumor progression is sensitive to *p27* dosage and support the hypothesis that downregulation of p27^{kip1} is an early event in prostatic neoplasia. Expression of these three genes is known to decrease in clinical samples, providing further evidence of their cooperation in tumor progression.

Telomeres

Telomeres are specialized structures composed of repeat DNA sequences at the ends of chromosomes that are complexed with binding proteins and required for maintenance of chromosomal integrity. In cells lacking sufficient levels of the enzyme telomerase, telomeres progressively shorten with cell division as a result of the end replication problem and/or oxidant stress. The enzyme telomerase can add new repeat sequences to the ends of chromosomes, which stabilizes the telomeres and ensures proper telomere length. Excessive shortening can lead to improper segregation of chromosomes during cell division, genomic instability, and the initiation of tumorigenesis. Mice double mutant for telomerase and *p53* display increased epithelial cancer incidence, suggesting a role for short telomeres in cancer initiation. The majority of high-grade PIN and prostate cancer cases have abnormally short telomeres exclusively in the luminal cells. This supports the notion that cells in the luminal compartment may be the target of neoplastic transformation. Oxidative damage can result in telomere shortening, and this may go along with the proposed inflammation-oxidative stress model of prostate cancer progression.

MicroRNAs

MicroRNAs (miRNAs) are small noncoding RNA molecules that negatively regulate gene expression by interfering with translation. They are initially generated from primary transcripts (pri-mRNAs) and processed by the RNase III endonucleases Drosha and Dicer to produce the mature miRNA molecule. In the cytoplasm, the mature miRNA associates with the RNA-induced silencing complex (RISC) and binds to the 3' UTR of its target mRNA, leading to degradation or transcriptional silencing. Since their discovery over a decade ago, miRNAs have been shown to play key roles in development, and there is increasing evidence of their widespread dysregulation in cancer. Recently, a limited number of studies have reported a predominant decrease in the levels of miRNAs in prostate carcinoma, including let-7, miR-26a, miR-99, and miR-125-a-b. Although a prostate cancer-specific miRNA signature has not emerged from these small-scale studies, greater knowledge of miRNA expression patterns and target genes may reveal the role of miRNAs in the etiology of prostate cancer. MiRNAs that are consistently dysregulated in prostate tumorigenesis and have a strong relationship with disease progression may potentially serve as novel biomarkers for prognosis.

EPIGENETICS

Epigenetic alterations in prostate carcinogenesis include changes in chromosome structure through abnormal deoxycytidine methylation of wild-type DNA sequences and histone modifications (acetylation, methylation). Epigenetic events occur earlier in prostate tumor progression and more consistently than recurring genetic changes. However, the mechanisms by which these changes arise are poorly understood. Silencing of genes through aberrant methylation may occur as a result of altered DNA methyltransferase (DNMT) activity. DNMTs establish and maintain the patterns of methylation in the genome by catalyzing the transfer of methyl groups to deoxycytidine in CpG dinucleotides. Several genes that are silenced by epigenetic alterations have been identified.

GSTP1

Glutathione S-transferases are enzymes responsible for the detoxification of reactive chemical species through conjugation to reduced glutathione. The *GSTP1* gene, encoding the pi class glutathione S-transferase, was the first hypermethylated gene to be characterized in prostate cancer. GSTP1 is thought to protect prostate epithelial cells from carcinogen-associated and/or oxidative stress-induced DNA damage, and loss of GSTP1 may render prostate cells unprotected from such genomic insults. As an example, in LNCaP prostate cancer cells, devoid of GSTP1 as a result of epigenetic gene silencing, restoration of GSTP1 function affords protection against metabolic activation of the dietary heterocyclic amine carcinogen 2-amino-1-methyl-6-phenylimidazo [4,5–*β*] pyridine (PhIP), known to cause prostate cancer when ingested by rats. The loss of enzymatic defenses against reactive chemical species encountered as part of dietary exposures or arising endogenously associated with epigenetic silencing of GSTP1 provides a plausible mechanistic explanation for the marked impact of the diet and of inflammatory processes in the pathogenesis of human prostate cancer.

The most common genomic DNA mark accompanying epigenetic gene silencing in cancer cells is an accumulation of 5-meC bases in CpG dinucleotides clustered into CpG islands encompassing transcriptional regulatory regions. Hypermethylation of these CpG island sequences directs the formation of repressive heterochromatin that prevents loading of RNA polymerase and transcription of hnRNA that can be processed for translation into protein. For *GSTP1*, the 5-meCpG dinucleotides begin to appear in the gene promoter region of rare cells in PIA lesions, with more dense CpG island methylation changes emerging as PIA lesions progress to PIN and carcinoma. The proliferative expansion of cells with hypermethylated *GSTP1* CpG island sequences as PIA lesions progress hints at some sort of selective growth advantage, though how loss of GSTP1 can be selected during prostatic carcinogenesis has not been established. In addition to *GSTP1*, many other critical genes undergo epigenetic silencing during the pathogenesis of prostate cancer. The mechanisms by which epigenetic defects arise in prostate cancer cells, or in other human cancer cells, have not been discerned. However, the consistent appearance of such changes in inflammatory precancerous lesions and conditions, including PIA lesions, inflammatory bowel disease, chronic active hepatitis, and others, supports the contribution of inflammatory processes to some sort of epigenetic or DNA methylation catastrophe.

APC

The Adenomatous polyposis coli (APC) protein is a component of the Wnt/β-catenin signaling pathway that negatively regulates cell growth. APC is typically found in a complex with Glycogen synthase kinase-3 (GSK3) and Axin. This complex is responsible for targeting free cytosolic β-catenin for ubiquitin-mediated degradation. The pathway is activated by the binding of the Wnt protein to the Frizzled family of seven trans-membrane receptors and LRP5/6, followed by down-regulation of GSK3β, which allows accumulation of β-catenin and subsequent translocation of β-catenin to the nucleus. Once inside the nucleus, β-catenin is able to activate transcription of Wnt target genes.

Indirect support of the concept that APC inactivation may be mechanistically tied to prostate cancer progression comes from studies showing frequent hypermethylation of its promoter region and that the extent of methylation correlates with stage, grade, and biochemical recurrence. Activating mutations occur in approximately 5% of prostate cancers, and aberrant nuclear localization of β-catenin appears to occur only somewhat more frequently. The latter finding suggests that if APC is functioning as a tumor suppressor in prostate adenocarcinoma, then its primary role in the prostate may not relate to nuclear translocation of β-catenin. More direct support for the notion that APC may be a tumor suppressor in prostate cancer comes from a mouse model in which prostate-specific deletion of *APC* in adult mice resulted in carcinoma induction. Whether APC is involved in prostate cancer formation or progression or not, the methylation of its promoter may become a useful biomarker in prostate cancer diagnosis since this methylation may be detectable in bodily fluids such as urine or blood.

CONCLUSION

The pathogenesis of human prostate cancer features histological progression from normal prostate glandular epithelium to PIA, emerging as a consequence of procarcinogenic epithelial damage, to PIN, and then to prostatic adenocarcinoma. This histological evolution is accompanied by the accumulation of somatic genetic and epigenetic alterations that lead to dysregulated over- or underexpression of key genes, such as *MYC*, *NKX3.1*, *PTEN*, *p27*, *AR*, *GSTP1*, and *APC*, and to expression of abnormal fusion genes, formed from androgen-regulated genes such as *TMPRSS2* and *ETS* family oncogenes. The dysregulated genes contribute to the prostate cancer cell phenotype, creating a reliance on AR signaling for cell growth and survival, and a vulnerability to reactive oxygen and nitrogen species and dietary carcinogens. Various assays of gene and protein targets of somatic genetic and epigenetic changes are under scrutiny as molecular biomarkers of the disease, perhaps useful for prostate cancer screening, detection, diagnosis, and treatment monitoring.

ACKNOWLEDGMENTS

NIH/NCI Specialized Program in Research Excellence (SPORE) in Prostate Cancer #P50CA58236 (Johns Hopkins), NIH/NCI #R01 CA070196, and the Patrick C. Walsh Foundation.

KEY CONCEPTS

- Prostatic adenocarcinomas arise from precursor lesions termed proliferative inflammatory atrophy (PIA) and prostatic intraepithelial neoplasia (PIN).
- PIA lesions, which are characterized by epithelial damage, regeneration, and inflammatory cell infiltration, appear in response to a variety of procarcinogenic stresses.
- Heredity contributes significantly to the risk of prostate cancer development. Inherited prostate cancer susceptibility genes/loci discovered to date include *RNASEL*, encoding an enzyme that acts as a defense against viral infection, *MSR1*, encoding a receptor for bacteria on macrophages, and several sites at 8q24.
- Consistent alterations in the expression of key genes, with overexpression of *MYC* and *AR*, and underexpression of *NKX3.1*, *PTEN*, *p27*, *GSTP1*, and *APC*, accompany the pathogenesis of prostate cancer. Mouse modeling studies have implicated each of these genes in the development or progression of prostatic neoplasia.
- Shortening of telomere sequences ubiquitously appears early during prostate cancer development.
- Somatic epigenetic alterations, with hypermethylation and transcriptional silencing of several key genes, are the most abundant and earliest genome abnormalities, evident in PIA and PIN as well as in prostate cancer.
- Acquired genetic defects in prostate cancer include translocations and deletions that give rise to fusion transcripts between androgen-regulated genes, such as *TMPRSS2*, and genes encoding ETS family transcription factors.

SUGGESTED READINGS

1. Abate-Shen C, Banach-Petrosky WA, Sun X, et al. NKX3.1; PTEN mutant mice develop invasive prostate adenocarcinoma and lymph node metastases. *Cancer Res.* 2003;63:3886–3890.
2. De Marzo AM, Marchi VL, Epstein JI, et al. Proliferative inflammatory atrophy of the prostate: Implications for prostatic carcinogenesis. *Am J Pathol.* 1999;155:1985–1992.
3. De Marzo AM, Nelson WG, Isaacs WB, et al. Pathological and molecular aspects of prostate cancer. *Lancet.* 2003;361:955–964.
4. De Marzo AM, Platz EA, Sutcliffe S, et al. Inflammation in prostate carcinogenesis. *Nat Rev Cancer.* 2007;7:256–269.
5. Di Cristofano A, Pandolfi PP. The multiple roles of PTEN in tumor suppression. *Cell.* 2000;100:387–390.
6. Dong JT. Chromosomal deletions and tumor suppressor genes in prostate cancer. *Cancer Metastasis Rev.* 2001;20:173–193.
7. Kaarbo M, Klokk TI, Saatcioglu F. Androgen signaling and its interactions with other signaling pathways in prostate cancer. *Bioessays.* 2007;29:1227–1238.
8. Lee WH, Morton RA, Epstein JI, et al. Cytidine methylation of regulatory sequences near the pi-class glutathione S-transferase gene accompanies human prostatic carcinogenesis. *Proc Natl Acad Sci USA.* 1994;91:11733–11737.
9. McNeal JE, Bostwick DG. Intraductal dysplasia: A premalignant lesion of the prostate. *Hum Pathol.* 1986;17:64–71.
10. Meeker AK, Hicks JL, Platz EA, et al. Telomere shortening is an early somatic DNA alteration in human prostate tumorigenesis. *Cancer Res.* 2002;6405–6409.
11. Nelson WG, De Marzo AM, Isaacs WB. Prostate cancer. *N Engl J Med.* 2003;349:366–381.
12. Nelson CP, Kidd LC, Sauvageot J, et al. Protection against 2-hydroxyamino-1-methyl-6-phenylimidazo[4,5-b]pyridine cytotoxicity and DNA adduct formation in human prostate by glutathione S-transferase P1. *Cancer Res.* 2001;61:103–109.
13. Nelson WG, Yegnasubramanian S, Agoston AT, et al. Abnormal DNA methylation, epigenetics, and prostate cancer. *Front Biosci.* 2007;12:4254–4266.
14. Porkka KP, Pfeiffer MJ, Waltering KK, et al. MicroRNA expression profiling in prostate cancer. *Cancer Res.* 2007;67:6130–6135.
15. Tomlins SA, Rhodes DR, Perner S, et al. Recurrent fusion of TMPRSS2 and ETS transcription factor genes in prostate cancer. *Science.* 2005;310:644–648.
16. Wang S, Gao J, Lei Q, et al. Prostate-specific deletion of the murine PTEN tumor suppressor gene leads to metastatic prostate cancer. *Cancer Cell.* 2003;4:209–221.
17. Xu J, Zheng SL, Komiya A, et al. Germline mutations and sequence variants of the macrophage scavenger receptor 1 gene are associated with prostate cancer risk. *Nat Genet.* 2002;32:321–325.
18. Zheng SL, Sun J, Wiklund F, et al. Cumulative association of five genetic variants with prostate cancer. *N Engl J Med.* 2008;358:910–919.

Molecular Biology of Breast Cancer

Natasa Snoj . Phuong Dinh . Philippe Bedard . Christos Sotiriou

INTRODUCTION

Breast cancer is the most common cancer among women with a yearly incidence of 109.8 per 100,000 individuals. It is also the second leading cause of cancer-related death. Since the late 1980s, breast cancer-related deaths have significantly declined, partly due to screening and prevention and partly due to improvements in systemic therapy. With the widespread implementation of screening programs, the increasing frequency of screen-detected invasive cancers and noninvasive lesions has shifted the profile of tumor characteristics seen in breast cancer today toward smaller tumors. Breast cancer is increasingly being recognized as a biologically heterogeneous disease entity, with distinct subtypes demonstrating different natural history and clinical outcomes. Many of these differences are attributable to the heterogeneity that exists at the molecular level. As a result, the discovery and development of molecular markers have recently taken center stage. These markers are being investigated for their potential to serve as better predictors of prognosis and response to treatment. Clinically useful biomarkers that can better select patients for tailored clinical trials, according to molecular characteristics, may ultimately lead to individualized treatment for breast cancer patients.

TRADITIONAL BREAST CANCER CLASSIFICATION

Breast cancer has traditionally been classified according to histopathological type, and its anatomical extent reflected by the TNM classification. For many years, this breast cancer classification was the only tool available to asses the risk of relapse after local therapy and to base decisions on whether or not chemotherapy should be given.

Histopathological Features of Breast Cancer

The histopathological features of breast cancer are the foundation of traditional breast cancer pathology. They consist of (i) histological type, (ii) grade, and (iii) vascular invasion. These features provide clinicians with essential information for decision making regarding prognosis and treatment.

Histological Type

In the past, breast cancer was divided into lobular and ductal carcinoma according to the common belief that lobular carcinomas emanated from the lobules and ductal carcinomas from the ducts. Today, after demonstration that most breast cancers arise from the same location—the terminal duct lobular unit—these histomorphological differences are regarded as a manifestation of their distinct molecular profiles. The predominant type of breast cancer is ductal carcinoma *in situ* (DCIS). Some breast lesions not classified as breast cancer are also in this category as they are closely related to breast cancer in terms of histopathology, expression of molecular markers, and clinical presentation.

Lobular Neoplasia Lobular neoplasia can be classified as (i) lobular intraepithelial neoplasia, (ii) lobular carcinoma *in situ*, and (iii) invasive lobular carcinoma. Lobular intraepithelial neoplasia and lobular carcinoma are commonly referred to as *in situ* lobular neoplasia. Lobular neoplasia (*in situ* lobular neoplasia and invasive lobular carcinoma) is characterized by a population of small aberrant cells with small nuclei, individual private acini, and a lack of cohesion between cells. The distinctive molecular feature of lobular neoplasia is the loss of E-cadherin (*CDH1*) which can be demonstrated by immunohistochemistry. Since ductal carcinomas can also be E-cadherin negative, it should not be considered pathognomonic for lobular neoplasia. Other molecular characteristics of lobular carcinomas are epidermal growth factor receptor 1 (EGFR-1) and HER-2 negativity and positivity for antibody 34b E12 (cytokeratins 1, 5, 10, and 14), ER, and PR. These markers, while helpful, are also not pathognomonic, with the pleomorphic variant often described as ER, PR negative, and HER-2 positive. Therefore, while these molecular markers provide additional information, they should be correlated with the classical histopathological findings to ensure accurate diagnosis.

The diagnosis of *in situ* lobular neoplasia is often made following surgical excision of a suspicious nodule. It is often bilateral and in 50–70% of cases is multicentric. There is a high frequency (10–15% over a 10-year period and lifetime risk of up to 50%) of subsequent invasive ductal carcinoma in patients with a prior *in situ* lobular neoplasia, for reasons that are not well understood. Invasive lobular carcinoma is the second most common invasive breast cancer and represents 10–20% of all invasive breast cancers.

Ductal Carcinoma *in situ* DCIS is characterized by the proliferation of malignant ductal cells in without invasion of the surrounding stromal tissue. The *European Organisation for Research and Treatment of Cancer* (EORTC) grading system, based on cytonuclear pattern, distinguishes well-differentiated from intermediately differentiated and poorly differentiated DCIS. For the purpose of screening programs, this has been modified into groups based on low, intermediate, and high grade. Compared with normal epithelium, the growth potential of DCIS is 10 times greater and its apoptosis rate is increased 15-fold. Comedo-type DCIS is typically necrotic in the center of the ducts and has a higher risk of recurrence. Most ductal carcinomas are diffusely positive for luminal cell markers (CK8, CK18, CK19), but negative for basal cell markers (CK5/6 and CK14). In contrast, benign ductal hyperplasia may show a mosaic staining pattern for any of these markers, indicating a heterogeneous underlying cell population.

DCIS is multifocal in 30% of the cases, often in the same breast. In 2–6% of cases, axillary lymph node invasion is found at pathological examination, and this is thought to occur as a result of an unidentified invasive component. In the era of widespread mammographic screening, DCIS is more frequently detected and represents 25% of screen-detected breast cancers.

The treatment of DCIS is based on the extent of disease within the breast. In patients with extensive or multifocal DCIS, mastectomy with sentinel node biopsy is indicated, with the possibility of reconstruction. Sentinel lymph node biopsy is also indicated when microinvasion is found within DCIS. If breast-conserving surgery is undertaken, additional local irradiation has documented benefit in terms of lowering the incidence of subsequent noninvasive relapse and the progression to frankly invasive disease. Those patients with estrogen receptor positive DCIS may also benefit from adjuvant systemic treatment with tamoxifen in addition to radiotherapy after breast-conserving surgery to reduce the risk of relapse. However, our understanding of the biology of DCIS remains limited, and further clinical trials are necessary to optimize local and systemic treatment modalities.

Invasive Ductal Carcinoma Invasive ductal carcinoma is the most common invasive carcinoma of the breast, as it represents roughly 70% of all cases. Invasive lobular carcinoma is the second most common histopathological type representing 10–20% of all breast malignancies. The other 10–20% consists mostly of (i) medullary carcinoma (characterized by sharp tumor borders and lymphoid infiltration), (ii) mucoid carcinoma (with large amounts of extracellular mucous), (iii) papillary carcinoma (with well-differentiated papillary structure), and (iv) inflammatory carcinoma which occurs in 1–2% of all cases. Inflammatory carcinomas typically mimic the clinical presentation of benign inflammatory disease although they are characterized by extensive invasion of the lymphatic vessels within the dermis on histological examination. Other types of rare invasive breast cancers include (i) adenoid cystic carcinoma, (ii) apocrine carcinoma, and (iii) carcinoma with squamous metaplasia.

Another distinct histopathological entity of breast cancer is Paget's disease of the nipple. It is a variant of DCIS, where malignant cells infiltrate the ducts without invasion of the baseline membrane and extend to the major ducts of the nipple and the epidermis of the areola. An invasive component is present in proximally 90% of all cases. There are other malignant diseases of the breast that can resemble breast cancer, including phyllodes tumor, sarcoma, and malignant lymphoma.

Often the distinction between invasive and noninvasive carcinoma is difficult, as carcinoma cells can be found in both lesions and there are no reliable markers that differentiate between invasive and noninvasive cells. E-cadherin is often helpful in distinguishing malignant disease, as it is mostly positive in ductal carcinomas, but negative in lobular carcinomas. Invasive ductal carcinoma may express a variety of markers that have been extensively evaluated.

Other breast lesions, such as benign papillomas, are characterized by the expression of myoepithelial markers (alpha-SMA, myosin, calponin, p63, CD10), although myoepithelial markers may also be present in intraductal papillary carcinomas. Preservation of the myoepithelial layer is the distinguishing characteristic of benign sclerosing lesions, including carcinoma with pseudoinvasive structures. Given their overlapping molecular profiles, comparison with histopathological findings using hematoxylin and eosin is essential to make a proper diagnosis.

Vascular Invasion

The term vascular invasion refers to the invasion of lymphatic and blood vessels with tumor cells. It is assessed in hematoxylin-eosin-stained slides. Lymphovascular invasion is routinely included in the pathological evaluation and reporting of all breast cancers, although its interpretation is often difficult. In breast cancer, vascular invasion is a poor prognostic factor and predicts for increased local failure and reduced overall survival. Whether lymphatic or hematogenous invasion is more closely linked to prognosis is still unknown. Lymphovascular invasion by tumor cells is found in approximately 15% of invasive ductal carcinoma of the breast and is present in 10% of tumors without axillary lymph node invasion.

Perineural Invasion

The independent prognostic significance of perineural invasion is not clear, as it has only been assessed in combination with lymphovascular invasion.

Histological Grading

The histological grading of malignancy is the classical evaluation method for prognostication in breast cancer patients, and it is the simplest method, requiring only hematoxylin-eosin staining. Histological grading of breast cancer typically consists of three factors: tubule formation, mitotic count, and nuclear atypia. These three factors are all important in identifying patients at high risk of recurrence, although the relative importance of each feature is unclear. In addition, intratumoral heterogeneity and interobserver variation add to the difficulty of accurate prognostication based upon grading. In a review of the prognostic significance of histological grading, only nuclear atypia was correlated with the rate of recurrence. A grading system consisting of mitotic count and nuclear atypia (without tubule formation) was more strongly related to the risk of recurrence than a system using all three factors. Therefore, tubule formation is often excluded from the overall grading system even though it has the lowest heterogeneity within the primary tumor, compared with the other two.

Cells in the mitotic phase can be counted using light microscopy. However, the duration of the mitotic phase is often variable, especially in aneuploid tumors, so the number of mitoses is not linearly correlated with cellular proliferation. The recommendation for standardized counting of the mitotic activity index (MAI) includes assessment at $\times 400$ magnification in an area of $1.6\,\mathrm{mm}^2$ in the highest proliferative invasive area at the periphery of the tumor. MAI is not affected by fixation delay; it may impair morphological assessment. Ideally, mitoses should be counted before chemotherapy, but the mitotic index retains prognostic value even after chemotherapy. Mitoses should preferably be counted on excisional biopsies or mastectomies to avoid sampling error. Often underestimated in core biopsies, the MAI has a limited role in the decision of whether neoadjuvant chemotherapy should be given or not. Nonetheless, in several

retrospective and prospective studies, the MAI has proven to be a very strong, independent prognostic factor and, subsequently, is regarded as a category I prognostic factor for breast cancer by the College of American Pathologists.

TNM

Prognosis estimation has traditionally been based on grouping patients according to the anatomical extent of their disease. Incorporating primary tumor size, lymph node involvement, and distant metastases the TNM system combines these three factors into a stage grouping, with clinical and pathological variants. The TNM system is the most widely used tumor staging system, adopted by both the *Union International Contre le Cancer* (UICC) and the *American Joint Committee* (AJC). Although still clinically relevant, this system has limitations because it does not incorporate other important biological features that can influence the overall prognosis of the patient. Other important biological features include steroid hormone receptor content, HER-2 overexpression, tumor grade, indices of tumor proliferation, the presence or absence of vascular or lymphatic vessel invasion.

BIOMARKERS

Molecular analyses of cancer have led to the discovery of oncogenes and tumor suppressor genes. In practice, pathologists look for protein products (biomarkers) of the genes mainly through immunohistochemistry or immunocytochemistry. Biomarkers are potentially helpful in distinguishing between various histopathological types of breast cancers, as well as in assessing prognosis and predicting for a response to specific systemic therapy.

Biomarkers typically have continuous values, and cutoff points are arbitrarily established and used to simplify the clinical decision making.

Estrogen Receptor

Estrogen receptors (ER) are members of a large family of nuclear transcriptional regulators that are activated by steroid hormones, such as estrogen. ERs exist as two isoforms, α and β, that are encoded by two different genes. Although both isoforms are expressed in the normal mammary gland, it appears that only ER-α is critical for normal gland development. However, there is growing evidence that ER-β may antagonize the function of ER-α, and that high levels of ER-β are associated with a more favorable response to tamoxifen treatment.

Most research on the biology of ER has focused on its function as a nuclear transcription factor [classical ER-α function, or nuclear-initiated steroid signaling (NISS)], but there is also mounting evidence that estrogen can bind to ER located in or near the plasma membrane to activate other signaling pathways [nongenomic function, or membrane-initiated steroid signaling (MISS)].

With NISS, ER functions as a hormone-regulated nuclear transcription factor that can induce expression of specific genes in the nucleus. Upon estrogen ligand binding, ER binds to estrogen response elements (ERE) in target genes, recruits a coregulator complex, and regulates specific gene transcription. This ER action is controlled by a number of coregulatory proteins, termed coactivators and corepressors, that recruit enzymes to modulate chromatin structure to facilitate or repress gene transcription. It has been postulated that the cellular environment of coregulators may indeed influence whether selective estrogen receptor modulators (SERMS) are agonistic or antagonistic in action. Whether levels of coregulators are associated with prognosis and hormone resistance in breast cancer is unknown, but has been the focus of several studies.

Estrogen can therefore affect many processes, such as cell proliferation, the inhibition of apoptosis, invasion, and angiogenesis. c-Myc, vascular endothelial growth factor, bcl-2, insulin-like growth factor (IGF-1), insulin receptor substrate-1, transforming growth factor-α, cyclin-D1, and IGF-2 have all been shown to be regulated by estrogen.

It has long been recognized that estrogen is also capable of inducing other cellular effects with a much more rapid onset than would be possible via changes imposed by gene transcription through NISS. Such nongenomic effects have been attributed to MISS, involving membrane-bound or cytoplasmic ER. ER has been localized outside of the nucleus by biochemical analyses and by direct visualization using immunocytochemistry or more sophisticated microscopy. In fact, non-nuclear ER has been shown to exist in complexes with such signaling molecules as the IGF-1 receptor (IGF-1R) and the p85 subunit of phosphatidylinositol-3-OH kinase. While these membrane and nonnuclear ER functions have been well described in experimental model systems, they must still be confirmed in clinical breast cancer.

Targeting the ER—Tamoxifen Therapy

There is a significant body of evidence to suggest that steroid hormones, particularly estrogen, play a major role in the development of breast cancer. Approximately 75% of all breast cancers express estrogen receptors (ER+). ER expression has a high negative predictive value but a suboptimal positive predictive value for the efficacy of endocrine therapy. Patients with ER negative (ER−) tumors derive no benefit from endocrine therapy, while for ER+ tumors, estrogen signaling blockade is an important therapeutic strategy that can lead to improved outcomes, including increasing cure rates in early breast cancer, improving response rates and disease control in advanced disease, and reducing breast cancer incidence with prevention.

Tamoxifen is a selective modulator of ER that competitively impedes the binding of estradiol disrupts

mechanisms that regulate cellular replication and proliferation. Tamoxifen has been used in clinical trials since the 1970s, with initial nonselective use across the entire breast cancer population. In these unselected patients, tamoxifen can produce responses of up to 30%. In ER+ disease, responses of up to 80% have been observed, while in ER− disease, less than 10% may derive benefit.

Testing for ER

The introduction of routine ER testing in the early 1990s changed the way in which endocrine therapy is prescribed, particularly in preventing unnecessary tamoxifen-related toxicity for patients with ER− disease. However, the treatment of individual ER+ patients is less clearcut, largely because of the ambiguity created by varying methodologies and cutoffs employed by different laboratories for ER testing. Indeed, responses have been demonstrated in tumors with as few as 1–10% of cells positive for ER by immunohistochemistry (IHC). It is therefore more clinically important to distinguish between ER disease from measurable, but low-level ER expression, variably reported as either ER+ or ER−, depending on which cutoff values are used.

Quantitative ER as a continuous variable has been demonstrated in metastatic studies to be proportionally correlated to the response to hormonal therapy. In the adjuvant setting, a large meta-analysis has shown a similar relationship between tumoral ER expression and tamoxifen efficacy, and in the neoadjuvant setting, for tumor response to both tamoxifen and letrozole.

Nevertheless, quantitative ER is still an imprecise predictive tool because even tumors in the highest strata of ER expression can be endocrine-resistant (de novo resistance). Furthermore, a significant proportion of ER+ patients who initially respond to endocrine therapy will eventually fail treatment (acquired resistance). Although this resistance is not fully understood, several mechanisms have been hypothesized to be responsible for the development of resistance, including (i) loss of ER in the tumor, (ii) selection clones with ER mutations, (iii) deregulation of cellcycle components including ER-regulatory proteins, and (iv) cross-talk between ER and other growth factor receptor pathways. Given the limitations of ER assessment, the development of other molecular tools, reflecting the complex biology of these tumors, is needed to improve treatment for patients with ER+ disease.

Progesterone Receptor

The progesterone receptor (PR) gene is an estrogenregulated gene. PR mediates the effect of progesterone in the development of the mammary gland and breast cancer. In the 75% of breast cancers that express ER, more than half of these tumors also express PR. It has been hypothesized that PR levels in breast cancer may be a marker of an intact ER signal transduction pathway and that PR levels may therefore add independent predictive information. Emerging laboratory and clinical data also suggest that among ER+ tumors, PR status may predict differential sensitivity to antiestrogen therapy.

Although the etiology of ER+PR− disease some studies have shown that this phenotype may evolve through the loss of PR, while other studies suggest that they reflect a distinct molecular origin, with unique epidemiologic risk factors when compared with ER+PR+ disease.

Whether patients with ER+PR+ disease derive greater benefit from adjuvant tamoxifen than patients with ER+/PR− is controversial. Some studies suggested differential benefit according to PR status, although this was not confirmed in the recent Early Breast Cancer Trialists' Collaborative Group meta-analysis of individual patient data from randomized clinical trials.

Loss of PR expression has also been suggested to be predictive of greater benefit from aromatase inhibitors (AI) over tamoxifen for post-menopausal women with ER+ early breast cancer. More recent data from central pathological review of the large adjuvant AI trials, however, has confirmed that PR loss is a negative prognostic factor but does not predict differential sensitivity for an AI over tamoxifen.

The HER-2 Receptor

Following the discovery of ER, HER-2 has more recently emerged as an important molecular target in the treatment of breast cancer. Much of the recent success with anti-HER-2 therapy has been a direct result of being able to properly select the right patient subpopulation for targeted treatment.

HER-2 belongs to the human epidermal growth factor receptor family of tyrosine kinases consisting of EGFR, HER-2, HER-3, and HER-4. All of these receptors have an extracellular ligand-binding region, a single membrane-spanning region, and a cytoplasmic tyrosine-kinase-containing domain. This intracellular kinase domain is non-functional in the HER-3 receptor. Ligand binding to the extracellular region results in homodimer and heterodimer activation of the cytoplasmic kinase domain and phosphorylation of a specific tyrosine kinase. This leads to the activation of various intracellular signaling pathways, such as the mitogen-activated protein kinase (MAPK) and the phosphatidylinositol 3-kinase (P13K)-AKT pathways involved in cell proliferation and survival.

HER family signaling can become dysregulated via a number of mechanisms, including overexpression of a ligand; overexpression of the normal HER receptor; overexpression of a constitutively activated mutated HER receptor; and defective HER receptor internalization, recycling, or degradation.

HER-2 was first identified as an oncogene activated by a point mutation in chemically induced rat neuroblastomas. Soon afterward, it was found to be amplified in breast cancer cell lines. Early studies suggested that 30% of breast cancers overexpressed HER-2 which is

associated with a more aggressive phenotype and a poorer disease-free survival. HER-2 positivity associated with a relative, but not an absolute, resistance to endocrine therapy and resistance to certain chemotherapeutic agents. Most importantly, HER-2 status is predictive for benefit from anti-HER-2 therapies, such as trastuzumab.

Identifying the HER-2 Receptor

For the various reasons described above, accurate determination of HER-2 status is critically important for clinical decision-making. Currently, two different methods of HER-2 evaluation are routinely used: (i) immunohistochemistry (IHC) and (ii) fluorescence *in situ* hybridization (FISH).

IHC is a semiquantitative method that identifies HER-2 receptor expression on the cell surface using a grading system (0, 1+, 2+, and 3+), where an IHC result of 3+ is regarded as a marker of HER-2 overexpression. It is the most widely used technique, performed on paraffin tumor blocks. While fairly easy to perform and relatively inexpensive, results can vary depending on different fixation protocols, assay methods, scoring systems, as well as the selected antibodies.

FISH is a quantitative method measuring the number of copies of the HER-2 gene present in each tumor cell and is reported as either positive or negative. It is a more reproducible test but is comparatively more time-consuming and expensive.

Concordance between IHC and FISH has been extensively studied. In a study of 2963 samples using FISH as the standard method, the positive predictive value of an IHC 3+ result was 91.6%, and the negative predictive value of an IHC 0 or 1+ result was 97.2%. They also found that FISH had a significantly higher failure rate (5% versus 0.08%), was more costly, and required more time for testing (36 versus 4 hours) and interpretation (7 minutes versus 45 seconds) than IHC.

Another problem with HER-2 testing is the poor reproducibility between laboratories, even when the same technique is used. When the concordance between local and central evaluation in the NCCTG-N9831 adjuvant trastuzumab trial was examined, FISH concordance was 88.1% compared to an IHC concordance of 81.6%. This high rate of discordance was attributed to methodologic factors, such as inadequate quality-control procedures.

In recognizing the importance of accurate HER-2 testing, guidelines from the *American Society of Clinical Oncology* (ASCO)/*College of American Pathologists* (CAP) for HER-2 testing have recently been published. In these guidelines, recommendations for HER-2 evaluation are provided using an algorithm for positive, equivocal, and negative results:

1. HER-2 positive—IHC staining of 3+ (uniform, intense membrane staining of >30% of invasive tumor cells, a FISH result of more than 6.0 HER-2

gene copies per nucleus, or a FISH ratio (HER-2 gene signals to chromosome 17 signals) of more than 2.2;

2. HER-2 negative—IHC staining of 0 or 1+ FISH result of less than 4.0 HER-2 gene copies per nucleus, or FISH ratio of less than 1.8; and

3. HER-2 equivocal—IHC 3+ staining of 30% or less of invasive tumor cells or 2+ staining, a FISH result of 4 to 6 HER-2 gene copies per nucleus, or FISH ratio between 1.8 and 2.2.

Importantly, the guidelines strongly recommended "validation of laboratory assay or modifications, use of standardized operating procedures, and compliance with new testing criteria to be monitored with the use of stringent laboratory accreditation standards, proficiency testing, and competency assessment."

Targeting HER-2: Trastuzumab

The HER family of receptors is an ideal target for anti-cancer therapy. To date, two main therapeutic strategies have been developed to target the HER-2 receptor: monoclonal antibodies and small molecule kinase inhibitors.

Trastuzumab (Herceptin; Genentech, South San Francisco) is a recombinant, humanized anti-HER-2 monoclonal antibody and was the first clinically active anti-HER-2 therapy to be developed. Trastuzumab exerts its action through several mechanisms, including (i) induction of receptor downregulation/degradation, (ii) prevention of HER-2 ectodomain cleavage, (iii) inhibition of HER-2 kinase signal transduction via ADCC, and (iv) inhibition of angiogenesis.

In metastatic breast cancer (MBC), trastuzumab monotherapy produces response rates ranging from 12% to 34% with a median duration of disease control of 9 months. Preclinical studies have also shown additive or synergistic interactions between trastuzumab and multiple cytotoxic agents, including platinum analogues, taxanes, anthracyclines, vinorelbine, gemcitabine, capecitabine, and cyclophosphamide. The use of trastuzumab combined with chemotherapy can further increase response rates (RR), time to progression (TTP), and overall survival (OS).

Encouraged by the highly reproducible antitumor activity of trastuzumab in the metastatic setting, four major international studies with enrollment of over 13,000 women were launched in 2000–2001 to investigate the role of trastuzumab in the adjuvant setting: HERA, the combined North American trials NSABP-B31 and NCCTG/N9831, and BCIRG 006. In 2005, the initial results of these four trials, alongside a smaller Finnish trial, demonstrated that adjuvant trastuzumab produced significant benefit in reducing recurrence and mortality. Updated analyses for most of these trials have recently been presented, including that of another small trial PACS-04 which, in contrast, failed to show a benefit for adjuvant trastuzumab for reasons that are not well understood.

Despite differences in patient population and trial design, including chemotherapy regimen, the timing of trastuzumab initiation, and the schedule and duration of trastuzumab administration, remarkably consistent results were reported across these studies: a 33–58% reduction in the recurrence rate and a 30% reduction in mortality. This degree of benefit in early breast cancer is the largest reported since the introduction of tamoxifen for ER+ disease.

There are many other novel drugs being developed for use following failure of trastuzumab, either because of *de novo* or acquired resistance.

GENE EXPRESSION PROFILING

Historically, the classification of breast cancers has been based on histological type, grade, and expression of hormone receptors, but the advent of microarray technology has since demonstrated significant heterogeneity occurring also at the transcriptome level. Through their ability to interrogate thousands of genes simultaneously, microarray studies have allowed for a comprehensive molecular and genetic profiling of tumors. Not only have these studies changed the way in which we have traditionally classified breast cancer, their results have also yielded molecular signatures with the potential to significantly impact clinical care by providing a molecular basis for treatment tailoring.

Microarray Technology

Gene expression profiling, using microarray technology, relies on the accurate binding, or hybridization, of DNA strands with their precise complementary copies where one sequence is bound onto a solid-state substrate. These are hybridized to probes of fluorescent cDNAs or genomic sequences from normal or tumor tissue. Through analysis of the intensity of the fluorescence on the microarray chip, a direct comparison of the expression of all genes in normal and tumor cells can be made. At present, there are multiple microarray platforms that use either cDNA-based or oligonucleotide-based microarrays.

cDNA-based microarrays have double-stranded PCR products amplified from expressed sequence tag (EST) clones and then spotted onto glass slides. These have inherent problems with frequent hybridization among homologous genes, alternative splice variants, and antisense RNA. Oligonucleotide-based microarrays are shorter probes with uniform length. Shorter oligonucleotides (25 bases) may be synthesized directly onto a solid matrix using photolithographic technology (Affymetrix), and for longer oligonucleotides (55–70 bases), they may be either deposited by an inkjet process or spotted by a robotic printing process onto glass slides.

In the past, there has been much skepticism regarding the reliability and reproducibility of microarray technology. However, the overall reproducibility of the microarray technology has been found to be acceptable, as shown by the MicroArray Quality Control (MAQC) project conducted by the U.S. Food and Drug

Administration (FDA). They found that similar changes in gene abundance were detected, despite using different platforms. Furthermore, this reproducibility appears comparable to that of other diagnostic techniques, for example, with immunohistochemical analysis for hormone receptors in breast cancer.

Molecular Classification of Breast Cancer

One of the most important discoveries stemming directly from microarray studies has been the reclassification of breast cancer into molecular subtypes. This new classification has not only furthered our understanding of tumor biology, but it has also altered the way that physicians and clinical investigators conceptually regard breast cancer—not as one disease, but a collection of several biologically different diseases.

Four main molecular classes of breast cancers have been consistently distinguished by gene expression profiling. Based upon the original classification system, these subtypes are (i) basal-like, (ii) HER-2+, (iii) luminal-A, and (iv) luminal-B breast cancers.

In the basal-like subtype, there is a high expression of basal cytokeratins 5/6 and 17 and proliferation-related genes, as well as laminin and fatty-acid binding protein 7. In the HER-2+ subtype, there is a high expression of genes in the erbb2 amplicon, such as GRB7. The luminal cancers are ER+. Luminal A is characterized by a higher expression of ER, GATA3, and X-box binding protein trefoil factor 3, hepatocyte nuclear factor 3 alpha, and LIV-1. Luminal B cancers are generally characterized by a lower expression of luminal genes.

Beyond differing gene expression profiles, these molecular subtypes appear to have distinct clinical outcomes and responses to therapy that seem reproducible from one study to the next. The basal-like and HER-2+ subtypes are more aggressive with a higher proportion of *TP53* mutations and a markedly higher likelihood of being grade III than luminal A tumors. Despite a poorer prognosis, they tend to respond better to chemotherapy including higher pathologic complete response rates after neo-adjuvant therapy.

Fewer than 20% of luminal subtypes have mutations in *TP53* and these tumors are often grade I. They are more sensitive to endocrine therapy, less responsive to conventional chemotherapy, and demonstrate overall clinical outcomes.

Despite differences in testing platforms, the collective results of these studies suggest that ER expression is the most important discriminator. Beyond ER expression, distinct expression patterns can also be differentiated, perhaps reflecting distinct cell types of origin. However, this molecular classification is not without its inherent limitations, with up to 30% of breast cancers not falling into any of the four molecular categories. Exactly how many true molecular subclasses of breast cancer exist remains uncertain, and it is plausible that the molecular classification will evolve with new technological platforms, with the availability of larger data sets, as well as with improved understanding of tumor biology.

Gene Expression Signatures to Predict Prognosis

Traditional prognostic factors based on clinical and pathological variables are unable to fully capture the heterogeneity of breast cancer. Guidelines like the *National Comprehensive Cancer Network* (NCCN) used in the United States and the *International St. Gallen Expert Consensus* used in Europe to guide treatment decisions take into account relapses-risk-based traditional anatomical and pathological assessment. These guidelines cannot accommodate for the substantial variability that can exist between patients with similar stages and grades of disease. Using microarray technology, several independent groups have conducted comprehensive gene expression profiling studies with the aim of improving on traditional prognostic markers used in the clinic.

Using the top-down approach, where gene expression data are correlated with clinical outcome without a prior biological assumption, a group of researchers from Amsterdam identified a 70-gene prognostic signature, using the Agilent platform, in a series of 78 systemically untreated node-negative breast cancer patients under 55 years of age. This signature included mainly genes involved in the cell cycle, invasion, metastasis, angiogenesis, and signal transduction. Validated on a larger set of 295 young patients, including both node-negative and node-positive disease as well as treated and untreated patients, the 70-gene signature was found to be the strongest predictor for distant metastasis-free survival, independent of adjuvant treatment, tumor size, histological grade, and age.

Using the same top-down approach, another group in Rotterdam identified a 76-gene signature, using the Affymetrix technology which considered ER-positive patients separately from ER-negative patients. These 76 genes were mainly associated with cell cycle and cell death, DNA replication and repair, and immune response. In a training set of 115 patients and a multicentric validation set of 180 patients, they were able to demonstrate comparable discriminative power in predicting the development of distant metastases in untreated patients in all age groups with node-negative breast cancer.

Both the Amsterdam and Rotterdam gene signatures have been independently validated by TRANS-BIG, the translational research network of the *Breast International Group* (BIG). Despite having only three genes in common, both signatures were able to outperform the best validated tools to assess clinical risk. In particular, these signatures were both superior in correctly identifying the low-risk patients but were limited in identifying the high-risk patients, as half of those identified in this category did not, in fact, relapse. This suggests that the highest clinical utility for these molecular signatures may be in potentially reducing overtreatment of low-risk patients.

Using the top-down approach, investigators in collaboration with Genomic Health Inc., developed a recurrence score (RS) based on 21 genes that appeared to accurately predict the likelihood of distant recurrence in tamoxifen-treated patients with node-negative, ER+ breast cancer. A final panel of 16 cancer-related genes and 5 reference genes forms the basis for the Oncotype DX™ Breast Cancer Assay.

The RS classifies patients into three risk groups, based on cutoff points from the results of the NSABP trial B-20: high risk of recurrence is assigned if RS >31; intermediate risk if RS is 18–30; and low risk if RS <18. Retrospective validation of this predictor in 675 archival samples of the NSABP trial B-14 showed that the RS was significantly correlated with distant recurrence, relapse-free interval, and overall survival, independent of age and tumor size.

Using a different approach that is hypothesis-driven, otherwise known as the bottom-up approach, investigators examined whether gene expression patterns associated with histologic grade could improve prognostic capabilities especially within the class of intermediate-grade tumor. Accounting for 30–60% of all breast cancers, these intermediate-grade tumors display the most heterogeneity in both phenotype and outcome.

Of the unique 97 genes that formed the gene-expression grade index (GGI), most were associated with cell-cycle progression and differentiation. These genes were differentially expressed between low-grade and high-grade breast tumors, without a distinct gene-expression pattern to distinguish the intermediate group. Instead, the intermediate tumors showed expression patterns and clinical outcomes matching those of either low-grade or high-grade cases. The GGI, therefore, could potentially improve treatment decision making for these otherwise problematic patients with intermediate grade by reclassifying them into two distinct and clinically relevant subtypes.

In an examination of genomic grade with ER status, ER– tumors with poor clinical outcome were found to be mainly associated with high GGI, although ER+ tumors were more heterogeneous with a mixture of GGI levels. Thus, these two variables are not entirely independent from each other, with tumor genomic grading capable of providing an extra level of information when stratifying the ER+ group.

Other prognostic signatures derived from the bottom-up approach include the wound response signature, mutant/wild p53 signature, invasive gene signature (IGS), and the cancer stem cell signature. These and other prognostic signatures, whether derived with the bottom-up or top-down approach, have only a few genes in common but seem to offer similar prognostic information, with proliferation-related genes being the major driving force. Furthermore, it appears that the prognostic power of many of these signatures is limited.

CONCLUSION

Breast cancer is a clinically heterogeneous disease, and for many years, this heterogeneity was explained by the differing histopathological characteristics identified mainly by the microscope. Current practice guidelines based on histopathology have been important as risk stratification tools for therapy selection, but these guidelines are limited in the tailoring of treatment for the individual patient. It has long been observed that among patients with anatomic and pathological risk profiles, there can be substantial variability in both the natural history and response to treatment.

Novel molecular technologies and a better understanding of the tumor biology of breast cancer have resulted in significant advances in recent years. Previous emphasis on refining the traditional histopathologic criteria, on developing indices of proliferation (S-phase fraction, Ki67), on understanding genomic instability (e.g., DNA ploidy), and on analyzing single gene expression profiles (p53 expression) has shifted dramatically to molecular profiling, reflecting the expression of many thousands of genes.

With new knowledge, new challenges emerge. The excitement that comes with each new biomarker must be matched by the scientific rigor of prospective validation, with each biomarker being assessed for clinical and economic utility. The challenge also remains how to best integrate multiple new sources and levels of prognostic data with the existing histopathological model. Therefore, it should be with cautious optimism that we move forward in this era of unraveling the mystery behind the many molecules and mutations of this disease.

KEY CONCEPTS

- Breast cancer is a biologically heterogeneous disease, with distinct molecular subtypes associated with recurrence risk and patterns of relapse.
- The estrogen receptor (ER) and human epidermal growth factor-2 (HER2) receptor are predictive biomarkers that are routinely used in clinical practice to select patients for anti-hormonal and anti-HER2 therapy.
- There is considerable interlaboratory variation in ER and HER2 testing, highlighting the importance of compliance with standardized testing criteria.
- There are several commercially available gene expression profiles for early breast cancer that provide additional prognostic information beyond traditional clinical and pathological characteristics.
- The prognostic performance of these prognostic gene expression profiles is similar, driven largely by improved quantification of proliferation in the ER+/HER2– subtype.

SUGGESTED READINGS

1. Berry DA, Cirrincione C, Henderson C, et al. Estrogen-receptor status and outcomes of modern chemotherapy for patients with node-positive breast cancer. *J Am Med Assoc.* 2006;295(14): 1658–1667.
2. Early Breast Cancer Trialists' Collaborative Group. Effects of chemotherapy and hormonal therapy for early breast cancer on recurrence and 15-year survival: An overview of the randomised trials. *Lancet.* 2005;365(9472):1687–1717.
3. Goldhirsch A, Ingle JN, Gelber RD, et al. Thresholds for therapies: highlights of the St Gallen International Expert Consensus on the Primary Therapy of Early Breast Cancer 2009. *Ann Oncol.* 2009.

4. Harvey JM, Clark GM, Osborne CK, et al. Estrogen receptor status by immunohistochemistry is superior to the ligand-binding assay for predicting response to adjuvant endocrine therapy in breast cancer. *J Clin Oncol.* 1999;17(5):1474–1481.

5. Horowitz KB, McGuire WL. Predicting response to endocrine therapy in human breast cancer: A hypothesis. *Science.* 1975;189 (4204):726–727.

6. Kuerer HM, Albarracin CT, Yang WT, et al. Ductal carcinoma *in situ:* State of the science and roadmap to advance the field. *J Clin Oncol.* 2009;27(2):279–288.

7. Paik S, Shak S, Tang G, et al. A multigene assay to predict recurrence of tamoxifen-treated, node-negative breast cancer. *N Engl J Med.* 2004;351(27):2817–2826.

8. Paik S, Shak S, Tang G, et al. Expression of the 21 genes in the Recurrence Score assay and tamoxifen clinical benefit in the NSABP study B-14 of node negative, estrogen receptor positive breast cancer [abstract]. *J Clin Oncol.* (Meeting Abstracts) 2005;23(suppl 16):510.

9. Perou CM, Sorlie T, Eisen MB, et al. Molecular portraits of human breast tumours. *Nature.* 2000;406:747–752.

10. Pestalozzi BC, Zahrieh D, Mallon E, et al. Distinct clinical and prognostic features of infiltrating lobular carcinoma of the breast: Combined results of 15 International Breast Cancer Study Group clinical trials. *J Clin Oncol.* 2008;26(18):3006–3014.

11. Piccart-Gebhart MJ, Procter M, Leyland-Jones B, et al. Trastuzumab after adjuvant chemotherapy in HER2-positive breast cancer. *N Engl J Med.* 2005;353(16):1659–1672.

12. Rakha EA, El-Sayed ME, Lee AH, et al. Prognostic significance of Nottingham histologic grade in invasive breast carcinoma. *J Clin Oncol.* 2008;26(19):3153–3158.

13. Romond EH, Perez EA, Bryant J, et al. Trastuzumab plus adjuvant chemotherapy for operable HER2-positive breast cancer. *N Engl J Med.* 2005;353(16):1673–1684.

14. Saphner T, Tormey DC, Gray R. Annual hazard rates of recurrence for breast cancer after primary therapy. *J Clin Oncol.* 1996;14 (10):2738–2746.

15. Sotiriou C, Wirapati P, Loi S, et al. Gene expression profiling in breast cancer: Understanding the molecular basis of histologic grade to improve prognosis. *J Natl Cancer Inst.* 2006;98(4): 262–272.

16. van't Veer LJ. Gene expression profiling predicts clinical outcome of breast cancer. *Nature.* 2002;415:530–536.

17. van de Vijver MJ. A gene-expression signature as a predictor of survival in breast cancer. *N Engl J Med.* 2002;347:1999–2009.

18. Wirapati P, Sotiriou C, Kunkel S, et al. Meta-analysis of gene expression profiles in breast cancer: Toward a unified understanding of breast cancer subtyping and prognosis signatures. *Breast Cancer Res.* 2008;10(4):R65.

19. Wolff AC, Hammond MH, Scwartz JN, et al. American Society of Clinical Oncology/College of American Pathologists Guideline Recommendations for Human epidermal growth factor receptor 2 testing in breast cancer. *J Clin Oncol.* 2007;25(1):118–145.

27

Molecular Basis of Skin Disease

Vesarat Wessagowit

MOLECULAR BASIS OF HEALTHY SKIN

A key role of skin is to provide a mechanical barrier against the external environment. The cornified cell envelope and the stratum corneum restrict water loss from the skin and keratinocyte-derived endogenous antibiotics provide innate immune defense against bacteria viruses and fungi.

Normal skin also has been shown to have a very effective defense system against microbes. In the stratum corneum there is an effective chemical barrier maintained by the expression of S100A7 (psoriasin). This antimicrobial substance is very effective at killing *E. coli*. Subjacent to this in the skin there is another class of antimicrobial peptides such as RNASE7, which is effective against the broad spectrum of microorganisms, especially *Enterococci*. RNASE7 serves as a protective minefield in the superficial skin layers and helps to destroy invading organisms. Below this in the living layers of the skin are other antimicrobial peptides such as a β-defensins. The antimicrobial activity of most peptides occurs as a result of unique structural properties to enable them to destroy the microbial membrane while leaving human cell membranes intact. Some may play a specific role against certain microbes in normal skin whereas others act only when the skin is injured and the physical barrier is disrupted.

Certain antimicrobial peptides can also influence host cell responses in specific ways. The human catheli-cidin peptide LL-37 can activate mitogen-activated protein kinase (MAPK), an extracellular signal-related kinase in epithelial cells, and blocking antibodies to LL-37 hinder wound repair in human skin equivalents. Defensins and cathelicidins have immuno-stimulatory and immuno-modulatory capacities as catalysts for secondary host defense mechanisms, and can be che-motactic for distinct subpopulations of leukocytes as well as other inflammatory cells. Human β-defensins (hBDs) 1–3 are chemotactic for memory T-cells and immature dendritic cells—hBD2 attracts mast cells and activated neutrophils whereas hBD3–4 is also chemotactic for monocytes and macrophages. Cathe-licidins are chemotactic for neutrophils, monocytes/macrophages, and CD4 T-lymphocytes.

Skin immunity also is provided by a distinct population of antigen presenting cells in the epidermis known as Langerhans cells. Langerhans cells function as intraepidermal macrophages. Langerhans cells then leave the epidermis and migrate via lymphatics to regional lymph nodes. In the paracortical region of lymph nodes the Langerhans cell (or interdigitating reticulum cell as it is then known) expresses protein on its surface to present to a T-lymphocyte that can then undergo clonal proliferation. Langerhans cells contribute to several skin pathologies, including infections, inflammation, and cancer, and thus they play a pivotal role in regulating the balance between immunity and peripheral tolerance. Langerhans cells have characteristics that are different from other dendritic cells, in that they are more likely to induce Th-2 responses than the Th-1 responses that are usually necessary for cellular immune responses against pathogens. Langerhans cells, or a subset thereof, may also have immuno-regulatory properties that counteract the pro-inflammatory activity of surrounding keratinocytes.

Besides the antigen detection and processing role of epidermal Langerhans cells, cutaneous immune surveillance also is carried out in the dermis by an array of macrophages, T-cells, and dendritic cells (Figure 27.1). These immune sentinel and effector cells can provide rapid and efficient immunological back-up to restore tissue homeostasis if the epidermis is breached. The dermis contains a very large number of resident T-cells. Indeed, there are approximately 2×10^{10} resident T-cells, which is twice the number of T-cells in the circulating blood. Dermal dendritic cells vary in their functionality. Some have potent antigen-presenting capacities whereas others have the potential to develop into CD1a-positive and Langerin-positive cells, and some are pro-inflammatory. A recent addition to the family of skin immune sentinels is type 1 interferon-producing plasmacytoid predendritic

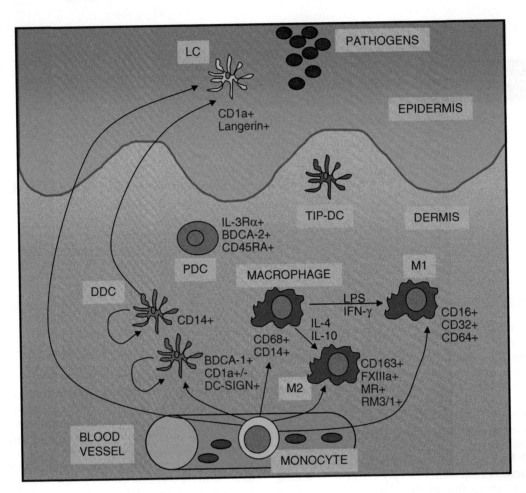

Figure 27.1   Diversity of immune sentinels in human skin. These include CD1a+ Langerin+ Langerhans cells located in the epidermis and various subtypes of dendritic cells and macrophages in the dermis. This figure illustrates some of the recent immunophenotypic and functional findings of these immune sentinels. The macrophage population expressing CD68 and CD14 can be further subdivided into classically activated macrophages (M1) and alternatively activated macrophages (M2), which develop under the influence of IL-4 and IL-10. Several cells have self-renewing potential under conditions of tissue homeostasis. Under inflammatory conditions, circulating blood-derived monocytes are potential precursors of Langerhans cells, dermal dendritic cells, and macrophages. Adapted from Nestle FO, Nickoloff BJ. Deepening our understanding of immune sentinels in the skin. *J Clin Invest.* 2007;117:2382–2385.

cells, which are rare in normal skin but which can accumulate in inflamed skin. A further component of the dermal immune system is the dermal macrophage. Dermal immune sentinels exhibit flexibility or plasticity in function. Depending on microenvironmental factors and cues they may acquire an antigen-presenting mode, a migratory mode, or a tissue resident phagocytic mode.

SKIN DEVELOPMENT AND MAINTENANCE PROVIDE NEW INSIGHT INTO MOLECULAR MECHANISMS OF DISEASE

The development of a stratified epithelium such as the skin requires a detailed architecture of maintaining an inner layer of proliferating cells, but that can also give rise to multiple layers of terminally differentiating cells

that extend to the body surface and are subsequently shed. A detailed understanding of this process, which generates a self-perpetuating barrier to keep microbes out and essential body fluids in, is becoming clearer and this improved understanding is providing new insights into skin maintenance as well as the pathogenesis and molecular mechanisms underlying certain developmental disorders.

One fundamental issue has been trying to provide an explanation for how epithelial progenitor cells retain a self renewing capacity. In 1999, it was shown that mice lacking the transcription factor p63 had thin skin and abnormal skin renewal. p63 is an evolutionary predecessor to the *p53* protein, part of a family of transcriptional regulators of cell growth differentiation and apoptosis. Although *p53* is a major player in tumorigenesis, p63 and another family member, p73, appear to have pivotal roles in embryonic development (Figure 27.2). p73-deficient mice have neurological and inflammatory pathology whereas p63 knockout mice

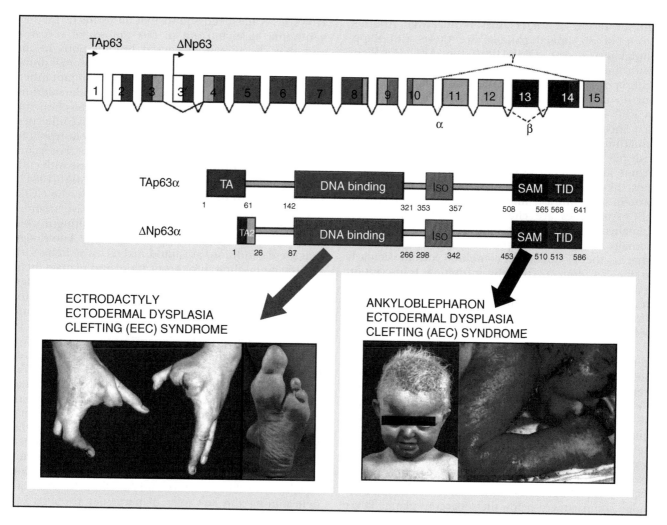

Figure 27.2 Genomic and functional domain organization of the transcription factor _p63_. At least six different isoforms can be generated by use of alternative translation initiation sites or alternative splicing. The main isoform expressed in human skin is ΔNp63α. Autosomal dominant mutations in the DNA binding domain of the _p63_ gene lead to ectrodactyly, ectodermal dysplasia, and clefting (EEC) syndrome. In contrast, autosomal dominant mutations in the SAM domain result in ankyloblepharon, ectodermal dysplasia, and clefting (AEC) syndrome. A number of other ectodermal dysplasia syndromes may also result from mutations in the _p63_ gene.

have major defects in epithelial limb and craniofacial development. These observations suggest that p63 has a crucial role in tissue morphogenesis and maintenance of epithelial stem cell compartments. Lack of p63 compromises skin formation either by creating an absence of lineage commitment and an early block in epithelial differentiation or by causing skin failure through a defect in epithelial stem cell renewal. p63 has been linked to several important signaling pathways such as epidermal growth factor (EGF), fibroblast growth factor (FGF), bone morphogenic protein (BMP), and Notch/Wnt/hedgehog signaling. p63 also directly regulates expression of extracellular matrix adhesion molecules, including the $\alpha_6\beta_4$ basal integrins and desmosomal proteins such as PERP, all of which are essential for epithelial integrity.

The human _p63_ gene consists of 16 exons, located on chromosome 3q28. There are two different promoter sites and three different splicing routes, which create at least six different protein isoforms. Several functional domains have been identified (Figure 27.2).

A further understanding of the role of p63 in skin development has been gleaned from studies on naturally occurring mutations in this gene. Heterozygous mutations cause developmental disorders, displaying various combinations of ectodermal dysplasia, limb malformations, and orofacial clefting. Thus far seven different disorders have been linked to mutations in the _p63_ gene. These conditions may have overlapping genotypic features, but there are some distinct genotype/phenotype correlations. The most common p63 ectodermal dysplasia is EEC syndrome (ectrodactyly, ectodermal dysplasia, and cleft/lip palate (OMIM604292)). It is characterized by three major clinical signs of cleft lip and/or palate, ectodermal dysplasia (abnormal teeth, skin, hair, nails, and sweat glands, or combinations thereof), and limb malformations in the form of split hand/foot (ectrodactyly)

353

and/or fusion of fingers/toes (syndactyly). Another group of p63-linked patients are those with Rapp-Hodgkin syndrome (OMIM129400) or AEC/Hay-Wells syndrome (OMIM106260). These syndromes fulfill the criteria of ectodermal dysplasia and orofacial clefting, but do not have the severe limb malformation seen in EEC syndrome. The features of these syndromes may include eyelid fusion (ankyloblepharon filiform adnatum), severe erosions at birth, and abnormal hair with pili torti or pili canaliculi. Mutations in EEC syndrome are clustered in the DNA-binding domain and most likely alter the DNA-binding properties of the protein. By contrast, mutations in Rapp-Hodgkin or AEC syndromes are clustered in the SAM and TI domains in the carboxy-terminus of p63α. The SAM domain is involved in protein–protein interactions whereas the TI domain is combined intramolecularly to the TA domain, thereby inhibiting transcription activation. All p63-associated disorders are inherited in an autosomal dominant manner and mutations are thought to have either dominant-negative or gain-of-function effects.

The epidermis contains a population of epidermal stem cells that reside in the basal layer, although it is not clear how many cells within the basal layer have a stem cell capacity. Stem cells are proposed to express elevated levels of β_1 and α_6 integrins and differentiate by delamination and upward movement to form the spinous layer, a granular layer, and the stratum corneum. The proliferation of epidermal stem cells is regulated positively by β_1 integrin and transforming growth factor α and negatively by transforming growth factor β signaling. Hair follicle stem cells reside in the bulge compartment below the sebaceous gland. These stem cells express the cell surface molecules CD34 and VdR as well as the transcription factors TCF3, Sox9, Lhx2, and NFATc1. These bulge area stem cells generate cells of the outer root sheath, which drive the highly proliferative matrix cells next to the mesenchymal papillae. After proliferating, matrix cells differentiate to form the hair channel, the inner root sheath, and the hair shaft. The mechanisms that control the stem cell proliferation and differentiation in the skin are now providing new insight into skin homeostasis.

MOLECULAR PATHOLOGY OF MENDELIAN GENETIC SKIN DISORDERS

There are approximately 5000 single gene disorders, of which nearly 600 have a distinct skin phenotype. Many of these disorders also have been characterized at a molecular level. Most inherited skin disorders are transmitted either by autosomal dominant, autosomal recessive, X-linked dominant, or X-linked recessive modes of inheritance. Understanding the precise pattern of inheritance, however, is essential for accurate genetic counseling. For example, the connective tissue disorder pseudoxanthoma elasticum (OMIM264800) for many years was thought to reflect a mixture of autosomal dominant and autosomal recessive genotypes.

However, it has been shown that all forms of pseudoxanthoma elasticum are in fact autosomal recessive and that the disease is caused by mutations in the ABCC6 gene, as the earlier reports of autosomal dominant inheritance were actually pseudo-dominant inheritance in consanguineous pedigrees. Understanding the molecular pathology of pseudoxanthoma elasticum also has identified similar autosomal recessive inherited skin diseases that have an overlapping phenotype, but in which there are subtle clinical differences. For example, a pseudoxanthoma elasticum-like disease with cutis laxa skin changes and coagulopathy (OMIM610842) has been shown to result from mutations in the GGCX gene. In both conditions, there is progressive accumulation of calcium phosphate in tissue, resulting in ocular and cardiovascular complications. The knowledge that these conditions are autosomal recessive helps with genetic counseling, prognostication, and management of families with affected members. Moreover, understanding the precise molecular pathology allows for more careful clinical monitoring and patient follow-up.

One of the principal functions of human skin is to provide a mechanical barrier against the external environment. The structural integrity of the skin depends on several proteins. These include intermediate filaments inside keratinocytes, intercellular junctional proteins between keratinocytes, and a network of adhesive macromolecules at the dermal-epidermal junction (Figure 27.3). Over the last 20 years, several Mendelian genetic disorders resulting from autosomal dominant or autosomal recessive mutations in structural proteins in the skin have provided fascinating insights into skin structure and function. In addition, determining the key roles of particular proteins has provided a plethora of new clinically and biologically relevant data.

One of the best characterized groups of disorders is epidermolysis bullosa (EB), a group of skin fragility disorders characterized by blister formation of the skin and mucous membranes that occurs following mild trauma. EB simplex is the commonest form of inherited EB. Transmission is mainly autosomal dominant. Ultrastructurally, the level of split is through the cytoplasm of basal cells, often close to the inner hemidesmosomal plaque. In dominant forms of EB simplex, there may be disruption of keratin tonofilaments or aggregation of keratin filaments into bundles (Figure 27.4). The molecular defects that cause EB simplex lie in either the keratin 5 gene (KRT5) or the keratin 14 gene (KRT14), or, in cases of the autosomal recessive EB simplex-muscular dystrophy, in the plectin gene (PLEC1). The molecular pathology of keratin gene mutations provides some insight into genotype/phenotype correlation. Notably mutations in the helix initiation/helix termination motifs in helices 1A and 2B of the keratin genes result in more severe subtypes. Clinically, EB simplex patients have the mildest skin lesions and scarring is not frequent. Extracutaneous involvement is rare, apart from cases with plectin pathology. The mildest clinical subtype of EB simplex is the Weber-Cockayne variant (OMIM131800) in which blistering occurs mainly on the palms and

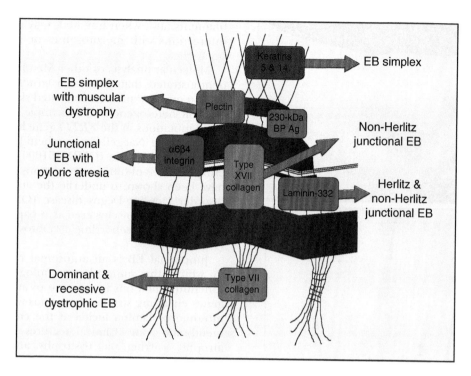

Figure 27.3   Illustration of the integral structural macromolecules present within hemidesmosome-anchoring filament complexes and the associated forms of clinical epidermolysis bullosa that result from autosomal dominant or autosomal recessive mutations in the genes encoding these proteins.

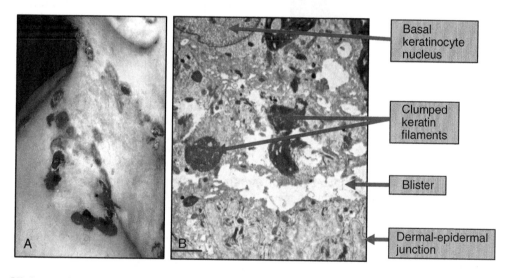

Figure 27.4   Clinico-pathological consequences of mutations in the gene encoding keratin 14 (*KRT14*), the major intermediate filament protein in basal keratinocytes. (A) The clinical picture shows autosomal dominant Dowling-Meara epidermolysis bullosa simplex. **(B)** The electron micrograph shows keratin filament clumping and basal keratinocyte cytolysis (bar = 1 μm).

soles (Figure 27.5). The molecular pathology involves autosomal dominant mutations that occur mainly outside the critical helix boundary motifs. In some forms of EB simplex, such as the Köbner subtype, the disease is more generalized than the Weber-Cockayne variant, although there is considerable overlap. The most severe form of EB simplex is the Dowling-Meara type

(OMIM131760). This often presents with generalized blister formation shortly after birth and it can be fatal in neonates. A characteristic of this condition in later childhood is the grouping of lesions in a herpetiform clustering arrangement (Figure 27.4). This form of EB simplex is associated with the greatest disruption of keratin tonofilaments. Another autosomal dominant

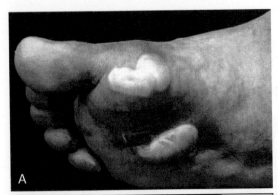

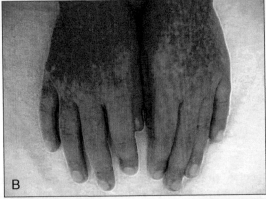

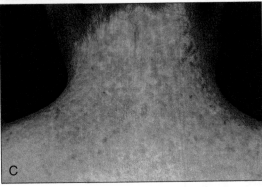

Figure 27.5 Spectrum of clinical abnormalities associated with dominant mutations in keratin 5 (KRT5). (A) Missense mutations in the nonhelical end domains result in the most common form of EB simplex, which is localized to the hands and feet (Weber-Cockayne variant). (B) A specific mutation in keratin 5, p.P25L, is the molecular cause of epidermolysis bullosa simplex associated with mottled pigmentation. (C) Heterozygous nonsense or frame shift mutations in the KRT5 gene leads to Dowling-Degos disease.

form of EB simplex is one associated with mottled skin pigmentation (OMIM131960). Apart from blisters, there is diffuse speckled hyperpigmentation as well as keratoderma of the palms and soles. The hypermelanotic macules are most evident in the axillae, limbs, and lower abdomen. The underlying molecular pathology in all reports is the same heterozygous proline to leucine substitution at codon 25 in the nonhelical V1 domain of the KRT5 gene. This proline residue is expressed on the outer part of polymerized keratin

filaments and, when mutated, may result in abnormal interactions with melanosomes or other keratinocyte organelles.

Molecular analysis of other Mendelian disorders has demonstrated that these skin structural proteins may also be mutated in other inherited skin diseases, known as allelic heterogeneity. For example, heterozygous nonsense mutations in the KRT14 gene have been shown to result in the Naegeli-Franceschetti-Jadassohn form of ectodermal dysplasia (OMIM161000). In addition, heterozygous loss-of-function mutations in the KRT5 gene have been shown to underlie the autosomal dominant disorder Dowling-Degos disease (OMIM179850). This is characterized by clustered skin papules, the histology of which shows seborrhoeic keratosis-like morphology (Figure 27.5).

Junctional EB is an autosomal recessive condition in which the molecular pathology involves loss-of-function mutations in any one of at least six different genes encoding structural proteins within the hemidesmosome or lamina lucida at the cutaneous basement membrane zone. Clinical features include blistering, atrophic scarring, nail dystrophy, and defective dental enamel as well as other abnormalities affecting the hair, eyes, and genito urinary tract. Ultrastructurally, the level of split is mainly through the lamina lucida of the basement membrane zone. The most severe type of junctional EB is the Herlitz subtype (OMIM226700). The underlying molecular pathology involves homozygous or compound heterozygous loss-of-function mutations in any of the three genes that encode the laminin-332 polypeptide; that is, the LAMA3, LAMB3, or LAMC2 genes (Figure 27.6)

Some forms of junctional EB are termed non-Herlitz (OMIM226650). In these cases there may be extensive blisters at birth, but the disease typically lessens in severity with time. This type of junctional EB is genetically heterogeneous. It may result from mutations in either the LAMA3, LAMB3, or LAMC2 genes (i.e., the subcomponents of laminin-332) or alternatively due to loss-of-function mutations on both alleles of the gene encoding type XVII collagen, COL17A1. The range of COL17A1 gene mutations includes missense, nonsense, frame-shift, or splice-site mutations, but usually there is total ablation of type XVII collagen protein. Another subtype of junctional EB is associated with a further extracutaneous abnormality, namely, pyloric atresia. The molecular pathology involves abnormalities in the $\alpha_6\beta_4$ integrin complex. Most patients have mutations in the gene encoding β_4 integrin, ITGB4, with the most severe cases usually having nonsense or frame-shift mutations. Mutations in the ITGA6 gene are seen less frequently.

Dystrophic EB represents a third type of inherited skin blistering (Figure 27.7). Ultrastructurally, the level of split is below the lamina densa and the underlying molecular defect in all types of dystrophic EB involves mutations in the type VII collagen gene (COL7A1). Transmission electron microscopy reveals abnormalities in the number and/or morphology of anchoring fibrils, which are composed principally of type VII collagen. Dominant forms of dystrophic EB

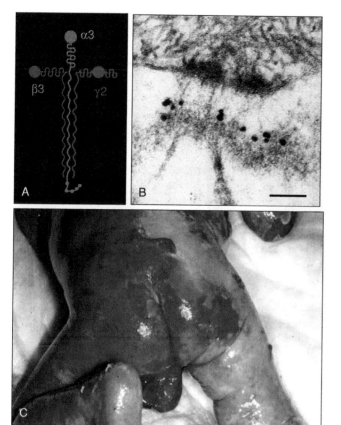

Figure 27.6 Laminin-332 mutations result in junctional epidermolysis bullosa. (A) Laminin-332 consists of three polypeptide chains: α3, β3, and γ2. **(B)** Immunogold electron microscopy shows laminin-332 staining at the interface between the lamina lucida and lamina densa subjacent to a hemidesmosome (bar = 50 nm). **(C)** Loss-of-function mutations in any one of these genes encoding the three polypeptides chains results in Herlitz junctional epidermolysis bullosa, which is associated with a poor prognosis, usually with death in early infancy.

usually are caused by heterozygous glycine substitution mutations within the type VII collagen triple helix. These result in dominant-negative interference and disruption of anchoring fibril formation. Affected individuals have mild trauma-induced blisters, mainly on skin overlying bony prominences such as the knees, ankles, or fingers. Blistering is followed by scarring and milia formation. Nail dystrophy is common.

It is clear, however, that the *COL7A1* pathology alone cannot account for the phenotypic variability and that other modifying genes, or environmental factors, may play a role. The most severe form of dystrophic EB is the autosomal recessive Hallopeau-Siemens subtype (OMIM226600). Blister formation in affected individuals starts from birth or early infancy and the skin is very fragile. Wound healing is often poor, leading to chronic ulcer formation with exuberant granulation tissue formation, repeated secondary infection, and frequent scar formation. Mucous membranes are extensively affected and esophageal involvement causes dysphagia and obstruction due to stricture. Affected

individuals also have an increased risk of squamous cell carcinoma. The molecular pathology involves loss-of-function mutations on both alleles of the *COL7A1* gene, leading to markedly reduced or completely absent type VII collagen expression at the dermal-epidermal junction. Overall, genotype/phenotype correlation suggests that in recessive dystrophic EB the amount of type VII collagen that is expressed at the dermal-epidermal junction is inversely proportional to clinical severity in terms of scarring and extent of blistering.

To maintain the structural function of the epidermis a number of intercellular junctions exist, including desmosomes, tight junctions, gap junctions, and adherens junctions. Mutations in these junctional complexes result in several Mendelian inherited skin diseases. Desmosomes are important cell–cell adhesion junctions found predominantly in the epidermis and the heart. They consist of three families of proteins: the armadillo proteins, cadherins, and plakins (Figure 27.8). Mutations in desmosomal proteins result in skin, hair, and heart phenotypes (Figure 27.9). Armadillo proteins contain several 42 amino acid repeat domains and are homologous to the drosophila armadillo protein. They bind to other proteins through their armadillo domains and play a variety of roles in the cell, including signal transduction, regulation of desmosome assembly, and cell adhesion. Mutations in the plakophilin 1 gene (*PKP1*) result in autosomal recessive ectodermal dysplasia-skin fragility syndrome (OMIM604536). Affected individuals have a combination of skin fragility and inflammation and abnormalities of ectodermal development, such as scanty hair, keratoderma, and nail dystrophy. Pathogenic *PKP1* mutations are typically splice-site or nonsense mutations. Autosomal dominant and recessive mutations also have been described in the plakoglobin gene (*JUB*). A recessive mutation results in Naxos disease (OMIM601214), a genodermatosis frequently seen on Naxos Island where approximately in 1:1000 individuals are affected with clinical features of arrhythmogenic right ventricular dysplasia, diffuse palmar keratoderma, and woolly hair. Autosomal dominant mutations in plakoglobin can also result in cardiomyopathy.

The cadherins comprise a group of desmogleins and desmocollins. Heterozygous mutations in the desmoglein 1 gene (*DSG1*) result in autosomal dominant striate palmoplantar keratoderma (OMIM148700). The molecular pathology in this disorder results from desmoglein 1 haploinsufficiency. Mutations in the desmoglein 4 gene (*DSG4*) result in localized autosomal recessive hypotrichosis (OMIM607903) in which affected individuals have hypotrichosis restricted to the scalp, chest, arms, and legs, but sparing of axillary or pubic hair. Papules on the scalp show atrophic curled up hair follicles and shafts with marked swelling of the precortical region. Recessive mutations in desmoglein 4 may also underlie some cases of autosomal recessive monilethrix. Plakins comprise a family of proteins that cross-link the cytoskeleton to desmosomes. They include desmoplakin, envoplakin, periplakin, plectin, bullous pemphigoid antigen 1, corneodesmosin, and microtubule actin crosslinking factor.

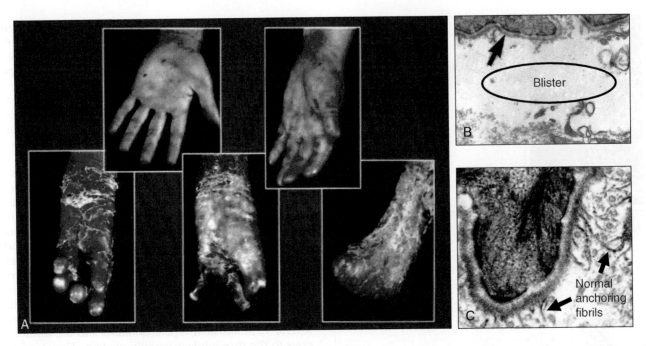

Figure 27.7   Clinico-pathological abnormalities in the dystrophic forms of epidermolysis bullosa. (A) This form of epidermolysis bullosa is associated with variable blistering and flexion contraction deformities, here illustrated in the hands. **(B)** The disorder results from mutations in type VII collagen (*COL7A1* gene), the major component of anchoring fibrils at the dermal-epidermal junction. This leads to blister formation below the lamina densa (lamina densa indicated by arrow). **(C)** In contrast, in normal human skin there is no blistering and the sublamina densa region is characterized by a network of anchoring fibrils.

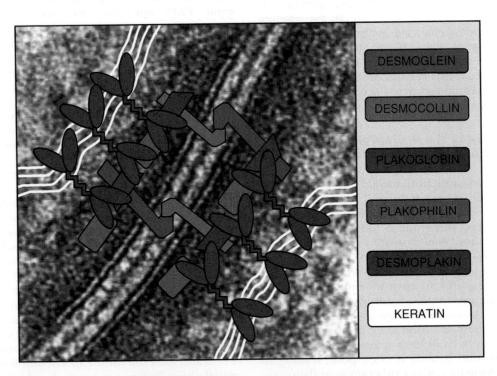

Figure 27.8   Protein composition of the desmosome linking two adjacent keratinocytes. The major transmembranous proteins are the desmogleins and the desmocollins. Several desmosomal plaque proteins, including desmoplakin, plakophilin, and plakoglobin, provide a bridge that links binding between the transmembranous cadherins and the keratin filament network within keratinocytes.

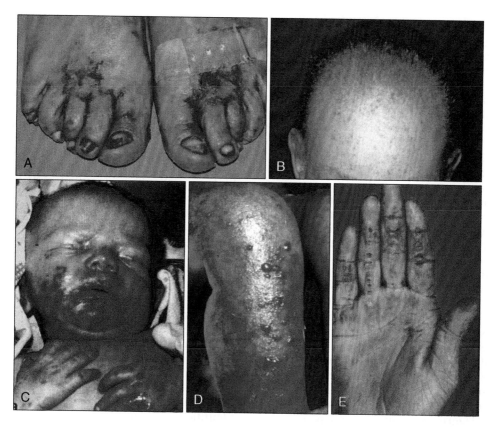

Figure 27.9   Clinical abnormalities associated with inherited gene mutations in desmosome proteins. **(A)** Recessive mutations in plakophilin 1 result in nail dystrophy and skin erosions. **(B)** Woolly hair is associated with several desmosomal gene abnormalities, particularly mutations in desmoplakin. **(C)** Recessive mutations in plakophilin 1 can result in extensive neonatal skin erosions, particularly on the lower face. **(D)** Recessive mutations in desmoplakin can lead to skin blistering. **(E)** Autosomal dominant mutations in desmoplakin do not result in blistering, but can lead to striate palmoplantar keratoderma.

Mutations in the desmoplakin gene (*DSP*) result in a variable combination of skin, hair, and cardiac abnormalities. These can be autosomal dominant or recessive. The phenotype in autosomal dominant cases ranges from cutaneous striate palmoplantar keratoderma to cardiac arrhythmogenic ventricular dysplasia 8 (OMIM607450). The phenotype in autosomal recessive patients ranges from skin fragility-woolly hair syndrome (OMIM607655) to other syndromes in which the skin and heart may be abnormal, such as dilated cardiomyopathy with woolly hair and keratoderma (OMIM605676). Compound heterozygous mutations that almost completely ablate the desmoplakin tail recently have been demonstrated to produce the most severe clinical subtype known as lethal acantholytic epidermolysis bullosa. Autosomal dominant mutations in the corneodesmosin gene (*CDSN*) result in hypotrichosis of the scalp (OMIM146520).

MOLECULAR PATHOLOGY OF COMMON INFLAMMATORY SKIN DISEASES

Atopic dermatitis and psoriasis represent two of the most common inflammatory skin dermatoses. Recent molecular insights are now providing new ideas about disease susceptibility and pathogenesis.

Atopic dermatitis is a chronic itching skin disease that results from a complex interplay between strong genetic and environmental factors. There are two forms of atopic dermatitis—the extrinsic and intrinsic forms. The former is associated with IgE-mediated sensitization whereas the latter is characterized by a normal total serum IgE level and the absence of specific IgE responses to aero-allergens and food-derived allergens.

To dermatologists, the association between atopic dermatitis and the monogenic disorder ichthyosis vulgaris (OMIM146700) has been evident for many years, given that several patients with ichthyosis vulgaris also have atopic dermatitis. Histopathologically, many cases of ichthyosis vulgaris are associated with abnormal (diminished) keratohyalin granules within the granular layer. Filaggrin is the major component of keratohyalin granules. Filaggrin is a composite phrase for <u>fil</u>ament <u>aggr</u>egating prote<u>ins</u>, repeat units of complex polypeptides derived from profilaggrin that help aggregate keratin filaments in the formation of the epidermal barrier. In 2006, it was shown that ichthyosis vulgaris resulted from loss-of-function mutations in the *FLG* gene. Ichthyosis vulgaris is a semidominant condition with heterozygotes displaying no phenotype or just mild ichthyosis whereas homozygotes or compound heterozygotes have a more severe form. Filaggrin mutations are very common in the general population, occurring in approximately 10% of the European population. Subsequently, it has been shown that filaggrin gene mutations are a major primary predisposing risk factor for atopic dermatitis. It is evident that approximately 50% of all cases of severe atopic dermatitis harbor mutations in the filaggrin gene. It has also been shown that the presence of filaggrin mutations is also a risk factor for asthma, but only for asthma in combination with atopic dermatitis, and not for asthma alone. This finding suggests that asthma in individuals with atopic dermatitis is secondary to allergic sensitization, which develops because of the defective epidermal barrier that allows allergens to penetrate the skin to make contact with antigen presenting cells. Filaggrin is not expressed in respiratory epithelium and therefore the new data on filaggrin mutations offer an intriguing new concept that atopic asthma may be initiated as a result of a primary cutaneous (rather than respiratory) abnormality in some individuals. The hypothesis of a defective epidermal barrier underlying asthma, and indeed allergic sensitization, has been verified in several studies, which have shown an association between filaggrin gene mutations and extrinsic atopic dermatitis associated with high total serum IgE levels and concomitant allergic sensitizations.

Primary defects in filaggrin, however, are not the entire basis of the molecular pathology of atopic dermatitis. It is likely that genes in several other factors, particularly in milder cases of atopic dermatitis, are also important.

Recent research focus on abnormalities of the epidermal barrier in atopic dermatitis is providing fascinating new insights into understanding the nature and etiology and perhaps is leading to novel treatments of atopic dermatitis. The presence of filaggrin gene mutations is set to influence and accelerate the design of new treatments that restore filaggrin expression and skin barrier function, given the new evidence that restoration of an intact epidermis may prevent both atopic dermatitis and cases of atopic dermatitis-associated asthma as well as systemic allergies. One of the common mutations in filaggrin is a nonsense mutation (p.R501X), which may represent an attractive drug target for small molecule approaches that modify post-transcriptional mechanisms designed to increase read-through of nonsense mutations and thereby stabilize mRNA expression. Other approaches that involve drug library screening or *in silico* methods to identify compounds capable of increasing filaggrin expression in the epidermis are also likely to lead to new evidence-based topical preparations suitable for the treatment of atopic dermatitis and ichthyosis vulgaris.

Psoriasis is a common and complex disease. It may manifest with inflammation in the skin as well as the nails and joints of patients. In recent years, considerable evidence has emerged for specific patterns of immunological abnormalities and for the critical role of certain cytokines and chemokine networks in psoriasis, findings that may be relevant to the design of new molecular therapies (Figure 27.10).

There is overwhelming evidence that psoriasis has an important genetic component, in that there is a higher incidence among first- and second-degree relatives of patients than unaffected control subjects and there is a higher concordance in monozygotic compared to dizygotic twins. The disease has a bimodal distribution of age of onset, with an early peak between 16 and 22 years and a later one between 57 and 60 years. Linkage studies have identified several genetic loci.

Aside from genetic risk factors, environmental risk factors include trauma (the Köbner phenomenon at sites of injury is well known), infection (including *Streptococcal* bacteria and human immunodeficiency virus infections), drugs (lithium, anti-malarials, β-blockers, and angiotensin converting enzymes), sunlight (in a minority), and metabolic factors (such as high-dose estrogen therapy and hypocalcemia). Other factors including psychogenic stress, alcohol, and smoking also have been implicated.

Psoriasis traditionally has been considered as a Th-1 disease, based on the identification of IFN-γ and TNF-α in the lesions with little or no detection of IL-4, IL-5, or IL-10. More recently other cytokines, including IL-18, IL-19, IL-22, and IL-23 have been identified as being up-regulated in psoriatic lesions. With relevance to the molecular pathology of psoriasis, one of the first clues came from the observation of clearing of psoriasis in renal transplant patients given the immunosuppressant cyclosporine A. In psoriatic skin that resolved in these individuals there were reductions in levels of several cytokines, indicating a connection between immunocytes, cytokines, and maintenance of psoriatic plaques. The effects of neutralizing a single cytokine can be extremely helpful, particularly with the use of anti-TNF-α therapy, a treatment that initially was developed for patients with sepsis. TNF-α blockers or receptor blockers are beneficial in patients with psoriasis, but it is unclear if the improvement of psoriatic plaques and arthritis is due to a local or a systemic effect.

Intriguing new ideas, however, recently have been gleaned from specific mouse models, in which prepsoriatic skin is grafted onto an AGR mouse (which lacks type I or IFN-α receptors as well as lacking type II or IFN-γ receptors, and is RAG deficient). In such mice, psoriatic plaques develop spontaneously. However, if these mice receive injections of anti-TNF-α agents, psoriasis does not develop. This demonstrates that resident immunocytes, contained within prepsoriatic skin, are necessary and sufficient to trigger psoriasis and that the local production of TNF-α is critically important in the generation of skin lesions.

A new area of investigation in the field of chronic inflammation centers around a possible inflammatory axis in which IL-12/IL-23 influences levels of a cytokine known as IL-17. This has led to a new paradigm through which Th-17-type T-cells contribute to autoimmunity and chronic inflammation. Thus, besides a Th-1- or Th-2-type immune system, there is also a

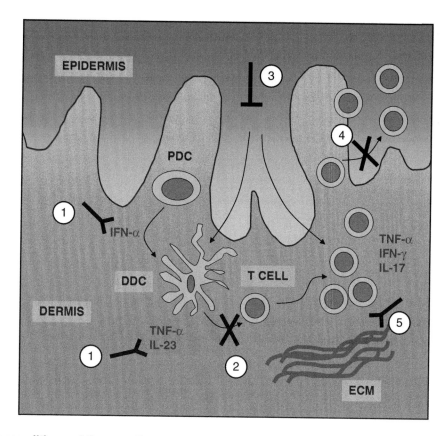

Figure 27.10 Abnormalities and therapeutic potential for inflamed skin in psoriasis. There is increasing evidence for a role of tissue resident immune cells in the immunopathology of psoriasis. New therapies may be developed by **(1)** antagonizing local cytokines and chemokines, such as IFN-α; **(2)** blocking of adhesion molecules (e.g., integrins) and costimulatory molecules within the tissue; **(3)** modification of keratinocyte proliferation and differentiation (e.g., use of corticosteroids or vitamin D preparations); **(4)** blocking of entry of dermal T-cells into the epidermis; and **(5)** modification of the micro-environment, including the extracellular matrix. Adapted from Boyman O, Conrad C, Tonel G, et al. The pathogenic role of tissue-resident immune cells in psoriasis. *Trends Immunol.* 2007;28:51–57.

cytokine network dominated by a Th-17-type response. IL-17 may be important in psoriasis because it can promote accumulation of neutrophils and can affect skin barrier function by inducing release of pro-inflammatory mediators by keratinocytes. Understanding the molecular of IL-17 in psoriatic skin may provide new insights for developing novel therapies.

SKIN PROTEINS AS TARGETS FOR INHERITED AND ACQUIRED DISORDERS

The integrity of the skin as a mechanical barrier depends, in part, on adhesive complexes that link cell-to-cell and cell-to-basement membrane. Two key junctional complexes in this task are the hemidesmosomes and the desmosomes. In addition to genetic diseases, however, further clues to the function of hemidesmosomal and desmosomal proteins have been derived from animal models or human diseases, in which the same structural proteins can be targeted and disrupted by auto-antibodies (Figure 27.11). Thus several hemidesmosomal and desmosomal proteins

may serve as target antigens for both inherited and acquired disorders (Figure 27.12).

One of the main transmembranous hemidesmosomal proteins is type XVII collagen, also known as the 180-kDa bullous pemphigoid antigen. Auto-antibodies against this protein typically result in bullous pemphigoid, a chronic vesiculo-bullous disease that usually affects the elderly. Histology shows subepidermal blisters with eosinophils and the pathogenic antibodies usually are directed against a particular epitope (within the NC16A domain) in the first noncollagenous extracellular part of type XVII collagen. IgG1 subclass auto-antibodies are found in patients with active skin lesions whereas IgG4 auto-antibodies are found in patients in remission. Besides bullous pemphigoid, other blistering conditions may also have auto-antibodies against type XVII collagen—for example, pemphigoid gestationis, which is an acute, pruritic vesiculo-bullous eruption in pregnant women. Patients produce auto-antibodies against the same epitope of type XVII collagen. IgA auto-antibodies against the NC16A domain of type XVII collagen give rise to two different skin diseases: chronic bullous disease of childhood and

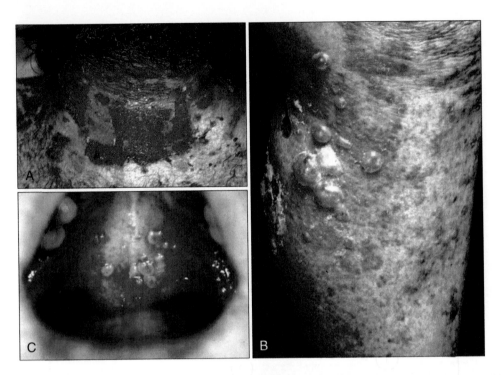

Figure 27.11   Clinical pathology resulting from auto-antibodies against desmosomes or hemidesmosomes.
(A) Pemphigus vulgaris resulting from antibodies against desmoglein 3. **(B)** Bullous pemphigoid associated with antibodies against type XVII collagen. **(C)** Mucous membrane pemphigoid associated with antibodies to laminin-332.

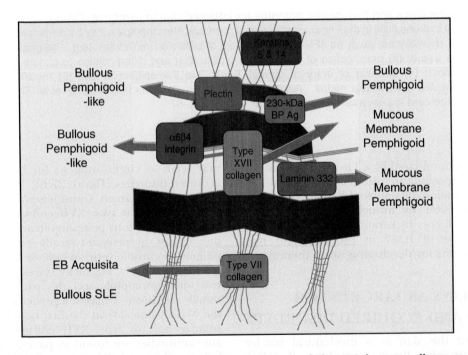

Figure 27.12   Illustration of hemidesmosomal structural proteins and the autoimmune diseases associated with antibodies directed against these individual protein components.

linear IgA bullous dermatosis. Chronic bullous disease of childhood usually presents in young children with clustered tense bullae, often in the perioral or perineal regions. Clinical manifestations can mimic bullous pemphigoid or dermatitis herpetiformis. A further disease with auto-antibodies against type XVII collagen is lichen planus pemphigoides. Patients typically have lichen planus lesions mixed with tense blisters, either on lichen planus lesions or on normal skin. Auto-antibodies react with epitopes within the NC16A domain, but the precise epitope is different from bullous pemphigoid.

Auto-antibodies to type VII collagen, the major component of anchoring fibrils, give rise to two different conditions: epidermolysis bullosa acquisita and bullous systemic lupus erythematosus. In both epidermolysis bullosa acquisita and bullous systemic lupus erythematosus there is linear deposition of IgG at the dermal-epidermal junction. A useful technique in immunodermatology to distinguish epidermolysis bullosa acquisita from bullous pemphigoid, however, is indirect immunofluorescence using 1 M sodium chloride split-skin (Figure 27.13). Incubation of normal skin in saline results in a split within the lamina lucida. Thus, certain skin antigens such as type XVII collagen map to the roof of the split whereas others such as type VII collagen map to the base. This means that in bullous pemphigoid and epidermolysis bullosa acquisita, labeling with sera from patients with these conditions on salt-split skin can permit an accurate diagnosis to be made. Bullous systemic lupus erythematosus reflects nonspecific bullous changes in patients with active disease. The histology resembles dermatitis herpetiformis, but the immunofluorescence and immunological analyses are typical of lupus erythematosus patients, and not all cases have antibodies to type VII collagen.

Another autoimmune blistering condition that targets hemidesmosomal components is mucous membrane pemphigoid. This is a chronic progressive autoimmune subepithelial disease characterized by erosive lesions of the skin and mucous membranes that result in scarring. Lesions commonly affect the ocular and oral mucosae. Direct immunofluorescence studies show IgG and/or IgA auto-antibodies at the dermal-epidermal junction. Salt-split indirect immunofluorescence can show epidermal, dermal, or roof and base labeling patterns, which reflects the different auto-antigens seen in this disease. In most cases, the target epitope is part of the extracellular domain of type XVII collagen, although this differs from the epitope associated with bullous pemphigoid. A minority of patients with mucous membrane pemphigoid display antibodies against laminin-332. This is an important subset of patients to identify since there is association with certain solid tumors, particularly malignancies of the upper aero-digestive tract. Cases of mucous membrane pemphigoid with antibodies to β_4 integrin tend to have only ocular involvement and may therefore be referred to as ocular mucous membrane pemphigoid.

Although many of the immunobullous diseases that target hemidesmosomal proteins involve immunopathology in which antibodies initially are directed against critical epitopes on individual proteins, it has been shown that in many cases there may be several antigens targeted by humoral immunity. This phenomenon is known as "epitope spreading" and reflects chronic inflammation that may alter the immunogenicity of other neighboring proteins involved in adhesion. In some patients, epitope spreading can lead to transition from one autoimmune disease to another. Published examples include transition from mucosal dominant pemphigus vulgaris to mucocutaneous pemphigus vulgaris; pemphigus foliaceus to pemphigus vulgaris; bullous pemphigoid to epidermolysis bullosa acquisita; dermatitis herpetiformis to bullous pemphigoid; pemphigus foliaceus to bullous pemphigoid; and concurrent bullous pemphigoid with pemphigus foliaceus. Antibodies to the plakin protein,

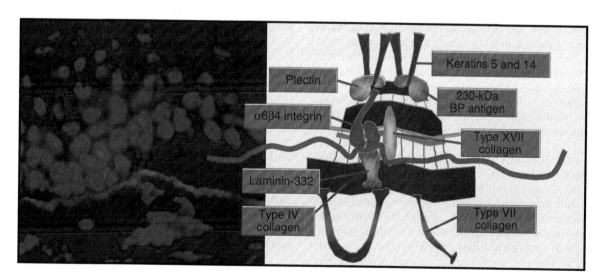

Figure 27.13 Salt-split skin technique to diagnose immunobullous disease. Incubation of normal human skin in 1 M NaCl overnight at 4°C results in cleavage through the lamina lucida. This results in separation of some proteins to the roof of the split and some to the base (above and below pink line on the schematic). In the skin labeling shown, immunoglobulin from a patient's serum binds to the base of salt-split skin. Further analysis revealed that the antibodies were directed against type VII collagen. This technique is useful in delineating bullous pemphigoid from epidermolysis bullosa acquisita, both of which are associated with linear IgG at the dermal-epidermal junction in intact skin.

desmoplakin, which are characteristically found in paraneoplastic pemphigus, also have been detected in pemphigus vulgaris as well as oral and genital lichenoid reactions. In epidermolysis bullosa acquisita, antibodies to laminin-332 may be found in addition to the typical type VII auto-antibodies. In bullous systemic lupus erythematosus, multiple auto-antibodies against type VII collagen, bullous pemphigoid antigen 1, laminin-332, and laminin-311 have been reported. The phenomenon of epitope spreading is perhaps best exemplified in paraneoplastic pemphigus, which is characteristically associated with auto-antibodies against many desmosomal and hemidesmosomal proteins, including desmogleins, plakins, bullous pemphigoid antigen 1, plectin, and plakoglobin.

In addition to hemidesmosomal proteins, the structural components of desmosomes may also be targets for autoimmune diseases. The skin blistering disease pemphigus is associated with auto-antibodies against desmosomal cadherins, principally desmoglein 3 and desmoglein 1. The intraepithelial expression patterns of desmoglein 1 and desmoglein 3, however, differ between the skin and mucous membranes, hence the different clinical manifestations when patients have antibodies against different desmogleins. In the skin, desmoglein 1 is expressed throughout the epidermis but more so in superficial part, as opposed to desmoglein 3, which is predominantly expressed in the basal epidermis. In the oral mucosa, both types of desmogleins are expressed throughout the epithelium, but desmoglein 1 is expressed at a much lower level than desmoglein 3. When there is coexpression of desmoglein 1 and desmoglein 3, these proteins may compensate for each other, but when expressed in isolation, specific pathology can arise when those proteins are targeted by auto-antibodies. For example, when there are only desmoglein 1 antibodies in the skin, blisters appear in the superficial epidermis, a site where there is no coexpression of desmoglein 3, whereas in the basal epidermis, the presence of desmoglein 3 can compensate for the loss of function of desmoglein 1. In the oral epithelium, keratinocytes express desmoglein 3 at a much higher level than desmoglein 1 and thus, despite the presence of desmoglein 1 antibodies, no blisters form. Therefore, sera containing only desmoglein 1 antibodies cause superficial blisters in the skin without mucosal involvement and the clinical consequences are pemphigus foliaceus and its localized form, pemphigus erythematosus. Desmoglein 1 is also the target antigen in endemic pemphigus, known as fogo selvagem. Desmoglein auto-antibodies result in skin with very superficial blisters, such that crust and scale are the predominant clinical features. Desmoglein 1 may also be targeted in a different manner—not by auto-antibodies, but by bacterial toxins (Figure 27.14). Specifically, exfoliative toxins A-D, which are produced by *Staphylococcus aureus*, specifically cleave the extracellular part of desmoglein 1. This leads to bullous impetigo and Staphylococcal scalded skin syndrome with superficial blistering in the epidermis, findings that are histologically similar to pemphigus foliaceus associated with auto-antibodies to the extracellular part of desmoglein 1.

When the patient's sera contains only desmoglein 3 auto-antibodies, however, coexpressed desmoglein 1 can compensate for the impaired function of desmoglein 3, resulting in no or only limited skin lesions. However, in the mucous membranes, oral erosions predominate as desmoglein 1 cannot compensate for the impaired desmoglein 3 function because of its low expression. Therefore, the patient typically has painful oral ulcers without much initial skin involvement. This accounts for the clinical phenotype in patients with the mucosal-dominant type of pemphigus vulgaris. However, when sera contain antibodies to both desmoglein 1 and desmoglein 3, as seen in the mucocutaneous type of pemphigus vulgaris, patients have extensive blisters, mucosal erosions, and skin blisters, because the function of both desmogleins is disrupted.

Intercellular auto-antibodies in pemphigus typically are composed of IgG isotypes, although some patients with superficial blistering may have intercellular auto-antibodies of the IgA subclass. There are two clinical subtypes of IgA pemphigus. One is a subcorneal pustular dermatosis subtype, and this is associated with flaccid vesico-pustules with clear fluid and pus arranged in an annular/polycyclic configuration. Indirect immunofluorescence studies show IgA auto-antibodies reacting against the superficial epidermis. These antibodies usually recognize epitopes within the desmosomal cadherin, desmocollin 1. In contrast, IgA pemphigus patients may have a different phenotype known as the intra-epidermal neutrophilic subtype, which presents with a sunflower-like configuration and pustules and vesicles with indirect immunofluorescence showing IgA intercellular auto-antibodies throughout the entire thickness of the epidermis. As well as spontaneous onset cases of pemphigus, certain drugs as such D-penicillamine can lead to drug-induced pemphigus. Most patients have auto-antibodies against the same target epitopes as pemphigus patients. In paraneoplastic pemphigus, the clinical presentation involves progressive blistering and erosions on the upper trunk, head, neck, and proximal limbs with an intractable stomatitis. Erythema multiforme-like lesions on the palms and soles can help distinguish the condition from ordinary pemphigus. Indirect immunofluorescent studies on rat bladder, an organ rich in transitional epithelial cells, demonstrate IgG intercellular auto-antibodies consistent with autoimmunity against plakin proteins, although numerous antigenic targets are usually present.

MOLECULAR PATHOLOGY OF SKIN CANCER

Skin cancer is very common. Nonmelanoma skin cancers, which include basal cell carcinomas (BCCs) and squamous cell carcinomas (SCCs), are the most common forms of human neoplasia.

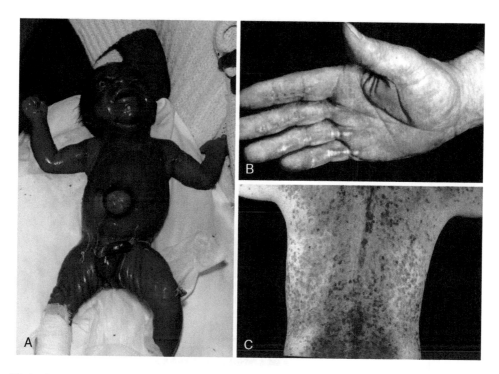

Figure 27.14 Clinical consequences of disruption of desmoglein 1 in human skin. (A) Staphylococcal toxins cleave the extracellular part of desmoglein 1 and result in staphylococcal scalded skin syndrome. **(B)** Inherited autosomal dominant mutations in desmoglein 1 can result in striate palmoplantar keratoderma. **(C)** Auto-antibodies against desmoglein 1 result in pemphigus foliaceus, which is associated with superficial blistering and crusting in human skin.

The main risk factor for skin neoplasia is environmental exposure to ultraviolet irradiation. The ultraviolet component of sunlight can be divided into three energy subtypes: UVC (100–280 nm), UVB (280–315 nm), and UVA (315–400 nm). The action spectra for UVB and UVC closely match the absorption spectrum of DNA, resulting in the formation of pyrimidine dimers, involving nucleotides, thiamine (T), and/or cytosine (C). In contrast, UVA is absorbed by other cellular chromophores, thereby generating oxygen reactive species such as hydroxyl radicals. These can result in DNA strand breaks and chromosome translocations. Of the ultraviolet light that reaches the earth's surface, UVA accounts for more than 90%, and UVB approximately 10%. Most UVB-induced mutations are located almost exclusively at dipyrimidine nucleotide sites (TT, CC, CT, and TC). Approximately 70% of observed mutations are C-to-T transitions and 10% are CC-to-TT, the latter representing an ultraviolet signature mutation. Cellular mechanisms of DNA repair are not always effective and these ultraviolet B-induced mutations can proceed unchecked, becoming permanent and subsequently inherited by all the progeny of the mutated cell, thereby allowing expression of the aberrant gene/protein function. Aside from ultraviolet radiation inducing DNA point mutations and small deletions, it may also result in gross chromosomal changes.

Collectively, these genetic/chromosomal changes initiate and promote cancer formation as well as the increased genomic instability and loss-of-heterozygosity frequently observed in nonmelanoma skin cancer. Cytogenic analysis has enabled the identification of a number of chromosomal abnormalities associated with nonmelanoma skin cancer and has implicated certain regions containing oncogenes and tumor suppressor genes that may be involved in their development. For example, in BCCs, early loss-of-heterozygosity studies identified regions on chromosome 9q22 as a common observation specific to these tumors. These regions harbor the Patched tumor suppressor gene (*PTCH*), which is a transmembrane receptor involved in the regulation of hedgehog signaling (Figure 27.15). Subsequently, the discovery of mutations that are known to activate hedgehog signaling pathways, including PTCH, Sonic hedgehog (Shh), and Smoothened (Smo), implicates hedgehog signaling as a fundamental transduction pathway in skin tumor development.

In the skin, the Shh pathway is crucial for maintaining the stem cell population and regulating the development of hair follicles and sebaceous glands. Although key embryonic developmental signaling pathways may be switched off during adulthood, aberrant activation of these pathways in adult tissue is often oncogenic. The Shh pathway may be activated in many neoplasms, and abnormalities in Shh signaling pathway components—such as Shh, PTCH1, Smo, GLI1, and GLI2—are major contributing factors in the development of BCCs. The function of PTCH1 is to repress Smo signaling. This function is impaired

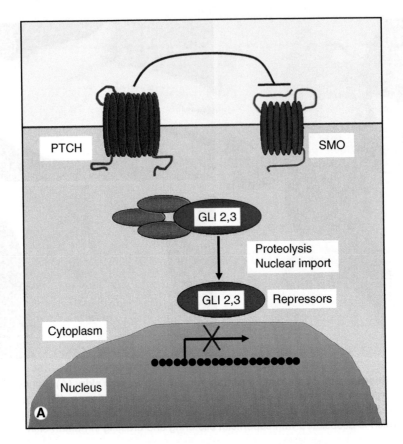

Figure 27.15 **The SHH signaling pathway. (A)** In the absence of SHH, PATCHED constitutively represses smoothened, a transducer of the SHH signal.

Continued

when PTCH1 is mutationally inactivated or when stimulated by Shh binding: both of these lead to uncontrolled Smo signaling. Downstream of Smo are the GLI transcription factors. Overexpression of GLI1 or GLI2 can lead to BCC development and GLI1 can also activate platelet-derived growth factor receptor-α, the expression of which may be increased in BCCs. The roles of other Smo target molecules such as the suppressor of fused Su(Fu) and protein kinase A (PKA) in the development of BCC are not fully understood. Other components of the pathway include a putative antagonist of Smo signaling, known as hedgehog interacting protein (HIP), and an actin binding protein, missing in metastases (MIM), which is a Shh-responsive gene. MIM is a part of the GLI/Su(Fu) complex and potentiates GLI-dependent transcription using domains distinct from those used for monomeric actin binding. Alterations in Shh regulate cell proliferation and associated cell cycle events. Shh overexpression leads to epidermal hyperplasia, accompanied by the proliferation of normally growth-arrested cells. Shh-expressing cells fail to exit the S and G2/M phases in response to calcium-induced differentiation signals and are unable to block the p21CIP1/WAF1-induced growth arrest in skin keratinocytes. Furthermore, PTCH1 protein interacts with phosphorylated cyclin B1 and blocks its translocation to the nucleus. The Shh/GLI pathway also up regulates

expression of the phosphatase CDC25b, which is involved in G2/M-transition. Thus, the loss of regulation of cell cyclin control is associated with the development of epithelial cancers, including BCCs.

Patients with the autosomal dominant disorder known as nevoid basal cell carcinoma syndrome or Gorlin syndrome (OMIM109400) have substantially increased susceptibility to BCCs and other tumors. They may also manifest jaw cysts and ectopic calcification, spina bifida, rib defects, palate abnormalities, coarse facies, hypertelorism, microcephaly, and skeletal abnormalities. Affected individuals develop BCCs on any part of the body. These patients have heterozygous germline mutations in the *PTCH* gene. The BCCs in these individuals retain a mutant germline and lose the wild-type allele as a second hit. Mutations in the *PTCH* gene have also been demonstrated in 50–60% of sporadic BCCs, emphasizing that *PTCH* gene mutations are important in the development of BCCs. Other sporadic BCCs that do not have mutations in *PTCH* may carry mutations in the *Smo* gene. An understanding of the mutations that lead to activation of hedgehog signaling has thus expanded our knowledge of the genetic basis of BCCs.

Insight into the molecular pathology of BCCs has led to development of a number of animal models that may be useful in developing chemoprevention strategies, as well as confirming specific contributions from

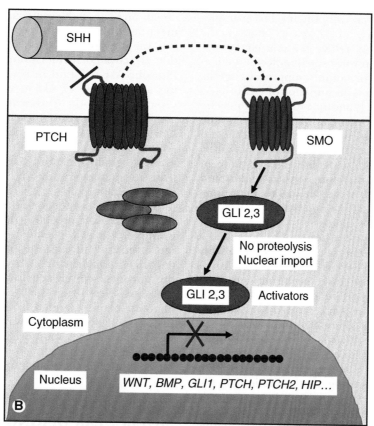

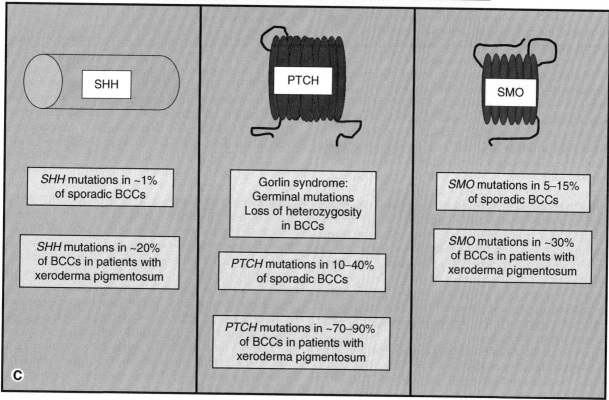

Figure 27.15–Cont'd. **(B)** Binding of the ligand SHH to PTCH relieves its inhibition of SMO and transcriptional activation occurs through the GLI family of proteins, resulting in activation of target genes. **(C)** Mutations in *SHH* or *PTCH* or *SMO* may be associated with basal cell carcinomas, both in sporadic tumors as well as in certain genodermatoses, such as xeroderma pigmentosum, that are associated with an increased risk of BCC. Germinal mutations in the *PTCH* gene underlie Gorlin's syndrome. Adapted from Singh M, Lin J, Hocker TL, et al. Genetics of melanoma tumorigenesis. *Br J Dermatol.* 2008;158:15–21.

hedgehog signaling pathway components. For example, drugs such as cyclopamine are known to be a specific inhibitor of Shh signaling. Other possible chemoprevention strategies might involve small molecule hedgehog signaling inhibitors or immunosuppressive agents such as rapamycin, which is an inhibitor of GLI1.

Like BCCs, melanoma is another tumor that may be entirely curable by early recognition and therapeutic intervention. Nevertheless, for patients with advanced metastatic melanoma, the 5-year survival is currently estimated at only 6% with a median survival time of 6 months. It is important, therefore, to try to understand the molecular pathology of melanoma to determine which patients are most at risk and which tumors will exhibit the most aggressive biology.

Dissection of the melanoma genome has revealed a number of *driver* mutations (i.e., those that probably influence the growth of tumors) and *passenger* mutations (i.e., those that do not specifically promote tumor growth). One important pathway is the RAS/Mitogen-activated protein kinase (MAPK) pathway, which regulates cell proliferation and survival in several cell types (Figure 27.16). Activating cell mutations in NRAS and BRAF have a combined prevalence of approximately 90% in melanoma and in benign melanocytic lesions, suggesting that activation of the MAPK pathway is an early essential step in melanocytic proliferation. Activating somatic NRAS mutations occur in 10% to 20% of melanomas. Of note, three highly recurrent NRAS missense changes represent over 80% of all mutations in this gene. NRAS mutations are more common on chronically sun-exposed sites and appear to occur early in tumorigenesis as well as being common in congenital nevi. Downstream of NRAS, activating mutations in BRAF have been identified, including a common missense mutation at valine 600. This mutation is equally prevalent in benign nevi and in melanoma, suggesting that BRAF activation is necessary for melanocytic proliferation, but not for tumorigenesis. To date, there has been no correlation between BRAF mutations, disease progression, and clinical outcome.

A proto-oncogene, KIT, encodes the stem cell factor receptor tyrosine-kinase found on numerous cell types, including melanocytes. c-KIT signals via the MAPK pathway and a downstream target is microphthalmia transcription factor (MITF), which is a critical regulator of melanocyte function. MITF regulates the development and differentiation of melanocytes and maintains melanocyte progenitor cells. MITF has a clear role in melanocyte survival and one of its transcriptional targets is the apoptosis antagonist and proto-oncogene BCL-2.

Apart from MAPK signaling, another pathway, the V-RAS/phosphatidyloinositol-3 kinase (PIK3) pathway, may also be activated in certain melanomas. Although PIK3 itself is not mutated in melanoma, several downstream components may have a role in melanoma tumorigenesis, including PTEN and Akt. PTEN has a tumor suppressor role in melanomas and its expression has been shown to be reduced or lost, through epigenetic inactivation, in some primary or metastatic melanomas. Restoration of PTEN in PTEN-deficient melanoma can reduce melanoma tumorigenicity and metastases. Akt is an important kinase in melanoma survival and progression. Akt3 is the main Akt isoform that is activated in melanomas. Strong expression of Phospho-Akt is found in several melanomas, suggesting a direct role of Akt in tumor progression. Phosphorylation of Akt is associated with increased PIK3 activity. Knowledge of these pathways is important if new treatments are to be developed for advanced melanoma since it is clear that alterations in survival and growth signaling pathways in melanoma tumor cells lead to increased tumorigenesis and resistance to chemotherapy.

A detailed knowledge of the molecular pathology of melanomas may also have therapeutic relevance. For example, it maybe possible to render melanoma cells more sensitive to existing forms of chemotherapy, to enhance apoptosis, or to restrict the proliferation of the oncogene-driven cells by targeting these aberrant pathways. New drugs such as Oblimersen (antisense oligonucleotide targeted to the anti-apoptotic protein BCL-2) and Sorafenib (a small molecule inhibitor of BRAF that induces apoptosis) are giving some encouraging results, especially when used in combination with traditional chemotherapy.

MOLECULAR DIAGNOSIS OF SKIN DISEASE

Understanding the molecular pathology of skin diseases has the potential to bring about several clinical and translational benefits for patients. Molecular data are helpful in diagnosing both inherited and acquired skin diseases and also contribute to improved disease classification, prognostication, clinical management, and the feasibility for designing and developing new treatments for patients. In many infectious diseases, the gold standard for diagnosis is identification of infectious agents by culture and species identification. However, material yield can be unsatisfactory and the process can take several weeks, or months, before a pathogen is identified. Indeed, certain diseases can be caused by different microorganisms and these may have different responses to antimicrobial agents. For example, subcutaneous mycosis with lymphangiotic spread is typically caused by the mold *Sporothrix schenkii*, but lymphatic sporotrichoid lesions can be caused by diverse pathogens. These include other nontuberculous mycobacteria or by bacteria such as *Nocardia* species; therefore precise identification of species is very important in the optimal management of such infections. Mycetoma is another chronic localized skin infection caused by different species of fungi or actinomycetes. Grains from the lesions contain pathogenic organisms, but final organism identification can be protracted. Molecular biology is able to help since sequence-based identification of large ribosomal subunits specific to particular organisms can be used to identify and characterize precise certain organisms that contribute directly to disease pathogenesis. PCR

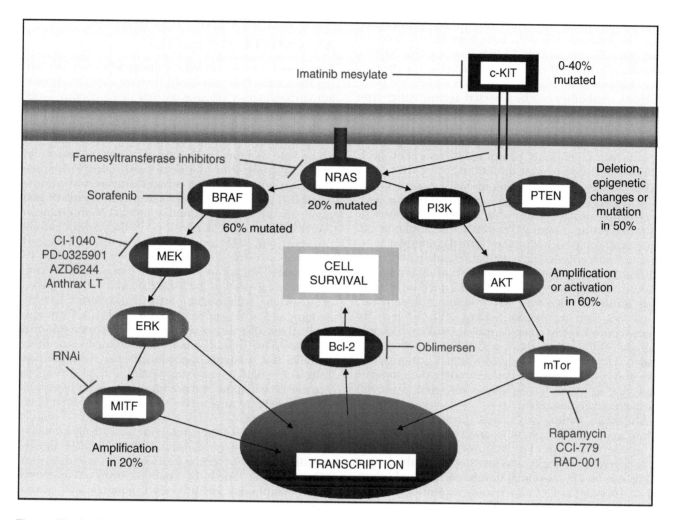

Figure 27.16 Potential for targeted therapies in melanoma. Recent improvement in defining the genetics of melanoma has led to the development of targeted therapeutic agents that are directed at specific molecular aberrations involved in tumor proliferation and resistance to chemotherapy. Adapted from Singh M, Lin J, Hocker TL, et al. Genetics of melanoma tumorigenesis. *Br J Dermatol.* 2008;158:15–21.

and sequencing approaches are also useful in identifying subsets of human papillomavirus. For example, epidermodysplasia verucciformis is a rare genodermatosis characterized by profound susceptibility to cutaneous infection with certain human papillomavirus subtypes. Approximately 50% of the patients develop nonmelanoma skin cancers on sun-exposed areas in the fourth or fifth decades of life. The cancer-associated viral subtypes are HPV-5, HPV-8, and HPV-14d; therefore molecular identification of these viruses in susceptible individuals can be helpful in promoting more vigorous surveillance of at-risk patients.

Molecular profiling also has proved to be useful in the identification and classification of cutaneous malignancies. Notably, T-cell receptor gene rearrangements, which arise as a result of variability in the V-J segment, provide important information relevant to the diagnosis and prognosis of cutaneous T-cell lymphoma. The presence of clonality usually is associated with the presence of malignancy. Thus, in many cases,

a T-cell receptor gene rearrangement can be helpful in distinguishing a malignant from a benign lymphocytic infiltrate, especially when clinical and histologic features are inconclusive. Clonality can also influence prognosis. For example, Sézary syndrome patients with T-cell clonality are more likely to die from lymphoma/leukemia than their nonclonal counterparts. DNA microarray profiling has also been used in patients with diffuse large B-cell lymphoma. Specific gene signatures have been linked to certain subtypes (e.g., prominent germinal center B-cell profile or activated B-cell profile), which can be directly associated with different prognoses.

The recent characterization of the molecular basis of the acquired immunobullous diseases also has led to new diagnostic tests. For example, antigen-specific ELISA kits for desmogleins 1 and 3 are now commercially available for the assessment of serum samples from patients with pemphigus. These ELISA tests are more sensitive and specific than immunofluorescence microscopy when used in diagnosing pemphigus. More importantly, the titers have been shown to correlate with disease activity. ELISA tests for the NC16A domain of type XVII collagen are useful in the diagnosis of bullous pemphigoid. In mucous membrane pemphigoid, antibodies can be directed against either type XVII collagen or laminin-332. Molecular approaches to determining the target antigen can be important since there is an increased risk of certain malignancies in mucous membrane pemphigoid associated with laminin-332 targeted auto-antibodies.

For inherited skin blistering diseases, several clinical subtypes can have similar phenotypes and routine light microscopy usually is not able to assist diagnosis or help determine prognosis. Nevertheless, antibodies to structural components at the dermal-epidermal junction are now commercially available and these can be helpful in diagnosing recessive forms of epidermolysis bullosa. Testing may involve labeling of skin sections with antibodies to type VII collagen, laminin-332, integrin-α_6 and β_4, plectin, or type XVII collagen. In many recessive forms of epidermolysis bullosa, immunostaining with one of these antibodies is reduced or undetectable. This immunohistochemical approach is therefore a useful prelude to determining the candidate gene that can then be sequenced and used to identify the pathogenic mutations.

Molecular pathology can also provide insight into other genodermatoses that conventional microscopy is unable to provide. For example, in some cases of hereditary leiomyomas, lesions may be multiple and there can be an autosomal dominant mode of inheritance. In such families, skin leiomyomas usually appear in adolescence or early adulthood. Some cases, however, may be complicated by the subsequent development of uterine leiomyomas, usually in the early 20s. The molecular pathology may involve mutations in the fumarase (*FH*) gene and identification of such cases through molecular diagnosis can have important clinical implications. For example, if an *FH* gene mutation is identified in a woman with hereditary multiple leiomyomas who plans to have children at some stage, it may be prudent to advise her to consider having her children at an early age before the onset of uterine fibroids. Diagnosis of *FH* gene mutations in hereditary multiple leiomyomas may also have other prognostic significance, in that a subset of patients may be at risk from developing aggressive type II papillary renal cell carcinomas.

Understanding the molecular pathology of inherited skin diseases also has changed clinical practice by allowing the development of newer techniques for prenatal diagnosis for certain disorders. Advances in fetal medicine have led to further advances in which the molecular pathology of inherited skin diseases can be used to develop new techniques for prenatal testing. These include preimplantation genetic diagnosis and preimplantation genetic haplotyping. Further technical advances are likely to lead to the development of less invasive forms of fetal screening, such as the analysis of free fetal DNA in the maternal circulation.

NEW MOLECULAR MECHANISMS AND NOVEL THERAPIES

Two groups of dermatological diseases that are benefiting from new molecular-based therapies are the chronic auto-inflammatory diseases as well as cutaneous T-cell lymphoma. The auto-inflammatory diseases are complex and clinically difficult to manage. Nevertheless, new insights into molecular mechanisms are leading to immediate therapeutic benefits. The cryopyrinopathies represent a spectrum of diseases associated with mutations in the cold-induced autoinflammatory syndrome 1 (*CIAS1*) gene that encodes cryopyrin. Cryopyrin and pyrin (the protein implicated in familial Mediterranean fever) belong to the family of pyrin domain-containing proteins. The *CIAS1* gene mutations result in increased cryopyrin activity. Cryopyrin regulates interleukin-1β production. Mutations in the gene that encodes for the CD2 binding protein 1 (*CD2BP1*), which binds pyrin, are associated with pyogenic arthritis, pyodermic gangrenosum, and acne syndrome (OMIM604416). Common to all these diseases is activation of the interleukin-1β pathway. This finding has been translated into direct benefits for patients. Specifically, the recombinant human interleukin-1 receptor antagonist, Anakinra, has proved to be helpful, and in some cases results in a quick and dramatic treatment for these chronic auto-inflammatory diseases.

Management of the cutaneous T-cell lymphomas is also set to improve following a more detailed understanding of the molecular pathology of these disorders (Figure 27.17). The most common forms of cutaneous T-cell lymphoma are mycosis fungoides and Sézary syndrome. Mycosis fungoides usually presents in a skin with erythematous patches, plaques, and sometimes tumors. Sézary syndrome is a triad of generalized erythroderma, lymphadenopathy, and the presence of circulating malignant T-cells with cerebriform nuclei (known as Sézary cells). Recent studies have identified a number of changes in various tumor suppressor and apoptosis-related genes in patients with mycosis fungoides and Sézary syndrome. Discovery of specific gene abnormalities and patterns of altered gene expression have value diagnostically and prognostically, but also offer new insight into developing treatments that transform a malignant disease into a less aggressive chronic illness. There are accumulating data regarding specific immune and genetic abnormalities in these diseases that might lead to new therapies that block the trafficking or proliferation of malignant T-cells. Of note, immunophenotyping suggests that mycosis fungoides and possibly Sézary cells are derived from CLA-positive effector memory cells. The malignant cells demonstrate altered cytokine profiles with interleukin-7 and -18 being up regulated in the plasma and skin of affected individuals. Loss of T-cell diversity is also a feature of these conditions and antigen presenting dendritic cells can have an important role in the pathogenesis of the disorders, especially in maintaining the survival of proliferating malignant T-cells. Specific chemokine receptors are associated with

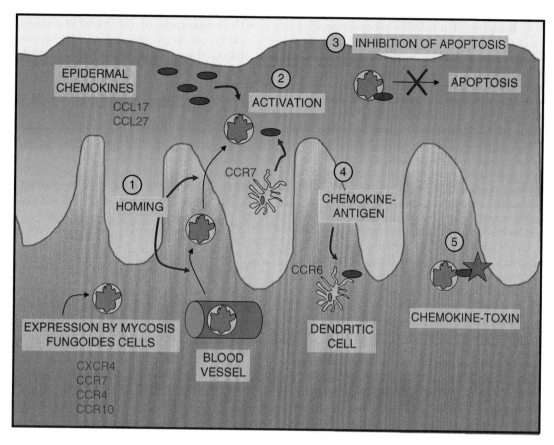

Figure 27.17 Roles for chemokine receptors and possible therapeutic manipulation in cutaneous T-cell lymphoma. Chemokine receptors may have important roles in enabling malignant T-cells to enter and survive in the skin. **(1)** Homing: Activation of T-cell integrins permits T-cell adhesion to endothelial cells in the skin and subsequent binding to extracellular matrix proteins. T-cells can then migrate along a gradient of chemokines (e.g., CCL17 and CCL27) to the epidermis. **(2)** Activation: chemokine receptors allow T-cells to interact with dendritic cells such as Langerhans cells, leading to T-cell activation and release of inflammatory cytokines. **(3)** Inhibition of apoptosis: chemokine receptor engagement can lead to up-regulation of PI3K and AKT, which are pro-survival kinases. T-cells can therefore survive and proliferate in the skin. **(4)** Chemokine-antigen fusion proteins can be used to target tumor antigens from cutaneous T-cell lymphoma cells to CCR6+ presenting dendritic cells that can stimulate host anti-tumor immunity. **(5)** Chemokine toxin molecules can also target specific chemokine receptors found on cutaneous T-cell lymphoma cells to mediate direct killing. Adapted from Hwang ST, Janik JE, Jaffe ES, et al. Mycosis fungoides and Sezary syndrome. *Lancet.* 2008;371:945–957.

mycosis fungoides and Sézary syndrome and these are directly relevant to skin tropism of malignant T-cells.

Given these abnormalities, new treatments are being developed that go beyond current combinations of nitrogen mustard, corticosteroids, and radiotherapy. The overproduction of Th-2 cytokines, such as interleukins 4, 5, and 10 in mycosis fungoides and Sézary syndrome, suggests that cytokines that promote a Th-1 phenotype might be clinically useful. Indeed interleukin-12 has shown clinical benefit. Interleukin-12 is a Th-1-promoting cytokine synthesized by phagocytic cells and antigen presenting cells. It enhances cytolytic T-cell and natural killer cell functions and is necessary for interferon-γ production by activated T-cells. Recombinant interferon-α and interferon-γ can also shift the balance from a Th-2 toward a Th-1 phenotype and may have clinical relevance.

Other approaches using antibodies to CD4 (Zanolimumab) and CD52 (Alemtuzumab) also broadly target T-cells and may help in relieving certain symptoms such as erythroderma and pruritus in Sézary syndrome. Vaccine therapy for mycosis fungoides and Sézary syndrome is another intriguing, but somewhat preliminary approach. This is due to the scarcity of target antigens, but identifying T-cell receptor sequences expressed by malignant lymphocytes should be technically feasible and may provide targets for immunotherapy. In similar fashion, loading autologous dendritic cells with tumor cells treated with Th-1-priming cytokines is another approach. Vaccination of patients with mimotopes (i.e., synthetic peptides that stimulate antitumor CDA-positive T-cells) is also a promising clinical option. Specific molecular targeting can also be attempted using histone deacetylase inhibitors such as Depsipeptide and Verinostat. These drugs induce growth arrest in conjunction with cell differentiation and apoptosis and have been shown to improve erythroderma and pruritus and a number of circulating Sézary cells in some patients with Sézary syndrome.

Drugs such as Imiquimod, a toll-like receptor agonist, can also benefit cutaneous T-cell lymphoma by inducing interferon-α. Other toll-like receptor agonists may have clinical utility, such as TLR9, which recognizes unmethylated CpG-containing nucleotide motifs that are present in most bacteria and DNA viruses. Treatment with CpG oligodeoxynucleotides leads to plasmacytoid dendritic cell up-regulation of costimulatory molecules and migratory receptors that subsequently generate type 1 interferons and promote a strong Th-1 immune response and enhance cellular immunity. These drugs may represent a useful adjuvant in immunotherapy or combined with cytotoxic drugs.

Another approved drug for use in cutaneous T-cell lymphoma is Bexarotene. This is a retinoid that modulates gene expression through selective binding to retinoid receptors. These receptors form either homodimers or heterodimers with other nuclear receptors that then act as transcription factors. Bexarotene therapy may also be combined with new treatments such as Denileukin diftitox, which is a fusion molecule containing the interleukin-2 receptor binding domain and the catalytically active fragment of diphtheria toxin. This targets the high-affinity interleukin-2 receptor present on activated T and B cells. A newer compound that probably works by the same mechanism is anti-Tac(Fv)-PE38 (LMB-2), which is an anti-CD25 recombinant immunotoxin.

Overall these new insights into the pathophysiology of cutaneous T-cell lymphoma are providing opportunities to therapeutically target specific aspects of the molecular pathology. These approaches, in which more selective therapy is the goal, are starting to have benefits for patients that result in better survival and fewer side effects.

KEY CONCEPTS

- A key role of skin is to protect the body against the external environment by providing a mechanical barrier and also a defense immune system.
- Integrity of this mechanical barrier property is provided by structural proteins in the hemidesmosomes and desmosomes.
- These structural proteins are targets for inherited and acquired skin disorders, many of which are mechanobullous.
- Defects in filaggrin cause ichthyosis vulgaris and is a major predisposing factor for its related condition, atopic dermatitis.
- Pathology of psoriasis involves tissue-resident immune cells.
- Sonic hedgehog signaling is a fundamental transduction pathway in skin tumor development.
- Molecular techniques are now being used in diagnosing inherited and acquired skin diseases, infectious diseases, identification and classification of cutaneous malignancies, and prenatal diagnosis.

- Understanding of the molecular mechanisms of skin pathology has led to the development of targeted molecular therapeutic agents.

SUGGESTED READINGS

1. Alizadeh AA, Eisen MB, Davis RE, et al. Distinct types of diffuse large B-cell lymphoma identified by gene expression profiling. *Nature.* 2000;403:503–511.
2. Amagai M, Komai A, Hashimoto T, et al. Usefulness of enzyme-linked immunosorbent assay using recombinant desmogleins 1 and 3 for serodiagnosis of pemphigus. *Br J Dermatol.* 1999; 140:351–357.
3. Borman AM, Linton CJ, Miles SJ, et al. Molecular identification of pathogenic fungi. *J Antimicrob Chemother.* 2008;61:(Suppl 1): i7–i12.
4. Boyman O, Conrad C, Tonel G, et al. The pathogenic role of tissue-resident immune cells in psoriasis. *Trends Immunol.* 2007; 28:52–57.
5. Chan YM, Yu QC, Fine JD, et al. The genetic basis of Weber-Cockayne epidermolysis bullosa simplex. *Proc Natl Acad Sci USA.* 1993;90:7414–7418.
6. Christiano AM, Greenspan DS, Hoffman GG, et al. A missense mutation in type VII collagen in two affected siblings with recessive dystrophic epidermolysis bullosa. *Nat Genet.* 1993;4: 62–66.
7. Daya-Grosjean L, Couve-Privat S. Sonic hedgehog signalling in basal cell carcinomas. *Cancer Lett.* 2005;225:181–192.
8. Fairley JA, Woodley DT, Chen M, et al. A patient with both bullous pemphigoid and epidermolysis bullosa acquisita: An example of intermolecular epitope spreading. *J Am Acad Dermatol.* 2004;51:118–122.
9. Fassihi H, Renwick PJ, Black C, et al. Single cell PCR amplification of microsatellites flanking the COL7A1 gene and suitability for preimplantation genetic diagnosis of Hallopeau-Siemens recessive dystrophic epidermolysis bullosa. *J Dermatol Sci.* 2006; 42:241–248.
10. Fine JD, Eady RA, Bauer EA, et al. Revised classification system for inherited epidermolysis bullosa: Report of the second international consensus meeting on diagnosis and classification of epidermolysis bullosa. *J Am Acad Dermatol.* 2000; 42:1051–1066.
11. Hanakawa Y, Schechter NM, Lin C, et al. Molecular mechanisms of blister formation in bullous impetigo and staphylococcal scalded skin syndrome. *J Clin Invest.* 2002;110:53–60.
12. Hashimoto T. Skin diseases related to abnormality in desmosomes and hemidesmosomes—Editorial review. *J Dermatol Sci.* 1999;20:81–84.
13. Hoffman HM, Rosengren S, Boyle DL, et al. Prevention of cold-associated acute inflammation in familial cold autoinflammatory syndrome by interleukin-1 receptor antagonist. *Lancet.* 2004;364: 1779–1785.
14. Hwang ST, Janik JE, Jaffe ES, et al. Mycosis fungoides and Sezary syndrome. *Lancet.* 2008;317:945–947.
15. Lowes MA, Bowcock AM, Krueger JG. Pathogenesis and therapy of psoriasis. *Nature.* 2007;445:866–873.
16. McGrath JA, McMillan JR, Shemanko CS, et al. Mutations in the plakophilin 1 gene result in ectodermal dysplasia/skin fragility syndrome. *Nat Genet.* 1997;17:240–244.
17. Pulkkinen L, Christiano AM, Gerecke D, et al. A Homozygous nonsense mutation in the beta 3 chain gene of laminin 5 (LAMB3) in Herlitz junctional epidermolysis bullosa. *Genomics.* 1994;24:357–360.
18. Rinne T, Hamel B, van BH, et al. Pattern of p63 mutations and their phenotypes—Update. *Am J Med Genet A.* 2006;140: 1396–1406.
19. Singh M, Lin J, Hocker TL, et al. Genetics of melanoma tumorigenesis. *Br J Dermatol.* 2008;158:15–21.
20. Smith FJ, Irvine AD, Terron-Kwiatkowski A, et al. Loss-of-function mutations in the gene encoding filaggrin cause ichthyosis vulgaris. *Nat Genet.* 2006;38:337–342.

28

Molecular Pathology: Neuropathology

Joshua A. Sonnen . C. Dirk Keene .
Robert F. Hevner . Thomas J. Montine

ANATOMY OF THE CENTRAL NERVOUS SYSTEM

The central nervous system (CNS) is composed of cellular components organized in a complex structure that is unlike other organ systems. Broadly, its cellular components can be divided into neuroepithelial and mesenchymal elements. Neuroepithelial elements derive from the primitive neural tube and include neurons and glia. The mesenchymal elements include blood vessels and microglia. Microglia are a bone marrow-derived population of scavenger cells that play a central role in CNS inflammation. At the macroscopic level, the parenchyma of the CNS can be categorized into two structurally and functionally unique components: gray and white matter. Gray matter is the location of most neurons and is the site of the integration of neural impulses by neurotransmitters across synapses between neurons. White matter functions to conduct these impulses efficiently and quickly between neurons in different gray matter regions.

Microscopic Anatomy

Gray Matter

Macroscopically, gray matter forms a ribbon of cortex in the human cerebrum and cerebellum as well as the mass of the deep nuclei. Microscopically, it is composed of cells forming and embedded in a finely interdigitating network of cellular processes. This network, referred to as neuropil, is sufficiently dense that individual cellular processes cannot be distinguished.

The cellular constituents of gray matter include neurons, glia (astrocytes and oligodendrocytes), endothelium, and microglia (Figure 28.1). Neurons are the electrically active cells of the brain. A neuron is composed of slender branching dendrites on which other neurons synapse and propagate action potentials, a body or soma where the metabolic and synthetic processes of the neuron are orchestrated, an axonal hillock where electrochemical impulses are integrated, an axon along which the integrated electrochemical impulse is conducted, and an axonal terminal where the electrochemical signal is passed to another neuron's dendrites or an effector cell across a synapse. The soma of neurons is easily identifiable by routine histologic stains. Neurons have generally round nuclei with a prominent nucleolus and open chromatin. The cytoplasm of neurons is remarkable in that it generally contains abundant rough endoplasmic reticulum called Nissl substance that is demonstrable by numerous histologic techniques. Loss of Nissl substance is a sign of early neuronal injury and is seen in a variety of conditions including axonal transection and hypoxia/ischemia. The neuron's axons and dendrites are major components of neuropil.

Neurons require a constant supply of oxygen and glucose, and even short interruptions can cause neuronal death. Neurons are the most susceptible to most forms of CNS injury and are the first cells lost to necrosis or apoptosis under stressful conditions. Further, for reasons that are not fully understood, populations of neurons have differential susceptibilities to different types of stress. Cells in one region of the hippocampus may become necrotic in response to hypoxia, while neurons in other regions are spared; this pattern of injury may be reversed in hypoglycemia.

Glia, from Latin for "glue," form the bulk of the CNS parenchyma and outnumber neurons on the order of 1000 to 1. The primary glial cell of gray matter is the star-shaped astrocyte. Astrocytes maintain a variety of supportive functions including structure, metabolic

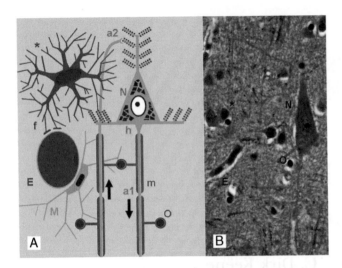

Figure 28.1   (A) Schematic of the microanatomy of gray matter. (B) Photomicrograph of the microanatomy of gray matter. (A) A neuron (N) contains a prominent nucleus with open chromatin, a conspicuous nucleolus and cytoplasmic Nissl substance that is composed of abundant rough endoplasmic reticulum. The neuron elaborates numerous apical and lateral dendrites that are decorated with many receptors. Electrochemical impulses are generated at dendrites and integrated across the neuron's body at the axonal hillock (h) and transmitted along the axon (a1). Oligodendrocytes (O) envelop the axon within a segmented myelin (m) sheath allowing more rapid and efficient conduction of impulses. An endothelial cell (E) forms a small capillary space surrounded by a resting microglial cell (M). A nearby astrocyte (*) has numerous cytoplasmic processes, some of which rest foot processes (f) on the vessel, helping to maintain the blood-brain barrier. An axon (a2) from a distant neuron forms an axonal terminal (t) on a dendrite of the pictured neuron, forming a synapse and releasing neurotransmitters to the receptors on the dendrite. **(B)** Section of frontal cortex stained with hematoxylin and eosin (H&E) and Luxol fast blue (LFB).

support for neurons, management of cellular waste products, uptake and release of neurotransmitters, regulation of extracellular ion concentration, and interactions with the vasculature including helping to maintain the blood-brain barrier and responding to injury. Normally, astrocytes have irregular, potato-shaped nuclei and numerous fine cellular processes that are indistinct within the neuropil. In response to noxious stimuli, astrocytes increase production of their characteristic intermediate filament, glial fibrillary acidic protein (GFAP). The astrocyte's soma swells and becomes prominent. This process is known as gliosis and is analogous to scar formation outside the CNS. A second population of glia in gray matter is the oligodendrocytes; these will be discussed in more detail later under white matter.

Microglia are a population of bone marrow-derived scavenger cells. Usually unobtrusive, the quiescent microglia have small rod-shaped nuclei and inapparent cytoplasm. Ramified (quiescent) microglia do not express major histocompatibility complex (MHC) I/II

antigens, unlike other scavenger cells. However, in response to injury, microglia replicate and migrate. They produce MHC I/II antigen, activate inflammatory and cytotoxic signaling, and can phagocytize material and process it for antigen presentation to T-cells.

White Matter

White matter is composed of numerous axonal processes (from neurons whose bodies reside in gray matter), glia, blood vessels, and microglia (Figure 28.2). The axons are insulated in segments by layers of myelin, and this insulation allows more rapid and efficient conduction of electrochemical signals along the axon. Myelin is composed of concentric proteolipid membranes which are extensions of cytoplasmic processes of the primary glia of white matter, the oligodendrocyte. Oligodendrocytes have small round nuclei with condensed chromatin and indistinct cytoplasm. A single oligodendrocyte myelinates numerous passing axons. Because oligodendrocytes are responsible for the maintenance of a large amount of myelin, they have relatively high metabolic demands and are consequently relatively sensitive to injury compared to other white matter elements. Injury to oligodendrocytes causes local loss of myelin; this is discussed further in the section on demyelination. Astrocytes and microglia are minority populations within white matter and are less sensitive to injury than oligodendrocytes. Tissue response to noxious stimuli is mediated through astrocytic gliosis and microglial activation as in gray matter.

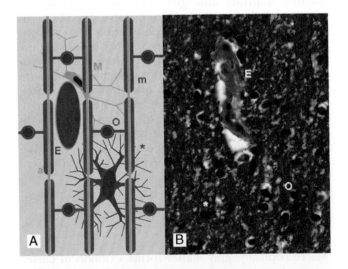

Figure 28.2   (A) Schematic of the microanatomy of white matter. (B) Photomicrograph of the microanatomy of white matter (H&E/LFB). Numerous axons (a) from distant neurons pass through white matter conducting nerve impulses. Oligodendroglia (O) are the most common cells and myelinate many adjacent segments of passing axons. The insulating myelin (m) allows rapid and efficient conduction of nerve impulses. Oligodendrocytes have high metabolic needs, and endothelium (E) form numerous capillaries. In the absence of disease, astrocytes (*) and microglia (M) are inconspicuous.

Cerebrospinal Fluid and the Ventricular System

Within the brain there are interconnected fluid-filled chambers called ventricles. The ventricles are filled with cerebrospinal fluid (CSF). Most CSF is produced by a specialized organ within the ventricles called the choroid plexus. The choroid plexus is a network of large capillaries and glia with a specialized epithelial lining. CSF surrounding the brain acts as a cushion for the brain and spinal cord. CSF is taken back up into the venous circulation and the entire volume of CSF (~150 mL) turns over 4 to 5 times a day.

Gross Anatomy

The CNS can be divided into a number of anatomic regions, each with specific neurologic or cognitive functions. Disease or damage to these regions produces neurologic or cognitive deficits that correlate with the anatomic location and extent of disease. The main divisions are the cerebrum, cerebellum, brainstem, and spinal cord. The cerebrum is covered by a folded layer of gray matter called the cortex and is generally believed to be the location of conscious thought. The folding allows greater surface area to fit within the confines of the skull. The folds themselves are called gyri and are characteristic of normal cortical development in humans. Lesions of the cortex cause deficits of cognition and conscious movement or sensation. The cortex can be further divided into lobes which subserve different cognitive domains or neurologic functions (Figure 28.3). The frontal lobes anteriorly are involved in executive function (self-control, planning) and personality. Damage to this region produces personality changes and socially inappropriate behavior. Posteriorly, the frontal lobe houses the primary motor cortex which controls

voluntary movement for the opposite side (contralateral) of the body. Damage causes contralateral weakness or paralysis. The parietal lobe contains the primary sensory cortex for the contralateral half of the body, and damage causes anesthesia and neglect. The temporal lobe is involved in the conscious processing of sound, but also contains the hippocampus that is integral to the formation of new memories. Damage to the hippocampus produces memory dysfunction. The occipital lobe contains the primary visual cortex for the field of vision on the opposite side of the body, and damage causes partial or complete loss of conscious perception of visual stimuli. The white matter underlying and connecting these cortical regions may also be damaged and produces deficits related to the regions connected. The cerebrum also contains deeply situated gray matter nuclei. The most often implicated of these in disease are the basal ganglia (Figure 28.4). These nuclei form important inhibitory circuits on the cerebrum, and disease causes movement disorders and well as disorders of cognition.

The cerebellum is involved in coordination and balance. The brainstem has three primary functional domains. First, it contains white matter tracks that connect the cerebrum to the spinal cord and carries information between these two structures. Second, numerous autonomic functions are coordinated by nuclei within the brainstem that serve as integrative centers. Third, the cranial nerve nuclei reside in the brainstem, and they control motor function of the head and neck and receive sensory input from these same structures. The spinal cord is primarily involved in the transmission of nerve impulses to and from the peripheral nervous system along its abundant white matter tracks, but it also contains central gray matter nuclei that regulate autonomic functions and reflexes.

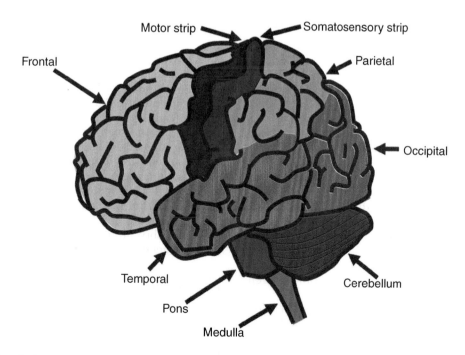

Figure 28.3   Lateral view of the central nervous system.

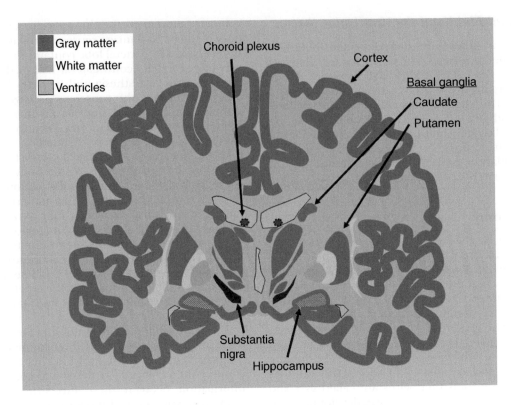

Figure 28.4   Coronal section of cerebrum through the deep gray matter nuclei.

Summary

The anatomy of the CNS is complex. The correlation between structure and function illuminates an underlying order. Knowledge of the location or cellular target of injury can give great insight into the deficits present and vice versa.

NEURODEVELOPMENTAL DISORDERS

Developmental neuropathology encompasses a broad variety of cerebral malformations and functional impairments caused by disturbances of brain development manifesting during ages from the embryonic period through adolescence and young adulthood. The neurologic and psychiatric manifestations of neurodevelopmental disorders range widely depending on the affected neural systems and include such diverse manifestations as epilepsy, mental retardation, cerebral palsy, breathing disorders, ataxia, autism, and schizophrenia. In terms of morbidity and mortality, the spectrum is extremely broad: the mildest neurodevelopmental disorders can be asymptomatic, while the worst malformations often lead to intrauterine or neonatal demise.

Current systems for classifying neurodevelopmental malformations emphasize the underlying developmental processes that are perturbed.

To understand the pathogenesis of neurodevelopmental disorders, one must first acquire at least a rudimentary knowledge of brain and spinal cord development, including underlying cellular mechanisms and interactions. A brief review of processes highlighting recent progress will be the starting point for subsequent consideration of selected specific brain malformations.

Organizing the Central Nervous System: Signaling Centers and Regional Patterning

The central nervous system begins at the neural plate, which folds along the midline and closes dorsally to form the neural tube. From its formation, the neural plate is patterned along the rostrocaudal and mediolateral axes in a grid-like fashion by gradients and boundaries of gene expression. *HOX* (human homeobox) genes, for example, are differentially expressed along the rostrocaudal axis and play an important role in specifying segmental organization of the spinal cord and hindbrain. Such differences of gene expression are programmed by both the intrinsic developmental history of each region and by extracellular factors produced in signaling centers that define positional information through interactions with developing neural tissue. Gene expression gradients, compartments, boundaries, and signaling centers continue to be important throughout the embryonic period until each brain subdivision has been generated and acquired its specific identity.

Neural Tube Closure and Wnt-PCP Signaling

Neural tube closure is a key early event in brain and spinal cord development in which the planar epithelium of the neural plate folds at the midline along the anteroposterior axis, and the lateral edges of the neural plate move dorsally, contact each other, and

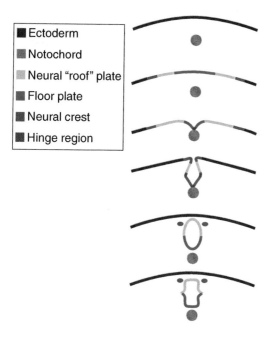

Figure 28.5 Neural tube formation. In cross-section, the neural tube first derives from a planar (two-dimensional) epithelium with mediolateral organization and transforms to a tubular (three-dimensional) with both mediolateral and dorsoventral axes.

fuse dorsally to form the neural tube (Figure 28.5). Dorsal fusion first occurs in the region of the cervical spinal cord primordium, followed by separate closure events at midbrain and rostral telencephalic points. The exact location and number of dorsal fusion events

vary between and within species. From each of these sites, fusion proceeds rostrally and caudally in a zipper-like mode until closure is complete (primary neurulation) from the rostral end (anterior neuropore) to the caudal end (posterior neuropore). Interestingly, this mode of primary neural tube closure does not quite extend the full caudal length of the spinal cord, but ends around the lumbosacral region. More caudal regions (mainly sacral) appear to develop by a distinct mechanism involving cavitation of the caudal eminence (tail bud) mesenchyme, known as secondary neurulation. A schematic of this process is shown in Figure 28.6.

Primary neurulation depends on deformation of the neural plate due to convergent extension, a process of cell migration within the plane of the neuroepithelium. At the molecular level, convergent extension utilizes the Wnt-planar cell polarity (Wnt-PCP) signaling pathway for cell adhesion and polarity. Many details remain to be elucidated, but this pathway is essential for cells to distinguish mediolateral and anteroposterior directions within the neuroepithelium, and thus migrate in the correct directions. Studies of mutant mouse strains, notably loop-tail (*Lp*), have shown that mutations of several Wnt-PCP genes (such as *Vangl2* in *Lp*) cause craniorachischisis, a severe form of open NTD extending from brain to spinal cord.

The incidence of neural tube defects (NTD) has recently declined in developed countries due to dietary supplementation with folate, but the molecular mechanism of this effect remains unknown. Polymorphisms in enzyme genes involved in folate metabolism are known to confer risk for NTD. One polymorphism of 5,10-methylene tetrahydrofolate reductase (MTHFR)

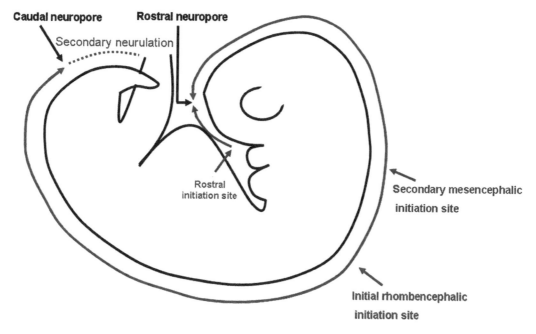

Figure 28.6 Simplified schematic of the sequence of neural tube fusion. After the neural tube begins closure from rostral and posterior cerebral sites, closure propagates anteriorly and posteriorly. The locations of the neuropores are potential sites for defects of neural tube closures. Although two initiation sites of primary neural tube fusion are shown here, this has not been rigorously proven in humans, and other mammals may have more initiation sites.

is thermolabile and, if present in fetus or mother, confers increased risk of NTD. Environmental and toxic exposures have also been implicated in the pathogenesis of NTD. A large number of physical agents ranging from X-irradiation to alcohol to folate antagonists have been associated with NTD in epidemiologic studies and can cause NTD in animal models. Maternal health status, including diabetes, obesity, infections, and toxic exposures, also is associated with NTD in humans.

Importantly, the process of neurulation transforms the axes of the developing CNS neuroepithelium from a planar to a tubular coordinate system. The planar (two-dimensional) neural plate is defined by rostrocaudal and mediolateral axes, while the tubular (three-dimensional) neural tube is defined by rostrocaudal, mediolateral, and dorsoventral axes. During this reorganization process, the lateral edge of the neural plate becomes the dorsal midline (roof plate) of the neural tube, and the medial edge (midline) of the neural plate becomes the ventral midline (floor plate) of the neural tube (Figure 28.5). The new axes become the substrate for further patterning and remain important throughout later development, although additional axial transformations occur locally in the developing brain.

Finally, neural tube closure is essential for subsequent development of posterior tissues including the vertebral arches and cranial vault, paraspinal muscles, and posterior skin of the head and back. In some cases, neural tube closure proceeds normally, but the overlying skin and mesodermal structures fail to cover the posterior neural tube. It is unknown whether these malformations (such as encephalocele, myelocele, and sacral agenesis) are caused by primary defects of neural tube closure (as frequently assumed in the neuropathology literature) or mesodermal and ectodermal development.

Human Neural Tube Defects

The most severe NTD characterized by the complete failure of neural tube closure along the entire craniospinal axis, is called craniorachischisis (Figure 28.7A). This lethal malformation corresponds to the severe form of NTD found in *Lp* mice, described previously, and is probably caused by defects of Wnt-PCP signaling during convergent extension. Closure defects limited to the cranial region result in anencephaly, a severe, lethal defect (Figure 28.7B). In anencephaly, extensive destruction of the brain tissue is secondary to direct exposure to amniotic fluid: initially, the cerebral tissue proliferates and grows outside the surrounding skull base, a transient stage known as exencephaly. The most common human NTD are limited to the lumbosacral region, sites of posterior neuropore closure and secondary neurulation. Typically, the spinal tissue has a ragged interface with mesodermal and ectodermal derivatives and is exposed externally. This appearance is called myelomeningocele (Figure 28.7C), somewhat erroneously because there is usually no cele or closed sac. Less often, malformations with a closed skin surface and sac are also seen. In the great majority of cases, lumbosacral myelomeningocele

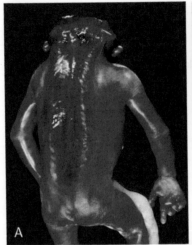

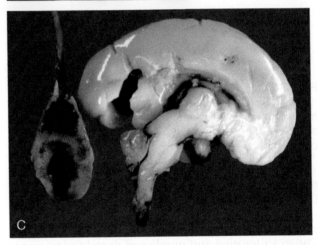

Figure 28.7 Neural tube defects. (A) Craniorachischisis. **(B)** Anencephaly. **(C)** Myelomeningocele with Chiari type II malformation. The spinal cord with open lumbosacral myelomeningocele is shown at left; the medial view of brain with hydrocephalus, tectal beaking, and herniation of the medulla over the spinal cord is shown at right.

is accompanied by a malformation of the midbrain, hindbrain, and skull base known as Chiari type II malformation, historically called the Arnold-Chiari malformation (Figure 28.7C).

Rostrocaudal and Dorsoventral Patterning of the Neural Tube

A program of segmental, compartmentalized gene expression begins to define rostrocaudal subdivisions of the CNS during neural plate stages and this process accelerates with neurulation. Morphologically, the spinal cord is partitioned into cervical, thoracic, lumbar, and sacral segments associated with mesodermal somites. Morphological development of the brain is more complicated (Figure 28.8), as it is initially divided into three vesicles (prosencephalon/forebrain, mesencephalon/midbrain, rhombencephalon/hindbrain), and then

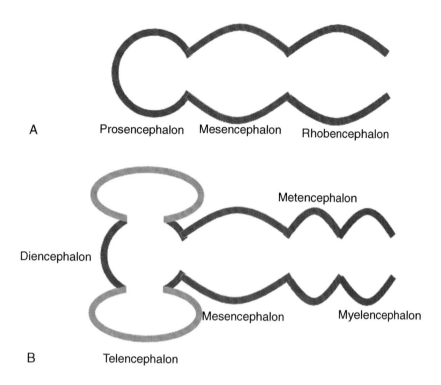

Figure 28.8 Segmentation and cleavage of the developing brain. (A) After fusion of the neural tube, the anterior-most portion forms swellings that will become the cerebrum (prosencephalon), the midbrain (mesencephalon), and hindbrain (rhombencephalon). **(B)** Following this, proliferation of the cortical neuroepithelium forms telencephalic vesicles and will become the majority of the brain's paired hemispheres, and the rhombencephalon divides into the metencephalon and myelencephalon, which will become the pons and cerebellum and medulla, respectively.

further subdivided into five vesicles (telencephalon, diencephalon, mesencephalon, metencephalon, myelencephalon). Key molecular players in spinal and hindbrain segmentation are the *HOX* genes, which encode transcription factors expressed in nested patterns. The *HOX* gene family and their functions are highly conserved in evolution: their roles in rostrocaudal segmentation were initially discovered in the fruit fly, *Drosophila melanogaster*, and have been confirmed in other vertebrate and invertebrate species. More rostral segments of the CNS (forebrain and midbrain) also show segment-like subdivisions and nested gene expression patterns, but *HOX* genes are not involved. Instead, these brain areas are patterned by distantly related transcription factors with homologous DNA-binding domains (homeobox), along with other families of transcription factors, expressed in complex tissue patterns that interact with several signaling centers in the embryonic brain. The latter include the isthmus at the midbrain-hindbrain junction and the zona limitans intrathalamica at the junction of rostral and caudal thalamic subdivisions.

Concurrently with rostrocaudal segmentation, the neural tube is patterned along the dorsoventral axis by a different set of molecules and signaling centers. Important ventral signaling centers include the notochord (primordium of vertebral body nucleus pulposus), a mesodermal structure located ventral to the spinal cord and hindbrain; the prechordal plate

mesendoderm, located ventral to the forebrain and midbrain; and the floor plate of the neural tube, a specialized neuroepithelial structure in the ventral midline of spinal cord and brain. All three of these ventral signaling centers produce Sonic hedgehog (Shh), a small, post-translationally modified, secreted protein with potent ventralizing activity on neural structures. The key dorsal signaling center is the roof plate, a specialized neuroepithelial structure in the dorsal midline that ultimately develops into choroid plexus. The roof plate, characterized by high-level expression of *ZIC* genes, produces secreted morphogens belonging to the bone morphogenetic protein (BMP) and Wnt families, with potent dorsalizing activity. The balance of antagonistic ventralizing and dorsalizing signals patterns the neural tube into basal and alar subdivisions associated with motor and sensory pathways, respectively. Accordingly, the ventral spinal primordium generates motor neurons, while the dorsal spinal primordium generates sensory relay neurons. Sensory and motor distinctions become more subtle in the midbrain and forebrain. For example, most of the forebrain is composed of alar (sensory) plate neuroepithelium, including such ventral (anatomically speaking) structures as the neural retina and optic pathways. In contrast, the neurohypophysis (ventral hypothalamus) may be considered a motor pathway, inasmuch as hormone-producing neurons generate somatic responses and behaviors.

Holoprosencephaly

The best characterized and most common defect of dorsoventral patterning in humans is holoprosencephaly, defined as complete or partial failure of cerebral hemispheric separation. Holoprosencephaly exhibits a spectrum of severity. The mildest forms (lobar holoprosencephaly) show some development of the interhemispheric fissure with continuity of cortex across the midline. Forms with intermediate severity (semilobar holoprosencephaly) show partial development of the interhemispheric fissure, while severe forms (alobar holoprosencephaly) show complete absence of the interhemispheric fissure and a single forebrain ventricle (Figure 28.9). The defective midline structures in holoprosencephaly result from abnormalities of ventral or dorsal patterning molecules, including Shh and Zic2. Dorsoventral patterning is essential for differential proliferation and apoptosis of dorsal midline and ventrolateral forebrain structures, the essential underlying mechanisms of hemispheric separation. Overall proliferation is severely reduced in holoprosencephaly and the brain is invariably small.

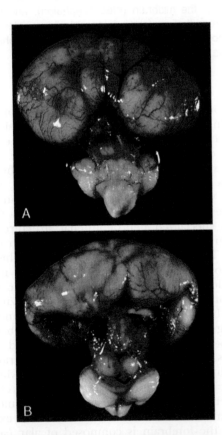

Figure 28.9 Alobar holoprosencephaly in a 23-week gestational age fetus. (A) Anterior view: the single "holosphere" exhibits shallow sulci, but no deep interhemispheric fissure. (B) Posterior view: cortex is continuous across the midline and the posteriorly open (due to disrupted delicate membranes) single forebrain ventricle is visible.

NEUROLOGICAL INJURY: STROKE, NEURODEGENERATION, AND TOXICANTS

Basic Mechanisms of Injury

Before we embark on a discussion of specific types of nervous system injuries, it is first worthwhile to consider a few basic mechanisms of damage to the nervous system and the nervous system's response. The following mechanisms are not exclusive to any neurologic disease, but are commonly proposed to contribute to varying degrees to the pathogenesis of stroke, neurodegeneration, and neurotoxicant injury.

The adult CNS has limited regenerative ability, as noted previously. The natural history of CNS damage is that cells vulnerable to injury become necrotic or apoptotic and their debris is removed by scavenger cells. The overall mass of the affected region is reduced and shrinks in a state called atrophy. The residual neuroepithelial cells, usually astrocytes and microglia, respond by gliosis.

Excitotoxicity

L-glutamate is the most abundant excitatory amino acid (EAA) neurotransmitter in the CNS. It and its cell surface ligand-activated ion channels, such as AMPA and NMDA receptors, participate in a wide array of neurological functions. Figure 28.10 depicts normal activity at an excitatory synapse. The glutamatergic presynaptic terminal releases glutamate upon depolarization into the synapse that then can bind to postsynaptic AMPA receptor to initiate influx of Na^+ into the postsynaptic element that has a potential gradient across its membrane generated primarily by the action of Na^+/K^+-ATPase. Glutamate also binds to the NMDA receptor, but unless the postsynaptic membrane is sufficiently depolarized, flow of cations, including Ca^{2+}, is blocked by Mg^{2+}. Glutamate is removed from the synapse by a number of transporters on neurons and astrocytes; astrocytic glutamate can enter the glutamine cycle to replenish neurotransmitter pools. Excitatory neurotransmission can be subverted to contribute to several neurologic diseases. Indeed, drugs that target excitatory neurotransmission are currently approved for use in patients with Alzheimer's disease and remain a focus of those interested in limiting damage from ischemia or multiple sclerosis. Excessive EAA receptor stimulation sets in motion a cascade of events that can contribute to neuronal injury through a process called excitotoxicity. Broadly, there are two types of excitotoxicity: direct and indirect.

Figure 28.10B depicts events in direct excitotoxicity. The key initiator in direct excitotoxicity is increased synaptic concentration of EAAs either by release from the presynaptic terminal (shown in red in Figure 28.10), decreased reuptake, or exposure to excitatory neurotoxicants like domoic acid in algal blooms or as occurs in lathyrism. Extensive activation of AMPA receptors

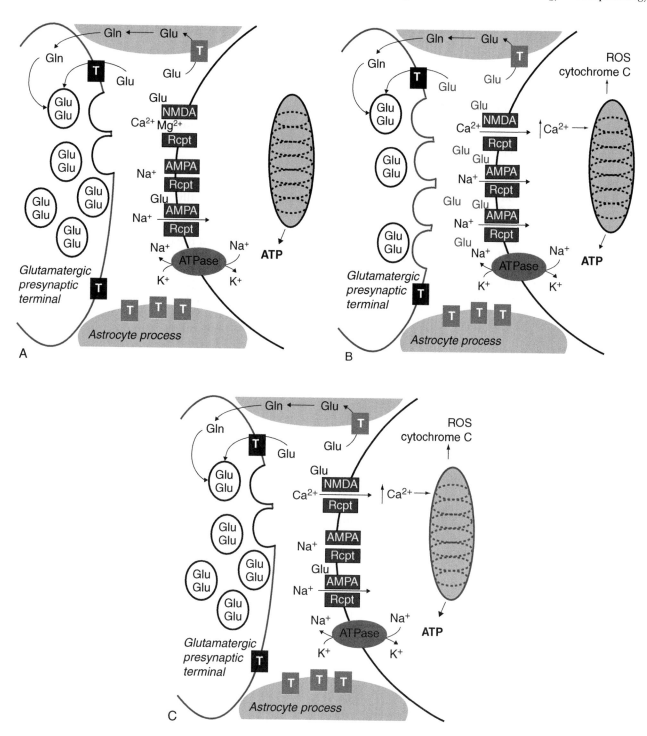

Figure 28.10 Excitotoxicity. Diagrams show excitatory neurotransmission **(A)**, direct excitotoxicity **(B)**, and indirect excitotoxicity **(C)**.

sufficiently depolarizes the postsynaptic membrane to relieve the Mg^{2+} block of the NMDA receptor. Rapid injury and lysis can follow this depolarization-dependent increase in ion influx. Delayed toxicity can develop even after EAA has been removed secondary to increased intraneuronal Ca^{2+} levels with subsequent inappropriate activation of calcium-dependent enzymes, mitochondrial damage, increased generation of free radicals, and transcription of proapoptotic genes.

Indirect excitotoxicity is depicted in Figure 28.10C. In distinction from direct excitotoxicity, the key event is impaired mitochondrial function with reduced ATP generation. One consequence is reduced activity of Na^+/K^+-ATPase with partial depolarization of the postsynaptic membrane that, when combined with normal levels of glutamate release, are sufficient to relieve the Mg^{2+} block of NMDA-mediated ion conductance. This again leads to increased intraneuronal

Ca^{2+} with consequences similar to those described for direct excitotoxicity.

Mitochondrial Dysfunction

That neuronal mitochondrial dysfunction leads to neuron damage and death is clearly established by toxicants, perhaps the best studied being 1-methyl-4-phenyl-tetrahydropyridine (MPTP). Interrupting electron transport, diminishing the proton gradient, or inhibition of complex V all can result in reduced ATP production and lead to stressors like those described previously. The contribution of mitochondrial dysfunction to neuron injury likely extends far beyond toxicants to include some unusual diseases caused by inherited mutations in the mitochondrial genome or genes encoded in the nucleus whose protein products are incorporated into mitochondria, and perhaps some neurodegenerative diseases like Huntington's disease and Parkinson's disease.

Free Radical Stress

Free radicals are normal components of second messenger signaling pathways. However, excess or uncontrolled free radical production is detrimental to cells either by direct damage to macromolecules or liberation of toxic byproducts. A number of sites for free radical generation exist in the nervous system. A major source appears to be oxidative phosphorylation in mitochondria. Other sources of free radicals may become significant during pathological states. These include excitotoxicity, enzymes such as cyclooxygenase, lipoxygenase, myeloperoxidase, NADPH oxidase, and monoamine oxidase, autoxidation of catechols such as dopamine, and amyloid β peptides.

Innate Immune Activation

Activation of innate immune response in brain, primarily mediated by microglia, is a feature shared by several neurodegenerative diseases. Moreover, microglial-mediated phagocytosis of potentially neurotoxic protein aggregates (*vide infra*) might be an important means of neuroprotection. Indeed, while some aspects of the innate immune activation are an appropriate response to the stressors presented by neurodegenerative diseases, other facets of glial activation, especially when protracted, may contribute to neuronal damage. This balance between beneficial and deleterious actions of immune activation is well appreciated by students of pathology of other organs and is currently an area of very active investigation among neuroscientists with the goal of promoting neurotrophic or neuroprotective actions while suppressing paracrine damage to neurons. Key elements in paracrine neurotoxicity from activated glia appear to include IL-1β, TNFα, and prostaglandin E_2, among others with activation of COX2, iNOS, and NADPH oxidase that interface with excitotoxicity and free radical stress.

Vascular Disease and Injury

Disease in the cerebrovasculature can lead to compromised perfusion of regions of the brain or spinal cord, a process called vascular brain injury. The most common types of vascular diseases are atherosclerosis and arteriolosclerosis. The most common forms of vascular brain injury are ischemia and hemorrhage.

Ischemia

The clinical consequences of acute ischemic stroke are dramatic and critically dependent on the precise regions of CNS involved. The pathologic consequences of ischemia assume one of three general forms depending on the severity of the insult (Figure 28.11). The first is a complete infarct with necrosis of all parenchymal elements. The second form of injury, an incomplete infarct, occurs when ischemia is less severe, producing necrosis of some cells, but not all tissue elements. Incomplete infarcts demonstrate that neurons and, to a lesser degree, oligodendroglia are more vulnerable than other cells in the CNS to ischemia. The ultimate tissue manifestation of an incomplete infarct resembles the edge of a complete infarct; that is, neuronal and oligodendroglial depopulation, myelin pallor, astrogliosis, and capillary prominence. The final form of ischemic injury results in damage and dysfunction, but without death of parenchymal elements.

Although necrotic brain is essentially irretrievably lost, it is important to realize that the zones with incomplete infarction or damage without necrosis hang in the balance between vulnerability to further injury and salvage or perhaps even regeneration. Indeed, some functional recovery typically occurs in the days, weeks, and even months following an ischemic stroke, in

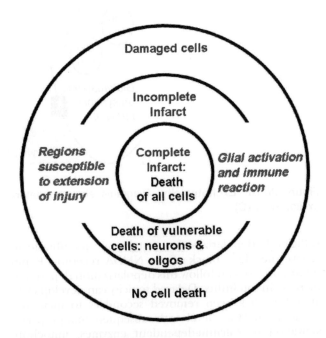

Figure 28.11   Diagram of varying levels of damage and reaction in vascular brain injury.

part from resolution of reversible stressors, post-stroke neurogenesis in at least some regions of the CNS, and functional reorganization of surviving elements. The balance of deleterious versus beneficial glial and immune responses in these surviving but damaged regions is thought to be key to optimal clinical outcome and is an area of intense investigation.

Regardless of the cause, CNS infarcts share a common evolution of coagulative necrosis, leading to liquefaction that culminates in cavity formation. Although death of cells occurs in CNS tissue within minutes of sufficient ischemia, the earliest structural sign of damage is not apparent until 12 to 24 hours following the ictus, when it takes the form of coagulative necrosis of neurons, or red neuron formation (Figure 28.12). The histologic hallmarks of this process are neuronal karyolysis and cytoplasmic hypereosinophilia. Infarcts become macroscopically apparent about 1 day after onset as a poorly delineated edematous lesion (Figure 28.13). Subsequent evolution of infarcts is dictated by the inflammatory cell infiltrate. Around 1 to 2 days after infarction, the lesion is characterized by disintegration of necrotic neurons, capillary prominence, endothelial cell hypertrophy, and a short-lived influx of neutrophils that are rapidly replaced by macrophages. Initially, macrophages may be difficult to discern in the inflamed and necrotic tissue, but within several days they accumulate to very high density and become enlarged with phagocytized material; this is the phase of liquefaction. These debris-laden macrophages ultimately return to the bloodstream. When their task is completed, the solid mass of necrotic tissue that characterized the acute infarct has been transformed into a contracted, fluid-filled cavity that is traversed by a fine mesh of atretic vessels, but does not develop a collagenous scar that is typical of other organs (Figure 28.14). Usually, the cavitated infarct is surrounded by a narrow zone of astrogliosis and neuron loss that abuts with

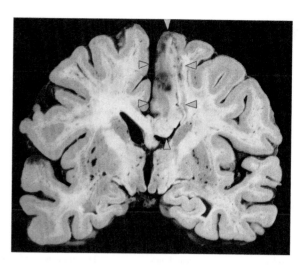

Figure 28.13   Coronal section of cerebrum shows acute infarct in the territory supplied by the left anterior cerebral artery (blue arrowheads).

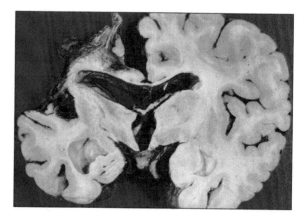

Figure 28.14   Coronal section of cerebrum shows remote infarct in the territory supplied by the left middle cerebral artery.

histologically normal brain parenchyma. Distal degeneration of fiber tracts is also a late manifestation of infarction in the CNS.

The type of vessel occluded leads to characteristic patterns of regional ischemic damage to CNS (Table 28.1). Territorial infarcts describe regions of necrotic tissue secondary to occlusion of an artery, such as the anterior cerebral artery (Figure 28.13). A common mechanism for obstruction of large arteries is complicated atherosclerosis with thrombus formation (Figure 28.15A and Figure 28.15B). Some will progress to hemorrhage as the obstruction is lysed and the necrotic regions reperfused. The intraparenchymal large arterioles that arise from arteries at the base of the brain are vulnerable to processes that produce arteriolosclerosis such as hypertension and diabetes mellitus (Figure 28.16). The two major complications of arteriolosclerosis in brain are smaller infarcts (<1 cm), called lacunar infarcts (Figure 28.17),

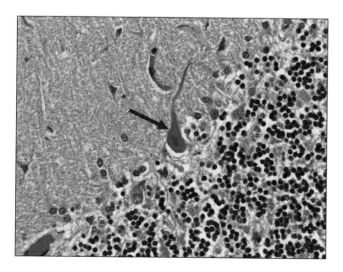

Figure 28.12   Red neuron. Photomicrograph of H&E-stained section of cerebellum shows Purkinje neuron with changes of coagulative necrosis, aka "red neuron" (black arrow).

Table 28.1	Characteristic CNS Injuries Related to Size of Vessel Occluded	

Vessel	Type of Lesion	Examples of Diseases
Arteries	Territorial infarct, sometimes hemorrhagic	Atherosclerosis, vasculitis, vasospasm, dissection
Large Arterioles	Lacunar infarct Microaneurysm	Arteriolosclerosis, diabetes, hypertension
Microvessels	Widespread injury of varying severity	Arteriolosclerosis, CADASIL, fat or air emboli
Veins	Hemorrhagic infarct, often bilateral	Thrombosis

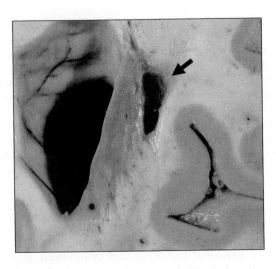

Figure 28.17 Lacunar infarct. Coronal section of right cerebrum shows remote lacunar infarct in the internal capsule adjacent to caudate head (black arrow).

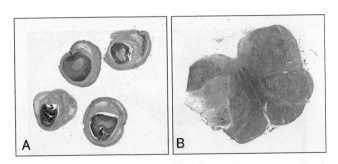

Figure 28.15 Thrombosis. Photomicrographs of H&E/LFB-stained sections show thrombus occluding the basilar artery that also has complicated atherosclerosis **(A)** and the corresponding subacute infarct of the lateral medullary plate **(B)**.

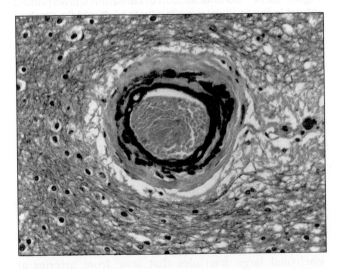

Figure 28.16 Arteriolosclerosis. Section of subcortical white matter demonstrates an arteriole with a thickened vascular wall with immunohistochemical reaction for anti-muscle-specific actin demonstrating continuous hypertrophy Gomori trichrome.

and formation of aneurysms that may rupture and lead to intracerebral hemorrhage.

Rather than focal arteriolar disease leading to discrete lacunar infarcts, widespread disease of small arterioles and capillaries, also known as microvessels, can produce more pervasive ischemic damage to brain that often presents clinically with global neurologic dysfunction. There are several causes of this form of ischemic injury to brain that typically vary in severity from diffuse degeneration with astrogliosis with or without microinfarcts (discernible only by microscopy) that are sometimes hemorrhagic; this variation in damage likely reflects a gradient of insults from oligemia to ischemia. One form of this type of injury thought to be secondary to widespread arteriolosclerosis is called subcortical arteriosclerotic encephalopathy or Binswanger's disease that is associated with hypertension. Another cause of this type of ischemic injury is Cerebral Autosomal Dominant Arteriopathy with Subcortical Infarcts and Leukoencephalopathy (CADASIL). CADASIL is associated with missense mutations in the *NOTCH3* gene in the vast majority of patients and is characterized by a microangiopathy with degeneration of vascular smooth muscle cells that leads to discontinuous immunoreactivity for smooth muscle actin in arterioles (Figure 28.18). Although CADASIL is relatively rare, knowledge gained from investigation of this disease has spurred the search for other genetic causes and risk factors for ischemic brain injury. Other causes of widespread small arteriole/capillary occlusion are thrombotic thrombocytopenic, rickettsial infections, and fat or air emboli.

Thrombosis of veins leads to stagnation of blood flow in the territories being drained and is observed most frequently as a complication of systemic dehydration, hypercoagulable states, or phlebitis. Edema and extravasation of erythrocytes are observed initially. However, venous thrombosis can lead to infarcts that are commonly bilateral and conspicuously hemorrhagic.

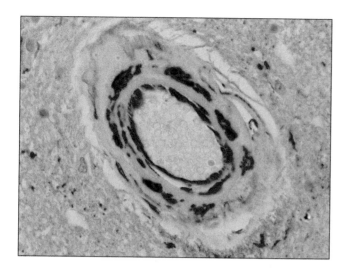

Figure 28.18   Small vessel in CADASIL. Photomicrograph of immunohistochemical reaction for antimuscle-specific actin shows discontinuous smooth muscle in a white matter arteriole from a patient with cerebral autosomal dominant arteriopathy with subcortical infarcts and leukoencephalopathy (CADASIL).

In contrast to regional ischemic damage from diseases that affect blood vessels, global ischemic damage is produced by profound reductions in cerebral perfusion pressure. There are different forms of global ischemic damage that vary with the severity of the insult. In its most extreme form, global ischemia culminates in total cerebral necrosis, or brain death. Although the patient's vital functions may be maintained artificially for some time, cerebral perfusion typically becomes blocked by the massively increased intracranial pressure, leading to necrosis of the entire brain. Global ischemia need not be so severe as to produce total cerebral necrosis. These instances are typified by sudden decrease in mean arterial pressure, such as shock or cardiac arrest, from which the patient is resuscitated. The pattern of tissue damage reflects an exaggerated or selective vulnerability of some groups of neurons to ischemic injury. The most susceptible are the pyramidal neurons of the hippocampus and Purkinje cells of the cerebellum. In more severe cases, arterial border zone, or watershed, infarcts occur at the distal extreme of arterial territories, regions of the brain and spinal cord where distal vascular territories overlap. While there are several arterial border zones in adults, incompletely developed anastomoses between the cerebral cortical long penetrating arteries and the basal penetrating arteries produce a periventricular arterial border zone in premature infants that is inordinately sensitive to global ischemia; the resulting lesion is called periventricular leukomalacia.

The pathophysiology of ischemic injury to the CNS is complex. Within seconds to minutes, there is failure of energy production, release of K^+ and glutamate, and massive increase in intraneuronal calcium (direct excitotoxicity). What follows over hours to weeks is an intricate balance of further damage and response to injury that critically involves elements of immune activation. Enormous effort has been spent investigating potential therapeutic targets and experimental interventions focused in the acute phase of ischemic stroke. Sadly, none has translated to general patient care. Indeed, the current therapeutic approach to ischemic stroke is recombinant tissue plasminogen activator to reverse vessel obstruction; however, this can be complicated by hemorrhage into injured brain.

Hemorrhage

Intracranial hemorrhage occurs when a vessel ruptures and releases blood from the intravascular compartment into some other compartment in the cranium. The key to appreciating the clinical significance of intracranial hemorrhages is an understanding of the type and location of the vessel involved (Table 28.2).

Rupture of an artery leads to release of blood under high pressure into the spaces surrounding the brain that can rapidly lead to fatal compression, while rupture of an arteriole releases blood under high pressure into brain parenchyma with equally devastating effects. Rupture of microvessels leads to lesions called microhemorrhages that conspire with microinfarcts to produce apparently progressive widespread neurologic dysfunction. Finally, rupture or tearing of veins that bridge from the subarachnoid space to the dural sinuses releases blood under venous pressure that produces subdural hematomas.

Intraparenchymal hemorrhages from arteriolar disease, which can be large and life-threatening, deserve further discussion. As described in Table 28.1, aneurysm formation in cerebral arterioles is a consequence in at least some individuals with hypertension. Rupture of these weakened regions of vessels is a common cause of intraparenchymal hemorrhage that more commonly occur centrally or deep in the cerebral hemisphere (Figure 28.19). Another common cause of intraparenchymal hemorrhage is rupture of arterioles sufficiently weakened by amyloid deposition, a condition called cerebral amyloid angiopathy (CAA). These more commonly occur in one of the cerebral

| Table 28.2 | Characteristic CNS Injuries Related to Size of Vessel Ruptured |

Vessel	Location of Ruptured Vessel	Outcome
Artery	Epidural Subarachnoid	Medical emergency
Arteriole Microvessel	Intraparenchymal	Cumulative effects can contribute to widespread progressive neurologic deficits
Vein	Subdural	Varies, often slowly progressive

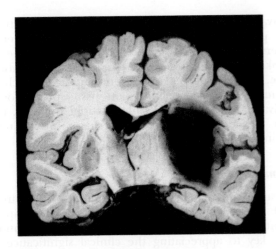

Figure 28.19   Hemorrhage. Coronal section of cerebrum shows intracerebral hemorrhage involving deep structures on the right.

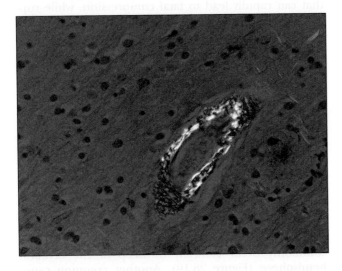

Figure 28.20   Congophilic amyloid angiopathy. Polarized light from Congo red-stained cerebral cortex shows birefringent parenchymal blood vessel. Note also adjacent parenchymal amyloid deposit from a patient with Alzheimer's disease.

lobes, such as the occipital lobe, and so are termed lobar hemorrhages (Figure 28.20). Amyloid means starch-like and describes a group of proteins that share common interrelated features such of high β sheet conformation, affinity for certain histochemical dyes, and the capacity to form fibrillar structures of limited solubility. CAA can result from the deposition of amyloid (A) β peptides, cystatin C, prion protein, ABri protein, transthyretin, or gelsolin. CAA with Aβ peptides can occur apparently sporadically with advancing age and in patients with Alzheimer's disease. Rarely, CAA is caused by autosomal dominant mutations and is called Hereditary Cerebral Hemorrhage with Amyloidosis (HCHWA). These include Dutch, Italian, and Flemish families with mutations in the amyloid precursor protein (*APP*) gene and the Icelandic type that is caused by mutations in the cystatin gene. It is

important to note the other mutations in *APP* are rare autosomal dominant causes of Alzheimer's disease. Furthermore, amyloid deposition in blood vessels is not always limited to the cerebrum, for example, HCHWA-Icelandic type typically shows more widespread involvement of vessels throughout the brain.

Degenerative Diseases

Neurodegenerative diseases range from very common illnesses like Alzheimer's disease to rare illnesses like prion diseases. Function is localized within the CNS, so the clinical presentation of neurodegenerative illnesses is dictated by the regions of brain affected. For example, Alzheimer's disease focuses initially in the hippocampus and closely related structures and later involves primarily frontal, temporal, and parietal lobes. Clinically, Alzheimer's disease is characterized by early impairment in declarative memory that is followed by impairments in other cognitive domains. Another example is amyotrophic lateral sclerosis (ALS) in which degeneration of Betz cells in the parietal lobe and anterior horn cells in the spinal cord produces a characteristic combination of weakness and paralysis. A fuller understanding of the human functional neuroanatomy than is possible to discuss here is needed to appreciate the correlations between affected regions and clinical presentation. However, Table 28.3 presents a very broad overview for selected diseases.

Etiology

The cause of each one of these illnesses is known to some extent; however, this mostly concerns forms that have patterns of highly penetrant autosomal dominant inheritance. For example, Huntington's disease is caused by inheritance of an abnormally expanded trinucleotide repeat (CAG) in the *HD* gene that is translated into an expanded glutamine repeat in the Huntintin protein. The situation is more complex for the other four neurodegenerative diseases that we are considering. Alzheimer's disease, Parkinson's disease, ALS, and Creutzfeldt-Jakob disease, the most common type of prion disease, all have an uncommon subset of patients with autosomal dominant forms of the disease, but also much more common sporadic forms that are not caused by inherited mutations, although some have been associated with inherited risk factors (Table 28.4). It is key to understand that identification of genetic causes and risk factors defines relevance, but not mechanism. For example, the mutation that leads to Huntington's disease has been known for 15 years, but the mechanisms of neurodegeneration remain an area of intense investigation. Mechanism of disease is important because it is the foundation for evidence-based therapeutic interventions.

Pathogenesis of Neurodegeneration

While there is evidence for each of the basic mechanisms of neuronal injury contributing to neurodegenerative

Table 28.3   Anatomic and Clinical Features of the Neurodegenerative Diseases

Disease	Affected Region	Corresponding Clinical Features
Alzheimer's disease	Hippocampus, regions of cerebral cortex	Dementia
Creutzfeldt-Jakob disease	Cerebral cortex, basal ganglia, cerebellum	Dementia, movement disorders
Parkinson's disease	Midbrain and basal ganglia	Bradykinesia, rigidity, tremor
Huntington's disease	Basal ganglia	Chorea
ALS	Primary motor neurons	Weakness and paralysis

Table 28.4   Autosomal Dominant and Sporadic Forms of the Neurodegenerative Diseases

Disease	Prevalence	Autosomal Dominant		Sporadic	
		Inherited Cause	Frequency	Inherited Risk Factor	Frequency
AD	~20% of people over 65	Mutation in *APP, PSEN1,* or *PSEN2*	Uncommon	APOE ε4 allele	Common
PD	3–5% of people over 65	Mutation in *SNCA, PARK2, UCHL1, DJ1,* or *PINK1*	Uncommon	SNCA polymorphisms	Common
ALS	~4 per million	Mutation in *SOD1*	Uncommon	Not yet identified	Common
CJD	~1 per million	Mutation in *PRNP*	Uncommon	*PRNP* polymorphisms	Common
HD	~3 per 100,000 in Western Europeans	Expanded CAG repeat in *HD*	All	Not applicable	None

diseases, a relatively specific hallmark feature of several neurodegenerative diseases, including all of those discussed in this section, is the accumulation of aggregates of misfolded protein that in some instances form amyloid. Although protein misfolding is not limited to neurologic disease, within neurologic disease it seems limited to neurodegenerative conditions; we will discuss this mechanism here. The focus on protein abnormalities in neurodegenerative diseases stems from the characteristic amyloid deposits known to neuropathologists for over a century: amyloid plaques in Creutzfeldt-Jakob disease that contain prion protein (PrP) amyloid, senile plaques of Alzheimer's disease that contain amyloid β peptides, and α-synuclein-containing Lewy bodies in Parkinson's disease (Figure 28.21). Our understanding of the role of protein misfolding in neurodegenerative disease has advanced greatly from identification of amyloid. Current ideas about this aspect of neurodegeneration are best demonstrated by prion diseases like Creutzfeldt-Jakob disease (Figure 28.22).

The *PRNP* gene normally is transcribed (step 1) and translated (step 2) to PrP-C (green circle), a glycosyl phosphatidyl inositol- (GPI-) anchored protein that is expressed by many cells but at high levels by neurons. In some forms of prion diseases, PrP-C misfolds from its normal conformation to a pathogenic form (red square), PrP-Sc (step 3), that is high in β-sheet (Sc is an abbreviation for scrapie, a form of prion diseases in sheep). Some of the prion disease-causing mutations in *PRNP* encode for proteins with increased susceptibility to misfold into these pathogenic forms. These abnormal conformers of PrP-Sc are thought to

promote further subversion of PrP-C folding (step 5) leading to apparently self-propagated generation of abnormal conformers that organize progressively into ordered complexes called fibrils (step 6) that can fracture to generate new seeds for further recruitment of PcP-C (step 7), and accumulate as amyloid in brain (step 8). Cellular defenses against protein misfolding, self-aggregation, and accumulation include chaperones, the ubiquitin-proteosome system, and autophagy, among others. Precisely how these abnormal conformers or higher ordered complexes lead to neuron death is not entirely clear, but likely involves activation of at least some of the pathogenic processes described previously.

Since similar pathologic features to those described so far for prion diseases are shared by several neurodegenerative diseases, many now propose that similar molecular mechanisms underlie the more common Alzheimer's disease where the misfolded and accumulating proteins are Aβ peptides. The amyloid precursor protein (APP) is the product of the *APP* gene mentioned previously that, when mutated, can cause a highly penetrant form of early-onset autosomal dominant Alzheimer's disease. APP is a single membrane-spanning protein that is expressed at high levels by neurons and that undergoes exclusive endoproteolytic cleavages by α-secretase to generate secreted and internalized segments, or by β-secretase and γ-secretase to generate secreted protein, amyloid β peptides, and internalized segments, all of which have biological activity; however, the major research focus has been on the neurotoxic properties of Aβ peptides. The

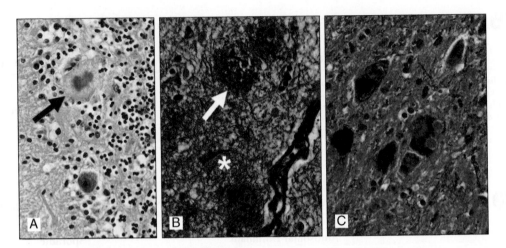

Figure 28.21   Proteinaceous inclusions of neurodegenerative diseases. Photomicrographs show **(A)** plaque (black arrow) in cerebellum from patient with CJD (H&E), **(B)** senile plaque (white arrow) and neurofibrillary tangle (white asterisk) in a patient with AD (modified Bielschowsky), and **(C)** a pair of Lewy bodies, one among several pigmented (dopaminergic) neurons of the substantia nigra from a patient with Parkinson's disease (H&E/LFB).

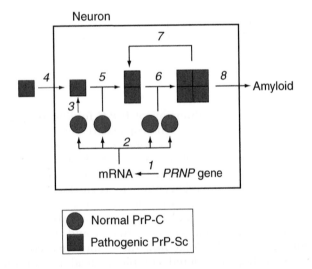

Figure 28.22   Diagram of prion disease pathogenesis. The *PRNP* gene is transcribed (1) and translated (2) to PrP-C (green circle). The pathogenic protein (red square) forms by misfolding of PrP-C (3) or can be transmitted (4). PrP-Sc promotes further recruitment of PrP-C (5) into fibrils (6) that can generate new seeds (7) and form amyloid (8).

studied, with $A\beta_{1-42}$ being more fibrillogenic and neurotoxic in model systems. Key to ultimately understanding Alzheimer's disease will be unraveling the mechanistic connections between accumulation of $A\beta$ peptides, not only in brain parenchyma as shown earlier, but also in cerebral blood vessels (Figure 28.20), and the accumulation of pathologic forms of the microtubule-associated protein tau in structures called neurofibrillary tangles (Figure 28.21B). While this connection may involve some of the processes described previously, it currently remains enigmatic.

It is critically important to emphasize that a unique feature of prion diseases is their transmissibility (step 4 in Figure 28.22). This is a real but rare clinical issue. However, it is achieved routinely in laboratory animals. Indeed, protease-resistant fragments of PrP-Sc are now widely viewed as the transmissible agent in prion diseases. This is not the case for $A\beta$ peptides or aggregated proteins characteristic of other neurodegenerative diseases. Indeed, no other neurodegenerative disease is transmissible. Perhaps this simply reflects varying potency for transmission. Alternatively, this may point to a fundamental difference in mechanism among these diseases.

Before we leave the topic of neurodegenerative diseases, it is important to remember the immense looming public health challenge posed by late-onset Alzheimer's disease (LOAD). LOAD describes those patients with Alzheimer's disease who have onset in later adult life, typically older than 65 years, and who have not inherited a causative mutation; sometimes this is referred to as sporadic Alzheimer's disease. While identification and investigation of autosomal dominant forms have provided invaluable insight into the etiology and pathogenesis of Alzheimer's disease, LOAD may present additional facets for investigation. One of these might be the possible intersection of Alzheimer's disease with vascular brain injury in older patients as diagrammed in Figure 28.23.

identity of β-secretase is now known and is called BACE1 (β-site APP-cleaving enzyme), while at least part of the multicomponent γ-secretase appears to be the protein products of *PSEN1* and *PSEN2*, mutations of which also cause highly penetrant forms of early-onset autosomal dominant Alzheimer's disease. Indeed, the clustering of mutations that cause early-onset Alzheimer's disease around the generation of $A\beta$ peptides is the foundation of the amyloid hypothesis for this disease. Promiscuity in the cleavage site by γ-secretase is responsible for generating $A\beta$ peptides of varying lengths. Of these, $A\beta$ that are 40 ($A\beta_{1-40}$) or 42 ($A\beta_{1-42}$) amino acids in length are the most intensely

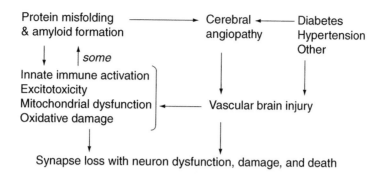

Figure 28.23   **Diagram of potential interactions between Alzheimer's disease and vascular brain injury in dementia.**

Neurotoxicants

Neurotoxicology has significance that reaches beyond the identification of xenobiotics (neurotoxicants) or endogenous agents (neurotoxins) that are deleterious to the nervous system. Although considered separate fields of study, neurotoxicology, neurodegeneration, stroke, trauma, and metabolic diseases of the nervous system inform each other about the mechanisms of neuronal dysfunction and death, as well as response to injury in the nervous system. Indeed, there are several examples of compounds first identified as neurotoxicants that subsequently came into use as models of human neurodegenerative disease; perhaps the most striking example is 1-methyl-4-phenyl-1,2,3,6-tetrahydropyridine (MPTP). Moreover, many compounds initially identified as neurotoxicants have become fundamental tools used by neuroscientists, such as tetrodotoxin, curare, kainic acid, and 6-hydroxydopamine. Conversely, progress in other fields of neuroscience continually advances understanding of the mechanisms of neurotoxicants.

Biochemical Mechanisms of Selected Neurotoxicants

Domoic Acid A large number of experiments and trials have provided indirect or pharmacological support for a role for excitotoxicity in neurological injury. Humans accidentally exposed to high doses of EAA receptor agonists and who subsequently developed neurologic disease underscore the importance of EAAs in disease. Perhaps the most striking example is the domoic acid intoxications that occurred in the Maritime Provinces of Canada in late 1987 (Figure 28.24). A total of 107 patients were identified who suffered an acute illness that most commonly presented as gastrointestinal disturbance, severe headache, and short-term memory loss within 24 to 48 hours after ingesting mussels. A subset of the more severely afflicted patients was subsequently shown to have chronic memory deficits, motor neuropathy, and decreased medial temporal lobe glucose metabolism by positron emission tomography (PET). Neuropathological investigation of patients who died within 4 months of intoxication disclosed neuronal loss with reactive gliosis that was most prominent in the hippocampus and amygdala, but also affected regions

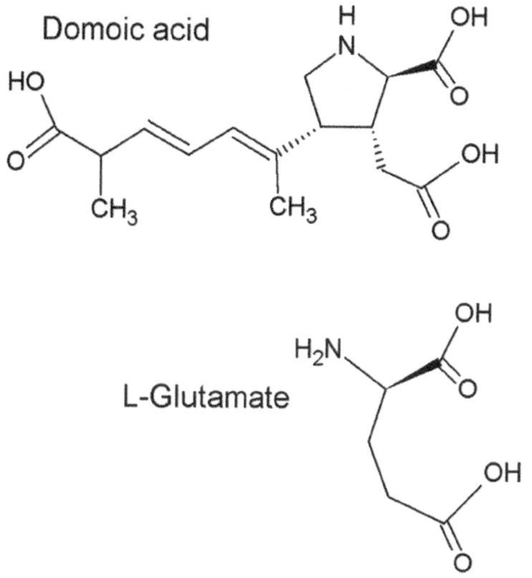

Figure 28.24   **Structures for domoic acid and L-glutamate.**

of the thalamus and cerebral cortex. The responsible agent was identified as domoic acid, a potent structural analog of L-glutamate that had been concentrated in cultivated mussels.

MPTP The history of MPTP-induced parkinsonism in young adults who inadvertently injected themselves with this compound is well known. MPTP is a protoxicant that, after crossing the blood-brain barrier, is metabolized by glial MAO-B to a pyridinium intermediate (MPDP$^+$) that undergoes further two-electron oxidation to yield the toxic metabolite methyl-phenyl-tetrahydropyridinium (MPP$^+$) that is then selectively transported into nigral neurons via the mesencephalic dopamine transporter (DAT) (Figure 28.25). Once inside these neurons, MPP$^+$ is thought to act primarily as a mitochondrial toxin by inhibiting complex I activity in the mitochondrial electron transport chain, thereby reducing ATP production and increasing ROS generation. Indeed, MPTP-induced dopaminergic neurodegeneration can be diminished by free radical scavengers, inhibitors of the inducible form of nitric oxide synthase, and by EAA receptor antagonists. Alternatively, transgenic mice lacking some elements of antioxidant defenses are significantly more vulnerable to MPTP-induced dopaminergic neurodegeneration.

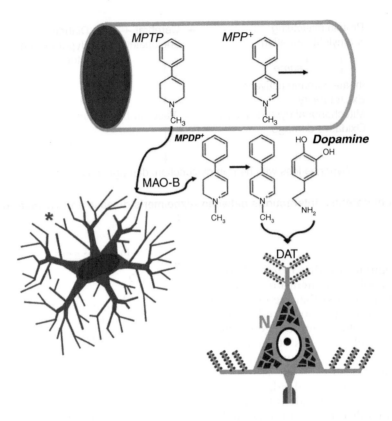

Figure 28.25 Diagram of steps in dopaminergic toxicity from MPTP exposure. MAO-B in astrocytes (*) catalyzes oxidation of MPTP to ultimately form MPP$^+$ which is selectively taken up by dopaminergic neurons (N) expressing the dopamine transporter (DAT).

So far, the search for xenobiotics that may act similarly to MPTP and could be potential environmental toxicants that promote PD has not yielded clear candidates.

NEOPLASIA

Primary CNS neoplasms can be defined as uncontrolled growth of cells derived from normal CNS tissues. The most common histologically resemble glia and are termed gliomas. Although it was previously believed that gliomas were derived from mature glial elements, such as astrocytes or oligodendrocytes, it is more likely that these tumors arise from glio-neuronal progenitor cells, so-called cancer stem cells.

Gliomagenesis results from a number of known genetic defects associated with control of cell cycle and proliferation, apoptosis pathways, cell motility, and invasive potential.

Unlike other organ systems that use the TNM (local [T]umor growth, regional lymph [N]ode spread, and distant [M]etastasis) staging system for establishing prognosis of tumors, primary CNS tumors are assigned a histologic grade that correlates with their predicted behaviors. The *World Health Organization* (WHO) classification of tumors (2007) identifies tumors as grade I if they are curable by resection alone and have >10 year median survival, grade II if they are not curable by resection alone and have 5–10 year median survival, grade III for 2–3 year median survival, and grade IV

if they have <2 year median survival. The histologic features underlying WHO grading include nuclear atypia, mitoses, vascular proliferation, and necrosis.

Diffuse Gliomas

A defining characteristic of the diffuse gliomas is their ability to infiltrate widely throughout the CNS parenchyma, causing them to have no clearly recognizable boarder with normal tissue (Figure 28.26). Because of this feature, they are not curable by resection alone. The natural history of diffuse gliomas is a tendency toward progression from low to high grade by the accumulation of additional genetic defects. De novo, grade IV gliomas also occur, but usually by a different set of genetic defects. Broadly, there are four types of genetic alterations involved in gliomagenesis: those which affect (i) cell survival, (ii) cell proliferation (cell cycle regulators), (iii) brain invasion, and (iv) neovascularization. Mutations in Rb, *p53*, receptor tyrosine kinase, integrin, and other signaling pathways critical to cell cycle regulation are important, as are mutations in cell death pathways (TNFR, TRAIL, CD95, Bcl-2, etc.). Transition from low to high grade is heralded by tumor enhancement by neuroimaging which correlates with microvascular proliferation (angiogenesis), in which pathways regulated by VEGF, HIF, and other molecules are critical. Finally, brain invasion, which is probably least understood of all the gliomagenic

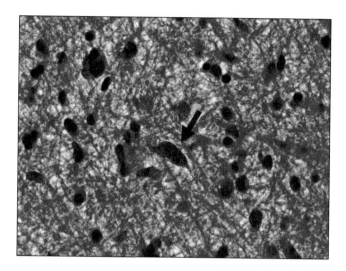

Figure 28.26 Photomicrograph of diffuse astrocytoma. White matter infiltrated by an enlarged, hyperchromatic (darkly stained), pleomorphic (irregularly shaped) neoplastic glial cell (arrow) in a diffuse astrocytoma (H&E 600× magnification).

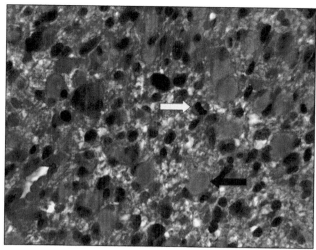

Figure 28.27 Photomicrograph of an anaplastic astrocytoma. Numerous enlarged, hyperchromatic (darkly stained) neoplastic glial nuclei have infiltrated brain parenchyma in this anaplastic astrocytoma. A mitotic figure (white arrow) is present. A globule of eosinophilic cytoplasm is seen within neoplastic astrocytes (black arrow) likely representing glial fibrillary acidic protein, the primary intermediate filament of astrocytes (H&E).

processes and the most critical to toxicity, is likely mediated by metalloproteinases and integrins, among others.

Diffuse Astrocytoma–Glioblastoma Sequence

Diffuse astrocytomas (WHO grade II–IV) are gliomas in which the neoplastic cells resemble microscopically normal or reactive astrocytes and are often characterized by their cytoplasmic expression of GFAP. Histologically, grade II astrocytomas are typified by mild nuclear atypia and hypercellularity and an absence of mitotic activity, vascular proliferation, or necrosis. Anaplastic astrocytomas (WHO grade III) are histologically characterized by increased hypercellularity with nuclear atypia and mitotic activity in the absence of vascular proliferation and necrosis (Figure 28.27).

Glioblastoma (GBM; WHO grade IV) is the most common primary glioma with an incidence of approximately 3/100,000 population per year. GBMs are characterized by rapid progression and poor survival. They can be classified, based on clinical and molecular characteristics, into those that arise *de novo* and those that progress from lower grade astrocytomas (Figure 28.28).

Primary or *de novo* GBMs account for 90% of cases, generally arise in older patients, and are more prone histologically to be of the small cell subtype. The characteristic abnormalities of signal transduction in primary GBM include overexpression or signal amplification of the epidermal growth factor receptor (*EGFR*) or loss-of-function mutations of *PTEN* (phosphatase and tensin homology). Deletion of *EGFR* exons 2–7 (EGFRvIII), the most common *EGFR* mutation (found in 20–30% of GBM and 50–60% of GBM with *EGFR* amplification), results in constitutive EGFR activation and insensitivity to EGF. EGFRvIII stimulates a different signal transduction pathway than full-length EGFR.

EGFR activation operates primarily through the Ras-Raf-MAPK and PI3K-Akt pathways. Constitutive activation of EGFRvIII mutants increases proliferation and survival by preferentially activating the (PI3K)/Akt pathway, and possibly through stimulation of a second messenger system not available to nonmutant EGFR. The characteristic genetic alterations involving cell cycle control in primary GBMs include overexpression of *MDM2* (murine double minute 2 protein), which suppresses *p53*, and deletions of *CDKN2A*, which encodes the tumor suppressor p16^{INK4A}, a potent regulator of retinoblastoma (*RB*) tumor suppressor gene, or, through an alternate reading frame (ARF) p14^{ARF}, an important accessory to *p53* activation. Other common findings include loss of heterozygosity in chromosome 10p and overexpression of Bcl2-like-12 protein, a potent antiapoptotic molecule. The genetic abnormalities occur together in a random distribution and are not progressive. Homozygous deletion of *CDKN2A* is associated with EGFR overexpression, higher proliferative activity, and may account for poorer overall survival of primary GBMs. It is interesting to note that the small cell phenotype common in primary GBMs appears to be most closely associated with EGFR overexpression, suggesting a molecular-histologic link.

Secondary GBMs progress from lower grade astrocytomas with stepwise accumulation of additional genetic defects and generally occur in younger individuals. Common early genetic abnormalities include direct mutations of the cell cycle suppressor genes *TP53* and *RB1*, or overexpression of platelet-derived growth factor ligand and/or receptors. Loss of heterozygosity (LOH) of chromosomes 11p and 19q is common in progression of these low-grade

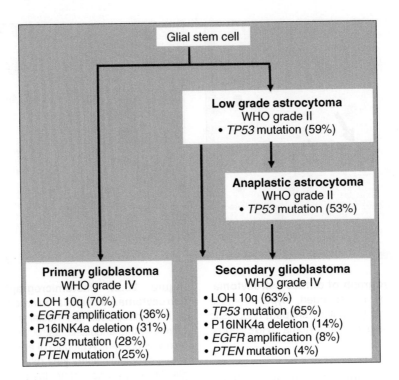

Figure 28.28   Genetic alterations associated with primary and secondary glioblastoma.

astrocytomas to anaplastic astrocytomas. The transition from anaplastic astrocytoma to GBM is less well characterized, but can involve generally LOH of chromosome 10q, mutations indirectly affecting EGFR/PTEN pathways, and alterations to the cell cycle inhibitory p16^{INK4a}/RB1 pathway. These pathways are summarized in Figure 28.29.

By definition, all GBMs are prone to spontaneous necrosis and bizarre microvascular proliferation (Figure 28.30). Aberrations in the genetic control of growth and cell cycle give rise to the hypoxia, necrosis, and angiogenesis that underlie these characteristic histologic features. A model of this process (Figure 28.31) begins with genetic alterations in

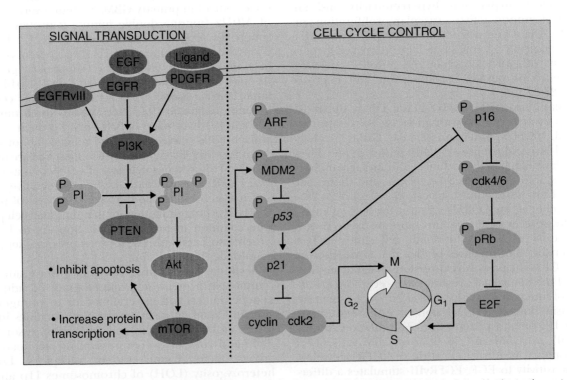

Figure 28.29   Molecular pathways implicated in gliomagenesis. Multiple molecules can be involved in the pathway to glioma.

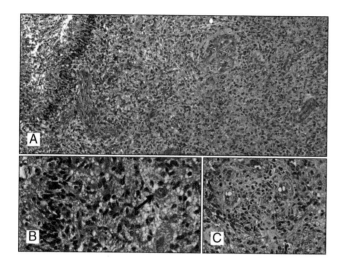

Figure 28.30 Photomicrograph of glioblastoma. (A) Sections from a brain infiltrated by glioblastoma stained with H&E at 100× magnification and **(B)** 600× magnification (below left). A line of parallel nuclei forming a pseudo-palisade can be seen extending from the panel above through the left lower panel. The pseudopalisade is adjacent to necrosis. A neuron can be seen surrounded by neoplastic cells (arrow). **(C)** The collagen network of a proliferative vessel is highlighted blue-green by a trichrome stain.

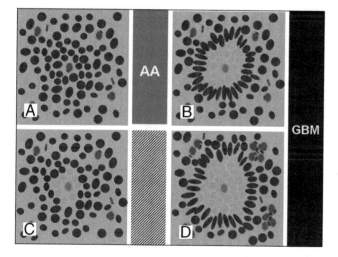

Figure 28.31 Progression from anaplastic astrocytoma (AA) to glioblastoma (GBM). (A) AA is mitotically active producing an area of hypercellularity, but no necrosis or vascular cell hyperplasia. **(B)** Neoplastic glia produce thrombophilic compounds such as thrombin and tissue factor causing thrombosis of a microvessel, focal ischemia and hypoxia, and local necrosis. This change is the hallmark of the transition from AA to GBM. **(C)** Neoplastic glia respond to hypoxia by migrating away, forming a "pseudopalisade" of nuclei surrounding the necrosis. **(D)** Migrating neoplastic glia also produce angiogenic factors such as VEGF and HIF, causing vascular cell hyperplasia and abnormal vessel formation.

tumor cells which cause a variety of downstream effects which alter the blood-brain barrier, damage endothelium, or directly promote intravascular thrombosis (such as secreting tissue factor). Small vessels in the tumor become occluded, and the resulting damaged tissue releases thrombin as part of the coagulative cascade, which in turn binds to its ligand the protease-activated receptor 1 (PAR-1). Local hypoxia leads to focal necrosis. Tumor cells react to PAR-1 by forming a wave of migration away from the area of hypoxia, the histologic counterpart of which is a pseudopalisade. These migrating cells are severely hypoxic and express high levels of hypoxia-inducible factor (HIF), which activates hypoxia-responsive element domains (HRE), of which VEGF is the most characteristic in GBM. High levels of VEGF cause the vascular proliferation characteristic of this tumor.

An epigenetic feature of GBM that may have therapeutic implications is the methylation status of the DNA repair gene *MGMT* (O^6-methylguanine–DNA methyltransferase). *MGMT* is a DNA-repair protein that removes alkyl groups from the O^6 position of guanine. Epigenetic silencing of this gene is identifiable in a little less than half of GBMs and predicts response to alkylating chemotherapy.

Oligodendroglioma

Oligodendrogliomas are diffusely infiltrating gliomas that represent the second most common parenchymal tumor of the adult CNS. The neoplastic cells in oligodendrogliomas histologically resemble mature oligodendrocytes. They generally have round nuclei with condensed chromatin and inconspicuous cytoplasm and few cytoplasmic processes. A classical appearance of the neoplastic cells in oligodendroglioma is a round hyperchromatic nucleus with a perinuclear clearing or halo, the so-called fried egg cells. They may express a small amount of cytoplasmic GFAP, but generally less than astrocytic neoplasms. Histologically, they form monotonous proliferations of infiltrating neoplastic cells with occasional hypercellular nodules. They have characteristic chicken-wire vasculature composed of a network of fine branching capillaries (Figure 28.32A). The neoplastic cells also have a tendency to cluster, or satellite, around infiltrated neurons and vascular structures (Figure 28.32B).

A large subset of oligodendrogliomas contain a chromosomal translocation-mediated loss of the short arm of chromosome 1 (1p) and the long arm of chromosome 19 (19q), commonly detected using *in situ* hybridization or DNA amplification techniques. Loss of 1p/19q has been associated with both classic histomorphology and better clinical behavior. Patients whose tumors have this relative co-deletion demonstrate improved disease-free survival, median survival, and may respond better to alkylating chemotherapeutics. Loss of 1p/19q is inversely correlated with mutations in *TP53*, loss of chromosomal arms 9p and 10q, and amplification of EGFR. Co-deletion of 1p/19q is also associated with lower

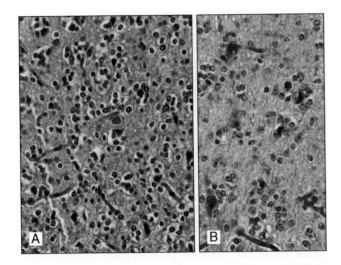

Figure 28.32   Photomicrograph of oligodendroglioma. (A) Oligodendrocytes-like neoplastic cells with round nuclei and perinuclear clearing give this oligodendroglioma a fried egg appearance. **(B)** Non-neoplastic neurons are clustered about by neoplastic cells. Fine capillaries are seen in the background of both panels (H&E).

MGMT expression and *MGMT* promoter hypermethylation and may explain the greater chemosensitivity of these tumors. Because of the robust correlation of this specific genetic aberration to both histomorphology and prognosis, oligodendrogliomas exist at the forefront of molecular diagnostics in gliomas. Molecular testing of tumors for loss of 1p/19q is already widely accepted as a prognostic marker and is commonly used by clinicians to guide therapy. Furthermore, although not currently accepted as a diagnostic test for oligodendroglioma, loss of 1p/19q has proven useful in differentiating oligodendrogliomas from histologic mimics and in cases of ambiguous histomorphology. The molecular mechanism(s) by which loss of 1p and 19q mediate their effects has not been elucidated. The chromosomal translocation that underlies most losses of 1p/19q occurs at pericentromeric sites, excluding the possibility of a transgene product driving this neoplasm. Attempts to identify the presumed tumor suppressor genes on 1p and 19q have been unsuccessful. Proteomic analysis of tumors with and without the paired deletion have demonstrated increased expression of proteins associated with malignant behavior in tumors without the deletion, but definitively decreased expression of proteins coded on the lost arms has not been demonstrated.

Oligodendrogliomas are classified in the WHO system into two grades. WHO grade II oligodendrogliomas have the same histomorphology as described previously (Figure 28.32). Grade III anaplastic oligodendrogliomas are histologically similar, but display additional nuclear atypia and mitotic activity, and in addition to the classic chicken wire vasculature, they often have vascular proliferation reminiscent of that seen in GBM. Oligodendrogliomas may progress to a histology that is indistinguishable from GBM, although

LOH of 1p/19q probably portends a more favorable prognosis than astrocytic GBM.

DISORDERS OF MYELIN

Myelin is composed of layers of a complex proteolipid membrane that allows rapid and efficient conduction of action potentials along the axons of neurons. Loss or dysfunction of the normal myelin sheath causes abnormalities of this normal electrical signaling. There are two broad categories that describe ways myelin can be damaged: dysmyelination and demyelination. Dysmyelination is the loss of abnormally formed myelin or loss of the oligodendrocytes that produce and maintain myelin and is due to biochemically deranged myelin processing. Demyelination is the loss of structurally and biochemically normal myelin either by immune-mediated attack against myelin or oligodendrocytes or due to metabolic derangement of oligodendrocytes.

Leukodystrophies

Leukodystrophies are genetic disorders that cause damage to white matter (*leuko* from the Greek for "white"). The leukodystrophies are autosomal recessive (AR) or X-linked disorders that result from loss-of-function mutations in enzymes involved in either production of myelin or normal myelin turnover and degradation (Figure 28.33). These enzymatic defects are not necessarily specific only to the CNS, and other organ systems may be affected. Clinically, these diseases typically present in early childhood with progressive loss of motor control, cognitive function, seizures, and eventual death.

Metachromatic leukodystrophy (MLD) is the most common leukodystrophy, is AR, and is caused by deficiency of the arylsulfatase A enzyme, which breaks down galactosyl-3-sulfatide to galactocerebroside in lysosomes. Galactosyl-3-sulfatide accumulates in many tissues, but the symptoms are related to destruction of myelin. MLD is so named because the accumulated material will change the color (metachromasia) of acidified cresyl violet stain from violet to brown in tissue. Krabbe's globoid cell leukodystrophy is another AR leukodystrophy caused by an enzyme deficiency in the sulfatide breakdown pathway. Here, there is a defect in the enzyme galactocerebroside ß-galactocerebrosidase, which breaks down galactocerebroside to galactose and ceramide.

Adrenoleukodystrophy (ALD) affects both the CNS myelin and adrenal glands. ALD is usually X-linked and is unusual among the leukodystrophies in that it is due to a defect in a peroxisomal transporter protein. ALD is most often caused by a defect in *ABCD1*, a member of the ATP-binding cassette transporter family. *ABCD1* encodes a transmembrane protein which is half of a heterodimeric transporter that transfers the enzyme peroxisomal acyl coenzyme A into peroxisomes, where it β-oxidizes long chain fatty acids. Deficiencies in this enzyme cause a buildup of very long chain fatty acids (VLCFA), especially hexacosanoic

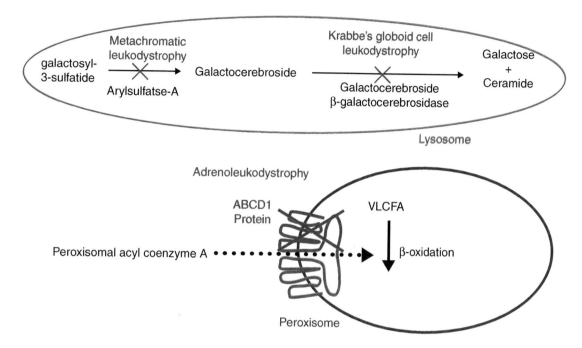

Figure 28.33 **Myelin metabolic pathway and loss of function mutations associated with dysmyelinating disease.**

(C26:0) and tetracosanoic (C24:0) acid. This form of ALD is X-linked. A more severe autosomal recessive form of the disease is due to absent or reduced peroxisome receptor-1, which causes more widespread peroxisomal dysfunction.

Numerous other leukodystrophies exist, and the molecular defects are known in some. Pelizaeus-Merzbacher leukodystrophy is X-linked and caused by abnormalities in a highly abundant myelin structural protein, proteolipid protein (PLP), which interferes with normal myelin formation. Interestingly, complete loss of the protein causes a relatively mild form of the disease, while missense mutations confer more severe changes, probably because the abnormal protein product interferes conformationally with the compact layering of normal myelin. Alexander's disease is a leukodystrophy in which the pathology is primary to astrocytes, not oligodendrocytes. It is caused by a dominant gain of toxic function mutation in the glial fibrillary acidic protein (GFAP) gene in which the protein forms abnormal aggregates. Large amounts of GFAP and other proteins accumulate in astrocyte processes. The mechanism of damage to oligodendrocytes or to myelin is not known.

Demyelination

Idiopathic Demyelinating Disease Multiple sclerosis (MS) describes a variety of related disorders characterized by relapsing and remitting, multifocal immune-mediated damage to oligodendrocytes and myelin that are separated by time and anatomic location. These diseases are further clinically subclassified by the rate of progression of disability and the presence or absence of partial recovery between attacks. Histologically, there is loss of

Figure 28.34 **Photomicrograph of old demyelinating lesion in multiple sclerosis.** Sections of cerebral white matter from the edge of demyelinating lesion in multiple sclerosis. A Luxol fast blue demonstrates loss of myelin (blue) in the left half of the figure, while a Holmes silver stain shows preserved axons in black.

myelin with relative preservation of axons (Figure 28.34). Although the patterns and mechanisms of injury in MS are increasingly well characterized, the ultimate cause(s) of MS is unknown. The incidence of the disease is characterized by a markedly uneven geographic distribution generally with higher latitudes having higher risk, perhaps suggesting a role for environmental factors. Genetic factors also clearly have a role, as the incidence of disease is elevated in family members of a proband, with highest risk in monozygotic twins, which

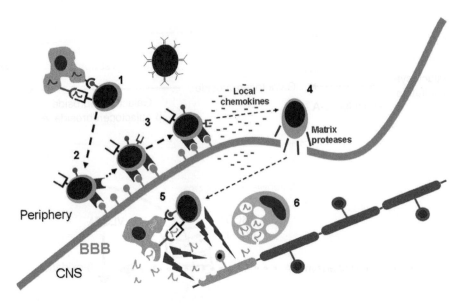

Figure 28.35   CNS directed autoimmunity in demyelinating disease. T-cells are activated in the periphery by antigen presenting cells (APCs) through the interaction of the T-cell receptor (TCR) and antigen bound to major histocompatibility complex (MHC) class II along with co-regulatory signals (1). Activated T-cells express molecule surface receptors for adhesion molecules on the luminal surface of the endothelium of the blood-brain barrier (BBB) (2). Activated T-cells also respond to local chemokines (3). The activated T-cells, under the influence of local chemokines, release matrix proteases that allow them to traverse the BBB (4). In the CNS, T-cells interact with microglia-derived or CNS infiltrating APCs through the interaction of the TCR and antigen-MHC and co-stimulatory pathways, thus mediating damage to myelin and oligodendrocytes (5). Macrophages remove debris (6).

reaches a probability of almost 1/3. There is a prevailing multistep hypothesis of MS (Figure 28.35): inflammatory cells outside the CNS are activated and upregulate adhesion molecules that guide them to the BBB. In this activated state, they are also more reactive to local chemokines and secrete matrix proteases to gain entry into the CNS, where they mediate damage. The disease mechanism of MS is generally understood to be T-cell-mediated autoimmune processes which require some as yet to be characterized trigger. T-cells outside the CNS are activated via the T-cell receptor (TCR) binding with antigen presented on the major histocompatibility complex (MHC) molecule and additional signals by antigen-presenting cells (APC). Additional signaling pathways also play a role in immune activation. One of these pathways that has been characterized involves the B7 signaling molecule on APCs. B7 interacts with signaling molecules on different T-cell subsets (most notably CD28 and CTLA-4) that modify the TCR response which can either cause activation and development of T-cell function, or unresponsiveness and apoptosis. APCs also release chemokines such as interleukin (IL)-12 and IL-23 that generate CD4-positive T-helper type 1 cells (Th$_1$). Th$_1$ cells are proinflammatory, interferon γ-producing cells that are part of the normal antiviral response. These and other proinflammatory T-cell subsets have been implicated in demyelinating disease.

Neuromyelitis optica (NMO), also known as Devic's disease, is a variant of MS for which an etiology has been elucidated. Clinically, NMO is characterized by generally monophasic demyelination of the optic nerve,

brainstem, and spinal cord without demyelination of the intervening cerebrum. Pathologically, the disease is characterized by a circulating autoantibody called the NMO-immunoglobulin (Ig)G. The NMO-IgG is directed against aquaporin 4, a water transporter in the foot processes of astrocytes where they abut CNS microvessels. The mechanism by which oligodendrocytes and glia are damaged is not fully understood.

Acquired Metabolic Demyelination   There are two primary osmotic demyelination syndromes. Central pontine myelinolysis (CPM) is a dire complication associated with rapid correction of systemic hyponatremia. Rapid replacement of sodium intravenously in persons with low serum sodium concentrations causes an acute loss of myelin that is usually confined to the middle of the base of the pons. The disease usually occurs in the setting of chronic alcoholism, liver or renal disease, post-transplant, or malnourishment. Although the disease mechanism is not well understood, hypotheses include rapid shifts of water in and out of glia, glial dehydration, myelin degradation, or apoptosis of oligodendrocytes.

Marchiafava-Bignami disease is a similar disease entity in which the white matter of the corpus callosum, the major tract that connects the left and right cerebral hemispheres, undergoes a similar demyelination to that seen in CPM, albeit with more necrosis. Initially described in poorly nourished Italian men who consumed large quantities of crude red wine, it is now known to be associated with alcoholism and to have a worldwide distribution.

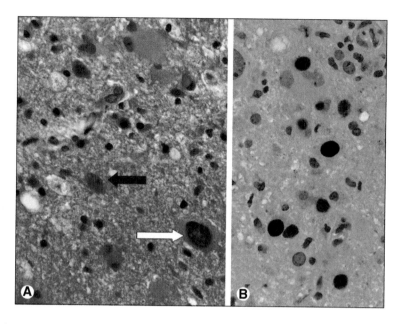

Figure 28.36   Photomicrograph of Progressive Multifocal Leukoencephalopathy (PML). Sections of cerebral white matter from a demyelinating lesion of PML. **(A)** H&E stains demonstrate enlarged oligodendrocyte nuclei with homogeneous "ground-glass" viral inclusions (black arrow) and bizarre enlarged astrocytes (white arrow) simulating a neoplasm. **(B)** Immunoperoxidase staining for antibodies against Simian virus 40 which cross-react with JC virus nuclear inclusions in oligodendroglia (brown).

Infectious Demyelination Progressive multifocal leukoencephalopathy (PML) is an acute, progressive demyelinating disorder caused by a polyomavirus. JC virus is a small (~50 nm), double-stranded, icosahedral DNA virus lacking a membranous envelope. Approximately 40% of individuals in the developed world are sero-positive for the virus but, are asymptomatic. Recrudescence of the JC virus causes disease in immunocompromised individuals, such as in AIDS, of which it is a defining illness. JC virus preferentially infects oligodendrocytes in the CNS. Histologically, PML is characterized by demyelination, lymphocytic inflammation, oligodendrocytes with enlarged nuclei with a ground-glass appearance, and bizarre reactive astrocytes (Figure 28.36). Electron microscopy shows oligodendrocyte nuclei packed with viral particles. The JC virus is oncogenic in rodents producing astrocytomas.

KEY CONCEPTS

- The CNS is composed of specialized cells that have unique functions and metabolic requirements. The specialized functions of the CNS are carried out by specific anatomic structures, and injury or dysfunction of these structures leads to characteristic neurologic disorders.
- Common basic mechanisms of injury to the CNS include excitotoxicity, mitochondrial dysfunction, free radical stress, and innate immune activation. Following injury, the CNS has very limited regenerative ability. Susceptible cell populations are lost and replaced with reactive glia in a process termed gliosis.

- Injury to the CNS caused by vascular disease is common and arises from cessation of blood flow (ischemia) and bleeding (hemorrhage). Because of their high metabolic demands, neurons, and to a lesser extent oligodendrocytes, are susceptible to even short periods of ischemia. The causes and consequences of vascular injury are specific to the types (arteries or veins) and sizes of vessels involved.
- Different neurodegenerative diseases differentially affect different CNS regions and produce syndromes characteristic of dysfunction in those regions. The causes of most neurodegenerative diseases remain unknown but many are characterized by aggregation of abnormal insoluble proteins. Model systems of neurodegenerative diseases using neurotoxicants act through the common basic mechanisms of injury to the CNS.
- Gliomas are primary CNS neoplasms that, similar to other organ systems, have genetic defects associated with cell cycle control and proliferation, apoptosis, cell motility, and invasion. Unlike most other organ systems, gliomas are generally diffusely infiltrative and cannot be cured by resection. Increasingly, specific genetic abnormalities are recognized to drive both the histologic appearance and behavior of gliomas.

SUGGESTED READINGS

1. Aldape K, Burger PC, Perry A. Clinicopathologic aspects of 1p/19q loss and the diagnosis of oligodendroglioma. *Arch Pathol Lab Med.* 2007;131:242–251.
2. Aloisi F. Immune function of microglia. *Glia.* 2001;36:165–179.

3. Bansal K, Liang ML, Rutka JT. Molecular biology of human gliomas. *Technol Cancer Res Treat.* 2006;5:185–194.

4. Bar-Or A. The immunology of multiple sclerosis. *Semin Neurol.* 2008;28:29–45.

5. Baumann P-D. Biology of oligodendrocyte and myelin in the mammalian central nervous system. *Physiol Rev.* 2001;18:871–927.

6. Benjamin R, Capparella J, Brown A. Classification of glioblastoma multiforme in adults by molecular genetics. *Cancer J.* 2003;9:82–90.

7. Copp AJ. Neurulation in the cranial region—Normal and abnormal. *J Anat.* 2005;207:623–635.

8. Ebers GC. Environmental factors and multiple sclerosis. *Lancet Neurol.* 2008;7:268–277.

9. Endres M, Engelhardt B, Koistinaho J, et al. Improving outcome after stroke: Overcoming the translational roadblock. *Cerebrovasc Dis.* 2008;25:268–278.

10. Fuller CE, Perry A. Molecular diagnostics in central nervous system tumors. *Adv Anat Pathol.* 2005;12:180–194.

11. Golden JA, Harding BN, eds. *Developmental Neuropathology: Pathology and Genetics.* Basel: ISN Neuropath Press; 2004:42–48.

12. Haines DE. *Neuroanatomy: An Atlas of Structures, Sections, and Systems.* 7th ed. Lippincott: Williams & Wilkins; 2008.

13. Langston JW. The etiology of Parkinson's disease with emphasis on the MPTP story. *Neurology.* 1996;47:S153–S160.

14. McLendon RE, Turner K, Perkinson K, et al. Second messenger systems in human gliomas. *Arch Pathol Lab Med.* 2007;131:1585–1590.

15. Ohgaki H, Kleihues P. Genetic pathways to primary and secondary glioblastoma. *Am J Pathol.* 2007;170:1445–1453.

16. Prusiner SB. Shattuck lecture—Neurodegenerative diseases and prions. *N Engl J Med.* 2001;344:1516–1526.

17. Rong Y, Durden DL, Van Meir EG, et al. Mechanisms of necrosis in GBM. *J Neuropathol Exp Neurol.* 2006;65:529–539.

18. Sadler TW. Embryology of neural tube development. *Am J Med Genet C Semin Med Genet.* 2005;135C:2–8.

19. World Health Organization. *Pathology and Genetics of Tumours of the Nervous System.* Geneva: WHO Press; 2007.

Practice of Molecular Medicine

Practice of
Molecular
Medicine

Part

29

Molecular Diagnosis of Human Disease

Lawrence M. Silverman . Grant C. Bullock

INTRODUCTION

Developments in molecular diagnostics primarily reflect technological breakthroughs in molecular biology. Examples of these breakthroughs include the Southern blot, the polymerase chain reaction (PCR), and high throughput DNA sequencing. The diagnostic utility of the information provided by these molecular tools drives, in part, basic scientific breakthroughs, such as the completion of the Human Genome Project. Superimposed on these developments in molecular diagnostics came changes in the regulatory environment, originated by existing regulatory agencies, professional organizations, and new advisory committees, all of whom strived to describe the good laboratory practices which all clinical laboratories should attempt to attain.

Initially, we will provide a brief background into the history of molecular diagnostics followed by illustrative examples of molecular testing in infectious diseases, oncology, drug metabolism (or pharmacogenetics) and heritable disorders. For each the preanalytical, analytical, and postanalytical considerations which comprise good laboratory practice will be described. The selected examples present new concepts or mechanisms regarding disease pathogenesis or treatment applied to (i) clinical diagnosis, (ii) population screening, (iii) selected screening, and (iv) tests associated with disease prediction and risk assessment.

HISTORY OF MOLECULAR DIAGNOSTICS

Regulatory Agencies and CLIA

The major regulatory force in clinical molecular diagnostics is the original Clinical Laboratory Improvement Amendments (CLIA) and the impact of subsequent changes. Since the original legislation (1988) and the publication of the final rule (1992), only CLIA-certified laboratories may provide testing of human specimens (other than for forensic and research purposes) which results in clinical decision making. Thus, any changes in CLIA regarding molecular testing will have tremendous impact on the clinical molecular laboratories. Over the years, these changes have been relatively minor, so that the issues of quality of testing and patient safety reflect the general aspects of CLIA rather than a specific area of application. While most of molecular testing is well covered by the existing CLIA mandates, genetic (heritable disorders) testing represents a challenge, particularly regarding preanalytical and postanalytical areas.

It was regarding the preanalytical and postanalytical areas of genetic testing that most of the public and media concern has focused. As early as 1994, the Institute of Medicine of the National Academy of Sciences convened a committee to deal with issues regarding genetic testing. Among the many concerns were the issues surrounding the use of genetic testing to predict future outcomes of individuals and their families. It is this unique aspect of genetic testing that differs from most other clinical laboratory tests and demonstrates the need for special attention in the preanalytical and postanalytical areas.

In 1995, a second committee was commissioned to deal with these issues. The Department of Energy convened a Task Force on Genetic Testing and emphasized the importance of restricting genetic testing to CLIA-certified clinical laboratories. It is important to note that the number of genetic tests and the number of laboratories providing these services was limited (while data are scarce prior to 2000, the number of tests available at this time was probably less than 100). In 2008 over 1500 genetic tests were listed in GeneTests (http://www.ncbi.nlm.nih.gov/sites/GeneTests/?db=GeneTests).

Finally, in 1998 the Department of Health and Human Services Secretary Donna Shalala established the Secretary's Advisory Committee on Genetic Testing (SACGT)—which was later renamed SACGHS (Secretary's Advisory Committee on Genetics Health and Society)—to take a more global view of genetic testing and the regulatory agencies who oversaw various aspects described by the previous committees. Over the past 10 years, this committee has been very active in assessing genetic testing and its wider implications, including genetic discrimination and patents, in addition to the roles of various regulatory agencies regarding genetic testing. However, molecular technology and applications continue to develop in previously unanticipated areas.

Because of the nature of molecular genetic testing many methods were developed by laboratories for use only in that laboratory. Thus, these laboratory-developed tests (LDTs) presented unusual challenges to traditional regulatory agencies, such as the Food and Drug Administration (FDA). To this day, LDTs are the life blood of molecular diagnostics, and differentiate this area from most other areas of the clinical laboratory which depend more on kit and instrument manufacturers. However, regulatory oversight for LDTs is more stringent, by necessity, than those having FDA approval. Thus, we enter the universe of ASR (analyte-specific reagents manufactured for LDTs), RUO (research use only kits), and IVDMIA (*in vitro* diagnostic multivariate index assays). Each of these designations imparts specific FDA guidelines and additional regulatory oversight. Typically, a clinical molecular laboratory uses a mixture of FDA-approved kits and LDTs, which include many ASRs. Occasionally, an RUO kit may be in use for research projects, and perhaps in the future labs may incorporate IVDMIAs, although currently these are primarily found in commercial laboratories.

In order to maintain CLIA certification, clinical laboratories must undergo inspections, usually every 2 years, by accrediting organizations recognized by the *Centers for Medicare and Medicaid Services* (CMS), such as the *College of American Pathologists* (CAP). A CAP inspection uses checklists to monitor laboratory performance and quality, including personnel qualifications, and maintains oversight over molecular testing. However, because of the mixed bag of tests performed and the various clinical applications, the inspection procedure requires multiple checklists to cover the various applications. This requires inspectors to have extensive inspection experience and training.

Quality Assurance, Quality Control, and External Proficiency Testing

An additional regulatory aspect involves quality assurance (QA) programs consisting of both external proficiency testing (PT) and the use of quality control (QC) materials. While QA is a part of all laboratory testing areas, unique problems surround the dearth of appropriate control materials and external PT programs, which are intimately related for the following reason. Molecular testing is confounded by many factors including (i) the rare nature of many heritable conditions, (ii) the multiplicity of mutations associated with a single disorder (for example, cystic fibrosis), and (iii) the level of tissue heterogeneity, particularly when dealing with malignant conditions. For molecular infectious disease testing, particular care must be given to ensure that lack of an amplified product signifies no measurable target as opposed to amplification failure. In this latter case, use of internal controls added to the amplification reaction assures the laboratory that a negative result is a true negative. Also, since nucleic acid is found in living and dead organisms, the presence of a positive signal may not represent active infectious agents.

Method Validation

Adding to this complexity are the difficulties associated with validating molecular tests. When validation is applied to LDTs, the FDA (via the ASR rules) states "clinical laboratories that develop tests are acting as manufacturers of medical devices and are subject to FDA jurisdiction." To underscore this, CLIA requirements for test validation vary by test complexity and by whether the test is FDA approved, FDA cleared, or laboratory developed. CLIA requires that each lab establish or verify the performance specifications of moderate-complexity and high-complexity test systems that are introduced for clinical use. For an FDA-approved test system without any modifications, the system is validated by the manufacturer; therefore, the laboratory need only verify (confirm) these performance specifications. In contrast, for a modified FDA-approved test or an in-house developed test, the laboratory must establish the performance specifications for a test, including accuracy, precision, reportable range, and reference range. The laboratory must also develop and plan procedures for calibration and control of the test system. In addition, for a modified FDA-approved test or a laboratory-developed test, the laboratory must establish analytic sensitivity and specificity and any other relevant indicators of test performance. How and to what extent performance specifications should be verified or established is ultimately under the purview of medical laboratory professionals and is overseen by the CLIA certification process, including laboratory inspectors.

Clinical Utility

In 1997, the Department of Energy Task Force on Genetic Testing proposed three criteria for the evaluation of genetic tests: (i) analytic validity, (ii) clinical validity, and (iii) clinical utility. By clinical utility, the report referred to "... the balance of benefits to risks. ..." However, more frequently, especially with new molecular genetic tests, clinical utility may not be readily available when a test is put into clinical use. In fact, it may take years to accumulate data on clinical utility.

MOLECULAR LABORATORY SUBSPECIALTIES

Heritable Disorders

Fragile X Syndrome

The most common form of heritable mental retardation is called the Fragile X Syndrome (FXS), so-called because of the cytogenetic appearance of increased fragility at a locus on the long arm of the X chromosome, under specific cell culture conditions.

This observation is frequently associated with characteristic clinical features, primarily in males, including mild to moderate mental retardation, coarse fascies with large low-set ears, and enlarged testes postpuberty. Affected females have a less well-defined phenotype, due to X-inactivation (also called lyonization). The inherent difficulties of cytogenetic testing for fragile X chromosomes led to the development of a molecular test once the gene was identified. The first molecular test was based on Southern blot analysis, coupled with specific restriction enzymes, which could identify the characteristic molecular defect.

The *FMR1* gene contains a CGG repeat region in the 5′ untranslated region (Figure 29.1). The number of CGG repeats is variable in the general population. When increased beyond 200 CGG repeats, the gene is hypermethylated, leading to absence of expression of the gene product, FMR1. In the normal population, the number of repeats varies from 5 to 44 repeats, with a median of 29. In individuals with FXS, the number of repeats usually exceeds 200. Premutation carriers have between 55 and 200 repeats, which has been associated with (i) increased risk of full expansion during female (but not male) meiosis leading to FXS in her offspring, (ii) increased risk of ovarian dysfunction (premature ovarian failure), and (iii) increased risk of FXTAS (fragile x-associated tremor/ataxia syndrome). Clearly, an accurate and precise method for determining the number of CGG repeats, and the associated hypermethylation status, has significant clinical implications. However, the Southern blot technique was able to assess only hypermethylation, but not accurately assess the number of CGG repeats (Figure 29.2).

Validation of Molecular Sizing of FMR1

The Southern blot procedure for FXS is well documented. The focus here is on the validation of a sizing assay for the CGG region of *FMR1* performed on a capillary electrophoresis platform. The LDT is based on the report by Fu, while an RUO assay is commercially available from Celera Diagnostics. Inherent in any method validation, well-characterized samples must be available. These can be patient samples analyzed by an independent laboratory, immortalized cell lines (Coriell Cell Repositories), or artificially prepared standards (National Institute of Standards and Technology). Each of these materials has advantages and disadvantages. Ideally, all three sources provide the most complete assessment of method validation. Conditions for both the LDT and the RUO are available from the literature.

PCR and Capillary Electrophoresis

Sizing of the *FMR1* 5′ UTR is performed by capillary electrophoresis following PCR of the critical region. Details of the PCR and capillary electrophoresis are discussed in the suggested readings.

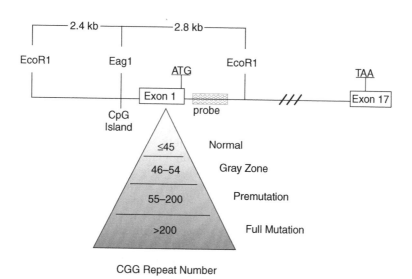

CGG Repeat Number

Figure 29.1   Schematic representation of the Ce\GG repeat in exon 1 of *FMR1* and associated alleles. A CGG-repeat number less than or equal to 45 is normal. A CGG-repeat number of 46 to 54 is in the gray zone and has been reported to expand to a full mutation in some families. A CGG-repeat number of 55 to 200 is considered a premutation allele and is prone to expansion to a full mutation during female meiosis. A CGG-repeat number in excess of 200 is considered a full mutation and is diagnostic of fragile X syndrome.

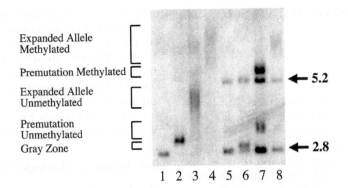

Expanded Allele
Methylated

Premutation Methylated

Expanded Allele
Unmethylated

Premutation
Unmethylated
Gray Zone

← 5.2

← 2.8

1 2 3 4 5 6 7 8

Figure 29.2   Southern blot analysis for the diagnosis of fragile X syndrome. Patient DNA is simultaneously digested with restriction endonucleases *EcoR*1 and *Eag*1, blotted to a nylon membrane, and hybridized with a ^{32}P-labeled probe adjacent to exon 1 of *FMR1* (see Figure 29.1). *Eag*1 is a methylation-sensitive restriction endonuclease that will not cleave methylated DNA. Normal male control DNA with a CGG-repeat number of 22 on his single X chromosome (*lane 1*) generates a band about 2.8 kb in length corresponding to *Eag*1-*EcoR*1 fragments (see Figure 29.1). Normal female control DNA with a CGG-repeat number of 20 on one X chromosome and a CGG-repeat number of 25 on her second X chromosome (*lane 5*) generates two bands, one at about 2.8 kb and a second at 5.2 kb. *EcoR*1-*EcoR*1 fragments approximately 5.2 kb in length represent methylated DNA sequences characteristic of the lyonized chromosome in each cell that is not digested with restriction endonuclease *Eag*1. DNA in *lane 2* contains an *FMR1* CGG-repeat number of 90 and is characteristic of a normal transmitting male. The pattern observed in *lane 3* is representative of a mosaic male with a single X chromosome with a full mutation (>200 repeats). However, the full mutation in some cells is unmethylated; in other cells, the full mutation is fully methylated, hence the term *mosaic*. In those cells in which the full mutation is unmethylated, digestion by both *Eag*1 and *EcoR*1 occurs, and in those cells in which the full mutation is fully methylated, digestion of the DNA by *Eag*1 is inhibited. The pattern observed in *lane 4* is diagnostic of a male with fragile X syndrome, illustrating the typical expanded allele fully methylated in all cells. *Lane 6* is characteristic of a female with one normal allele that has 29 CGG-repeats and a larger gray zone allele with a CGG-repeat number of 54. *Lane 7* is the pattern observed from a premutation carrier female with one normal allele having a CGG-repeat number of 23 (band at about 2.8 kb) and a second premutation allele with CGG repeats of 120 to about 200 (band at about 3.1 kb). In premutation carrier females, in cells in which the X chromosome with the premutation allele is lyonized, the normal 5.2 kb *EcoR*1-*EcoR*1 band is larger because of the increased CGG-repeat number and is about 5.5 kb in length. *Lane 8* is diagnostic of a female with fragile X syndrome with one fully expanded mutation allele that is completely methylated and a second normal allele with a CGG-repeat number of 33.

Statistical Analysis

Validation studies should determine reproducibility (imprecision) and accuracy, allowing the laboratory to report values with confidence intervals. These estimates of precision can be obtained by multiple determinations of the same sample during a single run of samples and by analyzing the same sample on multiple runs, followed by determining the mean and the standard deviation. These terms are referred to as within-run and between-run precision.

Postanalytical Considerations

Interpretation of *FMR1* sizing analysis depends on the clinical considerations for ordering the test. These may be (i) diagnosis, (ii) screening, (iii) predictive within a family with affected members, or (iv) risk assessment. For example, a diagnostic application would include determining whether an individual with the characteristic phenotype has FXS, and depends on the sex of the patient. A male, having a single X chromosome, should have a single product following PCR amplification, with a normal number of CGG repeats (5–44 repeats). One of the limitations of the PCR procedure is the size of the CGG repeat region. Depending on the PCR primers and amplification conditions, the maximum number of CGG repeats that can be amplified is ~100 repeats. Thus, a fully amplified CGG region of >200 repeats would not produce a discernable PCR product. However, a female, with two X chromosomes, with a full expansion would still have an unaffected X chromosome, resulting in a single product following PCR amplification, representing the normal-sized allele. Another limitation with females is that both alleles may be the same size, resulting in a single product following amplification. For females, any analysis which yields a single product following PCR, and with the absence of a product in a male, Southern blot analysis would be necessary, as this analysis is not limited by the constraints of PCR amplification. Thus, Southern blot analysis should detect fully expanded alleles.

If an affected family member has been identified, carrier determination can be performed, exemplifying a predictive use of this testing. However, another level of complexity involves risk assessment of female premutation carriers. These individuals have between 55 and 200 repeats, not large enough for full mutations, but sufficient for the associated conditions mentioned previously (premature ovarian failure and FXTAS), in addition to increased risk of passing on a fully expanded allele to offspring (the risk is proportional to the number of CGG repeats). Note that individual alleles between 45 and 54 repeats are considered intermediate and may carry an increased risk for being premutation carriers, but not for passing on full mutations. No data exist on the risk of these individuals for ovarian dysfunction or FXTAS. Obviously, knowing the precision and accuracy of the assay is paramount in interpreting the number of CGG repeats.

Benefit/Risk Assessment (Clinical Utility)

Since defining clinical utility takes considerable clinical experience with newer assays, it is difficult, if not impossible, to assess clinical utility before setting up an LDT, as advocated by some of the regulatory agencies. For example, *FMR1* sizing assays demonstrate the evolving nature of clinical utility. When Southern blot analysis was the only molecular method available, the clinical utility was based on the contribution of the *FMR1* assay to the diagnosis of FXS. However, by this method, limited information was available for assessing the premutation carrier status. With the development of *FMR1* sizing assays, more accurate assessment of the premutation allele could be made, and the clinical utility for ovarian dysfunction and FXTAS could be assessed. Thus, improvements in methodology can affect the clinical utility of an analyte.

Infectious Diseases

Enteroviral Disease

Enteroviruses (EV) are a general category of single-stranded, positive-sense, RNA viruses of the Picornaviridae family, which includes the subgroups enteroviruses, polioviruses, coxsackieviruses (A and B), and echoviruses. Rhinoviruses are closely related members of the Picornaviridae family that are a major cause of the common cold. Most EV infections are asymptomatic. Symptomatic EV infections affect many different organ systems. EV infections peak during the first month of life and cause the majority of aseptic meningitis worldwide.

Until recently, a diagnosis of EV infection was based on clinical presentation and the isolation of virus in cell culture from throat, stool, blood, or cerebrospinal fluid (CSF) specimens. EV detection by culture methods is insensitive and takes too long to be useful in clinical management decisions. By contrast, reverse transcription-PCR (RT-PCR) is more rapid and sensitive and has been used to detect EV RNA in CSF, throat swabs, serum, stool, and muscle biopsies. EV RNA detection by molecular methods is now considered the gold standard for the diagnosis of enteroviral meningitis.

Validation of LightCycler-Based, RT-PCR Assay for Enterovirus

For rapid detection of EV RNA in CSF specimens, we established a qualitative semi-nested LightCycler (LC-PCR) method based on EV RNA isolation and cDNA synthesis followed by PCR amplification with primers that target the highly conserved EV 5'-noncoding region (5'-NCR). To improve sensitivity and specificity, we used a second round of semi-nested PCR. The semi-nested PCR uses two different viral subgroup-specific forward primers (one for polioviruses and one for coxsackieviruses), and a common reverse primer that generates 85 bp to 87 bp amplicons depending on the viral subgroup. The amplicons bind the EV-TaqMan probe which allows for real-time fluorometric detection and excludes cross-hybridization with rhinoviral sequences. Viral RNA is isolated from infected CSF using the High Pure Viral RNA Kit (Roche Diagnostics). The extracted nucleic acids are subjected to RT-PCR in the LightCycler using the LC RNA Master Hybridization Probes Kit (Roche Diagnostics). Amplicons are diluted with sterile water, and subsequent semi-nested PCR is carried out in the LightCycler with the EV-TaqMan probe using the LC Fast Start DNA Master Hybridization Probes Kit (Roche Diagnostics).

QC materials include external positive and negative controls and a no template control (water blank). An appropriate internal positive control was not available at the time of assay development. Therefore, a technical failure cannot be ruled out when a patient sample is negative. The crossing thresholds for known positive samples were determined during the validation set by a limiting dilution approach using negative CSF samples spiked with known amounts of cell culture grown EVs. A minimum cycle threshold (Ct) of 30 cycles was established. Limits of detection for several EVs were determined and compared to results from a reference lab for the same specimen control.

Preanalytical Considerations

A semi-nested RT-PCR method is especially susceptible to contamination by nucleases or unrelated nucleic acids, which cause false negative or false positive results, respectively. Laboratory organization and equipment should be designed to avoid contamination, and specimens should be handled with gloves in a biosafety cabinet to prevent contamination with fingertip RNases and biohazard exposures. The quality of viral RNA in CSF depends on what happens to the specimen before RNA sample extraction occurs. Therefore, only 0.5 ml or greater volumes of CSF specimens collected in the appropriate sterile plastic containers are accepted. The EV virion is very stable and protects the viral RNA from degradation. Sample stabilization is not required as long as specimens are dispatched to the laboratory within hours of acquisition. If there will be a delay, then the specimen should be refrigerated or placed on ice. If the delay will be more than 24 hours, then the specimen should be stored frozen and kept frozen until RNA extraction. Samples that have been frozen before dispatch and arrive in the state of thawing should be rejected. Heparin inhibits both RT and PCR reactions and should be avoided.

Postanalytical Considerations

Fluoresence signal detected above background before the 30th cycle of the semi-nested EV LC-PCR is diagnostic of EV infection. Ct values that are equal to or greater than 30 cycles are negative. In a patient with the clinical signs and symptoms of viral meningitis, a positive result is consistent with a diagnosis of EV meningitis. Interpretation of a negative result is more difficult due to the absence of an internal positive control for the RT-PCR. A negative result could indicate a technical failure with the RT or PCR steps. Alternatively, the sample may not contain EV RNA. Thus, a negative result does not exclude EV infection and needs to be communicated in the report to the ordering physicians. Finally, because EV needs to be reported to the state health department, a positive result must be called to the physician and reported to hospital epidemiology.

Benefit/Risk Assessment (Clinical Utility)

The distinction between EV and bacterial infections in neonates can be difficult, and patients are usually treated empirically with antibiotics until an EV infection is confirmed, at which point supportive care is the treatment choice. Therefore, a rapid molecular method for detecting EV RNA has clinical utility if available daily. A positive molecular EV test helps to exclude bacterial or other etiologies that may require longer hospitalization for treatment. A negative EV test is not very useful clinically because it does not rule out EV meningitis.

Pharmacogenetics Hepatitis C Viral Genotyping

Millions of people in the United States are chronically infected with HCV, and many are unaware that they are infected. Tens of thousands are newly infected with HCV each year and the same magnitude dies annually in the United States from the sequelae of hepatitis C. HCV is also the most frequent indication for liver transplant in adults in the United States. Since the outcome of HCV infection depends on both the viral load and HCV genotype, the determination of HCV genotype provides important clinical information regarding the duration and type of antiviral therapy used for a given HCV-infected patient. In addition, the genotype is an independent predictor of the likelihood of sustained HCV clearance after therapy. There are more than 11 genotypes and more than 50 subtypes of HCV. In the United States, genotypes 1a, 1b, 2a/c, 2b, and 3a are most common. Genotypes 4, 5, 6, and 7 are endemic to Egypt, South Africa, China, and Thailand, respectively, and are rarely found in the United States. Since patients infected with HCV genotype 1 may benefit from a longer course of therapy, and genotypes 2 and 3 are more likely to respond to combination interferon-ribavirin therapy, the clinically relevant distinction among the affected population of the United States is between genotype 1 and nontype 1.

In addition to viral load, HCV genotype is important clinically. Line-probe and sequencing-based methods are available for HCV genotyping, but are expensive, time consuming, and provide more information than is currently needed by clinicians for patient management decisions. With this in mind, we developed a rapid HCV genotyping assay for the LightCycler to differentiate HCV type 1 from nontype 1 infections.

Validation of HCV Genotyping

The determination of HCV genotype is based on isolation of HCV viral RNA and cDNA synthesis followed by PCR amplification of the 5′-untranslated region (5′-UTR). Viral RNA is isolated from infected serum after HCV viral load determination using the MagNA Pure Total Nucleic Acids Extraction Kit or the Qiagen EZ1 BioRobot. The extracted nucleic acids are subjected to RT-PCR using primers specific for 5′-UTR sequence that is conserved among all known HCV genotypes. The RT-PCR step is performed using an Applied Biosystems GeneAmp 9600 PCR System (block cycler). A second PCR amplification using a semi-nested pair of primers is carried out in the LightCycler in the presence of a pair of FRET oligonucleotide probes. The FRET detection probe allows the discrimination of HCV genotypes 1a/b, 2a/c, 2b, 3a, and 4 during melting curve analysis (Figure 29.3), because it hybridizes with differential affinity to a region of the 5′-UTR that varies among the different HCV genotypes. Magnesium concentration was critical for optimal genotype discrimination. Primer and probe design and PCR details have been described. For method validation, patient samples analyzed by an independent reference laboratory were used. Positive QC materials were pooled patient serum samples of previously determined genotype. Negative QC materials were pooled patient serum samples previously determined to be HCV negative. A water blank is also always included with each run. A serial limiting dilution of positive patient samples with known viral titers was used to determine the limits of detection. A more sensitive fully nested PCR method was developed and is able to genotype specimens with 130 IU per mL. The fully nested approach is rarely needed because patients usually have high viral loads upon initial presentation, and this is the ideal time to assess the HCV genotype.

Statistical Analysis

In addition to determination of the accuracy of the genotyping assay, melting point precision was determined within run and between runs using QC materials. An analysis of assay performance after 18 months showed no assay failures for 70 runs with 411 patient samples using three different LightCyclers operated by six different technicians. The genotype-specific melting temperatures remained nonoverlapping with coefficients of variation that were less than 1% for all genotypes. Melting temperatures are continuously monitored to assure assay performance.

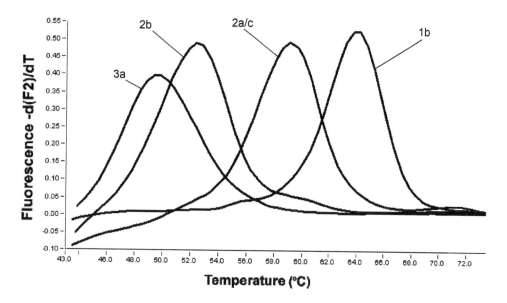

Figure 29.3 HCV genotyping assay. Comparison of samples from patients infected with HCV genotypes 1b, 2a/c, 2b, and 3a. Shown are genotype-specific melting transitions for four samples in 2 mM $MgCl_2$. Data were obtained by monitoring the fluorescence of the LCRed640-labeled FRET sensor probe during heating from 40 to 80°C at a temperature transition rate of 0.1°C/s.

Preanalytical Considerations

The semi-nested or nested RT-PCR method is especially susceptible to contamination by RNases, previously amplified DNA, or unrelated nucleic acids. Unidirectional flow of samples through physically separated preamplification and postamplification areas should be maintained. Equipment, reagents, plasticware, and specimens should be handled with gloves to prevent contamination with fingertip RNases or cross-contamination of specimens. The use of plastic microfuge tube cap openers instead of gloved fingertips is recommended to prevent cross-contamination. Samples should be added one at a time, first followed by positive QC, negative QC, and finally the water blank. The only known interference is heparin, which may inhibit the RT-PCR reaction. Evaluation of isolation of nucleic acid by the MagNA Pure system has shown that heparin is removed during the isolation of nucleic acid by this method. A minimum of 500 microliters of serum is the required sample type for this method. Whole blood samples should be centrifuged, and serum should be separated within 6 hours of collection. HCV is an enveloped RNA virus and is labile to repeated freeze-thaw cycles. Viral titers will drop upon storage at –20°C. Therefore, fresh samples are preferred. Since aqueous solutions of RNA are unstable and susceptible to spontaneous hydrolysis even at –20°C, it is best to perform the RT-PCR step on the same day as RNA extraction. MagNA Pure-extracted nucleic acid samples should be stored at –80°C if the RT-PCR step cannot be performed on the day of isolation.

Postanalytical Considerations

Interpretation of HCV genotype is reported as type 1, type 2, type 3, type 4, or none detected if negative. Both early viral load measurements and viral genotype are used to determine likelihood of clinical response and influence management decisions during combination antiviral therapy. Therefore, these tests should be ordered together during the initial workup of a patient with viral hepatitis. Reflexive genotyping of patients who are positive for HCV with sufficiently high viral loads and no prior genotype should be considered. Once the genotype is determined, there is no need for additional genotyping unless there is clinical suspicion of a second infection with a different HCV genotype. The HCV LC-PCR assay is able to detect two genotypes in one patient sample. One limitation with this assay is the inability to detect some rare genotypes that may occur in the United States, such as genotypes 5 or 6. Thus, the detection of a novel melting temperature could indicate an infection with a non-1, -2, -3, or -4 HCV genotype. This specimen would be sent out for sequencing by a reference laboratory. Similarly a specimen with a high viral load but no detectable genotype indicates a technical failure or a novel genotype and would also be sent to a reference laboratory for sequencing.

Benefit/Risk Assessment (Clinical Utility)

HCV genotyping assays are an essential component of an evolving clinical decision tree used to determine the duration of therapy and the likelihood of sustained virological response (SVR) to this therapy. Absolute viral load, log decline in viral load from baseline, and viral genotype are clinically useful in predicting SVR and nonsustained virological response (NSVR) to treatment. The *NIH Consensus Development Conference Statement, Management of Hepatitis C:2002*, stresses the importance of early prediction of SVR and NSVR in patient management in the first 12 weeks of therapy. Combination interferon plus ribavirin therapy is given for 24 or 48 weeks, depending on viral genotype and viral load. The therapy is expensive and difficult for patients to tolerate due to side effects. HCV genotype and viral loads are used to predict the likelihood of SVR or NSVR. The likelihood of SVR and NSVR is useful for clinicians and patients when making decisions about the duration of therapy. HCV genotype determination is clinically useful in risk assessment and making therapeutic decisions instead of making a diagnosis.

Oncology

B-Cell and T-Cell Clonality

B-cell and T-cell clonality assays are based on the detection of a predominant antigen receptor gene rearrangement that represents the outgrowth of a predominant lymphocyte clone. In the correct clinical and pathologic context, the presence of a predominant clonal lymphocyte population is consistent with lymphoma or lymphoid leukemia. Multiplex PCR-based B-cell and T-cell clonality assays have replaced Southern-blot-based assays because they are robust, faster to perform, and easier to interpret. Unlike Southern blots, PCR-based assays are amenable to formalin-fixed, paraffin-embedded (FFPE) tissues. PCR-based assays amplify the DNA between primers that target the conserved framework (FR) and joining (J) sequence regions within the lymphoid antigen receptor genes. These conserved regions flank the V-J gene segments that randomly rearrange during the maturation of B-lymphocytes and T-lymphocytes.

Random genetic rearrangement of the V-J segments occurs in the immunoglobulin heavy chain gene (*IGH*) and the kappa and lambda immunoglobulin light chain genes in B-cells. Random genetic rearrangement of the V-J segments also occurs in the alpha (*TCRA*), beta (*TCRB*), gamma (*TCRG*), and delta (*TCRD*) T-cell receptor genes during T-cell development. Antigen receptor gene rearrangement occurs in a sequential fashion one allele at a time. If the first allele fails to successfully recombine (produces a nonfunctional protein product), then the other allele will undergo rearrangement. It is important to remember that the nonfunctional rearrangement can still be detected by the clonality assay. Because of the sequential manner in which these genes rearrange, a mature lymphocyte may carry one (mono-allelic clone) or two (bi-allelic clone) rearranged antigen receptor genes of a given type. Each maturing B-cell or T-cell has either one or two gene-specific V-J rearrangements unique in both sequence and length. Therefore, when the V-J region is amplified, one or two unique

amplicons will be produced per cell. If the population of lymphocytes being examined is a lymphoma, then one or two unique amplicons will dominate the population. In contrast, if the population is a polyclonal lymphoid infiltrate, then there will be a Gaussian distribution of many different sized amplicons centered around a statistically favored, average-sized rearrangement (Figure 29.4).

Because the antigen receptors are polymorphic and subject to mutagenesis as part of the normal rearrangement process, multiple primer sets are used to increase the ability of the assay to detect the majority of possible V-J rearrangements. Each set of multiplex primers has a defined valid amplicon size range. After antigen exposure, germinal center B-cells undergo somatic hypermutation of the V regions, and this may affect the binding of the V-region specific (FR) primers and prevent the detection of lymphoma clones arising from the germinal center or postgerminal center B-cell compartments. Primer sets that target *IGH* D-J rearrangements are less susceptible to somatic hypermutation and allow the detection of more terminally differentiated B-cell malignancies.

Validation of B-Cell and T-Cell Clonality Assays

PCR-based clonality assays are routinely used by many laboratories. However, prior to 2004, referral lab PCR-based clonality assays used nonstandardized, lab-specific PCR primer sets and methods, making proficiency testing and lab performance comparisons difficult. In addition, many of these institution-specific B-cell and T-cell clonality assays are not comprehensive and target a limited number of possible V-J rearrangements. To address this concern, we used the optimized InVivoScribe (IVS) multiplex PCR primer master mixes. The IVS master mixes have been standardized and extensively validated by testing more than 400 clinical samples using the *World Health Organization* Classification scheme. The validation was done at more than 30 prominent independent testing centers throughout Europe in a collaborative study known as the BIOMED-2 Concerted Action.

The *TCRG* locus is on chromosome 7 (7q14). Although most mature T-cells express the alpha/beta TCR, the gamma/delta TCR genes rearrange first and are carried by almost all alpha/beta T-cells. The IVS *TCRG* assay uses only two master mixes and is able to detect 89% of all T-cell lymphomas tested in the BIOMED-2 Concerted Action. The positive and negative controls included in the IVS kit are used to assess each of the master mixes. The specimen control size standard is a separate master mix included with the kit that evaluates each patient DNA sample by targeting six different housekeeping genes potentially producing amplicons of 84, 96, 200, 300, 400, and 600 base pairs. Under ideal conditions and using capillary electrophoresis to analyze the amplicons, the *TCRG* assay can resolve a 1% clonal population in a polyclonal background.

The *IGH* locus is on chromosome 14 (14q32.3). The IVS *IGH* assay uses five master mixes and is able to detect 92% of all B-cell lymphomas tested in the BIOMED-2

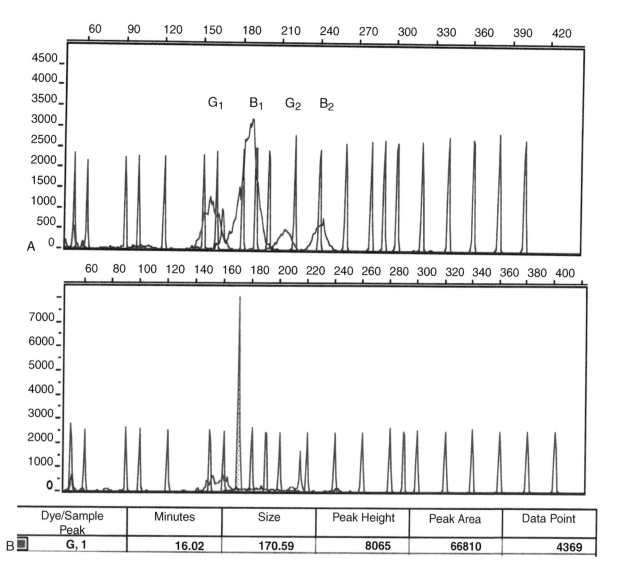

Dye/Sample Peak	Minutes	Size	Peak Height	Peak Area	Data Point
B ▣ G, 1	16.02	170.59	8065	66810	4369

Figure 29.4   T-cell receptor gamma chain PCR assay for clonality. (A) Polyclonal reactive T-cell proliferation pattern. A polyclonal population of T-cells with randomly rearranged T-cell receptor gamma chain genes produces a normal or Gaussian distribution of fluorescently labeled PCR products from each primer pair in the multiplex reaction. This produces four bell-shaped curves that represent the valid size range for an individual primer pair. Two of the valid size ranges are green and two are blue; G_1, B_1, G_2, B_2. The red peaks represent size standards. **(B)** Clonal T-cell proliferation pattern. A clonal T-cell proliferation results in a relative dominance of a single T-cell receptor gamma chain gene producing a predominant spike of a discrete size on the corresponding electropherogram. Data were obtained using an ABI 3100 capillary electrophoresis system and ABI Prism Software.

Concerted Action validation study. Master mixes A, B, and C target the framework regions (FR) 1, 2, and 3 of the V gene segments, respectively, and all J gene segments. Master mixes D and E target the D and J regions. The *IGH* locus rearranges the D and J segments first during maturation, followed by the addition of a random V segment to the previously rearranged D-J fusion product. In addition, the D segments are less likely to be mutated by somatic hypermutation during postgerminal center development. Because the primers in tubes D and E target D-J regions, these tubes are more likely to detect clonal *IGH* rearrangements in lymphoma cells that arose either very early or late in B-cell ontogeny. Positive, negative, and specimen extraction

controls are included and run on each master mix and specimen, respectively. The specimen control size standard is used to assess the integrity of the DNA extracted from the specimen and is identical to the specimen control size standard described previously for the *TCRG* kit. Capillary electrophoresis using the ABI 3100 Avant Genetic Analyzer was better at determining the size of predominant amplicons than both agarose and polyacrylamide gel systems during method development. The IGH assay can resolve a 1% clonal population in a polyclonal background.

Genomic DNA is isolated from whole blood, body fluids, fresh tissues, and FFPE tissue blocks, using the appropriate Qiagen DNA isolation protocol. A slightly modified Qiagen DNA isolation protocol for FFPE samples was selected after a comparison to three other DNA isolation protocols during validation. After DNA isolation, multiple dilutions are tested for beta-globin amplification by PCR to assess DNA quality prior to analysis using the IVS system. Samples that fail to amplify beta-globin at any concentration are quantified by UV spectrophotometry. Approximately 100 to 200 ng of input DNA are ideal for *TCRG* or *IGH* PCR reactions. Samples that are undetectable by UV spectrophotometry and fail beta-globin PCR should be reported as quantity insufficient and retested by increasing the amount of starting material by 2- to 3-fold if possible. Samples that fail beta-globin PCR but contain sufficient quantities of DNA are most likely nonamplifiable due to inhibitors and should be reported as nonamplifiable due to PCR inhibitors in the sample. Additional dilutions could be retested by beta-globin PCR if desired. The least dilute DNA that gives a strong beta-globin signal should be used for the InVivoScribe B-cell or T-cell clonality assay. Each of the products of the IVS PCR is combined with size standards, denatured, and subjected to capillary electrophoresis. After assay validity is confirmed by checking all controls, the results are analyzed and a report is generated.

After the method was established, a variety of clinically and histopathologically diagnostic cases of B-cell and T-cell lymphomas were used to evaluate the performance of the method. This laboratory-modified method was able to detect 92% of B-cell lymphomas and 85% of T-cell lymphomas used in the validation set. An ideal QC control material is not available for this assay. The ideal QC material would consist of a high and low positive (clonal) FFPE QC and a negative (polyclonal) FFPE QC. FFPE QC material would control for the DNA extraction in addition to the PCR and interpretation phases of this method. The currently available QC and PT control materials do not assess the DNA extraction phase of this test.

Preanalytical Considerations

Special considerations regarding specimen type and handling are important for this method. One of the most common specimens submitted for *IGH* or *TCRG* gene rearrangements is FFPE tissue blocks. PCR inhibitors are commonly found with this specimen type. Xylene, heme, and excessive divalent cation chelators are also inhibitory and need to be removed during DNA extraction procedures. The fixative used prior to paraffin embedment is critical for optimum results because formalin of inferior quality can introduce PCR inhibitors or prevent the extraction of amplifiable DNA. Fresh, high-quality, 10% neutral-buffered formalin is acceptable, and this should be communicated to the pathologists who will be submitting specimens for gene rearrangement studies. Fixatives that contain heavy metals such as Hg (Zenkers and B5) should not be used for PCR. Decalcified bone marrow biopsy specimens are unacceptable as are specimens that were fixed with strong acid-based fixatives like Bouin's

solution. Tissues should be fixed at room temperature using proper histologic standards. Fixation depends on formaldehyde-mediated cross-linking of proteins and nucleic acids. Excessive cross-linking will inhibit DNA extraction and shorten the average size of the recovered DNA molecules; therefore, overfixation should be avoided. Fixation using the approved fixatives at room temperature progresses at a rate of approximately 1 mm/hour. Typical 0.5 mm thick tissue sections should fix for a minimum of 6 hours and a maximum of 48 hours. After 48 hours in fixative, tissues can be stored in 50% ethanol for up to 2 weeks prior to processing and embedding. The volume of fixative used should be at least 10 times the volume of tissue to achieve optimal results. Some laboratories use razor blades to gouge tissue out of the block, but this destroys the block and prevents any additional studies that may be necessary in the future. The best way to obtain FFPE samples for PCR is to cut 4 micron thick sections on the microtome. Histotechnologists should cut the sections for PCR first thing in the morning with a fresh microtome blade and new water. Several initial sections are discarded to remove any possible contaminating DNA on the surface of the block. One to five sections are obtained, depending on the size and type of tissue sample and the relative proportion of the tissue that is involved by the suspicious lymphoid infiltrate. PCR of lymphoid infiltrates in FFPE skin or brain tissue blocks are somewhat refractory to PCR analysis and require more sections for analysis.

To snap freeze fresh tissues, one should cut them into 1 mm^3 to 3 mm^3 pieces, snap freeze them in liquid nitrogen, and then store at $-70°C$ until processed. Specimens snap frozen at a different location should be shipped on dry ice to arrive during normal business hours and be stored at $-70°C$ until processed. Whole blood or bone marrow aspirate specimens should be at least 200 microliters of EDTA or citrated material stored at $4°C$ for up to 3 days before DNA extraction.

CSF specimens should be at least 200 microliters depending on cellularity. CSF samples should be stored at $4°C$ for up to 24 hours or at $-20°C$ for long-term storage. Other sample types, such as stained or unstained tissue on slides, can be used.

Postanalytical Considerations

The results of this test must be interpreted in the context of the other clinicopathologic data for the specimen. The detection of a clonal lymphoid proliferation in a tissue sample does not necessarily indicate a diagnosis of lymphoma or leukemia. On the other hand, the failure to detect a clonal lymphoid proliferation does not necessarily rule out a lymphoma/leukemia. The test is one puzzle piece that must be integrated with other clinicopathologic data before rendering a diagnosis. The molecular B-cell and T-cell clonality assay is an adjunct ordered by pathologists for cases that are suspicious but not obviously lymphoma/leukemia, and the results of this test do not stand alone.

Although helpful guidelines for the interpretation of results are provided by the manufacturer, this assay is not an out-of-the-box assay. After method development, an important part of validation is the development of interpretive guidelines that are appropriate for the established method, specimen types, and the patient population. An important consideration is how the results will be interpreted by the ordering group of pathologists and oncologists. Specimen handling, DNA extraction procedures, and fragment analysis procedures will affect the final results from this assay, and this will affect the ability to detect a clonal proliferation in a polyclonal background. Our laboratory has established a set of interpretive guidelines that include a required minimum peak height for positive peaks and positive peak to background peak ratio values that are appropriate for our method and reduce the number of false positives. The details of our interpretive guidelines are beyond the scope of this chapter and are described elsewhere.

Results are interpreted and reported to the ordering pathologist in an interpretive report (CoPath) format. The results are interpreted as (i) positive for a clonal lymphoid proliferation; (ii) suspicious but not diagnostic for a clonal lymphoid proliferation (with a recommendation to repeat the assay if clinical concerns persist); (iii) oligoclonal reactive pattern; (iv) insufficient material/inadequate quality. The reports also contain information on the pertinent clinical and laboratory history, the specimen submitted for PCR, and the relevant surgical pathology case if appropriate. Suspicious results are usually repeated for confirmation.

Benefit/Risk Assessment (Clinical Utility)

In many cases of atypical lymphoid proliferations suspicious for lymphoma, the morphologic, immunohistochemical, flow cytometric, or clinical features are equivocal or contradictory. Detection of a predominant antigen receptor gene rearrangement (detection of a T-cell or B-cell clone) can support a diagnosis of lymphoma. Early diagnosis and appropriate treatment of lymphoma can prolong patient survival and in many cases result in a complete remission or cure. Because the morphology of many lymphomas is indistinguishable from benign lymphoid hyperplasia, ancillary tests that demonstrate an atypical clonal lymphocyte proliferation are essential in early diagnosis and subsequent treatment of many lymphoma cases. Multiplex PCR assays for clonality have become the standard of practice in lymphoma diagnosis and may eventually be used on all cases of atypical lymphoid proliferations to either rule in or rule out lymphoma earlier. For those cases that are obviously lymphoma, molecular characterization of the specific lymphoma cell clone may also prove useful in patient follow-up after treatment to screen for residual or recurrent disease. In addition many of the morphologic subtypes of lymphoma have not been extensively characterized from the molecular diagnostics standpoint. For example, for biopsy cases that are suspicious but not diagnostic for lymphoma, detection of a clonal B-cell or T-cell population predicts lymphoma behavior or patient outcome. The diagnostic

tools that allow these types of correlative studies have only recently become available, and this assay represents one of the assays that will allow these kinds of questions to be considered.

Another important consideration is the improved performance and diagnostic utility of the newer generation of multiplex PCR assays since 2004. Many descriptions on the utility of PCR-based clonality assays refer to pre-2004 multiplex PCR assays that cannot detect all possible *IGH* or *TCRG* gene rearrangements because of suboptimal primer sets. This can increase the number of false negatives because the PCR primers that could recognize the clonal gene rearrangement are missing. In addition, oligoclonal reactive proliferations may appear to be clonal because a limiting number of primers may not detect all of the members of the oligoclonal proliferation.

KEY CONCEPTS

- One way that basic scientific breakthroughs are translated from the "bench to the bedside" is by the clinical application of molecular assays to diagnose disease and portend prognosis.
- The clinical application of molecular testing is regulated by CLIA to insure that good laboratory practices and standards are maintained across all clinical laboratories.
- Testing for heritable disorders requires additional regulatory oversight steps because the test results may predict future outcomes of individuals and their families.
- CLIA requirements for test validation vary by test complexity and by whether the test is FDA approved, FDA cleared, or laboratory developed.
- The determination of the number of CGG repeats in the FMR1 gene is important for fragile X syndrome diagnosis and risk assessment. In addition, FMR1 repeat expansion has been correlated to other conditions suggesting additional clinical applications for the molecular FMR1 sizing assay.
- Rapidly detecting enterovirus RNA in CSF specimens has clinical utility by allowing the rapid confirmation of enteroviral meningitis.
- Molecular hepatitis C virus genotyping assays that discriminate type 1 from nontype 1 HCV infection are useful for making pharmacotherapeutic decisions.
- Molecular lymphocyte clonality assays are clinically useful and have become an essential part of the diagnostic work-up for many cases of suspected leukemia/lymphoma.

SUGGESTED READINGS

1. Bullock GC, Bruns DE, Haverstick DM. Hepatitis C genotype determination by melting curve analysis with a single set of fluorescence resonance energy transfer probes. *Clin Chem.* 2002; 48:2147–2154.

2. Chute DJ, Cousar JB, Mahadevan MS, et al. Detection of immunoglobulin heavy chain gene rearrangements in classic Hodgkin lymphoma using commercially available BIOMED-2 primers. *Diagn Mol Pathol.* 2008;17:65–72.

3. Fu YH, Kuhl DP, Pizzuti A, et al. Variation of the CGG repeat at the fragile X site results in genetic instability: Resolution of the Sherman paradox. *Cell.* 1991;67:1047–1058.

4. Haverstick DM, Bullock GC, Bruns DE. Genotyping of hepatitis C virus by melting curve analysis: Analytical characteristics and performance. *Clin Chem.* 2004;50:2405–2407.

5. Jennings L, Van Deerlin VM, Gulley MA. Principles and practice for validation of clinical molecular pathology tests. *Arch Pathol Lab Med.* 2009;133:743–755.

6. Mikesh LM, Crowe SE, Bullock GC, et al. Celiac disease refractory to a gluten-free diet? *Clin Chem.* 2008;54:441–444.

7. Neumaier M, Braun A, Wagener C. Fundamentals of quality assessment of molecular amplification methods in clinical diagnostics. International Federation of Clinical Chemistry Scientific Division Committee on Molecular Biology Techniques. *Clin Chem.* 1998;44:12–26.

8. Pozo F, Casas I, Tenorio A, et al. Evaluation of a commercially available reverse transcription-PCR assay for diagnosis of enteroviral infection in archival and prospectively collected cerebrospinal fluid specimens. *J Clin Microbiol.* 1998;36:1741–1745.

9. Romero JR. Reverse-transcription polymerase chain reaction detection of the enteroviruses. *Arch Pathol Lab Med.* 1999; 123:1161–1169.

10. van Dongen JJM, Langerak AW, Bruggemann M, et al. Design and standardization of PCR primers and protocols for detection of clonal immunoglobulin and T-cell receptor gene recombinations in suspect lymphoproliferations: Report of the BIOMED-2 Concerted Action BMH4-CT98–3936. *Leukemia.* 2003;17:2257–2317.

11. Watkins-Riedel T, Woegerbauer M, Hollemann D, et al. Rapid diagnosis of enterovirus infections by real-time PCR on the lightCycler using the TaqMan format. *Diagn Microbiol Infect Dis.* 2002;42:99–105.

12. Wilson JA, et al. Consensus characterization of 16 FMR1 Reference Materials: A consortium study. *J Mol Diagn.* 2008;10:2–12.

30

Molecular Assessment of Human Disease in the Clinical Laboratory

Joel A. Lefferts . Gregory J. Tsongalis

INTRODUCTION

New technologies that were once labeled *For Research Use Only* have made a rapid transition into the clinical laboratory as user-developed assays (UDAs), analyte-specific reagents (ASRs), or FDA-cleared/approved diagnostic assays. Molecular diagnostics has spanned the entire spectrum of applications that will be performed on a routine clinical basis in most hospital laboratories. While the ability for clinical laboratories to detect human genetic variation has historically been limited to a rather small number of traditional genetic diseases where no clinical laboratory testing was ever available, our current understanding of many disease processes at a molecular level has expanded our testing capabilities to both human and nonhuman applications. The identification of numerous new genes and disease-causing mutations, as well as benign polymorphisms that may influence a specific phenotype, has created a rapid demand for molecular diagnostic testing.

There is a growing need for clinical laboratories to provide high-quality nucleic acid-based tests that have significant clinical relevance, excellent performance characteristics, and shortened turnaround times. This need was initially driven by the completion of the Human Genome Project, which identified thousands of genes and millions of human single nucleotide polymorphisms (SNPs) and culminated in disease associations. An unprecedented demand has now been placed on the clinical laboratory community to provide increased diagnostic testing for rapid and accurate identification and interrogation of genomic targets.

Molecular technologies first entered the clinical laboratories in the early 1980s as manual, labor-intensive procedures that required a working knowledge of chemistry and molecular biology, as well as an exceptional skill set. Testing capabilities have moved very quickly away from labor intense, highly complex, and specialized procedures to more user-friendly, semiautomated procedures. This transition began with testing for relatively high-volume infectious diseases such as *Chlamydia trachomatis*, HPV, HIV-1, HCV, and others. Much of this was championed by the availability of FDA-cleared kits and higher throughput, semiautomated instruments such as the Abbott LCX and Roche Cobas Amplicor systems. These two instruments that were based on the ligase chain reaction (LCR) and polymerase chain reaction (PCR), respectively, fully automated the detection steps of the assays with manual specimen processing and amplification reactions.

In the early 1980s, the Southern blot was the routine method used in the few laboratories performing clinical testing for a variety of clinical applications, even though the turnaround time was in excess of 2–3 weeks. The commercial availability of restriction endonucleases and various agarose matrices helped in making this technique routine. The PCR revolutionized blotting technologies and detection limits by offering increased sensitivity and the much-needed shortened turnaround time for clinical result availability. Initial PCR thermal cycling occurred in different temperature waterbaths, until the discovery of Taq polymerase, the first commercially available thermostable polymerase. Programmable thermal cyclers with reaction vessels containing all of the amplification reagents have gone through several generations to current instrumentation. Many modifications to the PCR have been introduced, but none as significant as real-time capability. The elimination of post-PCR detection systems and the ability to perform the entire assay in a closed vessel had significant advantages for the clinical laboratory. Various detection chemistries for real-time PCR were rapidly introduced, and many of the older detection methods (including gel electrophoresis, ASO blots, and others) vanished from the laboratory.

More recently, real-time PCR has become a method of choice for most molecular diagnostics laboratories. This modification of the traditional PCR allows for the simultaneous amplification and detection of amplified nucleic acid targets for both qualitative and quantitative applications. The main advantages of real-time PCR are the speed with which samples can be analyzed, as there are no post-PCR processing steps required, and the closed-tube nature of the technology that helps to prevent contamination. The analysis of results via amplification curve and melt curve analysis is very simple and contributes to its being a much faster method for delivering PCR results.

DNA sequencing technologies, fragment sizing, and loss of heterozygosity studies have benefited from automated capillary electrophoresis instrumentation. Several forms of microarrays are currently available as post-PCR detection mechanisms for multiplexed analysis of SNPs, gene expression, and pathogen detection. Today, molecular diagnostic laboratories are well equipped with various instruments and technologies to meet the challenges set forth by molecular pathology needs (Table 30.1).

This chapter will introduce the new molecular diagnostic paradigm and several new directions for the molecular assessment of human diseases.

THE CURRENT MOLECULAR INFECTIOUS DISEASE PARADIGM

Since early molecular technologies began entering the clinical laboratory, clinical applications were few in number, techniques were labor intense, and turnaround times were in excess of 2 weeks. Nonetheless,

there was significant clinical utility for Southern blot transfer assays such as the IgH and TCR gene rearrangements, Fragile X Syndrome, and linkage analysis. These studies were qualitative interrogations of specific genetic sequences to determine the presence or absence of a disease-associated abnormality, directly or indirectly. It was not until molecular applications for infectious disease testing were being developed that the need for quantitative and resistance testing became apparent.

Molecular infectious disease testing posed the first comprehensive testing paradigm using molecular techniques. That is, clinical applications of molecular diagnostics to infectious diseases included the molecular trinity: qualitative testing, quantitative testing, and resistance genotyping. This paradigm spanned the spectrum of current molecular capabilities and has provided significant insight to the nuances of molecular testing for other disciplines (Table 30.2). Molecular diagnostic testing for infectious disease applications continues to be the highest volume of testing being performed in clinical laboratories (Table 30.3).

While accurate and timely diagnosis of infectious diseases is essential for proper patient management, traditional testing methods for many pathogens did not allow for rapid turnaround time. Prompt detection of the microbial pathogen would allow providers to institute adequate measures to interrupt transmission to a susceptible hospital or community population and/or to begin proper therapeutic management of the patient. Unlike traditional microbial techniques, molecular techniques offered higher sensitivity and specificity with turnaround times that were unprecedented once amplified technologies were introduced to the clinical laboratory. However, a qualitative test to identify the presence or absence of an organism did not meet all of the necessary clinical needs, especially when trying to monitor therapeutic efficacy. Clinically, it was more useful to identify how much of an organism was present versus if it was or was not present. This would not only allow for the use of the assay as a therapeutic monitoring tool, but also help distinguish a clinically significant infection from a latent or resolving infection. Quantitative assays, initially for viral load testing in HIV-1-infected patients, were developed to monitor the success of newly

Table 30.1 Current Technologies Being Used in the Molecular Pathology Laboratory at the Dartmouth Hitchcock Medical Center

Technology	Platform/Instrument
DNA/RNA extraction	Manual
	Spin column
	Qiagen EZ1 robot
Fluorescence *in situ* hybridization (FISH)	Hybrite oven
	Molecular imaging system
bDNA	Siemens System 340
Hybrid capture	Digene HCII System
Real-time PCR	Cepheid GeneXpert
	ABI 7500
	Cepheid Smartcycler
	Roche Taq48
Capillary electrophoresis	Beckman CEQ
	Beckman Vidiera
Traditional PCR	MJ Research
	DNA Engines
Microarray	Luminex
	Superarray
	Nanosphere

Table 30.2 Nuances of Molecular Diagnostic Testing

Preanalytical	Analytical	Postanalytical
1. Specimen type	1. Technology used	1. LIS
2. Specimen collection device	2. Type of assay: qualitative vs quantitative	2. Result reporting and interpretation
3. Specimen transport	3. Controls	3. Consultation
4. DNA vs RNA	4. Quality assurance	4. Standardization

Table 30.3   Molecular Infectious Disease Tests That Are Currently FDA Cleared and Being Performed in Clinical Laboratories

Bacteria	Viruses
Bacillus anthracis	Avian flu
Candida albicans	Cytomegalovirus
Chlamydia trachomatis	Enterovirus
Enterococcous faecalis	HBV (quantitative)
Francisella tularensis	Hepatitis C virus (qualitative and quantitative)
Gardnerella, Trichomonas, and *Candida* spp.	HIV drug resistance
Group A streptococci	HIV (quantitative)
Group B streptococci	HBV/HCV/HIV blood screening assay
Legionella pneumophila	Human papillomavirus
Methicillin-resistant *Staphylococcus aureus*	Respiratory viral panel
Mycobacterium tuberculosis	West Nile virus
Mycobacterium spp.	
Staphylococcus aureus	

introduced protease and reverse transcriptase inhibitors. This application for HIV-1 viral load testing also represented the first true need for a companion molecular diagnostic assay, not in the sense of prescribing a therapeutic but rather in monitoring its efficacy effectively. To do so, PCR became quantitative, and other quantitative chemistries such as branched DNA (bDNA), transcription-mediated amplification (TMA), and nucleic acid sequenced-based amplification (NASBA) were born. More recently, real-time PCR capabilities have been further developed and routinely include more accurate methods for quantification of target sequences.

HIV-1 once again became the poster child for a novel molecular-based application when it was realized that the virus often mutated in response to therapy. The need to detect viral genome mutations also represented the first major routine application of sequencing methods in the clinical laboratory. These assays would predict response to regimens of antiretroviral therapy and help guide the practitioner's choices of therapy. In this sense,

HIV-1 resistance testing also represented the first routine application of personalized medicine. This application resulted in the need to address the implementation and maintenance of accurate databases as well as deal with specimens co-infected with different mutant viruses. From a single disease moiety, AIDS, a molecular paradigm evolved that would set the stage and expectations for all future molecular diagnostic applications.

A NEW PARADIGM FOR MOLECULAR DIAGNOSTIC APPLICATIONS

Our ability to test individuals for many types of human and nonhuman genetic alterations in a clinical setting is expanding at an enormous rate. This is in part due to our better understanding of human disease processes, advanced technologies, and expansion of the molecular trinity—qualitative, quantitative, and resistance genotyping—for more common diseases such as cancer. Human cancer represents a complex set of abnormal cellular processes that culminates in widespread systemic disease. Add to this the rapid identification of new biomarkers, introduction of novel therapeutics and progressive management strategies, and human cancer now becomes a model for the new molecular diagnostic paradigm demanding the three phases of testing (qualitative, quantitative, and resistance genotyping) capabilities of the clinical laboratory.

Phase I: Qualitative Analysis

Qualitative testing results in the determination of presence or absence of a particular genetic sequence. This target can be associated with an infectious agent, can be some human genomic alteration associated with a clinical phenotype, or some benign polymorphism associated with personalized medicine traits. Using the numerous tools available to the Molecular Pathology Laboratory, performing this testing has become routine for many applications, including such targets as parvovirus B19, mutation screening for cystic fibrosis, and *CYP2D6* genotyping (Table 30.4).

One such oncology target is the *JAK2* V617F mutation. V617F is a somatic point mutation in a conserved

Table 30.4   Qualitative Molecular Diagnostic Tests That Are Routinely Performed in the Dartmouth Molecular Pathology Laboratory

Genetics	Hematology	Infectious	Oncology	Personalized Medicine
AAT	BCL-2	*B. holmesii*	BRAF	CYP2C9
ApoE	FLT3	*B. parapertussis*	MSI	CYP2D6
CFTR	IGH	*B. pertussis*	SLN	EGFR
Factor II	JAK2	*Brucella* spp.		GSTP1
Factor V	TCR	GBS		HCV Genotype
HFE		HHV6		*K-ras*
MTHFR		HPV		UGT1A1
SNCA Rep1		MRSA		VKORC1
		Parvo B19		

region of the autoinhibitory domain of the Janus kinase 2 tyrosine kinase that is associated with chronic myelo-proliferative diseases (CMPD). The G>T mutation results in a substitution of phenylalanine for valine at position 617. This single point mutation leads to loss of JAK2 autoinhibition and constitutive activation of the kinase.

CMPD is a group of hematopoietic proliferations characterized by increased circulating red blood cells, extramedullary hematopoiesis, hyperplasia of hemato-poietic cell lineages, and bone marrow dysplasia and fibrosis. CMPDs include chronic myelogenous leuke-mia, polycythemia vera, essential thrombocythemia, and chronic idiopathic myelofibrosis. The diagnosis of a CMPD is made from clinical and morphologic features which can often be confounded by the stage of disease. Numerous molecular methods have been developed for identifying this mutation including PCR-RFLP, allele-specific PCR, DNA sequencing, and real-time PCR (Figure 30.1). Identification of the *JAK2* V617F mutation in CMPD provides a significant molecular bio-marker as it is found in almost all cases of polycythemia vera and can be used to rule out other CMPDs. Due to the lack of accurate diagnostic tests for the CMPDs, the utilization of the *JAK2* mutation assay has increased sig-nificantly and has in many cases become part of a testing algorithm for the CMPDs.

Phase II: Quantitative Analysis

Quantitative molecular testing refers to our ability to enumerate the target sequence(s) present in any given tissue type. As with infectious diseases, this application has resulted mainly from the need to monitor therapy and determine eligibility for therapy. Commonly per-formed quantitative molecular assays are listed in Table 30.5. Typical of the diagnosis of a human cancer is the identification of numerous protein targets through immunohistochemical techniques that help pathologists in their diagnostic decision making. In addition, molecular markers have become part of this armamentarium not so much for the diagnosis of dis-ease, but more so for the selection and monitoring of therapies. The first example of this type of application

Table 30.5	Quantitative Molecular Diagnostic Tests That Are Routinely Performed in the Dartmouth Molecular Pathology Laboratory

Hematology	Infectious	Personalized Medicine
BCR-ABL	BKV	HER2
Chimerism	HCV	
t(15;17)	HIV	

was the quantitative detection of ERBB2 (v-erb-b2 erythroblastic leukemia viral oncogene homolog 2) in human breast cancers.

The human epidermal growth factor receptor 2 (*ERBB2*, commonly referred to as *HER2*) gene had been recognized as being amplified in breast cancers for many years. In fact, this gene is amplified in up to 25% of all breast cancers. Amplification of the *HER2* gene is the primary mechanism of HER2 overexpression that results in increased receptor tyrosine kinase activity (Figure 30.2). HER2-positive breast cancers have a worse prognosis and are often resistant to hormonal therapies and other chemotherapeutic agents. Trastuzu-mab (Herceptin), the first humanized monoclonal antibody against the HER2 receptor, was approved by the FDA in 1998. The availability of a therapeutic now made it necessary for laboratories to determine HER2 gene copy number or protein overexpression before patients would be eligible for treatment. Several FDA-cleared tests for immunohistochemical detection of HER2 protein and fluorescence *in situ* hybridization (FISH) detection of gene amplification are commercially available and approved as companion diagnostics for Herceptin (Figure 30.3). Recently, the American Society of Clinical Oncology and the College of American Pa-thologists published guidelines for performing and inter-preting *HER2* testing. These guidelines attempt to standardize *HER2* testing by addressing preanalytical, analytical, and postanalytical variables that could lessen

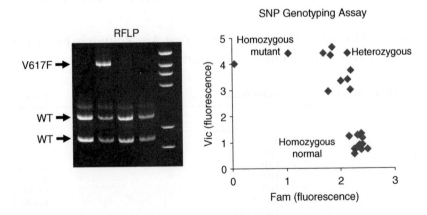

Figure 30.1   Detection of the *JAK2* V617F mutation using PCR-RFLP (left) and real-time PCR allelic discrimination (right).

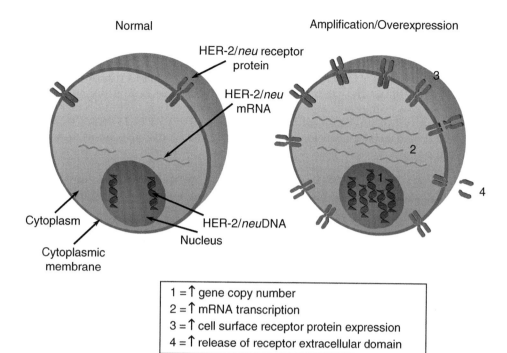

Normal   Amplification/Overexpression

HER-2/*neu* receptor protein

HER-2/*neu* mRNA

Cytoplasm

Cytoplasmic membrane

HER-2/*neu*DNA

Nucleus

1 = ↑ gene copy number
2 = ↑ mRNA transcription
3 = ↑ cell surface receptor protein expression
4 = ↑ release of receptor extracellular domain

Figure 30.2   Schematic diagram of the *HER2* amplification process in normal and tumor breast epithelial cells.

Figure 30.3   *HER2* amplification as detected by FISH analysis (orange signal, LSI HER2; green signal, CEP17).

result variability due to technical and interpretative subjectivity. While many laboratories have adopted an algorithm for screening breast cancer cases by immuno-histochemical staining and reflexing equivocal results for FISH analysis, others have gone to a primary FISH screen for *HER2* gene amplification.

Phase III: Resistance Testing

With respect to resistance genotyping, our infectious disease experience has taught us that this can be attrib-uted to identification of subtypes of organisms such as

in Hepatitis C virus (HCV) or in actual mutations of the viral genome as in the HIV-1 virus. It is well known that tumor cells develop resistance to therapeutic agents via multiple mechanisms and, more importantly, are sensi-tive to newer therapeutic modalities if certain genetic abnormalities are or are not present in the tumor cells. For example, significant improvements have been made in the response rates, progression-free survival, and overall survival for colorectal cancers, the second lead-ing cause of cancer death in the United States, in the past decade. This has been due to new combinations of chemotherapeutic agents and new small molecule targeted therapies.

The overall efficacy of traditional chemotherapeutic agents for treating cancer has been extremely low. More modern targeted therapies hope to improve on this. Selecting patients most likely to benefit from molecu-larly targeted therapies, however, has been challenging. Recently, K-*ras* mutations have been shown to be an independent prognostic factor in patients with advanced colorectal cancer who are treated with Cetux-imab. Cetuximab is known to be active in metastatic colorectal cancer patients which are resistant to irinote-can, oxaliplatin, and fluoropyrimidine. In these studies, K-*ras* mutation has also been identified as a marker of tumor cell resistance to Cetuximab. Some investigators associated the presence of wild-type K-*ras* sequences with panitumumab (a humanized monoclonal antibody against the epidermal growth factor receptor) efficacy in patients with metastatic disease. These data demon-strate the need for identifying resistant mutations in other genes, as well as identifying the presence of the target for these novel therapies.

K-*ras* mutations are present in >30% of colorectal cancers and play a critical role in the pathogenesis of this disease. K-*ras* is a member of the guanosine-5′-triphosphate (GTP)-binding protein superfamily. Stimulation of a growth factor receptor by ligand binding results in activated K-*ras* by binding to GTP. This results in downstream activation of pathways such as RAF/MAP kinase, STAT, and PI3K/AKT, that in turn results in cellular proliferation, adhesion, angiogenesis, migration, and survival. While the association of K-*ras* mutations and cancer had been known for quite some time, the association of mutated K-*ras* with a response to a novel and expensive therapeutic has made its interrogation a prerequisite to receiving therapy. It is now estimated that all colorectal cancer patients who will receive these therapies will first be tested for K-*ras* mutations.

BCR-ABL: A MODEL FOR THE NEW PARADIGM

Chronic myelogenous leukemia (CML) is considered one of the diseases in the group of myeloproliferative disorders. These are clonal hematopoietic malignancies characterized by the proliferation and survival of one or more of the myeloid cell lineages. CML has an estimated 5000 newly diagnosed cases per year, and it accounts for about 20% of all adult cases of leukemia. Typically, the disease progresses through three clinical phases: (i) a chronic phase (which may be clinically asymptomatic), (ii) an accelerated phase, and (iii) a blast phase or blast crisis. Most patients are diagnosed with the disease during the chronic phase by morphologic, cytogenetic, and molecular genetic techniques.

The (9;22)(q34;q11) translocation, which results in the *BCR-ABL* fusion gene, is a defining feature of this disease. *BCR-ABL* analysis warrants the three phases of the presented paradigm for qualitative diagnostic testing, quantitative monitoring, and resistance mutation analysis. CML has traditionally been diagnosed and monitored by cytogenetics, FISH, and Southern blot analysis (Figure 30.4). Conventional cytogenetics, such as bone marrow karyotyping that is performed on 30–50 cells, can be used to visualize the translocation by G-banding analysis. FISH, typically performed on 100–1000 cells in interphase or metaphase, allows the visualization of the translocation through the use of fluorescently labeled probes (Figure 30.4). These techniques are limited in sensitivity and cannot detect more than a 2-log reduction in residual CML cells and have increased turnaround times.

Real-time reverse transcriptase quantitative PCR (qPCR) has become the preferred method for

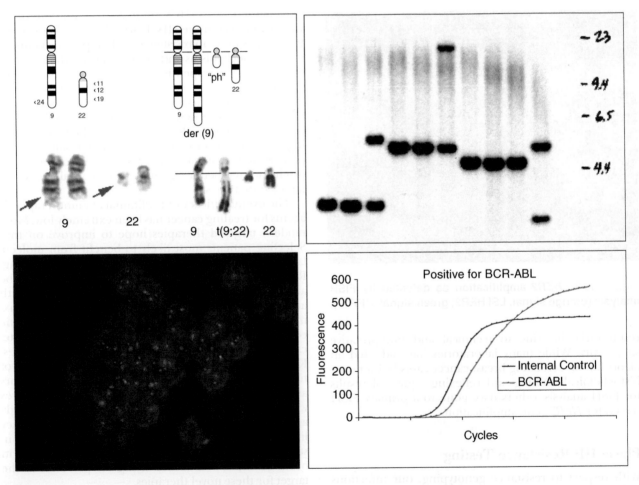

Figure 30.4 Testing methods for the detection of the *BCR-ABL* translocation. Traditional cytogenetic karyotyping, Southern blot transfer analysis, FISH, and real-time PCR.

monitoring the second phase of the paradigm, minimal residual disease in CML patients treated with imatinib (Figure 30.4). The major advantage of qPCR is that it has a sensitivity on the order of one CML cell per 10^5 normal cells, thus achieving a 3-log improvement in sensitivity compared to cytogenetics and FISH. Novel instrumentation now allows clinical laboratories the ability to provide quantitative analysis using real-time PCR in under 2 hours (Figure 30.5).

The detection of low levels of disease in CML patients is critical to the proper management of these patients. Studies have shown that a rapid 3-log quantitative reduction in *BCR-ABL* transcript predicts risk of progression in CML patients being treated with imatinib. Similarly, a significant increase in *BCR-ABL* transcript in a patient being treated with imatinib may indicate an acquired mutation which confers resistance to treatment. Molecular analysis of the transcript is performed to identify mutations that confer therapeutic resistance, the third phase of the paradigm. It has been recommended that patients who fail to achieve a 3-log reduction in *BCR-ABL* transcript levels after 18 months of imatinib therapy have an increase in their dose in an attempt to achieve a major molecular response.

The recognition of the role of BCR-ABL kinase activity in CML has provided an attractive target for the development of therapeutics. Although interferon therapy and stem cell transplantation were previously the first-line treatments of choice, tyrosine kinase inhibitors such as imatinib mesylate (Gleevec®; Novartis, Basel, Switzerland), and dasatinib (SPRYCEL®; Bristol-Myers Squibb, New York, NY) have now become first- and second-line therapies. Imatinib binds to the inactive form of the wild-type Bcr-Abl protein, preventing its activation. However, acquired mutations in the *BCR-ABL* kinase domains prevent imatinib binding. Dasatinib binds to both the inactive and active forms of the protein, and also binds to all imatinib-resistant mutants except the threonine to isoleucine mutation at amino acid residue 315 (T315I). The introduction of more efficacious therapeutics such as imatinib and dasatinib has redefined the therapeutic responses that are achievable in the CML patient (Table 30.6). CML now represents a molecular oncology paradigm similar to HIV-1 whereby qualitative, quantitative, and resistant genotyping can be performed routinely.

Table 30.6 Therapeutic Goals for the CML Patient

Therapeutic Goal	Therapeutic Response
Hematologic response	Normal peripheral blood valves, normal spleen size
Cytogenetic response	Reduction of Ph+ cells in blood or bone marrow
1. complete	1. 0% Ph+ cells
2. partial	2. 1–35% Ph+ cells
3. minor	3. 36–95% Ph+ cells
Molecular response	Reduction or elimination of BCR-ABL mRNA in peripheral blood or bone marrow

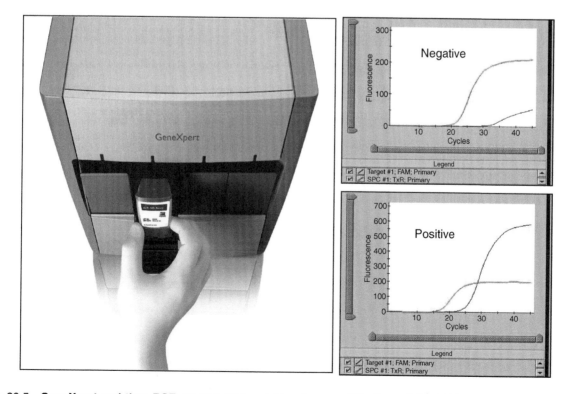

Figure 30.5 **GeneXpert real-time PCR system and cartridge for quantification of the *BCR-ABL* transcript.**

CONCLUSION

Numerous genes and mutations have been identified for many human diseases, including cancer. Yet little of this knowledge has made it into the clinical laboratory as validated diagnostic tests. This is beginning to change with our increased medical and biological knowledge of these diseases. Genomics and proteomics in the clinical context of diagnostic testing are progressing at record speeds. Technology, as is typical in the clinical laboratory, has far outpaced proven clinical utility. Routine applications of information from the "omics-era" are currently being performed for a new paradigm in molecular diagnostic testing that consist of qualitative, quantitative, and resistance genotyping for various clinical conditions.

KEY CONCEPTS

- Molecular pathology has matured to a level of diagnostics that is unprecedented by any other technology or clinical laboratory effort.
- The molecular diagnostic paradigm includes qualitative, quantitative, and "resistance genotyping" for many disease states.
- Real-time polymerase chain reaction (PCR) has become the method of choice for most molecular diagnostic applications.
- Molecular infectious disease applications represent the first comprehensive testing paradigm using molecular techniques. That is, clinical applications of molecular diagnostics to infectious diseases include qualitative testing, quantitative testing, and resistance genotyping.
- Qualitative testing results in the determination of presence or absence of a particular genetic sequence.
- Quantitative molecular testing refers to our ability to enumerate the target sequence(s) present in any given tissue type. This application has resulted mainly from the need to monitor therapy and determine eligibility for therapy.
- Resistance genotyping can be used to subtype organisms such as the Hepatitis C virus or identify mutations in the viral genome as in the HIV-1

virus. Tumor cells also develop resistance to therapeutic agents via multiple mechanisms.
- Chronic myelogenous leukemia (CML) represents a new model for molecular diagnostic testing using qualitative, quantitative, and resistance testing tools.

SUGGESTED READINGS

1. Amado RG, Wolf M, Peeters M, et al. Wild-type KRAS is required for panitumumab efficacy in patients with metastatic colorectal cancer. *J Clin Oncol.* 2008;26:1626–1634.
2. Erali M, Voelkerding KV, Wittwer CT. High-resolution melting applications for clinical laboratory medicine. *Exp Molec Pathol.* 2008;85:50–58.
3. Kralovics R, Passamonti F, Buser AS, et al. A gain of function mutation of JAK2 in myeloproliferative disorders. *N Engl J Med.* 2005;352:1779–1790.
4. Krause DS, Van Etten RA. Tyrosine kinases as targets for cancer therapy. *N Engl J Med.* 2005;353:172–187.
5. Levine RL, Gilliland DG. Myeloproliferative disorders. *Blood.* 2008;112:2190–2198.
6. Lièvre A, Bachet JB, Boige V, et al. KRAS mutations as an independent prognostic factor in patients with advanced colorectal cancer treated with cetuximab. *J Clin Oncol.* 2008;26:374–379.
7. Ma WW, Adjei AA. Novel agents on the horizon for cancer therapy. *CA Cancer J Clin.* 2009;59:111–137.
8. Martinelli G, Iacobucci I, Soverini S, et al. Monitoring minimal residual disease and controlling drug resistance in chronic myeloid leukaemia patients in treatment with imatinib as a guide to clinical management. *Hematol Oncol.* 2006;24:196–204.
9. Mendelsohn J, Baselga J. Epidermal growth factor receptor targeting in cancer. *Semin Oncol.* 2006;33:369–385.
10. Ratcliff RM, Chang G, Kok T, et al. Molecular diagnosis of medical viruses. *Curr Issues Mol Biol.* 2007;9:87–102.
11. Sauter G, Lee J, Bartlett JMS, et al. Guidelines for human epidermal growth factor receptor 2 testing: Biologic and methodologic considerations. *J Clin Oncol.* 2009;27:1323–1333.
12. Schiffer CA. BCR-ABL tyrosine kinase inhibitors for chronic myelogenous leukemia. *N Engl J Med.* 2007;357:258–265.
13. Tsongalis GJ, Coleman WB. Clinical genotyping: The need for interrogation of single nucleotide polymorphisms and mutations in the clinical laboratory. *Clin Chim Acta.* 2006;363:127–137.
14. van der velden VHJ, Hochhaus A, Cazzaniga G, et al. Detection of minimal residual disease in hematologic malignancies by real time quantitative PCR: Principles, approaches, and laboratory aspects. *Leukemia.* 2003;17:1013–1034.
15. Wolff AC, Hammond MEH, Schwartz JN, et al. American Society of Clinical Oncology/College of American Pathologists guideline recommendations for human epidermal growth factor receptor 2 testing in breast cancer. *J Clin Oncol.* 2007;25:118–145.

31

Pharmacogenomics and Personalized Medicine in the Treatment of Human Diseases

Hong Kee Lee . Gregory J. Tsongalis

INTRODUCTION

In 2001, the first draft of the Human Genome Project was simultaneously published by two groups. This was the beginning of an era that promised better diagnostic and prognostic testing that would lead to preventive medicine and more personalized therapy. As part of this promise, the ability to select and dose therapeutic drugs through the use of genetic testing has become a reality. Pharmacogenomics represents the role that genes play (pharmacogenetics) in the processing of drugs by the body (pharmacokinetics) and how these drugs interact with their targets to give the desired response (pharmacodynamics) (Figure 31.1). While other environmental factors also play a role in this selection, an individual's genetic makeup provides new insights into the metabolism and targeting of commonly prescribed as well as novel targeted therapies. The importance of implementing this knowledge into clinical practice is highlighted by the more than 2 million annual hospitalizations and greater than 100,000 annual deaths in the United States due to adverse drug reactions.

Current medical practices often utilize a trial-and-error approach to select the proper medication and dosage for a given patient (Figure 31.2). Pharmacogenetics (PGx) utilizes our knowledge of genetic variation to assess an individual's response to therapeutic drugs, and when interpreted in the context of the entire being, pharmacogenomics can result in better personalized medicine. This application of human genetics is our attempt to more accurately and efficaciously determine genomic sequence variations and expression patterns involved in response to the absorption, distribution, metabolism, and excretion of therapeutic drugs at a systemic level. A second component to this application of genomics is in the identification of target genes for novel therapies at the cellular level, as well as acquired mutations in genes of specific pathways that may result in a better or worse response to a novel therapy. The overall aim of PGx testing is to decrease adverse responses to therapy and increase efficacy by ensuring the appropriate selection and dose of therapy.

Numerous genetic variants or polymorphisms in genes that code for drug metabolizing enzymes, drug transporters, and drug targets that alter response to therapeutics have been identified (www.pharmgkb.org/) (Table 31.1). Classification of these enzymes includes specific nomenclature, CYP2D6*1, such that the name of the enzyme (CYP) is followed by the family (for example, 2), subfamily (for example, D), and gene (for example, 6) associated with the biotransformation. Allelic variants are indicated by an asterisk (*) followed by a number (for example, *1). The technologies that are currently available in clinical laboratories and are employed to routinely test for these variants are shown in Table 31.2.

Single nucleotide polymorphisms (SNPs) in these genes may result in no significant phenotypic effect, change drug metabolism by >10,000-fold, or alter protein binding by >20-fold. More than 3.5 million SNPs have been identified in the human genome (www.hapmap.org), making analysis quite challenging. As shown in Table 31.1, different genes can

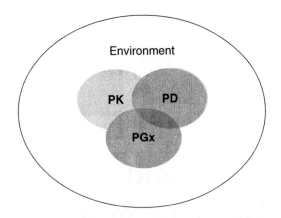

Table 31.1   Examples of Enzymes and Their Designated Polymorphic Alleles

Enzyme	Alleles
Cytochrome P450 2D6 (CYP2D6)	*2, *3, *4, *5, *6, *7, *8, *11, *12, *13, *14, *15, *16, *18, *19, *20, *21, *38, *40, *42, *10, *17, *36, *41
Cytochrome P450 2C9 (CYP2C9)	*1, *2, *3, *4, *5, *6
Cytochrome P450 2C19 (CYP2C19)	*2, *3
UGT1A1	*6, *28, *37, *60

Figure 31.1 Venn diagram of pharmacogenomics showing the interactions of pharmacogenetics (PGx), pharmacokinetics (PK), and pharmacodynamics (PD), along with other environmental factors that affect drug response.

contain various numbers of SNPs that alter the function of the coded protein. It is important to clearly differentiate these benign polymorphisms (present in greater than 1% of the general population) from rare disease-causing mutations that are present in less than 1% of the general population. The SNPs associated with pharmacogenetics do not typically result in any form of genetic disease. We do know that these SNPs may be used to evaluate individual risk for adverse drug reactions (ADRs) so that we may decrease ADRs, select optimal therapy, increase patient compliance, develop safer and more effective drugs, revive withdrawn drugs, and reduce time and cost of clinical trials.

In addition to the genotype predicting a phenotype, drug metabolizing enzyme activity can be induced or inhibited by various drugs. Induction leads to the production of more enzyme within 3 or more days of exposure to inducers. Enzyme inhibition by commonly prescribed drugs can be an issue in polypharmacy and is usually the result of competition between two drugs for metabolism by the same enzyme. A better understanding of drug metabolism has led to a number of amended labels for commonly prescribed drugs (Table 31.3).

Historical Perspective

Friedrich Vogel was the first to use the term pharmacogenetics in 1959. However, in 510 BC, Pythagoras recognized that some individuals developed hemolytic anemia with fava bean consumption. In 1914, Garrod expanded on these early observations to state enzymes detoxify foreign agents so that they may be excreted harmlessly, but some people lack these enzymes and experience adverse effects. Hemolytic anemia due to fava bean consumption was later determined to occur in glucose-6-phosphate dehydrogenase deficient individuals.

Through the early 1900s PGx evolved as investigators combined Mendelian genetics with observed phenotypes. In 1932, Snyder performed the first global study of ethnic variation and deduced that taste deficiency was inherited. As such, the phenylthiourea non-taster phenotype was an inherited recessive trait, and the frequency of occurrence differed between races. Similarly, polymorphisms in the N-acetyl transferase enzyme also segregate by ethnicity. Shortly thereafter other genetic differences such as aldehyde dehydrogenase and alcohol dehydrogenase deficiencies were discovered. In 1956, it was recognized that variants of glucose-6-phosphate dehydrogenase caused primaquine-induced hemolysis. It was shown that the metabolism of nortriptyline and desipramine was highly variable among individuals, and two phenotypes were identified in 1967. Subsequently, genetic deficiencies in other enzyme systems were documented

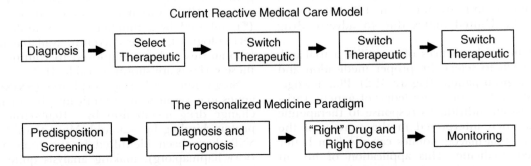

Figure 31.2 Schematic diagram illustrating current medical practice and the changes that occur in a personalized medicine model.

Table 31.2 Genotyping Methodologies Currently in Use in the Clinical Laboratory

Technology	Methodology*	Company*
Gel electrophoresis	Southern blot PCR-RFLP**	Generic equipment
Real-time PCR	Allele-specific PCR	Applied Biosystems
	Allelic discrimination	Biorad
		Cepheid
		Roche Molecular
DNA sequencing	Sanger—fluorescent detection	Applied Biosystems
		Beckman Coulter
		Pyrosequencing
Arrays	Liquid beads	Autogenomics
	Microarray	Luminex
	Microfluidic	Nanosphere
		Roche Molecular

*Representative examples.
**PCR-RFLP, polymerase chain reaction-restriction fragment length polymorphism analysis.

Table 31.3 Examples of Therapeutic Drugs That Include PGx Information or Have Had Labels Amended Due to New Knowledge of Metabolism

Atomoxetine	Strattera®
Thioridazine	Mellaril®
Voriconazole	Vfend®
6-mercaptopurine	Purinethol®
Azathioprine	Imuran®
Irinotecan	Camptosar®
Warfarin	Coumadin®

and shown that metabolism of drugs played a critical role in a patient's response as well as risk for development of ADRs. The central dogma for assessing human diseases at a molecular level, DNA → RNA → protein, became the model for moving the newly discovered knowledge base and technologies forward at the molecular level.

Genotyping Technologies

We currently have the means to successfully and accurately genotype patients for polymorphisms and mutations on a routine basis (Table 31.2). Genotyping involves the identification of defined genetic variants that give rise to the specific drug response phenotypes. Genotyping methods are easier to perform and more cost effective than the more traditional phenotyping methods. Because an individual's genotype is not expected to change over time, most genotyping applications need be performed only once in an individual's lifetime. The exception to this is for the acquired genetic variants that occur in certain disease conditions such as cancer where variants of tumor cells may change during the course of the disease. While a single genotype can determine responses to numerous therapeutic drugs, it is worth noting that patients and providers will need to refer back to genotype information on numerous occasions.

Early genotyping efforts utilized conventional procedures such as Southern blotting and the polymerase chain reaction (PCR) followed by restriction endonuclease digestion to interrogate human gene sequences for polymorphisms. The introduction of real-time PCR to clinical laboratories has resulted in assays which can be performed much faster and in a multiplexed fashion so that more variants are tested for at the same time (Figure 31.3). Direct sequencing reactions have been developed on automated capillary electrophoresis instruments to detect specific base changes that result in a variant allele.

More recently, platforms utilizing various array technologies for genotyping have been introduced into the clinical laboratory. Roche Molecular was the first to introduce an FDA-cleared microarray, Ampli-Chip CYP450 test, for PGx testing on an Affymetrix platform which detects 31 known polymorphisms in the CYP2D6 gene, including gene duplication and deletion, as well as two variations in the CYP2C19 gene. Other array platforms being introduced into

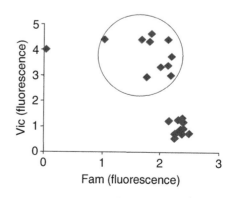

Figure 31.3 Allelic discrimination using real-time PCR and Taqman probes to identify individuals who are homozygous or heterozygous (circle) for a particular SNP.

the clinical laboratory are the AutoGenomics INFINITI™, Luminex 100 and 200 microbead-array technologies, and the Nanosphere Verigene system.

PGx and Drug Metabolism

Most of the enzymes involved in drug metabolism are members of the cytochrome P450 (CYP450) superfamily. CYP450 enzymes are mainly located in the liver and gastrointestinal tract and include greater than 30 isoforms. The most polymorphic of these enzymes responsible for the majority of biotransformations are the CYP3A, CYP2D6, CYP2C19, and CYP2C9. Benign genetic variants or polymorphisms in these genes can lead to the following phenotypes: poor, intermediate, extensive, and ultrarapid metabolizers. Poor metabolizers (PM) have no detectable enzymatic activity; intermediate metabolizers (IM) have decreased enzymatic activity; extensive metabolizers (EM) are considered normal and have at least one copy of an active gene; and ultrarapid metabolizers (UM) contain duplicated or amplified gene copies that result in increased drug metabolism. The following are several examples of polymorphic drug metabolizing enzymes that can affect response to therapy.

CYP2D6

CYP2D6 is an example of one of the most widely studied members of this enzyme family. It is highly polymorphic and contains 497 amino acids. The *CYP2D6* gene is localized on chromosome 22q13.1 with two neighboring pseudogenes, *CYP2D7* and *CYP2D8*. More than 50 alleles of *CYP2D6* have been described, of which alleles *3, *4, *5, *6, *7, *8, *11, *12, *13, *14, *15, *16, *18, *19, *20, *21, *38, *40, *42, and *44 were classified as nonfunctioning and alleles *9, *10, *17, *36, and *41 were reported to have substrate-dependent decreased activity. CYP2D6 alone is responsible for the metabolism of 20–25% of prescribed drugs (Table 31.4). Screening for CYP2D6*3, *4, and *5 alleles identifies at least 95% of poor metabolizers in the Caucasian population. Based on the type of metabolizer, an individual can determine the response to a therapeutic drug.

CYP2C9

A more specific example of CYP450 metabolizing enzyme polymorphisms and drug metabolism is demonstrated by *CYP2C9* and warfarin. In the past six years, there has been tremendous interest in studying the effect of genetics on warfarin dosing. *CYP2C9* polymorphisms (CYP2C9*2 and CYP2C9*3) were first found to reduce the metabolism of S-warfarin, which can lead to dosing differences. Soon after, the identification of a polymorphism in the vitamin K epoxide reductase (*VKORC1*) gene proved that drug target polymorphisms are also important in warfarin dosing.

Table 31.4   Some Therapeutic Drugs Metabolized by CYP2D6

Cytochrome P450 2D6	
Amitriptyline	Lidocaine
Aripiprazole	Metoclopramide
Atomoxetine	Metoprolol
Carvedilol	Nortriptyline
Chlorpromazine	Oxycodone
Clomipramine	Paroxetine
Codeine	Propafenone
Desipramine	Propranolol
Dextromethorphan	Risperidone
Duloxetine	Tamoxifen
Flecainide	Thioridazine
Fluoxetine	Timolol
Fluvoxamine	Tramadol
Haloperidol	Venlafaxine
Imipramine	Zuclopenthixol

Since then, several warfarin dosing algorithms that take into consideration patient demographics, *CYP2C9* and *VKORC1* polymorphisms, and concurrent drug use have been developed. In August 2007, the FDA announced an update to the warfarin package insert to include information on *CYP2C9* and *VKORC1* testing.

UGT1A1

The uridine diphosphate glucuronosyltransferase (UGT) superfamily of endoplasmic reticulum-bound enzymes is responsible for conjugating a glucuronic acid moiety to a variety of compounds, thus allowing these compounds to be more easily eliminated. It is a member of this family that catalyzes the glucuronidation of bilirubin, allowing it to be excreted in the bile. As irinotecan therapy for advanced colorectal cancers became more widely used, it was observed that patients who had Gilbert syndrome (defects in UGT leading to mild hyperbilirubinemia) suffered severe toxicity. Irinotecan is converted to SN-38 by carboxylesterase-2, and SN-38 inhibits DNA topoisomoerase I activity. SN-38 is glucuronidated by uridine diphosphate glucuronosyltransferase (UGT), forming a water-soluble metabolite, SN-38 glucuronide, which can then be eliminated (Figure 31.4). The observation made in Gilbert syndrome patients revealed that SN-38 shares a glucuronidation pathway with bilirubin. The decreased glucuronidation of bilirubin and SN-38 can be attributed to polymorphisms in the *UGT1A1* gene. The wild-type allele of this gene, *UGT1A1*1*, has six tandem TA repeats in the regulatory TATA box of the *UGT1A1* promoter. The most common polymorphism associated with low activity of *UGT1A1* is the *28 variant, which has seven TA repeats. In August 2005, the FDA amended the irinotecan (Camptosar®) package insert to recommend genotyping for the UGT1A1 polymorphism and suggested a dose reduction in patients homozygous for the *28 allele.

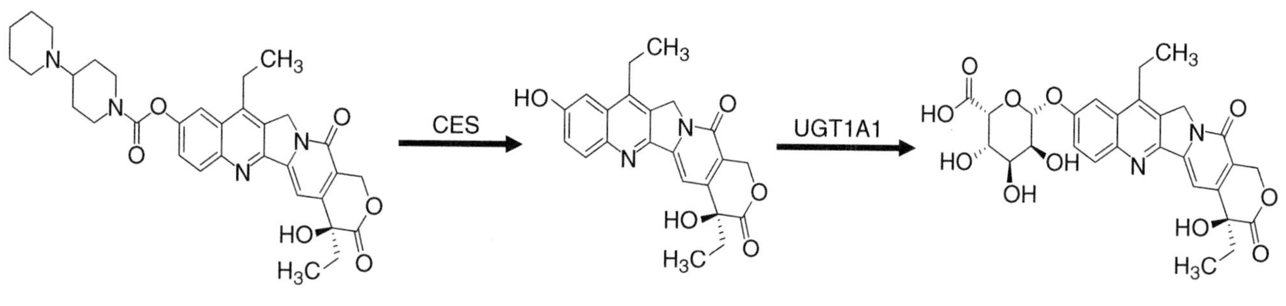

Figure 31.4   Schematic diagram of irinotecan metabolism.

PGx and Drug Transporters

Although the genes that code for drug metabolizing enzymes have received more attention in recent years as markers for PGx testing, the genes that code for proteins used to transport drugs across membranes also need to be considered when discussing PGx. These drug transporter proteins move substrates across cell membranes, bringing them into cells or removing them from cells. These proteins are essential in the absorption, distribution, and elimination of various endogenous and exogenous substances including pharmaceutical agents.

Several groups of drug transporters that may be significant in the field of pharmacogenomics exist, including multidrug resistance proteins (MDRs), multidrug resistance-related proteins (MRPs), organic anion transporters (OATs), organic anion transporting polypeptides (OATPs), organic cation transporters (OCTs), and peptide transporters (PepTs). ABCB1 (MDR1) is a member of the multidrug resistance protein family and is one example of a drug transporter protein important in the field of PGx.

ABCB1

ABCB1 is a member of the ATP-binding cassette (ABC) superfamily of proteins. Also known as P-glycoprotein (P-gp) or MDR1, it is a 170 kDa glycosylated membrane protein expressed in various locations including the liver, intestines, kidney, brain, and testis. Generally speaking, ABCB1 is located on the membrane of cells in these locations and serves to eliminate metabolites and a wide range of hydrophobic foreign substances, including drugs, from cells by acting as an efflux transporter. Due to the localization of ABCB1 on specific cells, ABCB1 aids in eliminating drugs into the urine or bile and helps maintain the blood-brain barrier.

Like other eukaryotic ABC proteins, the ABCB1 protein is composed of two similar halves, each half containing a hydrophobic membrane binding domain and a nucleotide binding domain. The membrane binding domains are each composed of six hydrophobic transmembrane helices, and the two hydrophilic nucleotide binding domains are located on the intracellular side of the membrane where they bind ATP.

ABCB1 was first identified in cancer cells that had developed a resistance to several anticancer drugs because of an overexpression of the transporter. When expressed at normal levels in noncancerous cells, ABCB1 has been shown to transport other classes of drugs out of cells including cardiac drugs (digoxin), antibiotics, steroids, HIV protease inhibitors, and immunosuppressants (cyclosporin A).

Genetic variations in the ABCB1 gene expressed in normal cells have been shown to have a role in interindividual variability in drug response. Although many polymorphisms have been detected in the ABCB1 gene, correlations between genotype and either protein expression or function have been described for only a few of the genetic variants. Most notable among these is the 3435 C>T polymorphism, found in exon 26, which has been found to result in decreased expression of ABCB1 in individuals homozygous for the T allele. These results, however, have been found to be slightly controversial. The possible importance of the 3435 C>T is particularly interesting considering the mRNA levels are not affected by this polymorphism, which is found within a coding exon. Although the correlation between the 3435 C>T polymorphism and ABCB1 protein levels may be due to linkage of this polymorphism to others in the ABCB1 gene, a recent study showed that this polymorphism alone does not affect ABCB1 mRNA or protein levels but does result in ABCB1 protein with an altered configuration. The altered configuration due to 3435 C>T is hypothesized to be caused by the usage of a rare codon which may affect proper folding or insertion of the protein into the membrane, affecting the function, but not the level of ABCB1. Genetic variations affecting the expression or function of drug transporter proteins, such as ABCB1, could drastically alter the pharmacokinetics and pharmacodynamics of a given drug.

PGx and Drug Targets

Most therapeutic drugs act on targets to elicit the desired effects. These targets include receptors, enzymes, or proteins involved in various cellular events such as signal transduction, cell replication, and others. Investigators have now identified polymorphisms in these targets that render them resistant to the particular therapeutic agent. While these polymorphisms in receptors may not show dramatic increases or decreases in drug activity as the drug metabolizing enzymes, biologically significant effects occur frequently. The human μ-opioid receptor gene (*OPRM1*) and *FLT3* are two examples of the effects of polymorphisms on drug targets.

OPRM1

The human μ-opioid receptor, coded by the *OPRM1* gene, is the major site of action for endogenous β-endorphin, and most exogenous opioids. Forty-three SNPs have been discovered in the *OPRM1* gene, but the 118A>G SNP is the most well studied. The allele frequency of *OPRM1* 118A>G SNP in Caucasians is 10–30%. This SNP causes an increased affinity of the μ-opioid receptor for β-endorphin, which improves pain tolerance in humans. However, the effect of this SNP is controversial in people taking exogenous opioids. Studies showed that 118G allele carriers required an increased dose of morphine in patients with acute and chronic pain, but a decreased dose of fentanyl in patients requiring epidural or intravenous fentanyl for pain control during labor. The exact mechanism of how this happens is still unknown.

FLT3

FLT3 is a receptor tyrosine kinase expressed and activated in most cases of acute myeloid leukemia (AML), which has a relatively high relapse rate due to acquired resistance to traditional chemotherapies. An internal tandem duplication (ITD) mutation in the FLT3 gene is found in up to 30% of AML patients, while point mutations have been shown to account for approximately 5% of refractory AML. The FLT3-ITD induces activation of this receptor and results in downstream constitutive phosphorylation in STAT5, AKT, and ERK pathways. This mutation in FLT3 is a negative prognostic factor in AML. Recently, a novel multitargeted receptor tyrosine kinase inhibitor, ABT-869, has been developed to suppress signal transduction from constitutively expressed kinases such as a mutated FLT3.

PGx Applied to Oncology

Cancer represents a complex set of deregulated cellular processes that are often the result of underlying molecular mechanisms. While there is recognized interindividual variability with respect to an observed chemotherapeutic response, it is also apparent that there is considerable cellular heterogeneity within a single tumor that may account for this lack of efficacy. It is becoming clear that molecular genetic variants significantly contribute to an individual's response to a particular therapy of which some is due to polymorphisms in genes coding for drug metabolizing enzymes. However, in the cancer patient the acquired genetics of the tumor cell must also be taken into account. Unlike traditional PGx testing, those tests for the cancer patient must be prepared to identify acquired or somatic genetic alterations that deviate from the underlying genome of the individual.

Cancer patients exhibit a heterogeneous response to chemotherapy with only 25–30% efficacy. PGx can improve on chemotherapeutic and targeted therapy responses by providing a more informative evaluation of the underlying molecular determinants associated with tumor cell heterogeneity. In the cancer patient, PGx can be used to integrate information on drug responsiveness with alterations in molecular biomarkers. Thus, as with previous examples of PGx testing, therapeutic management of the cancer patient can be tailored to the individual patient or tumor phenotype.

Estrogen Receptor

One of the first and most widely used parameters for a targeted therapy is the evaluation of breast cancers for expression of the estrogen receptor (ER) (Figure 31.5). ER-positive breast cancers are then treated with hormonal therapies that mimic estrogen. One of these estrogen analogues is Tamoxifen (TAM), which itself has now been shown to have altered metabolism due to CYP450 genetic polymorphisms. TAM is used to treat all stages of estrogen receptor-positive breast cancers. TAM and its metabolites compete with estradiol for occupancy of the estrogen receptor, and in doing so inhibit estrogen-mediated cellular proliferation. Conversion of TAM to its active metabolites occurs predominantly through the CYP450 system (Figure 31.6). Conversion of TAM to primary and secondary metabolites is important because these metabolites can have a greater affinity for the estrogen receptor than TAM itself. For example, 4-OH *N*-desmethyl-TAM (endoxifen) has approximately 100 times greater affinity for the estrogen receptor than tamoxifen. Activation of tamoxifen to endoxifen is primarily due to the action of CYP2D6. Therefore, patients with defective *CYP2D6* alleles derive less benefit from tamoxifen therapy than patients with functional copies of *CYP2D6*. The most common null allele among Caucasians is *CYP2D6*4*, a splice site mutation (G1934A) resulting in loss of enzyme activity and, therefore, lack of conversion of TAM to endoxifen. This would result in significantly

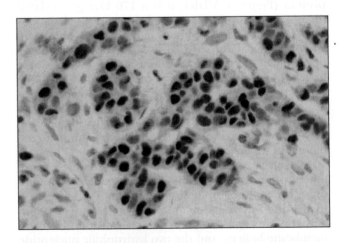

Figure 31.5 Immunohistochemical staining for the estrogen receptor in breast cancer.

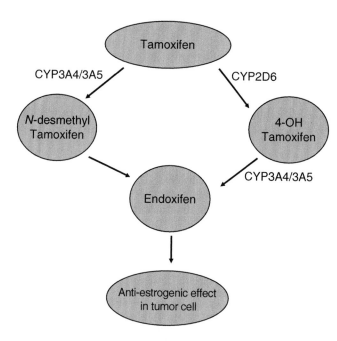

Figure 31.6 Simplified schematic diagram showing metabolism of tamoxifen by CYP450 enzymes with the resulting production of the active metabolite endoxifen.

decreased response to this commonly used antihormonal therapy. Common polymorphisms such as this make it feasible to accurately genotype patients so that treatments may be optimized.

HER2

The human epidermal growth factor receptor 2 (*ERBB2* or *HER2*) gene is amplified in up to 25% of all breast cancers. While this gene can be expressed in low levels in a variety of normal epithelia, amplification of the *HER2* gene is the primary mechanism of HER2 overexpression that results in increased receptor tyrosine kinase activity. HER2 status has been implemented as an indicator of both prognosis as well as a predictive marker for response to therapy such that HER2-positive breast cancers have a worse prognosis and are often resistant to hormonal therapies and other chemotherapeutic agents. As a predictive marker, HER2 status is utilized to determine sensitivity to anthracycline-based chemotherapy regimens. Determination of HER2 status is also recognized as the first FDA-approved companion diagnostic since Trastuzumab (Herceptin), the first humanized monoclonal antibody against the HER2 receptor, was approved by the FDA in 1998. Introduction of this therapeutic into routine use made it necessary for laboratories to determine the HER2 status in breast cancer cells before patients would be eligible for treatment. Several FDA-cleared tests for immunohistochemical detection of HER2 protein and fluorescence *in situ* hybridization (FISH) detection of gene amplification are commercially available and approved as companion diagnostics for Herceptin therapy. Recently, the American Society of Clinical Oncology and the College of American

Pathologists published guidelines for performing and interpreting HER2 testing. These guidelines attempt to standardize HER2 testing by addressing preanalytical, analytical, and postanalytical variables that could lessen result variability due to technical and interpretative subjectivity.

TPMT

Thiopurine S-methyltransferase (TPMT) is a cytosolic enzyme that inactivates thiopurine drugs such as 6-mercaptopurine and azathioprine through methylation. Thiopurines are frequently used to treat childhood acute lymphoblastic leukemia. Variability in activity levels of TPMT enzyme function exists between individuals, and it has been found that this variability can be attributed to polymorphisms of the *TPMT* gene. The most common variant alleles, *TPMT*2*, *TPMT*3A*, and *TPMT*3C*, account for 95% of TPMT deficiency. Molecular testing is a relatively convenient method for assessment of TPMT enzyme function in patients before treatment with thiopurines. Low TPMT activity levels could put a patient at risk for developing toxicity, since too much drug would be converted to 6-thioguanine nucleotides (6-TGNs), the cytotoxic active metabolite incorporated into DNA. On the other hand, a patient with high TPMT activity levels would need higher than standard doses of a thiopurine drug to respond well to the therapy, since a large amount of the drug is being inactivated before it can be converted to 6-TGNs.

EGFR

The epidermal growth factor receptor (*EGFR* or *HER1*) is a member of the ErbB family of tyrosine kinases that also includes HER2. Once the ligand binds to the receptor, the receptor undergoes homodimerization or heterodimerization. This activation via phosphorylation initiates signaling to downstream pathways such as PI3K/AKT and RAS/RAF/MAPK which in turn regulates cell proliferation and apoptosis.

EGFR is expressed in some lung cancers and is the target of newly developed small molecule drugs, including gefitinib and erlotinib. It has been shown that tumors that respond to these tyrosine kinase (TK) inhibitors (TKIs) contain somatic mutations in the EGFR TK domain. Up to 90% of mutations in nonsmall cell lung cancers (NSCLC) can be attributed to two mutations, an inframe deletion of exon 19 and a single point mutation in exon 21 (T2573G). While these mutations are associated with a favorable response to the new TKIs, other mutations are associated with a poor response and potential resistance to these therapeutics (for example, in-frame insertion at exon 20 confers resistance). Mutation analysis in the *EGFR* represents a new application for molecular diagnostics as some mutations confer a favorable response, while others are not so favorable. It becomes critical to the management of the NSCLC patient that the clinical molecular diagnostics laboratory be able to identify such mutations on a routine basis.

CONCLUSION

Our knowledge base of PGx and clinical applications of such testing are progressing at record speeds. Technology is allowing us to perform these tests routinely in the clinical laboratory. Currently, PGx testing can be performed to detect polymorphisms in the genes for metabolizing enzymes and some drug targets. Clearly, one growing application in cancer patients is to detect polymorphisms and mutations associated with responses to newly developed small molecule targeted therapies. The ultimate goal of these efforts is to truly provide a "personalized medicine" approach to patient management for the purpose of eliminating ADRs, selecting more efficacious therapeutics, and improving the overall well-being of the patient.

KEY CONCEPTS

- The application of genetic testing to predict how well or how poorly an individual will respond to a therapeutic drug has made its way into the clinical laboratory.
- Medical practices often utilize a trial-and-error approach to select the proper medication and dosage for a given patient, whereas pharmacogenetics (PGx) assesses an individual's response to therapeutic drugs.
- A second component to PGx is the application of genomics to the identification of target genes for novel therapies.
- Single nucleotide polymorphisms (SNPs) in PGx-associated genes may result in no significant phenotypic effect, change drug metabolism by more than 10,000-fold, or alter protein binding by more than 20-fold.
- In 510 BC, Pythagoras recognized that some individuals developed hemolytic anemia with fava bean consumption, which led to future discoveries of enzymes that detoxify foreign agents and to discoveries that some people lack these enzymes.
- Most of the enzymes involved in drug metabolism are members of the cytochrome P450 (CYP450) superfamily, and SNPs in these genes can lead to poor, intermediate, extensive, and ultrarapid metabolizers.
- In August 2005, the FDA amended the irinotecan (Camptosar®) package insert to recommend genotyping for the UGT1A1 polymorphism and suggested a dose reduction in patients homozygous for the *28 allele.
- Interindividual variability with respect to observed chemotherapeutic response in cancer patients is

well recognized. It is also apparent that there is considerable cellular heterogeneity within a single tumor that may account for this lack of efficacy.
- Determination of HER2 status is also recognized as the first FDA-approved companion diagnostic since Trastuzumab (Herceptin), the first humanized monoclonal antibody against the HER2 receptor, was approved by the FDA in 1998.

SUGGESTED READINGS

1. Evans WE, McLeod HL. Pharmacogenomics—Drug disposition, drug targets, and side effects. *N Engl J Med.* 2003;348(6): 538–549.
2. Gardiner SJ, Begg EJ. Pharmacogenetics, drug-metabolizing enzymes, and clinical practice. *Pharmacol Rev.* 2006;58(3): 521–590.
3. Goetz MP, Rae JM, Suman VJ, et al. Pharmacogenetics of tamoxifen biotransformation is associated with clinical outcomes of efficacy and hot flashes. *J Clin Oncol.* 2005;23(36):9312–9318.
4. Ingelman-Sundberg M. Pharmacogenetics of cytochrome P450 and its applications in drug therapy: The past, present and future. *Trends Pharmacol Sci.* 2004;25(4):193–200.
5. Jamroziak K, Robak T. Pharmacogenomics of MDR1/ABCB1 gene: The influence on risk and clinical outcome of haematological malignancies. *Hematology.* 2004;9(2):91–105.
6. Kirchheiner J, Fuhr U, Brockmoller J. Pharmacogenetics-based therapeutic recommendations—Ready for clinical practice? *Nat Rev Drug Discov.* 2005;4(8):639–647.
7. Klein TE, Chang JT, Cho MK, et al. Integrating genotype and phenotype information: An overview of the PharmGKB project. Pharmacogenetics Research Network and Knowledge Base. *Pharmacogenomics J.* 2001;1(3):167–170.
8. Laudadio J, Quigley DI, Tubbs R, et al. HER2 testing: A review of detection methodologies and their clinical performance. *Expert Rev Mol Diagn.* 2007;7(1):53–64.
9. Lynch TJ, Bell DW, Sordella R, et al. Activating mutations in the epidermal growth factor receptor underlying responsiveness of non-small-cell lung cancer to gefitinib. *N Engl J Med.* 2004;350 (21):2129–2139.
10. Maitland ML, Vasisht K, Ratain MJ. TPMT, UGT1A1 and DPYD: Genotyping to ensure safer cancer therapy? *Trends Pharmacol Sci.* 2006;27(8):432–437.
11. Oertel B, Lotsch J. Genetic mutations that prevent pain: Implications for future pain medication. *Pharmacogenomics.* 2008;9(2): 179–194.
12. Rollason V, Samer C, Piguet V, et al. Pharmacogenetics of analgesics: Toward the individualization of prescription. *Pharmacogenomics.* 2008;9(7):905–933.
13. Shankar DB, Li J, Tapang P, et al. ABT-869, a multitargeted receptor tyrosine kinase inhibitor: Inhibition of FLT3 phosphorylation and signaling in acute myeloid leukemia. *Blood.* 2007;109 (8):3400–3408.
14. Shastry BS. Pharmacogenetics and the concept of individualized medicine. *Pharmacogenomics J.* 2006;6(1):16–21.
15. Urquhart BL, Tirona RG, Kim RB. Nuclear receptors and the regulation of drug-metabolizing enzymes and drug transporters: Implications for interindividual variability in response to drugs. *J Clin Pharmacol.* 2007;47(5):566–578.
16. Weinshilboum R, Wang L. Pharmacogenomics: Bench to bedside. *Nat Rev Drug Discov.* 2004;3(9):739–748.

Index

Index

Thyroid-stimulating hormone (TSH)
  mutations, 293
  receptor
    activating mutations in familial
      nonautoimmune
      hyperthyroidism, 295–296
    inactivating mutations, 293
TK, *see* Thymidine kinase
TLR, *see* Toll-like receptor
TMPRSS2-ETS, gene fusion in prostate
  cancer, 337–338
TNF-α, *see* Tumor necrosis factor-α
Tolerance, autoimmune disease, 213–215
Toll-like receptor (TLR)
  chronic disease exacerbation, 21–22
  functional overview, 207
  inflammation role, 18, 18*f*
  types, 18, 31*t*
  virus recognition, 37
Topoisomerase II
  inhibitors and acute myeloid leukemia
    induction, 198
TPMT, *see* Thiopurine S-methyltransferase
TPO, *see* Thyroid peroxidase
Transcription factor p63, 357*f*
Transcriptomics, systems biology, 127
Transforming growth factor-β (TGF-β)
  anti-inflammatory activity, 19–20
  fibrosis mediation in disease, 23*t*
  tissue remodeling in inflammatory
    disease, 23
  valvular heart disease dysregulation,
    168–169
Trastuzumab, breast cancer
  management, 350
Treg, *see* Regulatory T-cell
Triple A syndrome, 301
Trophoblastic diseases, *see* Pregnancy
Trypanosomes
  immune system evasion strategies, 26–29
  surface glycoprotein diversity, 27–29, 29*f*, 30*f*
TSG, *see* Tumor suppressor gene
TSH, *see* Thyroid-stimulating hormone

TTF-1, *see* Thyroid transcription factor-1
TTP, *see* Thrombotic thrombocytopenic
  purpura
Tuberculosis, inherited susceptibility, 212
Tumor necrosis factor-α (TNF-α)
  cancer cachexia role, 59
  liver injury role, 268
  tissue remodeling in inflammatory
    disease, 23–24, 24*t*
Tumor suppressor genes (TSGs), 219
  hypermethylation in cancer, 108–110, 109*f*

U

UMPD, *see* Unclassified myeloproliferative
  disorder
uPA, *see* Urokinase plasminogen activator
Uridine diphosphate
  glucuronosyltransferases,
    polymorphisms and drug
    metabolism, 428
Uterine cancer
  classification, 313
  endometroid endometrial cancer, 314–315
  epidemiology, 313
  hyperplasia, 313–314
  mesenchymal tumors, 315–316
  Type II endometrial cancers, 315

V

Vaginal carcinoma, 319
Valvular heart disease
  calcified aortic stenosis, 161–162
  cells
    endothelial cell, 161–162
    interstitial cell, 162, 163*f*
    leukocytes, 163
  connective tissue disorders, 168
  mitral valve prolapse, 168
  pathogenesis, 169*f*

transforming growth factor-β
  dysregulation, 168–169
Vascular cell adhesion molecule-1 (VCAM-1),
  leukocyte–endothelial adhesion, 15
Vascular endothelial growth factor
  (VEGF), 220
VCAM-1, *see* Vascular cell adhesion
  molecule-1
VEGF, *see* Vascular endothelial growth factor
VIN, *see* Vulvar intraepithelial neoplasia
Vinculin, defects in cardiomyopathy, 171
VNTRs, *see* Variable number of tandem
  repeats
von Willebrand disease, 181–182, 182*t*
Vulvar intraepithelial neoplasia (VIN), 319

W

WG, *see* Wegener's granulomatosis
Wiskott-Aldrich syndrome, 183
Wnt
  familial adenomatous polyposis
    signaling, 258*f*

X

X-linked agammaglobulineia, *see* Bruton's
  disease
X-linked dominant inheritance, 68
X-linked dominant male lethal
  inheritance, 68
X-linked proliferative disease, *see* Duncan's
  syndrome
X-linked recessive inheritance, 68

Y

Y-linked inheritance, 68

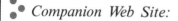

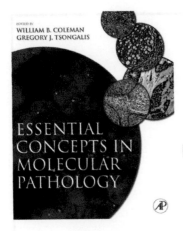

As a FREE special offer to book purchasers,
we are offering over 300 online study questions through

EXAM MASTER!

Please be sure to take advantage of this purchase bonus.

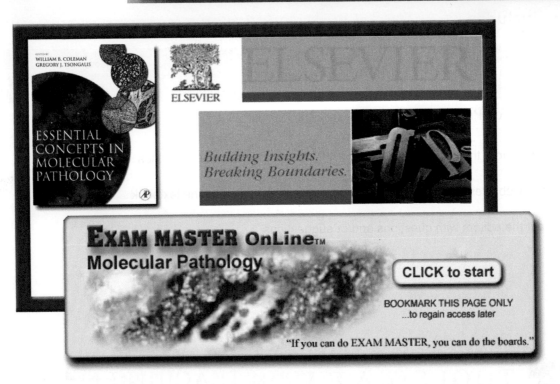

Gateway Code: http://www.exammaster2.com/wdsentry/elsevier.htm

EM OnLine Registration:

1. Go to your custom gateway: http://www.exammaster2.com/wdsentry/elsevier.htm
   - Click **"Click to Start"** and Click **"Not Registered Yet?"**
   - To continue, click **[I Accept]** on the Licensing Agreement page
2. Complete the registration form, click **[Submit Registration]**
3. Confirmation of registration and your temporary password will be e-mailed to you immediately
4. Log-in with your "User Name" and "Temporary Password"
5. The **Welcome Page** explains how to access the chapter questions

EXAM MASTER Corporation
www.exammaster.com

Printed and bound by CPI Group (UK) Ltd, Croydon, CR0 4YY

03/10/2024

01040728-0002